Bioactives in Fruit

Health Benefits and Functional Foods

Bioactives in Fruit

Health Benefits and Functional Foods

Editors

Margot Skinner
Professor in Food Science
School of Chemical Sciences and Institute of Plant and Food Science
University of Auckland
Auckland, New Zealand

Denise Hunter
Research Scientist in Food Innovation
The New Zealand Institute for Plant and Food Research Limited
Auckland, New Zealand

WILEY Blackwell

Registered office: John Wiley & Sons, Ltd, The Atrium, Southern Gate, Chichester, West Sussex, PO19 8SQ, UK

Editorial offices: 9600 Garsington Road, Oxford, OX4 2DQ, UK
The Atrium, Southern Gate, Chichester, West Sussex, PO19 8SQ, UK
111 River Street, Hoboken, NJ 07030-5774, USA

For details of our global editorial offices, for customer services and for information about how to apply for permission to reuse the copyright material in this book please see our website at www.wiley.com/wiley-blackwell.

Library of Congress Cataloging-in-Publication Data

Bioactives in fruit : health benefits and functional foods / editors,
Margot Skinner, Denise Hunter.
p. ; cm.
Includes bibliographical references and index.
ISBN 978-0-470-67497-0 (cloth : alk. paper) – ISBN 978-1-118-63555-1 –
ISBN 978-1-118-63556-8 (ePDF) – ISBN 978-1-118-63558-2 (emobi) – ISBN 978-1-118-63559-9 (epub)
I. Skinner, Margot. II. Hunter, Denise.
[DNLM: 1. Fruit–chemistry. 2. Biological Availability. 3 Functional Food.
4. Phytotherapy–methods. 5. Plant Extracts. WB 430]
RA784
613.2–dc23

2013006647

A catalogue record for this book is available from the British Library.

Wiley also publishes its books in a variety of electronic formats. Some content that appears in print may not be available in electronic books.

Cover image © Steven Allan/istockphoto.com
Cover design by Meaden Creative

Set in 10/12pt Times by Aptara® Inc., New Delhi, India
Printed and bound in Malaysia by Vivar Printing Sdn Bhd

1 2013

Contents

Contributors

Francisco A. Tomás-Barberán
CEBAS-CSIC
Department of Food Science and Technology
Murcia
Spain

Iris F.F. Benzie
Department of Health Technology and Informatics
The Hong Kong Polytechnic University
Kowloon
Hong Kong

David J. Burritt
Department of Botany
University of Otago
Dunedin
New Zealand

Antonio Camargo Garcia
Lipids and Atherosclerosis Unit
IMIBIC/Reina Sofia University Hospital/University of Cordoba, and CIBER Fisiopatologia Obesidad y Nutricion (CIBEROBN)
Instituto de Salud Carlos III
Spain

Paul F. Cancalon
Florida Department of Citrus
Lake Alfred
Florida
USA

Reinhold Carle
Hohenheim University
Institute of Food Science and Biotechnology
Stuttgart
Germany

William T. Cefalu
Diabetes and Nutrition Laboratory
Pennington Biomedical Research Center
Baton Rouge
Louisiana
USA

Suengmok Cho
Korea Food Research Institute
Seongnam
Republic of Korea

Alastair Currie
The New Zealand Institute for Plant and Food Research Ltd
Motueka
New Zealand

Gary M. Davenport
P&G Pet Care
Mason
Ohio
USA

Yoshiharu Deguchi
Laboratory of Drug Disposition and Pharmacokinetics
School of Pharmaceutical Sciences
Teikyo University
Japan

Michael Edwards
Department of Nutrition
Norwich Medical School
University of East Anglia
Norwich
UK

Carrie M. Elks
Nutritional Neuroscience and Aging Laboratory
Pennington Biomedical Research Center
Baton Rouge
Louisiana
USA

Richard Espley
The New Zealand Institute for Plant & Food Research
Auckland
New Zealand

Joseph Francis
Comparative Biomedical Sciences
Louisiana State University School of Veterinary Medicine
Baton Rouge
Louisiana
USA

María I. Gil
CEBAS-CSIC
Department of Food Science and Technology
Murcia
Spain

Anusooya Gnanavinthan
Food Science, School of Chemical Sciences
The University of Auckland
Auckland
New Zealand

Crystal F. Haskell
Brain, Performance and Nutrition Research Centre
School of Life Sciences
Northumbria University
Newcastle upon Tyne
UK

Denise Hunter
The New Zealand Institute for Plant & Food Research Ltd
Auckland
New Zealand

Donald K. Ingram
Nutritional Neuroscience and Aging Laboratory
Pennington Biomedical Research Center
Baton Rouge
Louisiana
USA

Kerrie L. Kaspar
Ocean Spray Cranberries, Inc.
Lakeville-Middleboro
Massachusetts
USA

Dietmar Rolf Kammerer
Hohenheim University
Institute of Food Science and Biotechnology
Stuttgart
Germany

Colin Kay
Department of Nutrition
Norwich Medical School
University of East Anglia
Norwich
UK

Christina Khoo
Ocean Spray Cranberries, Inc.
Lakeville-Middleboro
Massachusetts
USA

Carolyn E. Lister
The New Zealand Institute for Plant & Food Research Limited
Lincoln
New Zealand

Jose Lopez-Miranda
Lipids and Atherosclerosis Unit
IMIBIC/Reina Sofia University Hospital/University of Cordoba, and CIBER Fisiopatologia Obesidad y Nutricion (CIBEROBN)
Instituto de Salud Carlos III
Spain

Stefan Martens
Fondazione Edmund Mach, Istituto Agrario di San Michele all'Adige (IASMA)
Centro Ricerca e Innovazione, Dipartimento Qualita' Alimentare e Nutrizone, Biotecnologia dei Prodotti
Naturali
San Michele all'Adige
Italy

Bruno Mezzetti
Università Politecnica delle Marche
Ancona
Italy

Rosalind Miller
GlaxoSmithKline
Brentford
UK

Shingen Misaka
Department of Pharmacology, School of Medicine
Fukushima Medical University
Fukushima
Japan

Conchita Obón
Departamento de Ciencia y Tecnología de Alimentos
Universidad Miguel Hernández
Orihuela
Spain

Takashi Okura
Laboratory of Drug Disposition and Pharmacokinetics
School of Pharmaceutical Sciences
Teikyo University
Japan

Susan S. Percival
Food Science and Human Nutrition Department
University of Florida
Gainesville
Florida
USA

Francisco Perez-Jimenez
Lipid and Atherosclerosis Unit
IMIBIC/Reina Sofa University
Hospital/University of Cordoba
CIBER Fisiopatologa de la Obesidad y
Nutrición (CIBEROBN)
Instituto de Salud Carlos III
Cordoba, Spain

Sophie Putnam
GlaxoSmithKline
Brentford
UK

Diego Rivera
Departamento de Botánica
Universidad de Murcia
Murcia
Spain

George S. Roth
GeroScience, Inc.
Pylesville
Maryland
USA

David Ruiz
CEBAS-CSIC
Department of Food Science and Technology
Murcia
Spain

Zaid Saleh
The University of Auckland, Department of Chemical and Materials Engineering
The New Zealand Institute for Plant and Food Research Limited
Auckland
New Zealand

Catalina Sánchez-Roca
Departamento de Botánica
Universidad de Murcia
Murcia
Spain

Jessica Scalzo
The New Zealand Institute for Plant and Food Research Ltd
Hawkes Bay
New Zealand

Barbara Shukitt-Hale
Neuroscience Laboratory
USDA Human Nutrition Research Center on Aging
Tufts University
Boston
Massachusetts
USA

Margot Skinner
Food Science, School of Chemical Sciences
The University of Auckland
Auckland
New Zealand

Stephen J.M. Skinner
Department of Physiology
Faculty of Medicine and Health Sciences
The University of Auckland
Auckland
New Zealand

Bronwen G. Smith
Food Science, School of Chemical Sciences
The University of Auckland
Auckland
New Zealand

April J. Stull
Diabetes and Nutrition Laboratory
Pennington Biomedical Research Center
Baton Rouge
Louisiana
USA

Dongxiao Sun-Waterhouse
The New Zealand Institute for Plant & Food Research Limited
Auckland
New Zealand

Yuko Taki
Department of Pharmacokinetics and Pharmacodynamics
School of Pharmaceutical Sciences
University of Shizuoka
Japan

Keizo Umegaki
Information Center, National Institute of Health and Nutrition
Shinjuku-ku, Tokyo
Japan

Daniel Valero
Departamento de Ciencia y Tecnología de Alimentos
Universidad Miguel Hernández
Orihuela
Spain

Sissi Wachtel-Galor
Department of Health Technology and Informatics
The Hong Kong Polytechnic University
Kowloon
Hong Kong

Hiroshi Watanabe
Department of Clinical Pharmacology and Therapeutics
Hamamatsu University School of Medicine
Japan

Yasuo Watanabe
Yokohama College of Pharmacy
General Health Medical Center
Yokohama
Japan

Anthony W. Watson
Brain, Performance and Nutrition Research Centre
School of Life Sciences
Northumbria University
Newcastle upon Tyne
UK

Robin L. West
Department of Psychology
University of Florida
Gainesville
Florida
USA

Gary Woodward
Department of Clinical Biochemistry
Oxford University Hospitals
Oxford
UK

Shizuo Yamada
Department of Pharmacokinetics and Pharmacodynamics
School of Pharmaceutical Sciences
University of Shizuoka
Japan

Elena M. Yubero-Serrano
Brookdale Department of Geriatrics and Palliative Medicine
Division of Experimental Diabetes and Aging Division
Mount Sinai School of Medicine
New York, NY
USA

Preface

For centuries we have known that fruit is important for health but we are only just beginning to fully understand why. This book aims to summarize some of our current knowledge on the bioactive compounds that are associated with the health benefits of specific fruits with a strong emphasis on the validation of health benefits by human intervention trials. Because of the present interest in food and health, the book includes strategies to retain and enhance the bioactives in fruit through growing conditions, fruit storage, processing into ingredients and production of functional foods. To accomplish this task authors with expertise in biology, chemistry, pharmacology, food science, nutrition, medicine and horticulture have contributed. They come from universities, government and industry-funded research institutes and from biotechnology and food companies in Europe, the United States, Asia and New Zealand to give the book a broad perspective.

A brief introduction on the major classes of bioactives present in fruit is followed by a chapter on fruit fibre, exploring the many different chemical compositions that fibre from different sources, including commercial preparations, may contain. This is an important aspect to consider when attempting to attribute health benefits to fibre and when developing new fibre containing foods. The focus of the next chapter is the bioavailability of antioxidant compounds from fruit and includes vitamins C and E, as well as carotenoids and polyphenolic phytochemicals. It incorporates a discussion of the reasons why plasma antioxidant levels are maintained at low concentrations even when consumed at quite high levels, an important consideration as recommendations for fruit and vegetable consumption increase and supplements containing high amounts of these compounds are promoted by their manufacturers. Bioavailability and bioactivity of both foods and pharmaceuticals are related to their absorption, metabolism, distribution, excretion and transport across cell membranes, together with their ability to bind to specific receptors. The potential for functional food bioactives to interact with pharmaceuticals is considerable due to the large number of constituents that they contain. Some examples of the way in which selected functional foods and nutraceuticals influence the pharmacokinetics and pharmacodynamics of drugs are the subject of the next chapter, emphasizing the need to understand whether interactions of foods with medicines are clinically harmful or beneficial to drug therapy.

There then follow a number of chapters on the bioactives and health benefits of common and not-so-common fruits, which include apple and pear, citrus, pomegranate and stone fruit, grapes, blackcurrants, blueberry, cranberry and kiwifruit, feijoa and tamarillo. 'Vegetable' fruits, olives, avocado and cocoa, plant foods not commonly consumed as 'fruit', are not left out. The chapters are written from different perspectives depending on the interests and research focus of the authors. For example, the chapter on apples and pears includes an update on the biosynthesis and genetic regulation of bioactives within apples. The chapter on citrus includes attributes of citrus that have raised health concerns and again discusses the grapefruit–drug interactions alluded to in an earlier chapter, including ways that this may be overcome now and in the future. The chapter on blueberries includes a detailed examination of mechanistic and pre-clinical studies in models that range from longevity in fruit flies to myocardial ischemia in rats.

The health areas covered in the book span a wide range. These include basic antioxidant activity and all it may be associated with, immune modulation including positive effects associated with asthma, antibacterial properties and reduction of symptoms of infections such as cold and influenza, cardiovascular health, bone health, obesity, metabolic syndrome and diabetes, eye health and vision, oral and gastrointestinal health, urinary tract health, cognition and anxiety and, associated with a number of these areas, longevity. Supporting data, including results from mechanistic, *in vitro* and *in vivo* animal studies, are discussed together with validation from human intervention trials, and tentative or firm conclusions drawn, depending on the evidence available. It should be mentioned that epigenetic and nutrigenomic factors that may influence how individuals respond to the bioactive components in fruit is beyond the scope of this book. Once intervention trials are designed and conducted to provide more personalized outcomes, we may find that some of the equivocal results that are often obtained in intervention trials will be clarified and health benefits of functional foods be more commonly assigned to particular groups or genotypes.

The next two chapters are on the subject of horticulture. Using anthocyanins as an example, a chapter on breeding for enhanced levels of these bioactive compounds in blackcurrant and blueberry outlines gains that may be achieved through conventional breeding techniques, as well as the potential for using genetic transformation in strawberries as a proof of concept. As many of the bioactives in fruit are secondary metabolites, whose levels increase in plant cells after exposure to environmental stress, the potential of stress priming to enhance the bioactive contents of fresh fruit or produce new fruit-based products is considered in the chapter following.

The book finishes with two chapters that highlight different aspects of functional foods. In a chapter on recovery of bioactive compounds from entities such as peel and seeds, where they are present in high amounts, innovative process technologies are described that can be applied to purify, enrich and selectively fractionate the compounds from crude extracts recovered from food processing byproducts. The recovery of such components for use as bioactive compounds in functional or enriched foods, as well as dietary supplements, is presented as a promising strategy to produce ingredients and products from waste streams to help achieve sustainable agricultural food production. The last chapter discusses many of the interactions that may occur between fruit bioactives and other food components when they are incorporated into finished processed foods. It goes on to point out that it is the complexity of finished foods, coupled with the processing conditions, that influence the stability, bioaccessability and, ultimately, the bioavailability of the fruit bioactive compounds in a finished processed functional food product.

The information gathered in this book, describing fruit bioactives, their health benefits when consumed as a food and related topics regarding their development into fresh or processed functional foods, will be of use to postgraduate students, researchers, functional food product developers, food regulators and anyone who has curiosity about why 'fruit is good for you'. The information contained within may provide plant breeders with new targets for the development of value-added horticultural products. It may also provide nutritionists and dieticians with a useful resource for developing strategies to assist in preventing or slowing disease onset or severity. In many instances, the regular consumption of specific fruits in a quantity that is pleasurable and achievable within the context of the diet is all that may be required.

We extend our sincere thanks to the many individuals who have contributed so generously in accomplishing this project.

Margot Skinner
Denise Hunter

1 Introduction to the Major Classes of Bioactives Present in Fruit

Anusooya Gnanavinthan

BIOACTIVES

Bioactives are compounds that produce physiological effects when present in a living material, in other words they must exert physiological benefits related to promoting health and preventing effects of a disease (e.g. blood pressure reduction, blood glucose reduction etc.) (Aluko, 2011; Awika, 2011). When bioactives are taken orally, the compound must withstand the digestion that will destroy the active structure and render it physiologically inactive. But, in some cases, the inactive part of the compound becomes active once consumed, as a result of the action of digestive enzymes present in the gastrointestinal tract (Aluko, 2011). Also, bioactive compounds may exert their physiological effect within the digestive tract and may not be absorbed. However, in most cases, the compound must be absorbed from the gastrointestinal tract into the blood circulatory system, from where it is carried to target organs. Foods that contains bioactives, and that are consumed as part of a normal diet, are called functional foods (Shahidi, 2009).

Plants are a rich source of bioactive compounds. Plant-derived bioactive products such as fruits, vegetables and nuts are becoming popular because of their abundance and low cost (compared to animal products), and due to the wider acceptability of plant products as a result of religious, social or moral reasons that prevent many people consuming animal-derived products (Omaye *et al.*, 2000; Aluko, 2011). Bioactives from fruits show antimicrobial activity, anticancer activity, anti-inflammatory activity, immuno-stimulatory activity and antioxidant activity and so on (Hollman and Katan, 1999; Emilio, 2007; Weston, 2010).

CLASSIFICATION OF PLANT-DERIVED BIOCTIVES

Bioactives can be classified based on molecular identity or biopolymer type that includes polyphenolic compounds, indigestible carbohydrates (dietary fibres), functional lipids (mainly in cereals and seeds), proteins and peptides and carotenoids (Figure 1.1).

Phenolic compounds

Plant phenolics are a structurally diverse class of phytochemicals (Naczk and Shahidi, 2006). Phenolic compounds are defined by the presence of at least one aromatic ring bearing one

Bioactives in Fruit: Health Benefits and Functional Foods, First Edition. Edited by Margot Skinner and Denise Hunter.

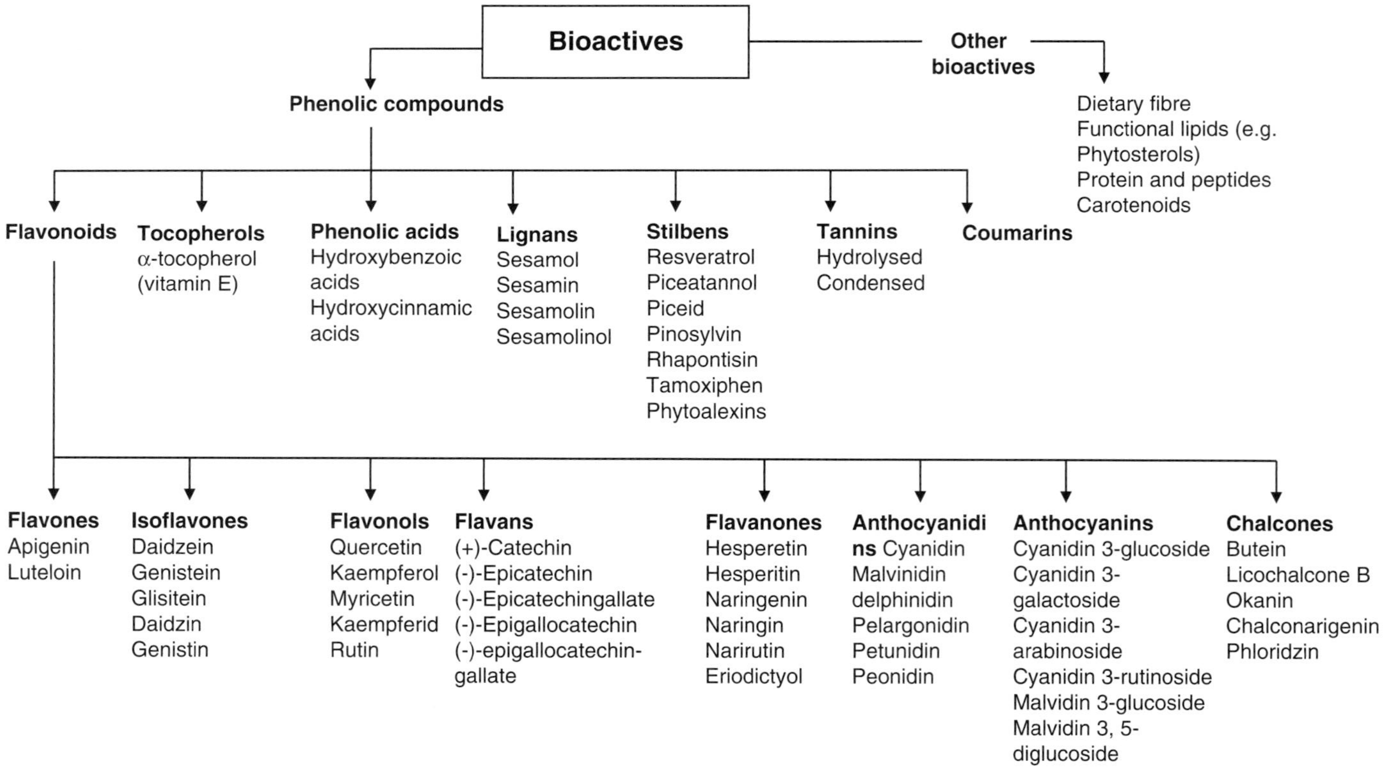

Fig. 1.1 Possible classification and examples of plant bioactive compounds (Von Elbe and Schwartz, 1996; Hollman and Katan, 1999; Tokuşoğlu and Hall, 2001; El Gharras, 2009; Naczk and Shahidi, 2006).

(phenol) or more (polyphenols) hydroxyl substituents, including their functional derivative (e.g. esters and glycosides) (Maffei, 2003). Polyphenols occur as plant secondary metabolites, are widely distributed in the plant kingdom and represent an abundant antioxidant component of the human diet (Tokuşoğlu, 2011). There has been an increased interest in the health benefits of polyphenols due to the corresponding antioxidant capacities (Wang *et al.*, 1996; Sun *et al.*, 2002).

Free radicals (reactive oxygen species) and antioxidant studies play a major role in medicine. Free radicals are produced by many biological reactions in the body and can damage crucial biomolecules. If these free radicals are not scavenged, they may lead to disease conditions (Fan *et al.*, 2007). Reactive oxygen species (ROS), such as superoxide (O_2^-), the hydroxyl radical ($^\bullet OH$), hydrogen peroxide (H_2O_2) and lipid peroxide radicals, have been implicated in playing an important role in chronic degenerative disease, such as cancer, inflammatory, cardiovascular and neurodegenerative diseases, and ageing (Mullen *et al.*, 2002; Price *et al.*, 2006; Sies, 2010). The harmful actions of free radicals can be blocked by antioxidants, by scavenging those free radicals and detoxifying the organisms (BeMiller and Huber, 1996; Fan *et al.*, 2007). Therefore, antioxidants are referred to as compounds that can counteract the damaging effects of oxygen in tissues, and the term is applied to molecules that protect from any free radical (molecule with unpaired electron) (Bilgiçli *et al.*, 2007). Antioxidants in food are defined as any substance that can relay, retard or prevent the development of food rancidity due to oxidation (Gordon, 2001). Natural antioxidants may inhibit lipid peroxidation in food and improve the quality and safety of the food (Fan *et al.*, 2007). Two inhibitory mechanisms may be involved: direct scavenging of free radicals (primary antioxidant, e.g. vitamin E – α-tocopherol) and indirect scavenging of free radicals (secondary antioxidant) (Gordon, 2001).

Types of antioxidants

1. Synthetic antioxidants: e.g. butylated hydroxyanisol (BHA), butylated hydroxytoluene (BHT), tertiary butylhydroquinone (TBHQ), ester of gallic acid and so on. BHA and BHT are heat stable and used for stabilization of fats in baked and fried foods (Gordon, 2001).
2. Natural antioxidants: the majority of natural antioxidants are phenolic compounds and grouped into **tocopherols, flavonoids, phenolic acids, cinnamic acid derivatives, lignans, stilbenes, tannins and coumarins** (Shahidi and Naczk, 1995; Von Elbe and Schwartz, 1996; Gordon, 2001; El Gharras, 2009). Natural antioxidants are more potent, efficient and safer than synthetic ones (Shi *et al.*, 2001).

Natural antioxidants

Flavonoids

Flavonoids are the main bioactive compounds found in fruits. They include a larger group of natural antioxidants (Haminiuk *et al.*, 2012). The structure of flavonoids contain a C_6–C_3–C_6 carbon skeleton (two aromatic rings linked by a three-carbon aliphatic chain which is condensed to form a pyran or a furan ring) (Shahidi and Naczk, 1995). Flavonoids comprise subclasses of flavones, isoflavones, flavonols, flavans, flavanones, anthocyanidins, anthocyanins and chalcones (Figures 1.1 and 1.2). Flavonoids that are linked to one or more sugar molecules are called flavonoid glycosides; when they are not connected to a sugar

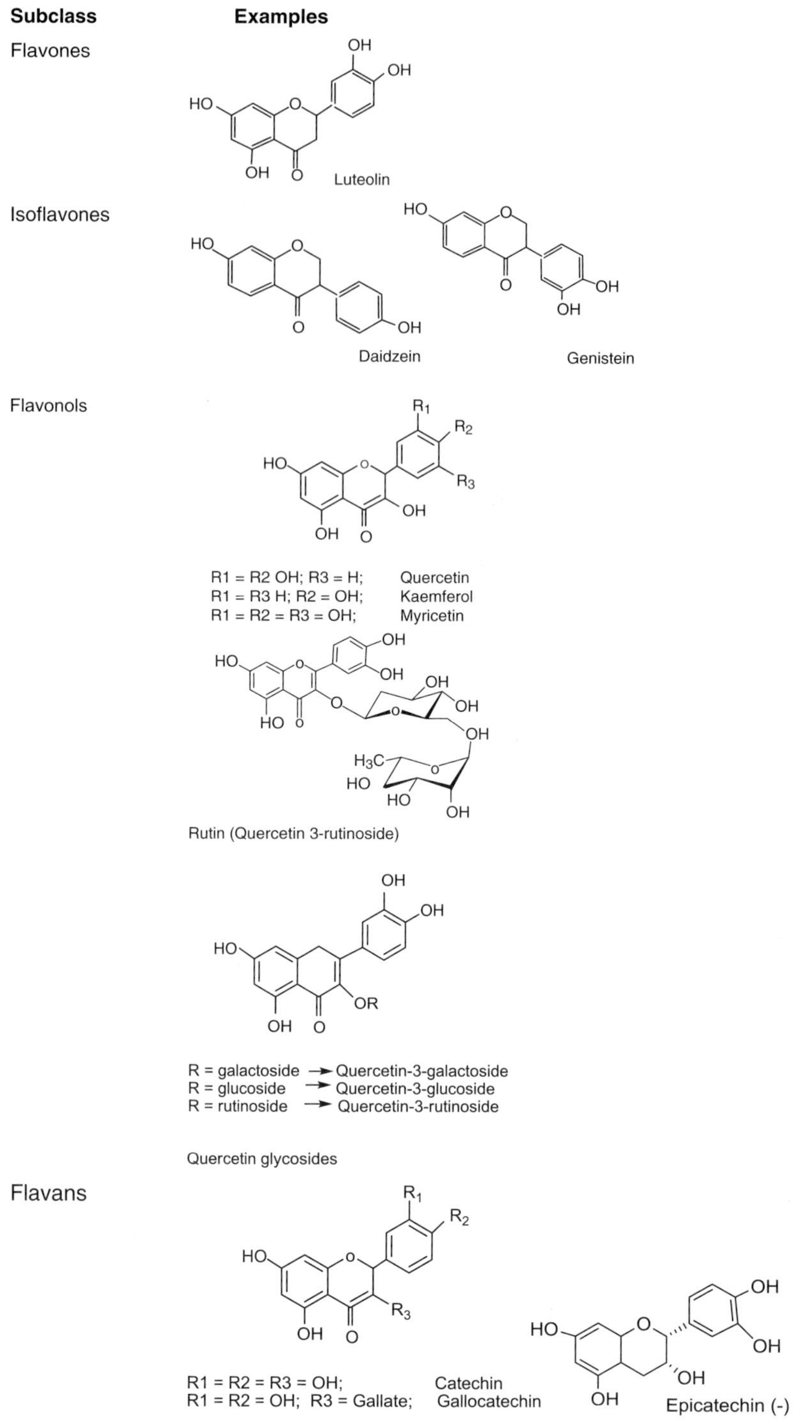

Fig. 1.2 Structures of different flavonoids (Berhow *et al.*, 1991; Hollman *et al.*, 1997a, 1997b; Sanoner *et al.*, 1999; Kähkönen and Heinonen, 2003; Robards, 2003).

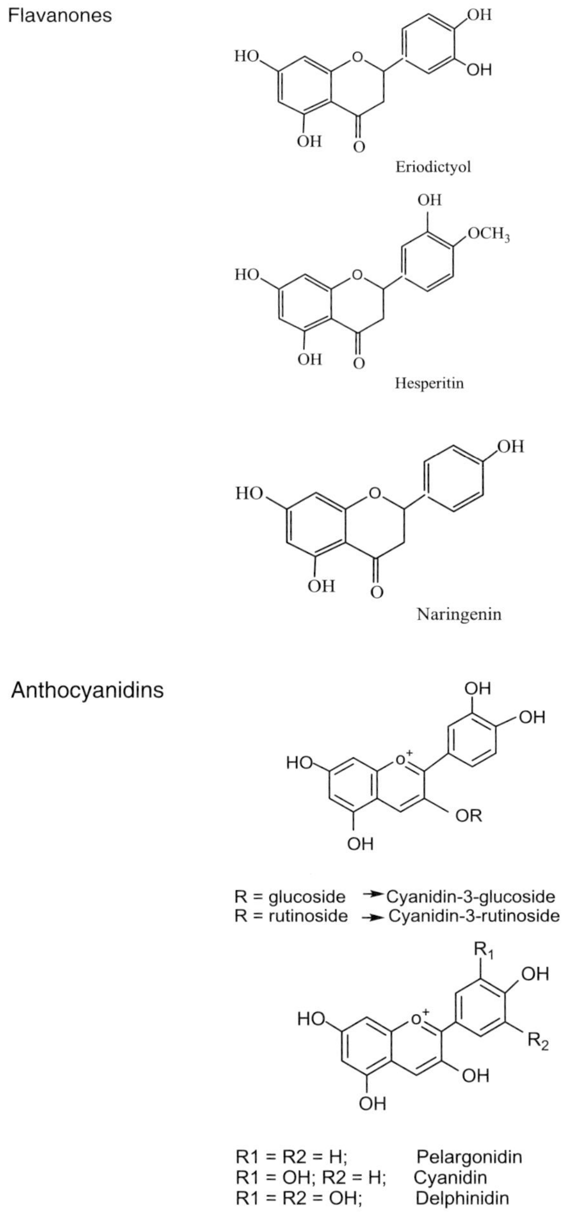

Fig. 1.2 (*Continued*) Structures of different flavonoids.

molecule are called aglycones (Haminiuk *et al.*, 2012). The flavonoids usually occur as glycosides which are less effective than aglycones. There is a relationship between structure and antioxidant activity. The groups, such as catechol moiety (Figure 1.3a – shown within the dotted ellipse), the 2, 3 double bond conjugation with a 4-oxofunction of a carbonyl group (Figure 1.3a – shown within the solid ellipse), and presence of hydroxyl groups at the 3 and 5 positions (Figure 1.3b – shown within the dotted ellipse), determine the free radical scavenging and oxidation potential (Shi *et al.*, 2001). Quercetin possesses all the

Anthocyanins

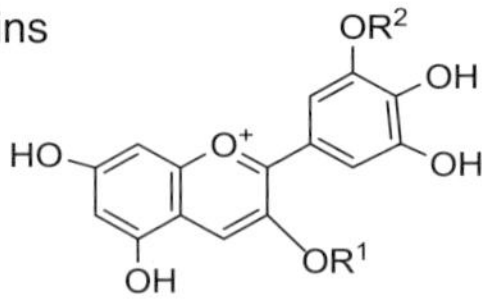

Anthocyanin compounds

R^1	R^2	Common name
β-D-glucopyranosyl	H	cyanidin 3-glucoside
6-*O*-α-L-rhanopyranosyl-b-D-glucopyranosyl	H	cyanidin 3-rutinoside
6-*O*-α-L-rhanopyranosyl-b-D-glucopyranosyl	OH	delphinidin 3-rutinoside
β-D-glucopyranosyl	OH	delphinidin 3-glucoside

Chalcones

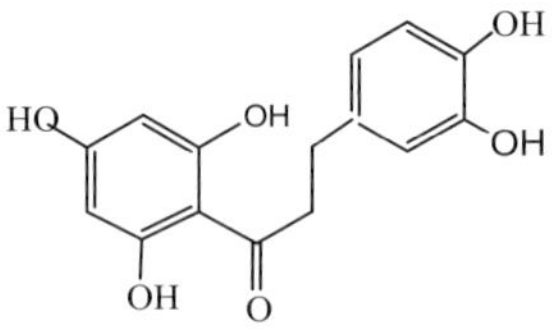

Butein

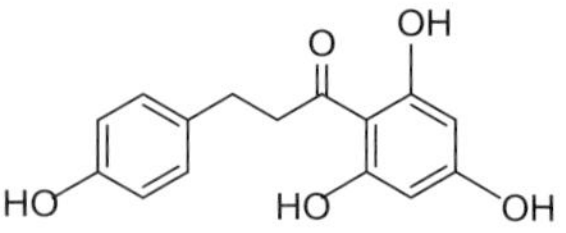

Phloretin

Fig. 1.2 (*Continued*) Structures of different flavonoids.

three structure groups, hence show higher antioxidant activity than kaempferol, which does not have catechol moiety (Shi *et al.*, 2001).

Flavonols are the most common flavanoids, occurring naturally as glycosylated forms (often glucose or rhamnose) (El Gharras, 2009). The most common flavonol aglycones are quercetin, kaempferol and myricetin (Figure 1.1) (Manach *et al.*, 2005; El Gharras, 2009). Flavones, on the other hand, consist of the glycosides of luteolin and apigenin (Manach *et al.*, 2004). Even though they are not very common in fruits, citrus fruits contain high concentrations of flavones. The main aglycones are naringenin in grapefruit, hesperetin in oranges and eriodictyol in lemons (El Gharras, 2009). Isoflavones can be aglycones or glycosides, occurring mainly in leguminous plants (e.g. soybeans). Flavanols exist in monomer (catechin) and polymer (proanthocyanidins) forms. These two flavanols are the main flavanols in fruits, whereas gallocatechin, epigallocatechin and epigallocatechin gallate are common in grapes (Manach *et al.*, 2004).

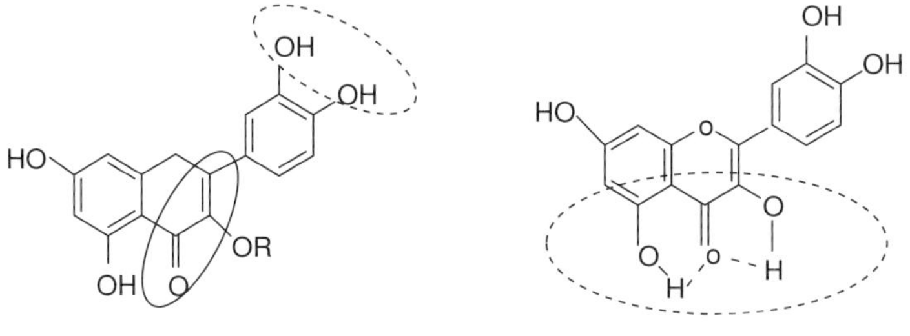

(a) (b)

Fig. 1.3 Antioxidant activity–structure relationships of flavonoids (Shi *et al.*, 2001).

Anthocyanins are pigments that impart red, blue, purple, violet and intermediate red–purple to berries and other fruits (Von Elbe and Schwartz, 1996). Similarly to other flavonoids, anthocyanins occur as glycosides. The de-glycosylated or aglycone forms of anthocyanins are known as anthocyanidins (Seeram, 2006). Hundreds of anthocyanins are known, varying in the basic anthocyanidin skeleton, such as cyanidin, delphinidin, pelargonidin, malvin, petunidin and peonidin (Table 1.1). Structural diversification is also achieved by the position at which glycosides (such as glucose, galactose, rhamnose and arabinose) and acyl groups (including phenolic acids such as caffeic, *p* coumaric, ferulic and sinapic) are attached to the skeleton (Naczk and Shahidi, 2006).

Tocopherols

Tocopherols are important biological antioxidants and widely used antioxidants. Tocopherols consist of two families, namely tocols and tocotrienols (Shahidi and Naczk, 1995). Depending on the number and position of methyl groups attached to the chromane rings, they are referred

Table 1.1 A summary of phytochemical present in different fruits (data from Nazck and Shahidi, 2006; Zadernowski *et al.*, 2009; Naczk *et al.*, 2011; Haminiuk *et al.*, 2012)

Phenolic compounds	Major Fruit source
Flavonols	
Quercetin	Blueberry, passion fruit, pomegranate
Kaempferol	Blackcurrant, fig
Myricetin	Apple, papaya
Rutin	Red grape, prunes, blueberry, apricot, apple
Flavones	
Apigeninn	Mango, durian
Luteolin	Lemon, pineapple, plum, watermelon, orange
Flavonones	
Hesperetin	Orange, grapefruit
Naringenin	Orange, grapefruit, lemon
Isoflavones	Soy beans
Flavanols	
Catechin	Grapes, cherry
Epicatechin	Apricot, cherry, grape, peach, blackberry, apple, avocado
Anthocyanins	
Cyanidin	Berry, pomegranate
Pelargonidin	Black currant, strawberry, raspberry, mangosteen
Peonidin	Blueberry, blackcurrant
Delphinidin	Black grape, blackcurrant, blueberry, grape fruit
Malvidin	Strawberry, plum, blueberry, grapefruit
Petunidin	Apple, blueberry
Hydrobenzoic acids	
Procatechuic acid	Blackberry
Gallic acid	Raspberry, banana, avocado
Vanillic acid	Avocado, strawberry
Syringic acid	Strawberry, grapes
Hydrocinnamic acids	
Caffeic acid	Papaya, peach, avocado
Chlorogenic acid	Kiwi fruit, passion fruit, blueberry, peaches
Coumaric acid	Cherry, orange, blackcurrant
Ferulic acid	Mango, orange, papaya, pineapple
Sinapic acid	Apple

to as α, β, γ, δ (Figure 1.4). Tocopherols also possess vitamin E activity; the most important antioxidant of this group is α-tocopherol (Shahidi and Naczk, 1995) (Figure 1.4). Fruits contain considerable amounts of tocopherols, whereas in animals they are only found in trace amounts. These are heat-stable and relatively weak antioxidants used in fats and oils (Gordon, 2001).

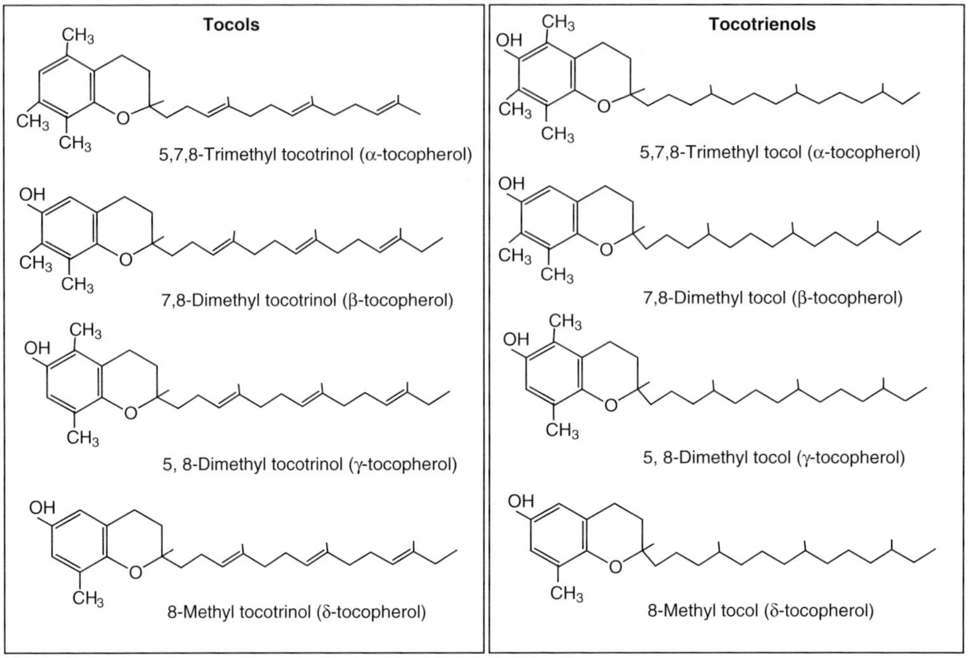

Fig. 1.4 Structures of tocopherols.

Phenolic acids

Two classes of phenolic acids have been identified – derivatives of benzoic acid and derivatives of cinnamic acid. Hydroxybenzoic acids are components of tannins (gallotannnins and ellagitannins), whereas hydroxycinnamic acids consist of *p*-coumaric, caffeic, ferulic and sinapic acids (Haard and Chism, 1996; Pontes, 2002; Manach *et al.*, 2004; Haminiuk *et al.*, 2012). Phenolic acids are rarely found as free forms but are commonly found in conjugates forms as esters and glycosides (Tiffany and Luke, 2007). Caffeic and quinic acid combine to form chlorogenic acid, which is rich in fruits (Clifford, 1999, 2000) (Figure 1.5). The antioxidant activity of phenolic acids and their esters depends on the number of hydroxyl group in the molecule. Hydroxylated cinnamic acids are more effective than their benzoic counterparts (Shahidi and Naczk, 1995; Tomás-Barberan and Clifford, 2000). Berries, plums, kiwifruit and apples contain a variety of phenolic acids which occur as derivatives of hydroybenzoic acid (e.g. gallic acid) and hydroxycinnamic acid (e.g. caffeic acid) (Figure 1.5). Concentrations typically decrease during the course of ripening (Seeram, 2006).

Other phenolic compounds

Other phenolic compounds, such as tannins, lignans, stilbenes and coumarins, are also present in plants (Figure 1.6). Lignans are formed from two phenylpropane units (Seeram,

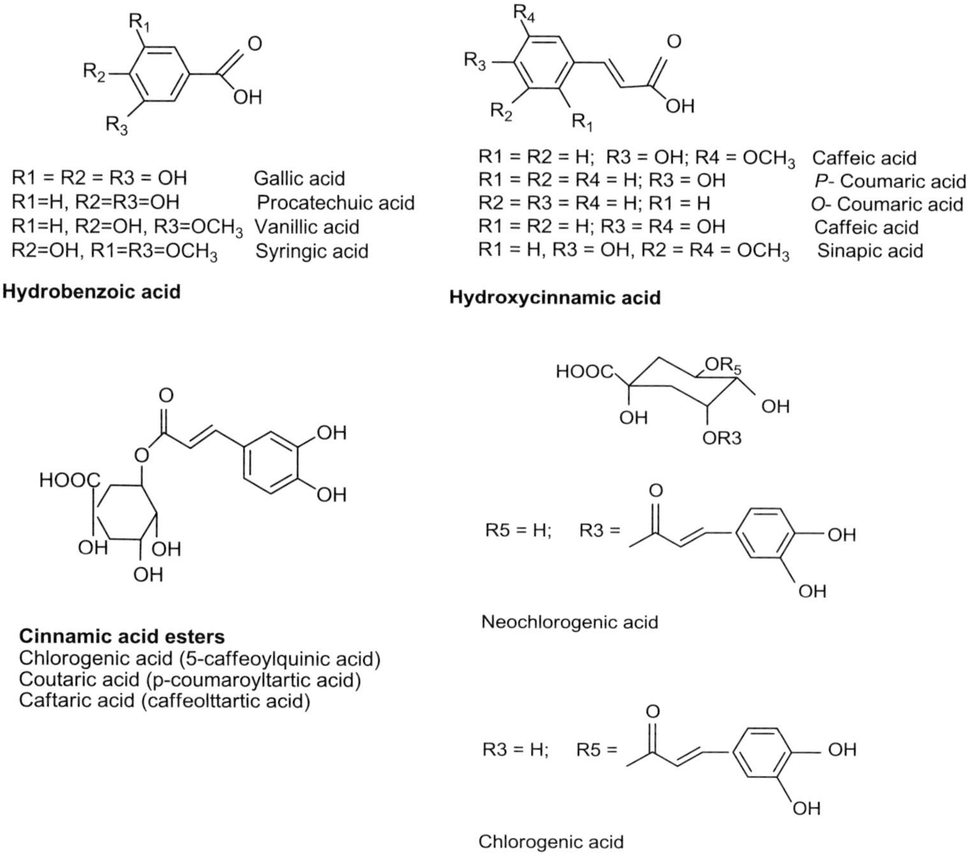

Fig. 1.5 Structures of phenolic acids.

2006; Haminiuk *et al.*, 2012). Stilbenes are phenolic based compounds such as resveratrol, which has anticancer effects and is found in small quantities in wine (Haminiuk *et al.*, 2012). Tannins, including different types of condensed tannins (e.g. proanthocyanidins, prodelphinidins etc.), and hydrolysable tannins are present in fruits. Hydrolysable tannins can be divided into gallotannins (esters of gallic acid) and ellagitannins (esters of hexahydroxydiphenic acid) (Seeram, 2006; Landete, 2011). Proanthocyanidins are dimers, oligomers and polymers of catechins that are bound together by links between C4 and C8 or C6 (Santos-Buelga and Scalbert, 2000). Proanthocyanidins are the major polypohenols in grapes and are responsible for the astringent character of fruits (El Gharras, 2009). Blueberries and cranberries contain high levels of proanthocyanidins.

Dietary fibre (DF)

Dietary fibre has been consumed for centuries and now has been recognized as having health benefits. The definition of dietary fibre was approved by the Codex Alimentarius Commission (ALINORM 09/32/A) at the FAO, Rome, Italy, on 29 July 2009, proposing that dietary fibre means carbohydrate polymers with 10 or more monomeric units which are neither digested nor absorbed in the small intestine (Cui *et al.*, 2011). Dietary fibres possess properties such as

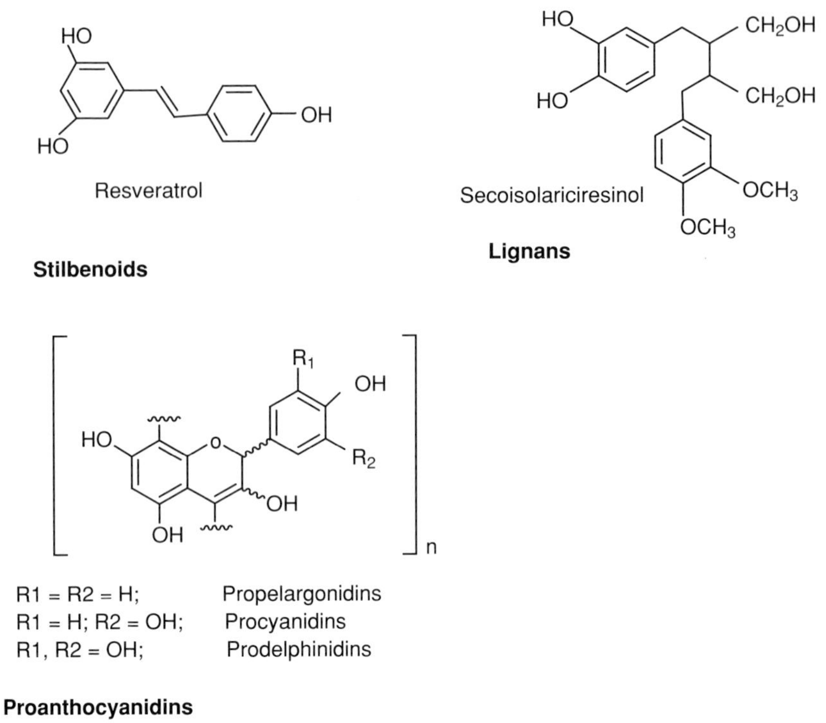

Fig. 1.6 Other phenolic compounds.

decreasing intestinal transit time and increasing stool bulk fermentable by colonic microflora, reducing blood total and LDL cholesterol levels and reducing postprandial blood glucose and insulin level, buffering the effect of excess acid in the stomach and also helping prevent constipation (Erkkila *et al.*, 1999; Brennan and Cleary, 2007; Filipovic *et al.*, 2007; Lunn and Buttriss, 2007). Low dietary fibre intake has been associated with a variety of diseases, such as diverticular disease, constipation, appendicitis, diabetes, obesity, coronary heart disease and bowel cancer (Tungland and Meyer, 2002; Slavin, 2003; Viuda-Martos *et al.*, 2010). A daily intake of approximately 25–38 g is encouraged to promote health benefits associated with fibre (King *et al.*, 2012).

Cell wall materials from fruits are good sources of dietary fibres. The major monosaccharides that form polysaccharides in the cell wall are rhamnose, fucose, arabinose, xylose, mannose, galactose, glucose, galacturonic acid and glucuronic acid (Harris and Smith, 2006). The main polysaccharides of primary cell walls are pectic, hemicellulosic (such as xyloglucans, glucomannans or galactoglucomannans and xylans) and cellulosic in varying proportions (Stevenson *et al.*, 1988; Carpita and McCann., 2000; Ridley *et al.*, 2001; Caffall and Mohnen, 2009). Different DFs from fruits such as apple, citrus as well as inulin and gums are incorporated into foods for their nutritional properties (Chau and Huang, 2004; O'Shea *et al.*, 2012).

Proteins and peptides

Bioactive peptides are protein fragments which have a positive impact on the function and condition of living beings (Pihlanto and Korhonen, 2003). Although the main role of proteins and peptides is to supply nitrogen and essential amino acids, in some instances they provide additional beneficial effects, such as antimicrobial, antioxidant, antithrombotic, antihypertensive and immunomodulatory activities (Perez Espitia *et al.*, 2012). Peptides with antimicrobial properties are produced by almost all species of life such as microorganisms, plants, animals and humans. Lunasin (found in soya bean), protease inhibitors, lupin conglutin-γ, resistant proteins, protein hydrolysates and peptides are some of the bioactives from plants having beneficial effects (Aluko, 2011).

Carotenoids

Carotenoids are lipid soluble compounds that are responsible for the yellow, orange, red and violet colours of various kinds of fruits (Von Elbe and Schwartz, 1996). Carotenoid bioactives are classified into carotenoid hydrocarbons (carotenes) and carotenoid alcohols (xanthophylls) (Fraser and Bramley, 2004). Xanthophylls consist of different derivatives, such as hydroxyl, epoxy, aldehyde and keto groups (Von Elbe and Schwartz, 1996). The carotenoid structural backbone consists of isoprene units linked together symmetrically by covalent bonds (either head-to-tail or tail-to-head) (Figure 1.7). Some structures may contain cyclic end groups (e.g. β-carotene), some contain either one or no cyclization (e.g. lycopene in tomatoes, papaya, watermelon, pink grape fruit and apricots), some contain shorter hydrocarbon skeletons (known as apocarotenal). The provitamin-A activity of some carotenoid bioactives helps in preventing chronic diseases such as cardiovascular disease and skin cancer (Mercke Odeberg *et al.*, 2003; Coyne *et al.*, 2005; Ramadan, 2011).

Functional lipids

Plant sterols (phytosterols) can occur either in the free form or esterified to fatty acids, sugar moieties or phenolic acids. They stabilize phospholipid bilayers in cell membranes (Kritchevsky and Chen, 2005). Though seeds are rich in phytosterols, fruits (e.g. berries, apricots) contain minor quantities (Kritchevsky and Chen, 2005; Seeram, 2006). Phytosterols are characterized by the number and location of double bonds and methylation at the C4

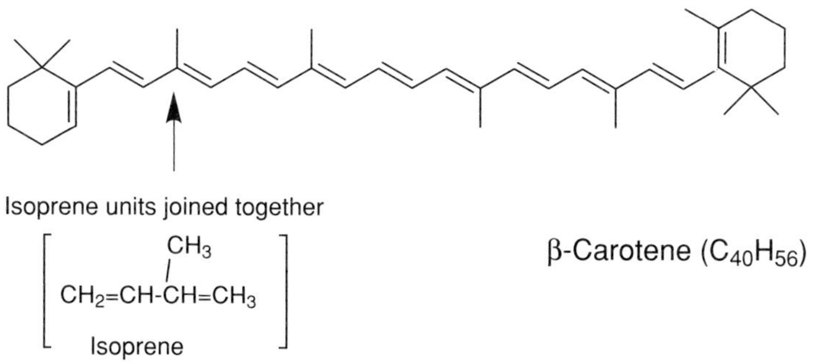

Fig. 1.7 Structure of β-carotene.

position on the ring system and alkylation and double bonds on the side chain. The common sterols found in fruits are sitosterol, campersterols, stigmasterol and avenasterols (Figure 1.8) (Tiffany and Luke, 2007; Plumb *et al.*, 2011; Sanclemente *et al.*, 2011). Phytosterols possess anticancer activity and help to reduce low-density cholesterol (Wolfreys and Hepburn, 2002; Kritchevsky and Chen, 2005; Seeram, 2006).

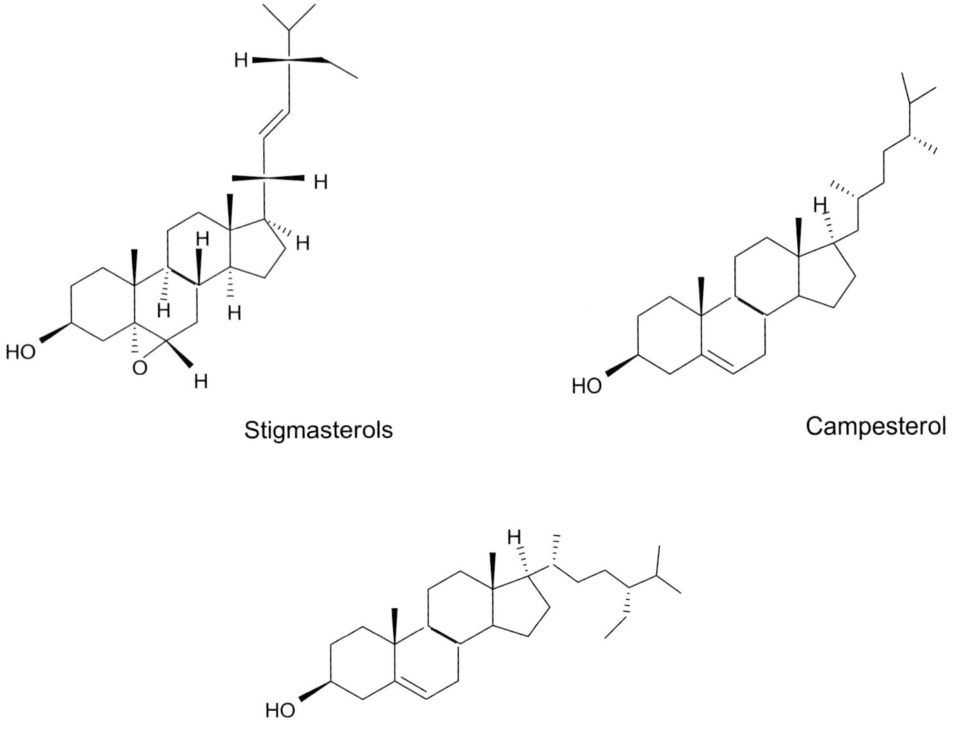

Sitosterol

Fig. 1.8 Structure of some common phytosterols.

BIOACTIVES IN FRUITS

Berries such as cranberries, blackberries, raspberries, black raspberries, blueberries, whortleberry, rowanberry, choke berry, bilberry and strawberries are consumed in fresh and processed forms such as beverages, yoghurts, jellies and jams (Seeram, 2008). Berries provide significant health benefits because of their high antioxidant, vitamin, mineral and fibre content (Kähkönen *et al.*, 2001; Ramadan, 2011). Laboratory and animal studies have shown that berries have anticancer properties due to the high content of bioactive phytochemicals (Tiffany and Luke, 2007; Seeram, 2008). Phenolics are the predominant phytochemicals present in berries (Tokuşoğlu and Stoner, 2011, Seeram *et al.*, 2006; Talcott, 2007; Wang, 2007). The bioactive phytochemicals are flavanoids (anthocyanins, flavonols and flavanols), condensed tannins, hydrolysed tannins, stilbenoids, phenolic acids and ligans (Seeram and

Nair, 2002; Manach *et al.*, 2004). The predominant phenolic acids in berries are hydroxybenzoic acids and hydroxyl cinnamic acids. Strawberries, raspberries and blackberries are high in ellagitannins. Blueberries and cranberries contain high levels of proanthocyanidins. Catechin and epicatechin are the main flavanols found in berry fruits (Seeram, 2006). Lignans are found in strawberry, blackberry, raspberry, cloudberry, cranberry, lingonberry and blueberry (Seeram, 2006). Although phenolic compounds are predominant in berries, nonpolar compounds such as ursolic acid, triterpene hydroxycinnamates and β-sitosterol are also reported to be found in berries (Murphy *et al.*, 2003).

Apricots are a rich source of bioactives, mainly polyphenols, carotenoids and vitamins (Tokuşoğlu, 2011). Chlorogenic acid (5-*O*-caffeoyquinic acid) is the most dominant phenolic compound in apricots (Erdogan-Orhan and Kartal, 2011). Some common phenolic acids found in apricots are neochlorogenic acid, caffeic acid, *p*-coumaric acid and their esters. Flavanols occur as glucosides and rutinosides of quercetin and kaempferol, but quercetin 3-rutinoside (rutin) is predominant (Dragovic-Uzelac *et al.*, 2005, 2007). Apricots are also a source of procyanidin B_1, procyanidin B_2, procyanidin B_3 and carotenoids (Dragovic-Uzelac *et al.*, 2007).

Plums are rich in bioactive compounds such as vitamins (A, C and E), anthocyanins and other phenols and carotenoids (Stacewicz-Sapuntzakis, 2012). Chlorogenic acid and its isomers are the major phenolic compounds in plums and prunes (Kim *et al.*, 2003). Neochlorogenic acid (3-*O*-caffeoylquinic acid) is a predominant polyphenol along with chlorogenic acid (5-*O*-caffeoylquinic acid), and cryptochlorogenic acid (4-*O*-caffeoylquinic acid) (Fang *et al.*, 2002; Kim *et al.*, 2003). Dried prunes contain higher amounts of phenolic compounds than fresh ones, since the dehydration process concentrates the constituents, although there is a partial degradation (Lombardi-Boccia *et al.*, 2003; Lombardi-Boccia, 2007). Phenolic acids, flavonols and flavans and small amounts of caffeic acids and coumaric acids are reported in plums and prunes. Cyanidin-3-glucosides and cyanidin 3-rutinosides are the major anthocyanins in plums (Chun *et al.*, 2003a, 2003b).

Apples are in rich in flavans (catechins, epicatechins), flavonols (rutin, quercetin and its derivatives), dihydrochalcones, hydrocinnamic acid derivatives and procyanidins (Lee *et al.*, 2003; McGhie *et al.*, 2005). Phlorizins (phloretin 2-β-D-glucoside) and phloretin 2-β-D-xylosyl-(1-6)-β-D-glucoside are the major hydrochalcones found in apples (Guyot *et al.*, 1998; Sanoner *et al.*, 1999).

Peaches and nectarines are rich in bioactives such as ascorbic acid (vitamin C), carotenoids (provitamin A) and phenolic compounds (Tomás-Barberan and Espín, 2001; Lavelli *et al.*, 2008; Cantín *et al.*, 2009; Tokuşoğlu and Hall 2011). The phenolic profiles in peaches and nectarines show similar trends. Peaches and nectarines are rich in chlorogenic acids and anthocyanin compounds (Gil *et al.*, 2002).

Palm dates possess antioxidant and antimutagenic activities, and are also considered as high energy fruit due to the sugar content (Tokuşoğlu, 2011). Dates contain anthocyanins, free phenolic acids (gallic acid, procatechuic acids, vanillic acid, syringic acid and ferulic acids), bound phenolic acids (gallic acid, procatechuic acid, *p*-hydrobenzoic acid, vanillic acid, caffeic aicd, syringic acid, *p*-coumaric acid, ferulic acid, *o*-coumaric acid), tannins and carotenoids (Al-Farsi *et al.*, 2005). Condensed tannins and phenolic acids are responsible for astringency in dates (Tokuşoğlu, 2011).

Pomegranates are a very rich source of anthocyanins (cyanidin 3, 5-di and 3-*O*-glucoside, delphinidin 3, 5-di and 3-*O*-glucoside, pelargonidin, 3, 5-di and 3-*O*-glucoside), ellagic acid, punicalagin isomers, flavanols (catechin, epicatechin, gallocatechin, epigallocatechin) and tannins (Gil *et al.*, 2000; Adams *et al.*, 2006; Nazck and Shahidi, 2006). Malic acid and citric acid are present in abundant amounts (Nazck and Shahidi, 2006; González-Molina *et al.*, 2009).

Citrus fruits are a rich source of flavonoids, cinnamic acid derivatives, coumarins, citric acid and vitamin C. Among the flavonoids, citrus fruits contain hesperidin and eriocitrin (flavanones), together with small amounts of flavones, minor flavonoids (quercetin and myricetin), and hydroxycinnamic acids (Manthey and Grohmann, 2001; González-Molina *et al.*, 2009). Grapefruit are rich in flavonones such as naringin and its derivatives; sweet oranges are rich in narirutin, hesperidin and hesperidin 7-rutinoside; sour oranges are rich in naringin, neohesperidin and hesperetin 7-neohesperidoside (Berhow *et al.*, 1991; Kanes *et al.*, 1993; El Gharras, 2009). A summary of phytochemical present in different fruits is given in Table 1.1.

CONCLUSION

Fruits are excellent sources of bioactives, which impart health benefits. The growing interest in bioactive substances is mainly because of their antioxidant potential and association between their consumption and the prevention of some diseases. Polyphenolic compounds are the most desirable food bioactives because of their antioxidant activity – their ability to scavenge oxygen radicals and other reactive species. These features make phenols a potentially interesting material for the development of functional foods.

REFERENCES

Adams L.S., Seeram N.P., Aggarwal B.B., Takada Y., Sand D. and Heber D. (2006) Pomegranate juice, total pomegranate ellagitannins, and punicalagin suppress inflammatory cell signaling in colon cancer cells. *Journal of Agricultural and Food Chemistry* **54**, 980–5.

Al-Farsi M., Alsalvar C., Morris A. *et al.* (2005) Comparison of antioxidant activity, anthocyanins, carotenoids, and phenolics of three native fresh and sun-dried date (Phoenix dactylifera L.) varieties grown in Oman. *Journal of Agricultural and Food Chemistry* **53**(19), 7592–9.

Aluko R.E. (2011) Plant derived bioactives. In: *Comprehensive Biotechnology*, 2nd edn (ed. Murray M.-Y.), pp. 501–15. Academic Press, Burlington.

Awika J. (2011) Effect of bioactive components on dough rheology, baking and extrusion. In: *Fruit and Cereal Bioactives* (eds Tokuşoğlu Ö. and Hall C.), pp. 337–45. CRC Press.

BeMiller J. and Huber K. (1996) Carbohydrates. In: *Food Chemistry* (ed. Fennema O.), pp. 83–154. Marcel Dekker, New York, USA.

Berhow M.A., Bennett R.D., Kanes K., Poling S.M. and Vandercook C.E. (1991) A malonic acid ester derivative of naringin in grapefruit. *Phytochemistry* **30**, 4198–200.

Bilgiçli N., Ibanoglu S. and Herken E. (2007) Effect of dietary fibre addition on the selected nutritional properties of cookies. *Journal of Food Engineering* **78**, 86–9.

Brennan C.S. and Cleary L.J. (2007) Utilisation Glucagel.5® in the [beta]-glucan enrichment of breads: A physicochemical and nutritional evaluation. *Food Research International*, **40**, 291–6.

Caffall K.H. and Mohnen D. (2009) The structure, function, and biosynthesis of plant cell wall pectic polysaccharides. *Carbohydrate Research* **344**, 1879–900.

Cantín C.M., Moreno M.a.A. and Gogorcena Y. (2009) Evaluation of the antioxidant capacity, phenolic compounds, and Vitamin C content of different peach and nectarine [*Prunus persica* (L.) Batsch] breeding progenies. *Journal of Agricultural and Food Chemistry* **57**, 4586–92.

Carpita N. and McCann M. (2000) The cell wall. In: *Biochemistry and Molecular Biology of Plants* (eds Buchanan B., Gruissem W. and Jones R.), pp. 52–97. American Society of Plant Physiologists, Rockville, Maryland, USA.

Chau C.F. and Huang Y.L. (2004) Characterization of passion fruit seed fibres – a potential fibre source. *Food Chemistry* **85**, 189–94.

Chun O.K., Kim D.-O. and Lee C.Y. (2003a) Superoxide radical scavenging activity of the major polyphenols in fresh plums. *Journal of Agricultural and Food Chemistry* **51**, 8067–72.

Chun O.K., Kim D.-O., Moon H.Y., Kang H.G. and Lee C.Y. (2003b) Contribution of individual polyphenolics to total antioxidant capacity of plums. *Journal of Agricultural and Food Chemistry* **51**, 7240–5.

Clifford M.N. (1999) Chlorogenic acids and other cinnamates – nature, occurrence and dietary burden. *Journal of the Science of Food and Agriculture* **79**, 362–72.

Clifford M.N. (2000) Chlorogenic acids and other cinnamates – nature, occurrence, dietary burden, absorption and metabolism. *Journal of the Science of Food and Agriculture* **80**, 1033–43.

Coyne T., Ibiebele T.I., Baade P.D., Dobson A., McClintock C., Dunn S., Leonard D. and Shaw J. (2005) Diabetes mellitus and serum carotenoids: findings of a population-based study in Queensland, Australia. *The American Journal of Clinical Nutrition* **82**, 685–93.

Cui S.W., Nie S. and Roberts K.T. (2011) Functional properties of dietary fiber. In: *Comprehensive Biotechnology*, 2nd edn (ed. Murray M.-Y.), pp. 517–25. Academic Press, Burlington.

Dragovic-Uzelac V., Pospišil J., Levaj B. and Delonga K. (2005) The study of phenolic profiles of raw apricots and apples and their purees by HPLC for the evaluation of apricot nectars and jams authenticity. *Food Chemistry* **91**, 373–83.

Dragovic-Uzelac V., Levaj B., Mrkic V., Bursac D. and Boras M. (2007) The content of polyphenols and carotenoids in three apricot cultivars depending on stage of maturity and geographical region. *Food Chemistry* **102**, 966–75.

El Gharras H. (2009) Polyphenols: food sources, properties and applications – a review. *International Journal of Food Science and Technology* **44**, 2512–8.

Emilio G. (2007) Detection and Isolation of bioactive natural products. In: *Bioactive Natural Products* (Colegate S.M. and Molyneux R.J.),pp. 11–76. CRC Press.

Erdogan-Orhan I. and Kartal M. (2011) Insights into research on phytochemistry and biological activities of *Prunus armeniaca* L. (apricot). *Food Research International* **44**, 1238–43.

Erkkila A., Sarkkinen E., Lehto S., Pyörälä K. and Uusitupa M. (1999) Dietary associates of serum total, LDL, and HDL cholesterol and triglycerides in patients with coronary heart disease. *Preventive Medecine* **28**, 558–65.

Fan L., Zhang S., Yu L. and Ma L. (2007) Evaluation of antioxidant property and quality of breads containing Auricularia auricula polysaccharide flour. *Food Chemistry* **101**, 1158–68.

Fang N., Yu S. and Prior R.L. (2002) LC/MS/MS characterization of phenolic constituents in dried plums. *Journal of Agricultural and Food Chemistry* **50**, 3579–85.

Filipovic N., Djuric M. and Gyura J. (2007) The effect of the type and quantity of sugar-beet fibers on bread characteristics. *Journal of Food Engineering* **78**, 1047–53.

Fraser P. and Bramley P. (2004) The biosynthesis and nutritional uses of carotenoids. *Progress in Lipid Research* **43**, 228–65.

Gil M.I., Tomás-Barberan F.A., Hess-Pierce B., Holcroft D.M. and Kader A.A. (2000) Antioxidant activity of pomegranate juice and its relationship with phenolic composition and processing. *Journal of Agricultural and Food Chemistry* **48**, 4581–9.

Gil M.I., Tomás-Barberán F.A., Hess-Pierce B. and Kader A.A. (2002) Antioxidant capacities, phenolic compounds, carotenoids, and vitamin c contents of nectarine, peach, and plum cultivars from California. *Journal of Agricultural and Food Chemistry* **50**, 4976–82.

González-Molina E., Moreno D.A. and García-Viguera C. (2009) A new drink rich in healthy bioactives combining lemon and pomegranate juices. *Food Chemistry* **115**, 1364–72.

Gordon M. (2001) The development of oxidative rancidity in foods. In: *Antioxidants in Food* (eds. Pokorny J., Yamashita M. and Gordon M.), pp. 7–21. Woodhead Publishing Limited, Florida, USA.

Guyot S., Marnet N., Laraba D., Sanoner P. and Drilleau J.F. (1998) Reversed-phase HPLC following thiolysis for quantitative estimation and characterization of the four main classes of phenolic compounds in different tissue zones of a french cider apple variety (*Malus domestica* Var. Kermerrien). *Journal of Agricultural and Food Chemistry* **46**, 1698–705.

Haard N. and Chism G. (1996) Characteristics of edible tissues. In: *Food Chemistry* (ed. b Fennema O.), pp. 943–1011. Marcel Dekker, Inc., New York, USA.

Haminiuk C.W.I., Maciel G.M., Plata-Oviedo M.S.V. and Peralta R.M. (2012) Phenolic compounds in fruits – an overview. *International Journal of Food Science and Technology* **47**, 2023–2044.

Harris P.J. and Smith B.G. (2006) Plant cell walls and cell-wall polysaccharides: structures, properties and uses in food products. *International Journal of Food Science and Technology* **41**, 129–43.

Hollman P.C.H. and Katan M.B. (1999) Dietary flavonoids: intake, health effects and bioavailability. *Food and Chemical Toxicology* **37**, 937–42.

Hollman P.C.H., van Trijp J.M.P., Buysman M.N.C.P., v.d. Gaag M.S., Mengelers M.J.B., de Vries J.H.M. and Katan M.B. (1997a) Relative bioavailability of the antioxidant flavonoid quercetin from various foods in man. *FEBS Letters* **418**, 152–6.

Hollman P.C.H., van Trijp J.M.P., Mengelers M.J.B., de Vries J.H.M. and Katan M.B. (1997b) Bioavailability of the dietary antioxidant flavonol quercetin in man. *Cancer Letters* **114**, 139–40.

Kähkönen M.P. and Heinonen M. (2003) Antioxidant activity of anthocyanins and their aglycons. *Journal of Agricultural and Food Chemistry* **51**, 628–33.

Kähkönen M.P., Hopia A.I. and Heinonen M. (2001) Berry phenolics and their antioxidant activity. *Journal of Agricultural and Food Chemistry* **49**, 4076–82.

Kanes K., Tisserat B., Berhow M. and Vandercook C. (1993) Phenolic composition of various tissues of rutaceae species. *Phytochemistry* **32**, 967–74.

Kim D.O., Chun O.K., Kim Y.J., Moon H.Y. and Lee C.Y. (2003) Quantification of polyphenolics and their antioxidant capacity in fresh plums. *Journal of Agricultural and Food Chemistry* **51**, 6509–15.

King D.E., Mainous A.G. and Lambourne C.A. (2012) Trends in dietary fiber intake in the United States, 1999–2008. *Journal of the Academy of Nutrition and Dietetics* **112**, 642–8.

Kritchevsky D. and Chen S.C. (2005) Phytosterols – health benefits and potential concerns: a review. *Nutrition Research* **25**, 413–28.

Landete J.M. (2011) Ellagitannins, ellagic acid and their derived metabolites: A review about source, metabolism, functions and health. *Food Research International* **44**, 1150–60.

Lavelli V., Pompei C. and Casadei M.A. (2008) Optimization of color and antioxidant activity of peach and nectarine puree: scale-up study from pilot to industrial plant. *Journal of Agricultural and Food Chemistry* **56**, 7091–9.

Lee K.W., Kim Y.J., Kim D.O., Lee H.J. and Lee C.Y. (2003) Major phenolics in apple and their contribution to the total antioxidant capacity. *Journal of Agricultural and Food Chemistry* **51**, 6516–20.

Lombardi-Boccia G. (2007) Rebuttal to comment on nutrients and antioxidant molecules in yellow plums (*Prunus domestica L.*) from conventional and organic productions: A comparative study. *Journal of Agricultural and Food Chemistry* **55**, 4279–80.

Lombardi-Boccia G., Lucarini M., Lanzi S., Aguzzi A. and Cappelloni M. (2003) Nutrients and antioxidant molecules in yellow plums (*Prunus domestica* L.) from conventional and organic productions: A comparative study. *Journal of Agricultural and Food Chemistry* **52**, 90–4.

Lunn J. and Buttriss J. (2007) Carbohydrates and dietary fibre. *British Nutrtion Foundation* **32**, 21–64.

Maffei M. (2003) Biochemistry, physiology and bioengineering of bioactive compounds from plants used as dietary supplements. In: *Dietary Supplements of Plant Origin* (ed. Maffei M.), pp. 105–70. CRC Press.

Manach C., Scalbert A., Morand C., Rémésy C. and Jiménez L. (2004) Polyphenols: food sources and bioavailability. *The American Journal of Clinical Nutrition* **79**, 727–47.

Manach C., Williamson G., Morand C., Scalbert A. and Rémésy C. (2005) Bioavailability and bioefficacy of polyphenols in humans. I. Review of 97 bioavailability studies. *The American Journal of Clinical Nutrition* **81**, 230S–42S.

Manthey J.A. and Grohmann K. (2001) Phenols in citrus peel byproducts. Concentrations of hydroxycinnamates and polymethoxylated flavones in citrus peel molasses. *Journal of Agricultural and Food Chemistry* **49**, 3268–73.

McGhie T.K., Hunt M. and Barnett L.E. (2005) Cultivar and growing region determine the antioxidant polyphenolic concentration and composition of apples grown in New Zealand. *Journal of Agricultural and Food Chemistry* **53**, 3065–70.

Mercke Odeberg J., Lignell Å., Pettersson A. and Höglund P. (2003) Oral bioavailability of the antioxidant astaxanthin in humans is enhanced by incorporation of lipid based formulations. *European Journal of Pharmaceutical Sciences* **19**, 299–304.

Mullen W., Graf B.A., Caldwell S.T., Hartley R.C., Duthie G.G., Edwards C.A., Lean M.E.J. and Crozier A. (2002) Determination of flavonol metabolites in plasma and tissues of rats by HPLC–radiocounting and tandem mass spectrometry following oral ingestion of [2-14C]quercetin-4′-glucoside. *Journal of Agricultural and Food Chemistry* **50**, 6902–9.

Murphy B.T., MacKinnon S.L., Yan X., Hammond G.B., Vaisberg A.J. and Neto C.C. (2003) Identification of triterpene hydroxycinnamates with *in vitro* antitumor activity from whole cranberry fruit (*Vaccinium macrocarpon*). *Journal of Agricultural and Food Chemistry* **51**, 3541–5.

Naczk M. and Shahidi F. (2006) Phenolics in cereals, fruits and vegetables: Occurrence, extraction and analysis. *Journal of Pharmaceutical and Biomedical Analysis* **41**, 1523–42.

Naczk M., Towsend M., Zadernowski, R. and Shahidi, F. (2011) Protein-binding and antioxidant potential of phenolics of mangosteen fruit (*Garcinia mangostana*). *Food Chemistry*, **128**(2), 292–8.
O'Shea N., Arendt E.K. and Gallagher E. (2012) Dietary fibre and phytochemical characteristics of fruit and vegetable by-products and their recent applications as novel ingredients in food products. *Innovative Food Science and Emerging Technologies* **15**, 1–13.
Omaye S., Bidlack W., Meskin M. and Yopham D. (2000) *Phytochemicals as Bioactive Agents*. Technomic Pub, Lancaster, PA, USA.
Perez Espitia P.J., de Fátima Ferreira Soares N., dos Reis Coimbra J.S., de Andrade N.J., Souza Cruz R. and Alves Medeiros E.A. (2012) Bioactive peptides: Synthesis, properties, and applications in the packaging and preservation of food. *Comprehensive Reviews in Food Science and Food Safety* **11**, 187–204.
Pihlanto A. and Korhonen H. (2003) Bioactive peptides and proteins. *Advances in Food and Nutrition Research* **47**, 175–276.
Plumb J.A., Rhodes M.J.C., Lampi A.M., Buchgraber M. and Kroon P.A. (2011) Phytosterols in plant foods: Exploring contents, data distribution and aggregated values using an online bioactives database. *Journal of Food Composition and Analysis* **24**, 1024–31.
Pontes P.V., Moreira R.F.A., Trugo L.C. and Maria C.A.B.D. (2002) The content of chlorogenic acids in tropical fruits. *Journal of the Science of Food and Agriculture* **82**, 1177–81.
Price J., Sanny C. and Shevlin D. (2006) Application of manual assessment of oxygen radical absorbent capacity (ORAC) for use in high throughput assay of 'total' antioxidant activity of drugs and natural products. *Journal of Pharmacological and Toxicological Methods* **54**, 56–61.
Ramadan M.F. (2011) Bioactive phytochemicals, nutritional value, and functional properties of cape gooseberry (*Physalis peruviana*): An overview. *Food Research International* **44**, 1830–6.
Ridley B.L., O'Neill M.A. and Mohnen D. (2001) Pectins: structure, biosynthesis, and oligogalacturonide-related signaling. *Phytochemistry* **57**, 929–67.
Robards K. (2003) Strategies for the determination of bioactive phenols in plants, fruit and vegetables. *Journal of Chromatography A* **1000**, 657–91.
Sanclemente T., Marques-Lopes I., Fajó-Pascual M., Cofán M., Jarauta E., Ros E., Puzo J. and García-Otín A.L. (2011) Naturally-occurring phytosterols in the usual diet influence cholesterol metabolism in healthy subjects. *Nutrition, Metabolism and Cardiovascular Diseases* **22**, 849–55.
Sanoner P., Guyot S., Marnet N., Molle D. and Drilleau J.F. (1999) Polyphenol profiles of french cider apple varieties (*Malus domestica* sp.). *Journal of Agricultural and Food Chemistry* **47**, 4847–53.
Santos-Buelga C. and Scalbert A. (2000) Proanthocyanidins and tannin-like compounds – nature, occurrence, dietary intake and effects on nutrition and health. *Journal of the Science of Food and Agriculture* **80**, 1094–117.
Seeram N.P. (2006) Berries. In: *Nutritional Oncology*, 2nd edn (eds David H., George L.B., Vay Liang W.G., Milner J. and Heber D.), pp. 615–28. Academic Press, Burlington.
Seeram N.P. (2008) Berry fruits for cancer prevention: Current status and future prospects. *Journal of Agricultural and Food Chemistry* **56**, 630–5.
Seeram N.P. and Nair M.G. (2002) Inhibition of lipid peroxidation and structure–activity-related studies of the dietary constituents anthocyanins, anthocyanidins, and catechins. *Journal of Agricultural and Food Chemistry* **50**, 5308–12.
Seeram N.P., Lee R., Scheuller H.S. and Heber D. (2006) Identification of phenolic compounds in strawberries by liquid chromatography electrospray ionization mass spectroscopy. *Food Chemistry* **97**, 1–11.
Shahidi F. (2009) Nutraceuticals and functional foods: Whole versus processed foods. *Trends in Food Science and Technology* **20**, 376–87.
Shahidi F. and Naczk M. (1995) Antioxidant properties of food phenolics. In: *Food Phenolics: Souces, Chemistry, Effects, Applications* (eds Shahidi F. and Naczk M.), pp 235–75. Technomic Publishing Company.
Shi H., Noguchi N. and Niki E. (2001) Introducing natural antioxidants. In: *Antioxdinats in food* (eds Pokorny J, Yanishlieva N and Gordon M), pp. 147–58. Woodhead Publishing Limited, Florida, USA.
Sies H. (2010) Polyphenols and health: Update and perspectives. *Archives of Biochemistry and Biophysics* **501**, 2–5.
Slavin J. (2003) Impact of the proposed definition of dietary fiber on nutrient databases. *Journal of Food Composition and Analysis* **16**, 287–91.
Stacewicz-Sapuntzakis M. (2012) Dried plums and their products: composition and health effects—an updated review. *Critical Reviews in Food Science and Nutrition*. doi: 10.1080/10408398.2011.563880.

Stevenson T.T., Darvill A.G. and Albersheim P. (1988) Structural features of the plant cell-wall polysaccharide rhamnogalacturonan-II. *Carbohydrate Research* **182**, 207–26.

Sun J., Chu Y.-F., Wu X. and Liu R.H. (2002) Antioxidant and antiproliferative activities of common fruits. *Journal of Agricultural and Food Chemistry* **50**, 7449–54.

Talcott S.T. (2007) Chemical components of berry fruits. In: *Berry Fruit* (ed. Zhao Y.), pp. 51–72. CRC Press.

Tiffany H. and Luke H. (2007) Berry fruit phytochemicals. In: *Berry Fruit* (ed. Zhao Y.), pp. 73–104. CRC Press.

Tokuşoğlu Ö. (2011) Phenolic and beneficial bioactives in drupe fruits. In: *Fruit and Cereal Bioactives* (eds Tokuşoğlu Ö. and Hall C.), pp. 83–105. CRC Press.

Tokuşoğlu Ö. and Hall, C. (2011) Introduction to bioactives in fruits and cereals. In: *Fruit and Cereal Bioactives* (eds Tokuşoğlu Ö. and Hall C.), pp. 3–8. CRC Press.

Tokuşoğlu Ö. and Stoner G. (2011) Phytochemical bioactives in berries. In: *Fruit and Cereal Bioactives* (eds Tokuşoğlu Ö. and Hall C.), pp. 143–70. CRC Press.

Tomás-Barberán F.A. and Clifford M.N. (2000) Dietary hydroxybenzoic acid derivatives – nature, occurrence and dietary burden. *Journal of the Science of Food and Agriculture* **80**, 1024–32.

Tomás-Barberán F.A. and Espín J.C. (2001) Phenolic compounds and related enzymes as determinants of quality in fruits and vegetables. *Journal of the Science of Food and Agriculture* **81**, 853–76.

Tungland B.C. and Meyer D. (2002) Nondigestible oligo- and polysaccharides (Dietary Fiber): Their physiology and role in human health and food. *Comprehensive Reviews in Food Science and Food Safety* **1**, 90–109.

Viuda-Martos M., López-Marcos M.C., Fernández-López J., Sendra E., López-Vargas J.H. and Pérez-Álvarez J.A. (2010) Role of fiber in cardiovascular diseases: A Review. *Comprehensive Reviews in Food Science and Food Safety* **9**, 240–58.

Von Elbe J. and Schwartz S. (1996) Colorants. In: *Food Chemistry* (ed. Fennema O.), pp. 655–722. Marcel Dekker, Inc., New York, USA.

Wang H., Cao G. and Prior R.L. (1996) Total antioxidant capacity of fruits. *Journal of Agricultural and Food Chemistry* **44**, 701–5.

Wang S. (2007) Antioxidant capacity and phenolic content of berry fruits as affected by genotype, preharvest conditions, maturity, and postharvest handling. In: *Berry Fruit* (ed. Zhao Y.), pp. 147–86. CRC Press.

Weston R.J. (2010) Bioactive products from fruit of the feijoa (*Feijoa sellowiana*, Myrtaceae): A review. *Food Chemistry* **121**(4), 923–926.

Wolfreys A.M. and Hepburn P.A. (2002) Safety evaluation of phytosterol esters. Part 7. Assessment of mutagenic activity of phytosterols, phytosterol esters and the cholesterol derivative, 4-cholesten-3-one. *Food and Chemical Toxicology* **40**, 461–70.

Zadernowski R., Czaplicki S. and Naczk M. (2009) Phenolic acid profiles of mangosteen fruits (*Garcinia mangostana*). *Food Chemistry* **112**, 685–9.

2 Fibre in Fruit

Bronwen G. Smith

INTRODUCTION

When we think of fruit, invariably it is the luscious fleshy forms, such as strawberries, cherries, apricots, apples and pears, that come to mind. Much of the fibre that we ingest when we consume fruit comes from the walls of the parenchyma cells in the tissues of these fleshy forms but not exclusively as, botanically, a fruit can take many forms (Grierson, 2002). Each cell in a plant is surrounded by a wall. New walls are formed at the end of cell division at a specialized region in the cytoplasm known as the phragmoplast (Hall *et al.*, 1984) with materials synthesised in the Golgi being rapidly deposited to separate the two daughter cells. When fully formed, plant cell walls contain an array of complex polysaccharides, including cellulose, pectic polysaccharides, xyloglucans, heteroxylans and galactoglucomannans that can differ in their fine structure and relative proportion among species, and undergo subtle structural changes during development, storage, and cooking. Commercial preparations of fibre used to confer textural attributes to manufactured foods, have usually been modified to give consistency of performance in a product and may be quite different to their native counterpart from which they have been extracted. Hence, it would seem prudent, when planning experiments to determine the health benefits of fibre and other food components, to be aware of these chemical differences so that outcomes can be causally related (Harris and Smith, 2006; Willats *et al.*, 2006; Jackson *et al.*, 2007; Ferguson *et al.*, 2010; Crowe *et al.*, 2012).

DIETARY FIBRE

Defining dietary fibre is not easy. The varying nature of the material has caused much debate over time, despite attempts by the *Codex Committee on Nutrition and Foods for Special Dietary Uses* to resolve the matter. A definition is important because it is the basis on which global food standards are set, labelling rules are determined and health claims are made. An 'agreed definition' (Cummings *et al.*, 2009) includes as components of dietary fibre: plant cell walls, resistant starch, other non-starch polysaccharides and some non-digestible oligosaccharides; all of which are resistant to digestion and absorption in the human small intestine but usually with complete or partial fermentation in the large intestine.

Bioactives in Fruit: Health Benefits and Functional Foods, First Edition. Edited by Margot Skinner and Denise Hunter.

Fibre is commonly referred to as being soluble or insoluble. However, such descriptions may not be entirely helpful when trying to understand potential health benefits of the various forms of fibre. This is because, with the exception of crystalline cellulose, all native cell wall polysaccharides are soluble in water to some extent, depending on the wall composition and structure, and processing (Harris and Smith, 2006). It may perhaps be more useful when assessing health benefits to understand its properties in saliva and gastric fluids rather than in solutions of chemicals used for fibre estimation. However, this very notion produces its own challenge, as saliva composition is complex and governed by an individual's physiology (Engelen *et al.*, 2007). Formulating a model saliva is also open to criticism (Preetha and Banerjee, 2005; Engelen *et al.*, 2007; de Almeida *et al.*, 2008). Furthermore, understanding the way in which fresh and cooked whole cell walls and their fragments, or their extracts, break down in the digestive system involves numerous challenges (Smith, 2004; Xu *et al.*, 2008; Tydeman *et al.*, 2010a, 2010b; Padayachee *et al.*, 2012). Hence, the whole prospect of studying the effects of fruit fibre on human well-being and disease mitigation is frustrated not only by the enigmas of cell wall composition and architecture but also by genomic differences among us. There is, therefore, a need to recognize the interrelationship between diet and genome (van Ommen, 2004; Kaput, 2008; Triggs *et al.*, 2010).

FRUIT

Fruit are living entities which respire and undergo physiological changes to facilitate their biological purpose of seed development and then distribution. A fruit results from fertilization of the egg to form the seed and evolvement of the ovule and associated carpellary tissues (Esau, 1961; Rudall, 1994). The new entities formed have widely differing morphologies among the angiosperms and are classified accordingly (Raven *et al.*, 1986; Grierson, 2002). Simple fruit include fleshy types, such as apples, pears, and grapes, as well as dry types, such as peas, beans and cereal grains or maize kernels. Aggregate fruit include raspberries and blackberries whereas pineapples and figs are examples of multiple fruits, having been formed from many flowers. Some vegetables contain seeds and, botanically, are therefore also fruit, for example tomatoes, cucumbers and squash.

Encouraging people to eat fruit, so that they obtain the various nutritional benefits from all of the components and potential protection against degenerative diseases (Hertog *et al.*, 1996; Steinmetz and Potter, 1996; Temple and Gladwin, 2003; Buttriss and Stokes, 2008; Kendall *et al.*, 2009; Soerjomataram *et al.*, 2010; Parkin and Boyd, 2011; Kaczmarczyk *et al.*, 2012), requires that they are provided with good quality fruit of their liking (Campbell *et al.*, 2009). One of the important quality attributes of fruit is texture, which arises from contributions from chemical components of the cell walls, wall architecture, tissue morphology, cell contents and turgor (Harker *et al.*, 1997; Waldron *et al.*, 1997). Therefore, understanding cell wall composition of fresh fruit during their maturation and storage is important in the development of cultivars that consumers like and which can be stored to satisfy market demands of a continued supply of high quality produce (Campbell *et al.*, 2009).

PLANT TISSUES AND TYPES OF CELL WALLS

The fleshy tissues of plant foods are largely composed of parenchyma cells, which have thin primary walls that have been formed during cell division and cell expansion. Other tissues, such as the sclereids (stone cells) found in pears, sclerenchyma fibres in pineapple and mango, and xylem tracheary elements, have thick secondary walls. These walls have been deposited over the primary wall at the end of cell expansion and are usually lignified (Bacic *et al.*, 1988; Bowes, 2004; Harris and Stone, 2008). But, most importantly, primary and secondary walls have quite different chemical compositions (Bacic *et al.*, 1988; Harris and Stone, 2008; Burton *et al.*, 2010). Some fruit may also have small amounts of thick-walled collenchyma tissue in locations where extra support is required (Matas *et al.*, 2011), for example under the skin. Hence, the physical form, as well as the chemical composition of the fibre from cell walls that we ingest, can be very different, depending on the anatomy of the plant material.

The angiosperms are divided into eudicotyledons, monocotyledons and basal angiosperms. The monocotyledon group includes the commelinid and non-commelinid monocotyledons (Harris and Hartley, 1980; Chase *et al.*, 2006; Givnish *et al.*, 2010; Harris, 2005; Harris and Trethewey, 2010; APG III 2009). Eudicotyledonous fruit include apples, pears, oranges, melons, grapes, raspberries and squash. Primary cell walls of this group typically contain cellulose, pectic polysaccharides (homogalacturonan, rhamnogalacturonans I and II, galactans, arabinans and arabinogalactans) and xyloglucans, as well as smaller amounts of galactoglucomannans and heteroxylans (O'Neill and York, 2003; Harris and Smith, 2006). In contrast, primary walls of commelinid monocotyledons studied, for example pineapples (Smith and Harris, 1995) and cereal grains, contain more heteroxylans (or arabinoxylans in cereal grain endosperm walls) (Fincher and Stone, 2004; Harris and Smith, 2006). However, there are some important differences in wall composition among members of this group. This largely relates to the presence of (1→3,1→4)-β-D-glucans, in the walls of some of order Poales (Smith and Harris, 1999; Trethewey *et al.* 2005), and the small amounts of cellulose in the endosperm cell walls of some cereal grains (Bacic *et al.*, 1988). The primary cell walls of the commelinid monocotyledons also contain ferulic acid (Harris, 2005; Harris and Trethewey, 2010), where it has been found ester-linked to heteroxylans (Ishii, 1991; Smith and Harris, 2001). Ferulic acid also occurs in the cell walls of the eudicotyledonous order Caryophyllales, linked to pectic polysaccharides (Fry, 1982; Ralet *et al.*, 1994; Quéméner and Ralet, 2004; Bunzel *et al.*, 2005). The primary walls of non-commelinid monocotyledons, for example onions (Redgwell and Selvendran, 1986), have a composition similar to that of eudicotyledons.

CELL WALL POLYSACCHARIDES

Cell wall polysaccharides are mainly constructed from seven neutral monosaccharides: rhamnose, fucose, arabinose, xylose, galactose, mannose and glucose, and the acidic monosaccharides galacturonic, glucuronic and 4-*O*-methylglucuronic acid. These monosaccharides are arranged in distinct ways to form six major families of polysaccharides, including cellulose, pectic polysaccharides, xyloglucans, heteroxylans, 1→3,1→4-β-D-glucans and galactoglucomannans. Some of these are homopolysaccharides, for example cellulose and 1→3,1→4-β-D-glucans, while the others are heteropolysaccharides, for example xyloglucans. There are some other unusual monosaccharides but these only occur in low concentrations in one particular polysaccharide, RG II, (Caffall and Mohnen, 2009). Some types of walls may also contain suberin, cutin, phenolic compounds, including lignin, and minerals as well as proteins and glycoproteins (Bacic *et al.*, 1988; Harris and Stone, 2008). Water is also present in primary walls.

→4)-β-D-Glc*p*-(1→4)-β-D-Glc*p*-(1→4)-β-D-Glc*p*-(1→4)-β-D-Glc*p*-(1→

Fig. 2.1 Generalized structure of cellulose. (The structure depicted is a single glucan chain.)

Cellulose is a linear polysaccharide of (1→4)-linked β-D-glucosyl residues (Figure 2.1), which rotate 180° around the axis forming a structural repeating unit of cellobiose (Bacic *et al.*, 1988). These molecules of (1→4)-β-D-glucans are stabilized by inter- and intramolecular hydrogen bonds and van der Waals forces, resulting in the formation of cellulose microfibrils approximately 2–3 nm in cross-section (Newman *et al.*, 1994; Smith *et al.*, 1998). One exception is strawberries, which have microfibrils with cross-sectional dimensions of less than 2 nm (Koh *et al.*, 1997). The relative proportion of cellulose in walls differs and may be as low as 3% in the endosperm walls of some cereal grains (Harris and Fincher, 2009) whereas in most parenchyma cell walls it may account for 30% or more (Mohnen, 2008). Although native cellulose is insoluble, some of the commercial forms, such as carboxymethylcellulose (E 466), methylcellulose (E 461) and hydroxypropylmethylcellulose (E 464), have improved solubilities and are widely used in manufactured foods. These commercial forms are usually produced from wood pulp through a series of chemical treatments resulting in modifications to the structure (Whistler and BeMiller, 1997) and, therefore, are not chemically the same as the native cellulose microfibrils found in fruit or other plant material.

(1→3,1→4)-β-D-Glucans are composed of three or four contiguous 1→4-linked β-glucosyl residues followed by individual 1→3-linked glucosyl residues. The two linkage types vary in relative proportion with 70% 1→4-linked and 30% 1→3-linked being reported. However, this may vary with tissue and cultivar and with growing conditions (Bacic *et al.*, 1988; Stone and Clarke, 1992; Fincher and Stone, 2004; Brennan and Cleary, 2005; Trethewey *et al.*, 2005; Harris and Fincher, 2009). The presence of the 1→3-linkage puts a kink in the molecule (Laziridou and Biliaderis, 2007; Burton *et al.*, 2010), which may help confer to the polysaccharide the viscosifying properties for which is it renowned.

Xyloglucans consist of a (1→4)-β-D-glucan backbone containing regions that are substituted with xylose residues linked through C(O)6 of the glucosyl residues, some of which may be further substituted with galactosyl and fucosyl residues (Hayashi, 1989; Fry *et al.*, 1993; Vincken *et al.*, 1997) (Figure 2.2). Xyloglucans vary in composition, depending on the source (Hsieh and Harris, 2009), and have important roles in the wall (Hayashi and Kaida, 2011), including cell expansion during fruit growth (Redgwell and Fry, 1993; Schröder *et al.*, 1998; Li *et al.*, 2009; Smith and Melton, 2012).

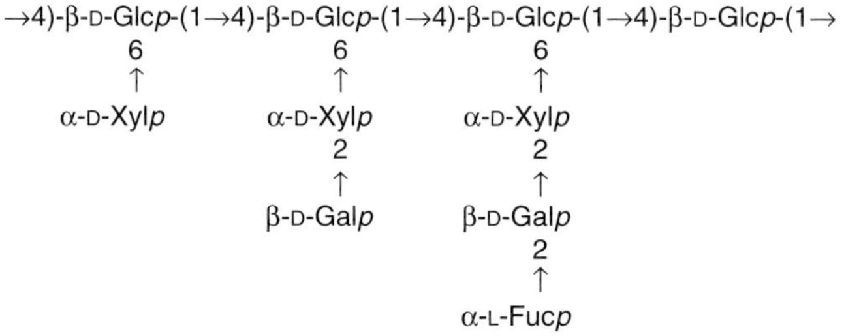

Fig. 2.2 Generalized structures of xyloglucan. (The structure depicted represents fucogalactoxyloglucans but other forms exist. Some galactose residues may be O-acetylated.)

Mannans, as galactoglucomannans in primary cell walls of fruit, are usually found in small amounts (Redgwell *et al.*, 1990; Schröder *et al.*, 2001; Melton *et al.*, 2009). These have a backbone of alternate (1→4)-β-D-mannosyl and (1→4)-β-D-glucosyl residues. Small side chains of galactosyl residues may be linked through C(O)6 of the mannosyl residues (Melton *et al.*, 2009).

Glucuronoarabinoxylans (GAXs) consist of (1→4)-β-D-xylose residues to which are linked single α-L-arabinofuranosyl and α-D-glucuronosyl (or the 4-*O*-methyl derivative) residues, linked at C(O)3 and C(O)2, respectively (Figure 2.3). In the primary cell walls of the endosperm of cereal grains this family is usually arabinoxylans (Fincher and Stone, 2004). In the primary walls of commelinid monocotyledons, the arabinose residues may be substituted with ferulic and *p*-coumaric acids (Harris, 2005). Pineapple is one example of a fruit which has ferulic acid ester linked to GAX in the unlignified cell walls (Smith and Harris, 2001).

```
→4)-β-D-Xylp-(1→4)-β-D-Xylp-(1→4)-β-D-Xylp-(1→4)-β-D-Xylp-(1→
                 3                                 2
                 ↑                                 ↑
             α-L-Araf                          α-D-GlcpA
                 5
                 ↑
            Ferulic acid
```

Fig. 2.3 Generalized structures of glucuronoarabinoxylan. (GlcA residues are much reduced and the arabinose residues more abundant in arabinoxylans from cereal grains. Arabinose residues may be feruloylated in commelinid monocotyledons.)

Pectic polysaccharides are a group of complex polysaccharides (Figure 2.4). They consist of residues of (1→4)-α-D-galacturonic acid arranged linearly but which incorporate a number of other monosaccharides, most of which are constituents of side chains. However, one monosaccharide, rhamnose, forms part of the distinct regions of the backbone (O'Neill *et al.*, 1990; Pilnik, 1990; Ridley *et al.*, 2001; Willats *et al.*, 2001a, 2001b; O'Neill and York, 2003; Willats *et al.*, 2006; Mohnen, 2008; Caffall and Mohnen, 2009; Voragen *et al.*, 2009; Ralet *et al.*, 2010). Pectic polysaccharides, are major components, representing up to 35% (Mohnen, 2008) of the primary cell walls of eudicotyledons and are distributed throughout the wall as well as in the middle lamella and junctions where cells adjoin (Liners *et al.*, 1992; Willats *et al.*, 2001a, 2001b; Guillemin *et al.*, 2005; Verhertbruggen *et al.*, 2009). The proportion of these polysaccharides in the primary walls of commelinid monocotyledons is usually much smaller, often less than 10% (Harris and Fincher, 2009). One exception is the palms, where concentrations similar to that in eudicotyledons are found (Carnachan and Harris, 2000). Three chemically distinct forms are recognised: homogalacturonan (HG), rhamnogalacturonan I (RG I) and rhamnogalacturonan II (RG II).

Homogalacturonan (HG) is a linear polymer of (1→4)-α-D-galacturonic acid residues which may be methyl esterified at C-6. It is these methyl groups which are removed by pectin methylesterase as fruit ripen (Brummell, 2006) and may, in part, give rise to changes in wall properties and subsequent softening of fruit during ripening and storage (Brummell and Harpster, 2001; Brummell, 2006). The galacturonic acid residues may also be acetylated at C-2 or C-3 (O'Neill *et al.*, 1990; O'Neill and York, 2003; Tanhatan-Nasseri *et al.*, 2011). These regions of non-methyl esterified galacturonic acid may associate with calcium forming junction zones known as the 'egg box structure' (Grant *et al.*, 1973). Hydrophobic interactions between the methyl groups on adjacent pectin molecules allow jams and jellies to set (Willats *et al.*, 2006; Smith and Melton, 2012).

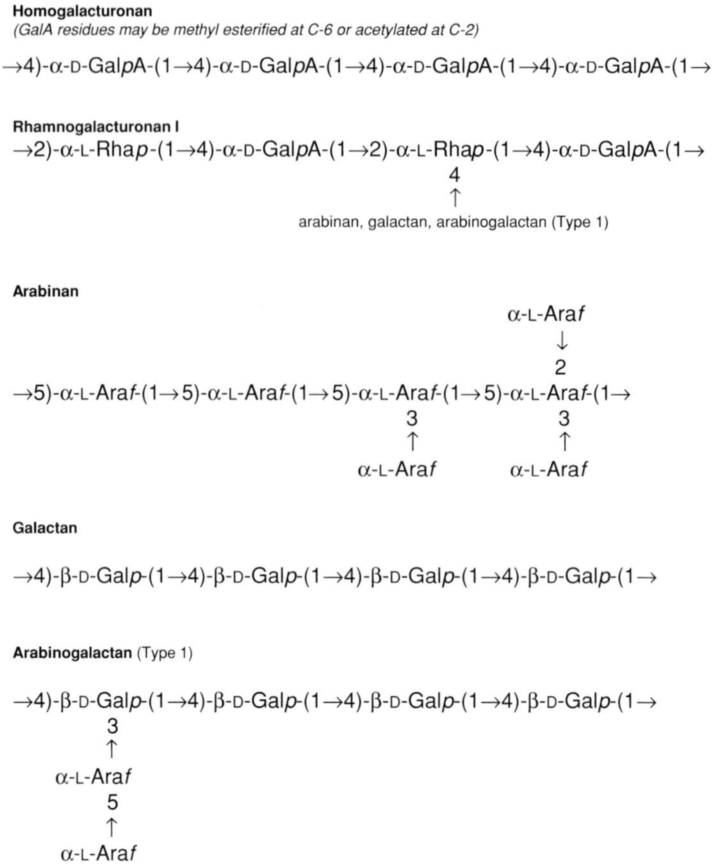

Fig. 2.4 Generalized structures of some pectic polysaccharides.

Rhamnogalacturonan (RG I) differs from HG in that the linear sequence of (1→4)-α-D-galacturonic acid residues of the backbone are interrupted by alternate residues of (1→2)-α-L-rhamnose units. These rhamnose residues may in turn be substituted at the C-4-position by side chain substituents of arabinans, galactans or arabinogalactans (Type I) (Bacic *et al.*, 1988; O'Neill *et al.*, 1990; O'Neill and York, 2003; Verhertbruggen and Knox, 2006; Mohnen, 2008; Ralet *et al.*, 2010).

Rhamnogalacturonan II is only present in small amounts in cell walls. While it shares some structural features with RG I, the arrangement of the side chains and inclusion of some unusual monosaccharides give it exceptional complexity (O'Neill *et al.*, 2004; Caffall and Mohnen, 2009).

Although these three forms of pectic polysaccharides are recognized, it must be noted that there can be variations in aspects of their structure and composition among species and in fruit as they ripen (Brummell *et al.*, 2004; Renard and Ginies, 2005; Billy *et al.*, 2008). As fruit develop from an immature state to a form we consider more palatable, there is often an accompanying loss of firmness in texture associated with changes in cell wall composition (Brummell, 2006). The side chains of RG I are subject to modification by endogenous enzymes resulting in a reduction in the number of residues. For example, in developing kiwifruit there is a loss of galactose residues (Redgwell *et al.*, 1997a, 1997b). Furthermore, the overall cell wall composition may differ among cultivars of a species, and even among individuals of a cultivar, in addition to there being changes during development and storage (de Vries *et al.*, 1982; Brummell *et al.*, 2004; Renard and Ginies, 2005; Brummell, 2006).

Pectic polysaccharides are commonly referred to as pectin. Commercial pectins, used as additives in food products, have been extracted from plant material, typically citrus peel and apple pomace. Extraction involves treatment of the material with solutions of acid and is followed by modification of the extracted polysaccharide to ensure consistency of performance in products. The great attraction of pectins is their capacity to form gels, and hence control texture in manufactured foods (Melton, 1988). Moreover, a wide range of pectins with differing degrees of methylation, acetylation and amidation is available to suit various food manufacturing applications (Pilnik, 1990; Whistler and BeMiller, 1997; Harris and Smith, 2006; Lopes da Silva and Rao, 2006; Willats *et al.*, 2006). Although choosing a suitable pectin for a particular food application might seem daunting, the task can be greatly eased by the wealth of information provided on the websites of major ingredient manufacturers. This information includes not only product specifications but also formulation and processing advice.

EFFECTS OF COOKING OR PROCESSING ON CELL WALL COMPOSITION

Some fruit, such as squash and quince, are unpalatable in the fresh state and must be cooked first. Others, for example peaches, apricots, tomatoes and apples, may find an excellent market as canned produce or are processed in some other manner to form a new product, for example sauces, purees, brined olives, pickles and dried fruit pieces. Such processed fruit often find its way into other manufactured goods, including breakfast cereals, snack bars, pasta sauces and tapenades, offering an additional source of cell wall fibre to that available as the fresh or original commodity. Many processing methods involve the use of heat both for pasteurisation and for the inactivation of enzymes, such as peroxidase, which might otherwise lower the quality of the food. Heat has the effect of changing the overall nature of the wall through the disruption of intermolecular bonds and the fine structure of the polysaccharides, in particular the pectic polysaccharides, which may be depolymerized (Jolie *et al.*, 2012; Renard and Maingonnat, 2012). These changes may be further dependent on the species and cultivar, and on the quality and age of the fruit used, and influence the overall composition of the fibre in the product in terms of quantity and solubility when compared to the unprocessed counterpart (Colin-Henrion *et al.*, 2009; Ella Missang *et al.*, 2012). In the production of some products, for example sauces and purees, the tissue may be macerated, resulting in fragmentation of the walls and, therefore, a different biological form being ingested by consumers. Additionally, there is potential for binding with components, such as anthocyanins found in vacuoles, which were previously separated from the walls by the plasma membrane and tonoplast (Padayachee *et al.*, 2012). Incidentally, such interactions may occur during mastication of fruit as well. Processing of olives also results in degradation of the cell walls as the alkali used to degrade the bitter oleuropein in the fruit also solubilizes some of the polysaccharides (Jimenez-Araujo *et al.*, 1994). This comes as no surprise to those working on cell walls, as we routinely treat walls with increasing concentrations of alkali to extract pectic polysaccharides, xyloglucans and heteroxylans for compositional analyses (Selvendran *et al.*, 1985; Fry, 1988; Smith and Harris, 1999; Quach *et al.*, 2001; Carnachan *et al.*, 2012).

HEALTH BENEFITS

Whilst we are encouraged to eat plenty of fresh fruit and vegetables, which on the whole contain large amounts of pectic polysaccharides, their complexity is such that determining the efficacy of a particular type of pectic polysaccharide is difficult. Even the use of commercial preparations of pectin is not straight forward and some intriguing results have ensued. For several years 'pectin' has been considered to have promise in limiting the progression of some cancers (Maxwell *et al.*, 2012). However, understanding its efficacy in doing so has been hindered by the sheer complexity of the polysaccharide in its native form, together with the modifications resulting from methods of extraction from cell walls and subsequent preparation for commercial applications. Although the chemical composition of the pectins or extracts used has not always been well characterized, there has been encouragement from more recent publications (Jackson *et al.*, 2007; Gunning *et al.*, 2009; Maxwell *et al.*, 2012) for this to be done.

Pectin-derived oligosaccharides have now been used in many studies, with particular interest being focussed on those that have efficacy in inhibiting the glycan-binding protein, galectin-3, which belongs to a family of highly evolutionary conserved proteins. Galectin-3 has been implicated in the advancement of some forms of cancer (Nangia-Makker *et al.*, 2002) and inflammatory diseases (Markov *et al.*, 2011). The means by which this is achieved is not completely understood, but small segments of RG I with galactan side chains have the ability to bind to the protein (Glinsky and Raz, 2009; Gunning *et al.*, 2009; Maxwell *et al.*, 2012). One modified citrus pectin, GCS 100, which appears to be a fragment of RG I with galactan side chains, can bind to and block galectin-3, and has shown success in promoting apoptosis of multiple myeloma cells (Chauhan *et al.*, 2005; Streetly *et al.*, 2010). The reader is referred to Maxwell *et al.* (2012) for an excellent review on the topic.

The use of human cancer cell lines is an attractive means of screening well characterized poly- and oligosaccharides for their ability to induce apoptosis. However, being able to determine the actual residues or structural form responsible is not always straightforward. Experiments designed to induce apoptosis in human prostate cancer cell lines using commercially prepared pectin extracts gave differing results, but a fractionated pectin powder showed promise (Jackson *et al.*, 2007). However, subsequent alkali treatment of the pectin to remove ester linkages reversed the positive outcome. Moreover, analysis of the treated material revealed an unexpected loss of arabinose side chain constituents from the preparation, which complicated the interpretation (Jackson *et al.*, 2007). These authors showed that by heat treating a commercially available citrus pectin they actually created an otherwise absent apoptotic capability. Heat, together with acid and alkali can modify galacturonic acid residues by removing the methyl ester groups on C-6. Once this has happened adjacent pectin molecules, each with extended sections of multiple negatively charged residues, can crosslink through divalent cations, for example Ca^{++} (or Mg^{++}) (Grant *et al.*, 1973; Morris *et al.*, 1982), resulting in the formation of a gel (Voragen *et al.*, 2009; Jolie *et al.*, 2012). Jackson *et al.*, (2007) postulated that some form of ester or other crosslinks may have been responsible for the observed apoptosis in the prostate cancer cell line. Hence, great caution should be observed when interpreting results, as unforeseen changes may occur during experimentation.

FUTURE CONSIDERATIONS FOR HEALTH FUNCTIONALITY

Eating fresh plant food does not necessarily mean that all of the components in the food make their way through our bodies unchanged from the native state. Cells maintain control by separation of substrates and enzymes by compartmentalization using membranes. When plant material is ruptured either by knives or teeth, components of the cell which are normally separated come into contact with each other and may react to form complexes. An example of this is the enzymic browning reaction which occurs in many fruit. When this mixing of cell contents occurs in the mouth with the addition of saliva the reactions may be even more complex. The mixing of enzymes and substrates held within the cell is one outcome but another is the interaction of polysaccharides from the cell wall with endogenous cell compounds, for example polyphenolics (Renard *et al.*, 2001; Palafox-Carlos *et al.*, 2011; Padayachee *et al.*, 2012). These aspects are worthy of greater investigation. An allied matter is the effect cell walls might have on the bioactivity of other cellular or food components. Sun-Waterhouse *et al.* (2008) showed that separately incubating mixtures of isolated apple cell walls or commercial pectin with either ascorbic acid or quercetin gave some protection of the antioxidant activity of ascorbic acid but not of quercetin.

The cell wall composition of fruit can vary in composition naturally because of biological determinants and effects of cooking and processing. Moreover, there is a difference between material ingested as whole cell walls and that which is merely an additive, for example pectin in jams and jellies. Assessing this situation is not altogether straightforward, as the commercial pectin is likely to be present along with all the other polysaccharides from the original fruit, whether remaining in the wall or extracted in the aqueous cooking matrix. Pectins are widely consumed in fresh fruit (and vegetables) and in their extracted form as additives, particularly in jams and jellies where, historically, they have been included in the formulation because of their gelling properties rather than anything to do with a perceived health outcome. Their application as food additives is becoming wider and they are now incorporated into dairy and other manufactured foods (as can be seen on websites of major ingredient manufacturers). In future, the use of an ingredient as an additive to achieve a particular structural form in a food product and adding an ingredient to give a health benefit may be overlapping aims of the manufacturer.

Plant cell walls might be considered as natural encapsulants for phytochemicals contained within the cell but to understand just how ‘protective’ the cell walls may be, their fate and that of their constitutive polysaccharides in the digestive system warrants fuller investigation. This was exemplified by Tydeman *et al.* (2010a, 2010b) who showed that a proportion of carrot cells in ingested tissue remained intact, slowing the release of the carotenoids and potentially lowered bioaccessibility and hence bioavailability. Nevertheless, we may not assume that cell walls are completely unchanged chemically during transit. Carnachan *et al.* (2012) showed that the degree of methylesterification on kiwifruit pectin was lowered, as was the molecular weight of pectins, during an *in vitro* digestion treatment. Importantly, the standard methods of measurement of dietary fibre would not have detected these subtle changes in composition (Carnachan *et al.*, 2012). Moreover, these changes may have as yet unknown physiological effects, suggesting that not only should we be aware of such changes when testing the functionality of these polysaccharides but that a more detailed chemical analysis of the fibre composition during transit is required.

The great capacity of polysaccharides to bind water from the hydroxyl groups of the monosaccharides, the ring oxygen and glycosidic linkages enables them to form gels and confer viscosity to foods (Whistler and BeMiller, 1997). Oats are particularly good sources of (1→3,1→4)-β-D-glucans and these polysaccharides generally have excellent viscosifying effects (Wood *et al.*, 1989). While some preparations of (1→3,1→4)-β-D-glucans have been shown to lower serum cholesterol and have a modulating effect on postprandial blood glucose (Braaten *et al.*, 1994; Hallfrisch *et al.*, 1995; Behall *et al.*, 1997; Wood, 2001) others have not (Keogh *et al.*, 2003) but the reasons are not clear (Harris and Smith, 2006). Moreover, different methods of extraction may result in preparations with differing degrees of success (Panahi *et al.*, 2007). The arabinoxylans present in endosperm walls are also good at forming gels (Izydorczyk and Biliaderis, 1992; Izydorczyk and Dexter, 2008).

Humans have co-evolved with their food and the fruit of certain plants have featured prominently in our various diets. When considering adding cell wall components to a food, one should consider the overall chemistry of the cell walls and the safe amounts to consume. The effects of various fibres may be different and transit time might be as important as chemical composition (Harris and Ferguson, 1993). The cell walls of fruit contain only a few different polysaccharides but in varying relative proportion and fine structure among the species and stage of development. Some of these polysaccharides can easily be solubilized in aqueous conditions and others not. Cooking changes the physical nature of the wall and the chemistry of the constituent polysaccharides, resulting in the increased solubility of some polysaccharides (Quach *et al.*, 2001; Ella Missang *et al.*, 2012; Jolie *et al.*, 2012; Renard and Maingonnat, 2012). It is possible that new interactions form among the solubilized material and with intercellular components, such as proteins and polyphenols, released following disruption of the plasma membrane and tonoplast. Such interactions may also occur as food is chewed. Hence, the fate of all components is worthy of increased attention in order to understand health benefits of whole cell walls, constitutive cell wall polysaccharides, which may leach out during transit, and that of modified polysaccharides used in manufactured foods. Most of us want to know the extent to which the polysaccharides in cell walls might act to protect us against life-threatening diseases, especially cancer. Since we intake these complex carbohydrates in a number of forms (as whole cell walls or as food additives) we must carefully tease out causality. We should not assume that all forms of fibre from cell walls suit all folk. Nutrigenomics, which examines the interaction of food components on the genome, may well help resolve the ambiguities found in earlier studies. Moreover, careful analysis of the cell walls and extracts during experimentation, whether with humans, models systems or cell lines, should be done to ensure all confounding variables, for example chemicals in reagents, and other food components are accounted for.

REFERENCES

APG III (2009) An update of the Angiosperm Phylogeny Group classification for the orders and families of flowering plants: APG III. *Botanical Journal of the Linnean Society* **161**, 105–121.

Bacic, A., Harris, P.J. and Stone, B.A. (1988) Structure and function of plant cell walls. In: *The Biochemistry of Plants* (eds P.K. Stumpf and E.E. Conn), pp. 297–371, Academic Press, New York.

Behall, K.M., Schofield, D.J. and Hallfrisch, J. (1997) Effect of beta glucan level in oat fiber extracts on blood lipids in men and women. *Journal of the American College of Nutrition* **16**, 46–51.

Billy, L., Mehinagic, E., Royer, G. *et al.* (2008) Relationship between texture and pectin composition of two apple cultivars during storage. *Postharvest Biology and Technology* **47**, 315–324.

Bowes, B.G. (2004) *A Colour Atlas of Plant Structure*. Manson Publishing Ltd, London.

Braaten, J.T., Wood, P.J., Scott, F.W. *et al.* (1994) Oat beta-glucan reduces blood cholesterol concentration in hypercholesterolemic subjects. *European Journal of Clinical Nutrition* **48**, 465–474.

Brennan, C.S. and Cleary, L.J. (2005) The potential use of 1→3,1→4-β-D-glucans as functional ingredients. *Journal of Cereal Science* **4**, 1–13.

Brummell, D.A. (2006) Cell wall disassembly in ripening fruit. *Functional Plant Biology* **33**, 103–119.

Brummell, D.A. and Harpster, M.H. (2001) Cell wall metabolism in fruit softening and quality and its manipulation in transgenic plants. *Plant Molecular Biology* **47**, 311–340.

Brummell, D.A. Dal Cin, V., Crisoto, C.H. and Labavitch, J.M. (2004) Cell wall metabolism during maturation, ripening and senescence of peach fruit. *Journal of Experimental Botany* **55**, 2029–2039.

Bunzel, M., Ralph, J. and Steinhart, H. (2005). Association of non-starch polysaccharides and ferulic acid in grain amaranth (*Amaranthus caudatus* L.) dietary fiber. *Molecular Nutrition and Food Research* **49**, 551–559.

Burton, R.A., Gidley, M.J. and Fincher, G.B. (2010) Heterogeneity in the chemistry, structure and function of plant cell walls. *Nature Chemical Biology* **6**, 724–732.

Buttriss, J.L. and Stokes, C.S. (2008) Dietary fibre and health: an overview. *Nutrition Bulletin* **33**, 186–200.

Caffall, K.H. and Mohnen, D. (2009) The structure, function, and biosynthesis of plant cell wall pectic polysaccharides. *Carbohydrate Research* **344**, 1879–1900.

Campbell, R.L., Smith, B.G., Jaeger, S.R. and Harker, F.R. (2009) Deterioration and disposal of fruit in the home: Consumer interviews and fruit quality assessments. *Journal of the Science of Food and Agriculture* **89**, 24–32.

Carnachan, S.M. and Harris, P.J. (2000) Polysaccharide compositions of primary cell walls of the palms *Phoenix canariensis* and *Rhopalostylis sapida*. *Plant Physiology and Biochemistry* **38**, 699–708.

Carnachan, S.M., Bootten, T.J., Mishra, S. *et al.* (2012) Effects of simulated digestion in vitro on cell walls polysaccharides from kiwifruit (*Actinidia* spp.). *Food Chemistry* **133**, 132–139.

Chase, M.W., Fay, M.F. Devey, D.S. *et al.* (2006) Multigene analyses of monocot relationships: a summary. *Aliso* **22**, 63–75.

Chauhan, D., Li, G., Podar, K. *et al.* (2005) A novel carbohydrate-based therapeutic GCS-100 overcomes bortezomib resistance and enhances dexamethasone-induced apoptosis in multiple myeloma cells. *Cancer Research* **65**, 8350–8358.

Colin-Henrion, M., Mehinagic, E., Renard, C.M.G.C. *et al.* (2009) From apple to applesauce: Processing effects on dietary fibres and cell wall polysaccharides. *Food Chemistry* **117**, 254–260.

Crowe, F.L., Key, T.J., Appleby, P.N. *et al.* (2012) Dietary fibre intake and ischaemic heart disease mortality: the European prospective investigation into cancer and nutrition-heart disease study. *European Journal of Clinical Nutrition* **66**, 950–956.

Cummings, J.H., Mann, J.I., Nishida, C. and Vorster, H.H. (2009) Dietary fibre: an agreed definition. *The Lancet* **373**, 365–366.

de Almeida, P.D.V., Grégio, A.M.T., Machado, M.Â.N. *et al.* (2008) Saliva composition and functions: A comprehensive review. *The Journal of Contemporary Dental Practice* **9**, 1–11.

de Vries, J.A., Voragen, A.G.J., Rombouts, F.M. and Pilnik, W. (1982) Changes in the structure of apple pectic substances during ripening and storage. *Carbohydrate Polymers* **2**, 25–33.

Ella Missang, C.E., Maingonnat, J-F., Renard, C.M.G.C. and Audergon, J-M. (2012) Apricot cell wall composition: Relation with the intra-fruit texture heterogeneity and impact of cooking. *Food Chemistry* **133**, 45–54.

Engelen, L., van den Keybus, P.A.M., de Wijk, R.A. *et al.* (2007) The effect of saliva composition on texture perception of semi-solids. *Archives for Oral Biology* **52**, 518–525.

Esau, K. (1961) *Anatomy of Seed Plants*. John Wiley & Sons, Inc., New York.

Ferguson, L.R., Smith, B.G. and James, B.J. (2010) Combining nutrition, food science and engineering in developing solutions to Inflammatory bowel diseases – omega-3 polyunsaturated fatty acids as an example. *Food and Function* **1**, 60–72.

Fincher, G.B. and Stone, B.A. (2004) Chemistry of non-starch polysaccharides. In: *Encyclopedia of Grain Science* (eds C. Wrigley, H. Corke and C.E. Walker), pp. 206–223, Elsevier, Oxford.

Fry, S.C. (1982) Phenolic components of the primary cell wall: Feruloylated disaccharides of D-galactose and L-arabinose from spinach polysaccharide. *Journal of Biochemistry* **203**, 493–504.

Fry, S.C. (1988) *The Growing Plant Cell Wall: Chemical and Metabolic Analysis*. Longman Scientific and Technical, Harlow.

Fry, S.C., York, W.S., Albersheim, P. *et al.* (1993) An unambiguous nomenclature for xyloglucan-derived oligosaccharides. *Physiologia Plantarum* **89**, 1–3.

Givnish, T.J., Ames, M., McNeal, J.R. *et al.* (2010) Assembling the tree of the monocotyledons: Plastome sequence phylogeny and evolution of Poales. *Annals of the Missouri Botanical Garden* **97**, 584–616.

Glinsky, V.V. and Raz, A. (2009) Modified citrus pectin anti-metastatic properties: one bullet, multiple targets. *Carbohydrate Research* **344**, 1788–1791.

Grant, G.T., Morris, E.R., Rees, D.A. *et al.* (1973) Biological interactions between polysaccharides and divalent cations: the egg-box model. *Febs Letters* **32**, 195–198.

Grierson, W. (2002) Fruit development, maturation and ripening. In: *Handbook of Plant and Crop Physiology* (ed. M. Pessarakli), Ch7, Marcel Dekker Inc. New York.

Guillemin, F., Guillon, F., Bonnin, E. *et al.* (2005) Distribution of pectic epitopes in cell walls of the sugar beet root. *Planta* **222**, 355–371,

Gunning, A. P., Bongaerts, R. J. M. and Morris, V. J. (2009) Recognition of galactan components of pectin by galectin-3. *The Federation of American Societies for Experimental Biology Journal* **23**, 415–424.

Hall, J.L., Flowers, T.J. and Roberts, R.M. (1984) *Plant cell Structure and Metabolism*. Longman Group Limited, Harlow.

Hallfrisch, J., Schofield, D.J. and Behall, K.M. (1995) Diets containing soluble oat extracts improve glucose and insulin response of moderately hypercholesterolemic men and women. *American Journal of Clinical Nutrition* **61**, 379–384.

Harker, F., Redgwell, R., Hallett, I. *et al.* (1997) Texture of fresh fruit. *Horticultural Reviews* **20**, 121–158.

Harris, P.J. (2005) Diversity in plant cell walls. In: *Plant Diversity and Evolution: Genotypic and Phenotypic Variation in Higher Plants* (ed. R.J. Henry), pp. 201–227, CAB International Publishing, Wallingford.

Harris, P.J. and Hartley, R.D. (1980) Phenolic constituents of the cell walls of monocotyledons. *Biochemical Systematics and Ecology* **8**, 153–160.

Harris, P.J. and Ferguson, L.R. (1993) Dietary fibre: its composition and role in protection against colorectal cancer. *Mutation Research* **290**, 97–110.

Harris, P.J. and Fincher, G.B. (2009) Distribution, fine structure and function of (1→3;1→4)-β-glucans in the grasses and other taxa. In: *Chemistry, Biochemistry, and Biology of (1→3)-β-Glucans and Related Polysaccharides* (eds A. Bacic, G.B. Fincher and B.A. Stone), pp. 621–654, Academic Press, San Diego, CA.

Harris, P.J. and Smith, B.G. (2006) Plant cell walls and cell-wall polysaccharides: structures properties and uses in food products. *International Journal of Food Science and Technology* **41** (S 2), 129–143.

Harris, P.J. and Stone, B.A. (2008) Chemistry and molecular organization of plant cell walls. In: *Biomass Recalcitrance* (ed. M.E. Himmel), pp. 60–93, Blackwell, Oxford, UK.

Harris, P.J. and Trethewey, J.A.K. (2010) The distribution of ester-linked ferulic acid in the cell walls of angiosperms. *Phytochemistry Reviews* **9**, 19–33.

Hayashi, T. (1989) Xyloglucans in the primary cell wall. *Annual Review of Plant Physiology Plant and Molecular Biology* **40**, 139–168.

Hayashi, T. and Kaida, R. (2011) Functions of xyloglucans in plant cells. *Molecular Plant* **4**, 17–24.

Hertog, M.G.L., Bueno-de-Mesquita, H.B., Fehily, A.M. *et al.* (1996) Fruit and vegetable consumption and cancer mortality in the Caerphilly study. *Cancer Epidemiology, Biomarkers and Prevention* **5**, 673–677.

Hsieh, Y.S.H. and Harris, P.J. (2009) Xyloglucans of monocotyledons have diverse structures. *Molecular Plant* **2**, 943–965.

Ishii, T. (1991) Isolation and characterization of a diferuloyl arabinoxylan hexasaccharide from bamboo shoot cell-walls. *Carbohydrate Research* **219**, 15–22.

Izydorczyk, M.S. and Biliaderis, C.A. (1992) Effect of molecular size and physical properties on wheat arbinoxylans. *Journal of Agricultural and Food Chemistry* **40**, 561–568.

Izydorczyk, M.S. and Dexter, J.E. (2008) Barley β-glucans and arabinoxylans: Molecular structure, physicochemical properties, and uses in food products – a review. *Food Research International* **41**, 850–868.

Jackson, C.L., Dreaden, T.M., Theobald, L.K. *et al.* (2007) Pectin induces apoptosis in human prostate cancer cells: correlation of apoptotic function with pectin structure. *Glycobiology* **17**, 805–819.

Jimenez-Araujo, A., Labavitch, J.M. and Heredia Morenot, A. (1994) Changes in the cell wall of olive fruit during processing. *Journal of Agricultural and Food Chemistry* **42,** 1194–1199.

Jolie, R.P., Christiaens, S., De Roeck, A. *et al.* (2012) Pectin conversions under high pressure: implications for the structure-related quality characteristics of plant-based foods. *Trends in Food Science and Technology* **24**, 1–16.

Kaczmarczyk, M.M., Miller, M.J. and Freund, G.G. (2012) The health benefits of dietary fiber: Beyond the usual suspects of type 2 diabetes mellitus, cardiovascular disease and colon cancer. *Metabolism Clinical and Experimental* **61**, 1058–1066.

Kaput, J. (2008) Nutrigenomics research for personalized nutrition and medicine. *Current Opinion in Biotechnology* **19**, 110–120.

Keogh, G.F., Cooper, G.J.S., Mulvey, T.B. *et al.* (2003). Randomized controlled crossover study of the effect of a highly β-glucan-enriched barley on cardiovascular disease risk factors in mildly hypercholesterolemic men. *American Journal of Clinical Nutrition* **78**, 711–718.

Kendall, C.W.C., Esfahani, A. and Jenkins, D.J.A. (2009) The link between dietary fibre and human health. *Food Hydrocolloids* **24**, 42–48.

Koh, T.H., Melton, L.D. and Newman, R.H. (1997) Solid-state ^{13}C NMR characterization of cell walls of ripening strawberries. *Canadian Journal of Botany* **75**, 1957–1964.

Lazaridou, A. and Biliaderis, C.G. (2007) Molecular aspects of cereal β-glucan functionality: Physical properties, technological applications and physiological effects. *Journal of Cereal Science* **46**, 101–118.

Li, X., Sakura, N. and Nevins, D.J. (2009) Characterization of kiwifruit xyloglucan. *Journal of Integrated Plant Biology* **51**, 1099–1104.

Liners, F., Thibault, J-F. and Van Cutsem, P. (1992) Influence of the degree of polymerization of oligogalacturonates and of esterification pattern of pectin on their recognition by monoclonal antibodies. *Plant Physiology* **99**, 1099–1104.

Lopes da Silva, J.A. and Rao, J.M. (2006) Pectins, structure functionality and uses. In: *Food Polysaccharides and their Applications*. 2nd edn (eds A.M. Stephen, G.O. Philips and P.A. Williams), pp. 354–410, CRC Press, Boca Raton, FL.

Markov, P.A., Popov, S.V., Nikitina, I.R. *et al.* (2011) Anti-inflammatory activity of pectins and their galacturonan backbone. *Russian Journal of Biorganic Chemistry* **37**, 817–821.

Matas, A.J., Yeats, T.H., Buda, G.J. *et al.* (2011) Tissue- and cell-specific transcriptome profiling of expanding tomato fruit provides insight into metabolic regulatory specialisation and cuticle formation. *The Plant Cell* **23**, 3893–3910.

Maxwell, E.G., Belshaw, N.J., Waldron, N.J. and Morris, V.J. (2012) Pectin – An emerging new bioactive food polysaccharide. *Trends in Food Science and Technology* **24**, 64–73.

Melton, L.D. (1988) Creating texture with polysaccharides. *Food Technology in New Zealand* **23**, 15–16, 18-19.

Melton, L.D., Smith, B.G., Ibrahim, R. and Schröder, R. (2009) Mannans in primary and secondary plant cell walls. *New Zealand Journal of Forestry* **39**, 153–160.

Mohnen, D. (2008) Pectin structure and synthesis. *Current Opinion in Plant Biology* **11**, 266–277.

Morris, E.R., Powell, D.A., Gidley, M.J. and Rees, D.A. (1982) Conformations and interactions of pectins I. Polymorphism between gel and solid states of calcium polygalacturonate. *Journal of Molecular Biology* **155**, 507–516.

Nangia-Makker, P., Hogan, V., Honjo, Y. *et al.* (2002) Inhibition of human cancer cell growth and metastasis in nude mice by oral intake of modified citrus pectin. *Journal of the National Cancer Institute* **94**, 1854–1862.

Newman, R.H., Ha, M.-A. and Melton, L.D. (1994) Solid-state ^{13}C NMR investigation of molecular ordering in the cellulose of apple cell walls. *Journal of Agricultural and Food Chemistry* **42**, 1402–1406.

O'Neill, M.A. and York, W.S. (2003) The composition and structure of primary cell walls. In: *The Plant Cell Wall* (ed. J.K.C. Rose), pp. 1–54, CRC Press, Boca Raton, FL.

O'Neill, M., Albersheim, P. and Darvill, A. (1990) The pectic polysaccharides of primary cell walls. In: *Methods in Plant Biochemistry*, 2nd edn (ed. P.M. Dey), pp. 415–441, Academic Press, London.

O'Neill, M.A., Ishii, T., Albersheim, P. and Darvill, A.G. (2004) Rhamnogalacturonan II: structure and function of a borate cross-linked cell wall pectic polysaccharide. *Annual Review Plant Biology* **55**, 109–139.

Padayachee, A., Netzel, G., Netzel, M. *et al.* (2012) Binding of polyphenols to plant cell wall analogues – Part 1: Anthocyanins. *Food Chemistry* **134**, 155–161.

Palfox-Carlos, H., Ayala-Zavala, J.F. and Gonzalez-Aguilar, G.A. (2011) The role of dietary fibre in the bioaccessibility and bioavailabiliy of fruit and vegetable antioxidants. *Journal of Food Science* **76**, R6–R15.

Panahi, S., Ezatagha, A., Temelli, F. *et al.* (2007) β-Glucan from two sources of oat concentrates affect postprandial glycemia in relation to the level of viscosity. *Journal of the American College of Nutrition* **26**, 639–644.

Parkin, D.M. and Boyd, L. (2011) Cancers attributable to dietary factors in the UK in 2010 1. Low consumption of fruit and vegetables. *British Journal of Cancer* **105**, S19–S23

Pilnik, W. (1990). Pectin – a many splendored thing. In: *Gums and Stabilizers for the Food Industry* (eds G.O. Phillips, P.A. Williams and D.J. Wedlock), pp. 313–326, Oxford University Press, Oxford.

Preetha, A. and Banerjee, R. (2005) Comparison of artificial saliva substitutes. *Trends in Biomaterials and Artificial Organs* **18**, 178–186.

Quach, ML., Melton, L.D., Harris, P.J. *et al.* (2001) Cell-wall compositions of raw and cooked corms of taro (*Colocasia esculenta). Journal of the Science of Food and Agriculture* **81**, 311–318.

Quéméner, B. and Ralet, M.-C. (2004) Evidence for linkage position determination in known feruloylated mono- and disaccharides using electrospray ion trap mass spectrometry. *Journal of Mass Spectrometry* **39**, 1153–1160.

Ralet, M.-C., Thibault, J.-F., Faulds, C. B. and Williamson, G. (1994) Isolation and purification of feruloylated oligosaccharides from cell walls of sugar-beet pulp. *Carbohydrate Research* **263**, 227–241.

Ralet, M.-C., Tranquet, O., Poulain, D. *et al.* (2010) Monoclonal antibodies to rhamnogalacturonan I backbone. *Planta* **231**, 1373–1383.

Raven, P.H., Evert, R.F. and Eichhorn, S.E. (1986) *Biology of Plants*. Worth, New York.

Redgwell, R.J. and Fry, S.C. (1993) Xyloglucan endotransglycosylase activity increases during kiwifruit (*Actinidia chinensis*) ripening. Implications for fruit softening. *Plant Physiology* **103**, 1399–1406.

Redgwell, R.J. and Selvendran, R.R. (1986) Structural features of cell-wall polysaccharides of onion *Allium cepa. Carbohydrate Research* **157**, 183–199.

Redgwell, R.J., Melton, L.D. and Brasch, D.J. (1990) Cell wall changes in kiwifruit following post harvest ethylene treatment. *Phytochemistry* **29**, 399–407.

Redgwell, R.J., MacRae, E.A., Hallet, I. *et al.* (1997a) In vitro and in vivo swelling of cell walls during softening. *Planta* **203**, 162–173.

Redgwell, R.J., Fischer, M., Kendal, E. and MacRae, E.A. (1997b) Galactose loss and fruit ripening: high molecular weight arabinogalactans in the pectic polysaccharides of fruit cell walls. *Planta* **203**, 174–181.

Renard, C.M.G.C. and Ginies, C. (2005) Comparison of the cell wall composition for flesh and skin of five different plums. *Food Chemistry* **114**, 1042–1049.

Renard, C.M.G.C. and Maingonnat, J.-F. (2012) Thermal processing of fruits and juices. In: *Thermal Food Processing. New Technologies and Quality Issues*, 2nd edn (ed. D.-W. Sun), pp. 413–438, CRC Press, Boca Raton, FL.

Renard, C.M.G.C., Baron, A., Guyot, S. and Drilleau, J.-F. (2001) Interactions between apple cell walls and native apple polyphenols: Quantification and some consequences. *International Journal of Biological Macromolecules* **29**, 115–125.

Ridley, B.L., O'Neill, M.A. and Mohnen, D. (2001) Pectins: structure, biosynthesis, and oligogalacturonide-related signaling. *Phytochemistry* **57**, 929–967.

Rudall, P. (1994) *Anatomy of Flowering Plants*. Cambridge University Press, Cambridge.

Schröder, R., Atkinson, R.G., Langkämper, G. and Redgwell, R.J. (1998) Biochemical and molecular characterisation of xyloglucan endotransglycosylase from ripe kiwifruit. *Plant Physiology* **204**, 1399–1406.

Schröder, R., Nicolas, P., Vincent, S.J.F. *et al.* (2001) Purification and characterisation of a galactoglucomannan from kiwifruit (*Actinidia deliciosa*). *Carbohydrate Research* **331**, 291–306.

Selvendran, R.R., Stevens, B.J.H. and O'Neill, M.A. (1985) Developments in the isolation and analysis of cell walls from edible plants. In: *Biochemistry of Plant Cell Walls* (eds C.T. Brett and J.R. Hillman), pp. 39–78. Cambridge University Press, Cambridge.

Smith, A.C. (2004) Texture and mastication. In: *Texture in Food, Vol 2 Solid Foods* (ed. D. Kilcast), pp. 53–81, Woodhead Publishing Ltd, Abington.

Smith, B.G. and Harris, P.J. (1995) Polysaccharide composition of unlignified cell walls of pineapples [*Ananas comosus* (L.) Merr.] Fruit. *Plant Physiology* **107**, 1399–1409.

Smith, B.G. and Harris, P.J. (1999) The polysaccharide composition of Poales cell walls: Poaceae walls are not unique. *Biochemical Systematics and Ecology* **27**, 33–53.

Smith, B.G. and Harris, P.J. (2001) Ferulic acid is esterified to glucuronoarabinoxylans in pineapple cell walls. *Phytochemistry* **56**, 513–519.

Smith, B.G. and Melton, L.D. (2012) Plant cell wall polysaccharides. In: *Food Carbohydrate Chemistry* (ed. R.E. Wrolstad), pp. 135–146, John Wiley & Sons, Ltd, Chichester.

Smith, B.G., Harris, P.J., Melton, L.D. and Newman, R.H. (1998) Crystalline cellulose in hydrated primary cell walls of three monocotyledons and one dicotyledon. *Plant and Cell Physiology* **39**, 711–720.

Soerjomataram, I., Oomen, D., Lemmens, V. *et al.* (2010) Increased consumption of fruit and vegetables and future cancer incidence in selected European countries. *European Journal of Cancer* **46**, 2563–2580.

Steinmetz, K.A. and Potter, D.A. (1996) Vegetables, fruit and cancer prevention: a review. *Journal of the American Dietetics Association* **96**, 1027–1039.

Stone, B. A. and Clarke, A.E. (1992) *Chemistry and Biology of (1→3)-β-Glucans*. La Trobe University Press, La Trobe University, Melbourne, Australia.

Streetly, M.J., Maharaj, L., Joel, S. *et al.* (2010) GCS-100, a novel galectin-3 antagonist, modulates MCL-1, NOXA and cell cycle to induce myeloma cell death. *Blood* **115**, 3939–3948.

Sun-Waterhouse, D., Melton, L.D., O'Connor, C.J. *et al.* (2008) Effect of apple cell walls and their extracts on the activity of dietary antioxidants. *Journal of Agricultural and Food Chemistry* **56**, 289–295.

Tanhatan-Nasseri, A., Crepeau, M.-J., Thibault, J-F. and Ralet, M.-C. (2011) Isolation and characterization of model homogalacturonans of tailored methylesterification patterns. *Carbohydrate Polymers* **86**, 1236–1243.

Temple, N.J. and Gladwin, K.K. (2003) Fruit, vegetables and prevention of cancer: Research challenges. *Nutrition* **19**, 467–470.

Trethewey, J.A.K., Campbell, L.M. and Harris, P.J. (2005) 1→3,1→4-β-D-Glucans in the cell walls of the Poales (*sensu lato*): an immunogold labelling study using a monoclonal antibody. *American Journal of Botany* **92**, 1669–1683.

Triggs, C.M., Munday, K., Hu, R. *et al.* (2010) Dietary factors in chronic inflammation: food tolerances and intolerances of a New Zealand Caucasian Crohn's disease population. *Mutation Research* **690**, 123–138.

Tydeman, E.A., Parker, M.L., Wickham, M.S.J. *et al.* (2010a) Effect of carrot (*Daucus carota*) microstructure on carotene bioaccessibility in the upper gastrointestinal tract. 1. *In vitro* simulations of carrot digestion. *Journal of Agricultural and Food Chemistry* **58**, 9847–9854.

Tydeman, E.A., Parker, M.L., Faulks, R.M. *et al.* (2010b) Effect of carrot (*Daucus carota*) microstructure on carotene bioaccessibility in the upper gastrointestinal tract. 2. *In vivo* digestions. *Journal of Agricultural and Food Chemistry* **58**, 9855–9860.

van Ommen, B. (2004) Nutrigenomics: Exploiting systems biology in the nutrition and health arenas. *Nutrition* **20**, 4–8.

Verhertbruggen, Y. and Knox, P.J. (2006) Pectic polysaccharides and expanding cell walls. *Plant Cell Monographs* **5**, 139–158.

Verhertbruggen, Y., Marcus, S.E., Haeger, A. *et al.* (2009) An extended set of monoclonal antibodies to pectic homogalacturonan. *Carbohydrate Research* **344**, 1858–1862.

Vincken, J.-P., York, W.S. Beldman, G. and Voragen, A.G.J. (1997). Two general branching patterns of xyloglucan: XXXG and XXGG. *Plant Physiology* **114**, 9–13.

Voragen, A.G.J., Coenen, G-J., Verhoef, R.P. and Schols, H.A. (2009) Pectin, a versatile polysaccharide present in plant cell walls. *Structural Chemistry* **20**, 263–275.

Waldron, K.W., Smith, A.C., Parr, A.J. *et al.* (1997) New approaches to understanding and controlling cell separation in relation to fruit and vegetable texture. *Trends in Food Science and Technology* **8**, 213–221.

Whistler, R.L. and BeMiller, J.N. (1997) *Carbohydrate Chemistry for Food Scientists*. Egan Press, St Paul, MN.

Willats, W.G.T., Orfila, C., Linberg, G. *et al.* (2001a) Modulation of the degree and pattern of methyl-esterification of pectin homogalacturonan in plant cell walls: Implications for pectin methyl esterase action, matrix properties, and cell adhesion. *Journal of Biological Chemistry* **276**, 19404–19413.

Willats, W.G.T., McCartney, L., Mackie, W. and Knox, J.P. (2001b) Pectin: cell biology and prospects for functional analysis. *Plant Molecular Biology* **47**, 9–27.

Willats, W.G.T., Knox, J.P. and Mikkelsen, J.D. (2006) Pectin: new insights into an old polymer are starting to gel. *Trends in Food Science and Technology* **17**, 97–104.

Wood, P.J. (2001) Cereal β-glucans: structure, properties and health claims. In: *Advanced Dietary Fibre Technology* (eds B.V. McCleary and L. Prosky), pp. 315–327, Blackwell Science, Oxford.

Wood, P.J., Weisz, J., Fedec, P. and Burrows, V.D. (1989) Large-scale preparation and properties of oat fractions enriched in (1→3)(1→4)-β-glucans. *Cereal Chemistry* **66**, 97–103.

Xu, W.L., Bronlund, J.E., Potgieter, J. *et al.* (2008) Review of the human masticatory system and masticatory robotics. *Mechanism and Machine Theory* **43**, 1353–1375.

3 Bioavailability of Antioxidant Compounds from Fruits

Iris F.F. Benzie and Sissi Wachtel-Galor

INTRODUCTION

A biological antioxidant can be defined as anything that prevents or opposes the oxidation of a biological substrate. Oxidation-induced damage to lipid, DNA and other important biomolecules is referred to as oxidative stress, which is understood to be a consequence of a relative or absolute lack of antioxidant defence. Oxidative stress is believed to underlie the development of deleterious age-related changes and various chronic diseases. There is a diverse array of antioxidants that constitutes the human antioxidant defence system. Many members of the system are endogenous, but plant-based foods and beverages are rich in various antioxidants that cannot be synthesized by the human body. Therefore, the diet is an important determinant of overall antioxidant defence and, by implication, susceptibility to oxidative stress. Epidemiological studies have provided convincing evidence that frequent consumption of antioxidant-rich fruits and vegetables is associated with better health and lower disease incidence, and fruits are promoted as forming an important part of healthy diets.

Fruits can be defined as the edible, usually sweet and fleshy form of the ripened ovary of a seed-bearing plant; they include apples, citrus fruits, kiwifruit, berries, red and green capsicum, banana, avocado and tomatoes. However, in regard to what we eat, the definition can be extended to cover a diverse range of dietary items, including seeds, nuts and pods, such as walnut, cocoa, coffee, and even mushroom, which is the fruiting body of a fungus. There is a huge variety of phytochemicals in fruits, many of which have antioxidant properties. These include ascorbic acid (vitamin C), vitamin E, the carotenoid family, which includes lycopene, β-carotene, lutein and zeaxanthin, and the flavonoid 'superfamily', which encompasses polyphenolic subfamilies such as the anthocyanins and catechins. None of these antioxidants are synthesized by the human body and many are highly effective antioxidants *in vitro*. However, it is noted that deficiency of only two dietary antioxidants, vitamin C and vitamin E, has been linked to specific biological syndromes, hence their vitamin status and the recommended daily allowances or intakes set for avoiding their deficiency. None of the other dietary antioxidants have a confirmed role in human biology, even though all show *in vitro* antioxidant activity and have potential benefit based on observational studies linking high dietary intake to improved health. The way in which they act as antioxidants varies across the different types and includes electron donating reductive action, iron binding and energy absorption. These different mechanisms prevent generation of or destroy or

Bioactives in Fruit: Health Benefits and Functional Foods, First Edition. Edited by Margot Skinner and Denise Hunter.

quench potentially damaging reactive oxygen species (ROS; 'free radicals'). The mechanisms of antioxidant action have been described well elsewhere and are not discussed further here. The focus here is on the antioxidant content of different fruits, the factors that affect this, the bioaccessibility, absorption and systemic distribution ('bioavailability') of fruit antioxidants, and how these can be modulated. The possible effects of higher intake or improved bioavailability of antioxidants on human health are also discussed briefly from the perspective of human evolution and dietary requirements and biological restrictions on antioxidant bioavailability.

ANTIOXIDANTS IN FRUITS

There is a huge number of phytochemicals with antioxidant properties in plant-based foods, and the individual antioxidants and their concentrations vary widely (Benzie and Strain, 2005). The main ROS scavenging antioxidants in fruits are vitamin C (ascorbic acid), which is colourless, and the carotenoid and flavonoid families of antioxidants that provide fruits with their orange, yellow, pink, red and purple colours. The vitamin E content of fruits is generally low, but seeds and nuts have high vitamin E content. Fruits also contain antioxidant enzymes, but these protein-based catalytic antioxidants are destroyed during digestion, and are not discussed further.

Antioxidants can facilitate photosynthesis and growth of plants, and provide plant cells with protection against reactive oxygen species (ROS). If eaten, absorbed and made systemically available, this antioxidant protection may extend to human tissues, augmenting endogenous defence (Benzie, 2003). This concept is supported by many epidemiological studies that show lower risk of various diseases in those who have a plant-rich diet or who have higher levels of plant-derived antioxidants in their plasma (Harding *et al.*, 2008; Benzie and Wachtel-Galor, 2009; Oude Griep *et al.*, 2010). For example, in a prospective study of over 20 000 apparently healthy subjects in England, a 20 μmol/l increase in plasma ascorbic acid was associated with a 9% decrease in risk of heart failure (Pfister *et al.*, 2011), and those with high ascorbic acid levels in plasma were found to have significantly lower odds ratios for subsequent development of hypertension, cardiovascular disease, diabetes and for all causes of mortality (Khaw *et al.*, 2001; Boekholdt *et al.*, 2006; Harding *et al.*, 2008; Myint *et al.*, 2011). In humans, all ascorbic acid is derived from the diet; these epidemiological findings imply strongly that ascorbic acid and/or other constituent(s) of foods rich in vitamin C prolong healthy life (Szeto *et al.*, 2004; Benzie and Wachtel-Galor, 2009).

There is strong link between oxidative stress and disease (Halliwell and Gutteridge, 2007), providing a plausible rationale for the benefits of antioxidants in plant-rich diets. Fruits have high antioxidant content but there are very large numbers of individual antioxidants that together make up the 'total' antioxidant content of fruits. While some specific dietary-derived antioxidants, such as ascorbic acid, α-tocopherol, epigallocatechin gallate (EGCG) and zeaxanthin can be been measured accurately, it is not possible to measure the entire array of individual antioxidants in a food or human blood plasma. Therefore, a widely adopted approach is to measure the 'total antioxidant content' of fruits and to measure post-ingestion changes in plasma total antioxidant content as a means to assess 'bioavailability' of antioxidants. This approach has generated a very large database on the 'total' antioxidant content (or capacity, activity or power) of various foods, including fruits, nuts and seeds (Szeto *et al.*, 2002; Guo *et al.*, 2003; Pellegrini *et al.*, 2006; Carlsen *et al.*, 2010; Fu *et al.*, 2011). This approach has also been used to investigate the effects of growing conditions, genetic manipulation, food processing and storage on the antioxidant content of foods, as well as antioxidant bioavailability. The following sections present brief descriptions of the main antioxidants and the 'total' antioxidant content of fruits.

Vitamin C (ascorbic acid)

Ascorbic acid is a colourless water soluble compound that is light and heat labile (Davey *et al.*, 2000). In human biology, ascorbic acid is a cofactor in the synthesis of collagen and is an important member in the human antioxidant network (Frei *et al.*, 1989). The human genome retains the gene that codes for the enzyme required in the final biosynthetic step of ascorbic acid production, but it is highly mutated and inactive (Benzie, 2003). Consequently, plasma and tissue ascorbic acid concentrations are maintained solely by dietary intake. In fasting plasma, ascorbic acid concentrations in apparently healthy adults cover quite a wide range (12–80 μmol/l). Deficiency is associated with plasma levels <10 μmol/l (Frei *et al.*, 1989; Chung *et al.*, 2001; Myint *et al.*, 2011). While rare nowadays in developed countries, the deficiency state known as scurvy is still found (Delanghe *et al.*, 2011). Plasma ascorbic acid concentration of >56 μmol/l has been suggested to be optimal in terms of maintaining and prolonging healthy life (Gey, 1998; Chung *et al.*, 2001; Khaw *et al.*, 2001; Boekholdt *et al.*, 2006: Harding *et al.*, 2008; Myint *et al.*, 2011).

As can be seen in Table 3.1, some fruits have a very high ascorbic acid content. Sweet peppers, lychee and kiwifruit have the highest content, and orange and cranberry juices are good sources. The recommended daily intake is 75 mg for adult females and 90 mg for adult males (Food and Nutrition Board and Institute of Medicine, 2000); this can be achieved easily by eating a few strawberries, kiwifruit or oranges per day (Szeto *et al.*, 2002). However, some commonly consumed fruits (especially those eaten most often by children), for example apples, grapes and bananas, are low in ascorbic acid and contribute little to the daily requirement. The type of cultivar, growing conditions and stage of ripeness affect the ascorbic acid content of fruit (Xu *et al.*, 2003; Roselló *et al.*, 2011). Furthermore, the vitamin is destroyed by heat, light and exposure to air (Davey *et al.*, 2000), affecting the content of fruits that have been pureed, juiced or cooked.

Vitamin E

Vitamin E is the collective name given to a group of eight lipophilic compounds, α-, β-, δ- and γ-tocopherol and α-, β-, δ- and γ-tocotrienol (Traber, 2007). The most studied member of the vitamin E family is α-tocopherol. It is in this form that most of the vitamin E in plasma and human tissues is found. This is due to biodiscrimination in absorption that favours absorption and retention of α-tocopherol over the others, despite there being higher quantities of γ-tocopherol in natural sources and significant amounts of tocotrienols in some seed oils (Traber, 2007). In humans, the vitamin is located in lipid-rich structures such as cell membranes and lipoproteins, where it protects polyunsaturated fatty acids from oxidation. Fasting plasma α-tocopherol concentrations are typically in the range 20–40 μmol/l but, because the vitamin is carried within lipoproteins, the concentration is best expressed relative to plasma lipid levels, which are usually in the range of 4.0–5.5 μmol/mmol total cholesterol plus triglycerides (Benzie *et al.*, 1998). Overt deficiency of vitamin E is associated with cerebellar ataxia but is rare, while low levels are probably common and have been linked to increased risk of cardiovascular disease and dementia (Traber 2007; Traber and Manor,

Table 3.1 Ascorbic acid content of common fruits, ranked in content per 100 g (source: USDA National Nutrient Database for Standard Reference, Release 20; http://www.ars.usda.gov/nutrientdata).

Fruit	Content: mg/100 g raw, edible fruit	Content: mg/serving of the raw fruit
Red capsicum	128	190
Kiwifruit	93	70
Green capsicum	80	190
Lychee	71	100
Papaya	62	86
Lemon (no skin)	60	31
Strawberries	59	98
Orange	53	70
Pineapple	48	74
Melon (cantaloupe)	37	59
Grapefruit	33	39
Mango	28	57
Tangerine	27	22
Raspberries	26	32
Blackberries	21	30
Melon (honeydew)	18	29
Tomatoes	13	16
Grapes	11	5
Blueberries	10	14
Plum	10	6
Banana	9	13
Watermelon	8	12
Cherries	7	5
Peach	7	7
Apples (with skin)	5	6
Cranberry juice (bottled)	*42*	*70*
Orange juice (fresh)	*50*	*100*

2012). The major dietary sources of vitamin E are seeds and seed oils, nuts and leafy green vegetables. The vitamin is light sensitive and the content of vitamin E in these foods is affected by storage and roasting/cooking. Due to their high water content, most fruits are very low in vitamin E, although avocado, red capsicum, kiwifruit, blackberries and mango (Table 3.2) can make a significant contribution toward the daily requirement of 15 mg/day for adults (Food and Nutrition Board and Institute of Medicine, 2000; Traber and Manor, 2012). Still, this intake is quite difficult to obtain without the use of supplements unless nuts and seeds are included in the diet.

Carotenoids

Carotenoids are a large family of lipophilic compounds that give the red and orange colours to watermelon, tomato, papaya and persimmon (Namitha and Negi, 2010). Carotenoids play an important role in light harvesting and in protection against photodamage in plants, but no specific physiological function or deficiency state has been revealed in humans. Therefore, although there is evidence that carotenoids can contribute to maintenance of cardiovascular health, there is no recommended daily allowance for carotenoids (Namitha

Table 3.2 α-Tocopherol content of some fruits, nuts and seeds, ranked in content/100 g (source: USDA National Nutrient Database for Standard Reference, Release 24; http://www.ars.usda.gov/nutrientdata).

Item	Content: mg/100 g as would be eaten	Content: mg/typical serving
Almonds	26	7,4
Sunflower seeds (roasted, salted kernels	26	7.4
Hazelnuts/filberts	15.2	4.3
Pine nuts (dried)	9.5	2.6
Peanuts (roasted, salted)	7.0	2.0
Brazil nuts (dried, unblanched)	5.8	1.6
Avocado	2.7	2.0
Red capsicum	1.6	2.3
Kiwifruit	1.5	1.1
Blackberries	1.2	1.7
Apricot	0.9	0.9
Mango	0.9	1.9
Raspberries	0.8	1.1
Nectarine	0.8	1.1
Blueberries	0.6	0.8
Peach	0.7	0.7
Tomato	0.5	1.0
Papaya	0.3	0.4
Strawberries	0.3	0.5
Grapes	0.2	0.3
Apple (with skin)	0.2	0.3
Orange	0.2	0.2

Fruits not mentioned above have <0.2 mg/100 g edible fruit

and Negi, 2010). Typical total carotenoid intake in the modern diet is a few milligrams per day (Benzie, 2003). The richest dietary source is carrot, but some fruits are good sources of specific carotenoids. The total carotenoid concentration in fasting plasma is low (<1 μmol/l), with individual carotenoids found at nanomolar levels. There are hundreds of individual carotenoids but the best known and studied are β-carotene, lycopene, lutein and zeaxanthin. Their distribution within the total carotenoid profile of human plasma has been estimated as: lutein (20%), lycopene (20%), β-carotene (6%) and zeaxanthin (3%) (Roberts *et al.*, 2009). Carotenoids, like many antioxidants, are prone to oxidation-induced destruction and the amount in food is affected by storage, processing and cooking, although cooking also increases their bioaccessibility (Namitha and Negi, 2010).

β-carotene is orange coloured, and found in relatively high amounts in apricots, cantaloupe melon and sweet red pepper (Table 3.3). It can be tranformed *in vivo* to active vitamin A and is sometimes referred to as provitamin A (Namitha and Negi, 2010). Lycopene is red in colour and is the most abundant carotenoid in tomatoes (Ford and Erdman, 2012). Watermelon, papaya and pink grapefruit also have high content of lycopene (Table 3.3). Several metabolites of lycopene ('lycopenoids') have been identified in human plasma and tissue; it is not clear if lycopene or the lycopenoids are the main bioactives responsible for lowering risk of chronic disease. The benefits of lycopene are mainly in regard to lower risk of prostate cancer, for which there is consistent and convincing evidence, especially for lycopene obtained from cooked tomatoes, and there is evidence also that lycopene

Table 3.3 β-Carotene and lycopene content of fruits (source: USDA National Nutrient Database for Standard Reference, Release 20; http://www.ars.usda.gov/nutrientdata).

Fruit	β-Carotene content: μg/100 g raw, edible fruit	Lycopene content: μg/100 g raw, edible fruit
Melon (cantaloupe)	2020	0
Capsicum red	1624	0
Apricot	1094	0
Pink grapefruit	686	1420
Mango	640	0
Tomato	486	2573
Watermelon	310	4532
Papaya	280	1852
Peach	162	0
Blackberries	128	0
Orange	72	0
Avocado	64	0
Kiwifruit	53	0

Fruits not mentioned contain <50 μg per 100 g fruit of β-carotene and no lycopene.

benefits cardiovascular health (Namitha and Negi, 2010; Riccione *et al.*, 2012). The health benefits of β-carotene are more controversial, though possible deleterious effects have been reported only for supplementation with pure β-carotene, highlighting the difference that may result from pure antioxidant supplements, as opposed to increased intake from whole foods (Padayatty and Levine, 2008).

Lutein and its isomer zeaxanthin are yellow coloured compounds of the xanthophyll subfamily of carotenoids. They are not converted to vitamin A *in vivo*. However, they have a role in maintenance of vision and lutein is reported to be cardioprotective (Namitha and Negi, 2010; Riccioni *et al.*, 2012). Lutein and zeaxanthin accumulate in the macula, where they form the macular pigment, or macula. The macula is believed to protect the lipid-rich photosensitive cells of the retina from photodamage. Zeaxanthin is concentrated within the central area of the fovea, or 'yellow spot', while lutein accumulates more peripherally in the retina (Sommerburg *et al.*, 1998). Therefore, zeaxanthin may be more important in relation to avoiding macular degeneration. Macular thinning and degeneration leads to loss of central vision and is an increasingly common cause of blindness in elderly people. Indeed, age-related macular degeneration (AMD) is the leading cause of blindness in people aged over 65 in developed countries. Eggs and green leafy vegetable are major of lutein but persimmon, kiwifruit, apricot and oranges also contain significant amounts (Table 3.4). Zeaxanthin is found in apricot, oranges, passion fruit and papaya, but yellow peppers and a small red berry (*Fructus barbarum L.)* commonly known as wolfberry, kei tze, Gou Qi Zi and the goji berry, are the richest natural sources of zeaxanthin (Table 3.4).

Flavonoids

Flavonoids are water soluble polyphenolic compounds that show strong antioxidant activities *in vitro*. They form the largest group of phytochemicals in the human diet, with more than 5000 identified to date. The flavonoid family can be divided into six different major classes: flavonols, flavanones, flavones, isoflavones, flavanols and anthocyanidins, based on differences in the number and arrangement of their phenol rings and type of attachments

Table 3.4 Lutein and zeaxanthin content of fruits (sources: data from Cheng *et al.*, 2005; Perry *et al.*, 2009; Namitha and Negi, 2010).

Fruit	Lutein content: μg/100 g raw, edible fruit	Zeaxanthin content: μg/100 g raw, edible fruit
Wolfberry (Kei Tze; Goji berry; *Fructus barbarum L.*)	580	19400
Capsicum red/orange/yellow/green	0/208/139/173	22/18/16665/0
Kiwifruit	70–170	11
Apricot	101	31
Orange	64	50
Passion fruit	15	44
Papaya	38	21
Tomato	32	0
Melon (cantaloupe)	12	0
Peach	11	3
Persimmon	9	7
Mango	5	5
Watermelon	4	0

(Egert and Rimbach, 2011). The flavonoid content of some fruits (mainly berries) is >100 mg/100 g fruit (Table 3.5). Strawberry, blueberry, blackberry and other berries are given their red, blue and purple colours by their constituent anthocyanidins (McGhie and Walton, 2007). Fruits (and chocolate) also contain catechins, of which epigallocatechin gallate (EGCG) is regarded as the most important. Catechins are more commonly associated with tea, which is very rich in these flavan-3-ols, but they are also found in some fruits, including grapes, some berries, peach and apple (http://flavo.vtt.fi/catechins.htm). Some fruits also contain the flavonol quercetin. Citrus fruits contain mainly flavanones. Daily intake of flavonoids can exceed 1 g, but plasma concentrations are very low and are unlikely to ever be above 3 μmol/l in total due to low bioavailability and rapid excretion. Nonetheless, flavonoids have powerful antioxidant properties and significant increases in the 'total antioxidant capacity' of plasma follow ingestion of flavonoid-rich foods or beverages (Benzie and Szeto, 1999; Serafini *et al.*, 2003; Wooten-Beard and Ryan, 2011).

'Total' antioxidant content of fruits

In fruits, and other plant-based foods, there are thousands of different antioxidant constituents. It is not feasible to attempt to measure all or even most of these in foods or in post-ingestion plasma, urine or biological fluids in the study of bioavailability and bioactivity. In addition, antioxidants may work in a cooperative or synergistic way to increase their effects. This has led to the concept of measuring the 'total' antioxidant content of a food or biological fluid, defined simply as the combined effects of all reductive (electron donating or scavenging type) antioxidants in the sample. Table 3.6 shows some examples of total antioxidant content of fruits from the now very extensive database available (Szeto *et al.*, 2002; Guo *et al.*, 2003; Halvorsen *et al.*, 2006; Pellegrini *et al.*, 2006; Thaipong *et al.*, 2006; Alothman *et al.*, 2009; Carlsen *et al.*, 2010; Mia *et al.*, 2010; Fu *et al.*, 2011; Wooten-Beard and Ryan, 2011). There are several methods available for measuring total antioxidant content but the data shown were generated using the Ferric Reducing/Antioxidant Power (FRAP) assay (Benzie & Strain, 1996a). The FRAP assay has high reproducibility, is simple, sensitive, quick, shows

Table 3.5 Flavonoid content of common fruits, mg/100 g edible portion (source: USDA database for the flavonoid content of selected food, Release 3 (2011); http://www.ars.usda.gov/nutrientdata).

Fruit	Total flavonoids	Flavan-3-ols	Anthocyanins	Quercetin
Elderberries	510	0	485 (mainly cyanidin)	27
Bilberries	290	0	285 (various)	3
Blueberries	158	7	141 (various)	8
Blackberries	137	42	90 (mainly cyanidin)	3
Cranberries	99	7	86 (mainly cyanidin and peonidin)	15
Plums Black/Red	86/9	34/0	40/5 (mainly cyanidin)	12/2
Grape Red/Green	50/7	4/6	44/0 (mainly malividin in red grapes)	1/1
Cherries Sweet	51	20	29 (mainly cyanidin and peonidin)	2
Raspberries	48	6	41 (mainly cyanidin)	1
Grapefruit/ Oranges	55/44 (flavanones naringenin/ hesperetin and naringenin	0/0	0/0	0.5/0
Strawberries	34	5	28 (mainly pelargonidin)	1
Apple Red Delicious/Gala	20/12	12/7	1/3.8 (mainly cyanidin)	4/4
Peach (white)	18	16	<1	0.5
Banana	14	6	7 (mainly delphinidin)	<0.1
Apricot	12	8	0	2
Nectarine	12	11	0	0.4

Lemon, pomelo and kumquat also contain mainly flavanones. Other fruits not mentioned (except for other berries) have flavonoid content <10 mg/100 g fresh fruit

high correlation with ascorbic acid and total phenolics content of food, is less affected by matrix and solvent than other methods, and has become widely used worldwide.

FACTORS THAT AFFECT ANTIOXIDANT RESPONSE TO INGESTION OF FRUIT: 'APPARENT' AND 'REAL' BIOAVAILABILITY

Different fruits have widely different antioxidant profiles and contain very different amounts of vitamin C, flavonols, carotenoids and 'total' antioxidants. Because higher intake of antioxidant-rich foods is associated with better health, dietary choices that include those fruits with higher antioxidant content may be helpful for maintenance and promotion of health (Haleem *et al.*, 2008). However, apart from the antioxidant content of raw, fresh fruits, there are several important considerations in relation to food production and dietary choices for health promotion through enhanced antioxidant intake and status. One relates to factors, such as genetic variation, growing conditions, storage, processing and cooking, that can increase or decrease the intrinsic antioxidant content of fruits. If the antioxidant content of a particular fruit is increased by manipulating the genome or growing conditions, then the

Table 3.6 Total antioxidant content of selected fruits measured using the FRAP assay (data from Szeto and Benzie, 2002; Fu *et al.*, 2011).

Fruit	FRAP value (μmol/100 g fresh fruit
Pomegranate	2557
Guava	2380
Strawberries	1594
Cherries	1456
Lemon (no skin)	1040
Orange	942
Kiwifruit	820
Persimmon	774
Peach	767
Lychee	722
Apple Green/Red	630/420
Papaya	519
Mango	506
Grape Red/Green	416/478
Banana	420
Watermelon	402
Pear Australian/Chinese	408/146

For a very extensive database on the FRAP value of fruits and other foods please refer to Halvorsen *et al.*, 2006, and Carlsen *et al.*, 2010.

amount of antioxidants ingested and absorbed per piece of fruit could be enhanced, increasing the 'apparent bioavailability' of antioxidants from this fruit. Other considerations relate to how much of the ingested dose is available for absorption. The cellulose walls of plant cells 'lock in' antioxidants. Antioxidants are released by disruption of the cell wall, making them more bioaccessible. Presence of fat, fibre and minerals, the pH, surface area and characteristics of the food matrix affect the stability and form in which antioxidants exist within the gastrointestinal tract and their bioaccessibility. True bioavailability of an antioxidant can be defined as how much of an ingested dose is absorbed and released into the circulating plasma for tissue uptake, metabolism and excretion. This is affected by various biological factors, such as speed of gastrointestinal transit, transport or 'harvesting' factors, and post-absorption metabolism. Content, bioaccessibility, absorption, metabolism and excretion of antioxidants in fruit can all be modulated, with subsequent effects on 'apparent' or 'true' bioavailability. This is discussed in the following section.

Intrinsic antioxidant content of fruit

Several factors affect the intrinsic antioxidant content of a fruit and those that enhance it could be regarded as making the fruit more desirable, of better value, or more beneficial to health. The stage of fruit development and ripeness affect antioxidant content. During ripening a 500-fold increase in the level of lycopene was seen in tomatoes (Bai and Lindhout, 2007). In pomegranates, the highest anthocyanin content (138 mg/100 g) was observed in 100 day old fruits, but between 20 and 100 days of development there was a 70% decrease in the ascorbic acid and total phenolics content of the fruits (Kulkarni and Aradhya, 2005). Another biological factor is genetic variation, which can be natural, through cross-breeding different varieties, or through *in vitro* techniques of genetic modification. As an example, the antioxidant and polyphenolic content of mango varied up to sixfold between eight different genotypes, and when five varieties of mango from four countries were compared, one variety (Ataulfo) showed higher beta-carotene, ascorbic acid, total phenolics and total antioxidant contents than the others (Manthey and Perkins-Veazie, 2009; Ma *et al.*, 2011)). Almost fourfold differences were seen in total antioxidant content across 10 Moroccan pomegranate varieties (Legua *et al.*, 2012). Another example is of apples, which are known to be a rich source of quercetin. Red Delicious apples contained 3.9 mg/100 g of quercetin in the edible portion, whereas the Granny Smith variety was lower, at 2.5 mg/100 g. Furthermore, when the skin was removed, the quercetin content of Red Delicious was only 0.41 mg/100 g, highlighting the different distribution of antioxidants within a fruit (USDA database). A study of 14 varieties of tomato showed variable content of ascorbic acid and β-carotene, and their lycopene content ranged from <5 to >200 mg/kg of fresh weight (Roselló *et al.*, 2011). This study also showed that the same variety grown in a different location within the same country (Italy) had different antioxidant content. Seasonal variation did not appear to have much effect on lycopene or β-carotene, but the ascorbic acid content of tomatoes grown in autumn/winter was markedly different (generally lower by 50–80%, but two varieties showed higher levels) compared to those grown in spring/summer (Roselló *et al.*, 2011).

The total antioxidant content of different varieties of apple, strawberry, apricot and peach were also studied and found to be different across fruits and varieties of the same fruit (Scalzo *et al.*, 2005). A marked effect of genotype on total antioxidant content and on total phenolics was also found in 27 apricot genotypes (Hegedűs *et al.*, 2010). Therefore, antioxidant content per fruit (and thereby possible response to ingestion, or 'apparent bioavailability' of antioxidants from that fruit) can be modulated by genetic variation, whether this is natural or imposed. Genome sequencing and transgenic techniques offer exciting new opportunities to improve nutritional quality of fruit, and genetic engineering gives plants additional characteristics and can add value for cultivation, postharvest stability and nutrition (Mattoo *et al.*, 2010). Most genetic engineering to date has been done in staple foods such as rice, potatoes and corn, increasing the concentrations of a range of antioxidants such as β-carotene, lycopene, ascorbic acid and zeaxanthin. Different transgenic types of tomato show two to threefold increases in lutein, lycopene and flavonoid content (Bai and Lindhout, 2007; Mattoo *et al.*, 2010). Therefore, depending on genotype, stage of growth, season, growing area and conditions, and which parts are eaten (with or without skins, including seeds or not) two portions of what is apparently the same type of fruit can have quite different antioxidant content, affecting the apparent bioavailability of antioxidants from these fruits.

After fruit is harvested there are various factors that can affect its antioxidant content. Fruits are often eaten raw, but are usually transported long distances and stored for days, weeks or months after harvesting before ending up on a fruit plate. Fruits are also processed into juices, purees and jams, and are canned, baked, dried and frozen. The processing methods, storage time and conditions all affect antioxidant content and, potentially, post-ingestion bioresponse (Álvarez *et al.*, 2012). Processing can lower antioxidant content due to thermal degradation and oxidation of labile antioxidants but some forms of processing, such as drying, can concentrate the antioxidant content. For example, the lycopene content in dried tomatoes (1400 mg/100 g fruit) is more than tenfold that of fresh tomato (124 mg/100 g fresh fruit) (USDA database). However, the type of drying process affects antioxidant content. Air drying is reported to result in higher antioxidant content in dried fruit compared to freeze drying, and when grapes are processed into wine the phenolics and antioxidant content are greatly increased (Papageorgiou *et al.*, 2008).

Bioaccessibility and bioavailability of antioxidants in fruit

Cooking of food, especially steaming, can release intracellular antioxidants for absorption (Wachtel-Galor *et al.*, 2008; Namitha and Negi, 2010). In addition, cooking can cause chemical changes, such as isomerization of the trans to the cis form of lycopene, which enhances its absorption (Namitha and Negi, 2010). In the case of β-carotene, higher absorption occurs when cooked carrots are eaten as compared to raw carrots, and the same is likely to be true for fruits (Namitha and Negi, 2010). However, while increasing the content or bioaccessibility of antioxidants in a fruit can lead to an increase in post-ingestion response, true bioavailability of an ingested substance is usually understood to relate to its gastrointestinal absorption and release into the circulating plasma. For antioxidants this can be modulated by the speed of transit through the gastrointestinal track, the food matrix, the size or surface area of the food particles, and by the presence of co-antioxidants in the food mixture, as well biological transport factors. The presence of dietary fibre has been reported to affect bioaccessibility and bioavailability of antioxidants from fruit (Palafox-Carlos *et al.*, 2011). Nanotechnology offers potential for development of encapsulating methods that provide protection and sustained release of phytochemicals, increasing bioaccessibility and bioavailability (Siddiqui *et al.*, 2010). The overall amount absorbed from the diet is also affected by the dosing pattern (dose and frequency of ingestion), by post-absorption metabolism and colonic biotransformation of antioxidants. Some of these factors are discussed briefly here in relation to different antioxidants in fruit.

Ascorbic acid

Ascorbic acid is absorbed quickly after ingestion and plasma levels can increase three to fourfold following intake, but bioavailability is limited. Plasma levels peak at 1–2 hours post ingestion and regain baseline (fasting) levels within 8–10 hours (Davey *et al.*, 2000). Study of post-ingestion increases in plasma ascorbic acid after vitamin C intake show that the response is related to dose but that higher doses lead to progressively smaller incremental increases in plasma concentration. For example, healthy fasting subjects who ingested 0.25 g of pure vitamin C showed a mean (SEM) peak increase in plasma ascorbic acid of 33(7) μmol/l over fasting levels, while peak increases after ingestion of 0.5 g, 1 g and 2 g were, respectively, 56(3.6), 58(6.6) and 66 (7.0) μmol/l, and doses >0.5 g did not lead to a significantly greater response (Benzie and Strain, 1997). Therefore, increasing bioavailability of ascorbic acid cannot be achieved effectively by ingestion of megadoses because the absorption mechanism is saturable (Davey *et al.*, 2000). Increased frequency of intake or sustained release maybe more effective, creating repeated 'waves' of absorption. Spreading the 'five-a-day' intake over the course of the day and ingesting vitamin C along with other dietary components that stabilize it or increase its bioaccessibility may enhance post-ingestion response. Flavonoids are reported to affect bioavailability of vitamin C and a Japanese study that investigated plasma response and urinary excretion of ascorbic acid after ingestion of 50 mg of pure vitamin C in water compared to the same amount in acerola (*Malpighia emarginate*) juice reported better ascorbic acid absorption and lower excretion when vitamin C was taken in acerola juice (Uchida *et al.*, 2011). However, no difference in bioavailability of ascorbic acid was seen when vitamin C was ingested in whole orange juice compared to a fortified low calorie beverage (Carter *et al.*, 2010). Vitamin C deficient mice were reported to have up to fivefold better delivery of ascorbic acid to tissues when the vitamin was ingested in a preparation of fresh kiwifruit, compared to when given in drinking water (Vissers *et al.*, 2011).

Vitamin E

In plant-based foods containing mixed isomers, α-tocopherol is preferentially harvested and retained within lipid-rich components of the body, but absorption of even this preferred form of vitamin E is very restricted (Traber, 2007). Plasma concentrations peak at around 10 hours post ingestion, but response to ingestion is markedly affected by whether or not dietary fat is present. In a study comparing uptake of deuterium-labelled α-tocopherol taken with water, semi-skimmed milk, full fat milk, and butter, the greatest response was seen with butter and the response to ingestion with water or semi-skimmed milk was negligible (Lodge *et al.*, 2004). As noted, the vitamin E content of most fruits is very low. Avocado, and seeds and nuts contain significant amounts, but these fruits or fruit parts should be taken with other foods containing fat so as to enhance the bioavailability of their vitamin E content.

Carotenoids

Carotenoids are lipophilic and their bioavailability is low, with a reported figure of 14% from vegetables (Namitha and Negi, 2010). There is no published overall carotenoid bioavailability figure for fruits, although an *in vitro* method of determining bioaccessiblity and bioavailability of β-carotene reported similar figures for cantaloupe and honeydew melons as for carrot (Fleshman *et al.*, 2011). Carotenoid absorption within the intestine requires release from plant tissues and binding to co-ingested dietary lipid (Palafox-Carlos *et al.*, 2011). This is followed by passive absorption within mixed micelles. Soluble fibre can inhibit the absorption of dietary fat and, thereby, carotenoids. For example, the presence of pectin lowers bioavailability of β-carotene by 42% (Palafox-Carlos *et al.*, 2011). Most studies of bioavailability of β-carotene have been on carrot, as it is the richest dietary source of this carotenoid. The bioavailability of β-carotene ingested in carrot juice was 70% greater compared to when eaten in the form of raw carrots, and that from cooked carrots was 50% higher than from raw carrots (Namitha and Negi, 2010; Palafox-Carlos *et al.*, 2011). The same is likely to be true of β-carotene from fruit juice compared to whole fruit, and cooked as opposed to raw fruit, but published data on this are lacking.

Lycopene bioavailability is twice as high from cooked tomato products, such as tomato puree and tomato sauce, compared to raw tomatoes. This is because of lycopene release from disrupted cells during cooking and to heat-induced isomerization of the *trans* form to the more absorbable *cis* form of lycopene, as well as to the co-presence of dietary fat in cooked tomatoes (Namitha and Negi, 2010). Lycopene is more stable than β-carotene. Greater degrees of mechanical homogenization of canned tomatoes increased bioavailability of both these carotenoids, but additional heating also enhanced the bioavailability of lycopene (van het Hof *et al.*, 2000). Lycopene absorption is also reported to be enhanced by co-ingestion of whey protein (Richelle *et al.*, 2002).

Absorption of lutein and zeaxanthin is enhanced by plant cell disruption and the co-presence of dietary fat. Thermal treatment and protein also enhanced the bioavailability of zeaxanthin from the small red berry known as wolfberry, kei tze and goji (*Fructus barbarum L.*), as seen in a controlled intervention trial of crossover design. The response seen in the triglycerides-rich fraction of plasma from 12 healthy subjects was two to threefold higher following ingestion (along with a standard meal containing fat) of an extract of the berry prepared at 80°C in skimmed milk, compared to extracts prepared in 80°C water and 40°C skimmed milk (Benzie *et al.*, 2006).

Flavonoids

Absorption of these polyphenolic phytochemicals is very low and there is extensive conjugation and/or rapid excretion of most of the flavonoids that humans do manage to absorb (Manach *et al.*, 2005; Palafox-Carlos *et al.*, 2011). Most of the thousands of different flavonoids in plants are in the *O*-glycosidic form, but some exist naturally in their aglycone form. The lipophilic aglycones are found in the peel and buds of fruits, rather than the more commonly eaten fleshy parts of fruit that are rich in the water soluble glycosidic forms (Wollenweber and Dietz, 1981). The sugar conjugate of most of the *O*-glycosidic flavonoids is removed in the intestine by the action of hydrolyzing enzymes, producing the flavonoid aglycones, a very small amount of which is absorbed (Del Rio *et al.*, 2010a, 2010b). The best studied flavonoids found in fruits are the deeply coloured anthocyanins, found in high amounts in berries and pomegranate, the flavan-3-ols (catechins) found in apples, grapes, peaches, cherries and apricots (and chocolate, which is produced from the cocoa seed), and the flavonol quercetin, from apples and grapes. Berries and pomegranates also contain ellagitannins, which are very large polyphenolic molecules called hydrolysable tannins. There are few studies on these, although ellagitannins do contribute to the antioxidant content of fruits. The interested reader is referred to the work of Heber (2011) for further information.

Bioavailability of flavonoids is low, particularly for the anthocyanins. These large, deeply coloured and highly water soluble polyphenolic antioxidants consist of an anthocyanidin plus a sugar attachment. The main anthocyanidins are delphinidin, cyanidin, petunidin, malvidin, pelargonidin and peonidin, and the sugars can be glucose, galactose, arabinose or xylose, among others (Del Rio *et al.*, 2010a; Chu *et al.*, 2011). Bioaccessibility and absorption are affected by the food matrix and also by the nature of individual anthocyanidins and the type of sugar. Following ingestion of up to several hundred milligrams of anthocyanins from sources that included elderberry, blueberry, chokeberry, strawberry, red grape juice and red wine, peak plasma concentrations of total anthocyanins were generally less than 100 nmol/l (0.1 μmol/l). The highest amount of anthocyanins ingested (179 μmol of pelargonidin-3-glucoside ingested in 200 g of strawberries) that has been recovered from the urine of human subjects is reported to be 1.8%, though absorption from red wine has been reported to be as high as 5% (McGhie and Walton, 2007; Parades-López *et al.*, 2010). However, bioavailability of some anthocyanins appears to be very poor. For example, only 0.16% of blackberry anthocyanins (mainly delphinidin-3-rutinoside and cyanidin-3-rutinoside) was recovered in urine in one human study, and $<0.05\%$ was recovered in another (Manach *et al.*, 2005; McGhie and Walton, 2007; Parades-López *et al.*, 2010; Jin *et al.*, 2011). A few features are noted in relation to anthocyanin bioavailability. The anthocyanins (the naturally occurring anthocyanidin glycosides) are absorbed without prior deglycosylation, which is in contrast to other flavonoids, and anthocyanins are found in the circulating plasma, albeit in very small amounts, within 15 minutes of ingestion (Manach *et al.*, 2005; McGhie and Walton, 2007). Their rapid appearance in plasma indicates absorption from the stomach and jejunum.

The catechins are flavan-3-ols. This group of antioxidants includes epigallocatechin gallate (EGCG), epicatechin gallate (ECG), epigallocatechin (EGC) and epicatechin (EC). Bioavailability of catechins is higher than for anthocyanins, but is low nonetheless, and, unlike the anthocyanins, absorbed catechins undergo rapid post-absorption conjugation, especially the non-gallated catechins (Del Rio *et al.*, 2010b). Glucuronidation, sulfation and methylation of non-gallated catechins occur quickly after absorption, followed by rapid renal clearance. Consequently, the plasma concentration of free and conjugated catechins is very low, even following ingestion of a large dose, and their half-life is generally only a few hours. Gallated catechins (EGCG and ECG) are handled differently to the others, being excreted in the bile rather than the urine. The catechins, in particular EGCG, escape rapid conjugation and renal excretion, and are found in free form in circulating plasma, although conjugation of EGCG is found after several hours. Data on catechin levels in plasma after ingestion of fruit are difficult to find, most studies having monitored responses to ingestion of tea or pure EGCG, but there are some studies with chocolate and cocoa (Del Rio *et al.*, 2010b). Bioavailability of flavan-3-ols from chocolate may be as high as 30% but plasma levels of total catechins remain <120 nmol/l in the hours following ingestion. Free EGCG is the greatest contributor to this post-ingestion increase but levels are reported to remain below 100 nmol/l (Del Rio *et al.*, 2010b). Still, the presence of free EGCG and the longer half-life of EGCG in the circulating plasma indicate it is the most bioavailable of the catechins, and the one that may be responsible for the reported health benefits of berries, tea, chocolate and other foods rich in flavan-3-ols.

Quercetin is the aglycone form of flavonoid glycosides, such as quercitrin and rutin found in apple peel, pomegranate, red grapes, citrus fruits, some berries and tomatoes. The glycosides are hydrolysed, releasing the aglyone form for gastrointestinal absorption, but quercetin concentrations in plasma after ingestion are in the low nanomolar range, and there is extensive metabolism of absorbed quercetin. In a human study, the peak plasma quercetin concentration was about 10 nmol/l and occurred four hours after ingestion of tomato juice containing 176 μmol of quercetin-*O*-rutinoside. Quercetin bioavailability from tomatoes is estimated to be <5% and that from apples even lower (Del Rio *et al.*, 2010a). The peak plasma quercetin concentration in human subjects after ingestion of apple peel powder containing 100 mg of quercetin aglyone equivalents was reported to average 63.8 ng/ml (approximately 200 nmol/l), though it is noted that the same dose of quercetin in the form of onion powder gave a fourfold greater peak quercetin value in plasma (Lee and Mitchell, 2012).

Overall, the numerous bioavailability studies performed have shown that different flavonoids are handled somewhat differently in regard to absorption and post-absorption metabolism and excretion, but that bioavailability of all flavonoids is very low, and most of the dietary flavonoid load is left unabsorbed within the gastrointestinal tract. The post-ingestion increase in individual flavonoids is generally well below 100 nmol/l, though EGCG and quercetin may transiently reach somewhat higher values. The total flavonoid concentration in plasma is in the 1–3 μmol/l range, despite daily ingestion of up to several grams of total flavonoids. This compares with about 50–80 μmol/l ascorbic acid in plasma with a daily intake of >100 mg/day, and a plasma α-tocopherol level of about 30 μmol/l with a daily intake of about 10 mg. Clearly, flavonoids would appear to make a negligible contribution to the 'total antioxidant content' of plasma, which is about 1000 μmol/l (Benzie and Strain, 1996a; Chung *et al.*, 2001). Still, the strong antioxidant properties of these multihydroxylated polyphenolic compounds may cause a greater increase in the post-ingestion total antioxidant content of plasma than might be expected from the post-ingestion magnitude of changes in flavonoid concentration *per se.* Assessing changes in post-ingestion total antioxidant content of plasma is an approach that has been used quite extensively to assess the bioavailability of antioxidants in food, although it is not without its pitfalls, and results need to be interpreted with caution, as described in the next section.

'Total antioxidants'

Because of the huge number and variety of individual antioxidants and the difficulties associated with measuring very small (low nanomolar levels in most cases) post-ingestion changes in these, measuring increases in the 'total antioxidant content' of plasma is an approach that has been widely adopted. There are many studies that have assessed acute changes in plasma antioxidant content following ingestion of, for example, fruit juices, wine, teas, herbs, apples, chocolate, tomatoes, berries and the fruiting body of fungi (Benzie *et al.*, 1999; Lee *et al.*, 2000; Serafini *et al.*, 2003; Wachtel-Galor *et al.*, 2004; Fernandez-Pachon *et al.*, 2005; Garcia-Alonso *et al.*, 2006; Hukkanen *et al.*, 2006; Parades-López *et al.*, 2010; Wooten-Beard and Ryan, 2011). Different assay methods, sampling times, as well as different food types and doses have been used, but studies performed with healthy human subjects demonstrate significant increases in plasma antioxidant content (or activity, capacity or power) shortly after ingestion of fruits or fruit juices. For example, 240 g strawberries, 200 g grapes and 400 ml of berry juice were reported to increase the plasma antioxidant capacity by, respectively, 14%, 5%, and 30% (Parades-López *et al.*, 2010). The greater response with berry juice likely reflects the better bioaccessibility of antioxidants from homogenized fruits, and also the fact that some berries, especially strawberry and cranberry, are very rich in ascorbic acid, which has the highest bioavailability among all the many thousands of dietary antioxidants.

In 12 subjects given 100 g of dark chocolate, the plasma antioxidant content (as the FRAP value) had increased by an average of 18% at one hour post ingestion (Serafini *et al.*, 2003). Interestingly, this increase was not seen with milk chocolate of much lower coca content, nor was a significant response seen with 100 g of dark chocolate taken with 200 ml full fat milk. This highlights the effects of other dietary constituent in antioxidant absorption. In a small study conducted in our laboratory, changes in the total antioxidant content of plasma and urine (as the FRAP value) after ingestion of cocoa and coffee were investigated in a single-blinded study of multiple crossover design involving six healthy, consenting subjects. The results (Figure 3.1) showed that after ingestion of a typical cup of coffee, plasma antioxidant capacity increased by 2–3%, with slower and less pronounced changes after ingestion of a cup of cocoa.

However, factors other than antioxidant absorption can affect postprandial changes in plasma total antioxidant activity, and these should be considered. Uric acid, formed in the body from purine breakdown, has antioxidant activity and contributes about 60% to the measured 'total antioxidant content' of human plasma (Benzie and Strain, 1996a). Fruit has high fructose content and a bolus of fructose increases uric acid in plasma (Lotito and Frei, 2006). Metabolism of fructose by fructokinase transiently depletes hepatic ATP and this leads to an acute but short term increase in the degradation of AMP, with increased uric acid formation. As an example, there was an approximately 12% increase in plasma antioxidant content after ingestion of apples, which are rich in fructose, but most of this increase was due to uric acid (Lotito and Frei, 2006). Intake of sorbitol, which is also found in fruit, and other fruit components, such as theobromine and theophylline, can also elevate plasma uric acid. Increases in total antioxidant content and uric acid were also seen after ingestion of wine. However, with both apples and wine, the increases in uric acid did not account for

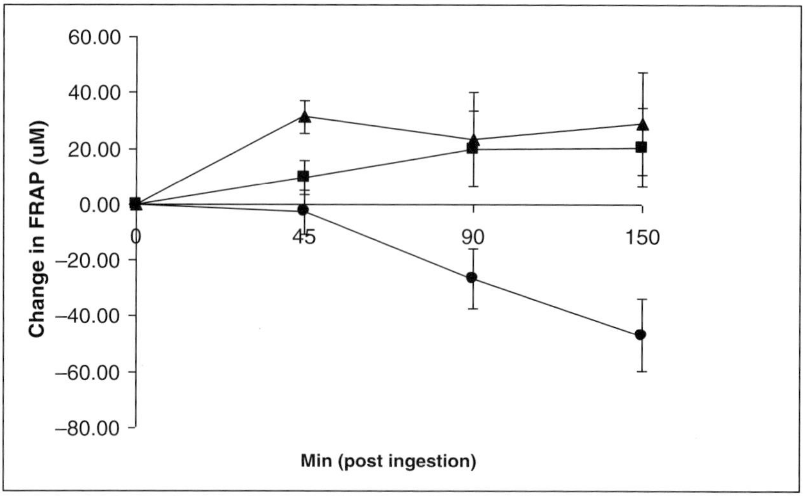

Fig. 3.1 Acute change in plasma antioxidant capacity, expressed as the ferric reducing/antioxidant power (FRAP) value after intake of 300 ml of instant coffee (triangles), cocoa drink (squares) and milk as control (circle). Results are Mean $\pm$SEM, $n = 6$; $* \ p < 0.05$ compared to baseline; # $p < 0.05$ compared to time matched value with control drink.

the entire post-ingestion changes in total antioxidant content of plasma, indicating some true absorption of antioxidants (Fernandez-Pachon *et al.*, 2005; Lotito and Frei, 2006). Therefore, while looking at acute or supplementation-related changes in plasma total antioxidant activity offers a simple and sensitive approach in agri-food and nutritional science, several points must be considered:

- The measurement is not specific for any individual antioxidant; it gives a composite result of the combined activity of a certain type of antioxidants.
- Different assays for total antioxidant activity are available, but they measure different things, some are affected by protein and lack sensitivity, and there are different ways to express results; data from different studies must take all this into account.
- The effect of uric acid should be corrected for; dietary factors can increase uric acid, leading to a falsely high impression of the impact of food ingestion on plasma total antioxidant content.

The influence of uric acid on plasma total antioxidant content can be corrected for very simply when using the FRAP assay. Stoichiometric factors are constant in the FRAP assay and the contribution of uric acid can be simply subtracted from the FRAP value (Benzie and Strain, 1996a, 1996b). The 'non-uric acid FRAP value' more specifically reveals true changes in the total activity of redox active antioxidants in the plasma following intake of fruits or other dietary agents. Uric acid is simple to measure using commercial uricase-linked kits. An alternative approach is to destroy uric acid in plasma by adding uricase, and then measuring the total antioxidant content remaining. A version of this approach using ascorbate oxidase is employed in a modification of the FRAP assay that can be used to measure ascorbic acid specifically and simultaneously with the total antioxidant content (Benzie and Strain 1999; Chung *et al.*, 2001).

GETTING TO THE END: A NOTE ON COLONIC METABOLITES

Antioxidant bioavailability testing focuses on measuring changes in the concentration of the antioxidant of interest (or 'total antioxidant content') in plasma and urine following ingestion of a known dose of a purified antoxidant, such as vitamin C or EGCG, or after eating 'whole' foods and beverages such as berries, chocolate, fruit juice, wine and tea. Some of the methods used to investigate responses have lacked sensitivity and specificity, and likely miss antioxidant metabolites, thereby over- or underestimating bioavailability. With the advent of highly specific and sensitive analytical tools, such as liquid chromatography with tandem mass spectrometry (LC-MS/MS) and an expanding database for identification of metabolites, studies that can accurately measure nanomolar or lower concentrations of individual compounds and their metabolites are adding to knowledge in bioavailability. These analytical tools are also supporting an emerging area of potential high importance: the identification and bioavailability of colonic metabolites of unabsorbed phytochemicals.

Most of the carotenoids and flavonoids that are ingested in food remain unabsorbed, and some absorbed antioxidant compounds re-enter the gastrointestinal tract via biliary excretion. These move into the large intestine, where they are available for biotransformation by the colonic microflora. Colonic metabolites include simple phenolic acids and ring fission products of polyphenolic compounds (Del Rio *et al.*, 2010b; Roowi *et al.*, 2010; Williamson and Clifford, 2010). The main benefits of unabsorbed and biotransformed phytochemicals may be through local action (antioxidant or otherwise) in the colon. However, it has become clear that many colonic metabolites of flavonoids appear in the plasma in a 'second wave' of absorption of compounds from flavonoid-rich foods such as berries, tea and wine (Del Rio *et al.* 2010b; Roowi *et al.*, 2010). Including these metabolites increases flavonoid bioavailability figures markedly, to 40% or more. Interestingly, some individuals were reported to have almost 100% bioavailability of flavon-3-ols when ring fission products were included in the calculation, while no colonic metabolites could be detected in the plasma of others (Del Rio *et al.*, 2010b; Roowi *et al.*, 2010). Further study is needed on the bioactivity, half-life in plasma, tissue uptake and physiological role of these colonic metabolites, as well as the effect of differences in microflora populations, probiotics and prebiotics on colonic biotransformation.

FINALLY, INCREASING BIOAVAILABILITY OF ANTIOXIDANTS: CAN IT AND SHOULD IT BE DONE?

There is strong supporting epidemiological evidence and a plausible scientific rationale for increasing intake of antioxidants from food as an effective means of prolonging healthy life. Higher intake of fruits and vegetables can achieve this. However, despite regular public health messages to include five or more serving of fruits and vegetables in the daily diet, few people achieve this. Increasing the content or bioavailability of antioxidants from foods such as fruit could achieve a better health-related outcome than repeated messages that have little impact on the dietary habits of the general public. The type and variety of fruit, where and how it is grown, stored and processed, and the form, amount and frequency in which it is eaten all help determine the dose ingested, the bioaccessibility and absorption ('bioavailability') of its intrinsic antioxidants. Colonic transformation of unabsorbed antioxidants leads to new compounds that may be bioavailable. Modulating any of these factors could help achieve enhanced antioxidant status by increased apparent or actual bioavailability. Possible strategies

Table 3.7 Summary of different approaches to increase 'apparent' or 'true' bioavailability of antioxidants from fruit.

Approach	Examples of how to achieve this	Rationale
Increase the antioxidant content intake	Increase intake of fruits rich in antioxidants; enhance synthesis of antioxidants in fruit by modulating genotype and growing conditions; select fruits at certain stages of growth cycle and ripeness	Supply greater amount of antioxidants per fruit serving → higher plasma response
Increase stability of antioxidants in the fruit	Avoid loss of antioxidants by improving storage of fresh fruit; for processed fruits, improve processing, storage and preservation methods	Supply greater amount of antioxidants per fruit serving → higher plasma response
Increase antioxidant bioaccessibility from the fruit	Cook fruits; modulate food matrix	Release intracellular antioxidants, increase antioxidant stability and surface area of the antioxidant-containing food matrix entering the gastrointestinal tract
Increase gastrointestinal absorption	Slow transit through the gastrointestinal tract; increase frequency of intake; selectively increase intake of fruits high in more absorbable antioxidants; manipulate antioxidant structure and food matrix; modulate colonic microflora	Allow more time for absorption; overcome absorption limitations with a 'small but often' dosing pattern; some antioxidant, e.g. ascorbic acid, are absorbed better than others, e.g polyphenols; some antioxidants in their natural form are non-absorbable until certain groups, e.g. carbohydrate residues, are removed; some antioxidants are absorbed better in the presence of co-antioxidants; vitamin E and other fat soluble antioxidants are absorbed better in the presence of dietary fat; dietary fibre can modulate absorption of antioxidants; colonic microflora can transform some unabsorbed antioxidants, e.g. catechins, into absorbable metabolites
Extend circulating half-life of absorbed and bioactive antioxidants	Lower metabolic inactivation and urinary excretion of absorbed antioxidants	*Theoretically,* by inhibiting glucuronidation and sulfation of polyphenols; lowering renal clearance of circulating antioxidants by complexing antioxidants or modulating their form

to achieve this are summarized in Table 3.7. However, there are important considerations to increasing our biological limitations in absorbing phytochemicals. Most dietary antioxidants are not absorbed, and that may be an evolved strategy to avoid toxic effects of what are, after all, xenobiotics.

Overcoming limits to absorb and retain dietary-derived antioxidants should be viewed with caution. Plasma levels of all dietary-derived antioxidants are kept low by restricted absorption, rapid post-ingestion metabolism and excretion. Even ascorbic acid, which has the highest bioavailability and for which we have a confirmed requirement, rarely reaches 100 μmol/l in plasma, even shortly after vitamin C ingestion, and there is rapid renal loss when plasma levels approach this concentration (Benzie and Strain, 1997; Davey *et al.*, 2000). Vitamin E (mainly α-tocopherol) is also a required element of the diet, but plasma levels rarely exceed 50 μmol/l due to our very limited biological harvesting of dietary vitamin E. Flavonoids and carotenoids can be ingested in very high quantities and still plasma concentrations remain in the nanomolar range. These observations raise intriguing questions, including:

- Why are plasma antioxidant concentrations kept so low, even when dietary intake is high?
- What can such small concentrations achieve in regard to human metabolism and health?
- Are all bioavailable antioxidants bioactive?
- Would it be beneficial to overcome normal mechanisms that limit bioavailability of dietary antioxidants, or is limited access an evolved biological defence against overload?

A few thoughts on these questions

Our current dietary needs are still rooted in what suited our ancient physiology (Benzie, 2003). The hominid diet of pre-agricultural times was rich in plant-based foods and the intake of flavonoids, carotenoids, as well as vitamin C and vitamin E, was likely to have been markedly higher than in the modern day diet. The loss of our ability to synthesize ascorbic acid was probably a metabolic saving that had no physiological downside because of the plentiful, regular supply in the habitual diet, and so was favoured and became an evolved feature of human biochemistry, although we retained an absolute need for vitamin C (Benzie 2003). So, for optimal physiological function the human body probably requires a fairly high intake of antioxidants in food. This aligns with the clear benefits of a well-planned vegetarian diet (Benzie and Wachtel-Galor, 2009, 2010). The cellular action, or bioactivity, of the great majority of phytochemicals is not known, but there could be specific molecular effects that require only the nano- or micromolar amounts that are normally achieved *in vivo*. Still, bioavailable does not equal bioactive and there is no known role for the flavonoids, many carotenoids and some members of the vitamin E family in human metabolism, even though they are found in plasma. Furthermore, increasing the amount of antioxidants absorbed by improving bioavailability is not necessarily desirable (Benzie and Wachtel-Galor, 2010, 2012). Moreover, bioactive does not necessarily mean biologically safe. ROS are potentially damaging, but many have important functions within the body, including in vasodilation, cell signalling and bacterial killing (Halliwell and Gutteridge, 2007). An overload of dietary antioxidants could interfere with these functions. Furthermore, many redox active antioxidants in plants, indeed probably all, can act as pro-oxidants *in vitro*. Whether this occurs *in vivo* is not known but pro-oxidant effects of dietary 'antioxidants' could be involved in their physiological action in activating gene transcription at nanomolar levels (Surh *et al.*, 2008; Benzie and Wachtel-Galor, 2010). However, such pro-oxidant properties could be damaging if the amount exceeds our evolved physiological safety zone and tips the redox balance in the direction of overt oxidative stress. This possibility provides cautionary food for thought in the planning of strategies to overcome our well evolved physiological limits of bioavailability (Benzie and Wachtel Galor, 2012).

CONCLUSIONS

Fruits contain many diverse antioxidants. Determining bioavailability of antioxidants from fruits and other foods is challenging but it can be done by looking at post-ingestion changes in specific or 'total' antioxidants in plasma and urine after ingestion. Many factors affect this and there are various strategies that can be adopted to increase bioavailability of antioxidants, the underlying rationale being that this will enhance antioxidant defence. Genetic manipulation, choice of growing conditions, processing methods and storage all affect the antioxidant content of fruit. The eating pattern, food composition and matrix and colonic microflora and biotransformation affect the bioaccessibility of dietary antioxidants. The focused and informed application of biotechnology, agricultural and food processing and dietary practices can enhance antioxidant content and bioaccessibility, increasing 'apparent' bioavailability. This can provide more 'antioxidant value' per fruit item while avoiding overcoming our evolved mechanisms of highly restricted bioavailability and rapid metabolism and excretion of most dietary-derived antioxidants, mechanisms that are likely to be important biological safeguards against phytochemical overload. There is no doubt that eating foods that are high in antioxidant content is a healthy life choice. Fruits offer a natural, attractive package of diverse antioxidants that contribute to a healthy life. Eat and enjoy!

ACKNOWLEDGEMENTS

The authors are grateful to The Hong Kong Polytechnic University for supporting this work.

DECLARATION OF INTEREST

The authors have no competing interests.

REFERENCES

Alothman, M., Bhat, R. and Karin, A.A. (2009) Antioxidant capacity and phenolic content of selected tropical fruits from Malaysia, extracted with different solvents. *Food Chem* **115**, 785–788.

Álvarez, R., Carvalho, C.P., Sierra, J. *et al.* (2012) Citrus fruit extraction systems: effect of chemical composition and antioxidant activity of clementine juice. *J Agric Food Chem* **60**, 774–781.

Bai, Y. and Lindhout, P. (2007) Domestication and breeding of tomatoes: what have we gained and what can we gain in the future? *Ann Bot* **100**, 1085–1094.

Benzie, I.F.F. (2003) Evolution of dietary antioxidants. *J Comp Biochem Physiol A* **136**, 113–126.

Benzie, I.F.F. and Strain, J.J. (1996a) The ferric reducing ability of plasma (FRAP) as a measure of 'antioxidant power': The FRAP Assay. *Anal Biochem* **239**, 70–76.

Benzie, I.F.F. and Strain, J.J. (1996b) Uric acid - friend or foe? *Redox Report* **2**, 231–234.

Benzie, I.F.F. and Strain, J.J. (1997) Acute post-ingestion changes in plasma ascorbic acid concentration: relationship to dose and to existing body stores. *Nutr Res* **17**, 187–190.

Benzie, I.F.F. and Strain, J.J. (1999) Ferric reducing (antioxidant) power as a measure of antioxidant capacity: the FRAP assay and its modification for measurement of ascorbic acid (FRASC). In: *Oxidants and Antioxidants, Volume 299 of 'Methods in Enzymology'* (ed. L. Packer), pp. 15–27. Academic Press, Orlando, FL.

Benzie, I.F.F. and Strain, J.J. (2005) Diet and Antioxidant Defence. In: *The Encyclopedia of Human Nutrition* (eds B. Caballero, L. Allen and A. Prentice), 2nd edn, pp. 131–137. Academic Press, London.

Benzie, I.F.F. and Szeto, Y.T. (1999) Total antioxidant capacity of teas by the ferric reducing / antioxidant power (FRAP) assay. *J Agric Food Chem* **47**, 633–636.

Benzie, I.F.F. and Wachtel-Galor, S. (2009) Biomarkers of long term vegetarian diets. *Adv Clin Chem* **47**, 171–222.

Benzie, I.F.F. and Wachtel-Galor, S. (2010) Vegetarian diets and public health, antioxidants, oxidative stress and redox balance: perspectives on connections, signals and needs. *Antiox Redox Signal* **13**, 1575–1591.

Benzie, I.F.F. and Wachtel-Galor, S. (2012) Increasing the antioxidant content of food: is it conceivable, possible or useful? *Int J Food Sci Nutr* **63**, 62–70.

Benzie, I.F.F., Janus, E.D. and Strain, J.J. (1998) Plasma vitamin C and E levels in Hong Kong Chinese. *Eur J Clin Nutr* **52**, 447–451.

Benzie, I.F.F., Szeto, Y.T., Strain, J.J. and Tomlinson, B. (1999) Consumption of green tea causes rapid increase in plasma antioxidant power in humans. *Nutr Cancer* **34**, 83–87.

Benzie, I.F., Chung, W.Y., Wang, J *et al.* (2006) Enhanced bioavailability of zeaxanthin in a milk-based formulation of wolfberry (Gou Qi Zi: *Fructus barbarum L.*). *Br J Nutr* **96**, 154–160.

Boekholdt, S.M., Meuwese, M.C., Day, N.E. *et al.* (2006) Plasma concentrations of ascorbic acid and C-reactive protein, and risk of future coronary artery disease, in apparently healthy men and women: the EPIC-Norfolk prospective population study. *Br J Nutr* **96**, 516–522.

Carlsen, M.H., Halvorsen, B.L., Holte, K. *et al.* (2010) The total antioxidant content of more than 3100 foods, beverages, spices, herbs and supplement used worldwide. *Nutr J* **22** (doi: 10.11.1186/1475-2891-9-3).

Carter, B., Monsivais, P, and Drewnowski, A. (2010) Absorption of folic acid and ascorbic acid from nutrient comparable beverages. *J Food Sci* **75**, H269–H293.

Cheng, C.Y., Chung, W.Y., Szeto, Y.T. and Benzie, I.F.F. (2005) Fasting plasma zeaxanthin response to *Fructus barbarum L.* (Wolfberry; Kei Tze) in a food-based human supplementation trial. *Br J Nutr* **93**, 123–130.

Chu, W., Cheung, S.C.M., Lau, R.A.W. and Benzie, I.F.F. (2011) Bilberry (*Vaccinium myrtillus L.*). In: *Herbal Medicine Biomolecular and Clinical Aspects* (eds I.F.F. Benzie and S. Wachtel-Galor), 2nd edn, pp. 55–71. Taylor and Francis, Boca Raton, FL.

Chung, W.Y., Chung, J.K.O., Szeto, Y.T. *et al.* (2001) Plasma ascorbic acid: measurement, stability and clinical utility revisited. *Clin Biochem* **34**, 623–627.

Davey, M.W., Van Montague, M., Samartin, M. *et al.* (2000) Plant L-ascorbic acid: chemistry, function, metabolism, bioavailability and processing. *J Sci Food Agric* **80**, 825–860.

Del Rio, D., Borges, G. and Crozier, A. (2010a) Berry flavonoids and phenolics: bioavailability and evidence of protective effects. *Br J Nutr* **104**, S67–S90.

Del Rio, D., Calani, L., Cordera, C. *et al.* (2010b) Bioavailability and catabolism of green tea flavan-3-ols in humans. *Nutrition* **26**, 1110–1116.

Delanghe, J.R., Langolois, M.R., De Buyzere, M.L. *et al.* (2011) Vitamin C deficiency: more than just a nutritional disorder. *Genes Nutr* **6**, 341–346.

Egert, S. and Rimbach, G. (2011) Which sources of flavonoids: complex diets or dietary supplements? *Adv Nutr* **2**, 8–14.

Fernandez-Pachon, M.S., Villano, D., Troncoso, A.M. and Garcia-Parrilla, M.C. (2005) Antioxidant capacity of plasma after red wine intake in human volunteers. *J Agric Food Chem* **53**, 5024–5029.

Fleshman, M.K., Lester, G.E., Riedl, K.M. *et al.* (2011) Carotene and novel apocarotenoid concentrations in orange-fleshed Cucumis melo melons: determinations of β carotene bioaccessibility and bioavailability. *J Agric. Food Chem* **59**, 4448–4454.

Food and Nutrition Board and Institute of Medicine (2000) Dietary reference intakes form vitamin C, vitamin E, selenium and carotenoids. National Academy Press, Washington, DC.

Ford, N.A. and Erdman, J.W., Jr. (2012) Are lycopene metabolites metabolically active? *Acta Biochim Pol* **59** (1), 1–4.

Frei, B., England, L. and Ames, B.N. (1989) Ascorbate is an outstanding antioxidant in human blood plasma. *Proc Natl Acad Sci* **86**, 6377–6381.

Fu, L., Xu, B.T., Xu, X.R. *et al.* (2011) Antioxidant capacities and total phenolic content of 62 fruits. *Food Chem* **129**, 345–350.

García-Alonso, J., Ros, G, Vidal-Guevara, M.L. and Peragio, M.J. (2006) Acute intake of phenolic-rich juice improves antioxidant status in healthy subjects. *Nutr Res* **26**, 330–339.

Gey, K.F. (1998) Vitamins E plus C and interacting conutrients required for optimal health. A critical and constructive review of epidemiology and supplementation data regarding cardiovascular disease and cancer. *Biofactors* **7**, 113–174.

Guo, C., Yang, J., Wei, J. *et al.* (2003) Antioxidant activities of peel, pulp and seed fractions of common fruits as determined by the FRAP assay. *Nutr Res* **23**, 1719–1726.

Haleem, M.A., Barton, K.L., Borges, G. *et al.* (2008) Increasing antioxidant intake from fruits and vegetables: practical strategies for the Scottish population. *J Hum Nutr Diet.* **21**, 539–546.

Halliwell, B. and Gutteridge, J.M.C. (eds) (2007) *Free Radicals in Biology and Medicine*. Oxford University Press, Oxford.

Halvorsen, B.L., Carlsen, M.H., Phillips, K.M. *et al.* (2006) Content of redox-active compounds (i.e. antioxidants) in foods consumed in the United States. *Am J Clin Nutr* **84**, 95–135.

Harding, A.H., Wareham, N.J., Bingham, S.A. *et al.* (2008) Plasma vitamin C level, fruit and vegetable consumption, and the risk of new-onset type 2 diabetes mellitus: the European prospective investigation of cancer – Norfolk prospective study. *Arch Intern Med* **168** (14), 1493–1499.

Heber, D. (2011) Pomegranate ellagitannins In: *Herbal Medicine Biomolecular and Clinical Aspects* (eds I.F.F. Benzie and S. Wachtel-Galor), 2nd edn, pp. 201–209. Taylor and Francis, Boca Raton, FL.

Hegedűs, A., Engel, R., Abrankó, L. *et al.* (2010). Antioxidant and antiradical capacities in apricot (*Prunus armeniaca L.*) fruits: variations from genotypes, years and analytical methods. *J Food Sci* **75**, C722–C730. http://flavo.vtt.fi/catechins.htm).

Hukkanen, A.T., Polonen, S.S., Karenlampi, S.O. and Kokko, H.I. (2006) Antioxidant capacity and phenolic content of sweet rowanberries. *J Agric Food Chem* **54**, 112–119.

Jin, Y., Alimbetov, D., George, T. *et al.* (2011) A randomized trial to investigate the effects of acute consumption of a blackcurrant juice drink on markers of vascular reactivity and bioavailability of anthocyanins in human subjects. *Eur J Clin Nutr* **65**, 849–856.

Khaw, K.T., Bingham, S., Welch, A. *et al.* (2001) Relation between plasma ascorbic acid and mortality in men and women in EPIC–Norfolk prospective study: a prospective population study. European Prospective Investigation into Cancer and Nutrition. *Lancet* **357**, 657–663.

Kulkarni, A.P. and Aradhya, S. M. (2005) Chemical changes and antioxidant activity in pomegranate arils during fruit development. *Food Chem* **93**, 319–324.

Lee, A., Thurnham, D.I. and Chopra, M. (2000) Consumption of tomato products with olive oil but not sunflower oil increases the antioxidant activity of plasma. *Free Radic Biol Med* **29**, 1051–1055.

Lee, J. and Mitchell, A.E. (2012) Pharmacokinetics of quercetin absorption from apples and onions in healthy humans. *J Agric Food Chem* **60**, 3874–3881.

Legua, P., Melgarejo, P., Abdelmajid, H. *et al.* (2012) Total phenols and antioxidant capacity in 10 Moroccan pomegranate varieties. *J Food Sci* **71**, doi: 101111/j.1750-3841.2011.02516x.

Lodge, J.K., Hall, W.L., Jeanes, Y.M. and Proteggente, A.R. (2004) Physiological factors influencing vitamin E biokinetics. *Ann NY Acad Sci.* **1031**, 60–73.

Lotito, S.B. and Frei, B. (2006) Consumption of flavonoid-rich foods and increased plasma antioxidant capacity in humans: Cause, consequence, or epiphenomenon? *Free Radic Biol Med* **41**, 1727–1746.

Ma, X., Wu, H., Liu, L. *et al.* (2011) Phenolic compounds and antioxidant properties in mango fruits. *Sci Hort* **129**, 102–107.

Manach, C., Williamson, G., Morand, C. *et al.* (2005) Bioavailability and bioefficiency of polyphenols in humans. 1. Review of 97 bioavailability studies. *Am J Clin Nutr* **8**, 230S–242S.

Manthey, J.A. and Perkins-Veazie, P. (2009) Influence of harvest date and local of beta-carotene, ascorbic acid, total phenols, the in vitro antioxidant capacity and phenolic profiles of five commercial varieties of mango (Mangifera indica L.). *J Agric Food Chem* **57**, 10825–10830.

Mattoo, A.K., Shukla, V., Fatima, T. *et al.* (2010) Genetic engineering to enhance crop-based phytonutrients (nutraceuticals) to alleviate diet-related diseases. *Adv Exp Med Biol* **698**, 122–143.

McGhie, T.K. and Walton, M.C. (2007) The bioavailability and absorption of anthocyanins: Towards a better understanding. *Mol Nutr Food Res* **51**, 702–713.

Mia, I., Lee, B.L., Lim, M.T. *et al.* (2010) Antioxidant activity and profiles of common fruits in Singapore. *Food Chem* **123**, 77–84.

Myint, P.K., Luben, R.N., Wareham, N.J. and Khaw, K.T. (2011) Association between plasma vitamin C concentrations and blood pressure in the European prospective investigation into cancer-Norfolk population-based study. *Hypertension* **58**, 372–379.

Namitha, K.K. and Negi, P.S. (2010) Chemistry and biotechnology of carotenoids. *Crit Rev Food Sci Nutr* **50**, 728–760.

Oude Griep, L.M., Geleijnse, J.M., Kromhout, D. *et al.* (2010) Raw and processed fruit and vegetable consumption and 10-year coronary heart disease incidence in a population-based cohort study in the Netherlands. *PLoS One* **5** (10), e13609.

Padayatty, S.J. and Levine, M. (2008) Fruit and vegetables: think variety, go ahead eat! *Am J Clin Nutr* **87**, 5–7.

Palafox-Carlos, H., Ayalaa-Zavala, J.F. and Gonzáalez-Aguilar, G.A. (2011) The role of dietary fibre in the bioaccessibility and bioavailability of fruit and vegetable antioxidants. *J Food Sci* **76**, R6–R15.

Papageorgiou, V., Mallouchos, A. and Komaitis, M. (2008) Investigation of the antioxidant behaviour of air- and freeze-dried aromatic plant materials in relation to their phenolic content and vegetative cycle. *J Agric Food Chem* **56**, 5743–5752.

Parades-López, O., Cervantes-Ceja, M.L., Vigna-Pérez, M. and Hernández-Pérez, T. (2010) Berries: improving human health and healthy aging, promoting quality of life – a review. *Plant Foods Human Nutr* **65**, 299–308.

Pellegrini, N., Serafini, M., Salvatore, S. *et al.*. (2006) Total antioxidant capacity of spices, dried fruits, nuts, pulses, cereals and sweets consumed in Italy by three different in vitro assays. *Mol Nutr Food Res* **50**, 1030–1038.

Perry, A., Rasmussen, H. and Johnson, E.J. (2009) Xanthophyll (lutein and zeaxanthin) content in fruits, vegetables and corn and egg products. *J Food Comp Anal* **22**, 9–15.

Pfister, R., Sharp, S.J., Luben, R. *et al.* (2011) Plasma vitamin C predicts incident heart failure in men and women in European Prospective Investigation into Cancer and Nutrition – Norfolk prospective study. *Am Heart J* **162**, 246–253.

Riccioni, G., Speranza, L., Pesce, M. *et al.* (2012) Novel phytonutrient contributors to antioxidant protection against cardiovascular disease. *Nutrition*. doi: 10.1016/j.nutr.2011.11.028.

Richelle, M., Bortlik, K., Liardet, S. *et al.* (2002) A food-based formulation provides lycopene with the same bioavailability to humans as that from tomato paste. *J Nutr* **132**, 404–408.

Roberts, R.L., Green, J. and Lewis, B. (2009). Lutein and zeaxanthin in eye and skin health. *Clinics Dermatol* **27**, 195–201.

Roowi, S., Stalmach, A., Mullen, W. *et al.* (2010) Green tea flavan-3-ols: colonic degradation and urinary excretion of catabolites by humans. *J Agric Food Chem* **58**, 1296–1304.

Roselló, S., Adalid, A.M., Cebolla-Cornejo, J. and Nuez, F. (2011) Evaluation of the genotype, environment and their interaction on carotenoid and ascorbic acid accumulation in tomato germplasm. *J Sci Food Agric* **91**, 1014–21.

Scalzo, J., Politi, A., Pellegrini, N. *et al.* (2005) Plant genotype affects total antioxidant capacity and phenolic contents in fruit. *Nutrition* **21**, 207–213.

Serafini, M., Bugianesi, R., Maiani, G. *et al.* (2003) Plasma antioxidants from chocolate. *Nature* **424**, 1013–1014.

Siddiqui, I.A., Adhami, V.M., Ahmad, N. and Mukhtar, H. (2010) Nanochemoprevention: sustained release of bioactive food components for cancer prevention. *Nutr Cancer* **62**, 883–890.

Sommerburg, O. Keunen, J.E.E., Bird, A.C. and van Kuijik, J.G.M. (1998) Fruits and vegetables that are sources for lutein and zeaxantin: the macular pigment in eyes. *Br J Opthalmol* **82**, 907–910.

Surh, Y.J., Kundu, J.K. and Na. H.K. (2008) Nrf2 as a master redox switch in turning on the cellular signaling involved in the induction of cytoprotective genes by some chemoprotective phytochemicals. *Planta Med* **74**, 1526–1539.

Szeto, Y.T. and Benzie, I.F.F. (2002) Effect of dietary antioxidants on human DNA *ex vivo*. *Free Radic Res* **36**,113–118.

Szeto, Y.T., Tomlinson, B. and Benzie, I.F.F. (2002) Total antioxidant power and ascorbic acid content of aqueous extracts of fresh fruits and vegetables: implications for dietary planning and food preservation. *Br J Nutr* **87**, 55–59.

Szeto, Y.T., Kwok, T.C.Y. and Benzie, I.F.F. (2004) Effect of long term vegetarian diet on biomarkers of antioxidant status and cardiovascular risk. *Nutrition* **20**, 863–866.

Thaipong, K., Boonprakob, U., Crosby, K. *et al.* (2006) Comparison of ABTS, DPPH, FRAP and ORAC assays for estimating antioxidant activity from guava fruit extracts. *J Food Comp Anal* **19**, 669–675.

Traber, M.G. (2007) Vitamin E regulatory mechanism. *Annu Rev Nutr* **27**, 347–362.

Traber, M.G. and Manor, D. (2012) Vitamin E. *Adv. Nutr.* **3**, 330–331.

Uchida, E., Kondo, Y., Amano, A. *et al.* (2011). Absorption and excretion of ascorbic acid alone and in acerola (Malpighia emarginata) juice: comparison in healthy Japanese subjects. *Biol Pharm Bull* **34**, 1744–1747.

USDA (US Department of Agriculture) DRI (Dietary Reference Intake) Tables; http://fnic.nal.usda.gov/dietary-guidance/dietary-reference-intakes/dri-tables (last accessed 14 February 2013).

USDA (US Department of Agriculture) Nutrient Data Laboratory (NDL) database of content of foods; http://www.ars.usda.gov/nutrientdata (last accessed 14 February 2013).

van het Hof, K.H., de Boer, B.C. and Tijburg, L.B. (2000) Carotenoid bioavailability in humans from tomatoes processed in different ways determined from the carotenoid response in the triglyceride-rich lipoprotein fraction of plasma after single consumption and in plasma after four days of consumption. *J Nutr* **130**, 1189–1196.

Vissers, M.C., Bozonet, S.M., Pearson, J.F. and Braithwaite, L.J. (2011) Dietary ascorbate intake affects steady state tissue concentrations in vitamin C-deficient mice: tissue deficiency after suboptimal intake and superior bioavailability from a food source (kiwifruit). *Am J Clin Nutr* **93**, 292–301.

Wachtel-Galor, S., Szeto, Y.T., Tomlinson, B. and Benzie, I.F.F. (2004), Ganoderma lucidum ('Lingzhi'); acute and short-term biomarkers response to supplementation. *Int J Food Sci Nutr* **55**, 75–83.

Wachtel-Galor, S., Wong, K.W. and Benzie, I.F.F. (2008) Effect of cooking on Brassica vegetables. *Food Chem* **110**, 706–710.

Williamson, G. and Clifford, M.N. (2010) Colonic metabolites of berry polyphenols: the missing link to biological activity? *Br J Nutr* **104**, S48–S66.

Wollenweber, E. and Dietz, V.H. (1981) Occurrence and distribution of free flavonoid aglycones in plants. *Phytochem* **20**, 869–932.

Wooten-Beard, P.C. and Ryan, L. (2011) Improving public health? The role of antioxidant-rich fruit and vegetable beverages. *Food Res Int* **44**, 3135–3148.

Xu, J., Yang, F., Chen, L. *et al.* (2003) Effect of selenium on increasing the antioxidant activity of tea leaves harvested during the early spring tea producing season. *J Agric Food Chem* **51**, 1081–1084.

4 Pharmacokinetic and Pharmacodynamic Interaction of Functional Foods with Medicines

Shizuo Yamada, Yuko Taki, Shingen Misaka, Takashi Okura, Yoshiharu Deguchi, Keizo Umegaki, Hiroshi Watanabe, Yasuo Watanabe and Margot Skinner

INTRODUCTION

Currently, the popularity of functional foods containing dietary supplements and botanical products and beverages is growing at a remarkable speed, in terms of the promotion of health, prevention and therapy of diseases. Herbal ingredients have received a great deal of attention in complementary and alternative medicine, and are used as dietary supplements or health products in many countries. It should be considered that much scientific knowledge of the efficacy and safety of herbal ingredients is still lacking. Elderly people frequently take dietary supplements with prescription drugs and such a tendency will further increase in the near future. A major concern is adverse events caused by a large excess intake or interactions with drugs. In fact, the potential for interaction with drugs is considerable because of the large number of constituents functional foods contain (Izzo and Ernst, 2001; Williamson, 2001). Possible changes could occur in the pharmacokinetics and pharmacodynamics of drugs: absorption in the small intestine, metabolism in the intestine and liver, distribution to target organs, transport across cell membrane and binding to specific receptors (Figure 4.1).

The induction and inhibition of hepatic drug-metabolizing enzymes by herbal ingredients or dietary compounds have been investigated. For example, St John's Wort, an herbal medicine used to treat mild depression, has been reported to decrease the blood concentrations of drugs by inducing hepatic cytochrome P450 (CYP) 3A4 activity, and thereby to attenuate the efficacy of drugs such as cyclosporin, indinavir and digoxin (Barone *et al.*, 2000; Durr *et al.*, 2000; Roby *et al.*, 2000). Furthermore, ginkgo biloba extract (GBE) and saw palmetto extract (SPE) are commonly prescribed for the treatment of cerebral insufficiency and peripheral vascular diseases (Kleijnen and Knipschild, 1992; McKenna *et al.*, 2001) and reduction of symptoms of benign prostatic hyperplasia (BPH) (Ernst, 2002), respectively, in some European countries. These herbs are used as a dietary supplement in the United States and Japan. Common beverages such as fruit juice, green tea and cranberry juice have been reported to affect the pharmacokinetics and pharmacodynamics of drugs (Takanaga *et al.*, 1998; Mohammed Abdul *et al.*, 2008; Okura *et al.*, 2008; Werba *et al.*, 2008). In fact, grapefruit juice has been shown to increase the bioavailability of 1,4-dihydropyridine calcium channel antagonists, benzodiazepines, antihistamines, and cyclosporin, due to significant inhibition of both metabolism by CYP3A4 enzymes and excretion through P-glycoprotein in the gastrointestinal tract (Ameer and Weintraub, 1997; Soldner *et al.*, 1999).

Bioactives in Fruit: Health Benefits and Functional Foods, First Edition. Edited by Margot Skinner and Denise Hunter.

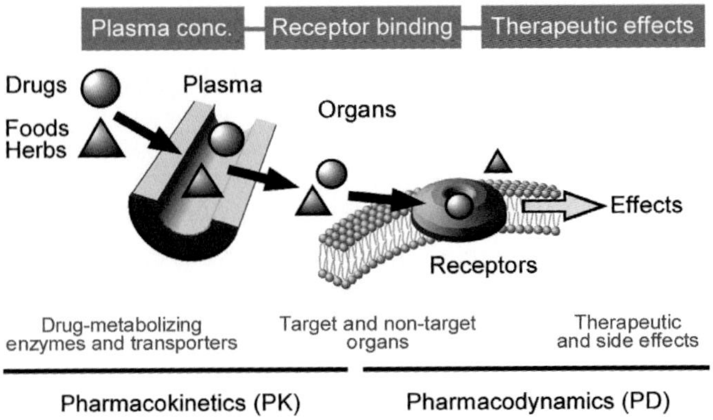

Fig. 4.1 Schematic representation of pharmacokinetic and pharmacodynamic interaction between ingredients of functional foods or herbs and drugs.

This chapter focuses on pharmacokinetic and pharmacodynamic interactions of fruit juices, ginkgo biloba extract (GBE), green tea, saw palmetto extract (SPE) and New Zealand fruit extracts with drugs.

FRUITS

Grapefruit juice

Grapefruit juice increases the bioavailability of several drugs including calcium channel blockers, benzodiazepins and immunosuppressants (Bailey *et al.*, 1991, 1993; Ducharme *et al.*, 1995; Fuhr *et al.*, 2002; Fukatsu *et al.*, 2006; Kupferschmidt *et al.*, 1995) (Figure 4.2). The main mechanism for the interaction is considered to be the irreversible

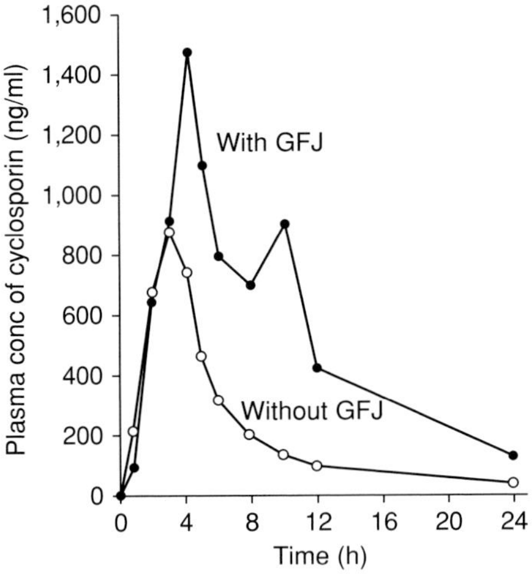

Fig. 4.2 Effect of ingestion of grapefruit juice (GFJ) on the plasma concentration of cyclosporine after on oral administration in healthy subjects (Reprinted by permission from Macmillan Publishers Limited on behalf of Cancer Research UK, *Clinical Pharmacology & Therapeutics*, © 1995, Ducharme *et al.*).

inhibition of CYP3A, a major drug-metabolizing enzyme in the small intestine, by furanocoumarins present in grapefruit juice (Paine *et al.*, 2006). Grapefruit juice inhibits not only metabolic enzymes but also drug transporters like P-glycoprotein (Marchetti *et al.*, 2007; Takanaga *et al.*, 1998; Dresser *et al.*, 2002), which play important roles in the intestinal barrier function in a coordinated manner with CYP3A (Synold *et al.*, 2001). The inhibitory effect of grapefruit juice on the intestinal barrier may enhance the oral bioavailability of drugs, which has been associated with a higher incidence of side effects.

Enhanced bioavailability can potentiate the therapeutic effect of drugs. Morphine is the most commonly used opioid analgesic for the treatment of cancer pain. A substrate of P-glycoprotein (Schinkel *et al.*, 1995), morphine's antinociceptive effect was enhanced by knockout of the P-glycoprotein gene in mice and the administration of a P-glycoprotein inhibitor in rats (Xie *et al.*, 1999; Zong and Pollack, 2000; Letrent *et al.*, 1999). In humans, the absorption of morphine is regulated by the intestinal P-glycoprotein (Kharasch *et al.*, 2003). Furthermore, P-glycoprotein may be partially associated with morphine tolerance (Aquilante *et al.*, 2000; Okura *et al.*, 2009), which limits the clinical use of morphine. The effects of grapefruit juice on oral morphine antinociception and pharmacokinetics in morphine-tolerant rats were examined (Okura *et al.*, 2008). Morphine tolerance was developed by repeated oral administration of morphine for five days, and grapefruit juice potentiated significantly the antinociceptive effect (Figure 4.3). Morphine concentrations in blood and intrathecal cerebrospinal fluid (CSF) measured by microdialysis were gradually decreased by the repeated treatment with morphine. The grapefruit juice significantly increased the blood concentration of morphine in morphine-tolerant rats (Figure 4.4). These results suggest that grapefruit juice enhances antinociception by increasing the intestinal absorption of morphine. The inhibition of intestinal P-glycoprotein by grapefruit juice may partly overcome morphine tolerance. However, little clinical evidence has been reported on the enhancement of effects of morphine by grapefruit juice. The inhibitory effect of grapefruit juice *in vivo* on P-glycoprotein activity in the human small intestine is not clear. Thus, further clinical studies will be important to examine the clinical effects of grapefruit juice on intestinal P-glycoprotein activity.

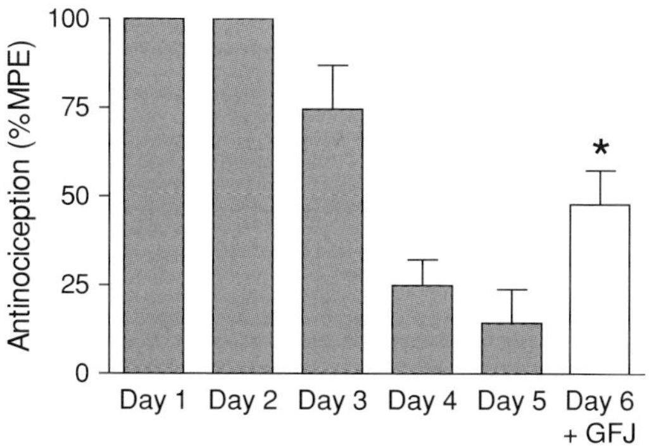

Fig. 4.3 Effects of repeated oral administration of morphine on the antinociceptive action in rats. Morphine (100 mg/kg) was given orally once a day for 1–5 days. After 120 min, rats were subjected to the tail-flick test. On the day after the 5-day-treatment with morphine (day 6), rats received grapefruit juice (GFJ) (2 ml/rat) 30 min before receiving morphine. The data are presented as % maximum possible effect (MPE). Each column represents the mean ± S.E. for four rats. *Significance versus Day 5, $p < 0.05$ (data from Okura *et al.*, 2008).

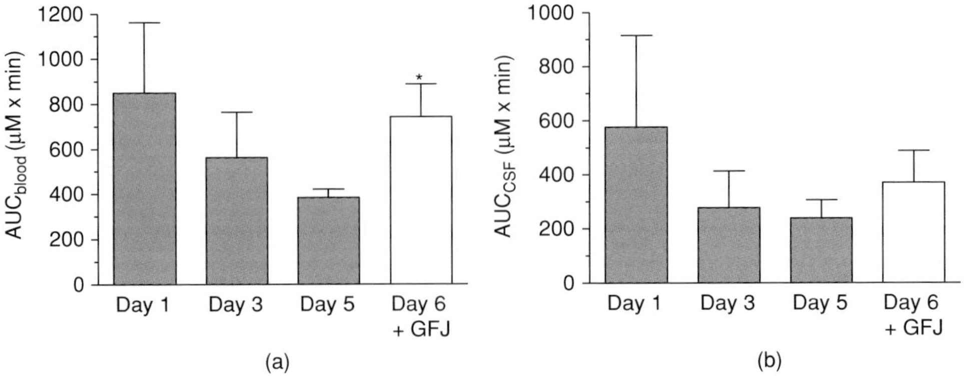

Fig. 4.4 Area under the concentration curve of morphine in the blood (AUC_{blood}) (a) and spinal CSF (AUC_{CSF}) (b) after oral administration for 1–5 days in rats. Microdialysis probes were implanted into the jugular vein and intrathecal space. Rats received orally morphine (100 mg/kg) once a day. On the day after the 5-day-treatment with morphine (day 6), rats received grapefruit juice (GFJ) (2 ml/rat) 30 min before receiving morphine. The concentrations of morphine in the dialysates were measured on day 1, 3, 5 and 6. Each column represents the mean ± S.E. for three to four rats. *Significance versus Day 5, $p < 0.05$ (data from Okura *et al.*, 2008).

Interactions mediated by drug transporters

Drug interactions caused by the inhibition of drug transporters have attracted increasing attention since the International Transporter Consortium published guidelines for transporter studies in drug development (Giacomini *et al.,* 2010). Numerous studies have suggested that transporters play a significant role in the pharmacokinetics and pharmcodynamics of therapeutic drugs *in vivo*. Dietary ingredients may have the greatest impact on drug disposition in the intestinal absorption process, as intestinal concentrations of ingredients are considered much higher than concentrations in other organs of the body. A number of *in vitro* and *in vivo* studies have shown that fruit juices and dietary ingredients inhibit drug transporters expressed in the intestinal epithelium including P-glycoprotein, breast cancer resistance protein (BCRP) and organic anion transporting polypeptides (OATPs). Grapefruit juice and its flavonoid, naringin, inhibited P-glycoprotein-mediated efflux across epithelial cell monolayers (Takanaga *et al.,* 1998; Dresser *et al.,* 2002). They also inhibited OATP-mediated uptake at much lower concentrations than those causing the equivalent P-glycoprotein inhibition (Dresser *et al.,* 2002). Orange juice and its flavonoid, hesperidin, are potent inhibitors of OATP *in vitro* (Dresser *et al.,* 2002). In addition to fruit juice, piperine, a major constituent of black pepper, capsaicin, a pungent component of chili pepper, and sesamin, a sesame lignin, have been shown to inhibit P-glycoprotein-mediated drug efflux in human intestinal cell lines (Bhardwaj *et al.,* 2002; Nabekura *et al.,* 2005; Okura *et al.,* 2010). Curcumin, a principal curcuminoid of turmeric, and daidzein and genistein, soybean isoflavones, have been reported to be inhibitors for BCRP (Shukla *et al.,* 2009; Enokizono *et al.,* 2007).

New Zealand fruit extracts

Recently, we have investigated the effect of New Zealand fruit extracts (feijoa, tamarillo, green kiwifruit, gold kiwifruit and blackcurrant) on CYP activities *in vitro*. Except for blackcurrant, extracts of the whole fruit, the skin and the flesh were obtained using methanol, and extract of the skin also was obtained with water. For blackcurrant, an extract available commercially as a powder was used. The extracts of tamarillo, green kiwifruit and gold kiwifruit had little effect on CYP activity. However, the whole extract and skin extract of feijoa significantly inhibited CYP1A2 activity. Also, the blackcurrant extract decreased all CYP activities, except CYP2E1, in a concentration-dependent manner. Blackcurrant contains many polyphenols such as anthocyanins, and anthocyanins have been reported to inhibit CYP (Dreiseitel *et al.,* 2008; Sand *et al.,* 2010). Therefore, anthocyanins might be the major inhibitor in blackcurrant. Also, extracts of whole feijoa and feijoa skin inhibited CYP1A2 activity, indicating that feijoa skin contains an inhibitory constituent of CYP enzymes. The skin contains many polyphenols that might be involved in CYP inhibitory activity.

In the *in vivo* experiments with mice, repeated oral administration of each extract (200 mg/kg, 5 days) of blackcurrant and feijoa tended to increase hepatic CYP activity. The extracts had little significant effect on CYP activity in the intestine. The extracts of green kiwifruit and gold kiwifruit did not affect the hepatic CYP activity and they decreased intestinal CYP activities. Additionally, the extract of tamarillo induced hepatic CYP3A activity. Generally, polyphenols including proanthocyanidins are poorly absorbed in the gut (Manach *et al.,* 2005). So, major constituents contained in blackcurrant and feijoa might have little effect on CYP activity *in vivo*. The proteolytic enzyme of actinidin contained in kiwifruit may be involved in the decrease of intestinal CYP. Furthermore, tamarillo contains a lot of vitamin E, and vitamin E induces hepatic CYP3A activity *in vivo*. Thus, vitamin E might be the major constituent contained in tamarillo that induces hepatic CYP3A activity.

GINKGO BILOBA EXTRACT (GBE)

Use of GBE and its constituents

Among herbal ingredients, GBE is one of the most popular, being used to improve cognitive function and peripheral arterial disease (Sierpina *et al.,* 2003). Recent randomized control trials failed to confirm the effectiveness of GBE in reducing the incidence of dementia in elderly individuals with normal cognition or with mild cognitive impairment (DeKosky *et al.,* 2008; Snitz *et al.,* 2009). Nevertheless, GBE is still popular among the elderly. As elderly people frequently take prescription drugs with dietary supplements (Nahin *et al.,* 2009), GBE–drug interactions would be a major concern.

GBE is a natural plant product containing many chemicals. Most commercially available GBE products are standardized according to amounts of ginkgo flavonol glycosides (glycosidic derivatives of quercetin, kaempferol, isorhamnetin) and terpenoids (ginkgolides A, B, C and bilobalide), which comprise 22–27% and approximately 5–7% of GBE, respectively, and less than 5 ppm of ginkgolic acid (Blumenthal, 1998). GBE products also contain 0.5–1% organic acids, such as vanillic acid and *p*-hydroxybenzoic acid. The exact constituents of GBE may vary among the products due to the time and place of harvest, and the extraction methods used.

Induction of CYPs

In pharmacokinetic studies with rats and mice, GBE induced the expression of hepatic drug-metabolizing enzymes, particularly cytochrome P450 (CYP), without causing hepatic

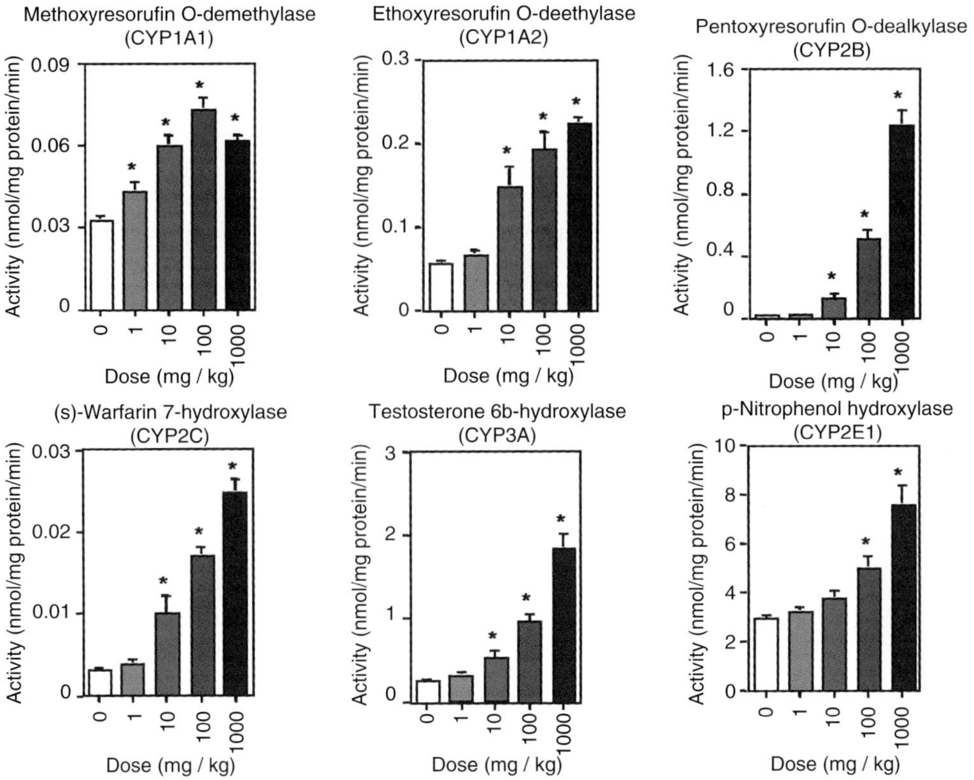

Fig. 4.5 Dose-dependent changes in hepatic CYP activities in rats given various doses of GBE. Rats were orally given GBE at doses of 0, 1, 10, 100 and 1000 mg/kg body weight for 5 days. The subtypes of CYP enzymes were determined by HPLC. Each column represents the mean ± S.E. *Significance versus untreated control, $p < 0.05$ (data from Umegaki *et al.*, 2002).

damage (Umegaki *et al.*, 2002, 2007). When rats were fed a GBE-containing diet or received GBE by intragastric gavage, the overall concentration of CYP and activity of various CYP enzymes in the liver were increased in a dose- and time-dependent manner (Figure 4.5). Significant increases in the concentration and activities of CYP enzymes were detected on day 1 of the feeding of a 0.5% GBE diet and after administration of 10 mg GBE/kg body weight for five days by intragastric gavage. The human equivalent dose, determined by the body surface normalization method (Reagan-Shaw *et al.*, 2008), is 1.62 mg/kg body weight, which is approximately 100 mg GBE/60 kg body weight and within the recommended dose range (up to 240 mg) taken from dietary supplements. Among the CYP enzymes, the activity of pentoxyresorufin O-dealkylase, a CYP2B enzyme, was markedly increased as confirmed by Western blot analysis and expression of m-RNA. GBE also increased levels of CYP2B1/2, CYP3A1 and CYP3A2 mRNA and related CYP activities in the rat liver (Shinozuka *et al.*, 2002; Umegaki *et al.*, 2002). A similar induction of hepatic CYPs by GBE in rats was observed with EGb761, a standardized GBE extract (Chatterjee *et al.*, 2005; Zhao *et al.*, 2006).

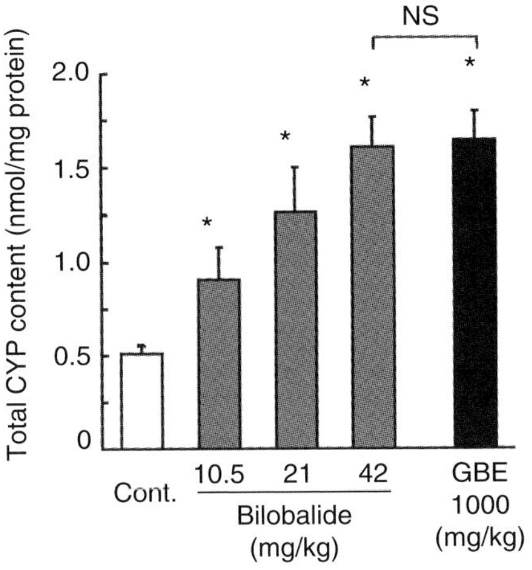

Fig. 4.6 Content of hepatic CYPs in mice given bilobalide or GBE containing the equivalent amount of bilobalide. Mice were administered either bilobalide (10.5, 21, 42 mg/kg) or GBE (1000 mg/k; 42 mg/kg as bilobalide) for 5 days. Each value is expressed as the mean ± S.D. for 5 mice. *Significance versus untreated control, $p < 0.05$. NS: not significant versus GBE (1000 mg/kg) (data from Umegaki *et al.*, 2007).

Mechanism of induction of hepatic drug-metabolizing enzymes

It is important to identify which substances in GBE are involved in the induction of CYPs. Studies performed in a cell-free enzyme system are very effective for screening for a substance in GBE that influences hepatic CYPs, but the results are not always applicable to the situation *in vivo*. In fact, there were discrepancies between the results of *in vitro* and *in vivo* studies (Gaudineau *et al.*, 2004; Mohutsky *et al.*, 2006; Sugiyama *et al.*, 2004c). To minimize the amount of test sample, substances that induced hepatic cytochrome P450 (CYP) within GBE were searched for in mice given either GBE or fractions of GBE. It was found that CYPs were induced markedly by a bilobalide-rich fraction, but not by flavonoid-rich fractions, and that bilobalide is the major substance in GBE that induces hepatic CYPs (Umegaki *et al.*, 2007) (Figure 4.6). In a rat study, quercetin, a major flavonoid in GBE, given for one week at a dose of 0.125% in the diet did not induce hepatic CYPs (Sugiyama *et al.*, 2004c). These findings indicate that the involvement of flavonoids in GBE in the induction of hepatic CYPs is unlikely. The three ginkgolides tested (A, B and C) enhanced hepatic CYP concentrations slightly, but did not significantly increase CYP activities (Taki *et al.*, 2012). Induction of hepatic CYP activities by GBE was compared in rats given two different GBE extracts, namely an acetone extract and an ethanol extract; the two extracts had same bilobalide content but different in flavonoid content. The extracts caused the induction of hepatic CYP activity to a similar extent (Sugiyama *et al.*, 2004c). This finding corroborates that bilibalide is a major substance inducing hepatic CYPs. The expression of hepatic CYPs induced by bilobalide was also reported for 7-methoxycoumarin O-demethylase activity in mice (Sasaki *et al.*, 1997), CYP2B1mRNA and CYP2B-mediated 7-benzyloxyresorufin O-dealkylation in primary cultured rat hepatocytes (Chang *et al.*, 2006) and CYP3A protein expression in human and rat primary hepatocyte study (Deng *et al.*, 2008). Although the contribution of bilobalide is unclear, GBE activated mouse and human PXR, a nuclear receptor involved in the transcriptional regulation of drug-metabolizing enzymes and transporters (Yeung *et al.*, 2008).

The reported half-life of bilobalide in blood is approximately two hours in rats and humans (Biber and Koch, 1999; Mauri *et al.*, 2001), indicating that it is eliminated easily from blood. When a single dose of GBE equivalent to approximately 40 mg /kg of bilobalide was given to rats, bilobalide in the liver was clearly detected at two hours but considerably decreased at 24 h and undetectable at 48 h. Thus, bilobalide is excreted rapidly and not accumulated in the liver. A single dose by gavage of bilobalide (30 mg/kg) in rats was found to produce a time-dependent induction of hepatic CYP activity and protein expression, and mRNA expression of CYP2B which was maximal at six hours and showed a similar response to that exhibited by the plasma and liver bilobalide concentration (Taki *et al.*, 2009). These findings suggest that bilobalide markedly induced CYPs, but the induction was turned off quickly due to bilobalide's rapid elimination from the liver. The rapid recovery of CYPs was confirmed in rats given excess GBE (Sugiyama *et al.*, 2004b); continuous and excess feeding of GBE (approximate dose: 500 mg GBE/kg and 21 mg bilobalide/kg) for one week to rats markedly induced hepatic CYPs, but discontinuation of the treatment led to normal levels of CYPs within one week. These findings suggest that interactions with drugs could be avoided by discounting GBE treatment.

Pharmacodynamic interaction

The induction of CYP by GBE suggested interaction with various drugs. In rats, GBE at 0.5% in the diet for two weeks increased hepatic CYPs and reduced the hypotensive effect of nicardipine, which is metabolized by CYP3A, with a decrease in the maximal nicardipine plasma concentration (Cmax) and the 23-h area under the curve ($AUC_{0–23}$) (Kubota *et al.*, 2003). Similarly, a 0.5 and 1.0% GBE diet given to rats for two weeks shortened the sleeping time of phenobarbital, which is known to be metabolized by CYP2B, with a reduction in the maximal phenobarbital plasma concentration (Cmax) and the 24-h area under the curve ($AUC_{0–24}$) (Kubota *et al.*, 2004). Interaction of GBE with tolbutamide, an oral antidiabetic agent, was also detected in aged and young rats, where a 5-day pretreatment with a 0.1% GBE diet attenuated the hypoglycemic action of tolbutamide, corresponding well to the enhanced activity of (S)-warfarin 7-hydroxylase, which is a CYP2C subtype and a major isoform metabolizing tolbutamide (Sugiyama *et al.*, 2004a) (Figure 4.7). It is noteworthy that the interaction of GBE with tolbutamide was clearly observed in the aged rats, which have a lower basal activity level of CYP subtypes in liver, while the induction by the GBE treatment was greater than that in young rats. The effects of GBE on the pharmacokinetics and pharmacodynamics of tolbutamide were significantly enhanced in rats maintained on a low-protein diet (Taki *et al.*, 2011). In mice, GBE interacted with (s)-warfarin, an active anticoagulant, through the induction of hepatic CYP2C by bilobalide, which resulted in increased warfarin metabolism, thereby decreasing the concentration of warfarin and its anticoagulant action (Taki *et al.*, 2012)(Figure 4.8).

Clinical study

Species differences exist for drug-metabolizing enzymes, thus it is important to know whether GBE has the potential to interact with drugs in humans at the current recommended doses. In contrast to studies with rats and mice, reports of GBE-drug interactions in clinical studies are

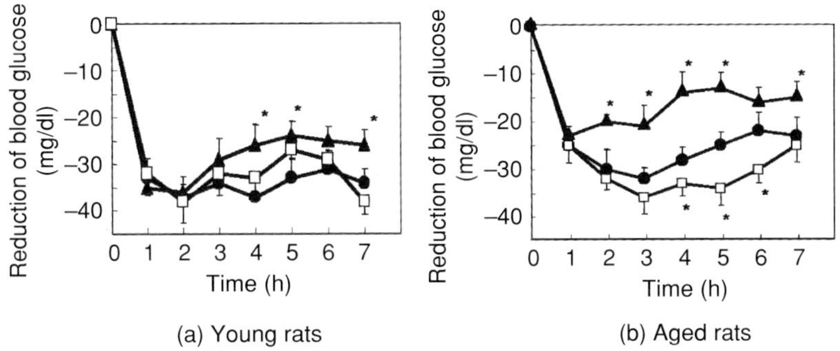

Fig. 4.7 Effects of simultaneous treatment and 5-day pretreatment with GBE on the hypoglycemic effect of tolbutamide in young (a) and old (aged) (b) rats. Young rats (7 weeks old) or old rats (19 months old) were administered tolbutamide (40 mg/kg, p.o.) with or without GBE treatment. The GBE-pretreated group was given feed containing 0.1% GBE for 5 days, and the simultaneous GBE-treated group was given a single dose of GBE (100 mg/kg, p.o.) with tolbutamide. After tolbutamide administration, blood was collected for the analysis of blood glucose concentrations. Each point represents the mean ± S.D. for 6 rats. •: control group; ▲: GBE-pretreated group; □: GBE-simultaneous-treated group. Each point is expressed as the mean ± S.D. for 5 rats. *Significance versus control, $p < 0.05$ (data from Sugiyama *et al.*, 2004a).

inconsistent, with some showing interactions (Uchida *et al.*, 2006; Robertson *et al.*, 2008) and others not (Duche *et al.*, 1989; Gurley *et al.*, 2002; Jiang *et al.*, 2005; Lei *et al.*, 2009; Zadoyan *et al.*, 2012). Intake of GBE (240 mg/day for 28 days) slightly decreased midazolam's $AUC_{0\text{–infinity}}$ and Cmax in 14 healthy subjects, indicating interaction of GBE with CYP3A4 drugs (Robertson *et al.*, 2008). Intake of GBE at 360 mg/day for 28 days slightly lowered the area under the concentration versus time curve ($AUC_{0\text{–infinity}}$) of tolbutamide and the blood glucose-lowering effect of tolbutamide in healthy male volunteers (Uchida *et al.*, 2006). On the other hand, GBE at 400 mg/day for 13 days did not influence the elimination half-life of antipyrine in a human study (Duche *et al.*, 1989). The administration of GBE at 240 mg/day for 28 days to healthy subjects caused no alteration in the activities of CYP3A4, CYP1A2, CYP2E1 or CYP2D6 assessed using a cocktail of specific substrates for

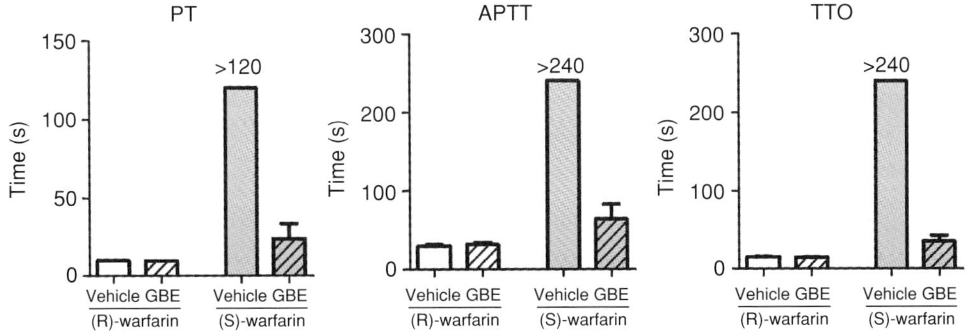

Fig. 4.8 Effect of GBE pretreatment on changes in anticoagulation parameters induced by (S)-warfarin- or (R)-warfarin. Mice were orally given GBE (100 mg/kg) for 5 days and (S)-warfarin or (R)-warfarin at a dose of 0.75 mg/kg for the last 3 days of the 5-day regimen. The coagulation parameters shown are prothrombin time (PT), activated partial thromboplastin time (APTT) and thrombotest (Owren) (TTO). Each column represents the mean ±S.E. for 5–6 mice (data from Taki *et al.*, 2012).

individual cytochromes (Gurley *et al.*, 2002). Seven days' pretreatment with recommended doses of ginkgo did not influence the pharmacokinetics and pharmacodynamics of warfarin in 12 healthy male subjects (Jiang *et al.*, 2005). The intake of Ginkgo biloba of 240 mg/day for 12 days did not affect the pharmacokinetics of voriconazole, a substrate of CYP2C19, in Chinese volunteers genotyped as either CYP2C19 extensive or poor metabolizers (Lei *et al.*, 2009). No relevant effect of GBE for major CYP enzymes (CYP1A2, CYP2C9, CYP2C19, CYP2D6, CYP3A) was observed following eight days' pretreatment with GBE at 240 mg/day in 18 healthy men and women (Zadoyan *et al.*, 2012).

According to the above reports, the interaction of GBE with drugs would be slight. In a study of GBE–atorvastatin interaction, treatment with 360 mg of GBE daily for 14 days slightly decreased plasma atorvastatin concentrations but had little significant effect on its cholesterol-lowering efficacy (Guo *et al.*, 2012). The different influences of GBE–drug interaction between humans and rats may be due to species differences in CYPs, dose of and intake periods of GBE, and the amount of active substance that induces CYPs.

Generally, GBE causes few adverse reactions, including headaches, gastric symptoms, diarrhoea and allergic skin reactions (De Smet, 2002). One of the most concerning adverse events associated with GBE is bleeding, which has been reported in those simultaneously taking GBE and anticoagulant drugs such as aspirin and warfarin (Ulbricht *et al.*, 2008). The PAF antagonistic action of ginkgolide B, which has been confirmed *in vitro* (Chung *et al.*, 1987; Lamant *et al.*, 1987), is an attractive explanation of bleeding episodes associated with GBE. However, an *in vivo* mouse study showed that ginkgolide B did not affect coagulation parameters, even at extremely high doses (up to 140 mg/kg) (Taki *et al.*, 2012). Also, clinical studies showed no influence of GBE intake on PAF-mediated platelet aggregation (Kudolo *et al.*, 2002) and prothrombin times (Kohler *et al.*, 2004). The half-maximal inhibitory concentration (IC50) of ginkgolide B for PAF-mediated human platelet aggregation was reported to be about 100 times that attained after the oral intake of standardized GBE at recommended doses (Koch, 2005). It is, therefore, unlikely that GBE and ginkgolide B, when used at the doses present in GBE products, induce spontaneous bleeding *in vivo* via antagonism of PAF. In animal studies, GBE attenuated rather than promoted the anticoagulant action of warfarin through the induction of hepatic CYPs by bilobalide (Taki *et al.*, 2012). Nonetheless, careful observation for bleeding and haemorrhage and interaction with drugs related to GBE-containing products will be needed in clinical practice, because of individual differences in sensitivity.

GREEN TEA

Green tea catechins

Flavonoids are part of a family of naturally occurring polyphenols contained in foods; they are divided into several classes, including flavonols, flavanols, flavanones and isoflavones, according to the degree of the oxidation of the oxygen heterocycle (Balentine *et al.*, 1997; Scalbert and Williamson, 2000). Catechins are categorized as flavanol compounds and found in various foods and beverages, such as green tea, chocolate, cocoa, pomegranate juice and red wine (De Pascual-Teresa *et al.*, 2000). Green tea (*Camellia sinensis*) is the most abundant source of catechins consisting of (–)–epigallocatechin-3-gallate (EGCG), (–)–epigallocatechin (EGC), (–)–epicatechin (EC) and (–)–epicatechin-3-gallate (ECG) (Graham, 1992), accounting for 30–42% of the dry weight of the solids in brewed green tea

(Khan and Mukhtar, 2007). In general, a cup of infused green tea (150 ml) supplies 30–40 mg of EGCG (Imai *et al.,* 1997). After the ingestion of a green tea extract containing 375 mg of EGCG, the plasma concentration of EGCG reached approximately 4 μM (Nakagawa *et al.,* 1997). Over recent decades, green tea has been recognized as a functional food for the prevention of cancer, cardiovascular disorders and infectious diseases (Khan and Mukhtar, 2007; Yang *et al.,* 2009), and consumed by a large proportion of the world's population. Accordingly, it is anticipated that the opportunities for the concomitant use of various drugs with green tea are increasing. Thus, the evaluation of scientific evidence on the possible drug interaction with green tea catechins is of great importance to reduce the risks of unwanted side effects. The green tea catechin–drug interactions mediated by the inhibition or induction of drug-metabolizing enzymes and drug transporters or other enzymes are considered here.

Cytochrome P450 enzyme-mediated interactions

Many drug interactions are attributed to the inhibition or induction of cytochrome P450 (CYP) enzymes (Wilkinson, 2005). The effects of green tea on the activity of CYP enzymes were first reported in rodents: drinking green tea significantly increased rat Cyp1a1, 1a2 and 2b1 activities, but not Cyp2e1 and 3a activities (Sohn *et al.,* 1994). Correspondingly, Park *et al.* recently demonstrated that repeated treatment with a green tea extract up-regulated Cyp2b1 and down-regulated Cyp3a mRNA expression in rat liver (Park *et al.,* 2009). Pharmacokinetic interactions between green tea catechins and Cyp substrate drugs including clozapine (Jang *et al.,* 2005), diltiazem (Li and Choi, 2008), midazolam (Nishikawa *et al.,* 2004), nicardipine (Choi and Burm, 2009), tamoxifen (Shin and Choi, 2009), and verapamil (Chung *et al.,* 2009) have been studied in rats. The main findings are summarized in Table 4.1. Those animal studies suggest that green tea catechins substantially inhibit Cyp3a activity in the liver or intestine and increase plasma concentrations of its substrates, whereas they may have an inductive effect on Cyp1a2 and reduce the plasma concentration of the substrate drugs. With respect to human CYP enzymes, Muto *et al.* reported that ECG and EGCG inhibited CYP1A1, 1A2, 2A6, 2C9, 2E1 and 3A4 activities depending on concentration in a human CYP-expressing membrane fraction (Muto *et al.,* 2001). Contrary to the inhibitory effects, the treatment with green tea extract for six hours induced CYP1A1 and 1A2 expression and increased mRNA levels for CYP2E1, 2D6 and 2C isoforms in human tongue cells (Yang and Raner, 2005).

In a clinical study, chronic consumption of decaffeinated green tea extract did not alter either the pharmacokinetics of alprazolam, a CYP3A4 probe drug, or the metabolic ratio of dextromethorphan, an index of CYP2D6 activity, in healthy volunteers (Donovan *et al.,* 2004). Chow and colleagues conducted a clinical trial to determine the effect of repeated green tea catechin administration on *in vivo* CYP activities using a drug cocktail containing caffeine (CYP1A2), dextromethorphan (CYP2D6), losartan (CYP2C19) and buspirone (CYP3A) (Chow *et al.,* 2006). Among the phenotypic indices investigated, only the area under the plasma concentration–time curve (AUC) of buspirone was significantly increased (by 1.2-fold) compared with the baseline value by green tea catechins, suggesting that green tea caused a small reduction in CYP3A activity but had no effects on CYP1A2, 2D6 and 2C19.

A recent case report has shown that green tea intake doubled the AUC of simvastatin, a cholesterol-lowering agent, and led to intense leg muscle cramps and pain in a hypercholesterolemic patient (Werba *et al.,* 2008). Because simvastatin is mainly metabolized by CYP3A

Table 4.1 Effects of green tea catechins on cytochrome P450 (CYP) enzymes in rats and humans *in vivo*.

CYP	Drug	Subject	Treatment	Effects	Comment	Reference
Cyp1a2	Clozapine, *p.o.*	Rats	GTE (175 mg/kg), 4 days	Cyp1a2 protein level 2.0-fold↑, C_{max} 0.6-fold↓, AUC 0.5-fold↓		(Jang *et al.*, 2005)
CYP1A2	Caffeine, *p.o.*	Human	EGCG (800 mg/day), 28 days	No effect		(Chow *et al.*, 2006)
Cyp2b		Rats	GTE (100 mg/kg), 7 days	Cyp2b1 mRNA levels 2.2-fold↑		(Park *et al.*, 2009)
CYP2C9	Losartan, p.o.	Human	EGCG (800 mg/day), 28 days	No effect		(Chow *et al.*, 2006)
CYP2D6	Dextromethorphan, *p.o.*	Human	Decaffeinated GTE, 14 days	No effect		(Donovan *et al.*, 2004)
CYP2D6	Dextromethorphan, p.o.	Human	EGCG (800 mg/day), 28 days	No effect		(Chow *et al.*, 2006)
Cyp3a		Rats	GTE (100 mg/kg), 7 days	Cyp3a mRNA level 0.5-fold↓		(Park *et al.*, 2009)
Cyp3a	Diltiazem, *p.o.*	Rats	EGCG (12 mg/kg)	AUC 1.8-fold↑, BA 1.8-fold↑	P-gp inhibition might be involved	(Li and Choi, 2008)
Cyp3a	Midazolam, *p.o.*	Rats	GTE (400 mg/kg), 7 days	C_{max} 2.1-fold↑, AUC 3.0-fold↑	Intestinal Cyp3a expression was decreased	(Nishikawa *et al.*, 2004)
Cyp3a	Nicardipine, *i.v.*	Rats	EGCG (10 mg/kg)	No effect		(Choi and Burm, 2009)
Cyp3a	Nicardipien, *p.o.*	Rats	EGCG (10 mg/kg)	C_{max} 1.5-fold↑, AUC 1.8-fold↑		(Choi and Burm, 2009)
Cyp3a	Tamoxifen, *p.o.*	Rats	EGCG (10 mg/kg)	C_{max} 1.9-fold↑, AUC 1.8-fold↑	P-gp inhibition might be involved	(Shin and Choi, 2009)
Cyp3a	Verapamil, *p.o.*	Rats	EGCG (10 mg/kg)	C_{max} 2.3-fold↑, AUC 2.1-fold↑	P-gp inhibition might be involved	(Chung *et al.*, 2009)
CYP3A	Alprazolam, *p.o.*	Human	Decaffeinated GTE, 14 days	No effect		(Donovan *et al.*, 2004)
CYP3A	Buspirone, p.o.	Human	EGCG (800 mg/day), 28 days	AUC 1.2-fold↑		(Chow *et al.*, 2006)

AUC, area under the plasma concentration–time curve; BA, bioavailability; GTE, green tea extract.

(Neuvonen *et al.*, 2006), this interaction may be due to the reduction of CYP3A activity by green tea. Further study will be required to clarify whether the interaction stems from the modulation of CYP3A activity. Collectively, green tea catechins probably have an inhibitory effect on CYP3A in humans. However, considering the low bioavailability of catechins (Lee *et al.*, 2002), pharmacokinetic interactions of green tea catechins with CYP3A substrate drugs may occur only in the gastrointestinal tract and not in the liver, just as for grapefruit juice (Mertens-Talcott *et al.*, 2006). It also cannot be excluded that green tea catechins have an inductive effect on some CYP subtypes, especially during the chronic consumption of green tea. Further investigations of the interaction between green tea catechins and CYP substrates, including its underlying mechanism, will help toward optimal pharmacotherapy in patients who drink green tea.

Transporter-mediated interaction

Recent evidence from *in vitro* and *in vivo* studies suggests that drug transporter-mediated drug interactions are clinically important (Dubuske, 2005). Recognition of drug interactions which lead to negative clinical outcomes, that is decreased effectiveness or tolerability, should support better medication and improve patient care. Several dietary flavonoids were found to be able to modulate an efflux transporter, P-glycoprotein (Castro and Altenberg, 1997; Shapiro and Ling, 1997). Concerning green tea catechins, it has been reported that EGCG inhibited the efflux of drugs mediated by P-glycoprotein *in vitro* (Jodoin *et al.*, 2002). Qian *et al.* also demonstrated that EGCG modulated the function of P-glycoprotein and reversed P-glycoprotein-mediated multidrug resistance in human cancer cells (Qian *et al.*, 2005). To date, there is little *in vivo* evidence to support the inhibitory effects of green tea catechins on P-glycoprotein. A recent study showed that intravenous administration of EGCG (20 mg/kg) to rats inhibited the transport of irinotecan and its active metabolite, SN-38, into the biliary tract, and prolonged their half-lives in plasma, possibly by modulating P-glycoprotein activity (Lin *et al.*, 2008). BCRP is another efflux transporter involved in cross-resistance to chemotherapeutic agents (Robey *et al.*, 2009). Some flavonoids have been identified as potent inhibitors of BCRP, however EGC and EGCG did not exhibit such inhibitory activity *in vitro* (Zhang *et al.*, 2004).

Uptake carriers such as OATP represent another class of drug transporters (Kim, 2006). Although no data are available regarding the *in vivo* impact of green tea catechins on OATP activity, there are a few *in vitro* studies which may facilitate further investigations of OTAP-mediated drug interaction with green tea catechins. Using OATP1B1-expressing HeLa cells, Wang *et al.* found that EGCG, but not EGC, was a potential inhibitor of OATP1B1 with an IC_{50} of 14.1 μM (Wang *et al.*, 2005). As for OATP2B1, which is expressed on human intestinal epithelium, green tea itself and green tea catechins including EC, ECG and EGCG significantly inhibited OATP2B1-mediated transport of estrone-3-sulfate in human embryonic kidney 293 cells at concentrations considered likely to be attainable in the human intestine (Fuchikami *et al.*, 2006). More recently, Roth *et al.* showed that ECG and EGCG inhibited uptake activities of OATP1A2, 1B1 and 2B1 in a concentration-dependent manner, while EC and EGC had minimal effects on OATPs (Roth *et al.*, 2011). Interestingly, EGCG was found to be a potent stimulator of OATP1B3-mediated uptake for one substrate examined, whereas for another substrate EGCG behaved as an inhibitor of OATP1B3 (Roth *et al.*, 2011). The authors proposed that potential inhibitors should be examined by using multiple and clinically relevant substrates when screening for OATP-mediated drug interactions.

In summary, more *in vivo* evidence will be needed for a better understanding of drug transporter-mediated drug interaction between green tea catechins and drugs. We also note that such interactions may have beneficial properties. For example, given that green tea catechins have potent inhibitory effects on efflux transporters such as P-glycoprotein *in vivo*, drinking green tea may be valuable for patients receiving cancer chemotherapy, because P-glycoprotein inhibition by catechins could suppress multidrug resistance in cancer cells.

Interactions through other mechanisms

Interactions mediated by other drug-metabolizing enzymes, such as phase II conjugating enzymes, have received less attention as compared to CYP enzymes (Kiang *et al.*, 2005). However, because many drugs and their metabolites undergo conjugation reactions, it is important to enhance our understanding of phase II enzyme-mediated drug interactions. Since green tea catechins undergo conjugate metabolism by UDP-glucuronosyltransferase (UGT), sulfotransferase (SULT) and catechol-*O*-methyltransferase (COMT) in rodents and humans (Feng, 2006), co-existence of green tea catechins and their substrates may cause drug interaction thorough the inhibition of those enzymes. Zhu *et al.* reported that a green tea catechin mixture and EGCG inhibited glucuronidation of estrone in a concentration-dependent manner with IC_{50} values of 12.5 and 10 μg/ml, respectively, in rat liver microsomes (Zhu *et al.*, 1998). Moreover, recent studies revealed that UGT1A1 and UGT1A4 activities were markedly inhibited by EGCG with IC_{50} values of 7.8 and 34.4 μg/ml in human liver microsomes (Mohamed *et al.*, 2010; Mohamed and Frye, 2011). EGCG showed weak inhibitory activities toward UGT1A6 and UGT1A9 (Mohamed and Frye, 2011). On the other hand, it has been reported that EGCG did not affect mRNA expression of UGT1A1 (Bothe *et al.*, 2011). To our knowledge, there is no clinical evidence regarding UGT-mediated drug interactions with green tea catechins. As for the other phase II enzymes, EGCG has been reported to inhibit COMT activity with IC_{50} values ranging from 0.07 to 0.2 μM in human liver cytosol (Lu *et al.*, 2003; Nagai *et al.*, 2004). In experiments *in vitro* using human recombinant SULT1A1 and SULT1A3, green tea catechins, particularly ECC and EGCG, inhibited SULT1A1 and SULT1A3 activities at around 10 μM (Nishimuta *et al.*, 2007). In addition to the inhibition of phase II enzymes, Golden and colleagues reported that EGCG directly reacted with bortezomib, an anticancer drug, and blocked its antiproliferative function in preclinical *in vitro* and *in vivo* models (Golden *et al.*, 2009). Presumably, this interaction arose as a result of a physicochemical interaction leading to the formation of a covalent cyclic boronate between EGCG and bortezomib.

In summary, the findings described highlight a possibility of green tea catechin–drug interactions through the modulation of not only phase II enzymes but also the chemical structure of co-administered drugs. More studies are warranted to evaluate the clinical consequences of such interactions.

SAW PALMETTO EXTRACT (SPE)

The ripe berries of the American dwarf palm (*Serenoa repens*) have been traditionally used to treat genitourinary problems; to enhance sperm production, breast size or libido; and as a mild diuretic (Ernst, 2002). SPE is almost exclusively used to treat benign prostatic hyperplasia (BPH). Fujino *et al.* (2007) have shown that repeated oral administration of SPE in rats had little significant influence on the content and activities of hepatic drug-metabolizing enzymes.

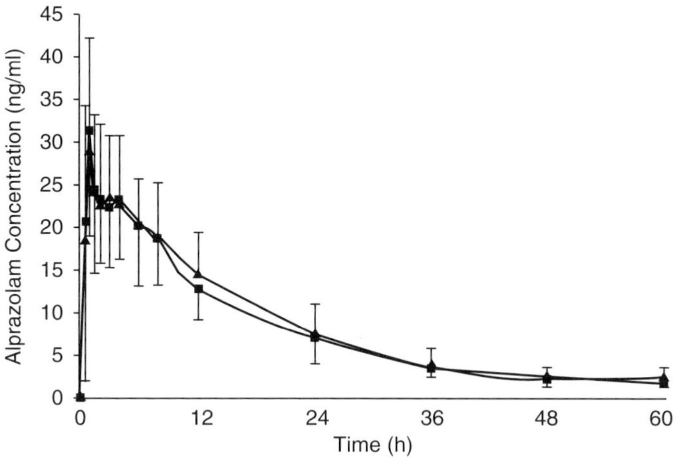

Fig. 4.9 Effect of repeated treatment (320 mg/day, 14 days) with saw palmetto on the plasma concentration of alprazolam (CYP3A4 activity) in normal volunteers (cited from Markowitz *et al.* (2003) *Clin Pharmacol Ther* **74**, 536–542).

Markowitz *et al.* (2003) reported that SPE (320 mg/day for 14 days) for the treatment of lower urinary tract symptoms (LUTS) suggestive of BPH did not alter plasma concentrations of probe drugs for CYP2D6 and CYP3A4 activity in normal volunteers (Figure 4.9). No effect of repeated treatment with SPE (160 mg/day for 28 days) was shown by using each probe drug for CYP1A2, CYP2D6, CYP2E1 and CYP3A4 (Gurley *et al.*, 2004). Therefore, it is unlikely that SPE at generally recommended doses alters the disposition of co-administered drugs. Also, repeated administration of SPE in rats had no significant effect on blood biochemical parameters except for a slight increase in the albumin value, suggestive of relative safety even with long-term intake (Fujino *et al.*, 2007: Suzuki *et al.*, 2009).

CONCLUSIONS

The oral intake of some functional foods has been shown to influence significantly the pharmacokinetics and pharmacodynamics of co-administered drugs. Such interactions may be in part mediated through significant inhibition or induction of drug-metabolizing enzymes and transporters in the small intestine, liver, kidney and brain. It must be clarified whether interactions of foods with medicines have clinically harmful or beneficial effects for drug therapy. Furthermore, some ingredients of functional foods may affect directly pharmacological targets of medicines, thereby causing the significant augmentation or diminution of therapeutic effects. For example, SPE has been shown to improve significantly urinary dysfunction possibly through the direct action of drug targets such as pharmacological α_1-adrenoceptors and muscarinic cholinoceptors in the prostate and bladder (Oki *et al.*, 2005; Suzuki *et al.*, 2007), and thus the combination of SPE and medicines may be advantageous in terms of a reduction in the dosage and cost and adverse effects of drugs. Further scientific and clinical evaluation of pharmacokinetic and pharmacodynamic interactions of functional foods with prescription drugs is prospectively encouraged to establish their proper uses in clinical settings. Finally, it should be kept in mind that combinations of functional foods (including beverages) and medicines should be viewed cautiously in terms of potential adverse interactions in patients treated with drugs having relatively narrow therapeutic windows.

ACKNOWLEDGEMENTS

The authors would like to express their appreciation to Miss Etsuko Sano for secretarial assistance in preparing this manuscript.

REFERENCES

Ameer, B. and Weintraub, R. A. (1997) Drug interactions with grapefruit juice. *Clin Pharmacokinet*, **33**, 103–21.

Aquilante, C. L., Letrent, S. P., Pollack, G. M. and Brouwer, K. L. (2000) Increased brain P-glycoprotein in morphine tolerant rats. *Life Sci*, **66**, PL47–51.

Bailey, D. G., Spence, J. D., Munoz, C. and Arnold, J. M. (1991) Interaction of citrus juices with felodipine and nifedipine. *Lancet*, **337**, 268–9.

Bailey, D. G., Arnold, J. M., Strong, H. A. *et al.* (1993) Effect of grapefruit juice and naringin on nisoldipine pharmacokinetics. *Clin Pharmacol Ther*, **54**, 589–94.

Balentine, D. A., Wiseman, S. A. and Bouwens, L. C. (1997) The chemistry of tea flavonoids. *Crit Rev Food Sci Nutr*, **37**, 693–704.

Barone, G. W., Gurley, B. J., Ketel, B. L. *et al.* (2000) Drug interaction between St. John's wort and cyclosporine. *Ann Pharmacother*, **34**, 1013–6.

Bhardwaj, R. K., Glaeser, H., Becquemont, L. *et al.* (2002) Piperine, a major constituent of black pepper, inhibits human P-glycoprotein and CYP3A4. *J Pharmacol Exp Ther*, **302**, 645–50.

Biber, A. and Koch, E. (1999) Bioavailability of ginkgolides and bilobalide from extracts of ginkgo biloba using GC/MS. *Planta Med*, **65**, 192–3.

Blumenthal, M. (1998) *The Complete German Commission E Monographs – Therapeutic Guide to Herbal Medicines*. The American Botanical Council, Austin, TX.

Bothe, H., Gassmann, K., Gotz, C. *et al.* (2011) Epigallocatechin-3-gallate does not affect the activity of enzymes involved in metabolic activation and cellular excretion of benzo[a]pyrene in human colon carcinoma cells. *Toxicol Lett*, **203**, 258–64.

Castro, A. F. and Altenberg, G. A. (1997) Inhibition of drug transport by genistein in multidrug-resistant cells expressing P-glycoprotein. *Biochem Pharmacol*, **53**, 89–93.

Chang, T. K., Chen, J. and Teng, X. W. (2006) Distinct role of bilobalide and ginkgolide A in the modulation of rat CYP2B1 and CYP3A23 gene expression by Ginkgo biloba extract in cultured hepatocytes. *Drug Metab Dispos*, **34**, 234–42.

Chatterjee, S. S., Doelman, C. J., Noldner, M. *et al.* (2005) Influence of the Ginkgo extract EGb 761 on rat liver cytochrome P450 and steroid metabolism and excretion in rats and man. *J Pharm Pharmacol*, **57**, 641–50.

Choi, J. S. and Burm, J. P. (2009) Effects of oral epigallocatechin gallate on the pharmacokinetics of nicardipine in rats. *Arch Pharm Res*, **32**, 1721–5.

Chow, H. H., Hakim, I. A., Vining, D. R. *et al.* (2006) Effects of repeated green tea catechin administration on human cytochrome P450 activity. *Cancer Epideml Biomarkers Prev*, **15**, 2473–6.

Chung, K. F., Dent, G., Mccusker, M. *et al.* (1987) Effect of a ginkgolide mixture (BN 52063) in antagonising skin and platelet responses to platelet activating factor in man. *Lancet*, **1**, 248–51.

Chung, J. H., Choi, D. H. and Choi, J. S. (2009) Effects of oral epigallocatechin gallate on the oral pharmacokinetics of verapamil in rats. *Biopharm Drug Dispos*, **30**, 90–3.

De Pascual-Teresa, S., Santos-Buelga, C. and Rivas-Gonzalo, J. C. (2000) Quantitative analysis of flavan-3-ols in Spanish foodstuffs and beverages. *J Agric Food Chem*, **48**, 5331–7.

De Smet, P. A. (2002) Herbal remedies. *New Eng J Med*, **347**, 2046–56.

Dekosky, S. T., Williamson, J. D., Fitzpatrick, A. L. *et al.* (2008) Ginkgo biloba for prevention of dementia: a randomized controlled trial. *JAMA*, **300**, 2253–62.

Deng, Y., Bi, H. C., Zhao, L. Z. *et al.* (2008) Induction of cytochrome P450 3A by the Ginkgo biloba extract and bilobalides in human and rat primary hepatocytes. *Drug Metab Lett*, **2**, 60–6.

Donovan, J. L., Chavin, K. D., Devane, C. L. *et al.* (2004) Green tea (*Camellia sinensis*) extract does not alter cytochrome p450 3A4 or 2D6 activity in healthy volunteers. *Drug Metab Dispos*, **32**, 906–8.

Dreiseitel, A., Schreier, P., Oehme, A. *et al.* (2008) Anthocyanins and their metabolites are weak inhibitors of cytochrome P450 3A4. *Mol Nutr Food Res*, **52**, 1428–33.

Dresser, G. K., Bailey, D. G., Leake, B. F. *et al.* (2002) Fruit juices inhibit organic anion transporting polypeptide-mediated drug uptake to decrease the oral availability of fexofenadine. *Clin Pharmacol Ther*, **71**, 11–20.

Dubuske, L. M. (2005) The role of P-glycoprotein and organic anion-transporting polypeptides in drug interactions. *Drug Saf*, **28**, 789–801.

Ducharme, M. P., Warbasse, L. H. and Edwards, D. J. (1995) Disposition of intravenous and oral cyclosporine after administration with grapefruit juice. *Clin Pharmacol Ther*, **57**, 485–91.

Duche, J. C., Barre, J., Guinot, P. *et al.* (1989) Effect of Ginkgo biloba extract on microsomal enzyme induction. *Int J Clin Pharmacol Res*, **9**, 165–8.

Durr, D., Stieger, B., Kullak-Ublick, G. A. *et al.* (2000) St John's Wort induces intestinal P-glycoprotein/MDR1 and intestinal and hepatic CYP3A4. *Clin Pharmacol Ther*, **68**, 598–604.

Enokizono, J., Kusuhara, H. and Sugiyama, Y. (2007) Effect of breast cancer resistance protein (Bcrp/Abcg2) on the disposition of phytoestrogens. *Mol Pharmacol*, **72**, 967–75.

Ernst, E. (2002) The risk-benefit profile of commonly used herbal therapies: Ginkgo, St. John's Wort, Ginseng, Echinacea, Saw Palmetto, and Kava. *Ann Intern Med*, **136**, 42–53.

Feng, W. Y. (2006) Metabolism of green tea catechins: an overview. *Curr Drug Metab*, **7**, 755–809.

Fuchikami, H., Satoh, H., Tsujimoto, M. *et al.* (2006) Effects of herbal extracts on the function of human organic anion-transporting polypeptide OATP-B. *Drug Metab Dispos*, **34**, 577–82.

Fuhr, U., Muller-Peltzer, H., Kern, R. *et al.* (2002) Effects of grapefruit juice and smoking on verapamil concentrations in steady state. *Eur J Clin Pharmacol*, **58**, 45–53.

Fujino (Oki), T., Suzuki, M. and Yamada, S. (2007) Effects of saw palmetto extract on urodynamic function and receptors in the lower urinary tract. *Jap J Complem Alter Med*, **4**, 41–50.

Fukatsu, S., Fukudo, M., Masuda, S. *et al.* (2006) Delayed effect of grapefruit juice on pharmacokinetics and pharmacodynamics of tacrolimus in a living-donor liver transplant recipient. *Drug Metab Pharmacokinet*, **21**, 122–5.

Gaudineau, C., Beckerman, R., Welbourn, S. and Auclair, K. (2004) Inhibition of human P450 enzymes by multiple constituents of the Ginkgo biloba extract. *Biochem Biophys Res Comm*, **318**, 1072–8.

Giacomini, K. M., Huang, S. M., Tweedie, D. J. *et al.* (2010) Membrane transporters in drug development. *Nat Rev Drug Discov*, **9**, 215–36.

Golden, E. B., Lam, P. Y., Kardosh, A. *et al.* (2009) Green tea polyphenols block the anticancer effects of bortezomib and other boronic acid-based proteasome inhibitors. *Blood*, **113**, 5927–37.

Graham, H. N. (1992) Green tea composition, consumption, and polyphenol chemistry. *Prev Med*, **21**, 334–50.

Guo, C. X., Pei, Q., Yin, J. Y. *et al.* (2012) Effects of Ginkgo biloba extracts on pharmacokinetics and efficacy of atorvastatin based on plasma indices. *Xenobiotica*, **42**, 784–90.

Gurley, B. J., Gardner, S. F., Hubbard, M. A. *et al.* (2002) Cytochrome P450 phenotypic ratios for predicting herb-drug interactions in humans. *Clin Pharmacol Ther*, **72**, 276–87.

Gurley, B. J., Gardner, S. F., Hubbard, M. A. *et al.* (2004) *In vivo* assessment of botanical supplementation on human cytochrome P450 phenotypes: Citrus aurantium, Echinacea purpurea, milk thistle, and saw palmetto. *Clin Pharmacol Ther*, **76**, 428–40.

Imai, K., Suga, K. and Nakachi, K. (1997) Cancer-preventive effects of drinking green tea among a Japanese population. *Prev Med*, **26**, 769–75.

Izzo, A. A. and Ernst, E. (2001) Interactions between herbal medicines and prescribed drugs: a systematic review. *Drugs*, **61**, 2163–75.

Jang, E. H., Choi, J. Y., Park, C. S. *et al.* (2005) Effects of green tea extract administration on the pharmacokinetics of clozapine in rats. *J Pharm Pharmacol*, **57**, 311–6.

Jiang, X., Williams, K. M., Liauw, W. S. *et al.* (2005) Effect of ginkgo and ginger on the pharmacokinetics and pharmacodynamics of warfarin in healthy subjects. *Br J Clin Pharmacol*, **59**, 425–32.

Jodoin, J., Demeule, M. and Beliveau, R. (2002) Inhibition of the multidrug resistance P-glycoprotein activity by green tea polyphenols. *Biochim Biophys Acta*, **1542**, 149–59.

Khan, N. and Mukhtar, H. (2007) Tea polyphenols for health promotion. *Life Sci*, **81**, 519–33.

Kharasch, E. D., Hoffer, C., Whittington, D. and Sheffels, P. (2003) Role of P-glycoprotein in the intestinal absorption and clinical effects of morphine. *Clin Pharmacol Ther*, **74**, 543–54.

Kiang, T. K., Ensom, M. H. and Chang, T. K. (2005) UDP-glucuronosyltransferases and clinical drug–drug interactions. *Pharmacol Ther*, **106**, 97–132.

Kim, R. B. (2006) Transporters and drug discovery: why, when, and how. *Mol Pharmacol*, **3**, 26–32.

Kleijnen, J. and Knipschild, P. (1992) Ginkgo biloba. *Lancet*, **340**, 1136–9.

Koch, E. (2005) Inhibition of platelet activating factor (PAF)-induced aggregation of human thrombocytes by ginkgolides: considerations on possible bleeding complications after oral intake of Ginkgo biloba extracts. *Phytomedicine*, **12**, 10–6.

Kohler, S., Funk, P. and Kieser, M. (2004) Influence of a 7-day treatment with Ginkgo biloba special extract EGb 761 on bleeding time and coagulation: a randomized, placebo-controlled, double-blind study in healthy volunteers. *Blood Coagul Fibrinolysis*, **15**, 303–9.

Kubota, Y., Kobayashi, K., Tanaka, N. *et al.* (2003) Interaction of Ginkgo biloba extract (GBE) with hypotensive agent, nicardipine, in rats. *In Vivo*, **17**, 409–12.

Kubota, Y., Kobayashi, K., Tanaka, N. *et al.* (2004) Pretreatment with Ginkgo biloba extract weakens the hypnosis action of phenobarbital and its plasma concentration in rats. *J Pharm Pharmacol*, **56**, 401–5.

Kudolo, G. B., Dorsey, S. and Blodgett, J. (2002) Effect of the ingestion of Ginkgo biloba extract on platelet aggregation and urinary prostanoid excretion in healthy and Type 2 diabetic subjects. *Thromb Res*, **108**, 151–60.

Kupferschmidt, H. H., Ha, H. R., Ziegler, W. H. *et al.* (1995) Interaction between grapefruit juice and midazolam in humans. *Clin Pharmacol Ther*, **58**, 20–8.

Lamant, V., Mauco, G., Braquet, P. *et al.* (1987) Inhibition of the metabolism of platelet activating factor (PAF-acether) by three specific antagonists from Ginkgo biloba. *Biochem Pharmacol*, **36**, 2749–52.

Lee, M. J., Maliakal, P., Chen, L. *et al.* (2002) Pharmacokinetics of tea catechins after ingestion of green tea and (−)-epigallocatechin-3-gallate by humans: formation of different metabolites and individual variability. *Cancer Epidem Biomarkers Prev*, **11**, 1025–32.

Lei, H. P., Wang, G., Wang, L. S. *et al.* (2009) Lack of effect of Ginkgo biloba on voriconazole pharmacokinetics in Chinese volunteers identified as CYP2C19 poor and extensive metabolizers. *Ann Pharmacother*, **43**, 726–31.

Letrent, S. P., Pollack, G. M., Brouwer, K. R. and Brouwer, K. L. (1999) Effects of a potent and specific P-glycoprotein inhibitor on the blood–brain barrier distribution and antinociceptive effect of morphine in the rat. *Drug Metab Dispos*, **27**, 827–34.

Li, C. and Choi, J. S. (2008) Effects of epigallocatechin gallate on the bioavailability and pharmacokinetics of diltiazem in rats. *Pharmazie*, **63**, 815–8.

Lin, L. C., Wang, M. N. and Tsai, T. H. (2008) Food-drug interaction of (−)-epigallocatechin-3-gallate on the pharmacokinetics of irinotecan and the metabolite SN-38. *Chem Biol Interact*, **174**, 177–82.

Lu, H., Meng, X. and Yang, C. S. (2003) Enzymology of methylation of tea catechins and inhibition of catechol-*O*-methyltransferase by (−)-epigallocatechin gallate. *Drug Metab Dispos*, **31**, 572–9.

Manach, C., Williamson, G., Morand, C. *et al.* (2005) Bioavailability and bioefficacy of polyphenols in humans. I. Review of 97 bioavailability studies. *Am J Clin Nutr*, **81**, 230S–242S.

Marchetti, S., Mazzanti, R., Beijnen, J. H. and Schellens, J. H. (2007) Concise review: Clinical relevance of drug drug and herb drug interactions mediated by the ABC transporter ABCB1 (MDR1, P-glycoprotein). *Oncologist*, **12**, 927–41.

Markowitz, J. S., Donovan, C. L., Devane, C. L. *et al.* (2003) Multiple doses of saw palmetto (*Serenoa repens*) did not alter cytochrome P450 2D5 and 3A4 activity in normal volunteers. *Clin Pharmacol Ther*, **74**, 536–42.

Mauri, P., Simonetti, P., Gardana, C. *et al.* (2001) Liquid chromatography/atmospheric pressure chemical ionization mass spectrometry of terpene lactones in plasma of volunteers dosed with Ginkgo biloba L. extracts. *Rapid Commun Mass Spectrom*, **15**, 929–34.

Mckenna, D. J., Jones, K. and Hughes, K. (2001) Efficacy, safety, and use of ginkgo biloba in clinical and preclinical applications. *Altern Ther Health Med*, **7**, 70–86, 88–90.

Mertens-Talcott, S. U., Zadezensky, I., De Castro, W. V. *et al.* (2006) Grapefruit–drug interactions: can interactions with drugs be avoided? *J Clin Pharmacol*, **46**, 1390–416.

Mohamed, M. E. and Frye, R. F. (2011) Inhibitory effects of commonly used herbal extracts on UDP-glucuronosyltransferase 1A4, 1A6, and 1A9 enzyme activities. *Drug Metab Dispos*, **39**, 1522–8.

Mohamed, M. F., Tseng, T. and Frye, R. F. (2010) Inhibitory effects of commonly used herbal extracts on UGT1A1 enzyme activity. *Xenobiotica*, **40**, 663–9.

Mohammed Abdul, M. I., Jiang, X., Williams, K. M. *et al.* (2008) Pharmacodynamic interaction of warfarin with cranberry but not with garlic in healthy subjects. *Br J Pharmacol*, **154**, 1691–700.
Mohutsky, M. A., Anderson, G. D., Miller, J. W. and Elmer, G. W. (2006) Ginkgo biloba: evaluation of CYP2C9 drug interactions in vitro and in vivo. *Amer J Ther*, **13**, 24–31.
Muto, S., Fujita, K., Yamazaki, Y. and Kamataki, T. (2001) Inhibition by green tea catechins of metabolic activation of procarcinogens by human cytochrome P450. *Mutat Res*, **479**, 197–206.
Nabekura, T., Kamiyama, S. and Kitagawa, S. (2005) Effects of dietary chemopreventive phytochemicals on P-glycoprotein function. *Biochem Biophys Res Commun*, **327**, 66–70.
Nagai, M., Conney, A. H. and Zhu, B. T. (2004) Strong inhibitory effects of common tea catechins and bioflavonoids on the *O*-methylation of catechol estrogens catalyzed by human liver cytosolic catechol-*O*-methyltransferase. *Drug Metab Dispos*, **32**, 497–504.
Nahin, R. L., Pecha, M., Welmerink, D. B. *et al.* (2009) Concomitant use of prescription drugs and dietary supplements in ambulatory elderly people. *J Am Geriatr Soc*, **57**, 1197–205.
Nakagawa, K., Okuda, S. and Miyazawa, T. (1997) Dose-dependent incorporation of tea catechins, (−)-epigallocatechin-3-gallate and (−)-epigallocatechin, into human plasma. *Biosci Biotechnol Biochem*, **61**, 1981–5.
Neuvonen, P. J., Niemi, M. and Backman, J. T. (2006) Drug interactions with lipid-lowering drugs: mechanisms and clinical relevance. *Clin Pharmacol Ther*, **80**, 565–81.
Nishikawa, M., Ariyoshi, N., Kotani, A. *et al.* (2004) Effects of continuous ingestion of green tea or grape seed extracts on the pharmacokinetics of midazolam. *Drug Metab Pharmacokinet*, **19**, 280–9.
Nishimuta, H., Ohtani, H., Tsujimoto, M. *et al.* (2007) Inhibitory effects of various beverages on human recombinant sulfotransferase isoforms SULT1A1 and SULT1A3. *Biopharm Drug Dispos*, **28**, 491–500.
Oki, T., Suzuki, M., Nishioka, Y. *et al.* (2005) Effects of saw palmetto extract on micturition reflex of rats and its autonomic receptor binding activity. *J Urol*, **173**, 1395–9.
Okura, T., Ozawa, T., Ito, Y. *et al.* (2008) Enhancement by grapefruit juice of morphine antinociception. *Biol Pharm Bull*, **31**, 2338–41.
Okura, T., Ozawa, T., Ibe, M. *et al.* (2009) Effects of repeated morphine treatment on the antinociceptive effects, intestinal absorption, and efflux from intestinal epithelial cells of morphine. *Biol Pharm Bull*, **32**, 1571–5.
Okura, T., Ibe, M., Umegaki, K. *et al.* (2010) Effects of dietary ingredients on function and expression of P-glycoprotein in human intestinal epithelial cells. *Biol Pharm Bull*, **33**, 255–9.
Paine, M. F., Widmer, W. W., Hart, H. L. *et al.* (2006) A furanocoumarin-free grapefruit juice establishes furanocoumarins as the mediators of the grapefruit juice-felodipine interaction. *Am J Clin Nutr*, **83**, 1097–105.
Park, D., Jeon, J. H., Shin, S. *et al.* (2009) Green tea extract increases cyclophosphamide-induced teratogenesis by modulating the expression of cytochrome P-450 mRNA. *Reprod Toxicol*, **27**, 79–84.
Qian, F., Wei, D., Zhang, Q. and Yang, S. (2005) Modulation of P-glycoprotein function and reversal of multidrug resistance by (−)-epigallocatechin gallate in human cancer cells. *Biomed Pharmacother*, **59**, 64–9.
Reagan-Shaw, S., Nihal, M. and Ahmad, N. (2008) Dose translation from animal to human studies revisited. *FASEB J*, **22**, 659–61.
Robertson, S. M., Davey, R. T., Voell, J. *et al.* (2008) Effect of Ginkgo biloba extract on lopinavir, midazolam and fexofenadine pharmacokinetics in healthy subjects. *Curr Med Res Opin*, **24**, 591–9.
Robey, R. W., To, K. K., Polgar, O. *et al.* (2009) ABCG2: a perspective. *Adv Drug Deliv Rev*, **61**, 3–13.
Roby, C. A., Anderson, G. D., Kantor, E. *et al.* (2000) St John's Wort: effect on CYP3A4 activity. *Clin Pharmacol Ther*, **67**, 451–7.
Roth, M., Timmermann, B. N. and Hagenbuch, B. (2011) Interactions of green tea catechins with organic anion-transporting polypeptides. *Drug Metab Dispos*, **39**, 920–6.
Sand, P. G., Dreiseitel, A., Stang, M. *et al.* (2010) Cytochrome P450 2C19 inhibitory activity of common berry constituents. *Phytother Res*, **24**, 304–7.
Sasaki, K., Wada, K., Hatta, S. *et al.* (1997) Bilobalide, a constituent of Ginkgo biloba L., potentiates drug-metabolizing enzyme activities in mice: possible mechanism for anticonvulsant activity against 4-O-methylpyridoxine-induced convulsions. *Res Comm Mol Pathol Pharmacol*, **96**, 45–56.
Scalbert, A. and Williamson, G. (2000) Dietary intake and bioavailability of polyphenols. *J Nutr*, **130**, 2073S–85S.

Schinkel, A. H., Wagenaar, E., Van Deemter, L. *et al.* (1995) Absence of the mdr1a P-Glycoprotein in mice affects tissue distribution and pharmacokinetics of dexamethasone, digoxin, and cyclosporin A. *J Clin Invest*, **96**, 1698–705.

Shapiro, A. B. and Ling, V. (1997) Effect of quercetin on Hoechst 33342 transport by purified and reconstituted P-glycoprotein. *Biochem Pharmacol*, **53**, 587–96.

Shin, S. C. and Choi, J. S. (2009) Effects of epigallocatechin gallate on the oral bioavailability and pharmacokinetics of tamoxifen and its main metabolite, 4-hydroxytamoxifen, in rats. *Anticancer Drugs*, **20**, 584–8.

Shinozuka, K., Umegaki, K., Kubota, Y. *et al.* (2002) Feeding of Ginkgo biloba extract (GBE) enhances gene expression of hepatic cytochrome P-450 and attenuates the hypotensive effect of nicardipine in rats. *Life Sci*, **70**, 2783–92.

Shukla, S., Zaher, H., Hartz, A. *et al.* (2009) Curcumin inhibits the activity of ABCG2/BCRP1, a multidrug resistance-linked ABC drug transporter in mice. *Pharm Res*, **26**, 480–7.

Sierpina, V. S., Wollschlaeger, B. and Blumenthal, M. (2003) Ginkgo biloba. *Am Fam Physician*, **68**, 923–6.

Snitz, B. E., O'Meara, E. S., Carlson, M. C. *et al.* (2009) Ginkgo biloba for preventing cognitive decline in older adults: a randomized trial. *JAMA*, **302**, 2663–70.

Sohn, O. S., Surace, A., Fiala, E. S. *et al.* (1994) Effects of green and black tea on hepatic xenobiotic metabolizing systems in the male F344 rat. *Xenobiotica*, **24**, 119–27.

Soldner, A., Christians, U., Susanto, M. *et al.* (1999) Grapefruit juice activates P-glycoprotein-mediated drug transport. *Pharm Res*, **16**, 478–85.

Sugiyama, T., Kubota, Y., Shinozuka, K. *et al.* (2004a) Ginkgo biloba extract modifies hypoglycemic action of tolbutamide via hepatic cytochrome P450 mediated mechanism in aged rats. *Life Sci*, **75**, 1113–22.

Sugiyama, T., Kubota, Y., Shinozuka, K. *et al.* (2004b) Induction and recovery of hepatic drug metabolizing enzymes in rats treated with Ginkgo biloba extract. *Food and chemical toxicology*, **42**, 953–7.

Sugiyama, T., Shinozuka, K., Sano, A. *et al.* (2004c) Effects of various ginkgo biloba extracts and proanthocyanidin on hepatic cytochrome P450 activity in rats. *Shokuhin Eiseigaku Zasshi*, **45**, 295–301.

Suzuki, M., Oki, T., Sugiyama, T. *et al.* (2007) Muscarinic and alpha-1-adrenergic receptor binding characteristics of saw palmetto extract in rat lower urinary tract. *Urol*, **69**, 1216–20.

Suzuki, M., Ito, Y., Fujino, T. *et al.* (2009) Pharmacological effects of saw palmetto extract in the lower urinary tract. *Acta Pharmacol Sin*, **30**, 271–81.

Synold, T. W., Dussault, I. and Forman, B. M. (2001) The orphan nuclear receptor SXR coordinately regulates drug metabolism and efflux. *Nat Med*, **7**, 584–90.

Takanaga, H., Ohnishi, A., Matsuo, H. and Sawada, Y. (1998) Inhibition of vinblastine efflux mediated by P-glycoprotein by grapefruit juice components in caco-2 cells. *Biol Pharm Bull*, **21**, 1062–6.

Taki, Y., Yamazaki, Y., Shimura, F. *et al.* (2009) Time-dependent induction of hepatic cytochrome P450 enzyme activity and mRNA expression by bilobalide in rats. *J Pharmacol Sci*, **109**, 459–62.

Taki, Y., Hagiwara, E., Hirose, C. *et al.* (2011) Effects of Ginkgo biloba extract on the pharmacokinetics and pharmacodynamics of tolbutamide in protein-restricted rats. *J Pharm Pharmacol*, **63**, 1238–43.

Taki, Y., Yokotani, K., Yamada, S. *et al.* (2012) Ginkgo biloba extract attenuates warfarin-mediated anticoagulation through induction of hepatic cytochrome P450 enzymes by bilobalide in mice. *Phytomedicine*, **19**, 177–82.

Uchida, S., Yamada, H., Li, X. D. *et al.* (2006) Effects of ginkgo biloba extract on pharmacokinetics and pharmacodynamics of tolbutamide and midazolam in healthy volunteers. *J Clin Pharmacol*, **46**, 1290–8.

Ulbricht, C., Chao, W., Costa, D. *et al.* (2008) Clinical evidence of herb–drug interactions: a systematic review by the natural standard research collaboration. *Curr Drug Metab*, **9**, 1063–120.

Umegaki, K., Saito, K., Kubota, Y. *et al.* (2002) Ginkgo biloba extract markedly induces pentoxyresorufin O-dealkylase activity in rats. *Jpn J Pharmacol*, **90**, 345–51.

Umegaki, K., Taki, Y., Endoh, K. *et al.* (2007) Bilobalide in Ginkgo biloba extract is a major substance inducing hepatic CYPs. *J Pharm Pharmacol*, **59**, 871–7.

Wang, X., Wolkoff, A. W. and Morris, M. E. (2005) Flavonoids as a novel class of human organic anion-transporting polypeptide OATP1B1 (OATP-C) modulators. *Drug Metab Dispos*, **33**, 1666–72.

Werba, J. P., Giroli, M., Cavalca, V. *et al.* (2008) The effect of green tea on simvastatin tolerability. *Ann Intern Med*, **149**, 286–7.

Wilkinson, G. R. (2005) Drug metabolism and variability among patients in drug response. *N Engl J Med*, **352**, 2211–21.

Williamson, E. M. (2001) Synergy and other interactions in phytomedicines. *Phytomedicine*, **8**, 401–9.

Xie, R., Hammarlund-Udenaes, M., De Boer, A. G. and De Lange, E. C. (1999) The role of P-glycoprotein in blood-brain barrier transport of morphine: transcortical microdialysis studies in mdr1a (−/−) and mdr1a (+/+) mice. *Br J Pharmacol*, **128**, 563–8.

Yang, C. S., Wang, X., Lu, G. and Picinich, S. C. (2009) Cancer prevention by tea: animal studies, molecular mechanisms and human relevance. *Nat Rev Cancer*, **9**, 429–39.

Yang, S. P. and Raner, G. M. (2005) Cytochrome P450 expression and activities in human tongue cells and their modulation by green tea extract. *Toxicol Appl Pharmacol*, **202**, 140–50.

Yeung, E. Y., Sueyoshi, T., Negishi, M. and Chang, T. K. (2008) Identification of Ginkgo biloba as a novel activator of pregnane X receptor. *Drug Metab Dispos*, **36**, 2270–6.

Zadoyan, G., Rokitta, D., Klement, S. *et al.* (2012) Effect of Ginkgo biloba special extract EGb 761(R) on human cytochrome P450 activity: a cocktail interaction study in healthy volunteers. *Eur J Clin Pharmacol*, **68**, 553–60.

Zhang, S., Yang, X. and Morris, M. E. (2004) Flavonoids are inhibitors of breast cancer resistance protein (ABCG2)-mediated transport. *Mol Pharmacol*, **65**, 1208–16.

Zhao, L. Z., Huang, M., Chen, J. *et al.* (2006) Induction of propranolol metabolism by Ginkgo biloba extract EGb 761 in rats. *Curr Drug Metab*, **7**, 577–87.

Zhu, B. T., Taneja, N., Loder, D. P. *et al.* (1998) Effects of tea polyphenols and flavonoids on liver microsomal glucuronidation of estradiol and estrone. *J Steroid Biochem Mol Biol*, **64**, 207–15.

Zong, J. and Pollack, G. M. (2000) Morphine antinociception is enhanced in mdr1a gene-deficient mice. *Pharm Res*, **17**, 749–53.

5 Health Properties of Apple and Pear

Richard Espley and Stefan Martens

INTRODUCTION

The potential health effects of apple have long been recognized. The proverbial saying 'an apple a day keeps the doctor away' can be traced back to a published statement, 'A Pembrokeshire proverb: Eat an apple on going to bed, and you'll keep the doctor from earning his bread' (Phillips, 1866). While the science behind this statement may have been lacking at the time, apple fruit has for many years carried the expectation of providing dietary health benefits. It is still a considerable challenge to fully understand the role of bioactives in the foods that contribute to a healthy diet but it presents significant opportunities for an integrated approach from both plant and health researchers (Martin *et al.*, 2011). This chapter aims to review our understanding of the health properties of apple. Pears and other pipfruit have been the subject of less research but will also be referred to where appropriate.

THE ORIGIN OF APPLES

The eminent Russian plant scientist Nikolai Vavilov was instrumental in establishing the origin of many species (Vavilov, 1931) and his theory, concerning the geographical centre of origin of apple and pear, has provided the inspiration for others. Most notably the research of Aimak Dzhangaliev, who dedicated his career to the conservation of the wild pome species (Dzhangaliev, 2003). The region Vavilov identified as the origin was around the Tien Shan Mountains in central Kazakhstan. This region contains the centre of origin for apple (*Malus*) and related genera such as pear (*Pyrus*), rowan or mountain ash (*Sorbus*), as well as being the centre of origin for apricot (Karychev *et al.*, 2005). It is here that *Malus sieversii* is found and there is genetic evidence that this species is the progenitor of the cultivated apple (Harris *et al.*, 2002). Wild apples have been discovered in a region that stretches from Central Asia to Western China (Zhang *et al.*, 1993; Watkins, 1995). *Malus* is classified as a member of the Rosaceae family with around 55 wild species that constitute the genus *Malus* (Way *et al.*, 1991; Harris *et al.*, 2002; Forsline *et al.*, 2003). These species are extremely variable in tree form, fruit size and fruit colour (for collections in the United Kingdom and the United States of America see: www.nationalfruitcollection.org.uk/ and www.ars.usda.gov/main/main.htm). Hybridization of these species would have been

Bioactives in Fruit: Health Benefits and Functional Foods, First Edition. Edited by Margot Skinner and Denise Hunter.

possible by seed dispersal by birds as well as by ruminants such as deer and antelope and bear (Juniper and Mabberley, 2006). The proximity of the Old Silk Road to the centre of origin is likely to have played an important role in the East–West dispersal of *Malus* species into Europe and China.

DOMESTICATION OF APPLES

The process of apple domestication would have been accelerated by the increasing popularity of ancient trading routes, particularly the Old Silk Road. Horses, donkeys and mules would have eaten the fruit and distributed seed along the routes. This may have opened up hybridization possibilities for species in geographically isolated areas. It is thought that in the Central Asian region there are as many as 6000 varieties (Juniper and Mabberley, 2006). It is likely that apples for human consumption would have spread from Central Asia along these trade routes. Initially, domesticated apples would have been grown from seed and random hybridizations would have taken place, but with the advent of new technologies, such as grafting, it was possible to produce more uniform crops of the best selections (Juniper *et al.*, 1998). By 1826, The Royal Horticultural Society (UK) listed more than 1200 varieties and today there are more than 20 000 named varieties. However, despite this huge genetic resource, only a small number of varieties are currently under commercial production.

APPLE PRODUCTION

World fruit production stands at around 500 million tonnes, with apple one of the most important crops, accounting for around 12% of total fruit production. Apple production value is estimated at US$10 billion (Food and Agriculture Organization of the United Nations [www.fao.org]). Apple is the most versatile of the temperate fruit crops, capable of growing in diverse conditions and severe temperature extremes from high latitude regions to high elevations in the tropics (Forsline *et al.*, 2003).

In economic terms, apple is the most important of the Rosaceae family. China accounts for 36% of total production, far in excess of the next largest producer, the United States of America, with 7.5% (Figure 5.1). Turkey, France, Italy and Iran are all leading exporters for Northern Hemisphere production, whilst Southern Hemisphere producers and exporters are primarily Chile, Argentina, New Zealand and South Africa (O'Rourke, 2007).

Commercial production is limited to relatively few varieties which include 'Golden Delicious', 'Red Delicious', 'McIntosh' and 'Jonagold' (derived from the USA), 'Fuji' (Japan), 'Granny Smith' (Australia) and 'Royal Gala', 'Braeburn' and 'Jazz' (New Zealand).

MAJOR CLASSES OF APPLE BIOACTIVES

In general, most of the bioactive compounds found in apple vary greatly between different varieties and genotypes. The chemical profile is also strongly influenced by growth period, growing season, pre- and postharvest conditions, orchard practice, climate conditions/geographical locations and other biotic and abiotic factors (Ceymann *et al.*, 2012). The five predominant classes of polyphenols are represented by derivatives belonging to the subclasses of phenolic acids, flavan-3-ols, dihydrochalcones, flavonols and, to some extent

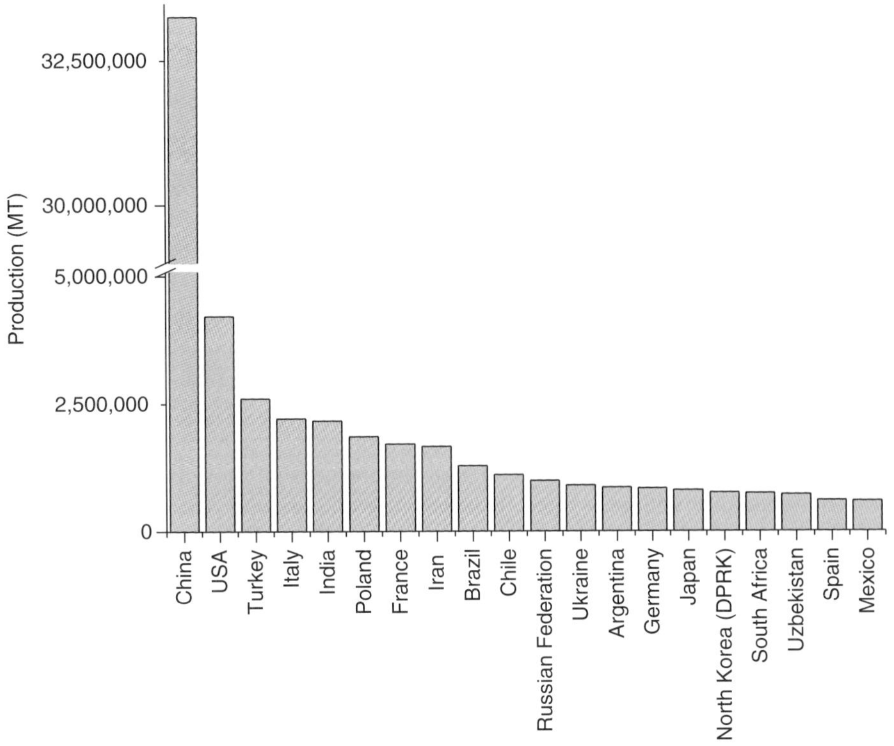

Fig. 5.1 Top twenty apple producers by country with estimated production in tonnes. Total world production is estimated at 55 million tonnes, updated June 2010 (data from www.fao.org).

also, anthocyanins; the first two are considered to be the major classes. Flavan-3-ols and its oligomers can make up to 90% of the total polyphenols in apple fruits and phenolic acid up to 31% (Wojdylo *et al.*, 2008). However, all five classes are directly or indirectly involved in the health effects of apple or other foodstuffs and have a large range of values in different cultivars.

Dihydrochalcones – phloridzin

Dihydrochalcone derivatives such as phloridzin (phloretin 2′-*O*-glucoside) and phloretin 2′-*O*-xylosylglucoside are generally accepted as characteristic compounds found in apple tissues, especially in *Malus* × *domestica* (*M. pumila*), and are found in both skin and cortex (Awad *et al.*, 2000; Lommen *et al.*, 2000). In apple fruits dihydrochalcone derivatives can reach up to 5% of total polyphenols (Wojdylo *et al.*, 2008). In skin their accumulation can reach more than 10-fold the concentration present in fruit flesh and is correlated with the occurrence of flavan-3-ols and flavonols (Chagné *et al.*, 2012a). In certain conditions or in specific cultivars the aglycon phloretin was described in traces (Wojdylo *et al.*, 2008). Other derivatives of this group were identified only in *Malus* spec., such as trilobatin in *Malus trilobata* or sieboldin in *Malus sieboldii* (Hunter, 1975; de Bernonville *et al.*, 2010). Due to the obvious absence in pear, dihydrochalcones are frequently used as a specific marker to control the authenticity of apple products, such as juices.

Flavonols – quercetin, kaempferol

Various vacuolar flavonols accumulate in the epidermal tissue of the fruit skin as UV-B absorbing pigments and are virtually absent or only detectable in trace amounts in the cortex (Chagné *et al.*, 2012a). However, they can still account for up to 10% of total polyphenols in fruits (Vrhovsek *et al.*, 2004; Wojdylo *et al.*, 2008). Major members of this group described in apple fruit tissues are quercetin (Qu) 3-*O*-glycosides, which were identified as the galactoside (isoquercitrin), -glucoside (isoquercetrin), -rhamnoside (quercitrin), -rutinoside (rutin), -xyloside (reynoutrin), -arabinopyranoside (guajaverin) and -arabinofuranoside (avicularin). The most abundant flavonol is Qu-3-*O*-galactoside while the aglycon is detectable only in few cases, suggesting a strong activity of the respective glycosyltransferases. A similar pattern was found in pear fruits with Qu 3-*O*-glucoside dominating over the respective 3-*O*-galactoside and 3-*O*-rutinoside. However, the most predominant Qu derivatives found in pear are the methylated isorhamnetin (isorhamnetin) glycosides. Isorha 3-*O*-glucoside was identified by Duggan (1969) and later described as omnipresent in pear cultivars by Schieber *et al.* (2001). The authors suggested the use of this metabolite as a marker to detect adulterations of apple juice with pear juice. This was revised only one year later when the same authors detected the compound to be present together with isorhamnetin 3-*O*-galactoside in the fruit skin of one apple cultivar (Schieber *et al.*, 2002). Additionally, Qu-7-*O*-xyloside, isorhamnetin 3-*O*-rhamnogalactoside, 3-*O*-galactoside and 3-*O*-rutinoside (narcissin) were identified in pear peels (Nortje, 1965; Duggan, 1969; Schieber *et al.*, 2001). Isorhamnetin diglycosides and acylated isorhamnetin glycosides are also described in pear fruits (Wald *et al.*, 1989).

Flavan-3-ols – catechins, epicatechins and its di-, oligo- and polymers

(–)–Epicatechin and procyanidin B2 were found to be the major polyphenolic compounds with a flavonoid structure in almost all apple cultivars (Vrhovsek *et al.*, 2004; Ceymann *et al.*, 2012; Chagné *et al.*, 2012a). High amounts of the monomer epicatechin were found to correlate significantly with the presence of procyanidin B2, which is an epicatechin dimer. Beside these two compounds (+)–catechin and procyanidin B1 were detectable in many cultivars but the concentration can range from below the limit of detection to several milligrams per 100 g FM in different cultivars. Procyanidin B1 is a dimer of both flavan-3-ol monomers, epicatechin and catechin, described in apple. Chagné *et al.* (2012a) showed the occurrence of five not yet identified proanthocyanidins found in both skin and cortex tissue of segregating population while Wojdylo *et al.* (2008) identified the trimer procyanidin C1. Furthermore, they quantified oligomeric procyanidins in old and new apple varieties and described a degree of polymerization between 3 and 29, which can reach 50 in cider apples. However, no tissue specific accumulation was detectable for this group of compounds. Henry-Kirk *et al.* (2012) described several apple cultivars whose epicatechin profile varied by as much as twofold from each other. Epicatechin was also reported in pear cultivars by Schieber *et al.* (2001).

Anthocyanins – cyanidins

Some apple cultivars are not able to produce anthocyanin at all or only under high-light conditions in minor amounts in the skin (e.g. 'Granny Smith', 'Golden Delicious') while others permanently express the respective pathway and thus impart red colour to the fruit (e.g. 'Summer Red') (Merzlyak *et al.*, 2008). Even within the same variety the concentration can vary due to environmental effects on the biosynthetic pathway. The predominant tissue for accumulation is the fruit skin but red-fleshed apple cultivars and species have been described (van Nocker *et al.*, 2012). The major anthocyanins follow the same pattern of accumulation as 3′,4′-hydroxylated compounds (quercetin, catechin, epicatechin), with up to four different cyanidin (Cy) derivatives detectable in red tissues. Among the Cy derivatives identified, the 3-*O*-galactoside accounts for the highest concentration. Other identified anthocyanins are Cy-3-*O*-arabinoside, Cy-3-*O*-glucoside and Cy-7-*O*-arabinoside (Vrhovsek *et al.*, 2004; Wojdylo *et al.*, 2008; Chagné *et al.*, 2012a). The aglycone Cy itself was only identified in minor amounts in some fruits and seems to be glycosylated very efficiently.

Carotenoids

Apple fruit is generally not considered a rich source of carotenoids and while levels are generally low in the flesh carotenoids do contribute to peel colour where varied levels of β-carotene, lutein, violaxanthin and antheraxanthin and others are often present (Knee, 1972; Felicetti and Schrader, 2009). Extrathylakoid carotenoids pigments are suggested to be involved in stress regulation in senescing leaves and also in ripening fruits. Ripening is generally accompanied by a degradation of chlorophylls and the build-up of xanthophylls (Merzlyak *et al.*, 2008). Ampomah-Dwamena *et al.* (2012) described the biosynthesis of carotenoids in genotypes of apple accumulating high levels of flesh and skin carotenoid. Fruit of commercial apple cultivars shows relatively low concentrations of carotenoids (<2.5 μg/g fresh weight [FW]) in comparison with fruit such as citrus (25 μg/g), papaya (60 μg/g) and persimmons (15 μg/g). In apple skin, this carotenoid is mainly lutein, while in flesh carotenoid can be beta-carotene, neoxanthin or violaxanthin.

Other phytochemicals/bioactives

Along with epicatechin and procyandin B2, the phenolic acid (hydroxycinnamate) chlorogenic acid (3-*O*-caffeoyl quinic acid) is one of the major individual polyphenolic compounds in apple varieties, with similar range as the two flavan-3-ols. Additionally, *p*-coumaroyl quinic acid was described in considerable amounts in whole apple extracts. Based on this finding Ceymann *et al.* (2012) suggested the division of apples into flavan-3-ol or phenolic acid predominant cultivars based on the obtained ratio of these groups. Generally, it is described that bitter varieties and cider apples contain higher concentrations of these phenolic acids than non-bitter genotypes and dessert apples (Wojdylo *et al.*, 2008). In contrast to the flavan-3-ols, phenolic acids show a more strict tissue specific accumulation, as they accumulate in the cortex and account for the main polyphenolic compound in this tissue. In skin they represent a small proportion of the total phenolics (Wojdylo *et al.*, 2008; Chagné *et al.*, 2012a). As with apple, the predominant phenolic acid compounds in pear are chlorogenic and *p*-coumaroyl quinic acid. Additionally, considerable amounts of caffeic, *p*-coumaric and ferulic acid are also present in both fruits (Schieber *et al.*, 2001). Chen *et al.* (2012) also identified gallic acid, a benzoic acid derivate, which was found together with ferulic, caffeic and *p*-coumaric acids in mainly the peel of three apple cultivars. He and Liu (2007) isolated and identified 13 triterpenoids as bioactive constituents in apple peel by bioactivity-guided fractionation. The major compound of this group was identified as ursolic acid, which was present in the largest quantity. Other structures identified are 2a-hydroxyursolic acid and the oleane-type triterpenoid maslinic acid plus their derivatives substituted with various *p*-coumaroyloxy units.

Arbutin, hydrochinon *O*-glucoside, was described as a unique marker of pear. However, it has become apparent that in some cultivars this compound is only found in very low amounts or is even absent. Therefore, the use as possible characteristic markers for the detection of admixtures in apple products is under discussion.

Besides the polyphenolic compounds thought to have antioxidant activities, apples are also known to contain a relatively high concentration of ascorbic acid. Due to this fact apples are an important source of ascorbic acid for the daily recommended dose (Vrhovsek *et al.*, 2004). Apple fruit varieties do vary in their vitamin C level. Total fruit L-ascorbate (L-AA) levels can range from 401 to 1448 nmol/g fresh weight (FW), a maximum 3.6-fold variation with commercial varieties such as 'Fuji' with 471 nmol/gFW and 'Braeburn' with 1283 nmol/gFW (Davey and Keulemans, 2004).

BIOSYNTHESIS AND GENETIC REGULATION OF APPLE BIOACTIVES

Dihydrochalcones

Although dihydrochalcones are closely related to flavonoid compounds from the biochemical point of view, knowledge of their biosynthetic pathways and the structural and regulatory genes is still limited. Recent enzymatic studies provide evidence that three steps lead to the formation of phloridzin the most abundant metabolite from this group found in apple. Starting from the common precursor of the competing flavonoid pathway, *p*-coumaroyl-CoA is reduced to *p*-dihydrocoumaroyl-CoA by the activity of a NADPH-dependent double bond reductase (DBR). This enzyme activity was detected by Gosch *et al.* (2009) in crude protein extract of apple leaves and also from other non-dihydrochalcone accumulating plants, which led to the assumption that the formation is not *Malus* specific and exhibited this functionality only as a side activity. They postulated that the absence of dihydrochalcone derivatives in *Pyrus* is due to the lack of a functional double bound reductase. However, recently Idbah *et al.* (personal communication) unambigiously identified an alkenal type double bond reductase specific for *p*-coumaroyl-CoA that catalysed the double bond reduction to the corresponding dihydrocompound. This enzyme was only active with *p*-coumaroyl-CoA and no other precursor of the pathway. Both crude enzyme extracts from apple leaves and isolates of four common recombinant chalcone synthases were found to be able to condense the previously formed CoA ester with three molecules of malonyl-CoA to phloretin without showing any substrate preference for the nature of CoA ester of dihydrochalcone or flavonoid pathway (Gosch *et al.*, 2009). The last step to form phloridzin is achieved by a highly substrate- and regio-specific phloretin 2′-glucosyltransferase recently cloned and functionally characterized by Judge *et al.* (2008). The overall regulation of this pathway branch is still not understood and requires further attention in the future.

Flavonoids

The regulation of non-anthocyanic flavonoids such as flavanols and proanthocyanidins (PAs, also referred to as condensed tannins) is less well understood in apple. Flavonols, including the quercetin glycosides, are generally restricted to the peel, as are anthocyanins. Other phenolics, for example chlorogenic acid, epicatechin, PAs and phlorizidin, can be present in both the peel and cortex (McGhie *et al.*, 2005). There are major cultivar differences for all bioactives, including phenolics, and these can be compounded by orchard management, geographical location and seasonal environmental variation (van der Sluis *et al.*, 2001; Tsao *et al.*, 2003; McGhie *et al.*, 2005). While it is likely that in apple and pear the regulation of flavonoids is controlled by transcription factors, as in other species (Nesi *et al.*, 2000, 2001), this has not been reported to date. Similarly, there are gaps in our knowledge on the biosynthesis of important polyphenols but many of these genes have now been successfully mapped to the apple genome and can be tracked in segregating populations, including the *HQT/HCT* genes involved in chlorogenic acid synthesis (Chagné *et al.*, 2012b). Other key biosynthetic enzymes for the production of catechin and epicatechin are well known, including LAR1/LAR2, which are involved in the synthesis of 2,3-*trans*-flavan-3-ol from luecoanthocyanidins, and ANR, which synthesises anthocyanidins into 2,3-*cis*-flavan-3-ols as well as FLS1 for the production of flavonols (Henry-Kirk *et al.*, 2012).

Anthocyanins

Anthocyanin levels in plants are modulated in response to environmental, developmental and stress signals. These include temperature, nutrient status, wounding, pathogen infection, water stress and light (Dixon and Paiva, 1995; Chalker-Scott, 1999). The major precursors for all flavonoids, including anthocyanins, are malonyl-CoA and *p*-coumaroyl-CoA (Dixon and Steele, 1999). From these precursors the enzyme chalcone synthase (*CHS*) forms chalcone, the first committed step in anthocyanin production and the establishment of the C_{15} backbone. Chalcone is then isomerized by chalcone isomerase (*CHI*) to produce chalcone naringenin, and from there a hydroxylation step, via flavanone 3β-hydroxylase (*F3H*), converts naringenin to dihydroflavonol. The reduction of dihydroflavonol by dihydroflavonol 4-reductase (*DFR*) produces leucoanthocyanidin and this is converted into the coloured compound anthocyanidin by leucoanthocyanidin dioxygenase (*LDOX*) (also known as anthocyanidin synthase, ANS). There follows a glycosylation step, usually mediated by uridine diphosphate (UDP)-glucose:flavonoid 3-*0*-glucosyltransferase (*UFGT*). Subsequent additions of sugars and acyl side groups may follow in some species. Colour variation may depend on the number of hydroxyl groups on the B-ring, the sugars and acyl side groups (Harborne, 1967), the environment of the vacuole including its pH (Koes *et al.*, 2005) or the accumulation of specific metal ions (Brouillard, 1988) or cellular ultrastructure (Noda *et al.*, 1994). One of the most common anthocyanin pigments is cyanidin, which, in the form of cyanidin 3-*0*-galactoside, is the pigment primarily responsible for red colouration in apple skin (Lancaster, 1992; Tsao *et al.*, 2003) and pear (Steyn *et al.*, 2004). The enzymes operating in this biosynthetic pathway in apple have been well characterized (Honda *et al.*, 2002; Kim *et al.*, 2003).

The biosynthesis of anthocyanin pigments has been well studied and is transcriptionally regulated by a combination of transcription factors, the MYB-bHLH-WD40 (MBW) complex (Allan *et al.*, 2008). In apple, the anthocyanin pathway is controlled by MYB transcription factors and genetic evidence suggests that these genes are the major controllers of colour in apple peel (Zhu *et al.*, 2010), flesh and foliage (Chagné *et al.*, 2007). The MYB10 TF interacts with bHLH3 (Espley *et al.*, 2007), which in turn interacts with the WD40 protein TTG1 (An *et al.*, 2012). This complex regulates the structural enzymes in the anthocyanin pathway to drive accumulation. This transcriptional complex is affected by environmental conditions, such as a repressive effect in hot temperatures (Lin-Wang *et al.*, 2011). A similar transcriptional complex is likely to perform this regulation in pear (Feng *et al.*, 2010; Lin-Wang *et al.*, 2010).

Carotenoids

Carotenoids are a diverse group of plant pigments that confer the yellow, orange and red colours seen in many flowers and fruits. They are made up of isoprenoid units and provide essential roles in phytohormone production, photoprotection and light harvesting support in plant tissues (Howitt and Pogson, 2006). Humans cannot synthesize carotenoids, so plant-derived carotenoids constitute an important part of dietary health and are responsible for provitamin A and promote antioxidant activity (Fraser and Bramley, 2004). Unlike anthocyanins, there does not appear to be a specific set of transcription factors that controls carotenoid accumulation. Rather, a number of different mechanisms, including rate limiting enzymes within the pathway that control flux, substrate diversion into other pathways and changes to source:sink ratios, are used to control carotenoid accumulation (Cazzonelli and Pogson, 2010).

In apple, the pathway flux and corresponding accumulation of carotenoids appears to be controlled by three enzymes within the pathway (Ampomah-Dwamena *et al.*, 2012). Using germplasm with varying levels of anthocyanin, the three enzymes, 15-cis-zeta-carotene isomerase (*ZISO*), carotenoid isomerase (*CRTISO*) and lycopene epsilon cyclase (*LYC*-ε), were shown to correlate with and predict levels of carotenoids in apple peel and flesh. To date, no similar studies have been reported for pear.

APPLE AND HUMAN HEALTH

In a comprehensive study of 10 000 individuals conducted in Finland, only one of the food types considered, apple, was associated with a positive reduction in the incidence of nearly all the chronic diseases reported (Knekt *et al.*, 2002). In this section, some of the evidence for apple and its association with important health properties are presented. There is a huge and diverse volume of literature related to apple and health so it is by no means complete, but some excellent reviews are cited for further information.

Bioavailability of apple bioactives

The link between plant bioactives and human health is highly complex and includes a wide array of non-nutritional secondary metabolites, including the three major classes: polyphenols, terpenoids and nitrogen-containing compounds, such as alkaloids (Harborne *et al.*, 1999). The bioavailability of these bioactives will undoubtedly affect any putative health benefits. In definition, bioavailability is the amount of compound processed or metabolized through normal metabolic pathways, generally after absorption through the gastrointestinal tract. Measurement of these processes presents significant technical issues for *in vivo* analysis and the use of model gut systems can help (McDougall *et al.*, 2005). For a bioactive molecule to be useful or bioavailable, it has to reach the target site in an effective form and at an appropriate level (Milner, 2006).

Just as evidence is emerging for the effects that bioactives have on endogenous antioxidant systems, it is possible that many of the other beneficial effects may be due to similar induction of cell signalling and/or the modulation of gut microbiota (van Duynhoven *et al.*, 2011). This may explain the dichotomy between the relatively low levels of bioavailability and the clear dietary benefits (Crozier *et al.*, 2009; Manach *et al.*, 2009).

Much of the research has been conducted on polyphenol absorption, metabolism or degradation and it has been shown to mainly occur in the small intestine and colon (Scalbert and Williamson, 2000). This may vary for both the individual consumer and for the specific bioactive and may partly be determined according to the food matrix and the presence of other bioactives within that matrix (Yang *et al.*, 2011). For apple, which has many bioactive compounds and is a rich source of sometimes hard to characterize polyphenols, elucidating bioavailability can be highly complex. The level of bioavailability will, naturally, be determined by the level of bioactives present and this varies according to apple variety and apple product (Biedrzycka and Amarowicz, 2008). Each bioactive may have a different potential for bioavailability according to its source. For example, a study measuring plasma concentrations of quercetin in healthy individuals showed that the bioavailability of apple quercetin was just 30% that of onion (Hollman *et al.*, 1997). It seems likely that many of these bioactives are heavily metabolized in the gastrointestinal tract; a study using apple juice suggested that around 67% were metabolized, degraded or absorbed in the small intestine (Kahle *et al.*, 2007). Again, the food matrix may determine the level of metabolism or absorption as, for example, apple smoothies may lead to proportionally higher levels of polyphenols reaching the colon than apple juice or cider (Hagl *et al.*, 2011).

Anthocyanins are likely to undergo similar metabolism and absorption in the stomach or jejunum (McGhie and Walton, 2007). Various anthocyanin derivatives, such as their aglycones, gallic acid, protocatechuic acid, and syringic acid, have been reported to be metabolized by gut microflora (Aura *et al.*, 2005; Avila *et al.*, 2009; Forester and Waterhouse, 2010). Anthocyanin stability may depend on the particular sugar moiety and the most common anthocyanin in apple, cyanidin galactoside, is likely to be relatively unstable (He *et al.*, 2005). However, the comparatively low level of anthocyanins in apple means that measurement of anthocyanic fate can be difficult or impossible to detect *in vivo* (Bouayed *et al.*, 2011). This may also be explained by the relative sensitivity of anthocyanins to pH, particularly with the transition from the acidic gastric environment to the alkaline intestinal environment. Given these factors, the recovery rate of anthocyanins from the gut is likely to be low (McDougall *et al.*, 2005).

Antioxidant activity of apple and apple products

A common theme for the putative health benefits of fruit, including apple, is the evidence for antioxidant activity. Epidemiological studies have linked apple consumption with a reduction in the risk of chronic diseases such as cancer, cardiovascular diseases (CVD), asthma and diabetes. Since it is widely known that oxidative damage can be the cause of certain chronic diseases, is it this antioxidant activity that provides the health component? It is unlikely to be due solely to vitamin C since, at least in apples, vitamin C contributes less than 0.4% of total antioxidant capacity (Eberhardt *et al.*, 2000). Conversely, the level of total phenolics in fruit does appear to be correlated with antioxidant activity (Sun *et al.*, 2002). Both apple and pear are relatively low in vitamin C concentration (0.057 and 0.040 mg/g) and have strong antioxidant activity, with apple out-performing many other fruits in *in vitro* trials (Sun *et al.*,

2002). Further studies have shown the antioxidant capacity for individual phytochemical components extracted from apple juice (Cefarelli *et al.*, 2006), with chlorogenic acid, caffeic acid and rutin having strong antioxidant effects (Schaefer *et al.*, 2006). In terms of tissue type, apple peel contains a significantly greater concentration of phytochemicals than apple flesh (cortex) and has a corresponding greater effect on antioxidant capacity (Wolfe *et al.*, 2003). *In vitro* trials have demonstrated the high antioxidant capacity for apple but are, of course, subject to the particular apple variety and numerous experimental conditions, such as pH, concentration and the particular method used. There has been an increasing emphasis on feeding trials to study the antioxidant effect in mammalian model species and humans.

In rats, intervention with apple juice induced higher levels of antioxidant response element genes (Soyalan *et al.*, 2011). A comparison of apple, peach and pear found that apple, particularly apple peel, was significantly higher in bioactive compounds and was the most effective at raising antioxidant capacity in rats fed with or without added dietary cholesterol (Leontowicz *et al.*, 2002). A study in pigs showed that in a comparison with strawberries and tomatoes, apple supplemented diets were the most effective at lowering oxidative stress and increasing the level of protection against DNA damage in mononuclear blood cells (Pajk *et al.*, 2006). In humans, a high antioxidant diet, such as one high in fruit phenolics, has been shown to produce biological samples with greater antioxidant concentration (Bianchi *et al.*, 2010). Numerous human intervention studies have shown that apple (or apple products) can improve antioxidant levels in serum (Vieira *et al.*, 2012) and plasma (Yuan *et al.*, 2011) and reduce the levels of oxidative stress and DNA damage (Ko *et al.*, 2005; Maffei *et al.*, 2007).

Recent studies suggest that increases in antioxidant levels, such as in plasma, are due to uric acid levels rather than polyphenols (Vieira *et al.*, 2012), as a result of a fructose-mediated increase in urate production (Godycki-Cwirko *et al.*, 2010).

Finally, it has been suggested that these dietary antioxidants are not so much functioning as hydrogen-donating antioxidants themselves, but that they activate endogenous antioxidant systems (Stevenson and Hurst, 2007). They confer their beneficial health-related properties by modulating endogenous cell signalling pathways (Williams *et al.*, 2004; Lotito and Frei, 2006).

Cancer

The relationship between apple consumption and decreased risks of developing chronic diseases has been well reviewed, particularly with reference to cancer (Boyer and Rui Hai, 2004; Gerhauser, 2008; Hyson, 2011). From a number of fruits tested in *in vitro* trials, apple was the second most effective bioactive in antiproliferation and antioxidant activity (shown as a bioactivity index) after cranberry (Sun *et al.*, 2002). In human studies, activity varies according to cancer type, although in one study all the cancer sites investigated (oral cavity and pharynx, oesophagus, larynx, colorectum, breast, ovary and prostate) consistently showed an inverse association with apple consumption (Gallus *et al.*, 2005). A lower incidence of lung cancer has been associated with apple (Knekt *et al.*, 2002), particularly for women (Feskanich *et al.*, 2000), and for both apple and pear in a large European study of nearly 500 000 participants (Linseisen *et al.*, 2007). Epidemiological studies also suggest that apples may be effective in preventing colorectal cancer (DeneoPellegrini *et al.*, 1996; Michels *et al.*, 2006).

While many studies conclude that flavonoids are the most likely bioactive in cancer prevention, other apple secondary metabolites, such as triperpenoids (e.g. ursolic acid), have also been implicated as antiproliferation agents (He and Liu, 2007).

Cardiovascular disease

There is strong evidence for a link between consumption of fruit and vegetables and reduced incidence of cardiovascular disease (CVD) (Goldberg, 2003). Numerous reports cite dietary apple as a contributing factor in this reduction and have been the subject of an excellent review (Weichselbaum *et al.*, 2010). A large survey of some 40 000 women showed that apple consumption lowered the risk of CVD disease by up to 22% (Sesso *et al.*, 2003) while another study of some 35 000 postmenopausal women found a similar association for apples and pears (Mink *et al.*, 2007). Women, but not men, were also at a lower risk of CVD in a Finnish study (Knekt *et al.*, 1996). The mechanism of this association is still not understood, although it seems that apples' bioactives may exert some effect on blood pressure and blood lipids (Weichselbaum *et al.*, 2010). Small mammal studies have demonstrated the role of apple or apple juice consumption on lowered total cholesterol, changes to cholesterol absorption and lipid modulation (Ogino *et al.*, 2007; Decorde *et al.*, 2008).

Diabetes

A rise in the incidence of type 2 diabetes has increased research into possible factors for reduction. A Finnish study showed an association between a lower risk of type 2 diabetes and apple consumption, with quercetin showing the highest association of all the flavonoids tested (Knekt *et al.*, 2002). A similar outcome was reported in an American study of 38 000 women and apple was the only food source associated with lower incidence levels (Song *et al.*, 2005), although in this study no association was found for the specific flavonoids considered, suggesting other apple bioactives may be causal. Anthocyanins have been associated with a lower risk of type 2 diabetes (Wedick *et al.*, 2012) and this study again confirmed apple and pear as having an association with a lower risk. Conversely, a study into the incidence of type 2 diabetes in postmenopausal women found no association between apple and incidence rate, or for any of the flavonoids considered (Nettleton *et al.*, 2006).

One explanation for the possible antidiabetic association of apple is the inhibitory effect on α-glucosidase, an enzyme implicated in the control of hyperglycemia and linked to type 2 diabetes (Adyanthaya *et al.*, 2010). Previous mouse model studies have shown that anthocyanins can help to reduce or inhibit elevation of blood glucose levels stimulated by a high fat diet (Tsuda *et al.*, 2003). Obesity has also been linked with insulin sensitivity and type 2 diabetes (Kahn *et al.*, 2006) and there is some evidence for an association with weight loss and apple and pears (de Oliveira *et al.*, 2003). Other fruits have also been shown to help inhibit weight gain, such as cherry and blood orange (Seymour *et al.*, 2009; Titta *et al.*, 2010), so it is possible that fruit, including apple, may play an important role in weight reduction.

Asthma

Evidence for the inverse association between apples and asthma has been well reviewed by Boyer and Rui Hai (2004) and more recently by Hyson (2011). Studies into both children and adults in various locations around the world have all added to this (Shaheen *et al.*, 2001; Romieu *et al.*, 2006; Chatzi *et al.*, 2007; Rosenlund *et al.*, 2011). Interestingly, one report showed a lower incidence of asthma symptoms amongst five-year-old children whose mothers had consumed apples during pregnancy, suggesting some protection for infants from maternal apple consumption (Willers *et al.*, 2007). However, the specific apple bioactives responsible for this association are not yet known. One study that selected three classes of apple flavonoids found no specific association with reduced incidence of asthma, suggesting that other non-flavonoid bioactives are involved or that function is based on synergy of a number of bioactives (Garcia *et al.*, 2005).

Cognitive processes

Combating age-related cognitive decline and neurodegeneration is of increasing importance to societies with lengthening life spans. The effects of flavanols, flavanones and anthocyanins from fruit have all been implicated in neural protection (Spencer, 2010). For apple, studies in normal aged mice have shown a neuroprotection effect with consumption of apple juice concentrate, particularly when presented with increased oxidative stress (Rogers *et al.*, 2004). While younger mice showed no difference, older mice were clearly aided in cognitive function and suffered reduced oxidative stress when apple was used to supplement a pro-oxidant diet deficient in folate and vitamin E (Tchantchou *et al.*, 2005). Similar methods were used to assess levels of the neurotransmitter acetylcholine and again apple was shown to prevent a decline when presented with a deficient diet (Chan *et al.*, 2006). A rat study using a diet supplemented with fresh apple showed that apple led to a decrease in anxiety levels and associated positive behavioural activity for aged animals but not at the expense of memory as long term potentiation was sustained. This may be attributed to the antioxidant activity of apple tissue, as the endogenous antioxidant marker superoxide dismutase (SOD) in old rats fed with apple was also reduced down to the level of young rats (Viggiano *et al.*, 2006). Whilst the mechanism of apple-derived bioactives on cognitive function remains unclear, the evidence from these model studies suggests that apple may prevent or slow age-related neurodegeneration.

Gastrointestinal health

Faecal output may also influence gut health. Studies have shown that anthocyanin-rich extracts in the diet of rats increased total faecal mass and moisture (Lala *et al.*, 2006). The authors suggested that this increase would reduce endogenous tumour-promoting compounds in the colon. In particular, faecal bile acids, which are known to promote colon cancer in rodent models (Magnuson *et al.*, 1993) were significantly reduced. These results may be due to modulation of the gut microbiota, changing levels of the overall bacterial population or the levels of individual bacterial species. In studies using bioactive extracts and *in vitro* model gut systems it has been shown, for example, that both flavanols and anthocyanins can modulate bacterial metabolism (Tzounis *et al.*, 2008; Hidalgo *et al.*, 2012), to a greater or lesser extent depending on the food matrix presented (Aprikian *et al.*, 2003; Bazzocco *et al.*, 2008), so it is quite possible that a phenolic-rich whole fruit such as apple may exert similar modulation. The composition of microbiota in the human gut influences the incidence of disease and general gut health and is largely determined by the complex interaction of gastrointestinal microbiota (He *et al.*, 2005). In effect, this could account for increased faecal output where these apple-derived bioactives are acting in a prebiotic manner. Previous studies have shown a relationship between a high antioxidant diet and faecal output, although this may be due to a proposed antibiotic effect of polyphenols (Bianchi *et al.*, 2010).

STORAGE AND PROCESSING

Apple fruit is subject to various processing techniques to produce pulp, juice and cider or to allow long term storage of fruit. A number of studies have considered the impact of storage and processing on the level of antioxidant capacity and bioactives, particularly the phenolic compounds. Notwithstanding the natural variation in levels between individual cultivars, in whole fruit it appears that these compounds are stable. In the peel of different cultivars, 'Granny Smith', 'Lady Williams', and 'Crofton', the level of phenolics remained relatively unchanged after nine months of storage at 0°C (Golding *et al.*, 2001). Neither flavonoid concentration nor antioxidant capacity was greatly effected in 'Jonagold', 'Golden Delicious', 'Cox's Orange' or 'Elstar' when subjected to 52 weeks in controlled atmosphere storage or 25 weeks in cold store, although there was a slight decrease in catechin (van der Sluis *et al.*, 2001). Treatment with gaseous 1-MCP, a commonly used postharvest treatment which inhibits ethylene activity and so delays ripening in storage, did not alter flavonoid levels but levels of chlorogenic acid were suppressed (MacLean *et al.*, 2006). A study involving flavonoid-enriched apple juice showed the relative stability of polyphenols during storage, with both polyphenolic levels and antioxidant activity relatively stable at ambient or cold storage temperature for up to six months (van der Sluis *et al.*, 2005).

Processing can severely affect bioactive levels in fruit. The commonly used practice of frying, for example to create apple slice snacks, can rapidly degrade anthocyanins (Patras *et al.*, 2010). Alternative methods may be less detrimental, such as vacuum drying, which significantly retained various bioactive compounds, including anthocyanins, vitamin C and antioxidant capacity, in red-fleshed apple slices (Joshi *et al.*, 2011). Similarly, ultra-high pressure homogenization (UHPH) has been shown to be effective at maintaining the levels of polyphenols, antioxidant capacity and vitamin C in apple juice processing, although β-carotene levels were reduced (Suarez-Jacobo *et al.*, 2011). These alternative methods are proving beneficial to bioactive levels in processed apple tissue and overcoming the potential sensory disadvantages of thermal processes (Choi and Nielsen, 2005).

Oxidation of apple fruit during crushing and pressing to produce juice can reduce levels of catechin and procyanidins and much of phenolic content is left behind in the pomace (Guyot *et al.*, 2003). Apple pomace is often treated as a waste product from the juicing process but the antioxidant capacity and polyphenol levels in pomace (Lu and Foo, 2000) suggests that it is a useful and relatively stable source of phytochemicals (Lavelli and Corti, 2011).

STRATEGIES TO IMPROVE THE BIOACTIVE CONTENT OF APPLES

Publication of the apple genome and further understanding of apple genetic diversity will assist in driving marker assisted breeding and gene-based technologies for the introduction of new varieties with increased levels of bioactive compounds (Velasco *et al.*, 2010; Micheletti *et al.*, 2011). New tools, such as single nucleotide polymorphisms (SNP) chips for apple, and highly developed apple genetic maps will further help this process (Costa *et al.*, 2010;

Chagné *et al.*, 2012a; Khan *et al.*, 2012). These maps are already proving useful for the development of high vitamin C apples (Davey *et al.*, 2006) and novel red-fleshed apples with high anthocyanin content, (Volz *et al.*, 2009; van Nocker *et al.*, 2012). Apple has also been the subject of genetic transformation (Bulley *et al.*, 2007). Much of this work has centred on horticultural traits, particularly disease resistance (Malnoy and Aldwinckle, 2007), but bioactive enhancement has also been achieved. One such example was apple transformation with a grape stillbene synthase gene, producing an increase in resveratrol glycoside levels (Ruhmann *et al.*, 2006). A transcription factor from maize, the basic helix loop helix (bHLH) *Lc*, was used to substantially increase anthocyanin (12-fold), flavan 3-ol epicatechin (14-fold), isomeric catechin (41-fold) and various proanthocyanidins (7 to 134-fold) in transgenic apple leaf although results for fruit were not reported (Li *et al.*, 2007). Anthocyanin levels have also been increased with the over-expression of the apple MYB transcription factor, *MYB10* (Espley *et al.*, 2007). This gene is autoregulatory, so that transformation with an allele from a wild red-fleshed apple accession into a commercial cultivar would increase anthocyanin levels in the plant and fruit (Espley *et al.*, 2009). This suggests a suitable candidate for cisgenic manipulation (Schouten and Jacobsen, 2008).

CONCLUSIONS

Apple represents one of the most versatile fruits, either eaten fresh (benefiting from its long storage life) or processed into pulp, juices or cider. In all its diverse forms it appears to maintain a diverse array of potential health bioactives. Considerable research into human and animal health has found an association with apple and a reduction in the incidence of many chronic diseases. As a dietary component, apple provides an excellent source for enhanced health protection.

REFERENCES

Adyanthaya I, Kwon YI, Apostolidis E, Shetty K (2010) Health benefits of apple phenolics from postharvest stages for potential type 2 diabetes management using *in vitro* models. *Journal of Food Biochemistry* **34**(1): 31–49.

Allan AC, Hellens RP, Laing WA (2008) MYB transcription factors that colour our fruit. *Trends in Plant Science* **13**(3): 99–102.

Ampomah-Dwamena C, Dejnoprat S, Lewis D, *et al.* (2012) Metabolic and gene expression analysis of apple (Malus × domestica) carotenogenesis. *Journal of Experimental Botany* **63**(12): 4497–4511.

An XH, Tian Y, Chen KQ, *et al.* (2012) The apple WD40 protein MdTTG1 interacts with bHLH but not MYB proteins to regulate anthocyanin accumulation. *Journal of Plant Physiology* **169**(7): 710–717.

Aprikian O, Duclos V, Guyot S, *et al.* (2003) Apple pectin and a polyphenol-rich apple concentrate are more effective together than separately on cecal fermentations and plasma lipids in rats. *The Journal of Nutrition* **133**(6): 1860–1865.

Aura AM, Martin-Lopez P, O'Leary KA, *et al.* (2005) *In vitro* metabolism of anthocyanins by human gut microflora. *European Journal of Nutrition* **44**(3): 133–142.

Avila M, Hidalgo M, Sanchez-Moreno C, *et al.* (2009) Bioconversion of anthocyanin glycosides by Bifidobacteria and Lactobacillus. *Food Research International* **42**(10): 1453–1461.

Awad MA, de Jager A, van Westing LM (2000) Flavonoid and chlorogenic acid levels in apple fruit: characterisation of variation. *Scientia Horticulturae* **83**(3–4): 249–263.

Bazzocco S, Mattila I, Guyot S, *et al.* (2008) Factors affecting the conversion of apple polyphenols to phenolic acids and fruit matrix to short-chain fatty acids by human faecal microbiota *in vitro*. *European Journal of Nutrition* **47**(8): 442–452.

Bianchi MA, Scazzina F, Del Rio D, *et al.* (2010) Ability of a high-total antioxidant capacity diet to increase stool weight and bowel antioxidant status in human subjects. *British Journal of Nutrition* **104**(10): 1500–1507.

Biedrzycka E, Amarowicz R (2008) Diet and health: Apple polyphenols as antioxidants. *Food Reviews International* **24**(2): 235–251.

Bouayed J, Hoffmann L, Bohn T (2011) Total phenolics, flavonoids, anthocyanins and antioxidant activity following simulated gastro-intestinal digestion and dialysis of apple varieties: Bioaccessibility and potential uptake. *Food Chemistry* **128**(1): 14–21.

Boyer J, Rui Hai L (2004) Apple phytochemicals and their health benefits. *Nutrition Journal* **3**: 5–15.

Brouillard R (1988) Flavonoids and flower colour. In: (Harborne JB, ed.) *The Flavonoids: Advances in Research since 1980*. Chapman and Hall, London, pp. 525–538.

Bulley SM, Malnoy M, Atkinson RG, Aldwinckle HS (2007) Transformed apples: traits of significance to growers and consumers. *Transgenic Plant Journal* **1**(2): 267–279.

Cazzonelli CI, Pogson BJ (2010) Source to sink: regulation of carotenoid biosynthesis in plants. *Trends in Plant Science* **15**(5): 266–274.

Cefarelli G, D'Abrosca B, Fiorentino A, *et al.* (2006) Free-radical-scavenging and antioxidant activities of secondary metabolites from reddened Cv. Annurca apple fruits. *Journal of Agricultural and Food Chemistry* **54**(3): 803–809.

Ceymann M, Arrigoni E, Scharer H, *et al.* (2012) Identification of apples rich in health-promoting flavan-3-ols and phenolic acids by measuring the polyphenol profile. *Journal of Food Composition and Analysis* **26**(1/2): 128–135.

Chagné D, Carlisle C, Blond C, *et al.* (2007) Mapping a candidate gene (*MdMYB10*) for red flesh and foliage colour in apple. *BMC Genomics* **8**(1): 212.

Chagné D, Crowhurst RN, Troggio M, *et al.* (2012a) Genome-wide SNP detection, validation, and development of an 8K SNP array for apple. *Plos One* **7**(2).

Chagné D, Krieger C, Rassam M, *et al.* (2012b) QTL and candidate gene mapping for polyphenolic composition in apple fruit. *BMC Plant Biology* **12**: 12.

Chalker-Scott L (1999) Environmental significance of anthocyanins in plant stress responses. *Photochemistry and Photobiology* **70**(1): 1–9.

Chan A, Graves V, Shea TB (2006) Apple juice concentrate maintains acetylcholine levels following dietary compromise. *Journal of Alzheimers Disease* **9**(3): 287–291.

Chatzi L, Apostolaki G, Bibakis I, *et al.* (2007) Protective effect of fruits, vegetables and the Mediterranean diet on asthma and allergies among children in. *Thorax* **62**(8): 677–683.

Chen C-S, Zhang D, Wang Y-Q, *et al.* (2012) Effects of fruit bagging on the contents of phenolic compounds in the peel and flesh of 'Golden Delicious', 'Red Delicious', and 'Royal Gala' apples. *Scientia Horticulturae* **142**(0): 68–73.

Choi LH, Nielsen SS (2005) The effects of thermal and nonthermal processing methods on apple cider quality and consumer acceptability. *Journal of Food Quality* **28**(1): 13–29.

Costa F, Longhi S, Magnago P, *et al.* (2010) Novel possibilities for marker-assisted breeding exploiting the apple genome. In: (Bassil NVMR, Martin R, eds) Proceedings of the International Symposium on Molecular Markers in Horticulture. Acta Horticulturae 859, International Society for Horticultural Science, pp. 357–360.

Crozier A, Jaganath IB, Clifford MN (2009) Dietary phenolics: chemistry, bioavailability and effects on health. *Natural Product Reports* **26**(8): 1001–1043.

Davey MW, Kenis K, Keulemans J (2006) Genetic control of fruit vitamin C contents. *Plant Physiology* **142**(1): 343–351.

Davey MW, Keulemans J (2004) Determining the potential to breed for enhanced antioxidant status in *Malus*: Mean inter- and intravarietal fruit vitamin C and glutathione contents at harvest and their evolution during storage. *Journal of Agricultural and Food Chemistry* **52**(26): 8031–8038.

de Bernonville TD, Guyot S, Paulin JP, *et al.* (2010) Dihydrochalcones: Implication in resistance to oxidative stress and bioactivities against advanced glycation end-products and vasoconstriction. *Phytochemistry* **71**(4): 443–452.

de Oliveira MC, Sichieri R, Moura AS (2003) Weight loss associated with a daily intake of three apples or three pears among overweight women. *Nutrition* **19**(3): 253–256.

Decorde K, Teissedre P-L, Auger C, *et al.* (2008) Phenolics from purple grape, apple, purple grape juice and apple juice prevent early atherosclerosis induced by an atherogenic diet in hamsters. *Molecular Nutrition and Food Research* **52**(4): 400–407.

DeneoPellegrini H, DeStefani E, Ronco A (1996) Vegetables, fruits, and risk of colorectal cancer: A case-control study from Uruguay. *Nutrition and Cancer – an International Journal* **25**(3): 297–304.
Dixon RA, Paiva NL (1995) Stress-Induced Phenylpropanoid Metabolism. *Plant Cell* **7**(7): 1085–1097.
Dixon RA, Steele CL (1999) Flavonoids and isoflavonoids – a gold mine for metabolic engineering. *Trends in Plant Science* **4**(10): 394–400.
Duggan MB (1969) Identity and occurrence of certain flavonol glycosides in four varieties of pears. *Journal of Agricultural and Food Chemistry* **17**(5): 1098–101.
Dzhangaliev AD (2003) The wild apple tree of Kazakhstan. In: (Janick J, ed.) *Wild Apple and Fruit Trees of Central Asia*. John Wiley & Sons, Inc., New York, pp. 63–303.
Eberhardt MV, Lee CY, Liu RH (2000) Nutrition: Antioxidant activity of fresh apples. *Nature* **405**(6789): 903–904.
Espley RV, Hellens RP, Putterill J, *et al.* (2007) Red colouration in apple fruit is due to the activity of the MYB transcription factor, MdMYB10. *The Plant Journal* **49**(3): 414–427.
Espley RV, Brendolise C, Chagné D, *et al.* (2009) Multiple Repeats of a Promoter Segment Causes Transcription Factor Autoregulation in Red Apples. *Plant Cell* **21**(1):168–183.
Felicetti DA, Schrader LE (2009) Changes in pigment concentrations associated with sunburn browning of five apple cultivars. I. Chlorophylls and carotenoids. *Plant Science* **176**(1): 78–83.
Feng SQ, Wang YL, Yang S, *et al.* (2010) Anthocyanin biosynthesis in pears is regulated by a R2R3-MYB transcription factor PyMYB10. *Planta* **232**(1): 245–255.
Feskanich D, Ziegler RG, Michaud DS, *et al.* (2000) Prospective study of fruit and vegetable consumption and risk of lung cancer among men and women. *Journal of the National Cancer Institute* **92**(22): 1812–1823.
Forester SC, Waterhouse AL (2010) Gut metabolites of anthocyanins, gallic acid, 3-O-methylgallic acid, and 2,4,6-trihydroxybenzaldehyde, inhibit cell proliferation of caco-2 cells. *Journal of Agricultural and Food Chemistry* **58**(9): 5320–5327.
Forsline PL, Aldwinckle HS, Dickson EE, *et al.* (2003) Collection, maintenance, characterization, and utilization of wild apples of Central Asia. In: (Janick J, ed.) *Wild Apple and Fruit Trees of Central Asia*. John Wiley & Sons, Inc., New York, pp. 1–62.
Fraser PD, Bramley PM (2004) The biosynthesis and nutritional uses of carotenoids. *Progress in Lipid Research* **43**(3): 228–265.
Gallus S, Talamini R, Giacosa A, *et al.* (2005) Does an apple a day keep the oncologist away? *Annals of Oncology* **16**(11): 1841–1844.
Garcia V, Arts ICW, Sterne JAC, *et al.* (2005) Dietary intake of flavonoids and asthma in adults. *European Respiratory Journal* **26**(3): 449–452.
Gerhauser C (2008) Cancer chemopreventive potential of apples, apple juice, and apple components. *Planta Medica* **74**(13): 1608–1624.
Godycki-Cwirko M, Krol M, Krol B, *et al.* (2010) Uric acid but not apple polyphenols is responsible for the rise of plasma antioxidant activity after apple juice consumption in healthy subjects. *Journal of the American College of Nutrition* **29**(4): 397–406.
Goldberg G (2003) Plants: diet and health. The report of a British Nutrition Foundation Task Force.
Golding JB, McGlasson WB, Wyllie SG, Leach DN (2001) Fate of apple peel phenolics during cool storage. *Journal of Agricultural and Food Chemistry* **49**(5): 2283–2289.
Gosch C, Halbwirth H, Kuhn J, *et al.* (2009) Biosynthesis of phloridzin in apple (Malus domestica Borkh.). *Plant Science* **176**, 223–231.
Guyot S, Marnet N, Sanoner P, Drilleau JF (2003) Variability of the polyphenolic composition of cider apple (Malus domestica) fruits and juices. *Journal of Agricultural and Food Chemistry* **51**(21): 6240–6247.
Hagl S, Deusser H, Soyalan B, *et al.* (2011) Colonic availability of polyphenols and D-(−)-quinic acid after apple smoothie consumption. *Molecular Nutrition and Food Research* **55**(3): 368–377.
Harborne J (1967) *Comparative biochemistry of the flavonoids*. Academic Press, London.
Harborne JB, Baxter H, Moss GP (1999) *Phytochemical dictionary: Handbook of bioactive compounds from plants*.Taylor and Francis, London.
Harris SA, Robinson JP, Juniper BE (2002) Genetic clues to the origin of the apple. *Trends in Genetics* **18**(8): 426–430.
He J, Magnuson BA, Giusti MM (2005) Analysis of anthocyanins in rat intestinal contentsimpact of anthocyanin chemical structure on fecal excretion. *Journal of Agricultural and Food Chemistry* **53**(8): 2859–2866.
He XJ, Liu RH (2007) Triterpenoids isolated from apple peels have potent antiproliferative activity and may be partially responsible for apple's anticancer activity. *Journal of Agricultural and Food Chemistry* **55**(11): 4366–4370.

Henry-Kirk RA, McGhie TK, Andre CM, *et al.* (2012) Transcriptional analysis of apple fruit proanthocyanidin biosynthesis. *Journal of Experimental Botany* **163**(15): 5437–5450.

Hidalgo M, Oruna-Concha MJ, Kolida S, *et al.* (2012) Metabolism of anthocyanins by human gut microflora and their influence on gut bacterial growth. *Journal of Agricultural and Food Chemistry* **60**(15): 3882–3890.

Hollman PCH, vanTrijp JMP, Buysman M, *et al.* 1997. Relative bioavailability of the antioxidant flavonoid quercetin from various foods in man. *FEBS Letters* **418**(1–2): 152–156.

Honda C, Kotoda N, Wada M, *et al.* (2002) Anthocyanin biosynthetic genes are coordinately expressed during red coloration in apple skin. *Plant Physiology and Biochemistry* **40**(11): 955–962.

Howitt CA, Pogson BJ (2006) Carotenoid accumulation and function in seeds and non-green tissues. *Plant Cell and Environment* **29**(3): 435–445.

Hunter LD 1975. Phloridzin and apple scab. *Phytochemistry* **14**(7): 1519–1522.

Hyson DA (2011) A comprehensive review of apples and apple components and their relationship to human health. *Advances in Nutrition* **2**(5): 408–420.

Joshi APK, Rupasinghe HPV, Khanizadeh S (2011) Impact of drying processes on bioactive phenolics, vitamin C and antioxidant capacity of red-fleshed apple slices. *Journal of Food Processing and Preservation* **35**(4): 453–457.

Judge H, Nguy D, Moller I, *et al.* (2008) Isolation and characterization of a novel glycosyltransferase that converts phloretin to phlorizin, a potent antioxidant in apple. *Febs Journal* **275**, 3804–3814.

Juniper BE, Mabberley DJ (2006) *The story of the Apple*. Timber Press, Portland.

Juniper BE, Watkins R, Harris SA (1998) The origin of the apple. *Acta Horticulturae* **484**: 27–34.

Kahle K, Huemmer W, Kempf M, *et al.* (2007) Polyphenols are intensively metabolized in the human gastrointestinal tract after apple juice consumption. *Journal of Agricultural and Food Chemistry* **55**(26): 10605–10614.

Kahn SE, Hull RL, Utzschneider KM (2006) Mechanisms linking obesity to insulin resistance and type 2 diabetes. *Nature* **444**(7121): 840–846.

Karychev RK, Salnikov Y, Nurtazin MT, Doud Miller D (2005) Tree fruit growing in Kazakhstan. *Chronica Horticulturae* **45**(4): 21–23.

Khan SA, Chibon P-Y, de Vos RCH, *et al.* (2012) Genetic analysis of metabolites in apple fruits indicates an mQTL hotspot for phenolic compounds on linkage group 16. *Journal of Experimental Botany* **63**(8): 2895–2908.

Kim S-H, Lee J-R, Hong S-T, *et al.* (2003) Molecular cloning and analysis of anthocyanin biosynthesis genes preferentially expressed in apple skin. *Plant Science* **165**(2): 403–413.

Knee M 1972. Anthocyanin, carotenoid, and chlorophyll changes in the peel of Cox's Orange Pippin apples during ripening on and off the tree. *Journal of Experimental Botany* **23**(1): 184–196.

Knekt P, Jarvinen R, Reunanen A, Maatela J (1996) Flavonoid intake and coronary mortality in Finland: A cohort study. *British Medical Journal* **312**(7029): 478–481.

Knekt P, Kumpulainen J, Jarvinen R, *et al.* (2002) Flavonoid intake and risk of chronic diseases. *American Journal of Clinical Nutrition* **76**(3): 560–568.

Ko SH, Choi SW, Ye SK, *et al.* (2005) Comparison of the antioxidant activities of nine different fruits in human plasma. *Journal of Medicinal Food* **8**(1): 41–46.

Koes R, Verweij W, Quattrocchio F (2005) Flavonoids: a colorful model for the regulation and evolution of biochemical pathways. *Trends in Plant Science* **10**(5): 236–242.

Lala G, Malik M, Cuiwei Z, *et al.* (2006) Anthocyanin-rich extracts inhibit multiple biomarkers of colon cancer in rats. *Nutrition and Cancer* **54**(1): 94–101.

Lancaster J (1992) Regulation of skin color in apples. *Critical Reviews in Plant Sciences* **10**: 487–502.

Lavelli V, Corti S (2011) Phloridzin and other phytochemicals in apple pomace: Stability evaluation upon dehydration and storage of dried product. *Food Chemistry* **129**(4): 1578–1583.

Leontowicz H, Gorinstein S, Lojek A, *et al.* (2002) Comparative content of some bioactive compounds in apples, peaches and pears and their influence on lipids and antioxidant capacity in rats. *The Journal of Nutritional Biochemistry* **13**(10): 603–610.

Li H, Flachowsky H, Fischer TC, *et al.* (2007) Maize Lc transcription factor enhances biosynthesis of anthocyanins, distinct proanthocyanidins and phenylpropanoids in apple (*Malus domestica Borkh.*). *Planta* **226**(5): 1243–1254.

Lin-Wang K, Bolitho K, Grafton K, *et al.* (2010) An R2R3 MYB transcription factor associated with regulation of the anthocyanin biosynthetic pathway in Rosaceae. *BMC Plant Biology* **10**: 50.

Lin-Wang KUI, Micheletti D, Palmer J, *et al.* (2011) High temperature reduces apple fruit colour via modulation of the anthocyanin regulatory complex. *Plant, Cell & Environment* **34**, 1176–1190.

Linseisen J, Rohrmann S, Miller AB, *et al.* (2007) Fruit and vegetable consumption and lung cancer risk: Updated information from the European Prospective Investigation into Cancer and Nutrition (EPIC). *International Journal of Cancer* **121**(5): 1103–1114.

Lommen A, Godejohann M, Venema DP, *et al.* (2000) Application of directly coupled HPLC-NMR-MS to the identification and confirmation of quercetin glycosides and phloretin glycosides in apple peel. *Analytical Chemistry* **72**(8): 1793–1797.

Lotito SB, Frei B (2006) Consumption of flavonoid-rich foods and increased plasma antioxidant capacity in humans: Cause, consequence, or epiphenomenon? *Free Radical Biology & Medicine* **41**(12): 1727–1746.

Lu YR, Foo LY (2000) Antioxidant and radical scavenging activities of polyphenols from apple pomace. *Food Chemistry* **68**(1): 81–85.

MacLean DD, Murr DP, DeEll JR, Horvath CR (2006) Postharvest variation in apple (*Malus* × *domestica Borkh.*) flavonoids following harvest, storage, and 1-MCP Treatment. *Journal of Agricultural and Food Chemistry* **54**(3): 870–878.

Maffei F, Tarozzi A, Carbone F, *et al.* (2007) Relevance of apple consumption for protection against oxidative damage induced by hydrogen peroxide in human lymphocytes. *British Journal of Nutrition* **97**(5): 921–927.

Magnuson BA, Carr I, Bird RP (1993) Ability of aberrant crypt foci characteristics to predict colonic tumor incidence in rats fed cholic acid. *Cancer Research* **53**(19): 4499–4504.

Malnoy M, Aldwinckle HS (2007) Development of fire blight resistance by recombinant DNA technology. *Plant Breeding Reviews* **29**: 315–358.

Manach C, Hubert J, Llorach R, Scalbert A (2009) The complex links between dietary phytochemicals and human health deciphered by metabolomics. *Molecular Nutrition and Food Research* **53**(10): 1303–1315.

Martin C, Butelli E, Petroni K, Tonelli C (2011) How can research on plants contribute to promoting human health? *The Plant Cell Online* **23**(5): 1685–1699.

McDougall GJ, Dobson P, Smith P, *et al.* (2005) Assessing potential bioavallability of raspberry anthocyanins using an in vitro digestion system. *Journal of Agricultural and Food Chemistry* **53**(15): 5896–5904.

McGhie TK, Hunt M, Barnett LE (2005) Cultivar and growing region determine the antioxidant polyphenolic concentration and composition of apples grown in New Zealand. *Journal of Agricultural and Food Chemistry* **53**(8): 3065–3070.

McGhie TK, Walton MC (2007) The bioavailability and absorption of anthocyanins: Towards a better understanding. *Molecular Nutrition and Food Research* **51**(6): 702–713.

Merzlyak MN, Melo TB, Naqvi KR (2008) Effect of anthocyanins, carotenoids, and flavonols on chlorophyll fluorescence excitation spectra in apple fruit: signature analysis, assessment, modelling, and relevance to photoprotection. *Journal of Experimental Botany* **59**(2): 349–359.

Micheletti D, Troggio M, Zharkikh A, *et al.* (2011) Genetic diversity of the genus Malus and implications for linkage mapping with SNPs. *Tree Genetics and Genomes* **7**(4): 857–868.

Michels KB, Giovannucci E, Chan AT, *et al.* (2006) Fruit and vegetable consumption and colorectal adenomas in the nurses' health study. *Cancer Research* **66**(7): 3942–3953.

Milner JA (2006) Diet and cancer: Facts and controversies. *Nutrition and Cancer – an International Journal* **56**(2): 216–224.

Mink PJ, Scrafford CG, Barraj LM, *et al.* (2007) Flavonoid intake and cardiovascular disease mortality: a prospective study in postmenopausal women. *American Journal of Clinical Nutrition* **85**(3): 895–909.

Nesi N, Debeaujon I, Jond C, *et al.* (2000) The *TT8* gene encodes a basic helix-loop-helix domain protein required for expression of *DFR* and *BAN* genes in *Arabidopsis* siliques. *Plant Cell* **12**(10): 1863–1878.

Nesi N, Jond C, Debeaujon I, *et al.* (2001) The *Arabidopsis TT2* gene encodes an R2R3 MYB domain protein that acts as a key determainant for proanthocyanidin accumulation in developing seed. *Plant Cell* **13**(9): 2099–2114.

Nettleton JA, Harnack LJ, Scrafford CG, *et al.* (2006) Dietary flavonoids and flavonoid-rich foods are not associated with risk of type 2 diabetes in postmenopausal women. *Journal of Nutrition* **136**(12): 3039–3045.

Noda K-i, Glover BJ, Linstead P, Martin C (1994) Flower colour intensity depends on specialized cell shape controlled by a Myb-related transcription factor. *Nature* **369**(6482): 661–664.

Nortje BK (1965) The flavonol glycosides in the fruit of Pyrus communis L. cultivar Bon Chretien. *The Biochemical Journal* **97**(1): 209–213.

O'Rourke D (2007) *Annual World Reviews*. Belrose Inc., Pullman, WA.

Ogino Y, Osada K, Nakamura S, *et al.* (2007) Absorption of dietary cholesterol oxidation products and their downstream metabolic effects are reduced by dietary apple polyphenols. *Lipids* **42**(2): 151–161.

Pajk T, Rezar V, Levart A, Salobir J (2006) Efficiency of apples, strawberries, and tomatoes for reduction of oxidative stress in pigs as a model for humans. *Nutrition* **22**(4): 376–384.

Patras A, Brunton NP, O'Donnell C, Tiwari BK (2010) Effect of thermal processing on anthocyanin stability in foods; mechanisms and kinetics of degradation. *Trends in Food Science and Technology* **21**(1): 3–11.

Phillips JP (1866) *A Pembrokeshire Proverb. Notes and Queries.* Oxford University Press, p. 153.

Rogers EJ, Milhalik S, Orthiz D, Shea TB (2004) Apple juice prevents oxidative stress and impaired cognitive performance caused by genetic and dietary deficiencies in mice. *The Journal of Nutrition, Health and Aging* **8**(2): 92–97.

Romieu I, Varraso R, Avenel V, *et al.* (2006) Fruit and vegetable intakes and asthma in the E3N study. *Thorax* **61**(3): 209–215.

Rosenlund H, Kull I, Pershagen G, *et al.* (2011) Fruit and vegetable consumption in relation to allergy: Disease-related modification of consumption? *Journal of Allergy and Clinical Immunology* **127**(5): 1219–1225.

Ruhmann S, Treutter D, Fritsche S, *et al.* (2006) Piceid (resveratrol glucoside) synthesis in stilbene synthase transgenic apple fruit. *Journal of Agricultural and Food Chemistry* **54**(13): 4633–4640.

Scalbert A, Williamson G (2000) Dietary intake and bioavailability of polyphenols. *Journal of Nutrition* **130**(8): 2073S–2085S.

Schaefer S, Baum M, Eisenbrand G, *et al.* (2006) Polyphenolic apple juice extracts and their major constituents reduce oxidative damage in human colon cell lines. *Molecular Nutrition and Food Research* **50**(1): 24–33.

Schieber A, Keller P, Carle R (2001) Determination of phenolic acids and flavonoids of apple and pear by high-performance liquid chromatography. *Journal of Chromatography A* **910**(2): 265–273.

Schieber A, Keller P, Streker P, *et al.* (2002) Detection of isorhamnetin glycosides in extracts of apples (Malus domestica cv. "Brettacher") by HPLC-PDA and HPLC-APCI-MS/MS. *Phytochemical Analysis* **13**(2): 87–94.

Schouten HJ, Jacobsen E (2008) Cisgenesis and intragenesis, sisters in innovative plant breeding. *Trends in Plant Science* **13**(6): 260–261.

Sesso HD, Gaziano JM, Liu S, Buring JE (2003) Flavonoid intake and the risk of cardiovascular disease in women. *American Journal of Clinical Nutrition* **77**(6): 1400–1408.

Seymour EM, Lewis SK, Urcuyo-Llanes DE, *et al.* (2009) Regular tart cherry intake alters abdominal adiposity, adipose gene transcription, and inflammation in obesity-prone rats fed a high fat diet. *Journal of Medicinal Food* **12**(5): 935–942.

Shaheen SO, Sterne JAC, Thompson RL, *et al.* (2001) Dietary antioxidants and asthma in adults – Population-based case-control study. *American Journal of Respiratory and Critical Care Medicine* **164**(10): 1823–1828.

Song YQ, Manson JE, Buring JE, *et al.* (2005) Associations of dietary flavonoids with risk of type 2 diabetes, and markers of insulin resistance and systemic inflammation in women: A prospective study and cross-sectional analysis. *Journal of the American College of Nutrition* **24**(5): 376–384.

Soyalan B, Minn J, Schmitz HJ, *et al.* (2011) Apple juice intervention modulates expression of ARE-dependent genes in rat colon and liver. *European Journal of Nutrition* **50**(2): 135–143.

Spencer JPE (2010) The impact of fruit flavonoids on memory and cognition. *British Journal of Nutrition* **104**: S40–S47.

Stevenson DE, Hurst RD (2007) Polyphenolic phytochemicals – just antioxidants or much more? *Cellular and Molecular Life Sciences* **64**(22): 2900–2916.

Steyn WJ, Holcroft DM, Wand SJE, Jacobs G (2004) Regulation of pear color development in relation to activity of flavonoid enzymes. *Journal of the American Society for Horticultural Science* **129**(1): 6–12.

Suarez-Jacobo A, Rufer CE, Gervilla R, *et al.* (2011) Influence of ultra-high pressure homogenisation on antioxidant capacity, polyphenol and vitamin content of clear apple juice. *Food Chemistry* **127**(2): 447–454.

Sun J, Chu YF, Wu XZ, Liu RH (2002) Antioxidant and anti proliferative activities of common fruits. *Journal of Agricultural and Food Chemistry* **50**(25): 7449–7454.

Tchantchou F, Chan A, Kifle L, *et al.* (2005) Apple juice concentrate prevents oxidative damage and impaired maze performance in aged mice. *Journal of Alzheimers Disease* **8**(3): 283–287.

Titta L, Trinei M, Stendardo M, *et al.* (2010) Blood orange juice inhibits fat accumulation in mice. *International Journal of Obesity* **34**(3): 578–588.

Tsao R, Yang R, Young JC, Zhu H (2003) Polyphenolic profiles in eight apple cultivars using high-performance liquid chromatography (HPLC). *Journal of Agricultural and Food Chemistry* **51**(21): 6347–53.

Tsuda T, Horio F, Uchida K, *et al.* (2003) Dietary cyanidin 3-O-{beta}-D-glucoside-rich purple corn color prevents obesity and ameliorates hyperglycemia in mice. *Journal of Nutrition* **133**(7): 2125–2130.

Tzounis X, Vulevic J, Kuhnle GGC, *et al.* (2008) Flavanol monomer-induced changes to the human faecal microflora. *The British Journal of Nutrition* **99**(4): 782–92.

van der Sluis AA, Dekker M, de Jager A, Jongen WMF (2001) Activity and concentration of polyphenolic antioxidants in apple: Effect of cultivar, harvest year, and storage conditions. *Journal of Agricultural and Food Chemistry* **49**(8): 3606–3613.

van der Sluis AA, Dekker M, van Boekel M (2005) Activity and concentration of polyphenolic antioxidants in apple juice. 3. Stability during storage. *Journal of Agricultural and Food Chemistry* **53**(4): 1073–1080.

van Duynhoven J, Vaughan EE, Jacobs DM, *et al.* (2011) Metabolic fate of polyphenols in the human superorganism. *Proceedings of the National Academy of Sciences, USA* **108**: 4531–4538.

van Nocker S, Berry G, Najdowski J, *et al.* (2012) Genetic diversity of red-fleshed apples (Malus). *Euphytica* **185**(2): 281–293.

Vavilov NI (1931) The wild relatives of fruit trees of the Asian part of the USSR and Caucasus and the problem of the origin of fruit trees. *Trans. Applied Bot. Genet. Breed.* **26**(3): 85–107.

Velasco R, Zharkikh A, Affourtit J, *et al.* (2010) The genome of the domesticated apple (Malus x domestica Borkh.). *Nature Genetics* **42**(10): 833–839.

Vieira FGK, Di Pietro PF, da Silva EL, *et al.* (2012) Improvement of serum antioxidant status in humans after the acute intake of apple juices. *Nutrition Research* **32**(3): 229–232.

Viggiano A, Viggiano A, Monda M, *et al.* (2006) Annurca apple-rich diet restores long-term potentiation and induces behavioral modifications in aged rats. *Experimental Neurology* **199**(2): 354–361.

Volz R, Oraguzie N, Whitworth C, *et al.* (2009) Red flesh breeding in apple – progress and challenges. *Acta Horticultureae* **814**: 337–342.

Vrhovsek U, Rigo A, Tonon D, Mattivi F (2004) Quantitation of polyphenols in different apple varieties. *Journal of Agricultural and Food Chemistry* **52**(21): 6532–6538.

Wald B, Wray V, Galensa R, Herrmann K (1989) Malonated flavonol glycosides and 3,5-dicaffeoylquinic acid from pears. *Phytochemistry* **28**(2): 663–664.

Watkins R (1995) Apple and pear. In: (Smartt J, Simmonds NW, eds) *Evolution of Crop Plants*. Longman, London, pp. 418–422.

Way RD, Aldwinckle HS, Lamb RC, *et al.* (1991) Apples (*Malus*). In: (Moore JN, Ballington JR, eds) *Genetic Resources of Temperate Fruit and Nut Crops*. International Society for Horticultural Science (ISHS), Wageningen, The Netherlands, pp. 1–62.

Wedick NM, Pan A, Cassidy A, *et al.* (2012) Dietary flavonoid intakes and risk of type 2 diabetes in US men and women. *American Journal of Clinical Nutrition* **95**(4): 925–933.

Weichselbaum E, Wyness L, Stanner S (2010) Apple polyphenols and cardiovascular disease – a review of the evidence. *Nutrition Bulletin* **35**(2): 92–101.

Willers SM, Devereux G, Craig LCA, *et al.* (2007) Maternal food consumption during pregnancy and asthma, respiratory and atopic symptoms in 5-year-old children. *Thorax* **62**(9): 773–779.

Williams RJ, Spencer JPE, Rice-Evans C (2004) Flavonoids: antioxidants or signalling molecules? *Free Radical Biology & Medicine* **36**(7): 838–849.

Wojdylo A, Oszmianski J, Laskowski P (2008) Polyphenolic compounds and antioxidant activity of new and old apple varieties. *Journal of Agricultural and Food Chemistry* **56**(15): 6520–6530.

Wolfe K, Wu XZ, Liu RH (2003) Antioxidant activity of apple peels. *Journal of Agricultural and Food Chemistry* **51**(3): 609–614.

Yang M, Koo SI, Song WO, Chun OK (2011) Food matrix affecting anthocyanin bioavailability: Review. *Current Medicinal Chemistry* **18**(2): 291–300.

Yuan LH, Meng LP, Ma WW, *et al.* (2011) Impact of apple and grape juice consumption on the antioxidant status in healthy subjects. *International Journal of Food Sciences and Nutrition* **62**(8): 844–850.

Zhang W, Zhang J, Hu X (1993) Distribution and diversity of *malus* germplasm resources in Yunnan, China. *Horticultural Science* **28**(10): 978–980.

Zhu Y, Evans K, Peace C (2010) Utility testing of an apple skin color MdMYB1 marker in two progenies. *Molecular Breeding* **27**: 525–532.

6 Orange and Grapefruit Bioactive Compounds, Health Benefits and Other Attributes

Paul F. Cancalon

INTRODUCTION

Initial studies on the relations between fruits and health emphasized macro and micronutrients, particularly vitamins and minerals (Rampersaud, 2007). It has now been established that a series of phytochemicals, among them polyphenols and carotenoids, are playing a major role in the health virtues of fruits. Early studies on the health benefits of fruit juices had focused on the free radical scavenging properties of phytochemicals known as the antioxidant theory. Hypotheses on the antioxidant mode of action were mainly derived from their physicochemical properties tested *in vitro*. Fruits were classified according to their antioxidant content measured *in vitro* and it was implied that antioxidant potentials and health benefits were correlated (Prior and Cao, 1999). It was rapidly shown that there was no link between the level of antioxidants measured in a container and health properties. It is not the phytochemicals present in the juice that have an effect but their metabolites produced by the body or by microorganisms. Once ingested, a significant fraction of the polyphenols is broken down by microorganisms into phenolic acids whose role is still unclear. Another fraction is absorbed as aglycones which are then glycosylated and sulfated, eliminating the major part of the antioxidant activity. Those metabolites have now been identified for the major fruits, particularly citrus, and their bioavailability, and concentrations in the blood and urine, have been assessed. Accumulated evidence shows that phytochemical metabolites act by modulating complex biochemical pathways. In many cases, they act by up- or down-regulating genes which, in turn, modulate enzymatic activities and at the end of the pathway generate a physiological change such as blood pressure (Williams *et al.*, 2004, Milenkovic *et al.*, 2011). Recent evidence shows that the majority of the beneficial effects of phytochemicals are aimed at several points within the chronic inflammation process. Inflammation is believed to be at the root of most non-communicable diseases: cardiovascular diseases (CVD), metabolic syndrome, insulin resistance and type 2 diabetes, osteoporosis, dementia and some forms of cancer (Pan *et al.*, 2010). So far, the most compelling results have been in the area of CVD, insulin resistance and cognition. However, today, the knowledge of the effects of fruits and juices on health is still very fragmented. Studies are often conducted under different conditions making comparison difficult. It should also be kept in mind that fruits and juices are not drugs and should not be expected to produce results of the same magnitude as medications. Discussions on how best to determine the most efficient way of

Bioactives in Fruit: Health Benefits and Functional Foods, First Edition. Edited by Margot Skinner and Denise Hunter.

uncovering the health benefits of citrus and foods in general, have opposed pharmacologists and nutritionists (Visioli, 2012). This question has not been resolved and should be actively debated.

COMPOSITION OF CITRUS

Orange and grapefruit contain macronutrients, micronutrients, minerals and vitamins that have been shown previously to have nutritional benefits of citrus, including low sodium, high potassium, absence of lipids, presence of folic acid and high level of vitamin C (Decuypere, 2000; Rampersaud, 2007). Vitamin C is essential to human survival and is involved in many physiological processes. However, approximately 13% of the United States population has been reported to be vitamin C deficient (Schleicher *et al.*, 2009). Folic acid has been shown to prevent birth defects, including spinal cord malformations such as spina bifida (Hasenau and Covington, 2002). These nutritional qualities of citrus juices are well known and have given rise to the main claims that are now allowed. However, they cannot explain all the positive effects reported. Recently, the attention has shifted toward the health properties of other compounds. Analyses of oranges have revealed the presence of 224 phytochemicals, including 23 monoterpenoids, 15 sesquiterpenoids, 13 diterpenoids, 32 flavones, 13 flavanones, 6 flavanols, 9 anthocyanins, 3 chalcones, 4 phenolic acids, 15 carotenoids and 4 coumarins (Berhow *et al.*, 1998; Manach, 2009; Neveu *et al.*, 2010). Research has focused mainly on the flavanones hesperidin and narirutin, the carotenoid beta cryptoxanthine and, to a lesser extent, the polymethoxylated flavones tangeretin and nobeletin. The specific effects of many minor components will probably never be examined; however, there is evidence that these compounds have synergistic activities, since studies have shown that a juice is more effective than an individual component. (Brand *et al.*, 2010; Milenkovic *et al.*, 2011; Morand *et al.*, 2011). Similarly, analyses of grapefruit have revealed the presence of many phytochemicals, including 13 polyphenols, naringin, narirutin, neoeriocitrin, neohesperidin, poncirin, didymin, eriocitrin, hesperidin (Berhow *et al.*, 1998; Neveu *et al.*, 2010), 20 carotenoids, particularly beta-carotene and lycopene, and, to a lesser extent, zeta-carotene and phytofluene (Khan and Mackinney, 1953). Grapefruits contain limonoids such as limonin and nomilin, which have been shown to have various potential health benefits (Manners, 2007). Other compounds found in grapefruit, the furanocoumarins, have been extensively examined because of their interaction with some drugs (Bailey *et al.*, 1991). The main furanocoumarins include 6', 7' dihydroxy bergamottin, bergamottin, paradisins and bergaptol. These compounds have been shown to interact with the cytochrome P450, CYP3A4 and increase the absorption of some drugs such as some calcium channel blockers and statins. The interaction is limited to selected drugs interacting with CYP3A4 and in each class of drug some are not affected by grapefruit (http://www.druginteractioncenter.org/; Mertens-Talcott *et al.*, 2006).

BIOAVAILABILITY AND METABOLIZATION OF CITRUS PHYTOCHEMICALS

It is now known that phytochemicals act mainly through metabolites generated in the digestive tract and various organs. It is therefore essential to know the fate of any compound under investigation (Stevenson and Hurst, 2007; Scalbert and Knasmüller, 2008). Studies have shown that orange juice polyphenols are mainly metabolized by bacteria then absorbed in the large intestine as aglycones (Crozier *et al.*, 2011; Del Rio *et al.*, 2012). Some flavonoids are poorly soluble in water and the question has been raised that water solubility may play an important role in their bioavailibility (Vallejo *et al.*, 2010). Furthermore, the bioavailability of hesperetin (hesperidin aglycone) can be increased fourfold after removal of the hesperidin rhamnose moiety and provide patients with hesperetin-7-glucoside (Nielsen *et al.*, 2006). The aglycones themselves are further transformed in various organs and it is these final metabolites that act on several pathways to promote health. The major orange juice phytochemicals are two flavanones: hesperidin and narirutin. Manach *et al.* (2003) examined the bioavailability of these two flavanones: five volunteers ingested one litre of a commercial orange juice providing 444 mg/l of hesperidin and 96.4mg/l narirutin. The peak plasma concentration of hesperetin was 1.28 μmol/l and 0.20 μmol/l for naringenin. Two main conclusions were drawn by the authors. The concentrations of metabolites in the blood are very low and the circulating forms for hesperidin were glucuronides (87%) and sulfoglucuronides (13%). More recently, the same group (Manach *et al.*, 2009) identified 61 metabolites following orange juice absorption. To this effect, the major metabolites have now been synthesized and are used in *in vitro* studies to determine their exact mode of action (Khan *et al.*, 2010; Quintana *et al.*, 2011).

In grapefruit, two flavanones, naringin and narirutin, which both are absorbed as the aglycone naringenin are the most studied. Erlund *et al.* (2001) had healthy volunteers ingest orange juice (five women and three men) or grapefruit juice (two women and three men) once (8 ml/kg). Blood and urine samples were collected between 0 and 24 hours after juice administration. C_{max} were 0.6 ± 0.4 and 6.0 ± 5.4 μmol/l for naringenin from orange juice and grapefruit juice, respectively, and 2.2 ± 1.6 μmol/l for hesperetin from orange juice. Similarly, Kanaze *et al.* (2007) gave orally 135 mg of hesperetin and naringenin to six healthy volunteers under fasting conditions. C_{max} for hesperetin and naringenin were 2731.8 ± 1358.4 nmol/l and 7386.6 ± 2833.4 nmol/l. Krogholm *et al.* (2012) evaluated flavonoids and enterolactone in overnight urine. A total of 191 individuals were examined. Concentrations of nine urinary flavonoid aglycones (quercetin, isorhamnetin, tamarixetin, kaempferol, hesperetin, naringenin, eriodictyol, phloretin and apigenin) and enterolactone were determined in overnight and 24-h urine samples; it was concluded that flavonoids and enterolactone in overnight urine samples may be used as a more feasible biomarker than 24-h urine for the assessment of fruit, juice, vegetable and tea intakes in epidemiological studies.

It can be concluded that the oral administration of the citrus juice leads to the formation of their flavanone conjugated forms and, because of the low bioavailability, any action must take place at the catalytic level. The biological effects of phytochemicals have been shown to be more diverse and involve cell-mediated responses and the modulation of various cell-signalling pathways (Williams *et al.*, 2004).

CITRUS AND SPECIFIC AILMENTS

From the hypotheses developed in the last few years, it is generally believed that the inflammation process is at the root of most ailments which seems to be positively affected by fruits (Pan *et al.*, 2010). The concept of inflammation can be defined as the reaction of the body to an injury, when the body reacts to repair itself (acute inflammation). This initial beneficial phase can become detrimental if repairs cannot be achieved. The constant low grade inflammation that ensues (chronic inflammation) is believed to be at the origin of many non-communicable diseases and has even been coined as the silent killer (Kassi *et al.*, 2011). The major diseases that may be initiated by an inflammation process include cardiovascular diseases (CVD), metabolic syndrome, insulin resistance and type 2 diabetes, osteoporosis, dementia and some forms of cancer. The influence of fruits and vegetables and the polyphenols they contain in attenuating inflammation was reviewed by Wu and Schauss (2012) and Recio *et al.* (2012).

Progress has also been made in understanding the biochemical events responsible for the biological effects of phytochemicals. They have been shown to be quite complex and to involve cell-mediated responses and the modulation of various cell-signalling pathways (Stevenson and Hurst, 2007). Several new disciplines grouped under the neologism 'omics', which aim to determine the effects of metabolites of the food phytochemicals on specific tissue and organs, are now being used to clarify these mechanisms (Scalbert and Knasmuller, 2008). They include methods to examine gene expression patterns (genomics, transcriptonomics) and also post-translational modifications (proteonomics), as well as metabolite profiles (metabolomics).

Citrus and cardiovascular health

Cardiovascular diseases (CVD) are one of the major health problems affecting humanity. According to recent data from the American Heart Association, CVD account for about 35% of annual deaths in the United States (Roger *et al.*, 2012). The effects of fruits and juices on cardiovascular diseases (CVD) including blood pressure, atherosclerosis, blood lipids, and endothelial reactivity have been the most extensively studies (Chanet *et al.*, 2012).

Epidemiological studies

The beneficial effects of fruits, vegetables and fruit juices were initially revealed by epidemiological studies. They have the advantage of exposing effects of a food type on a large number of individuals and they provide leads for further more specific studies, particularly human controlled trials. As early as 1999, Joshipura *et al.* showed that the lowest risks of ischemic stroke were observed for high consumption of cruciferous vegetables, citrus fruits and fruits rich in vitamin C. These results lead research to focus on the Mediterranean diet, which was implicated in reducing the risk of heart diseases. Other groups reported similar results (Fung *et al.*, 2009; Sofi, 2009; Estruch, 2010). Johnsen *et al.* (2003) followed a cohort of 54 506 men and women who were included in the Danish Diet, Cancer, and Health study from 1993 to 1997. During that time 266 cases of ischemic stroke were recorded. Individuals in the top quintile of fruit and vegetable intake (673 g/d) had a lower risk ratio of ischemic stroke than persons in the bottom quintile of intake (147 g/d). The authors found the lowest relative risk in the highest quintiles of intake of citrus fruit. The protective effect of citrus fruit was attributed by the authors to the high level of vitamin C. In a similar study, Mink *et al.* (2007) showed that the intake of flavonoids was inversely associated with coronary heart disease mortality. The authors concluded that foods rich in flavonoids were associated with reduced risk of death due to coronary heart disease and cardio vascular disease. Men aged 50–59 years, free of coronary heart disease, were recruited in France and Northern Ireland by Dauchet *et al.* (2004), who concluded that citrus intake, but not other fruits, was associated with lower rates of acute coronary events. Conversely, after reviewing the literature the same authors (Dauchet *et al.*, 2009) concluded that the evidence supporting a cardio protective effect of fruit and vegetable consumption was not very strong. The interpretation of the data may be quite complex, since in a recent review of randomized clinical trials (Kay *et al.*,

2012) suggested that the dose-response between flavonoids and markers of cardiovascular risk may be non-linear.

Biomarkers and pathways

Citrus components have been shown to affect several biochemical pathways involved in regulating cardiovascular events. Among those, the ability to alter blood lipids has been explored. Since the 1990s, studies have shown that citrus flavonoids could have the ability to modulate the synthesis and transport of cholesterol. Two enzymes were particularly examined: 3-hydroxy-3-methyl-glutaryl-CoA reductase (HMG-CoA reductase), an enzyme involved in cholesterol biosynthesis and targeted by statins, and acyl CoA: cholesterol transferase (ACAT), which transfers fatty acids to the lipoproteins. Several studies have shown that the activity of those two enzymes is significantly reduced by both hesperetin and naringenin. Bok *et al.* (1999) observed reduced activities of HMG-CoA reductase and ACAT in livers of rats fed a mixture of naringin and hesperidin. The same group (Lee *et al.*, 2003) showed that the supplementation of hesperetin and its metabolites to high-cholesterol fed rats significantly lowered the plasma total cholesterol and triglyceride concentrations compared to the control group. The hepatic HMG-CoA reductase and ACAT activities were significantly lower in the hesperetin and its metabolite supplemented groups than in the control group. They further showed (Jung *et al.*, 2006) that the two flavonoids lead to a decrease in the plasma and hepatic cholesterol levels, which may have been partly due to the decreased hepatic HMG-CoA reductase and ACAT activities in type 2 diabetic mice.

Kim *et al.* (2003, 2006) examined the effect of naringin on rats fed a high-fat and high-cholesterol diet for three or six weeks. Supplementation with naringin resulted in a significant decrease in the plasma cholesterol and triglyceride concentrations in the six-week trial. The hepatic cholesterol content was also lowered by naringin supplementation only in the six-week trial. The hepatic HMG-CoA reductase activity was lower in the rats supplemented with naringin for six weeks, while the hepatic acyl-coenzyme A: cholesterol acyltransferase activity was lower in both the three-week and six-week trials. Very similar results were obtained in rats by Rajadurai *et al.* (2006), in diabetic mice by Jung *et al.* (2006), and in insulin-resistant mice by Mulvihill *et al.* (2009). Goldwasser *et al.* (2010) showed in human hepatocytes that naringenin activated genes involved in the genesis and catabolism of lipids. They concluded that naringenin increases lipid metabolism in the liver. The cholesterol lowering effect of naringin and the statin lovostatin were compared by Kim *et al.* (2004) in low density lipoproteins (LDL) receptor knockout mice fed a cholesterol diet. The naringin and lovastatin supplementation significantly lowered plasma and hepatic cholesterol levels compared to the control group. The hepatic HMG-CoA reductase activity was significantly lower in the naringin and lovastatin supplemented groups than in the control group, whereas the ACAT activity was unaffected.

Nahmias *et al.* (2008) measured ApoB secretion in primary human hepatocytes following 24 hours of stimulation with naringenin. They demonstrated a dose-dependent decrease in ApoB secretion following naringenin stimulation. At 200 μM naringenin, ApoB secretion was reduced by 60% ± 7% and microsomal triglyceride transfer protein (MTP) activity by 58% ± 8%. The authors also demonstrated by polymerase chain reaction (PCR) that naringenin induced significant changes in hepatic gene transcription: HMGR transcription was reduced by 57% ± 3% while ACAT2 was reduced by 55% ± 7%. By contrast, the mRNA levels of actin, MTP, ACAT1 remained unchanged. Furthermore, the authors showed that intraperitoneal injection of naringin in mice at concentrations of 60 mg/kg results in a decrease by almost half of the total plasma triglycerides. Another possible mechanism was put forwards by Milenkovic *et al.* (2012), who examined the impact of dietary supplementation with polyphenols at nutritional doses on miRNA expression in the livers of apolipoprotein E-deficient mice (apo$E^{-}/^{-}$). In ApoE knock-out mice, polyphenols significantly reduced atherosclerosis lesion development via a lower monocyte infiltration. The authors concluded that polyphenols at nutritional doses modulate the expression of miRNA in the liver.

Another area where possible mechanisms have been examined is the effect of flavonoids on blood vessel walls, particularly the endothelium. In a nutrigenomic study, Liu *et al.* (2008) examined the effect of the citrus flavonoids hesperetin and naringenin on human umbilical vein endothelial cells. It was reported that hesperetin, but not naringenin, increased nitric oxide (NO) released from human umbilical vein endothelial cells in a dose-dependent manner. Real-time reverse transcription PCR and western-blotting analysis revealed that hesperetin up-regulated endothelium nitric oxide synthase (eNOS) expression. These results suggested that hesperetin exerts an anti-atherogenic effect, in part, via ER-mediated eNOS expression and subsequent increase of endothelial NO production. The increase in NO production following a four-week orange juice regimen was confirmed in a clinical study (Morand *et al.*, 2011). An inhibition of adipogenesis and adipocyte differentiation in 3T3-L1 cells by *Citrus aurantium* flavonoids was recently reported by Kim *et al.* (2012a).

These studies showed that orange and grapefruit juice components, mainly hesperidin and naringin, or more likely some of their metabolites, act to up- or down-regulate many genes associated with the maintenance of the blood circulation process. These complex mechanisms are not completely elucidated and will require more studies to be fully understood.

Animal and clinical studies

Work on the mechanistic aspect of citrus and cardiovascular health was complemented by animal and human studies. Gorinstein *et al.* (2004a) examined the effect of grapefruit and orange juice on plasma lipids in rats fed cholesterol and cholesterol-free diets; they concluded that diets supplemented with orange and, to a lesser degree, with grapefruit juices improved plasma lipid metabolism only in rats fed added cholesterol. These authors (Gorinstein *et al.*, 2005a) performed several similar studies with grapefruit and grapefruit juice. They compared the influence of naringin versus red grapefruit juice on plasma lipid levels in rats fed cholesterol-containing and cholesterol-free diets. After 30 days, it was found that, in rats fed cholesterol diets, the rise of triglycerides was less among those drinking red grapefruit juice than in the control, but HDL-C was not affected. Naringin had a comparable effect but to a lesser degree. Similar results were obtained in rats by the same group (Gorinstein *et al.*, 2005b) with red Star Ruby and blond Jaffa grapefruits. It is important to note that citrus appears to improve blood lipids only in animals receiving an abnormally high level of cholesterol. Conversely, Demonty *et al.* (2010) evaluated the LDL-C-lowering efficacy of pure hesperidin and naringin in 204 moderately hypercholesterolemic healthy men and women and showed that pure hesperidin and naringin had little effect on blood cholesterol. This shows the importance of examining the whole juice rather than a single phytochemical. The effects of orange and grapefruit juice were examined by Deyhim *et al.* (2006) on sham (control) and orchidectomized rats. Serum cholesterol and liver triglycerides increased in the orchidectomized group, however grapefruit but not orange juice produced a decrease in liver and in serum cholesterol as compared with the sham group, but under those conditions triglycerides levels were unaffected. The authors concluded that the potential benefits of citrus juices may be mediated by inhibition of the activity of HMG CoA reductase and ACAT. These studies on rats showed that they prevent the rise of noxious blood lipids in animals fed a high cholesterol diet.

Many problems linked to blood circulation have been shown to arise from the inflammation process. Ghanim *et al.* (2010) examined the effects of orange juice on the generation of inflammation due to high fat meals. Three groups of 10 subjects were given water or 300 kcal glucose or orange juice in combination with a 900-kcal high fat, high carbohydrate meal (HFHC meal). Blood samples were obtained before and 1, 3, and 5 hours after the drinks and meal combinations were consumed. Results indicated that protein expression of the toll-like receptors TLR2 and TLR4 messenger RNA (mRNA), and mRNA expression of matrix metalloproteinase (MMP)-9 increased significantly after glucose or water were consumed with the meal but not when orange juice was consumed with the meal. The authors concluded that orange juice intake with the HFHC meal prevented meal-induced oxidative and inflammatory stress, including the increase in TLR expression. The same group (Deopurkar *et al.*, 2010) further studied healthy normal-weight subjects who were given either glucose, cream and an equicaloric amount of orange juice or water after an overnight fast. The authors concluded that orange juice intake does not induce oxidative stress and inflammation, and that by curtailing the inflammation process orange juice contributes to limiting the development of circulatory problems.

In 2000, Kurowska *et al.* examined the effect of orange juice on high density lipoproteins (HDL) and low density lipoproteins (LDL). Sixteen healthy men and nine healthy women with elevated plasma total and LDL-cholesterol and normal plasma triacylglycerol concentrations consumed the AHA Step I diet during the 17-week study. Consumption of 750 ml but not of 250 or 500 ml orange juice daily increased HDL-cholesterol concentrations by 21%, triacylglycerol concentrations increased by 21% and 30%, respectively, whereas the LDL-HDL cholesterol ratio decreased by 16% during the same period. The authors reported a reduction in the LDL-HDL cholesterol ratio with the highest dose of orange juice was due to changes in HDL cholesterol concentrations.

The influence of the pummelo-grapefruit hybrid juice Oroblanco (also known as **Sweetie**) on humans suffering from hypercholesterolemia was studied by Gorinstein *et al.* (2004b). Seventy-two hypercholesterolemic patients were supplemented daily with 100 or 200 ml of fresh grapefruit juice. After 30 consecutive days of grapefruit juice supplemented diets, improvements in serum lipids levels were found in the two grapefruit juice groups; total cholesterol decreased by 9.5% and 16.1%, LDL cholesterol by 11.6% and 21.0%, and total glycerides by 11.5% and 24.7%, respectively. Similar results were obtained in a clinical study with red and white grapefruit (Gorinstein, 2006); fifty-seven hyperlipidemic patients who had coronary bypass surgery were randomly divided into three groups: two experimental (red and white grapefruit groups) and one control group. After 30 days, the authors reported that serum lipid levels in the grapefruit groups were decreased compared to the control and that fresh red grapefruit contains higher quantities of bioactive compounds than white grapefruit. Diet supplemented with fresh red grapefruit positively influenced serum lipid levels of all fractions, especially serum triglycerides. More recently, Diaz-Juarez *et al.* (2009) performed a double-blind crossover study to examine the effects of grapefruit and orange juice on forty volunteers (20 normotensive and 20 hypertensive). Patients received 250 ml of freshly prepared *Citrus paradise* juice, *Citrus sinensis* juice, cow's milk or a vitamin C supplemented beverage. In normotensive and hypertensive subjects, the consumption of *C. paradise* juice showed a statistically significant decrease in diastolic arterial pressure and systolic arterial pressure. The decrease in mean arterial pressure was more pronounced in patients who received *Citrus paradisi* juice, rather than *Citrus sinensis* juice. The authors concluded that the hypotensive effect could be due to naringenin and that this vasodilator effect could be mediated by modulation of the nitric oxide pathway.

The effect of orange juice on lipid oxidation was examined by Sanchez-Moreno *et al.* (2003). Six men and six women consumed 500 ml orange juice/d for 14 days. Blood was collected every hour for six hours on the first day and again on days 7 and 14. Levels of 8-*epi*-prostaglandin $F_{2\gamma}$ (8-*epi*-$PGF_{2\gamma}$), a marker for hypercholesterolemia, were measured. The authors reported that the concentrations of 8-*epi*-$PGF_{2\gamma}$ were significantly higher among the men than among the women. However, a significant decrease in 8-*epi*-$PGF_{2\gamma}$ was noted among men on day 14 but no significant changes were observed among the women. They also showed a significant decrease in plasma uric acid, a marker of oxidative stress and possibly cardiovascular diseases, in men after 14 days. Aptekmann and Cesar (2010), in a three-month study, examined the effect of the daily ingestion of 500 ml of orange juice in thirteen women who did one hour of aerobic training three times a week. The control group consisted of another 13 women who did the same aerobic training programme but did not consume orange juice. Results showed that in the experimental group serum LDL-C decreased 15% and HDL-C increased 18%. The same group (Cesar *et al.*, 2010) examined fourteen adults who consumed 750 ml/day of orange juice for 60 days and reported decreased low density lipoprotein cholesterol. Grapefruit consumption for six weeks was reported by Dow *et al.* (2011) to show a modest weight loss, a significant reduction in waist circumference and a significant reduction in systolic blood pressure.

A comprehensive study of the effect of orange juice on humans has been recently completed. The study followed the effect of orange juice from the genes to the biomarkers and finally to the endpoint results in the volunteers. Milenkovic *et al.* (2011) and Morand *et al.* (2011) examined twenty-four healthy men (50–65 years old) in a randomized controlled crossover study. In the chronic part of the study, made of three, four-week periods, volunteers consumed daily: 500 ml orange juice, 500 ml control drink and hesperidin or 500 ml of placebo. An acute postprandial study was conducted at the beginning of each experimental period. Results revealed that the diastolic blood pressure, an indicator of the peripheral blood vessel resistance, was lower after four weeks in orange juice group (80.4 $\pm$ 2.0 mm/Hg) compared to the placebo (84.9 $\pm$ 2.1) but systolic blood pressure was not affected. The possible mechanisms may involve an increase in the production of nitric oxide (NO) by the vascular endothelium. Indeed, the authors showed that orange juice tended to increase NO plasma concentration; furthermore, in the acute postprandial study an enhancement of the endothelial plasticity was observed. This result indicates that the impact of the juice on the microvascular function was focused on the endothelium-dependent vasorelaxation. It should also be mentioned that after the chronic study the concentrations of uric acid in plasma were significantly lower in the orange juice group (57.9 $\pm$ 3.1 umol/l) compared to control (64.3 $\pm$ 3.0 umol/l), confirming an effect of orange juice on oxidative stress as reported by Sanchez-Moreno *et al.* (2003). Part of this study was devoted to a human nutrigenomics study focused on the impact consumption of orange juice and pure hesperidin had on the expression of genes in leukocytes (Milenkovic *et al.*, 2011). Using micro array techniques, the authors reported that 3422 genes were differentially expressed by orange juice, 1819 by hesperidin and 1582 genes were common to both groups. This again shows the importance of examining the whole juice rather than a single phytochemical (Demonty *et al.*, 2010). The analysis showed that the most represented processes were transcription, signal transduction, cell adhesion, cell proliferation, chemotaxis, cell differentiation and lipid metabolic process. Many of the down-regulated genes are involved with the formation of the atherosclerotic plaque. The expression of these genes by orange juice consumption is associated with a decrease in the expression of genes coding for the adhesion molecules, while the expression is known to increase in the patients with coronary atherosclerosis. Other genes affected are known to be linked to cardiovascular disease development and, particularly, the nitric oxide pathways. This should result in a decrease in the inhibition of NO synthase that will improve the concentration of plasma NO. It should be pointed out that after this four-week study a very significant and beneficial reaction of the genome was observed; however, the number of significantly positive physiological changes was more limited. It can be expected that longer studies would have unveiled a larger array of significant results.

Studies have shown that citrus may have a beneficial effect on the cardiovascular system that can persist even when citrus metabolites have been eliminated from the body. Improvements have been reported in blood lipids, blood pressure, endothelial function, as well as down-regulation of the events leading to the formation of atherosclerosis. Although very encouraging, further, longer term, clinical studies are necessary to confirm and possibly quantify those benefits.

Citrus and cognition

An effect of polyphenols on brain activity and cognition is only possible because these compounds and their metabolites have been shown to cross the blood–brain barrier (BBB) that shields this organ from many chemicals. Milbury and Kalt (2010) reported that anthocyanin glucuronides can cross the BBB and access the neurons. The permeability of naringenin and quercetin across the BBB was revealed *in vitro* and *in situ* by Youdim *et al.* (2003). Several animal studies showed that naringin may protect rats against neuro-inflammation induced by chemicals such as kainic acid (Golechha *et al.*, 2011) or colchicine (Kumar *et al.*, 2010). It has also been suggested that citrus compounds may help maintain brain function. Unno *et al.* (2011) showed that beta-cryptoxanthin, a carotenoid found in citrus, may slow senescence in mice. However, Butchart *et al.* (2011) did not see that flavonoids from citrus fruits, apple and tea preserved IQ during an entire human life span. Recently, Ebrahimi and Schluesener (2012) summarized the protective effects of polyphenols against neurodegeneration and discussed some of the difficulties in incorporating these biochemical compounds into relevant therapeutics for neurodegenerative diseases. Nones *et al.* (2012) showed that treatment of astrocytes with hesperidin leads to an increase in cell proliferation. These results suggest that hesperidin modulate neuroglial interactions in the central nervous system. For those reasons Vafeiadou *et al.* (2009) also concluded that polyphenols prevent neuronal deterioration or even improve cognitive performance. Nurk *et al.* (2010) examined more than 2000 elderly subjects and reported positive cognitive effects with various plant foods, finding that citrus fruits and high-fibre bread were the most effective. Spencer's group (Spencer, 2010a, 2010b; Rendeiro *et al.*, 2012) reviewed the neuroprotective properties of dietary flavonoids and concluded, firstly, that they promote cerebral vascular blood flow and, secondly, that they interact with neuronal signalling cascades leading to an inhibition of cell death and to a promotion of neuronal differentiation. Improvement of cognition has been demonstrated for many fruits and there are indications that they may prevent neuronal deterioration.

Citrus and cancer

A large number of epidemiological studies have examined the influence of fruit and vegetable intake on the incidence of cancer. After examining epidemiologic surveys and 144-item food frequency questionnaires, Jansen *et al.* (2011) concluded that lower consumption of fruits, vegetables, whole grains and fibre was associated with pancreatic cancer. Li *et al.* (2010) suggested that citrus consumption is associated with reduced all-cancer incidence among 42 470 Japanese adults. García-Tirado *et al.* (2012) reviewed the literature for the effects of flavonoids and lung cancer. They identified a meta-analysis, a systematic review, 11 prospective cohort studies and five case-control studies and concluded that there was a small protective effect of flavonoid consumption (especially at high doses) against the development of lung cancer. A systematic review of prospective studies conducted by Martínez-González *et al.* (2011) was aimed at assessing the relationship between fruit and vegetable consumption and chronic diseases. The authors assessed 18 457 university graduates and reported that consumption of fruits and vegetables was inversely associated with CVD incidence and mortality but that the association with cancer was not so clear. Foschi *et al.* (2010) analysed data from a series of case-control studies to assess the effect of citrus on various forms of cancer. They studied a large number of patients with various forms of cancer, concluding that citrus have a protective role against cancers of the digestive and upper respiratory tract but that no consistent association was found with breast, endometrial, ovarian, prostate and renal cell cancer. Similarly, Freedman *et al.* (2007) reported a significant protective role of citrus consumption on oesophageal squamous cell carcinoma. Overall, epidemiological studies indicate an inverse relation between the consumption of fruits, fruit juices and vegetables, particularly citrus, and many inflammation-related diseases but definitive proof would require other types of evidence.

Citrus fruits, such as oranges, tangerines, grapefruits, lemons and limes, are rich in flavonoids, carotenoids and limonoids. Many studies have examined the effect of pure citrus compounds on cancer. The effect of naringin was studied by Camargo *et al.* (2012) in rats after intraperitoneally administering different doses of naringin (10, 25 and 35 mg/kg), for 50 days. At 25 mg/kg, naringin inhibited tumour growth. However, the use of very high levels of naringin raised questions about the possible effect of juices and fruits. β-Cryptoxanthin and hesperidin may also have an effect on induced rat colon, rat tongue and mouse lung tumorigenesis (Tanaka *et al.*, 2012). Similar results were reported by Vanamala *et al.* (2006) with grapefruit limonin. Nalini *et al.* (2012) reported that hesperetin supplementation showed an inhibition of cell proliferation markers, angiogenic growth factors and induction of apoptosis in rat colon cancer. Orally administered naringenin was shown by Qin *et al.* (2011) to inhibit the outgrowth of metastases after breast cancer surgery. Many studies have been based on examining the effect of isolated phytochemicals or fruit extracts on carcinoma cell cultures. Han *et al.* (2012) investigated the antiproliferative activity of the peels of *Citrus aurantium* extract on apoptosis in U937 human leukemia cells and reported a dose-dependent inhibition of the growth of the cells. Extracts from lemon seed were investigated by Kim *et al.* (2012b) for apoptotic effects in human breast adenocarcinoma (MCF-7) cells. The results suggested that aglycones and glucosides of the limonoids and flavonoids extract may potentially serve as a chemopreventive agent for breast cancer. Tian *et al.* (2001) also showed that grapefruit limonoids may prevent cancer cell proliferation *in vitro*. Similarly, citrus pectins may interfere with cancer cells' signal transduction (Liu *et al.*, 2001).

The interpretation of these results is difficult since the compounds that are placed in contact with the cancer cells are most often significantly modified before entering the body and could not reach the cells *in vivo*. Furthermore, the amounts of phytochemical used is very often far above physiological concentrations of the circulating metabolites. It can be concluded that if fruits and vegetables appear to have a preventive effect on cancer, more specific activity may be very hard to assess and it may be ethically difficult to perform significant clinical studies on cancer patients.

Bone health

Bone health is largely the result of equilibrium between osteoclast cells destroying bone and osteoblasts building it. Several fruit phytochemicals, mainly polyphenols and carotenoids, have been shown to have a positive influence on bone health and, particularly, the bone mineral density of postmenopausal women (Trzeciakiewicz *et al.*, 2009; Sacco *et al.*, 2013). An interesting hypothesis was raised by Uehara (2006) who pointed out that hesperidin regulates hepatic cholesterol synthesis by inhibiting the activity of 3-hydroxy-3-methlyglutaryl coenzyme A (HMG-CoA) reductase. Statins, cholesterol lowering agents, also induce bone formation and inhibit bone resorption both *in vitro* and *in vivo*. The author suggested that hesperidin may act on bone by the same mechanism as that of statins. Fruit juice carotenoids, β-cryptoxanthin, β-carotene and lycopene, and citrus flavonoids may also improve bone health by preventing bone destruction by osteoclasts (Habauzit *et al.*, 2011). Horcajada and Offord (2012) reviewed the effect of phytochemicals on bone health and concluded that phytonutrients and, especially, polyphenols can act on both osteoblasts and osteoclasts to modulate bone metabolism, since a balance between the activity of both cell types is required for bone health maintenance. Similarly, Yamaguchi (2012) reported that β-cryptoxanthin has stimulatory effects on osteoblastic bone formation and inhibitory effects on osteoclastic bone resorption *in vitro*. The ability of naringin to improve the decrease in bone weight coefficient, the length and the diameter of the bone as well as the mineral content was described by Wei *et al.* (2007). Bone resorption was significantly attenuated in male senescent rats by hesperidin and naringin, the two major citrus flavanones (Habauzit *et al.*, 2011). The same group examined the actions of a hesperidin metabolite, hesperetin-7-O-glucuronide, which is more bioavailable than hesperidin and as a result may have more effect on bone health than the original flavanone (Habauzit *et al.*, 2009; Trzeciakiewicz *et al.*, 2010). They suggested that hesperetin-7-O-glucuronide may regulate osteoblast differentiation and might be implicated in the regulation of osteoblast/osteoclast communication. The furanocoumarins, imperatorin and bergapten were reported by Tang *et al.* (2008) to enhance alkaline phosphatase (ALP) activity, type I collagen synthesis and bone nodule formation in primary cultured osteoblasts. Tang *et al.* (2011) showed that another furanocoumarin, psoralen, promoted osteoblast differentiation of mouse osteoblasts in a dose-dependent manner and up-regulated the expression of bone morphogenetic protein (BMP) genes. In a different area, it has also been suggested that potassium citrate (present in citrus) may contribute to the neutralization of the protein-induced metabolic acidosis that induces loss of bone calcium (Mardon *et al.*, 2008). Results accumulated so far have provided good indications that, at this time, citrus are the fruits that may play the most active role on bone health.

ATTRIBUTES OF CITRUS HAVING RAISED DISPROPORTIONATE CONCERNS

Grapefruit–drug interaction

In recent years, the interaction of grapefruit with various drugs has been widely reported in the press and has significantly impacted the sale of grapefruit juice. In 1991, Bailey *et al.* discovered that the blood level of the drug felodipine was significantly higher in patients who had consumed grapefruit juice. The same phenomenon was later found to occur with various drugs and, particularly, some of the calcium channel blockers and some of the cholesterol lowering drugs (statins). Further studies showed that the increase in drug bioavailability was due the elimination from the intestine of an enzyme of the P450 cytochrome family, CYP3A4. Cytochrome P450 enzymes are monoxygenases that play a major role in the metabolism of a wide array of xenobiotics, including drugs and various pollutants (Lewis, 2003). Cytochrome P450s catalyse an oxidation of the xenobiotics and result in mostly beneficial detoxification. The compounds responsible for the interaction have been identified as furanocoumarins (FCs), which are compounds synthesized in various plants, mainly the Apiaceae and Rutaceae, and found in fruits and vegetables ranging from grapefruit to parsnip. They have been shown to affect a wide variety of organisms, particularly insects (Li *et al.*, 2004). The interaction between plant FCs and insects has been extensively examined, particularly the co-evolution of FC and P450s in wild parsnip (Li *et al.*, 2004). The FCs found in wild parsnip (*Pastinaca sativa*) play a dominant role in the resistance of this plant to its principal enemy the parsnip webworm (*Depressaria pastinacella*) (Zangerl and Berenbaum, 2004).

The entire mechanism of grapefruit drug interaction, known as mechanism-based interaction, has now been elucidated. Once ingested, the FCs enter the enterocytes lining the intestinal walls. They react with CYP3A4 in normal enzymatic reactions affecting the furan ring. They generate various metabolites, particularly epoxyfurans and ketoenals. If the process was to stop here, the drug interaction problem would not exist. However, the released FC metabolites, in turn, bind irreversibly CYP3A4. The enzyme is then tagged for destruction and is rapidly catabolized (Li *et al.*, 2004). Not all FCs are equally effective, they have been ranked by order of inhibitory potency: paradisin C > 6', 7'-dihydroxybergamottin > bergamottin > bergapten > bergaptol with comparative strength of 260, 44, 20, 20, 1 (Ohnishi *et al.*, 2000). Following grapefruit juice ingestion, CYP3A4 is not replaced in the affected enterocytes and it is necessary to wait about three days for new enterocytes containing CYP3A4 to be formed. During that time drugs metabolized by CYP3A4 are no longer partially destroyed by intestinal first-pass clearance and may reach the bloodstream at increased concentrations; in the most extreme cases, a tenfold rise in bioavailability was reported. If grapefruit keeps being ingested in a regular basis, then the cytochrome population does not regenerate.

Several factors however can modify the physiological effects of the grapefruit drug interaction process. Among all the medications taken by mouth only a limited number are metabolized by CYP3A4, other drugs follow different pathways and are unaffected. Even among the drugs metabolized by CYP3A4, the effect of grapefruit can be highly variable. Detailed information of the effect of grapefruit on most drugs can be obtained on a web site established by Tufts University and the University of Florida (http://www.druginteractioncenter.org). The site reviews the drugs affected by grapefruit and classifies them as having a low or no interaction, a medium or a high interaction. For example, among statins: pravastatin is not or mildly affected, atovastatin is moderately influenced whereas a major effect was reported with simvastatin. Therefore, for each class of drugs there are some unaffected by grapefruit. The problem is further complicated by the fact that the level of cytochrome P450 varies greatly between individuals (Lin and Lu, 2001). In twenty patients, Paine *et al.* (1997) reported a 64-fold variation in the intra- and inter-intestinal level of CYP3A, ranging from 4.1 to 262 pmol/mg of protein. Under those conditions only patients with a high level of intestinal P450 would be affected by grapefruit drug interaction. There are also indications that P450 enzyme levels decrease with age, although the question is controversial (Sotaniemi *et al.*, 1997) and possibly older people would be less affected than the young.

The level of furanocoumarins in grapefruit juice is very variable and can be affected by various parameters, ranging from cultivar to storage. Study on the effects of processing on grapefruit juice furanocoumarins (FC) showed that the concentration of several FC decreases during storage. The mechanism responsible for FC degradation was investigated by HPLC and LC/MS (Cancalon *et al.*, 2011). It appears to be a pH, temperature dependent, non-enzymatic hydrolysis involving the removal of the side chain. Paridisin C, 6'7'dihydroxybergamottin (DHB) and, to a lesser extent, bergamottin have been identified so far as the compounds most affected. As expected, the reaction is associated with an increase in the amount of bergaptol present in the juice. Linalool oxide has been tentatively identified as the degradation product; this molecule is formed by the reorganization of the side chain once severed from the FC (Cancalon and Haun, 2006). Because of this hydrolysis process significant differences can be found in shelf-stable and refrigerated grapefruit juices (Cancalon *et al.*, 2011). Degradation of DHB and paradisin C has implications for the drug interaction potential of a juice, since they are the most active FCs and bergaptol generated in the reaction is a weak inhibitor. As a consequence, shelf-stable and heat-abused juices will have less inhibitory potential compared to equivalent fresh or refrigerated juices. The geographical location of the trees has also been reported as having a significant influence on FC levels in the fruits. Grapefruits and pummelos from Florida, California and Israel were examined for their FC content; for the same cultivars FCs were found to vary in the decreasing order California > Israel > Florida. The reason for these variations is still unclear but furanocoumarins are stress-induced compounds and climatic differences, such as rainfall and temperature variations, may be involved (Cancalon *et al.*, 2008).

Genetic differences have been shown to play a major role in the FC content of grapefruit and pummelo cultivars, with concentrations ranging from up to 800 mg/l of DHB to only traces or even no detectable amount. Two populations of fruits have been revealed: one has the better known composition, rich in DHB and some paradisin C, and another with little or no DHB but containing a set of weak inhibitors, mainly isoimperatorin, and a small amount of cnidicin (Cancalon *et al.*, 2009). It is significant to point out that in each of the two groups, including among seedless triploid hybrids, some have almost no measurable FCs and exhibit very limited interaction as measured by the microsome method (Greenblatt, personal communication). It is hoped that the breeding programme underway will develop grapefruit types without FCs, or with FCs having minimal drug interaction activities if any (Chen *et al.*, 2011).

The problem of grapefruit–drug interaction cannot be ignored but it is important to put it in the proper perspective. Its potential negative effect is limited to a small number of drugs and in most cases, such as statins, there are alternative drugs unaffected by grapefruit. The level of CYP3A4 varies so much among individuals that only patients with high levels of intestinal P450 could be affected by grapefruit. Therefore, patients consuming grapefruit and their physicians should, in most cases, be able to find a drug that can treat their specific disease.

Citrus body weight and sugar metabolism

Obesity is taking epidemic proportion in many countries. In the United States, the Center for Disease Control and Prevention (CDC) reported that about 65% of the US adult population aged 20–74 years is currently overweight (DHHS, 2004). Organizations and researchers have been investigating the origins of this problem. It is believed that excess consumption of calorie-rich foods is one of the main causes, sugar-rich drinks have been especially stigmatized and, by extension, the blame also reached fruit juices. Hess *et al.* (2012) reviewed recommendations for sugar intake and pointed out contradictions and discrepancies. As early as 1997, Dennisson *et al.* examined the effect of juice ingestion by 168 children (ninety-four 2-year-old children and seventy-four 5-year-old children). They concluded that consumption of 12 fl oz (355 ml)/day or more of fruit juice by young children was associated with short stature and obesity. In 2006, the same group (Faith *et al.*, 2006) added that among children initially either at risk for overweight or overweight, increased fruit juice intake was associated with excess adiposity gain, whereas parental offerings of whole fruits were associated with reduced gain in adiposity. However, Nicklas *et al.* (2008) reported that the mean daily juice consumption was 4.1 fl oz (121 ml), which contributed a mean intake of 58 kcal (3.3% of total energy intake). On average, children consumed less than the maximum amounts of 100% juice recommended by the American Academy of Pediatrics. The Academy recommends limiting fruit juice consumption to 4–6 ounces/day for children 1–6 years old and 8–12 ounces or 2 servings/day for children 7–18 years old (American Academy of Pediatrics, 2001).

The polemic was exacerbated by the question of high fructose corn syrup (HFCS), one of the main sweeteners of sodas. Bray *et al.* (2004) reported that the consumption of HFCS increased by more than a 1000% between 1970 and 1990, and that HFCS now represents more than 40% of caloric sweeteners added to foods and that the increased use of HFCS in the United States mirrors the rapid increase in obesity. It is believed that the effect of HFCS is due to the presence of fructose. The metabolism of fructose differs from glucose; in particular, fructose favours *de novo* lipogenesis and is thought to be responsible for an increased level of blood triglycerides (Stanhope *et al.*, 2011). In addition, unlike glucose, fructose does not stimulate insulin secretion, which is involved in the regulation of food intake. However, many studies on the deleterious effects of fructose were conducted on mice. Sievenpiper *et al.* (2011, 2012a) pointed out that the sugar metabolism of rodents and men are different, and that conclusion from one species to the other should be done carefully and that fructose *per se* is still healthy. It should also be kept in mind that fructose will need to be transformed into glucose, spreading overtime the level of blood glucose. The same group pointed out that small amounts of fructose decrease the glycemic response to high-glycemic index meals in humans and concluded from a meta-analysis that fructose doses ($\leq$36 g/d) may improve glycemic control without adverse effects on body weight, TAG, insulin and uric acid (Sievenpiper *et al.*, 2012b, 2012c). It should be pointed out that Morand *et al.* (2011) also showed that orange juice has a positive effect on blood uric acid. Similarly, Livesey and Taylor (2008) showed that fructose intake of 90 g/d or less significantly improved HbA(1c), a marker of dysglycemia. The authors reported no significant change in body weight at intakes less than 100 g fructose/d and that fructose intakes of less than 50 g/d had no significant postprandial effect on triglycerides.

The confusion was compounded by studies which pooled all beverages in the same category (Bremer and Lustig, 2012). Palmer *et al.* (2008) artificially defined a category as 'fruit drinks' that includes fruit juices other than orange or grapefruit, powdered drinks and fortified fruit drinks. Wang *et al.* (2008) stated that soda contributed approximately 67% of all sugar-sweetened beverage calories among the adolescents, whereas fruit drinks provided more than half of the sugar-sweetened beverage calories and added that young people derive 10–15% of total calories from sugar-sweetened beverages and 100% fruit juice. Studies that examine non-standardized data provide skewed and confusing conclusions. It is likely that excessive consumption of any calorie source is likely to lead to obesity. For example, Faith *et al.* (2006) showed that the consumption of very high levels of fruit juice could have a negative effect (24–30 fl oz, 709–889 ml). Small children given unrealistic amounts of juice contributing to 25–60% of their daily energy intake exhibited weight gain and failure to thrive (Smith and Lifshitz, 1994). These questions were reviewed by Nicklas *et al.* (2008). O'Neil and Nicklas (2008) evaluated the relationship between consumption of 100% fruit juice and bodyweight among children and adolescents and found that there is no systematic association between consumption of 100% fruit juice and overweight in children or adolescents. In a recent study, Morand *et al.* (2011) showed that ingestion of up to 500 ml of orange juice a day for four weeks does not lead to gain weight. O'Neil *et al.* (2011a, 2012) reported that usual intake of 100% fruit juice consumption for children of 2–5 years was associated with a better diet. The same group (O'Neil *et al.*, 2011b) showed that 100% orange juice consumption by children 2–18 years of age did not induce weight or body mass index changes and that juice intake does not significantly affect the risk of obesity. The findings of these analyses support previous studies by Skinner and Carruth (2001) and Newby *et al.* (2004) that showed no significant correlation between weight change and the consumption of 100% fruit juice, fruit drinks, milk, soda or diet soda in preschool-aged children. The impact of 100% orange juice on body composition was evaluated by Wang *et al.* (2012), who stated that adults but not children had lower BMI, waist circumference and percentage body fat but also lower odds ratio for overweight and obesity than non-consumers. Daily consumption of 250 ml of orange juice for three months was reported by Simpson *et al.* (2012) to not adversely affect insulin sensitivity, body composition or other indices of the metabolic syndrome.

Study of fruit juice intake based only on calorie calculations ignores several important points. It has been shown that fruit juices contain compounds that limit or prevent insulin resistance by fruit juices. Yoshida *et al.* (2007) showed that fruit juice consumption was inversely associated with fasting plasma glucose. A reduction of insulin resistance, oxidative stress and inflammation were reported after ingestion of several fruit juices including grapefruit juice (Yao *et al.*, 2004), grapefruit naringin (Kannappan and Anuradha, 2010) and orange juice (Ghanim *et al.*, 2007). Alcada *et al.* (2009) also showed that orange juice may increase the production of somatostatin, an inhibitor of insulin secretion. The authors concluded that fruit juices are a good source of the sugars needed by the body and also provide phytochemicals with a wide range of health benefits. Rampersaud (2007) classified fruit juices according to their nutrition density and showed that citrus juices, particularly pink grapefruit and orange juice, were more nutrient dense compared to the other non-fortified 100% juices. However, the Estimated Average Requirement for several vitamins and minerals was higher in juice consumers.

Cell biology studies have revealed mechanisms that could explain the beneficial effects of citrus juices on sugar metabolism. In mammals but not in birds, glucose is absorbed and transported across membranes by mediated translocation processes. Skopec *et al.* (2010) showed that flavonoids effectively decrease glucose absorption in rats, which rely on mediated absorption of glucose, but do not have an effect in robins, which rely on non-mediated absorption of glucose. At a physiologically attainable dose of 10 μM, grapefruit naringenin was shown by Harmon and Patel (2004) to reduce insulin-stimulated glucose uptake by nearly 25% in MCF-7 breast cancer cells. Li *et al.* (2006) showed that other citrus phytochemicals, the polymethoxylated flavones (PMFs), could ameliorate the fructose-induced hypertriglyceridemia in hamsters. After four weeks, PMF-treated groups showed a statistically significant decrease in serum triglyceride (TG) and cholesterol levels compared to the fructose-fed control group, as well as a decrease in insulin level and an improvement in glucose tolerance. Claussnitzer *et al.* (2011) showed that dietary flavonoids may interfere with glucose transport GLUT4 processes and that even under physiological conditions naringenin could impair glucose transport. The effect of polyphenols on uptake and apical to basolateral transport of glucose was investigated by Manzano and Williamson (2010) using Caco-2 intestinal cell monolayers. Substantial inhibition on both uptake and transport was observed by extracts. The authors concluded that polyphenols inhibit glucose transport from the intestinal lumen into cells. On the other hand, naringenin was shown by Zygmunt *et al.* (2010) to stimulate glucose uptake in muscle L6 myotubes in a dose- and time-dependent manner. Similarly, the PMF tangeretin may stimulate glucose uptake and improve glucose tolerance and, therefore, exerts antidiabetic effects in both cell culture and mouse models (Kim *et al.*, 2012c). Although still incomplete, these data appear to indicate that citrus compounds may improve sugar metabolism in humans by modulating its cellular transport.

The understanding of the effect of fruit juices, and particularly citrus juices, on sugar uptake and weight gain has been obscured by confusing definitions and experiments. Recent results indicate that at moderate doses citrus juices do not have deleterious consequences on sugar and lipid metabolism and, as a result, do not promote weight gain. Furthermore, they appear to be able to modulate sugar uptake and distribution and may have some positive outcome on the development of metabolic syndrome and insulin resistance.

Dental health

Several reports have postulated that fruit juices can affect dental health, promote caries and dissolve enamel. Reports of these effects appeared as early as 1954 (Thomas). These studies have been performed either *in vitro* or under long, non-natural conditions. Many studies are based on keeping isolated teeth or teeth fragments in a drink for various amounts of time (Willershausen *et al.*, 2008; Ren *et al.*, 2011). Lim *et al.* (2008) reported that children with a high consumption of soft drinks are at a higher risk of developing dental caries relative to those consuming milk and 100% fruit juice, and that damage can be avoided by regular use of fluoridated toothpaste. Issa *et al.* (2011) claimed that both juice and whole fruits could have a detrimental effect on teeth. However, no measurable association of intake of 100% fruit juice, milk, tea and some other drinks with the prevalence of tooth loss were reported by Tanaka *et al.* (2008). Other studies also concluded that fluoride or other additives may prevent acid food damage (Ren *et al.*, 2011; Scaramucci *et al.*, 2011). Davis *et al.* (2007) proposed that calcium added to juice can provide another beneficial effect. Besides those negative aspects that have extensively reviewed, recent studies have revealed that juice polyphenols have a beneficial effect on dental health (Hannig *et al.*, 2009; Ferrazzano *et al.*, 2009). Wood (2007) examined the ability of bioflavonoids to reduce dental caries and cariogenic bacteria and concluded that in rats naringenin supplementation reduced dental caries formation, possibly because of a decrease in dental plaque accumulation. Varoni *et al.* (2012) reported that plant polyphenols prevent oral diseases. Hesperidin was shown by Hiraishi *et al.* (2011) to preserve collagen and inhibit demineralization on bovine root dentine, and to enhance remineralization. These studies, performed mostly on artificial preparations, have shown that any damaging effect of citrus juices would only occur if there was prolonged contact between the juice and the teeth. *In vivo*, saliva will rapidly remove the juice from the mouth; furthermore, citrus flavonoids appear to have beneficial effect of their own.

CONCLUSIONS

Some beneficial properties of citrus have been known for a long time, among those, low sodium, high potassium, absence of lipids and the presence of vitamins, particularly folic acid, and vitamin C. These attributes have given rise to several health claims. In recent years, research has focused on the health properties of phytochemicals. Although our knowledge of the mechanism of action of those compounds is still limited, it is believed that they may be mostly acting through the control of chronic inflammation, a process that seems to promote the development of cardiovascular diseases, metabolic syndrome, insulin resistance, type2 diabetes, osteoporosis, dementia and some forms of cancer. Published research has provided insights on the role of citrus in human health. However, *in vitro* studies are often done on cell cultures with abnormally high levels of the original phytochemicals and not with the metabolites actually reaching the organ *in vivo*. Animal studies, such as those reported on the effects of fructose in rats, are difficult to extrapolate to humans. Even with clinical studies providing significant evidence, this support must be tempered with the understanding that fruits and fruit juices are not drugs and that the beneficial effects reported are usually mild and difficult to observe readily in very healthy volunteers. The most striking benefits reported in clinical studies have often been associated with patients who are at mild to significant risk for developing various health problems, such as pre-hypertension, overweight/obesity or cholesterolemia. Presently it cannot be distinguished whether beneficial effects in very healthy people are insufficient to be reflected by currently available clinical methodology or if clinical benefits from phytochemical consumption depend upon regulation of slightly different pathways in people with various degrees of health (Shao *et al.*, 2012; Almoosawi *et al.*, 2012). It is, therefore, necessary to obtain a consensus on the type and level of proof that should be expected from clinical studies, and if the pharmacological protocols used so far should be adapted for fruits, vegetables and other foodstuffs.

REFERENCES

Alcada, M.N., Monteiro, R., Calhau, C. and Azevedo, I. (2009) Orange juice vs soft drinks. *Arch Intern Med* **169**:315.

Almoosawi S., Tsang C., Ostertag L.M. *et al.* (2012) Differential effect of polyphenol-rich dark chocolate on biomarkers of glucose metabolism and cardiovascular risk factors in healthy, overweight and obese subjects: a randomized clinical trial. *Food Funct.* DOI: 10.1039/c2fo30060e.

American Academy of Pediatrics (2001) The use and misuse of fruit juice in pediatrics. Committee on Nutrition. *Pediatrics.* **107**:1210–1213.

Aptekmann, N.P. and Cesar, T.B. (2010) Orange juice improved lipid profile and blood lactate of overweight middle-aged women subjected to aerobic training. *Maturitas* **67**:343–347.

Bailey, D.G., Spence, J.D., Munoz, C. and Arnold, J.M. (1991) Interaction of citrus juices with felodipine and nifedipine. *Lancet* **337**:268–269.

Berhow, M., Tisserat, B., Kanes, K. and Vandercook, C. (1998) Survey of Phenolic Compounds Produced in Citrus. Agricultural Research Service Technical Bulletin Number 1856. United States Department of Agriculture.

Bok, S.-H., Lee, S.-H., Park, Y.-B. *et al.* (1999) Plasma and hepatic cholesterol and hepaticactivities for 3-hydroxy-3-methyl-glutaryl-CoA reductase and acy1CoA: cholesterol transferase are lower in rats fed citrus peel extract or a mixture of citrus bioflavonoids. *J Nutr* **129**:1182–1185.

Brand, W., Padilla, B., Van Bladeren, P.J. *et al.* (2010) The effect of co-administered flavonoids on the metabolism of hesperetin and the disposition of its metabolites in Caco-2 cell monolayers. *Mol Nutr Food Res* **95**:1089–1095.

Bray, G.A., Nielsen, S.J., and Popkin, B.M. (2004) Consumption of high-fructose corn syrup in beverages may play a role in the epidemic of obesity. *Am J Clin Nutr* **79**:537–543.

Bremer, A.A. and Lustig, R.H. (2012) Effects of sugar-sweetened beverages on children. *Pediatr Ann* **41**:26–30.

Butchart, C., Kyle, J., McNeill, G. *et al.* (2011) Flavonoid intake in relation to cognitive function in later life in the Lothian Birth Chort 1936. *Br J Nutr* **106**:141–148.

Cancalon P.F. and Haun, C. (2006) Grapefruit juice furanocoumarins and P450 eytochrome CYP3A4. *Proc Fl State Hort Soc* **119**:358–360.

Cancalon, P.F., Haun, C., Gmitter, F. *et al.* (2008) Differences in the furanocoumarin composition and drug interaction potential of pummelos and grapefruit from Florida, California, and Israel. Proceedings of the 59th Citrus Processors' and Subtropical Technology Conference, pp. 1–2.

Cancalon, P., Gmitter, F., Chen, C., and Greenblatt, D.J. (2009) Variations in the furanocoumarin compostion and drug interaction potential of pummelos and grapefruits (abstract). 4th *International Conference on Polyphenol and Health*, p. 116.

Cancalon, P.F., Barros, Santiago M. *et al.* (2011) Effect of maturity, processing and storage on the furanocoumarin composition of grapefruit and grapefruit juice (JFS-2010-1057.R1). *J Food Sci* **76**:C543–548.

Camargo, C.A., Gomes-Marcondes, M.C., Wutzki, N.C. and Aoyama, H. (2012) Naringin inhibits tumor growth and reduces interleukin-6 and tumor necrosis factor α levels in rats with Walker 256 carcinosarcoma. *Anticancer Res* **32**:129–133.

Cesar, T.B., Aptekmann, N.P., Araujo, M.P. *et al.* (2010) Orange juice decreases low-density lipoprotein cholesterol in hypercholesterolemic subjects and improves lipid transfer to high-density liporprotein in normal and hypercholesterolemic subjects. *Nutr Res* **30**:689–694.

Chanet, A., Milenkovic, D., Manach, C. *et al.* (2012) Citrus flavanones: what is their role in cardiovascular protection?. *J Agric Food Chem* **60**:8809–8822.

Chen, C., Cancalon, P.F., Haun, C. and Gmitter, F. (2011) Characterization of furanocoumarin profile and inheritance toward selection of low furanocoumarin seedless grapefruit cultivars. *J Amer Soc Hort Sci* **136**:1–6.

Claussnitzer, M., Skurk, T., Hauner, H. *et al.* (2011) Effect of flavonoids on basal and insulin-stimulated 2-deoxyglucose uptake in adipocytes. *Mol Nutr Food Res* **55**(Suppl 1):S26–34.

Crozier A., Del Rio, D. and Clifford, M.N. (2011) Bioavailability of dietary flavonoids and phenolic compounds. *Mol Aspects Med* **31**:446–467.

Dauchet, L., Ferrieres, J. and Arveiler, D. (2004) Frequency of fruit and vegetable consumption and coronary heart disease in France and Northern Ireland: the PRIME study. *Br J Nutr* **92**:963–972.

Dauchet, L., Amouyel, P. and Dallongevile, J. (2009) Fruits, vegetables and coronary heart disease. *Medscape Nat Rev Cardiol* **6**:599–608.

Davis, R.E., Marshall, T.A., Qian, F. *et al.* (2007) *In vitro* protection against dental erosion afforded by commercially available, calcium-fortified 100 percent juices. *J Am Dent Assoc* **138**:1593–1698.

Decuypere, J.D. (2000) Dr Decuypere's Nutrient Charts^{TM}: Fruit Chart. http://www.health alternatives2000.com/fruit-nutrition-chart.html (last accessed 23 February 2013).

Del Rio, D., Rodriguez-Mateos, A., Spencer, J.P. *et al.* (2012) Dietary (poly)phenolics in human health: structures, bioavailability, and evidence of protective effects against chronic diseases. *Antioxid Redox Signal*. doi: 10.1089/ars.2012.4581.

Demonty, I., Lin, Y., Zebregs, Y.E. *et al.* (2010) The citrus flavonoids hesperidin and naringin do not affect serum cholesterol in moderately hypercholesterolemic men and women. *J Nutr* **140**:1615–1620.

Dennison, B., Rockwell, H.L. and Baker, S.L. (1997) Excess fruit juice consumption by preschool-aged children is associated with short stature and obesity. *Pediatrics* **99**:15–22.

Deopurkar, R., Ghanim, H., Friedman, J. *et al.* (2010) Differential effects of cream, glucose, and orange juice on inflammation, endotoxin, and the expression of Toll-like receptor-4 and suppressor of cytokine signaling-3. *Diabetes Care* **33**:991–997.

Deyhim, F., Lopez, E., Gonzalez, J. *et al.* (2006) Citrus juice modulates antioxidant enzymes and lipid profiles in orchidectomized rats. *J Med Food* **9**:422–426.

DHHS (US Department of Health and Human Services, Centers for Disease Control and Prevention, National Center for Health Statistics) (2004) Health, United States, 2004 with Chartbook on Trends in the Health of Americans. US Government Printing Office, Washington, DC). [Library of Congress Catalog no. 76-641496.]

Díaz-Juárez, J.A., Tenorio-López, F.A., Zarco-Olvera, G. *et al.* (2009) Effect of Citrus paradise extract and juice on arterial pressure both in vitro and *in vivo*. *Phytother Res* **23**:948–954.

Dow, C.A., Going, S.B., Chow, H.H. *et al.* (2012) The effects of daily consumption of grapefruit on body weight, lipids, and blood pressure in healthy, overweight adults. *Metabolism* **61**(7):1026–1035.

Ebrahimi, A. and Schluesener, H. (2010) Natural polyphenols against neurodegenerative disorders: potentials and pitfalls. *Ageing Res Rev* **11**:329–345.

Erlund, I., Meririnne, E., Alfthan, G. and Aro, A. (2001) Plasma kinetics and urinary excretion of the flavanones naringenin and hesperetin in humans after ingestion of orange juice and grapefruit juice. *J Nutr* **131**:235–241.

Estruch, R. (2010) Anti-inflammatory effects of the Mediterranean diet: the experience of the PREDIMED study. *Proc Nutr Soc* **69**:333–340.

Faith, M.S., Dennison, B.A, Edmunds, L.S. and Stratton, H.H. (2006) Fruit juice intake predicts increased adiposity gain in children from low-income families:weight status-by-environment interaction. *Pediatrics* **118**:2066–2075.

Ferrazzano, G.F., Amato, I., Ingenito, A. *et al.* (2009) Anti-cariogenic effects of polyphenols from plant stimulant beverages (cocoa, coffee, tea). *Fitoterapia* **80**:255–262.

Foschi, R., Pelucchi, C., Dal Maso, L. *et al.* (2010) Citrus fruit and cancer risk in a network of case-control studies. *Cancer Causes Control* **21**:237–242.

Freedman, N.D., Park, Y., Subar A.F. *et al.* (2007) Fruit and vegetable intake and esophageal cancer in a large prospective cohort study. *Int J Cancer* **121**:2753–2760.

Fung, T.T, Rexrode, K.M., Mantzoros, C.S. *et al.* (2009) Mediterranean diet and incidence of and mortality from coronary heart disease and stroke in women. *Circulation* **119**:1093–1100.

García-Tirado, J., Rieger-Reyes, C. and Saz-Peiró, P. (2012) Efecto de los flavonoides en la prevención del cancer de pulmón: revision sistemática. *Med Clin (Barc)* **139**:358–363.

Ghanim, H., Mohanty, P., Pathak, R. *et al.* (2007) Orange juice or fructose intake does not induce oxicative and inflammatory response. *Diabetes Care* **30**:1406–1411.

Ghanim, H., Sia, C.L., Upadhyay, M. *et al.* (2010) Orange juice neutralizes the proinflammatory effect of a high-fat, high-carbohydrate meal and prevents endotoxin increase and Toll-like receptor expression. *Am J Clin Nutr* **1**:940–949.

Goldwasser, J., Cohen, P.Y., Yang, E. *et al.* (2010) transcriptional regulation of human and rat hepatic lipid metabolism by the grapefruit flavonoid naringenin: Role of PPARa, PPARc and LXRa. *PLoS ONE* **5**:e12399.

Golechha, M., Chaudhry, U., Bhatia, J. *et al.* (2011) Naringin protects against kainic acid-induced status epilepticus in rats: evidence for an antioxidant, anti-inflammatory and neuroprotective intervention. *Biol Pharm Bull* **34**:360–365.

Gorinstein, S., Leontowicz, H., Leontowicz, M. *et al.* (2004a) Fresh Israeli Jaffa blond (Shamouti) orange and Israeli Jaffa Red Star Ruby (Sunrise) grapefruit juices affect plasma lipid metabolism and antioxidant capacity in rats fed added cholesterol. *J Agric Food Chem* **52**:4853–4859.

Gorinstein, S., Caspi, A., Libman, I. *et al.* (2004b) Fresh israeli jaffa sweetie juice consumption improves lipid metabolism and increases antioxidant capacity in hypercholesterolemic patients suffering from coronary artery disease: studies in vitro and in humans and positive changes in albumin and fibrinogen fractions. *J Agric Food Chem* **52**:5215–5222.

Gorinstein, S., Leontowicz, H., Leontowicz, M. *et al.* (2005a) Changes in plasma lipid and antioxidant activity in rats as a result of naringin and red grapefruit supplementation. *J Agric Food Chem* **53**:3223–3228.

Gorinstein, S., Leontowicz, H., Leontowicz, M. *et al.* (2005b) Red Star Ruby (Sunrise) and blond qualities of Jaffa grapefruits and their influence on plasma lipid levels and plasma antioxidant activity in rats fed with cholesterol-containing and cholesterol-free diets. *Life Sci* **77**:2384–2397.

Gorinstein, S., Caspi, A., Libman, I. *et al.* (2006) Red grapefruit positively influences serum triglyceride level in patients suffering from coronary atherosclerosis: studies in vitro and in humans. *J Agric Food Chem* **54**:1887–1892.

Habauzit, V., Nielsen, I.L., Gil-Izquierdo, A. *et al.* (2009) Increased bioavailability of hesperetin-7-glucoside compared with hesperidin results in more efficient prevention of bone loss in adult ovariectomised rats. *Br J Nutr* **102**:976–984.

Habauzit, V., Sacco, S.M., Gil-Izquierdo, A. *et al.* (2011) Differential effects of two citrus flavanones on bone quality in senescent male rats in relation to their bioavailability and metabolism. *Bone* **49**:1108–1116.

Han, M.H., Lee, W.S., Lu, J.N. *et al.* (2012) *Citrus aurantium L.* exhibits apoptotic effects on U937 human leukemia cells partly through inhibition of Akt. *Int J Oncol* **40**:2090–2096.

Hannig, C., Sorg, J., Spitzmüller, B. *et al.* (2009) Polyphenolic beverages reduce initial bacterial adherence to enamel in situ. *J Dent* **37**:560–566.

Harmon, A.W. and Patel, Y.M. (2004) Naringenin inhibits glucose uptake in MCF-7 breast cancer cells: a mechanism for impaired cellular proliferation. *Breast Cancer Res Treat* **85**:103–110.

Hasenau, S.M. and Covington, C. (2002) Neural tube defects. *MCN Am J Matern Child Nurs* **27**:87–91.

Hess, J., Latulippe, M.E., Ayoob, K. and Slavin, J. (2012) The confusing world of dietary sugars: definitions, intakes, food sources and international dietary recommendations. *Food and Funct* **3**:477–486.

Hiraishi, N., Sono, R., Islam, M.S. *et al.* (2011) Effect of hesperidin in vitro on root dentine collagen and demineralization. *J Dent* **39**:391–396.

Horcajada, M.N. and Offord, E. (2012) Naturally plant-derived compounds: role in bone anabolism. *Curr Mol Pharmacol* **5**:205–218.

Issa, A.I., Toumba, K.J., Preston, A.J. and Duggal, M.S. (2011) Comparison of the effects of whole and juiced fruits and vegetables on enamel demineralisation in situ. *Caries Res* **45**:448–452.

Jansen, R.J., Robinson, D.P., Stolzenberg-Solomon, R.Z. *et al.* (2011) Fruit and vegetable consumption is inversely associated with having pancreatic cancer. *Cancer Causes Control* **22**:1613–1625.

Johnsen, S.P., Overvad, K., Stripp, C. *et al.* (2003) Intake of fruit and vegetables and the risk of ischemic stroke in a cohort of Danish men and women. *Am J Clin Nutr* **78**:57–64.

Joshipura, K.J., Ascherio, A., Manson, J.E. *et al.* (1999) Fruit and vegetable intake in relation to risk of ischemic stroke. *JAMA* **282:**1233–1239.

Jung, U.J., Lee, M.K., Park, Y.B. *et al.* (2006) Effect of citrus flavonoids on lipid metabolism and glucose-regulating enzyme mRNA levels in type-2 diabetic mice. *Int J Biochem Cell Biol* **38**:1134–1145.

Kanaze, F.I., Bounartzi, M.I., Georgarakis, M. and Niopas, I. (2007) Pharmacokinetics of the citrus flavanone aglycones hesperetin and naringenin after single oral administration in human subjects. *Eur J Clin Nutr* **61**:472–477.

Kannappan, S. and Anuradha, C.V. (2010) Naringenin enhances insulin-stimulated tyrosine phosphorylation and improves the cellular actions of insulin in a dietary model of metabolic syndrome. *Eur J Nutr* **2**:101–109.

Kassi, E., Pervanidou, P., Kaltsas, G. and Chrousos, G. (2011) Metabolic syndrome: definitions and controversies. *BMC Med* **9**:48–61.

Kay, C.D., Hooper, L., Kroon, P.A. *et al.* (2012) Relative impact of flavonoid composition, dose and structure on vascular function: A systematic review of randomised controlled trials of flavonoid-rich food products. *Mol Nutr Food Res* **56**:1605–1616.

Khan, M.U. and Mackinney, G. (1953) Carotenoids in grapefruit, Citrus paradisi. *Plant Physiol* **28**:550–552.

Khan, M.K., Rakotomanomana, N., Loonis, M. and Dangles, O. (2010) Chemical synthesis of citrus flavanone glucuronides. *J Agric Food Chem* **58**:8437–8443.

Kim, H.K., Jeong, T.S., Lee, M.K. *et al.* (2003) Lipid-lowering efficacy of hesperetin metabolites in high-cholesterol fed rats. *Clin Chim Acta* **327**:129–137.

Kim, H.J., Oh, G.T., Park Y.B. *et al.* (2004) Naringin alters the cholesterol biosynthesis and antioxidant enzyme activities in LDL receptor-knockout mice under cholesterol fed condition. *Life Sci* **74**:1621–1634.

Kim, S.Y., Kim, H.J., Lee, M.K. *et al.* (2006) Naringin time-dependently lowers hepatic cholesterol biosynthesis and plasma cholesterol in rats fed high-fat and high-cholesterol diet. *J Med Food* **9**:582–586.

Kim, G.S., Park, H.J., Woo, J.H. *et al.* (2012a) Citrus aurantium flavonoids inhibit adipogenesis through the Akt signaling pathway in 3T3-L1 cells. *Complement Altern Med* **12**:31–41.

Kim, J., Jayaprakasha, G.K., Uckoo, R.M. and Patil, B.S. (2012b) Evaluation of chemopreventive and cytotoxic effect of lemon seed extracts on human breast cancer (MCF-7) cells. *Food Chem Toxicol* **50:**423–430.

Kim, M.S., Hur, H.J., Kwon, D.Y. and Hwang, J.T. (2012c) Tangeretin stimulates glucose uptake via regulation of AMPK signaling pathways in C2C12 myotubes and improves glucose tolerance in high-fat diet-induced obese mice. *Mol Cell Endocrinol* **358**:127–134.

Krogholm, K.S., Bysted, A., Brantsaeter, A.L. *et al.* (2012) Evaluation of flavonoids and enterolactone in overnight urine as intake biomarkers of fruits, vegetables and 41:beverages in the Inter99 cohort study using the method of triads. *Br J Nutr* **27**:1–9.

Kumar A., Dogra, S. and Prakash, A. (2010) Protective effect of naringin, a citrus flavonoid, against colchicines-induced cognitive dysfunction and oxidative damage in rats. *J Med Food* **13**:976–984.

Kurowska, E.M., Spence, J.D., Jordan, J. *et al.* (2000) HDL-cholesterol-raising effect of orange juice in subjects with hypercholesterolemia. *Am J Clin Nutr* **72**:1095–1100.

Lee, M.K., Moon, S.S., Lee, S.E. *et al.* (2003) Naringenin 7-O-cetyl ether as inhibitor of HMG-CoA reductase and modulator of plasma and hepatic lipids in high cholesterol-fed rats. *Bioorg Med Chem* **11**:393–398.

Lewis, D.F. (2003) P450 structures and oxidative metabolism of xenobiotics. *Pharmacogenomics* **4**:387–395.

Li, W., Zangerl, A.R., Schuler, M.A. and Berenbaum, M.R. (2004) Characterization and evolution of furanocumarin-inducible cytochrome P450s in the parsnip webworm, *Depressaria pastinacella. Insect Mol Biol* **13**:603–613.

Li, R.W., Theriault, A.G., Au, K. *et al.* (2006) Citrus polymethoxylated flavones improve lipid and glucose homeostasis and modulate adipocytokines in fructose-induced insulin resistant hamsters. *Life Sci* **79**:365–373.

Li, W.Q., Kuriyama, S., Li, Q. *et al.*(2010) Citrus consumption and cancer incidence: the Ohsaki cohort study. *Int J Cancer* **127**:1913–1922.

Lim, S., Sohn, W., Burt, B.A. *et al.* (2008) Cariogenicity of soft drinks, milk and fruit juice in low-income African-american children: a longitudinal study. *J Am Dent Assoc* **139**:959–967.

Lin, J.H. and Lu, Y.H. (2001) Interindividual variability in inhibition and induction of cytochrome P450 enzymes. *Annu. Rev. Pharmacol. Toxicol* **41**:535–535.

Liu, Y., Ahmad, H., Luo, Y. *et al.* (2001) Citrus pectin: characterization and inhibitory effect on fibroblast growth factor-receptor interaction. *J Agric Food Chem* **49**:3051–3057.

Liu, L., Xu, D.M., and Cheng, Y.Y. (2008) Distinct effects of naringenin and hesperetin on nitric oxide production from endothelial cell. *J Agric Food Chem* **56**:824–829.

Livesey, G. and Taylor, R. (2008) Fructose consumption and consequences for glycation, plasma triacyglycerol, and body weight: meta-analyses and meta-regression models of intervention studies. *Am J Clin Nutr* **88**:1419–1437.

Manach, C., Morand, C., Gil-Izquierdo, A. *et al.* (2003) Bioavailability in humans of the flavanones hesperidin and naritutin after the ingestion of two dodses of orange juice. *Eur J Clin Nutr* **57**:235–2342.

Manach, C., Hubert, J., Llorach, R. and Scalbert, A. (2009) The complex links between dietary phytochemicals and human health deciphered by metabolomics. *Mol Nutr Food Res* **53**:1303–1315.

Manners, G.D. (2007) Citrus limonoids: analysis, bioactivity, and biomedical prospects. *J Agric Food Chem* **55**:8285–8294.

Manzano, S. and Williamson, G. (2010) Polyphenols and phenolic acids from strawberry and apple decrease glucose uptake and transport by human intestinal Caco-2 cells. *Mol Nutr Food Res* **54**:1773–1780.

Mardon J, Habauzit V, Trzeciakiewicz A *et al.*(2008) Long-term intake of a high-protein diet with or without potassium citrate modulates acid-base metabolism, but not bone status, in male rats. *J Nutr* **138**:718–724.

Martínez-González, M. Á., dela Fuente-Arrillaga, C., López-Del-Burgo, C. *et al.* (2011) Low consumption of fruit and vegetables and risk of chronic disease: a review of the epidemiological evidence and temporal trends among Spanish graduates. *Public Health Nutr* **14**:2309–2315.

Mertens-Talcott, S.U., Zadezensky, I, De Castro, W.V. *et al.* (2006) Grapefruit–drug interactions: can interactions with drugs be avoided? *J Clin Pharmacol* **46**:1390–1416.

Milbury, P.E. and Kalt, W. (2010) Xenobiotic metabolism and berry flavonoid transport across the blood–brain barrier. *J Agric Food Chem* **14**:3950–3956.

Milenkovic, D., Deval, C., Mazur, A. and Morand, C. (2011) Hesperidin displays relevant role in the nutrigenomic effect of orange juice on blood leukocytes in human volunteers: a randomized controlled cross-over study. *PLoS One* **6**:e26669.

Milenkovic, D., Deval, C., Gouranton, E. *et al.* (2012) Modulation of miRNA expression by dietary polyphenols in apoE deficient mice: a new mechanism of the action of polyphenols. *PLoS One* **7**:e29837.

Mink, P.J., Scrafford, C.G., Barraj, L.M. *et al.* (2007) Flavonoid intake and cardiovascular disease mortality: a prospective study in postmenopausal women. *Am J Clin Nutr* **85**:895–909.

Morand, C., Dubray, C., Milenkovic, D. *et al.* (2011) Hesperidin contributes to the vascular protective effects of orange juice: A randomized cross-over study on healthy volunteers. *Am J Clin Nutr* **93**:73–80.

Mulvihill, E.E., Allister, E.M., Sutherland, B.G. *et al.* (2009) Naringenin prevents dyslipidemia, apolipoprotein B overproduction, and hyperinsulinemia in LDL receptor-null mice with diet-induced insulin resistance. *Diabetes* **58**:2198–2210.

Nalini, N., Aranganathan, S. and Kabalimurthy, J. (2012) Chemopreventive efficacy of hesperetin (citrus flavonone against 1,2-dimethylhydrazine-induced rat colon carcinogenesis. *Toxicol Mech Methods* **22**:397–408.

Nahmias, Y., Goldwasser, J., Casali, M. *et al.* (2008) Apolipoprotein B-dependent hepatitis C virus secretion is inhibited by the grapefruit flavonoid naringenin. *Hepatology* **47**:1437–45.

Neveu V, Perez-Jiménez J, Vos F (2010) Phenol-Explorer: an online comprehensive database on polyphenol contents in foods. doi: 10.1093/database/bap024. (www.phenol-explorer.eu/; last accessed 23 February 2013)

Newby, P.K., Peterson, K.E., Berkey, C.S. *et al.* (2004) Beverage consumption is not associated with changes in weight and body mass index among low-income preschool children in North Dakota. *J Am Diet Assoc* **104**:1086–1094.

Nicklas, T.A., O'Neil, C.E. and Kleinman, R. (2008) Association between 100% juice consumption and nutrient intake and weight of children aged 2 to 11 years. *Arch Pediatr Adolesc Med* **162**:557–565.

Nielsen, I.L., Chee, W.S., Poulsen, L. *et al.*(2006) Bioavailability is improved by enzymatic modification of the citrus flavonoid hesperidin in humans: a randomized, double-blind, crossover trial. *J Nutr* **136**:404–408.

Nones, J., Spohr, T.C. and Gomes, F.C. (2012) Effects of the flavonoid hesperidin in cerebral cortical progenitors in vitro: indirect action through astrocytes. *Int J Dev Neurosci.* dx.doi.org/10.1016/j.ijdevneu.2012.01.008.

Nurk, E., Refsum, H., Drevon, C.A. *et al.* (2010) Cognitive performance among the elderly in relation to the intake of plant foods. *The Hordaland Health Study. Br J Nutr* **104**:1190–1201.

O'Neil, C.E. and Nicklas, T.A. (2008) A review of the relationship between 100% fruit juice consumption and weight in children and adolescents. *American Journal of Lifestyle Medicine* **2**: 315–354.

O'Neil, C.E., Nicklas, T.A., Rampersaud, G.C. and Fuloni, V.L. (2011a) One hundred percent orange juice consumption is associated with better diet quality, improved nutrient adequacy, and no increased risk for overweight/obesity in children. *Nutr Res* **31**:673–682.

O'Neil, C.E., Nicklas, T.A, Zanovec, M. and Fulgoni, V.L. (2011b) Diet quality is positively associated with 100% fruit juice consumption in children and adults in the United States: NHANES 2003–2006. *Nutr J* **10**:17.

O'Neil, C.E, Nicklas, T.A., Zanovec, M *et al.* (2012) Fruit juice consumption is associated with improved nutrient adequacy in children and adolescents: the National Health and Nutrition Examination Survey (NHANES) 2003–2006. *Public Health Nutr* **23**:1–8.

Ohnishi, A. Matsuo, H., Yamada, S. *et al.* (2000) Effect of furanocoumarin derivatives in grapefruit juice on the uptake of vinblastine by467 Caco-2 cells and on the activity of cytochrome P450 3A4. *Br J Pharmacol* **130**:1369–1377.

Paine, M.F., Khalighi, M., Fisher, J.M. *et al.* (1997) Characterization of interintestinal and intraintestinal variations in human CYP3A-dependent metabolism. *J Pharmacol Exp Ther* **283**:1552–1562.

Palmer, J.R., Boggs, D.A., Krishnan, S. *et al.* (2008) Sugar-sweetened beverages and incidence of type 2 diabetes mellitus in African American women. *Arch Int Med* **168**:1487–1492.

Pan, M. H., Lai, C.S. and Ho, C.T. (2010) Anti-inflammatory activity of natural dietary flavonoids. *Food and Funct* **1**:15–31.

Prior, R.L. and Cao, G. (1999) antioxidant capacity and plyphenolic compoentof teas: implications for altering *in vivo* antioxidant status. *Proc Soc Exp Biol Med* **220**:255–261.

Qin, L., Jin, L., Lu, L. *et al.* (2011) Naringenin reduces lung metastasis in a breast cancer resection model. *Protein Cell* **2**:507–516.

Quintana, M., Martin, J.F., Lyan, B. *et al.* (2011) Effects of consumption of hesperidin or orange juice on the plasma and urine metabolomic profiles. 5th *International Conference on Polyphenol and Health*, 17–20th October, Sitges (Barcelona), Spain.

Rajadurai, M. and Stanely Mainzen Prince, P. (2006) Preventive effect of naringin on lipids, lipoproteins and lipid metabolic enzymes in isoproterenol-induced myocardial infarction in Wistar rats. *J Biochem Mol Toxicol* **20**:191–197.

Rampersaud, G.C. (2007) A comparison of nutrient density scores for 100% fruit juices. *J Food Sci* **72**:S261–266.

Recio, M.C., Andújar, I. and Ríos, J.L. (2012) Anti-inflammatory agents from plants: progress and potential. *Curr Med Chem* **11**:823–835.

Ren, Y.F., Liu, X., Fadel, N. *et al.* (2011) Preventive effects of dentifrice containing 5000 ppm fluoride against dental erosion *in situ*. *J Dent* **39**:672–678.

Rendeiro, C., Guerreiro, J.D., Williams, C.M. and Spencer, J.P. (2012) Flavonoids as modulators of memory and learning: molecular interactions resulting in behavioural effects. *Proc Nutr Soc* **14**:1–17.

Roger, V.L., Go, A.S., Lloyd-Jones, D.M. *et al.* (2012) Heart disease and stroke statistics 2012 update: a report from the American Heart Association. *Circulation* **125**:e2–e220.

Sacco, S.M., Horcajada, M.N. and Offord, E. (2013) Phytonutrients for bone health during ageing. *Br J Clin Pharmacol* **75**:697–707.

Sanchez-Moreno, C., Cano, P.P., de Ancos, B. *et al.* (2003) Effect of orange juice intake on vitamin C concentrations and biomarkers of antioxidant status in humans. *Am J Clin Nutr* **78**:454–460.

Scalbert, A. and Knasmüller, S. (2008) Genomic effects of phytochemicals and their implication in the maintenance of health. *Br J Nutr* **1**:ES1–2.

Scaramucci, T., Hara, A.T., Zero, D.T. *et al.* (2011) *In vitro* evaluation of the erosive potential of orange juice modified by food additives in enamel and dentine. *J. Dent* **39**:841–8.

Schleicher, R.L., Carroll, M.D., Ford, E.S. and Lacher, D.A.(2009) Serum vitamin C and the prevalence of vitamin C deficiency in the United States: 2003–2004 National Health and Nutrition Examination Survey (NHANES). *Am J Clin Nutr* **90**:1252–1263.

Shao L, Liu K, Huang F *et al.* (2012) Opposite effects of quercetin, luteolin, and epigallocatechin gallate on insulin sensitivity under normal and inflammatory conditions in mice. inflammation. doi: 10.1007/s10753-012-9514-x

Sievenpiper, J.L., de Souza, R.J., Kendall, C.W. and Jenkins, D.J. (2011) Is fructose a story of mice but not men? *J Am diet Assoc* **111**:219–220.

Sievenpiper, J.L., de Souza, R.J. and Jenkins, D.J. (2012a) Sugar: fruit fructose is still healthy. *Nature* **482**:470.

Sievenpiper, J.L., Chiavaroli, L., de Souza, R.J. *et al.* (2012b) 'Catalytic' doses of fructose may benefit glycaemic control without harming cardiometabloic risk factors: a small meta-analysis of randomized controlled feeding trials. *Br J Nutr* **21**:1–6.

Sievenpiper, J.L., de Souza, R.J., Mirrahimi, A. *et al.* (2012c) Effect of fructose on body weight in controlled feeding trials: a systematic review and meta-analysis. *Ann Intern Med* **156**:291–304.

Simpson, E.J., Brown, S.J., Mendis, B. *et al.* (2012) The effect of daily orange juice consumption on insulin sensitivity and indices of the metabolic syndrome. *Proc Nutr Soc* **71**:E182.

Skinner, J.D., and Carruth, B.R. (2001) A longitudinal study of children's juice intake and growth: the juice controversy revisited. *J Am Diet Assoc* **101**:432–437.

Skopec, M.M., Green, A.K. and Karasov, W.H. (2010) Flavonoids have differential effects on glucose absorption in rats (*Rattus norvegicus*) and American robins (*Turdis migratorius*). *J Chem Ecol* **36**:236–243.

Smith, M.M. and Lifshitz, F. (1994) Excess fruit juice consumption as a contributing factor in nonorganic failure to thrive. *Pediatrics* **93**:438–443.

Sofi, F. (2009) The Mediterranean diet revisited: evidence of its effectiveness grows. *Curr Opin Cardiol* **24**:442–446.

Sotaniemi, E.A., Arranto, A.J., Pelkonen, O. and Pasanen, M. (1997) Age and cytochrome P450-linked drug metabolism in humans: an analysis of 226 subjects with equal histopathologic conditions. *Clin Pharmacol Ther* **61**:331–339.

Spencer, J.P. (2010a) Beyond antioxidants: the cellular and molecular interactions of flavonoids and how these underpin their actions on the brain. *Proc Nutr Soc* **69**:244–260.

Spencer, J.P. (2010b) The impact of fruit flavonoids on memory and cognition. *Br J Nutr* **3**:S40–47.

Stanhope, K.L., Bremer, A.A., Medici, V. *et al.* (2011) consumption of fructose and high fructose corn syrup increase postprandial triglycerides, LDL-cholesterol, and apolipoprotein-B in young men and women. *J Clin Endocrinol Metab* **96**:E1596–605.

Stevenson, D.E. and Hurst, R.D. (2007) Polyphenolic phytochemicals – Just antioxidants or much more? *Cell Mol Life Sci* **64**:2900–2916.

Tanaka, K., Miyake, Y., Sasaki, S. *et al.* (2008). Beberage consumption and the prevalence of tooth loss in pregnant Japanese women: the Osaka Maternal and Child Health Study. Osaka Materal and Child Health Study Group. *Fukuoka Igaku Zasshi* **99**:80–89.

Tanaka, T. Tanaka, T., Tanaka, M. and Kuno, T. (2012) Cancer chemoprevention by citrus pulp and juices containing high amounts of β-cryptoxanthin and hesperidin. *J Biomed Biotechnol* **2012**:516981–516991.

Tang CH, Yang RS, Chien MY *et al.* (2008) Enhancement of bone morphogenetic protein-2 expression and bone formation by coumarin derivatives via p38 and ERK-dependent pathway in osteoblasts. *Eur J Pharmacol* **579**:40–49.

Tang DZ, Yang F, Yang Z *et al.* (2011) Psoralen stimulates osteoblast differentiation through activation of BMP signaling. *Biochem Biophys Res Commun* **405**:256–261.

Thomas, A.E. (1954) Some observations on the influence of orange juice ingestion on the teeth and supporting structures. *Oral Surg Oral Med Oral Pathol* **7**:741–749.

Tian, Q., Miller, E.G., Ahmad, H. *et al.* (2001) Differential inhibition of human cancer cell proliferation by citrus limonoids. *Nutr Cancer* **40**:180–184.

Trzeciakiewicz, A., Habauzit, V. and Horcajada, M.N. (2009) When nutrition interacts with osteoblast function: molecular mechanisms of polyphenols. *Nutr Res Rev* **22**:68–81.

Trzeciakiewicz, A., Habauzit, V., Mercier, S. *et al.* (2010) Molecular mechanism of hesperetin-7-O-glucuronide, the main circulating metabolite of hesperidin, involved in osteoblast differentiation. *J Agric Food Chem* **58**:668–675.

Uehara, M. (2006) Prevention of osteoporosis by foods and dietary supplements. Hesperidin and bone metabolism. *Clin Calcium* **16**:1669–7166.

Unno, L., Sugiura, M., Ogawa, K. *et al.* (2011) Beta-cryptoxanthin, plentiful in Japanese mandarin orange, prevents age-related cognitive dysfunction and oxidative damage in senescence-accelerated mouse brain. *Biol Pharm Bull* **34**:311–317.

Vafeiadou, K., Vauzour, D., Lee, H.Y. *et al.* (2009) The citrus flavanone naringenin inhibits inflammatory signaling in glial cells and protects against neuroinflammatory injury. *Arch Biochem Biophys* **484**:100–109.

Vallejo F, Larrosa M, Escudero E *et al.* (2010) Concentration and solubility of flavanones in orange beverages affect their bioavailability in humans. *J Agric Food Chem* **58**:6516–6524.
Vanamala, J., Leonardi, T., Patil, B.S. *et al.* (2006) suppression of colon carcinogenesis by bioactive compounds in grapefruit. *Carcinogenesis* **27**:1257–1265.
Varoni, E.M., Lodi, G., Sardella, A. *et al.* (2012) Plant polyphenols and oral health: old phytochemicals for new fields. *Curr Med Chem* **19**:1706–1720.
Visioli F. (2012) Can experimental pharmacology be always applied to human nutrition? *Int J Food Sci Nutr* **63**:10–13.
Wang, Y.C., Bleich, S.N. and Gortmaker, S.L. (2008) Increasing caloric contribution from sugar-sweetened beverages and 100% fruit juices among US children and adolescents, 1988–2004. *Pediatrics* **121**:e1604–1614.
Wang Y, Lloyd B, Yang M *et al.* (2012) Impact of orange juice consumption on macronutrient and energy intakes and body composition in the US population. *Public Health Nutr* **20**:1–8
Wei, M., Yang, Z., Li, P. *et al.* (2007) Anti-osteoporosis activity of naringin in the retinoic acid-induced osteoporosis model. *Am J Chin Med* **35**:663–667.
Willershausen, B., Callaway, A., Azrak, B. and Duschner, H. (2008) Influence of apple juice on human enamel surfaces of the and second dentition – an *in vitro* study. *Eur J Med Res* **13**:349–54.
Williams, R.J., Spencer, J.P. and Rice-Evans, C. (2004) Flavonoids: antioxidants or signaling molecules? *Free Radic Biol Med* **36**:838–849.
Wood, N. (2007) the effects of selected dietary bioflavonoid supplementation on dental caries in young rats fed a high-sucrose diet. *J Med Food* **10**:694–701.
Wu, X. and Schauss, A.G. (2012) Mitigation of Inflammation with Foods. *J Agric Food Chem* [Epub ahead of print].
Yamaguchi, M. (2012) Beta-cryptoxanthin in bone homeostasis. Role of carotenoid. *J Biomed Sci* **19**:36–49.
Yao, F., Shu, G., Yang, Y. and Shi, B. (2004) Grapefruit and diet of the type 2 diabetes. *Diabetes* **53**:A591.
Yoshida, M., McKeown, N.M., Rogers, G. *et al.* (2007) Surrogate markers of insulin resistance are associated with consumption of sugar-sweetened drinks and fruit juice in middle and older-aged adults. *J Nutr* **137**:2121–2127.
Youdim, K.A., Dobbie, M.S., Kuhnle, G. *et al.* (2003) Interaction between flavonoids and the blood–brain barrier: *in vitro* studies. *J Neurochem* **85**:180–192.
Zangerl, A.R. and Berenbaum, M.R. (2004) Genetic variation in primary metabolites of *Pastinaca sativa*; can herbivores act as selective agents? *J Chem. Ecol* **30**:1985–2002.
Zygmunt, K., Faubert, B., MacNeil, J. and Tsiani, E. (2010) Naringenin, a citrus flavonoid, increases muscle cell glucose uptake via AMPK. *Biochem Biophys Res Commun* **398**:178–183.

7 Health Benefits from Pomegranates and Stone Fruit, Including Plums, Peaches, Apricots and Cherries

Francisco A. Tomás-Barberán, David Ruiz, Daniel Valero, Diego Rivera, Conchita Obón, Catalina Sánchez-Roca and María I. Gil

INTRODUCTION

The regular consumption of fruit and vegetables as part of a balanced diet has been associated with health benefits in different epidemiological studies. The health effects have been linked to a lower risk of cardiovascular diseases and some types of cancer. These effects have been linked to the free radical scavenging and antioxidant properties of many phytochemical constituents present in these food products, together with the contribution to the intake of dietary fibre, minerals, some vitamins like vitamin C, and a replacement effect decreasing the intake of other food products, and increasing satiety.

Recent studies have shown that none of these effects can be considered independently, as they can interact in many different ways, either during food intake as part of a complete meal, changes in the GI tract, interaction with gut microbiota and interaction with other food constituents (lipids, proteins, carbohydrates, vitamins, etc.).

There has been active research to evaluate the effect of these food constituents on health. Summarized in this chapter are the state of the art of the research on the health effects, bioactive compounds and changes with processing and storage in the main phytochemicals present in pomegranates and stone fruit.

THE ORIGIN, DIVERSITY AND TRADITIONAL USES OF THE POMEGRANATE (*PUNICA GRANATUM* L.)

Pomegranate (*Punica granatum* L.) is a diploid shrub or small tree, with 2n = 16, 18 and belongs to the genus *Punica* of the *Lythraceae* family (formerly included in *Punicaceae*), which consists of two species. The other species of the genus (*Punica protopunica* Balf.f.) is endemic to the island of Socotra (Yemen) in the Indian Ocean. The small bitter fruits of this last species are not appealing as human or livestock food (Miller, 2004).

Wild forms of *P. granatum* grow in masses in the south Caspian belt, in north-east Turkey and in Albania and Montenegro but it is impossible to distinguish between wild and feral populations. Domestication brought about an increase in fruit size and a shift from sexual reproduction to clonal propagation (Gabrielian and Zohary, 2004). Genetic diversity in cultivated populations of pomegranate is high from the Mediterranean to China (Yuan *et al.*, 2007; Verma *et al.*, 2010).

Bioactives in Fruit: Health Benefits and Functional Foods, First Edition. Edited by Margot Skinner and Denise Hunter.

The archaeological evidence for pomegranate is abundant in the Near East in the form of plant remains as well as in the form of images (Rivera *et al.*, 2012). Pomegranates are recorded in several ancient Mesopotamian cuneiform sources, from the second half of the third millennium BC (Zohary and Hopf, 1994). It is without doubt cited in the Bible under the Hebrew names *rimmôn*, or the Greek *roa, roia* or *kodon* (Moldenke and Moldenke, 1952). In Greece, between the fifth and third centuries BC, Hippocrates recorded numerous medicinal uses of pomegranate to treat heart pain, leucorrhoea, sterility, uterus, pleuritis and lung diseases, phthisis, and as an astringent, vulnerary, febrifuge, antiemetic, a panacea for women in gynaecology, postpartum, for cholera (Alliota *et al.*, 2003); Theophrastus thoroughly treated pomegranate morphology, cultivation and uses (Hort, 1980). Later the pomegranate medicinal uses of are treated by Dioscorides (first century AD), the Armenian Ananias of Shirak (seventh century AD), Al-Biruni and Avicenna (Rivera *et al.*, 2012).

Pomegranate is an appreciated minor crop in traditional horticulture of the Mediterranean and the Near East. The numerous seeds are covered with a juicy flesh which can be eaten fresh, or whose juice can be extracted, and can also be fermented into wine (Zohary and Hopf, 1994).

The fruits are consumed in the Bodrum area (Turkey) and the bark is used to treat parasites and diabetes in Edremit Gulf (Balikesir, Turkey); herbalists sell the flowers to treat diarrhoea in Istanbul, molasses for the same purpose in Gaziantep and the roots to expel ascaris and tapeworm in Bilecik. The branches are used as a circulatory tonic, the rind to treat diarrhoea, dysentery and hookworm, and the roots as an astringent and anthelmintic in the West Bank (Palestine); the rind is used to treat obesity, ulcer, urinary tract inflammation, stomach acidity, kidney stones and burns in Qalqilia (Palestine).

Pomegranate is widely cultivated in orchards and gardens in Iraq for its fruit, which can be eaten or used to make cooling drinks and, locally grown at Koysinciq, it is sold as an ingredient in herbal mixtures in the Qaysari Market, Erbil, Kurdish Autonomous Region (Iraq). One variety has small black fruits known as *Imm–harmal*, which are not edible but are used to make a local medicine. The pomegranate is sometimes grown as an ornamental shrub, or as a hedge, and the fruits are used to obtain the pomegranate sour molasses known as *Dibbs al–rumman* (Arabic) in Iraq. *Robb–e Anâr* (Persian) is thick, dark and tart pomegranate syrup used to enrich soups and stews in Iran. The pomegranates used are tart ruby–red, not the sweet variety from Lebanon or Spain (Rivera *et al.*, 2012).

Within the rites for the soul of the dead, a Mandaean sacred food is eaten which includes pomegranate; as this is one of the five most sacred foods of the Mandaeans. *Haoma* (Avestan) twigs (*Ephedra* sp.), dried and washed, with fresh pomegranate twigs, are pounded in the mortar together with a little water. The resultant juice is strained through a strainer and used in Zoroastrian rituals. *Hadhânaêpata* (Avestan) 'the evergreen', the fragrant, is the pomegranate tree which is considered to be an emblem of the immortality of the soul. Every Fire-temple of the Zoroastrian must have a date tree and a pomegranate tree. A few drops of the consecrated *Haoma* juice are poured into the mouth of the dying person as a symbol of the immortality of soul; if it is not available, the juice of a few grains of pomegranate is dropped into the mouth of the dying person (Rivera *et al.*, 2012).

THE ORIGIN, DIVERSITY AND TRADITIONAL USES OF THE STONE FRUITS (*PRUNUS* L.)

Stone fruit is a general term for cherries, plums, peaches, apricots and related species (Table 7.1) and interspecific hybrids (Table 7.2).

Table 7.1 Main types of stone fruits (Pérez *et al.*, 2008; USDA, 2012).

Group	Origin	Features	Representative cultivars, cultivar groups or trade marks
Apricots	Western Asia to Central Asia	Diploid (*Prunus armeniaca* L.)	Bergeron, Búlida, Galta Rocha, Harmat, Moniquí, Murciana
Sweet Cherries	Central Europe and Western Asia	Diploid (*Prunus avium* L.)	Ambrunés, Burlat, Jarandilla, Mollar, Napoleon, Picota, Reverchon
Dwarf cherries, Maraschino cherries	Western Asia	Tetraploid (*Cerasus vulgaris* Mill. and C. *fruticosa* (Pall.) Woronow). *Cerasus vulgaris* is an allotetraploid species supposed to result from natural hybridization between *P. fruticosa* and *P. avium*	Guindo, Kántorjánosi, Köröser, Weichsel
Peaches	Central Asia	Diploid (*Prunus persica* (L.) Stokes)	Catherina, Maygold
Nectarines	Central Asia	Diploid (*Prunus persica* (L.) Stokes)	Fantasia, Snow Queen, Summergrand
Cherry Plums, Myrobalan Plums	Western and Central Asia	Diploid (*Prunus cerasifera* Ehrh.)	Atropurpurea (Pisardii), Bur Alooche, Ghermez, Kok Sultan, Mayeras, Plastic, Thundercloud
Asian plum, California plum, Japanese plum	China. But most are descendant of the plum trees imported, inter-crossed and grown by Luther Burbank	Diploid (*Prunus salicina* Lindl.)	Blackamber, Black Beaut(y), Fortune, Friar, Golden Japan, President, Santa Rosa, Suplum Six (Angeleno)
Mirabelle plums	Western Asia to Central Asia	Hexaploid (*Prunus domestica L.* subsp. *syriaca (Borkh.) Janch. ex Mansf*)	Drap d'Or, Late Mirabelle
Bullaces, Damsons	Central Europe	Hexaploid (*Prunus domestica* L. subsp. *insititia* (L.) C.K.Schneid.)	Black Bullace, Crittenden, Pollizo, Saint Julian, Shropshire
Greengages	Central Europe	Hexaploid (*Prunus domestica* L. subsp. *italica* (Borkh.) Gams ex Hegi)	Jefferson, Reine Claude
Plums, Gages	Central Europe	Hexaploid (*Prunus domestica* L. subsp. *domestica*)	Pozegaka, Prune d'Agen
Mei, Mume, Japanese Apricot	China, Japan	Diploid (*Prunus mume* (Siebold) Siebold and Zucc.)	Chinese: Hongmei, Taigemei, Zhaoshuimei. Japanese: Bungo, Hibai, Yabai

Table 7.2 Main interspecific hybrid types of stone-fruits (Okie, 2006; Pérez *et al.*, 2008; Zhivondov and Uzundzhalieva, 2012).

Group	Origin	Features	Representative cultivars, cultivar groups or trade marks
Plumcots, Apriplums	Plumcot is a name created by Luther Burbank and these hybrids occur naturally	First-generation hybrids between a plum parent (*P. salicina* or *P. cerasifera* or their hybrids) and an apricot (*P. armeniaca*)	Flavorosa, King Kong, Plum Parfait, Sweet Treat
Pluots	Developed in the late 20th century by Floyd Zaiger	Later-generations that show more plum than apricot characteristics. The skin is smooth resembling that of a plum	Dapple Dandy, Flavorosa, Geo Pride, Raspberry Jewel, Red Ray, Splash
Apriums	Developed in the late 1980s by Floyd Zaiger	Complex plum–apricot hybrids that show more apricot traits. The flesh is usually dense and the skin is similar to the skin of apricots	Cot-N-Candy, Flavor-Delight
Nectaplums	Developed in the late 20th century by Floyd Zaiger	Hybrids of nectarines (*P. persica*) and plums. Both nectarine and plum traits are easily detectable	Spice Zee
Peacotums, Nectacotum Pluots	Developed in the late 20th century by Floyd Zaiger	Hybrids of peach (*P. persica*), apricot (*P. armeniaca*) and plum	Nectacotum
Duke cherries	Naturally occurring	Hybrids of sweet and sour cherries (*P.* x *gondouinii* (Poit. and Turpin) Rehder)	Guindo Garrafal, Guindo Tomatillo

Plums have been proven to have effective antioxidant activity that results from polyphenols (Donovan *et al.*, 1998; Gil *et al.*, 2002; Kim *et al.*, 2003). A prune is any one of various plum cultivars, mostly *Prunus domestica* or European Plum, sold as fresh or dried fruit. The dried fruit is also referred to as a dried plum or prune. In general, fresh prunes are freestone cultivars (the pit is easy to remove), whereas most other plums grown for fresh consumption are clingstone (the pit is more difficult to remove) (Wikipedia, 2012).

PHYTOCHEMICAL COMPOSITION

Pomegranate

Pomegranate fruit phytochemicals have been studied over the last 15 years, due to the high antioxidant capacity of pomegranate fruit juices and the large number of studies supporting their biological activity. Pomegranates are rich in polyphenols but their seeds, and particularly the seed oil, also contain some characteristic fatty acids such as punicic acid, a conjugated tri-unsaturated fatty acid of 18 carbon atoms. Some steroids and triterpenes have also been reported in seeds.

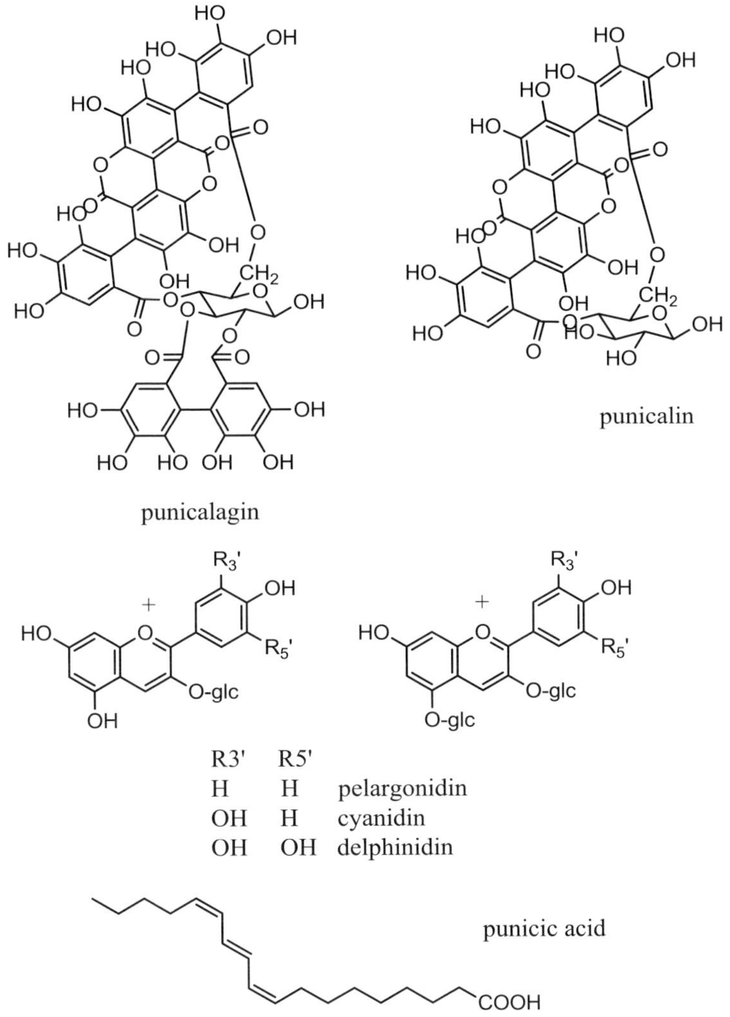

Fig. 7.1 Characteristic phytochemicals of pomegranate fruits.

Polyphenols

Anthocyanins

The characteristic red pigmentation of pomegranates is due to a combination of six anthocyanins, including the 3-glucosides and the 3,5-diglucosides of delphinidin, cyanidin and pelargonidin (Figure 7.1). The different hues observed in different cultivars with more scarlet or purple colours are due to the prevalence of pelargonidin or delphinidin pigments, respectively. Changes have been reported with ripening and with the position of pomegranates in the tree. In general, those fruit with less pigmentation in the husk show a more intense pigmentation in the arils, and correspondingly in the juice. Cultivars like 'Mollar', which is characteristically from Spain with soft seeds and good quality for fresh eating, are less pigmented and acidic than cultivars as 'Wonderful', which is more acidic, with a hard seed and with a an intense colour, and that are more convenient for juice production.

Ellagitannins
The main polyphenols, and more characteristically from pomegranates, are the ellagitannins. Punicalagin and punicalin isomers are the most characteristic ellagitannins of pomegranate, and are combinations of gallagic acid which esterifies glucose. In the case of punicalagin, hexahydroxy-diphenic acid and gallagic are linked to the glucose residue (Figure 7.1). Punicalagin oligomers have also been reported using advanced HPLC-MS-MS methodologies, although these are minor compounds.

Ellagic acid derivatives
Free ellagic acid and glycosidic combinations with glucose, rhamnose and pentoses (arabinose, and possibly xylose) are also present in pomegranates. Valoneic acid bilactone has also been reported in pomegranates, although at smaller concentrations. Ellagic acid dimethyl and trimethyl ether were also described in the seeds.

Gallotannins
This type of tannin is present in this fruit, at different concentrations, although these tannins are less relevant and less characteristic than the ellagitannins. The simplest is monogalloyl glucose but di-galloyl and trigalloyl glucose are also present. Often they combine with hexahydroxydiphenic acid to produce ellagitannins that also contain gallic acid. Some examples are granatin A, casuarinin, corilagin, pedunculagin and tellimagrandin I.

Proanthocyanidins
These are rare compounds in pomegranates, although they have been reported in the fruit membranes (Plumb *et al.*, 2002). Gallocatechin, gallocatechin-(4,8)-catechin, catechin-(4,8)-gallocatechin and gallocatechin-(4,8)-gallocatechin have been reported.

Flavonols
These are minor compounds in the fruit, although they are present in the fruit husk. Quercetin, myricetin, kaempferol and luteolin and the glycosides rutin (quercetin 3-rutinoside) and quercetin 3-glucoside, have been reported in fruits.

Organic acids
Both aliphatic and aromatic acids occur in pomegrantes. The aromatics caffeic, chlorogenic, cinnamic, o-coumaric, *p*-coumaric, ferulic, gallic acid protocatechuic acids have been reported. In addition, the aliphatic acids citric, malic, oxalic, quinic, succinic and tartaric acids have been detected.

Fatty acids and triglycerides

Pomegranate seed oils contain different unsaturated fatty acids, including eicosenoic, linoleic, linolenic, oleic and punicic acids. This last acid is characteristic of pomegranates and is an omega 5 tri-unsaturated conjugated fatty acid. It is present in pomegranate seed oil as tri-O-punicyl-glycerol and is also present in juice.

Sterols and triterpenes

The seeds and oil also contain a number of triterpenes, including β-sitosterol, stigmasterol, ursolic acid, betulinic and asiatic acids, as well as sterols including estrone, estradiol, estriol and testosterone (Seeram *et al.*, 2006), although other studies have failed to demonstrate the occurrence of these hormones in pomegranates.

Vitamin C

Pomegranates contain only a very small amount of L-ascorbic acid. This has been reported to be about 6 mg/100 g (FW) (Davey *et al.*, 2000).

Apricot

Apricot (*Prunus armeniaca* L.) fruits are a rich source of bioactive compounds such as vitamins, carotenoids and polyphenols, which contribute significantly to their taste, colour and nutritive values, but also have significant health effects in humans. Apricot fruit contains three major types of antioxidant molecules: polyphenols, comprising both hydro- and lipophilic components such as proanthocyanidins, hydroxycinnamic acid derivatives, flavonols and anthocyanins (Radi *et al.*, 1997; Ruiz *et al.*, 2005a; Dragovic-Uzelac *et al.*, 2007; Sochor *et al.*, 2010; Schmitzer *et al.*, 2011); lipid-soluble carotenoids, such as β-carotene, β-cryptoxanthin, γ-carotene and lycopene (Radi *et al.*, 1997; De Rigal *et al.*, 2000; Ruiz *et al.*, 2005b; Akin *et al.*, 2008); and water-soluble vitamin C (Munzuroglu *et al.*, 2003; Hegedüs *et al.*, 2010). In addition, the apricot is an important nutritional source of mono- and polysaccharides (Aubert and Chanforan, 2007; Bassi *et al.*, 1996; Gurrieri *et al.*, 2001; Ruiz and Egea, 2008), organic acids (Souty *et al.*, 1990; Audergon *et al.*, 1991; Mehlenbacher *et al.*, 1991; Gurrieri *et al.*, 2001; Ruiz and Egea, 2008), fibre and vitamins such as thiamin, riboflavin, niacin, pantothenic acid, vitamin B6 and folic acid (Marlett and Vollendorf, 1994; Voi *et al.*, 1995), and mineral elements such as potassium, calcium and magnesium (Drogoudi *et al.*, 2008). Apricot kernel oil is rich in mono- and polyunsaturated fatty acids, with oleic and linoleic acids being the major constituents, and a number of minor components, such as tocopherols and phenolic compounds (Turan *et al.*, 2007).

Polyphenols

The phenolic content of apricots has been measured by some authors and large variations in phenolic composition were found among cultivars. Phenolic characterization (achieved by HPLC) of nine apricot cultivars coming from different genetic origins led to the identification of chlorogenic and neochlorogenic acids, (+)-catechin and (–)–epicatechin and rutin (quercetin-3-rutinoside) as the major phenolic compounds in apricot. In addition to these compounds, other quercetin-3-glycosides and proanthocyanidins were detected, and two procyanidins were tentatively identified as procyanidins B2 and C1 (Radi *et al.*, 1997). The presence of the two hydroxycinnamics (chlorogenic and neochlorogenic acids) in high amounts in apricot fruits has been previously reported (Moller and Herrmann, 1983). More recently, the HPLC-MS/MS analysis of the extracts obtained from apricot peel and flesh in a wide set of apricot genotypes showed the occurrence of four main groups of phenolic compounds: proanthocyanidins, hydroxycinnamic acid derivatives, flavonols and anthocyanins (Ruiz *et al.*, 2005a). These groups coincide with those previously found in other stone fruits (peach and nectarine) (Tomás-Barberán *et al.*, 2001). Among these compounds, chlorogenic and neochlorogenic acids, procyanidins B1, B2 and B4, and some procyanidin trimers, quercetin 3-rutinoside, kaempferol 3-rhamnosyl-hexoside and quercetin 3-acetyl- hexoside, cyanidin 3-rutinoside, and 3-glucoside, were detected in the skin and flesh of the different cultivars (Ruiz *et al.*, 2005a). Other phenolic compounds that have also been identified in apricot were some phenolic acids and their derivatives, such as gallic acid, caffeic acid, *p*-coumaric acid and ferulic acid, as well as quercetin-3-glucoside, quercetin-3-galactoside and procyanidin B3 (Dragovic-Uzelac, *et al.*, 2007).

A wide range of variability was found among the apricot varieties with regard to the content of phenolic compounds. The total phenolics content, quantified as the addition of the individual compounds quantified by HPLC, ranged between 32.6 and 160.0 mg/100 g of edible tissue (Ruiz *et al.*, 2005a). Different studies have also reported high levels of total polyphenols in apricot (Dragovic-Uzelac *et al.*, 2007; Brat *et al.*, 2007; Drogoudi *et al.*, 2008; Sochor *et al.*, 2010, 2011; Hegedüs *et al.*, 2010; Schmitzer *et al.*, 2011). The phenolics concentration was greater in the peel than in the flesh of the different varieties (Ruiz *et al.*, 2005a; Schmitzer *et al.*, 2011). In addition, the apricot kernels were found to be rich in total phenolic content, ranging from 92.2 to 162.1 mg gallic acid equivalent/100 g (Korekar *et al.*, 2011).

Carotenoids

Apricots have been described as one of the most important dietary sources of provitamin A carotenoids because 250 g of fresh or 30 g of dried apricots provide 100% of the recommended daily allowance (Bolin and Stafford, 1974). The recommended daily intake is either 2 mg of β-carotene (recommended by the Deutsche Gesellschaft für Ernährung, Germany) or 5 ± 6 mg of β-carotene (recommended by the National Cancer Institute, USA). Apricots, with a carotenoid content ranging from 1.5 to 16.5 mg/100 g (Ruiz *et al.*, 2005b) belong to one of the most carotenoid-rich fruits (with grapefruit, papaya and nectarine).

In 1933, Brockmann conducted one of the first studies on the carotenoid characterization of apricot (*Prunus armeniaca* L.), and β-carotene was found to be the principal pigment (Curl, 1960). β-carotene represents more than 50% of total carotenoid content in apricot (Radi *et al.*, 1997; De Rigal *et al.*, 2000; Ruiz *et al.*, 2005b; Dragovic-Uzelac *et al.*, 2007). Numerous other carotenoids, such as γ-carotene, β-cryptoxanthin, phytoene, phytofluene, α-carotene, zeaxanthin, lycopene, and lutein, are present in apricots but in smaller amounts (Curl, 1960; Radi *et al.*, 1997; De Rigal *et al.*, 2000; Fraser and Bramley, 2004; Ruiz *et al.*, 2005b; Dragovic-Uzelac *et al.*, 2007). Carotenoid content varies significantly depending on the variety. In nine apricot cultivars collected in France, β-carotene was the major carotenoid and its concentration varied from 0.5 mg/100 g dry mass ('Bebeco') to 11.5 mg/100 g dry mass ('Henderson'), while γ-carotene and lycopene only were detected in three cultivars (Radi *et al.*, 1997). More recently, thirty seven apricot varieties were analysed and the total carotenoid content ranged from 1.5 to 16.5 mg/100 g fresh weight (FW) of edible portion, with β-carotene as the main pigment followed by β-cryptoxanthin and γ-carotene (Ruiz *et al.*, 2005b). Twenty nine apricot cultivars of Greek and American origin and their hybrids were evaluated (Drogoudi *et al.*, 2008) and mean values of total carotenoids (0.9–3.8 mg/100 g FW) were similar to those reported by Sass-Kiss *et al.* (2005) (2.0–5.7 mg/100 g FW) for 11 apricot. In three apricot cultivars from Croatia, carotenoid contents were around 1.0 mg/100 g FW (Dragovic-Uzelac *et al.*, 2007).

Vitamin C

Although very few studies have been conducted in relation to the determination of vitamin C in apricot, the data show that the apricot is a rich source of vitamin C. Fifteen apricot genotypes were analysed for vitamin C contents; it ranged from 3.04 to 16.17 mg/100 g FW in 'Goldrich' and 'Preventa' cultivars, respectively (Hegedüs *et al.*, 2010). In general, the tested apricots contained similar levels of vitamin C as white- and yellow-flesh peaches and plums (Gil *et al.*, 2002), except 'Preventa' which contained a larger amount. Similar values were reported in the vitamin C content of 11 Malatya (Turkey) apricot cultivars, with values ranging from 4.9 to 11.5 mg/100 g, if recalculated on a fresh weight basis (Akin *et al.*, 2008). The levels of vitamin C were measured in different apricot cultivars from Turkey and the vitamin C content ranged from 4.8 to 7.1 mg/100 g of fresh apricot (Munzuroglu *et al.*, 2003). Finally, an average value of 4.0 mg/100 g dry weight was found in apricot by Leong and Oey (2012).

Others

Apricot is an important nutritional source of mono- and polysaccharides (Aubert and Chanforan, 2007; Bassi *et al.*, 1996; Gurrieri *et al.*, 2001; Ruiz and Egea, 2008) and organic acids (Souty *et al.*, 1990; Audergon *et al.*, 1991; Mehlenbacher *et al.*, 1991; Gurrieri *et al.*, 2001; Ruiz and Egea, 2008). The dominant sugar in fruit tissue is sucrose, followed secondly by glucose and thirdly by sorbitol and fructose-inositol (Bassi and Selli, 1990; Voi *et al.*, 1995; Radi *et al.*, 2003; Drogoudi *et al.*, 2008). Other sugars, such as xylose, mannose, maltose (Bassi and Selli, 1990) and raffinose (Ledbetter *et al.*, 2006), were also found present in apricot fruit in varying concentrations. The apricot is also an important nutritional source of fibre and vitamins, such as thiamin, riboflavin, niacin, pantothenic acid, vitamin B6 and folic acid (Marlett and Vollendorf, 1994; Voi *et al.*, 1995). In addition, the contents of thirteen biogenic alpha-L-amino acids (arginine, asparagine, isoleucine, lysine, serine, threonine, valine, leucine, phenylalanine, tryptophan, tyrosine, proline and alanine) were determined in apricot using ion exchange chromatography with UV-Vis spectrometry detection (Sochor *et al.*, 2011). Potassium, calcium and magnesium are considered as major minerals of the apricot fruit. Twenty nine apricot cultivars of Greek and American origin and their hybrids were evaluated (Drogoudi *et al.*, 2008) and the highest content was found for potassium (mean values ranged between 1.327 and 2.087%), followed by calcium (0.060–0.260%) and then magnesium (0.040–0.075%). Mean values were in the range of those reported for apricot cultivars from Turkey (Haciseferogullari *et al.*, 2007; Akin *et al.*, 2008), although values were lower for potassium and magnesium, respectively.

Apricot kernels contain a wide variety of bioactive components. Apricot kernel oil is rich in mono- and polyunsaturated fatty acids, with oleic and linoleic acids being the major constituents, and a number of minor components, such as tocopherols and phenolic compounds (Turan *et al.*, 2007). The apricot kernel contains 40% oil, which is composed of 30% linoleic acid (C18:2) and 60% oleic acid (C18:1) (Al-Khalifa, 1996). Apricot oil is also a good source of vitamin E (78 mg/100 g) and phytosterols (Sizova and Andreeva, 2007).

Peach and nectarine

The number of peach and nectarine cultivars has increased significantly during recent years, to introduce appealing fruits with white or yellow flesh, with reddish or yellowish skins, trying to increase the fruit flavour and their attributes for postharvest storage. These fruits are mainly consumed fresh and only a small part is processed to produce canned peaches in syrup, fruit nectars and jams.

Polyphenols

Peaches and nectarines contain significant amounts of hydroxycinnamates, flavonols, proanthocyanidins and anthocyanins. These compounds are more abundant in the peel than in the flesh. Recent studies have shown that the non-extractable phenolics, and particularly flavonoids, quantified using degradative techniques such as hydrolysis with phloroglucinol or thiolysis, are a very relevant part of the phenolics present in peaches and nectarines; studies using these methodologies should be carried out to complete the databases on food composition that are currently incomplete.

Previous studies have shown that the peel tissues contained higher amounts of phenolics and anthocyanins, and flavonols were almost exclusively located in this tissue. No clear differences were found between the phenolic content of nectarines and peaches, or between white and yellow flesh cultivars (Tomás-Barberán *et al.*, 2001). In addition, no clear trend on the phenolic content of the different cultivars was observed with ripening when soluble phenolics were analysed. When cultivars from California were studied, some cultivars showed very high soluble phenolic content, as was the case of the white flesh nectarine cultivar 'Brite Pearl' (about 450 mg/kg hydroxycinnamates and 500 mg/kg soluble proanthocyanidins in the flesh) and the yellow flesh nectarine cultivar 'Red Jim' (about 200 mg/kg hydroxycinnamates and 300 mg/kg soluble proanthocyanidins in flesh), that contained 10 times more soluble phenolics than cultivars such as 'Fire Pearl' (50 mg/kg hydroxycinnamates and 30 mg/kg soluble proanthocyanidins in the flesh) (Tomás-Barberán *et al.*, 2001). Regarding peaches, the white flesh cultivars 'Snow King' (300 mg/kg hydroxycinnamates, 700 mg/kg soluble proanthocyanidins in the flesh) and 'Snow Giant' (130 mg/kg hydroxycinnamates and 550 mg/kg soluble proanthocyanidins in the flesh) showed the highest content (Tomás-Barberán *et al.*, 2001). The different phenolics were characterized by HPLC-MS-MS and quantified by HPLC-UV detection at the appropriate wavelengths. Recent studies using phloroglucinolysis degradation methods show that the proanthocyanidin content of peaches and nectarines can be much higher when the oligomeric and polymeric proanthocyanidins are quantified. Thus, nectarine cultivar 'Caldesi 2000' contained 200 mg proanthocyanidins per 100 g fruit (fresh weight) (Pliakoni *et al.*, 2010).

Carotenoids

The main carotenoids present in peaches and nectarines were β-carotene and β-cryptoxanthin. In all cases carotenoids were more abundant in the peel tissues than in the flesh. In all cases β-carotene was more abundant than β-cryptoxanthin (Gil *et al.*, 2002), and as could be expected the yellow cultivars contained close to ten times more carotenoids than the white flesh cultivars. Per serving (100 mg comprised of 85% flesh and 15% peel), white flesh peaches provided between 8 and 18 mg/serving, while yellow flesh cultivars reached between 95 and 197 mg/serving.

Vitamin C

Peaches and nectarines contain only a very small amount of L-ascorbic acid. This has been reported to be in the range of 3–14 mg/100 g (fresh weight) (Gil *et al.*, 2002).

Plums

Stacewicz-Sapuntzakis *et al.*, 2001 have shown the list of components that have been quantified in fresh plums, dried prunes and prune juice per 100 g of their ready-to-eat weight:

water, carbohydrates, protein, fat, sugars, total dietary fibre, amino acids, minerals, vitamins, carotenoids, organic acids and phenolic compounds. The quantity of each component has been studied in different samples of different varieties of *Prunus domestica* and *Prunus salicina*.

Polyphenols

Plums are a rich source of phenolic compounds and provide 42.0–109.2 mg/100 g (*P. salicina*) (Gil *et al.*, 2002); these are higher in red-fleshed plums than in yellow-fleshed plums (Gil *et al.*, 2002) and are more important in determining antioxidant activity than were the anthocyanins or carotenoids (Vizzotto *et al.*, 2007).

The total phenolic contents of various cultivars of plums were in a range of 174 to 375 mg/100 g, expressed as gallic acid equivalents (GAE), on a fresh weight basis (Kim *et al.*, 2003).

The skin contained 3–9 times higher concentration of phenolic compounds than the flesh, but the flesh provided about 70% of the total because of its greater fraction (Gil *et al.*, 2002; Cevallos *et al.*, 2006). It is known to contain various kinds of phenolic compounds, including hydroxycinnamic acids, flavonols and anthocyanins.

There is a strong correlation between total phenolics and antioxidant activity in *P. salicina* and *P. domestica* (Gil *et al.*, 2002) and the amount of total flavonoids and total antioxidant capacity also show a good correlation (Gil *et al.*, 2002). Donovan *et al.* (1998) have reported for dried plums or prunes (*Prunus domestica*) a good antioxidant capacity related to their high phenolic content.

Carotenoids

Fresh prune plums contain a provitamin A carotenoid (β-carotene) and oxycarotenoids (xanthophylls, lutein is most important because it is well absorbed by humans) without vitamin A activity (Stacewicz-Sapontzakis *et al.*, 2001). Carotenoids mainly occurr in the peel (Gil *et al.*, 2002). The USDA Carotenoid Database shows a content of β-carotene of 430 μg/100 g fruit and 240 μg of lutein/100 g fruit in fresh plums. Stacewicz-Sapontzakis *et al.* (2001) have reported 326 μg of lutein/100 g fruit and 393 μg of β-carotene/100 g fruit and 2100 μg of total carotenoids/100 g fruit in Italian prune plums and 77 μg of lutein/100 g fruit and 180 μg of β-carotene/100 g fruit and 750 μg of total carotenoids/100 g fruit in French plums cv 'Saghiv'. Another study with Californian plums quantified 83–231 μg of total carotenoids/100 g (Gil *et al.*, 2002).

According to the USDA Food Composition Database, prune juice contains only traces of Vitamin A activity.

Vitamin C

Stacewicz-Sapontzakis *et al.* (2001) have found that fresh prune plum d'Agen contained between 5.1 and 5.7 mg ascorbate/100 g fruit, in plums from Yugoslavia an average of 8.8 mg vitamin C per 100 g of plums was found and in Turkish cultivars 16.5 mg/100 g for the Krikon Damson cultivar was reported. In dried plums, vitamin C is degraded during the dehydration processing at high temperatures. A vitamin C content of 3.3 mg/100 g of fruit was reported in USA samples and 4.0 mg/100 g in European samples. Prune juice often contains more vitamin C because this may be added during processing.

The amount of vitamin C reported in *Prunus salicina* is 2.5–10.2 mg/100 g (Gil *et al.*, 2002) and this is in accordance with the values reported in the USDA Food Composition Database of 3–10 mg/100 g of fresh weight (http://ndb.nal.usda.gov/ndb/foods/show/2454).

Cherry

Sweet cherry is an important fruit with high commercial importance worldwide, although given its perishable nature the application of cold storage is a necessary postharvest tool to maintain fruit quality until consumption. Among the factors determining the consumer's acceptability total soluble solids, acidity and colour are the most important (Crisosto *et al.*, 2003). For this reason, producers use a number of parameters to establish the optimum time for harvesting, the most reliable being skin colour (Romano *et al.*, 2006). Red colour development in sweet cherry is used as indicator of quality and ripening, and is due to accumulation and profile of anthocyanins (Díaz-Mula *et al.*, 2008). Nowadays, inverse associations between fruit and vegetable intake and chronic diseases, such as different types of cancer and cardiovascular disease, have been demonstrated in numerous epidemiological studies, in which phytochemicals have been indicated to be responsible for this observed protective effect (Schreiner and Huyskens-Keil, 2006). Among these compounds, special interest has been focused on anthocyanins and other polyphenolics, carotenoids and vitamins C and E. In cherry, antioxidant potential has been associated with ascorbic acid, phenolics and anthocyanins, which are found at lower concentrations (Chaovanalikit and Wrolstad, 2004; Serrano *et al.*, 2005, 2009).

Polyphenols

Phenolic compounds are important components of many fruits, vegetables and beverages, and contribute to their colour and sensory properties. In the case of cherries, they are known for having an excellent source of polyphenols in the human diet, although differences exist among cultivars, ranging in between 98 and 200 mg/100 g (Table 7.3). Since phenolic compounds contribute to fruit quality in terms of modifying colour, taste, aroma and flavour

Table 7.3 Total anthocyanins, total phenolics and total antioxidant activity in the hydrophilic and lipophilic fractions in sweet cherry cultivars at commercial ripening stage (Díaz-Mula *et al.*, 2008; Serrano *et al.*, 2009).

Cherry cultivar	Total anthocyanins (mg/100 g)	Total phenolics (mg/100 g)	Hydrophilic antioxidant activity (mg/100 g)	Lipophilic antioxidant activity (mg/100 g)
Cristalina	225 ± 5.57	163 ± 4.63	128 ± 3.98	35 ± 1.71
Sonata	198 ± 7.46	192 ± 6.23	135 ± 3.29	75 ± 2.68
Santina	129 ± 6.21	167 ± 3.68	119 ± 2.68	35 ± 0.95
Newstar	124 ± 7.20	155 ± 5.51	92 ± 4.93	57 ± 3.69
Sunburst	97 ± 8.27	164 ± 7.68	106 ± 2.37	41 ± 1.43
Sweetheart	97 ± 6.13	118 ± 5.63	89 ± 7.66	45 ± 2.79
Prime Giant	87 ± 4.87	98 ± 5.16	75 ± 4.45	54 ± 1.05
Somerset	66 ± 6.8	145 ± 6.96	91 ± 4.31	44 ± 2.82
Brooks	40 ± 2.58	98 ± 4.65	69 ± 2.50	33 ± 1.79

(Tomás-Baberán and Espín, 2001), those cultivars with higher phenolics content will have higher quality. Anthocyanins are the phytochemical compounds that confer the visual quality of fruits and vegetables, contributing to the red, blue and purple pigments in plant tissues (Gao and Mazza, 1995). The main phenolic compounds in sweet cherry fruits are anthocyanins, which also differ in concentration depending on cultivar. Thus, 'Brooks' cherry has the lowest anthocyanin content (40 mg/100 g) while 'Cristalina' shows the highest (225 mg/100 g). Those cultivars with the lowest anthocyanins ('Brooks', 'Somerset', 'Prime Giant' and 'Sweetheart') are considered as light-coloured cultivars (score 3 on the CTIFL colour chart) while those with the highest anthocyanin content ('Crsitalina and 'Sonata') are classified as dark-coloured (score 5), showing a direct relationship between colour parameters and anthocyanin concentration (Díaz-Mula *et al.*, 2008). The predominant anthocyanins in cherry are cyanidin-3-rutinoside and cyanidin-3-glucoside, while peonidin- and pelargonidin- 3-glucoside and 3-rutinoside have been found at very low concentrations (Gonçalves *et al.*, 2004; Mozetiĉ *et al.*, 2006; Usenik *et al.*, 2008a).

Sweet cherries are also rich in phenolic acids. The most abundant among them are derivatives of the hydroxycinnamic acids, such as caffeic acid and *p*-coumaric acid. The most common colourless phenolics in sweet cherries are neochlorogenic acid (3′-caffeilquinic acid) and 3′-*p*-coumaroylquinic acid (Mozetiĉ *et al.*, 2002, Chaovanalikit and Wolstad, 2004). The hydroxycinnamates are increasingly receiving attention for their potential health-promoting effects through their potent antioxidant action, ability to inhibit low density lipoprotein (LDL) oxidation and their chemopreventive properties (e.g. inhibitory effects on tumour promotion and the ability to block the formation of mutagenic compounds such as nitrosamines) as demonstrated by *in vitro* studies (McCune *et al.*, 2011).

Sweet cherries also contain quercetin, a phenolic bioactive compound belonging to a class of flavonoids that are widely distributed in a plant-based diet with a potent antioxidant activity (TAA) (Boots *et al.*, 2008). The ability of quercetin to act as a free radical scavenger suggests it could play a beneficial role in reducing reactive oxygen species (ROS) (i.e. hydrogen peroxide, superoxide anion) associated with chronic diseases such as cardiovascular disease and cancer (Wilms *et al.*, 2005). As shown in Table 7.3, cherries exhibit antioxidant activity in both hydrophilic and lipophilic extracts by measuring the scavenging capacity of $ABTS^{\bullet+}$ radicals. For all cultivars, hydrophilic-TAA is higher than lipophilic-TAA (about 80% of TTA in 'Cristalina' and approximately 50% in 'Prime Giant'), showing that the major contributors to antioxidant activity are hydrophilic compounds, such as polyphenols and anthocyanins. Antioxidant vitamins, such as tocopherols, and carotenoids are lipophilic compounds that might contribute to lipophilic-TAA.

Carotenoids

Similar to anthocyanins, carotenoids also exist as plant pigments, responsible for red, yellow and orange colour, and also have health-promoting effects. Carotenoids are a group of lipid-soluble natural pigments (C40 tetraterpenoids) present in fruits and vegetables that impart colours from yellow to red and are originated from eight C5 isoprenoid units joined head to tail resulting in a symmetrical molecule located in the chromoplasts. The hydrocarbon carotenoids are known as carotenes (ß-carotene, lycopene etc.) while xanthophylls are oxygenated derivatives containing at least one hydroxyl group and, therefore, are more polar than carotenes. On the other hand, carotenoids can be acyclic (e.g. lycopene), monocyclic (γ-carotene) or dicyclic (α- and ß-carotene).

The most studied carotenoids have been ß-carotene, lycopene, lutein and zeaxanthin. Lycopene, a carotenoid with non-provitamin A activity, has been found to have both greater antioxidant capacity and stronger inhibition of cancer cell proliferation than other carotenoids (Omoni and Aluko, 2005). The antioxidant ability of several carotenoids (carotenes and xanthophylls) follows the sequence, from high to low: lycopene > ß-cryptoxanthin ≈ ß-carotene > lutein ≈ zeaxanthin > α-carotene > canthaxanthin. However, the main physiological effect of carotenoids in humans has been classically attributed to their role as provitamin A, since those carotenes with a ß-ring end group are converted to vitamin A (retinol) by the action of an intestinal monooxygenase. In addition, in the last decade several epidemiological studies suggest that carotenoids have also important roles in a range of diseases including age-related macular degradation, cataract, cardiovascular diseases and some types of cancer, due to their function in cell differentiation and proliferation regulators or cell-to-cell communication stimulators (Krinsky and Johnson, 2005; Voutilainen *et al.,* 2006).

According to National Cancer Institute (2004), sweet cherry contains important amounts of carotenoids, mainly ß-carotene (38 μg/100 g, on average) and lutein/zeaxanthin (85 μg/100 g, on average). Although carotenoids are other important bioactive constituents in fruits (Valero and Serrano, 2010), almost no evidence exists on their occurrence in sweet cherry. In a recent paper (Valero *et al.*, 2011) on two sweet cherry cultivars ('Prime Giant' and 'Cristalina'), carotenoids have been found but at different concentrations in both cultivars, with 'Prime Giant' having significantly higher total carotenoids (1.06 ± 0.07 mg/100 g) than 'Cristalina' (0.61 ± 0.04 mg/100 g). Also, Leong and Oey (2012) reported individual content in sweet cherry (unknown cultivar) with a concentration of 0.02 mg/g (dry weight) for ß-carotene, ß-cryptoxanthin and α-carotene, and 0.01 mg/g (dry weight) for lycopene and lutein.

Vitamins

Vitamins are a class of nutrients that are essentially required by the human body for its biochemical and physiological functions. Vitamins are subdivided into fat soluble and water soluble, the vitamins A, D, E and K being lipophilic, while C and B are hydrophilic. Tocopherols (vitamin E) and carotenoids (provitamin A) are the major lipid-soluble antioxidant vitamins in fruits and vegetables, while vitamin C is the major hydrophilic antioxidant vitamin, although recent evidence indicates that vitamin D could also have a role as antioxidant (Asensi-Fabado and Munné-Bosh, 2010). In sweet cherry, the main vitamins are vitamin C (7-50 mg/100 g) followed by vitamin E (0.07 mg/100 g) and vitamin K (2 μg/100 g). With respect to vitamin C, differences in concentration at time of harvest have been reported. Thus, '4-70' cultivars had 28 mg/100 g (Serrano *et al.*, 2005), while 'Souvenir', 'Samba' and 'Prime Giant' showed ascorbic acid values of 4, 2 and 6 mg/100 g, respectively (Schmitz-Eiberger and Blanke, 2012).

EFFECT OF PROCESSING AND STORAGE ON POMEGRANATE AND STONE FRUIT BIOACTIVES

Pomegranates

Pomegranates can be consumed fresh, after preparing the arils as ready-to-eat fresh products stored refrigerated and under modified atmospheres, or after processing as juice, the main processed product, or as wine, vinegar, sauces or nutritional extracts and nutraceutical preparations.

Agronomic practices

Punica granatum is quite resistant to arid conditions and even to salty soils, making it possible to grow this species in marginal cultivation lands where other crops cannot be cultivated. It needs mild environmental conditions with temperatures that do not go below 0°C. A difference between day and night temperatures during the development of the fruit, and particularly those days that are close to harvest, is essential for the development of aril and juice pigmentation (Gil *et al.*, 1995b). In some cultivars that are not rich in pigments, this temperature difference is essential to obtain fruits of sufficient quality. The position of the fruit in the tree has also some influence on polyphenols and fruit quality. For the 'Mollar' cultivar, those fruits that are located externally in the tree, and therefore more exposed to the sun and wind, show a higher husk pigmentation, which under extreme environmental conditions can develop fruit scald, and lower arils pigmentation, while those that are located internally in the tree show a yellowing husk but usually have a higher internal pigmentation (Gil *et al.*, 1995a).

Ripening also affects pomegranate juice pigmentation both quantitatively and qualitatively. Five pomegranate clones, cultivated under homogeneous conditions, were studied to evaluate the changes in juice anthocyanin content during ripening. The main six anthocyanin pigments described previously were detected as responsible of the red colour of pomegranate juice. These were quantitatively and qualitatively analysed by high-performance liquid chromatography and identified as delphinidin 3-glucoside and 3,5-diglucoside, cyanidin 3-glucoside and 3,5-diglucoside and pelargonidin 3-glucoside and 3,5-diglucoside. An increase in juice pigmentation during fruit ripening was generally observed. In the early fruit-ripening stages, delphinidin 3,5-diglucoside was the main pigment, followed by cyanidin 3,5-diglucoside, while in the later stages the monoglucoside derivatives cyanidin 3-glucoside and delphinidin 3-glucoside increased considerably. The pelargonidin derivatives were always present in small amounts (Hernandez *et al.*, 1999).

Postharvest handling and storage

Pomegranates increase their anthocyanin content and pigmentation during postharvest storage in air at the recommended temperatures (5–10°C). 'Mollar' pomegranates increased the anthocyanin content during storage in air and changes in the anthocyanin pigments (diglycosides and monoglycosides) were observed (Gil *et al.*, 1996a; Artes *et al.*, 1998). 'Wonderful' pomegranates were stored in air and in air enriched with 10 or 20 kPa CO_2 at 10°C for six weeks. The colour and anthocyanins content of the arils stored in air was higher than those of the initial control and those stored in CO_2-enriched atmospheres. The higher the CO_2 concentration, the lower the anthocyanin pigmentation. This was associated with a decrease in anthocyanin biosynthesis and Phenylalanine Ammonia Lyase (PAL) activity (Holcroft *et al.*, 1998).

Processing

Minimal processing to obtain fresh arils is today a common technological practice and these ready-to-eat products are available in the market. The changes in anthocyanin pigments were evaluated during the processing and storage of the fresh arils (Gil *et al.,* 1996b). Slight decreases in anthocyanin content were observed during the washing of arils. During storage in air, a discrete increase in anthocyanins was observed. Delphinidin was the most sensitive anthocyanin, while cyanidin and pelargonidin derivatives increased. The 3,5-diglucosides also were more stable than the 3-glucosides (Gil *et al.,* 2006b).

Thermal processing to obtain pomegranate juice has also relevant effects on polyphenol extraction and it was shown that industrial processing of pomegranate extracts some of the hydrolysable tannins present in the fruit rind. This could account for the higher antioxidant activity of commercial juices compared to the experimental ones produced only from pomegranate arils (Gil *et al.,* 2000). Storage of thermally treated pomegranate juice decreases the content of anthocyanins and increases the presence of brown polymers. This is particularly evident in the juice from the cultivar 'Mollar', while the pigmentation is better preserved in juices produced from cultivar 'Wonderful', probably due to a higher anthocyanin content that could preserve the pigments from degradation by self-association of or to a lower pH of this juice that helps to stabilize the anthocyanins in solution.

Apricots

Around 40–45% of the total world apricot production is processed (Madrau *et al.,* 2009). To extend the shelf life of apricot, different preservation methods have been developed, including storage at low temperature, canning, freezing, drying and packing in controlled atmospheres (Jiménez *et al.,* 2008). Different works have investigated the effect of fruit postharvest management on the antioxidant properties of apricot fruits (De Rigal *et al.,* 2000; Kamisli and Karatas, 2009; Madrau *et al.,* 2009; Kan and Bostan, 2010; Erdogan and Erdemoglu, 2011; Igual *et al.,* 2012; Leong and Oey, 2012), as well as the influence of storage at low temperature (Egea *et al.,* 2006; Jiménez *et al.,* 2008; Leccese *et al.,* 2010). In addition, some agronomical practices may have an adverse effect on fruit bioactives content (Kan and Bostan, 2010; Leccese *et al.,* 2010; Erdogan and Erdemoglu, 2011; Roussos *et al.,* 2011).

Agronomic practices

The effect of cultivation systems on the antioxidant properties of apricot fruits has been investigated by different authors. Total antioxidant capacity, total phenol content and carotenoid content were monitored on trees of five cultivars 'Tyrinthos', 'Cafona', 'Bella d'Italia', 'Vitillo' and 'Pellecchiella' under integrated and organic systems (Leccese *et al.,* 2010). The results highlighted the importance of considering the interactions between cultivar and cultivation system, in order to evaluate apricot fruit quality under organic management and choose those genotypes with better performances under organic systems (Leccese *et al.,* 2010). Conventional and organically grown fresh and dried apricot cultivars from the east part of Turkey were analysed to determine their phenolic and vitamin A contents; the polyphenol and vitamin A contents of organically grown samples were found to be higher than the conventional cultivars in all samples (Kan and Bostan, 2010). In addition, Erdogan and Erdemoglu (2011) determined that concentrations of polyphenols in organic apricots were higher than pesticide-treated samples, and roughly the same in ripe-fresh and sun-dried apricots. Polyphenol concentrations of either organic or pesticide-treated sulfited-dried apricots were lower than the ripe-fresh apricots, and considering the organic ripe-fresh apricots, there was a decrease in some polyphenol concentration in the sulfited-dried apricots compared with the ripe-fresh apricot (Erdogan and Erdemoglu, 2011). The effect of thinning on phytochemical concentration in apricot was evaluated (Roussos *et al.,* 2011). Total phenol concentration increased with thinning, without any similar increase of the major individual phenolic compounds detected (neo-chlorogenic acid, chlorogenic acid, rutin, catechin, epicatechin, ferulic acid, *p*-coumaric acid and caffeic acid). The antioxidant capacity of the pulp was not influenced by thinning. Overall, thinning enhanced the pomological traits of apricot fruits as well as their phytochemical content (Roussos *et al.,* 2011).

Postharvest handling and storage

All the phenolic compounds are affected by oxidation in apricots. Chlorogenic acid, neochlorogenic acid and catechin decreased rapidly during enzymatic browning (Radi *et al.,* 1997). Various causes (physiological, mechanical, technological) may induce enzymatic browning in apricots with the development of undesirable colour, flavour and a loss of nutrients. In addition, an inverse correlation between carotenoid content and apricot susceptibility to browning was reported (Radi *et al.,* 1997; De Rigal *et al.,* 2000).

The influence of the storage at low temperature, with and without cold coadjuvants like 1-MCP, on different antioxidant systems of apricot var. 'Bulida' was evaluated (Egea *et al.,* 2006). When evaluating the antioxidant systems (total carotenoids, ascorbic acid, superoxide dismutase and peroxidase activities and the total Trolox equivalent antioxidant capacity (TEAC)) results showed a sharp decrease in ascorbic acid during storage, without significant differences between treatments, while, in the remaining antioxidant systems, a positive effect of 1-MCP was observed, in contrast to cold storage at 2°C and in modified atmospheres, which showed similar antioxidant evolution and levels. Fruits from five apricot cultivars were collected at full maturity stage and analysed either immediately or after storage at 4 ± 0.5°C and 85% of relative humidity for 7 and 14 days. Storage conditions did not affect antioxidant properties of apricot fruits whose storability is mostly related to pomological and aesthetic parameters, such as the flesh firmness and fruit colour rather than the other physicochemical and antioxidant properties (Leccese *et al.,* 2010). In addition, different methods of conservation (frozen and canned) on the antioxidant properties of raw apricot were evaluated, and antioxidant activity of both types of processed fruit was monitored during 150 days of storage (Jiménez *et al.,* 2008). The raw apricot exhibited the highest inhibition of oxidation according to the lipid peroxidation assay. The freezing process led to a slight loss of antioxidant activity, whereas canned apricots lost their antioxidant capacity. During storage in frozen and canned apricots no important changes were detected in the different antioxidant activities assayed from 1 to 150 days.

Processing

The effects of processing, that is heating (98°C, 10 min), freezing (−20°C) and freeze drying on anthocyanins, carotenoids, and vitamin C in summer fruits and vegetables, among them apricots, has been evaluated (Leong and Oey, 2012). An interesting discovery was that heating increased the content of the total vitamin C content, L-ascorbic acid content and the L-ascorbic acid/dehydroascorbic acid ratio. Compared to heating, freezing could maintain or slightly increase the content of phytochemicals and reduce sugar for most of the commodities. In contrast, freeze-dried samples mostly resulted in a lower amount of phytochemicals, as compared to fresh, heated and frozen commodities.

There are scarcely any data on the effect of drying on the phenolic content and antioxidant activity of apricot fruit. The changes of phenolics and antioxidant activity in apricots as influenced by hot air drying at different temperatures have been evaluated (Madrau *et al.*, 2009) and the results demonstrated that phenols degradation during hot air drying did not always result in the same effects. The decrease in chlorogenic and neochlorogenic acid in 'Cafona' cultivar was higher at the lower drying temperature. Catechin showed the same behaviour of hydroxycinnamic acids in both cultivars, while the decrease in the other compounds was significantly more marked in the sample dried at 75°C. The higher drying temperature gave a better product from the antioxidant and polyphenol content point of view (Madrau *et al.*, 2009). Loss of vitamin C after heat treatment has been widely reported (Ryley and Kayda, 1993).

To prevent both enzymatic and non-enzymatic browning, quality loss and microbial activity during drying and storage and to facilitate the drying process, sulfiting at low concentration is the most commonly used pretreatment (Rossello *et al.*, 1993; Lewicki, 2006; Karabulut *et al.*, 2007; Miranda *et al.*, 2009). Sulfuric acid solution was used as a sulfur source to sulfite apricots. It is concluded that while amounts of vitamins decrease, level of malondialdehyde increases with increasing immersion time and, thus, sulfur contents (Kamisli and Karatas, 2009). There are scarcely any data on alternative methods of drying apricot halves and their effect both on the kinetics and on a wide spectrum of bioactive compounds and their relationship with the product's antioxidant activity. Other techniques, such as hot air drying, microwave drying or its combination, may be efficient alternatives with even shorter processing times and, consequently, less impact on the nutritional value of the apricot (Karabulut *et al.*, 2007; Karatas and Kamisli, 2007). The industrial processing of dried apricots may be improved by using microwave energy, as the drying time is considerably reduced, and the fruit obtained had a higher phenolic content, particularly of chlorogenic acid, catequin and epicatequin (Igual *et al.*, 2012). Nevertheless, as the contribution of these phenols to antioxidant capacity was not significant, microwave-dried samples maintained the same antioxidant capacity as the air-dried ones. When sulfite is added prior to the drying processes, care should be taken with the total phenols and the antioxidant capacity quantified, as it may interfere with the results depending on the methodology used (Igual *et al.*, 2012). Microwave is much more efficient that infrared apricot drying in terms of speed, preservation of original colour of apricot samples and less loss of vitamins (A, C and E) and malondialdehyde values (Karatas and Kamisli, 2007). Treatment with ionizing radiation has emerged as an alternative and effective preservation method in food technology. However, studies of the effect of ionization on apricots have not always been satisfactory. Ionization at doses of 1.0 kGy affected negatively the resistance to oxidative stress of 'Búlida' apricots during storage, which could contribute to the decrease in their shelf life and to the reduction of their nutritive value (Egea *et al.*, 2007)

Peach and nectarines

Agronomic practices

The effect of different agronomic practices on peach and nectarine bioactives has been the objective of some recent studies. Fruit load per tree, the use of reflective mulch, irrigation and the type of rootstocks were evaluated. The influence of fruit load per tree was evaluated on nectarines (cultivar 'Stark Red Gold') from two orchards with different tree density (Andreotti *et al.*, 2010). Low crop load increased fruit size and weight in both orchards.

Low load fruits had a higher accumulation of phenolic compounds in skin and pulp tissues. The concentrations of cinnamic acids and flavan-3-ols consistently showed an inverse relationship to crop load. The effect of reflective mulching that modifies the average maximum temperature and the intensity of reflected light was also evaluated in two different studies. The mulching changed fruit quality at harvest by promoting ripening and generally increasing the concentration of the main classes of phenolic compounds, particularly cinnamic acid derivatives and proanthocyanidins. This was particularly relevant in fruits located in the lower part of the canopy (Pliakoni *et al.*, 2010). The position in the tree also has some influence on the phenolic content of nectarines, as fruit exposed to the sun (upper part of the canopy) showed a higher content than shaded fruit (lower part of canopy) (Pliakoni *et al.*, 2010). The authors of these studies state that the health-promoting value of nectarines could be enhanced by increasing phenolic content via crop load and reflective mulches.

The effect of irrigation regime was also studied by different groups. In a study on nectarines (cultivar 'Caldesi 2000'), it was shown that deficit irrigation increased the content of total phenolics, including proanthocyanidins and phenolic acids, reaching similar amounts in the two different years that the study was carried out (Pliakoni *et al.*, 2010). Another study with 'Suncrest' peaches showed that water stress induced an increase in total anthocyanins and the content of proanthocyanidins varied in different ways depending on the rootstock (Tavarini *et al.*, 2011). All these results suggest that water management can be a relevant factor in maintaining good quality fruits with an optimal content of phenolic bioactives.

Postharvest handling and storage

Once peaches and nectarines are harvested they can be stored under refrigeration to slow down the ripening process and extend the shelf life. Although there are many studies on the effect of postharvest storage and the different postharvest treatments on peach and nectarine quality, physiological disorders and decay, there is only limited information regarding the effects of these treatments on the bioactives content. Ripening at room temperature also shows relevant changes in phytochemicals, although a complete study has not been carried out. In peach (cultivar 'Rich Lady') stored at room temperature for up to 15 days, the carotenoids increased during the first four days, although phenolics and antioxidant capacity did not change over the ripening and over-ripening period (Puerta-Gómez and Cisneros-Zevallos, 2011). Storage of 'Caldesi 2000' nectarines at 2°C for two, four and six weeks did not produce significant changes in anthocyanins and proanthocyanidins, showing that refrigerated storage preserved these bioactives (Pliakoni *et al.*, 2010).

Processing

Processing to obtain fresh cut nectarines and peaches and storage at refrigerated temperatures for seven days did not affect significantly the polyphenol, vitamin C and carotenoid content, showing that fresh cut products preserve the nutritional quality during the shelf life in which they maintain the sensory quality (Steiner *et al.*, 2006; Gil, unpublished results). The effect of processing clingstone peaches on phenolics and procyanidins was evaluated (Asami *et al.*, 2003). Lye peeling, freezing and storage temperature, as well as the thermal treatment (several sterilization temperatures) were evaluated to establish the impact on phenolic composition of the processed peaches. The sterilization temperature and time are the main factors affecting phenolic composition. Storage of the processed products also induced losses around 30–45% of phenolic compounds. It was demonstrated that both storage and thermal processing conditions profoundly impact the levels of polyphenols in peaches. The impact of thermal processing and storage on peaches proanthocyanidins was also evaluated (Hong *et al.*, 2004). Thermal processing produced reductions of 11% in monomers, 9% in dimers, 12% in trimers, 6% in tetramers and 5% in pentamers. Hexamers and heptamers showed a higher degradation with a 30% loss. Some of these proanthocyanidins migrated from the fruit to the syrup and can account for the losses observed during the canning process. Additional degradation of proanthocyanidins was observed during the storage for three months of the canned peaches with percentages from 10 to 80% depending on the polymerization degree, tetramers being the most degraded during storage.

High pressure processing to obtain nectarine puree was compared with thermal processing to evaluate the effects on carotenoids and antioxidant activity (García-Parra *et al.*, 2011). High pressure treatments produced a higher content of carotenoids and antioxidant activity than thermal treatment, and had less significant changes in colour during storage and processing. These differences, however, disappeared during the refrigerated storage of the purees. It was concluded that hydrostatic high pressure application in the industrial line of nectarine puree presented some advantages compared to the thermal treatment: however, some of the changes found reduced during the storage period.

Plums

Agronomic practices

Successful plum production requires cultivars well adapted to the specific growing conditions in the area of production. Due to the interaction between environment and genotype, it is very important to evaluate preliminarily the agronomic and pomologic performance of all newly released cultivars in the areas in which they will be cultivated (Liverani *et al.*, 2010). The combined selection of optimal cultivars and manipulation of agricultural practices can optimize health benefits of plums.

Selected phenolic-rich genotypes of *Prunus salicina* and hybrids of *Prunus persica* showed high antioxidant activity, stable colour properties and good antimicrobial activity. Colorants prepared from an anthocyanin-rich plum genotype showed similar hue to that of synthetic colorant FD&C Red 3 and higher stability than a commercial red grape colorant with respect to time, temperature and pH. Additionally, a selected rich phenolic plum genotype exhibited strong antimicrobial activity against *Salmonella enteritidis* and *Escherichia coli* O157:H7. Selection of crops high in phenolic compounds can be related to enhanced functional properties and opens the possibility of breeding fruits with targeted functional properties for the fresh produce and processing market (Cevallos *et al.*, 2006).

The anthocyanin concentrations were higher in plum (*Prunus domestica* L.) 'Stanley' fruits of successive harvesting dates, meaning that anthocyanin accumulation seemed to occur constantly during fruit development and ripening. The total phenolic content was within the range 70–214 mg gallic acid equivalents per 100 g fresh weight. The highest concentrations of anthocyanins and phenolics were observed in plum fruits harvested in 2009, which is most likely due to the favourable weather conditions.

Ripening resulted in statistically increased fruit weight and soluble solids, decreased fruit firmness, darker colour of fruits, increased concentration of total sugars, decreased concentration of total acids and increased concentration of anthocyanins. There was no influence of ripening on the content of phenols. The results show significant influences of cultivar on fruit weight, soluble solids content, firmness, fruit colour, concentration of total acids, concentration of total phenols and anthocyanins in European plums (*Prunus domestica* L.) (Usenik *et al.*, 2008b).

Postharvest handling and storage

Storage conditions are strongly influential upon the organoleptic properties and health benefits of plums. Plum (*Prunus domestica* L.) cultivars 'Jubileum', 'Victoria' and 'Opal' were stored at 2°C in a controlled atmosphere (CA) for three weeks. Flavour was better in fruit from air than from a CA when sensory evaluation was conducted immediately after CA storage. A strong bicarbonate taste occurred, particularly in fruit evaluated immediately after storage in 25% CO_2. Exposure to 5, 15 or 25% CO_2 resulted in accumulation of ethanol. Cracking of 'Opal' plums was largely avoided by storage at 2°C in CA. Storage in controlled atmosphere had little effect on titratable acids (TA) and pH in the three cultivars. In general, pH increased with extended storage time. Effects of CA storage on soluble solids content (SSC) and fruit firmness were inconclusive. The optical density in plum juice was initially low and showed a decreasing trend during storage (Wang and Verstrheim, 2003).

Eight *Prunus salicina* plum cultivars (four dark-purple and four yellow) were harvested at the commercial ripening stage. At harvest, large variations in phytochemicals and antioxidant activity were found among cultivars in peel and pulp tissues, and phytochemicals concentration and antioxidant activity (TAA) were higher in the peel than in the pulp. During storage, increases in total phenolics for all cultivars (peel and pulp), in total anthocyanin content in the peel of the dark-purple plums and total carotenoids in the peel and pulp of the yellow cultivars were observed. No significant loss of bioactive compounds and TAA occurred during prolonged plum storage. Moreover, for a better evaluation of the antioxidant potential of plums, the contribution to carotenoids should not be overlooked (Diaz-Mula *et al.*, 2009b) and, for the same cultivars during cold storage, a general increase was found for total phenolics, anthocyanins and carotenoids (Valero and Serrano, 2010) both in the peel and the flesh of two purple and two yellow skin plum cultivars under modified atmosphere packaging conditions (MAP). Results revealed that, in all cultivars, total phenolics and TAA increased in the peel and flesh during storage, as well as anthocyanins: cyanidin-3-glucoside and cyanidin-3-rutinoside in the purple cultivars. These changes were significantly delayed in fruit stored under MAP conditions. Total carotenoids and TAA increased in the yellow cultivars (in both peel and flesh) while decreases were observed in the purple cultivars, these changes also being delayed by the use of MAP (Diaz Mula *et al.*, 2011).

Treatment with 1-methylcyclopropene suppresses and delays the climacteric ethylene production with acceptable fruit quality during ripening at ambient temperature and is effective in extending the storage life by up to six weeks with minimal loss of plum fruit quality (Martinez Romero *et al.*, 2003a, 2003b; Skog *et al.*, 2003; Valero *et al.*, 2003; Menniti *et al.*, 2004, 2006; Khan *et al.*, 2009; Kan *et al.*, 2011).

Processing

Processing, that is heating (98°C, 10 min), freezing (–20°C) and freeze drying, produced effects on anthocyanins, carotenoids, and vitamin C in summer fruits; in plums the release of membrane-bound anthocyanins was enhanced, resulting in a higher content after processing compared to fresh fruits. The stability of ascorbic acid was increased if ascorbic acid oxidase was inactivated for example by heating (Leong and Oey, 2012).

Cherry

The sweet cherry (*Prunus avium* L.) horticultural production chain involves several steps: production, harvesting, pre-cooling, cooling, selection, grading, packaging, transport, distribution and consumption. The quality of sweet cherries may have a different meaning, depending on which part of the chain is considered. Consumer acceptance appears to be the most important overall characteristic of cherries (Crisosto *et al.*, 2003). Consumer choice and preference of fruits and vegetables are majorly influenced by factors, such as convenience, culture, price, appearance and taste and, in recent years, the nutrient value and content of bioactive compounds are increasingly demanded characteristics.

Agronomic practices

The content of phytochemicals and the nutritional value in cherries may be affected by several preharvest factors, such as temperature, light intensity, fruit crops and maturity, the latter being the most studied. Light intensity increases levels of ascorbic acid and different growing temperatures (day/night) also affect total phenolic content. High temperature growing conditions (25/30°C) significantly enhance anthocyanin and total phenolic content (Wang, 2006). Recently, there has been an increasing interest in growing cherries under plastic greenhouses, especially in cold areas. This cultivation system can influence canopy and soil temperature, quantity and quality of transmitted, reflected or absorbed light.

Fruit ripening is a highly coordinated, genetically programmed process occurring at the later stages of fruit development and involving a series of physiological, biochemical and sensory changes, leading to an edible ripe fruit with desirable quality parameters. In sweet cherry, the ripening process is characterized by colour changes, from green to red, which can be followed by the evolution of L*, a* and b* parameters, and are due to accumulation and profile of anthocyanins. In fact, red colour development in sweet cherry is used as an indicator of quality and ripening of fresh cherry (Esti *et al.*, 2002; Serrano *et al.*, 2005; Mozetiĉ *et al.*, 2006). Harvesting is usually performed based on the attainment of acceptable fruit size, colour and concentration of soluble solids. However, there is little available information about the changes in the content of health-promoting compounds during sweet cherry development and ripening on the tree. Serrano *et al.* (2005) reported changes in concentrations and activities of antioxidants of sweet cherry at 14 different stages of ripeness, with total anthocyanins increasing exponentially from stage 8 to the maximum value at stage 14 (63.26 mg cyanidin equivalent activity per 100 g fresh sample). The total antioxidant activity (TAA) decreased from stage 1 to stage 8, and increased again from stage 8 to stage 14, coinciding with the total phenolic compound concentration and the accumulation of anthocyanins. TAA reached the maximum activity at stage 14, with average ascorbic acid equivalent activity of 50.03 mg per 100 g fresh sample. Thus, harvesting sweet cherries at stage 12 of ripening, when fruit reaches maximum size would support the development of the highest organoleptic, nutritional and functional quality attributes.

The agronomic requirements (chilling unit, flowering, fruit set, harvest date, productivity and resistance) and fruit quality (weight, colour, firmness, sweetness, taste) are traditionally the main aspects considered in a plant breeding programme. However, the content of functional compounds could be taken into account when parentals are chosen for a traditional or biotechnological breeding programme. In this sense, some functional properties, such as the content of anthocyanins, polyphenols and antioxidant activity, were studied in 11 sweet cherry cultivars harvested at commercial ripening stage. The results indicated that two cultivars ('Cristalina' and 'Sonata') had the highest levels in total anthocyanins (≈200 mg/100 g) while 'Brooks' and 'Sommerset' showed the lowest (<70 mg/100 g). On the other hand, 'Sonata', 'Cristalina', 'New Star' and 'Sunburst' had a total phenolic content over 90 mg/100 g, while in 'Prime Giant' and 'Brooks' this content was lower than 70 mg/100 g. With respect to total antioxidant activity, 'Sonata' and 'Cristalina' were the cultivars with the highest values, while 'Brooks' showed the lowest one. Considering these results, some sweet cherry cultivars with very good fruit quality parameters, like 'Brooks', do not reach the maximum functional properties. Other cultivars with high functional properties (like 'Sonata') could be considered as a good parental for a sweet cherry breeding programme with the aim to improve other characteristics (Díaz-Mula *et al.*, 2009a).

Postharvest handling and storage

The extension of the postharvest life of sweet cherry depends on three factors: (i) reduction in dehydration and weight loss, (ii) slowing down the physiological processes of maturation and senescence, and (iii) avoiding the onset and rate of microbial growth. To control these three factors, the main tools are refrigeration and controlling the relative humidity (RH). The optimum temperature for harvest and handling of cherries is between 10 and 20°C (out of this temperature range more pitting is observed), while the optimum storage temperature is 0°C, with a RH of 90–95% (Romano *et al.*, 2006). Thus, storage at low temperatures is the main postharvest treatment to reduce sweet cherry metabolism, to maintain quality and to prolong the storability in those perishable fruits and vegetables considered as non-chilling sensitive, like sweet cherry fruits. Some research has considered the changes in bioactive compounds and antioxidant activity during cold storage although no general tendency is found. Thus, loss of health-beneficial compounds (phenolics and ascorbic acid) has been found in table grapes, broccoli, pomegranate and apple (Serrano *et al.*, 2011), in which the loss of phenolics was highly dependent on cultivar. However, increases in phytochemicals were reported for sweet cherry during cold storage, although different behaviour has been reported depending on storage temperature. Gonçalves *et al.* (2004) studied the phenolic compounds hydroxycinnamates, anthocyanins, flavonols and flavan-3-ols of 'Burlat', 'Saco', 'Summit' and 'Van' sweet cherry cultivars harvested at two different ripening stages and stored under different cold conditions. Phenolic acid contents generally decreased with storage at 1–2°C and increased with storage at 15 ± 5°C. Anthocyanin levels increased at both storage temperatures, while flavonols and flavan-3-ol contents remained quite constant.

Maturity stage at harvest determines also the antioxidant potential after cold storage of sweet cherries. In a study on 11 cherry cultivars harvested at three ripening stages (S1, S2 and S3), significant increases in anthocyanin content were found during cold storage and subsequent shelf life at 20°C, the accumulation of anthocyanins during storage being attributed to normal sweet cherry ripening (Serrano *et al.*, 2009). The HPLC-DAD chromatograms revealed that in all cultivars the main anthocyanins were cyanidin-3-rutinoside, followed by cyanidin-3-glucoside and pelargonidin-3-rutinoside, which increased with ripening from S1 to S3. With respect to total phenolics, an increase in total phenolic compounds as maturity advanced was observed (from S1 to S3) for all cultivars. Neochlorogenic acid was the predominant hydroxycinnamic acid followed by 3-*p*-coumaroylquinic acid and both increased significantly from S1 to S3 and along storage.

In recent years, special attention has been made on the use of natural, safe compounds as postharvest treatments to improve the content of bioactive compounds during storage of sweet cherries. Thus, 'Cristalina' and 'Prime Giant' cherries harvested at commercial ripening stage treated with salicylic acid (SA), acetylsalicylic acid (ASA) or oxalic acid (OA) at 1 mM before storage under cold temperature showed beneficial effects in terms on maintenance of organoleptic quality by a delay of the postharvest ripening process, manifested by lower acidity, colour changes and firmness losses. This delay was also manifested by a delay in the accumulation of total phenolics, anthocyanins and antioxidant activity (Valero *et al.,* 2011).

Another postharvest treatment with beneficial effects on reducing postharvest ripening of sweet cherry has been the use of edible coatings. In this sense, 'Sweetheart' cherry coated with sodium alginate at several concentrations (1%, 3% or 5% w/v) delayed the evolution of the parameters related to postharvest ripening, such as colour, softening and loss of acidity, and reduced respiration rate. In addition, the edible coatings showed a positive effect on maintaining higher concentrations of total phenolics and total antioxidant activity, which decreased in control fruits associated with the over-ripening and senescence processes (Díaz-Mula *et al.,* 2012). Since the ingestion of fruits and vegetables with higher amounts of phenolics has antioxidant activity *in vivo* by increasing the plasma antioxidants (Fernández-Panchón *et al.,* 2008), the use of alginate as an edible coating led to fruits with a higher proportion of functional properties than control ones. However, no data exist on the bioavailability and bioconversion of phenolic compounds after the intake of sweet cherry, and thus more research is needed about this issue.

Processing

Most fruits and vegetables are consumed as a raw commodity, although for centuries humans have transformed these perishable products into more stable, delicious and safer dishes. The most used unit operations include blanching, freezing, canning and cooking (boiling, frying, baking or microwaving). These operations affect the nutritional content of fruits and vegetables, and on a general basis significant losses of nutrients as well as the bioactive compounds with antioxidant activity occur.

Fresh produce contains enzymes which cause loss of colour and nutrients, flavour changes and colour changes in canned and frozen fruits and vegetables. Blanching is the exposure of the vegetables to boiling water or steam for a brief period of time to inactivate naturally occurring enzymes but it also helps to destroy microorganisms on the surface. Most reports dealing with antioxidant compounds losses studied the combination of blanching plus freezing and blanching plus canning. Ascorbic acid (vitamin C) has been used as a model to evaluate the nutrient losses during fruit and vegetable processing by calculating its retention, since it is considered the least stable of the bioactive compounds. On a general basis, vitamin C content decreases during thermal processing conditions, although large variations exist among vegetable products (Rickman *et al.,* 2007).

Phenolic compounds declined after thermal processing mainly due to leaching into the brine or syrup in the canning process. Thus, in cherries, 50% of the total phenolics was lost during canning and was found in the syrup transferred from the fruit. During freezing the content of total phenolics is largely depended on fruit type. Thus, no losses were found in peaches while 40% was lost in cherries after three months of storage (Chaovanalikit and Wrolstad, 2004).

Anthocyanin and other flavonoid content tends to decline when cooking with water, as reported for several berries (Jones, 2007), although they remain in the boiling water. However, magnitude and duration of heating has a strong influence on anthocyanin degradation after thermal processing, with logarithmic course degradation related to arithmetic increase in temperature. In addition, during storage of thermally processed commodities, anthocyanin showed continuous degradation affecting colour and antioxidant capacity (Sadilova *et al.*, 2007). On the contrary, heating cherries (by dipping in hot water at 98°C for 10 minutes and afterwards immediately cooled in an ice bath) increased their anthocyanin content and maintained unchanged the carotenoid concentration (Leong and Oey, 2012). Heating results in enzyme inactivation, texture changes of fruits and vegetables and unavoidable leaching of water-soluble compounds, which could alter the entire phytochemical profile and content of fruit and vegetables. Accordingly, heating has been reported to increase the chemical extractability of phytochemical compounds, because of the release of phytochemicals from chromotoplasts leading to an increase of concentration (Howard *et al.*, 1999).

Similarly to heating, freezing increased (two- to threefold) the anthocyanin content of cherries and maintained the carotenoids. Freezing induces the formation of ice crystals that favours localized concentration of solutes and reallocation of water molecules in the cell structure. In contrast, freeze-dried cherries mostly resulted in a lower amount of phytochemicals, as compared to fresh, heated and frozen ones, especially the content of anthocyanins. With respect to vitamin C, increases have been found after freezing at –20°C, although the freeze-drying processing did not alter the vitamin C content in cherries (Leong and Oey, 2012).

In jams from both sweet and sour cherries, thermal processing of fruits led to statistical significant losses of monomeric anthocyanins (92–93% from the value recorded in frozen fruits), followed by vitamin C content (54–78%), FRAP values (30–41%) and total phenolics content (25–43%). Additionally, storing jams for a period of three months at 20°C led to a decrease in vitamin C content of 22–33% from the value recorded one day after processing and a loss of anthocyanins content (Poiana *et al.*, 2011).

HEALTH EFFECTS OF POMEGRANATES

Cardiovascular diseases

Atherosclerosis is the origin of most cardiovascular diseases. It involves interactions between the arterial cell wall, blood cells (monocytes) and plasma lipoproteins. All the events leading to atherosclerosis can be considered inflammatory. Other factors affecting cardiovascular disease are hypertension, platelet aggregation, the type of blood lipids and oxidative stress. As pomegranates (and pomegranate juice) is one of the most powerful free-radical scavenging fruits (Gil *et al.*, 2000), it was thought that it could have relevant effects in cardiovascular disease (Fuhrman and Aviram, 2006).

Studies have demonstrated that pomegranate juice ('Wonderful' cultivar) inhibits the atherosclerotic lesion development. It was thought that the polyphenols present in pomegranate juice could be very potent antioxidants against LDL oxidation, and was additionally shown to inhibit atherosclerosis development both in mice and in humans (Aviram *et al.*, 2002a, 2002b). The bioavailability and metabolism of pomegranate polyphenols, both in rats and humans, show that these antioxidant compounds are not absorbed and that they are metabolized by the gut microbiota into urolithins, which are absorbed and which have no antioxidant activity (Cerdá *et al.*, 2003, 2004, 2005). Therefore, other mechanisms than the direct antioxidant effect of pomegranate polyphenols should be envisaged to explain the effects observed preventing atherosclerosis formation.

In addition, pomegranate juice has also shown some effects on hypertension. After a one-year study in humans, it was shown that the intake of pomegranate juice decreased systolic blood pressure but had no effect on diastolic blood pressure (Aviram *et al.,* 2004). In this study, it was suggested that the observed affect could have been mediated by an inhibition of cythochrome P450 enzymes and an indirect effect on ACE (angiotensin converting enzyme) activity.

Pomegranate juice has also shown to produce some reduction in collagen-induced platelet aggregation both *in vitro* and *in vivo*, this being an additional mechanism for the cardiovascular health benefits observed (Aviram *et al.,* 2000).

Intake of pomegranate juice showed no effects on the serum lipid pattern of healthy humans, but significant changes were observed in type 2 diabetic patients, with a reduction in total cholesterol, LDL-cholesterol and in the ratio LDL-cholesterol:HDL-cholesterol (Esmaillzadeh *et al.,* 2004).

As a juice rich in antioxidants, the study of its effect on oxidative stress in serum and on macrophage atherogenicity have also been the objective of many studies. It has been suggested that there is a combination of effects on serum lipid peroxidation, LDL and HDL oxidation, and paraoxonase 1 (PON1). The effect on oxidative stress of macrophages and cholesterol metabolism have also been studied and show relevant potential effects of constituents in pomegranate juice (Fuhrman and Aviram, 2006), although the bioactives responsible for the observed effects have not been identified yet.

Cancer

There are many *in vitro* studies that show the potential effects of different pomegranate extracts against cancer cells. Among them the most remarkable are against breast cancer cells and colon cancer cells. The anticancer effects against breast cancer cells have been demonstrated for pomegranate polyphenols and an even better activity demonstrated for pomegranate seed oil (Metha and Lanski, 2004). This activity was linked to inhibition of the estrogen biosynthesis and aromatase activity (Kim *et al.,* 2002).

Inhibition of the growth of prostate cancer cells *in vitro* has also been demonstrated, inhibiting the proliferation, metastatic potential and phospholipase A2 expression of these prostate cells (Lansky *et al.,* 2005). An *in vivo* study with mice in which human prostate cancer cells were implanted, showed that the intake of pomegranate extracts resulted in a significant inhibition of tumour growth concomitant with a significant decrease in serum prostate-specific antigen levels (Malik *et al.,* 2005). The occurrence of pomegranate ellagitannins and gut microbiota metabolites known as urolithins in the prostate tissues of volunteers after the intake of pomegranate juice has recently been demonstrated (Gonzalez-Sarrías *et al.,* 2010b).

Pomegranate juice and isolated pomegranate polyphenols were also shown to induce apoptosis in HT-29 colon cancer cells (Seeram *et al.,* 2005). Some effects *in vitro* were also observed in human HL-60 promyelocytic leukemia cells (Kawaii and Lanski, 2004).

The mechanisms by which pomegranate extracts exert their potential anticarcinogenic properties have been linked with an effect on NF-κB activation, inhibition of production of COX-2, inhibition of angiogenesis and decrease in cell invasion (Shishodia *et al.,* 2006).

All of these *in vitro* studies, with the exception of those that use colon cancer cells, have limited correlation with the potential effects seen *in vivo* (González-Sarrías *et al.,* 2009a), as many of the anticarcinogenic compounds present in pomegranates are not absorbed and will not reach the breast, prostate or leukemia cells.

Studies using human colon cancer cells with the pomegranate extracts and the gut microbiota metabolites urolithins, and at the concentrations that can be found in the gut, have shown that they arrest cell growth of cancerous cells at the S- and G2/M phases

(Gonzalez-Sarrias *et al.*, 2009b). The effect on the gene expression of these cells was also evaluated; it was shown that this might contribute to colon cancer prevention by modulation of multiple genes in the epithelial cells lining the colon.

Anti-inflammatory

The anti-inflammatory effects of pomegranate extracts and urolithins, the gut microbiota metabolites of pomegranate ellagitannins, have been demonstrated in rats using a model of colon inflammation induced by dextran sodium sulfate (Larrosa *et al.*, 2010a). Both pomegranate extract and urolithins decreased inflammation markers (iNOS, cyclooxygenase-2, PTGES and PGE2 in colonic mucosa) and modulated the gut microbiota favourably, promoting the growth of lactobacilli and bifidobacteria. Pomegranate extract, but not urolithins, decreased oxidative stress in plasma and colon mucosa. Only urolithin A preserved colonic architecture preventing the epithelium loss. It was also shown that the normal formation of urolithins in rats treated with pomegranate extract was prevented during inflammation. It was suggested that urolithin A could be the most active anti-inflammatory compound derived from pomegranate ingestion in healthy subjects, whereas in colon inflammation the effects could be due to the non-metabolized ellagitannin-related fraction (Larrosa *et al.*, 2010b). The mechanism of action implicated seems to be via the inhibition of activation of NF-κB and MAPK, downregulation of COX-2 and mPGE-1 expression, and consequently via the reduction of PGE2 production (González-Sarrías *et al.*, 2010a).

HEALTH EFFECTS OF STONE FRUIT

Apricots

Recent studies indicated that an apricot-rich diet has potential health effects in humans, such as cardioprotective effects on the myocardial ischemia-reperfusion, protective effects on unsaturated fatty acids, amelioration of the detrimental effects of low-dose X-rays on testis tissue and even anti-amyloidogenic activity against Alzheimer's disease (Parlakpinar *et al.*, 2009; Ozsahin and Yilmaz, 2010; Ugras *et al.*, 2010; Katayama *et al.*, 2011; Zhang *et al.*, 2011). In dietary guidelines presented by the World Cancer Research Fund, an approximate daily consumption of 600–800 g fresh fruit was recommended in order to prevent cancer.

Prunus mume Siebold and Zucc is a variety of Japanese apricot and is known as Ume in Japan (Adachi *et al.*, 2007). It has been a traditional Japanese medicine for centuries and is a familiar and commonly consum ed food. The health benefits of Ume are now being widely recognized and have been strengthened by recent studies showing that MK615, an extract of compounds from Ume, has various pharmacological activities that include a potential source of free radical scavenging, antiviral, antimicrobial, anti-inflammatory and anticancer activities (Jeong *et al.*, 2006; Adachi *et al.*, 2007; Mori *et al.*, 2007; Nakagawa *et al.*, 2007; Okada *et al.*, 2008; Morimoto *et al.*, 2009; Enomoto *et al.*, 2010; Chung *et al.*, 2011; Morimoto-Yamashita *et al.*, 2011; Park *et al.*, 2011).

Cardiovascular diseases

The cardioprotective potential of apricot feeding in the ischemia-reperfusion model of rats *in vivo* has been demonstrated. Total phenolic content, DPPH radical scavenging and ferric-reducing power as *in vitro* antioxidant capacities of rat chow were significantly increased after supplementation with apricot. Copper, zinc superoxide dismutase (Cu,Zn SOD) and catalase (CAT) activities were increased, and lipid peroxidation decreased significantly in the hearts of 20% apricot-fed group after ischemia-reperfusion (Parlakpinar *et al.*, 2009). In addition, it was found that an ethanol-based (70%) extraction of dry apricot had *in vivo* cardioprotective effects in a myocardial ischemia-reperfusion injury rat model (Parlakpinar *et al.*, 2009). Apricot kernel oil has also been proven to have potent cardioprotective effects on the myocardial ischemia-reperfusion using a rat model *in vivo*. Myocardial catalase, superoxide dismutase, glutathione peroxidase, and constitutive nitric oxide synthase activities, as well as NO concentrations, were all increased, whereas malondialdehyde content and inducible nitric oxide synthase were decreased in apricot oil-treated rats (Zhang *et al.*, 2011). According to these results, apricot kernel oil could be developed as a nutriment for the treatment and prevention of myocardial infarcts (Zhang *et al.*, 2011).

Cancer

The Japanese apricot *Prunus mume* (Ume) possesses various pharmacological activities that include anticancer activities. The anticancer effects of the Ume were investigated on two established cancer cells, the Kato-III stomach cancer and the HL-60 promyelocytic leukaemia cell lines (Adachi *et al.*, 2007). Without Ume, the cancer cells grew and formed colonies. When the Ume extract was added, cancer cells were dose-dependently eliminated and at <5 μl/ml no cancer cells survived. Similarly, the condensed Ume showed strong antitumour effects on human pancreatic cancer and dog fibrosarcoma. The Ume preparation showed no toxic effect on normal human blood cells (Adachi *et al.*, 2007). An active compound that inhibits cancer cells was isolated from the fruit of *Prunus mume*, and its structure and *in vitro* activities were characterized. The *n*-hexane fraction obtained from methanol extracts exhibited the strongest inhibitory effect on the growth of cancer cells. From the *n*-hexane fraction, a new compound named B-1 was purified. B-1 has a highly specific inhibitory effect against cancer cells but little effect against normal cells (Jeong *et al.*, 2006). This compound has the potential to be developed as a nutraceutical. Pro-apoptotic effects of methanol extracts of Japanese apricots (*Prunus mume*) in human leukemia U937 cells have been reported (Chung *et al.*, 2011). Similar beneficial effects were found on ethanol extracts of Japanese apricots (Park *et al.*, 2011). Taken together, these findings suggest that methanol and ethanol extracts of Japanese apricots may be a potential chemotherapeutic agent for use in the control of human leukemia U937 cells.

Other

Recent work (Katayama *et al.*, 2011) has demonstrated that lutein in apricot fruits has potent anti-amyloidogenic activity *in vitro* which could be used against Alzheimer's disease. The anti-amyloidogenic activity was found to correlate with a hydrophobic region, including conjugated double-bond chains, and the number and portion of hydroxyl groups on each side of the molecule. These findings will contribute to the development of anti-amyloidogenic agents from naturally occurring materials. Another study has confirmed that the apricot fruit extracts decrease lipid peroxidation level in the Fenton reagent environment and protect markedly unsaturated fatty acids in the environment with radical sourced oxidation (Ozsahin and Yilmaz, 2010). In addition, apricot kernel oil is also useful for the treatment of hypercholesterolemia (Ramadan *et al.*, 2011). The rats in the apricot oil-treated group showed significantly lower levels of total cholesterol, total triglycerides, low density lipoprotein-cholesterol, alanine-aminotransferase and aspartate aminotransferase activities, as well as high levels of high density lipoprotein-cholesterol and total protein in comparison with the hypercholesterolemic group. A preventive effect of Japanese apricot intake on chronic atrophic gastritis by inhibiting *Helicobacter pylori* infection and reducing active mucosal inflammation has been reported (Enomoto *et al.*, 2010). Antioxidants in apricot may prevent ethanol-induced testicular damage but further studies, including tissue and blood testosterone levels, are necessary for humans (Kurus *et al.*, 2009). A diet supplemented with ground apricot kernel led to improved liver function, lipid peroxide, and liver CAT, SOD and GSH (Abdel-Rahman, 2011). Apricot kernel oil causes improvement in liver antioxidant status in rats in comparison to sunflower oil, which is a commonly consumed vegetable oil (Kutlu *et al.*, 2009). Morimoto *et al.* (2009) found that MK615, an extract of compounds from Ume (*Prunus mume*), significantly reduced the production of pro-inflammatory mediators (tumour necrosis factor-alpha and interleukin-6) induced by *Porphyromonas gingivalis* lipopolysaccharide (LPS), a major etiological agent in localized chronic periodontitis, in murine macrophage-like RAW264.7 cells. These results suggest that MK615 has potential as a therapeutic agent for treating inflammatory diseases such as periodontitis. A later study by the same authors (Morimoto-Yamashita *et al.*, 2011) showed that MK615 (*Prunus mume*) may be a potent inhibitor of the growth of some oral bacteria and an inhibitor of biofilm formation by *Streptococcus mutans*, the principal etiological agent of human dental caries. These findings suggest that MK615 has potential as a therapeutic agent for treating and preventing oral diseases such as dental caries and periodontitis.

Juice extracted from Japanese apricots (Mei-Gin) is a highly concentrated dark black syrup and a common health supplement in Asia. Chuda *et al.* (1999) found that the fruit-juice concentrate of the Japanese apricot markedly improved the fluidity of human blood. In addition, a study showed that the active component, 5-hydroxymethylfurfural (HMF), has a concentration-dependent effect on the inhibition of beta-hexosaminidase degranulation upon antigen-sensitized RBL-2H3 cells which has shown anti-allergic properties (Wu *et al.*, 2011).

Peach and nectarines

A literature survey reveals that there is no human intervention study in which peaches or nectarines have been used to evaluate their health effects. Their phytochemical composition, however, is very similar to those of other stone fruits and, therefore, the potential health effects can be similar. A study of the peach intake in European countries revealed that France was the country with the highest intake, in which 48% of the population declared a consumption of 3–5 fruits per week, this consumption being higher in females and in older people (Konopacka *et al.*, 2010).

A cohort study regarding the effect of fruit and vegetable intake on head and neck cancer risk in the United States showed that those volunteers placed in the first tertile of peach and nectarine consumption, together with other rosaceae fruits, had a decreased risk of head and neck cancer (Freedman *et al.*, 2008).

The potential health benefits of peaches and nectarines were surveyed by Byrne *et al.* (2009) and based on *in vitro* studies, which often show no correlation with the *in vivo* situation, the antiproliferative effects of stone fruit extracts against certain types of cancerous human cell lines was demonstrated, included breast cancer cells.

Other *in vitro* studies showed the ability of peach extracts to bind bile acids, with potential in the prevention of cardiovascular diseases and cancer (Kahlon and Smith, 2007). These studies, however, need confirmation *in vivo*.

A more recent study evaluated the effects of different fruits on *Drosophila melanogaster* life span and found that nectarine was one of the best fruits for expanding life span (Boyd *et al.*, 2011). Supplementation with 4% nectarine in the diet extended life span, increased fecundity and decreased the expression of some metabolic genes, including a key gluconeogenesis gene and oxidative stress damage in wild-type females having a high-fat diet. These results suggest the nectarines might promote longevity by modulating glucose metabolism and reducing oxidative damage.

Plums

Oxidative stress is linked to neurodegenerative diseases, cancer, cardiovascular diseases and to some behavioural conditions, such as anxiety and depression (Bouayed *et al.*, 2007). Following recent evidence that disturbances in oxidative metabolism are involved in anxiety disorders, high anxiety levels and depression, the use of antioxidants has been proposed as an approach for the prevention or treatment of these conditions. Polyphenols are naturally-occurring antioxidant substances which can have pharmacological actions on the central nervous system (Bouayed, 2010). Plums are considered a source of phytochemicals with beneficial health effects (Gonzalez-Flores *et al.*, 2011).

There was a strong correlation (0.93–0.96) between total phenolics and antioxidant activity of plums (Gil *et al.*, 2002; Byrne *et al.*, 2009). Red-flesh plums generally had higher anthocyanin and phenolic contents than the other plums but not necessarily greater antioxidant capacity (Vizzotto *et al.*, 2007).

Cardiovascular diseases

There are three major risk factors involved in cardiovascular diseases: hypertension, dyslipidemia and stress oxidative.

Dried prunes contain 6.1 g of dietary fibre per 100 g, while prune juice is devoid of fibre due to filtration before bottling. The laxative action of both prune and prune juice could be explained by their high sorbitol content (14.7 and 6.1 g/100 g, respectively). Prunes are a good source of energy in the form of simple sugars but do not mediate a rapid rise in blood sugar concentration, possibly because of high fibre, fructose and sorbitol content. Prunes contain large amounts of phenolic compounds (184 mg/100 g), mainly as neochlorogenic and chlorogenic acids, which may aid in the laxative action and delay glucose absorption. Phenolic compounds in prunes had been found to inhibit human LDL oxidation *in vitro*, and thus might serve as preventive agents against chronic diseases, such as heart disease and cancer. Additionally, the high potassium content of prunes (745 mg/100 g) might be beneficial for cardiovascular health (Stacewicz-Sapuntzakis *et al.*, 2001).

The data predict the cardioprotective effects of prunes. They are used for hypertension (Ahmed *et al.*, 2010b). This study was undertaken to see the cardioprotective effects of prunes. A placebo controlled clinical trial study was designed to see the effects of *Prunus domestica* on blood pressure in 259 pre-hypertensive (systolic BP = 120–139 mmHg, diastolic BP = 80–89 mmHg) volunteers. Treated groups drank prune juice and ate the whole fruit (dried plums) where either three (about 11.5 g) or six prunes were soaked overnight in a glass of water, whereas the control group took only a glass of plain water early in the morning on an empty stomach. Blood pressure was recorded fortnightly for eight weeks, and blood samples were taken at week 0 and week 8. There was significant reduction of blood pressure by a single dose of prunes daily group and the controls ($p < 0.05$). With the double dose of prunes, only systolic BP was reduced significantly ($p < 0.05$). The control group had significantly increased serum HDL whereas test groups had significantly reduced serum cholesterol and LDL ($p < 0.05$). Data were analysed by paired-sample t-test with 95% confidence interval. The data predict cardioprotective effects of prunes.

Negishi *et al.* (2007) found that prune extract may contribute to the primary prevention of cardiovascular diseases. After five weeks of prune extract treatment in hypertensive rats, the elevation of blood pressure in the prune extract-treated animals was suppressed in comparison with the control group.

The consumption of prunes has been recommended for hypercholesterolemia (Tinker *et al.,* 1991). Forty-one free-living adult men with mild hypercholesterolemia (5.2–7.5 mmol/l) voluntarily participated in an eight-week crossover study designed to determine the effect of prunes as a source of fibre on plasma cholesterol and on faecal output and bile acid concentration. During the prune period, subjects supplemented their usual diets with 12 prunes (100 g; 6 g dietary fibre) daily. Plasma low density lipoprotein cholesterol was significantly lower after the prune period (3.9 mmol/l). Faecal bile acid concentration of lithocholic acid was significantly lower after the prune period (0.95 mg bile acid/g dry wt stool). Both faecal wet and dry weights were approximately 20% higher after the prune period.

Cancer

Commercial varieties of plums have been reported to contain a mixture of phenolics that may exert anticancer activity. The fruit of the plum tree *(Prunus salicina* Lindl.) has been used as a traditional medicinal food in humans to enhance immunity against infectious agents and to treat cancers (Lee *et al.*, 2009). Phenolic compounds in prunes have been found to inhibit human LDL oxidation *in vitro*, and thus might serve as preventive agents against chronic diseases, such as heart disease and cancer (Stacewicz-Sapuntzakis *et al.,* 2001).

Phenolics in plums preferentially suppress the growth of estrogen-independent MDA-MB-435 breast cancer cells The plum 'Black Splendor' extracts showed greater effects on MDA-MB-435 cells compared to the other breast cancer or the normal breast cell lines (Noratto *et al.,* 2009). Antiproliferative assays with three breast cell lines indicated that plum phytochemicals inhibited the cell proliferation for estrogen-receptor negative MDA-MB-435 breast cancer cells but not the estrogen-positive breast cancer MCF-7 line or the normal breast cell MCF-10A line. Proliferation studies with MBA-MD-435 cells and assays to measure the LDL oxidation inhibition activity with the extracts from 26 peach, nectarine and plum cultivars indicated a range of effectiveness (Byrne *et al.,* 2009).

The immunostimulatory effects of a methanol extract of plum fruit following methanol evaporation and dissolving in phosphate buffered saline (PBS) were assessed by *in vitro* lymphocyte proliferation, tumour cell cytotoxicity and nitric oxide (NO) production. The crude methanol extract stimulated spleen lymphocyte proliferation and NO production by cultured macrophages, and inhibited the viability of tumour cells, significantly greater than media controls (Lee *et al.,* 2009).

The immature plum (*P. salicina* Lindl. cv. Soldam) and its active compound, (–)-epicatechin are expected to be a natural resource for developing novel therapeutic agents for cancer prevention and treatment of hepatocellular carcinoma. The growth inhibition exerted by immature plum extracts was greatest in HepG2 (Yu *et al*., 2009). Inhibition of growth and induction of differentiation of colon cancer cells by peach and plum phenolic compounds was studied by Lea *et al.* (2008). They concluded that several peach and plum phenolic molecules can influence growth and differentiation in human colon cancer cells. Fujii *et al.* (2006) demonstrated that the prune extract (*Prunus domestica* L.) suppressed proliferation and induced apoptosis of human colon carcinoma Caco-2 cells.

Other

Osteoporosis

Dried prunes are an important source of boron, which is postulated to play a role in the prevention of osteoporosis (Arjmandi, *et al*., 1998, 1999, 2002; Stacewicz-Sapuntzakis *et al*., 2001). A serving of prunes (100 g) fulfils the daily requirement for boron (2–3 mg). Also, they contain about 50 mg of calcium, a similar amount of magnesium and 80 mg of phosphorus in 100 g of fruit. Prunes are also rich in copper which is essential for bone building processes (Stacewicz-Sapuntzakis *et al*., 2001).

Dried plums have positive effects on bone structural and biomechanical properties that coincide with the restoration of certain bone marrow myeloid and lymphoid populations, and suppressed splenocyte activation occurring with ovarian hormone deficiency (Smith, 2009; Rendina *et al*., 2012). Deyhim *et al.* (2005) evaluated whether dried plum was able to restore bone mass in osteopenic ovariectomized rats. Loss of bone volume accompanied by loss of trabecular connectivity is generally believed to be an irreversible process but their observations suggest that dried plum improves the trabecular microstructure of tibia after losses have already occurred. Bu *et al.* (2007) have shown that dried plum has potent effects on bone in terms of bone mass, microarchitecture and strength in osteopenic male rats. Although these changes may be mediated through the suppression of bone resorption, the fact that the restoration in some of the bone structural and biomechanical parameter shares some similarities with parathyroid hormone should not be overlooked. Further investigation is needed on a mechanistic level to clarify the influence of dried plum on bone metabolism.

Anxiety disorders

A varied diet with plums, cherry and other vegetables that are rich in naturally-occurring polyphenols could be an effective means to prevent (or delay) anxiety, depression and other diseases linked to oxidative stress (Bouayed, 2010). Chlorogenic acid is one of the most abundant polyphenols in fruits. Bouayed *et al.* (2007) demonstrated *in vivo* and *in vitro* that chlorogenic acid has anxiolytic effects coupled with antioxidant activity. Thus, fruits such as plums (Mirabelle) may provide health-promoting advantages to consumers.

Hepatic disorders

Prunes are used by some as a remedy for various diseases, including hepatitis. A clinical trial was designed to see the effects of prunes (*Prunus domestica* L.) on liver function (Ahmed *et al*., 2010a). 166 healthy volunteers were divided into three groups randomly. Either three (about 12 g) or six (23 g approximately) prunes were soaked in a glass of water (250 ml) overnight. Each subject from two test groups was asked to drink prune juice and eat whole fruit (single or double dose of prunes) as well, early in the morning, daily for eight weeks; whereas each subject from control group was given a glass of water to drink. Blood samples were taken at week 0 and week 8 for chemical analysis. There was significant reduction of serum alanine transaminase (p 0.048) and serum alkaline phosphatase (p 0.017) by the lower dose of prunes. There was no change in serum aspartate transaminase and bilirubin. Alteration in liver function by use of prunes may have clinical relevance in appropriate cases and prunes might prove beneficial in hepatic disease.

Prunes have also been used for their hepatoprotective effects (Soni *et al.,* 2003). A methanol:ethanol (70:30) extract of *Prunus domestica* was prepared and tested for its hepatoprotective effect against Paracetamol and CCl_4-induced hepatitis in rats. Alteration in the levels of biochemical markers of hepatic damage such as SGPT, SGOT, ALP, total bilirubin, direct bilirubin and tissue LPO, GSH, catalase and SOD were tested in both treated and untreated groups. Paracetamol (2 g/kg) and carbon tetrachloride (CCl_4) (1.5 ml/kg) enhanced the SGPT, SGOT, ALP, total bilirubin and direct bilirubin and tissue level of GSH. Treatment with extract of *P. domestica* fruits (150 mg/kg and 300 mg/kg) reduced the altered levels of biochemical markers to the near normal levels in a dose dependent manner.

Other disorders

The ingestion of Japanese plums (*Prunus salicina* Lindl. cv. Crimsonglobe) increases the urinary 6-sulfatoxymelatonin and total antioxidant capacity levels in young, middle-aged and elderly humans. The participants consumed 195 g of plums twice a day for five days. Nutritionally, plums were shown to contain low amounts of saccharides. From a functional point of view, serotonin, melatonin, phenolic compounds and anthocyanins were detected (Gonzalez-Flores *et al.,* 2011).

Cherries

Foods of plant origin contain hundreds of non-nutrient constituents with significant biological activity, generally called 'bioactive compounds' or phytochemicals with antioxidant activity. Thus, fruit and vegetable consumption has shown protective effects against several chronic diseases associated with ageing, including atherosclerosis, cardiovascular diseases, cancer, cataracts, blood pressure increase, ulcerous, neurodegenerative diseases, brain and immune dysfunction and even against bacterial and viral diseases. These protective effects have been attributed to several antioxidant compounds, which vary widely in chemical structure and function in plant tissues and are grouped in vitamins (C and E), carotenoids, phenolic and thiol (SH) compounds (Asensi-Fabado and Munné-Bosh, 2010; Wang *et al.*, 2011).

Sweet and sour cherries contain several antioxidants and polyphenols that possess many biological activities, such as antioxidant, anticancer and anti-inflammation properties (Ferreti *et al.,* 2010; McCune *et al.,* 2011). The antioxidant activity of cherry extracts has been studied in several biological samples and results have demonstrated that cyanidin 3-glucoside exhibits the highest antioxidant activity, as reported by the use of liposomes as model membranes (Heinonen *et al.,* 1998), for which sweet cherries are more active compared to those of other berries (blackberries, red raspberries, blueberries or strawberries). The use of neuronal PC 12 cells exposed to oxidative stress revealed that cherry extracts induced a protective action due to their antioxidant properties (Kim *et al.,* 2005). Chen *et al.* (2005) reported cancer-protective effects of cyanidin glucosides in several studies using cancer cell lines, including apoptotic effects via G2/M growth cycle arrest. Studies *in vitro* using several cell culture systems including colon, endothelial, liver, breast and leukemic cells and keratinocytes have demonstrated that anthocyanins exhibit multiple anticarcinogenic effects (Wang and Stoner, 2008). Potential cancer chemopreventive activities of anthocyanins revealed from *in vitro* studies include also their ability to stimulate the expression of Phase II detoxification enzymes (glutathione reductase, glutathione peroxidase and NAD(P)H: quinone reductase) and to inhibit mutagenesis by environmental toxins and carcinogens.

Interestingly, cherry fruit has also shown antioxidant activity *in vivo*. In an animal model fed with sour cherry juices, an increased activity of the antioxidant enzymes SOD (liver, blood) and Gpx (liver) and a decrease of lipid peroxidation were observed (Sarić *et al.*, 2009). Studies in human subjects, reported that the consumption of 280 g of cherries (about 45 sweet Bing cherries) increased plasma lipophilic antioxidant capacity (Jacob *et al.*, 2003; Prior *et al.*, 2007). Some studies have demonstrated that cherry intake inhibits inflammatory pathways. It has been well recognized that low grade inflammation is a potential risk factor for a wide range of chronic illnesses including cancer, cardiovascular disease, obesity and arthritis. Seeram *et al.* (2003) reported that cherries inhibit the cyclooxygenase (COX) enzymes responsible for inflammatory response. In a cell culture study assessing COX-1 and COX-2 enzyme activity the anthocyanin cyanidin, along with malvidin, were shown to have the greatest inhibitory effects. The strong inhibitory potential of cyanidin is thought to be the result of the chemical structure which exhibits a hydroxyl group positioned in the B ring of the compound.

However, more epidemiological studies are needed to further assess the role of cherry bioactive compounds and specific health outcomes. Such feeding studies should include some assessment of dose-response under standardized cherry production methods in order to more fully understand the optimal dose of cherry intake necessary to promote modulation of disease specific biomarkers.

CONCLUSION

Stone fruit and pomegranates constitute a relevant dietary source of antioxidant phytochemicals, fibre, vitamins and minerals. These constituents can help decrease the risk of some oxidation-associated diseases, as is the case of cardiovascular diseases, neurodegenerative diseases and cancer. A direct and unambiguous correlation between a positive health effect and the intake of a specific fruit or their components has not been sufficiently demonstrated in clinical studies yet. The evidence available so far is indirect, although epidemiology studies have shown that consumption of these fruits can provide health benefits. The antioxidants present in these fruits are generally poorly absorbed during the digestion process but they can exert effects in the gastrointestinal tract, preventing the oxidation of nutrients and vitamins, inhibiting enzymes such as glucosidases and lipases and, therefore, decreasing the absorption of nutrients and decreasing obesity. In addition, they interact with intestinal cells decreasing inflammation, and modulating the microbiota exerting a prebiotic effect, as has been demonstrated for pomegranate and different proanthocyanidin-rich foods, as is the case of stone fruit. Many factors affect the antioxidant composition of these fruits, including environmental (agronomical), technological, storage and culinary factors; these should be taken into consideration in order to preserve the optimal content of these beneficial compounds in fruit. In addition, this opens new opportunities for producing fruits with an enhanced content in bioactive phytochemicals with the consequent benefits for the consumers, the farmers and the industry.

REFERENCES

Abdel-Rahman, M.K. (2011) Can apricot kernels fatty acids delay the atrophied hepatocytes from progression to fibrosis in dimethylnitrosamine (DMN)-induced liver injury in rats? *Lipids Health Dis.* **10**, 114–123.

Adachi, M.; Suzuki, Y.; Mizuta, T. *et al.* (2007) The "Prunus mume Sieb. et Zucc" (ume) is a rich natural source of novel anti-cancer substance. *Int. J. Food. Prop.* **10**, 375–384.

Ahmed, T.; Sadia, H.; Khalid, A. *et al.* (2010a) Prunes and liver function: A clinical trial. *Pakistan J. Pharm. Sci.* **23**,463–466.

Ahmed, T.; Sadia, H.; Batool, S. *et al.* (2010b) Use of prunes as a control of hypertension. *J. Ayub. Med. Col.* **22**, 28–31.

Akin, E.B.; Karabulut, I. and Topcu, A. (2008) Some compositional properties of main Malatya apricot (*Prunus armeniaca* L.) varieties. *Food Chem.* **107**, 939–948.

Al-Khalifa, A.S. (1996) Physicochemical characteristics, fatty acid composition, and lipoxygenase activity of crude pumpkin and melon seed oils. *J. Agric. Food Chem.* **44**, 964–966.

Alliotta, G.; Piomelli, D.; Pollio, A. and Touwaide, A. (2003) *Le Piante Medicinali del "Corpus Hippocraticum"*. Guerini e Associati, Napoli, Italy.

Andreotti, C.; Ravaglia, D. and Costa, G. (2010) Effects of fruit load and reflective mulch on phenolic compounds accumulation in nectarine fruit. *Eur. J. Hort. Sci.* **75**, 53–59.

Arjmandi, B.H.; Soliman, A.; Juma, S. *et al.* (1998) Prune prevents ovariectomy-induced bone loss. 7th Annual Functional Foods for Health Retreat. Urbana, IL.

Arjmandi, B.H.; Deyhim, F.; Lucas, E. *et al.* (1999) Prune dose-dependently reverses bone loss in ovarian hormone deficient rats. 8th Annual Functional Foods for Health Retreat. Chicago, IL.

Arjmandi, B.H.; Khalil, D.A.; Lucas, E.A. *et al.* (2002) Dried plums improve indices of bone formation in postmenopausal women. *J. Women's Health Gender-Based Med.* **11**, 61–68.

Artes, F.; Tudela, J.A. and Gil, M.I. (1998) Improving the keeping quality of pomegranate fruit by intermittent warming. *Z. Lebensm. Unters. Forsch.* **207**, 316–321.

Asami, D.K.; Hong, Y.J.; Barrett, D.M. and Mitchell, A.E. (2003) Processing-induced changes in total phenolics and procyanidins in clingstone peaches. *J. Sci. Food Agric.* **63**, 56–63.

Asensi-Fabado, M.A. and Munné-Bosh, S. (2010) Vitamins in plants: occurrence, biosynthesis and antioxidant function. *Trends Plant Sci.* **15**, 582–592.

Aubert, C. and Chanforan, C. 2007. Postharvest changes in physicochemical properties and volatile constituents of apricot (*Prunus armeniaca* L.). Characterization of 28 cultivars. *J Agric. Food Chem.* **55**: 3074–3082.

Audergon, J.M.; Reich, M. and Souty, M. (1991) Abricot. Les variations des critères de qualité. *L'Arboriculture fruitière* **436**: 35–46.

Aviram, M.; Dornfeld, L.; Rosenblat, M. *et al.* (2000) Pomegranate juice consumption reduces oxidative stress, atherogenic modifications to LDL, and platelet aggregation: studies in humans and in the atherosclerotic apolipoprotein E-deficient mice. *Am. J. Clin. Nut.* **71**, 1062–1076.

Aviram, M.; Fuhrman, B.; Rosenblat, M. *et al.* (2002a) Pomegranate juice polyphenols decreases oxidative stress, low density lipoprotein atherogenic modifications and atherosclerosis. *Free. Rad. Res.* **36**, 72–73.

Aviram, M.; Dornfeld, L.; Kaplan, M. *et al.* (2002b) Pomegranate juice flavonoids inhibit low density lipoprotein oxidation and cardiovascular diseases: studies in atherosclerotic mice and humans. *Drug. Ex. Clin. Res.* **28**, 49–62.

Aviram, M.; Rosenblat, M.; Gaitini, D. *et al.* (2004) Pomegranate juice consumption for 3 years by patients with carotid artery stenosis reduces common carotid intima-media thickness, blood pressure and LDL oxidation. *Clin. Nutr.* **23**, 423–433.

Bassi, D. and Selli, R. (1990) Evaluation of fruit quality in peach and apricot. *Adv. Hortic. Sci.* **4**, 107–112.

Bassi, D.; Bartolozzi, F. and Muzzi, E. (1996) Patterns and heritability of carboxylic acids and soluble sugars in fruits of apricot (*Prunus armeniaca* L.). *Plant Breeding* **115**: 67–70.

Bolin, H.R. and Stafford, A.E. (1974) Effect of processing and storage on provitamin A and vitamin C in apricots. *J. Food Sci.* **39**, 1034–1036.

Boots, A.W.; Haenen, G.R. and Bast, A. (2008) Health effects of quercetin: From antioxidant to nutraceutical. *Eur. J. Pharmacol.* **585**, 325–337.

Bouayed, J. (2010) Polyphenols: A potential new strategy for the prevention and treatment of anxiety and depression. *Curr. Nut. Food Sci.* **6**, 13–18.

Bouayed, J.; Rammal, H.; Dicko, A. *et al.* (2007) Chlorogenic acid, a polyphenol from *Prunus domestica* (Mirabelle), with coupled anxiolytic and antioxidant effects. *J. Neurol. Sci.* **262**, 77–84.
Boyd, O.; Weng, P.; Sun, X.P. *et al.* (2011) Nectarine promotes longevity in Drosophila melanogaster. *Free Rad. Biol. Med.* **50**, 1669–1678.
Brat, P.; Mennen, L.; Georgé, S. *et al.* (2007) Determination of the polyphenol content of fruits and vegetables. Establishment of a database and estimation of the polyphenol intake in the French diet. *Acta Hort.* **744**, 61–69.
Brokmann, H. (1933) Die carotinoide der aprikose (*Prunus armeniaca*). *Z. Physiol. Chem.* **216**, 45.
Bu, S.Y.; Lucas, E.A.; Franklin, M. *et al.* (2007) Comparison of dried plum supplementation and intermittent PTH in restoring bone in osteopenic orchidectomized rats. *Osteoporosis Int* **18**, 931–942.
Byrne, D.H.; Noratto, G.; Cisneros-Zevallos, L. *et al.* (2009) Health benefits of peach, nectarine and plums. *Acta Hort.* **841**, 267–273.
Cerdá, B.; Llorach, R.; Cerón, J.J. *et al.* (2003) Evaluation of the bioavailability and metabolism in the rat of punicalagin, and antioxidant polyphenol of pomegranate juice. *Eur. J. Nutrition.* **42**, 18–28.
Cerdá, B.; Espín, J.C.; Parra, S. *et al.* (2004) The potent in vitro antioxidant ellagitannins from pomegranate juice are metabolized into bioavailable but poor antioxidant hydroxy-6H-dibenzopyran-6-one derivatives by the colonic microflora in healthy humans. *Eur. J. Nutr.* **43**, 205–220.
Cerdá, B.; Tomás-Barberán, F.A. and Espín, J.C. (2005) Metabolism of antioxidant and chemopreventive ellagitannins from strawberries, raspberries, walnuts and oak-aged wine in humans: Identification of biomarkers and individual variability. *J. Agric. Food Chem.* **53**, 227–235.
Cevallos, B.A.; Byrne, D.; Okie, W.R. and Cisneros-Zevallos, L. (2006) Selecting new peach and plum genotypes rich in phenolic compounds and enhanced functional properties. *Food Chem.* **96**, 273–280.
Chaovanalikit, A. and Wrolstad, R.E. (2004) Total anthocyanin and total phenolics of fresh and processed cherries and their antioxidant properties. *J. Food Sci.* **69**, FCT67–FCT72.
Chen, P.N.; Chu, S.C.; Chiou, H.L. *et al.* (2005) Cyanidin 3-glucoside and peonidin 3-glucoside inhibit tumor cell growth and induce apoptosis in vitro and suppress tumor growth in vivo. *Nut. Cancer.* **53**, 230–243.
Chuda, Y.; Ono, H.; Ohnishi-Kameyama, M. *et al.* (1999) Mumefural, citric acid derivative improving blood fluidity from fruit-juice concentrate of Japanese apricot (*Prunus mume* Sieb. et Zucc). *J. Agric. Food Chem.* **47**, 828–831
Chung, Y.J.; Park, C.; Jeong, Y.K. and Choi, Y.H. (2011) Apoptosis induction by methanol extract of *Prunus mume* fruits in human leukemia U937 cells. *Korean J. Life Sci.* **21**, 1109–1119.
Crisosto, C.H.; Crisosto, G.M. and Metheney, P. (2003) Consumer acceptance of "Brooks" and "Bing" cherries is mainly dependent on fruit SSC and visual skin color. *Postharvest Biol. Technol.* **28**, 159–167.
Curl, A.L. (1960). The carotenoids of apricots. *Food Res.* **24**, 190–196.
Davey, M.W.; Van Montagu, M.; Inze, D. *et al.* (2000) Plant L-ascorbic acid: chemistry, function, metabolism, bioavailability, and effects of processing. *J. Sci. Food Agric.* **80**, 825–860.
De Rigal, D.; Gauillard, F. and Richard-Forget, F. (2000) Changes in the carotenoid content of apricot (*Prunus armeniaca.* var. Bergeron) during enzymatic browning: β-carotene inhibition of chlorogenic acid degradation. *J. Sci. Food Agric.* **80**, 763–768.
Deyhim, F.; Stoecker, B.J.; Brusewitz, G.H. *et al.* (2005) Dried plum reverses bone loss in an osteopenic rat model of osteoporosis, *Menopause* **12**, 755–762.
Díaz-Mula, H.M.; Castillo, S.; Martínez-Romero, D. *et al.* (2008) Sensory, nutritive and functional properties of sweet cherry as affected by cultivar and ripening stage. *Food Sci. Technol. Int.* **15**, 535–543.
Díaz-Mula, H.M.; Valero, D.; Zapata, P.J. *et al.* (2009a) The functional properties of sweet cherry as a new criterion in a breeding program. *Acta Hort.* **839**, 275–280.
Díaz-Mula, H.M.; Zapata, P.J.; Guillén, F. *et al.* (2009b) Changes in hydrophilic and lipophilic antioxidant activity and related bioactive compounds during postharvest storage of yellow and purple plum cultivars. *Postharvest Biol. Technol.* **51**, 354–363.
Díaz-Mula, H.M.; Zapata, P.J.; Guillén, F. *et al.* (2011) Modified atmosphere packaging of yellow and purple plum cultivars. 2. Effect on bioactive compounds and antioxidant activity. *Postharvest Biol. Technol.* **61**, 110–116.
Díaz-Mula, H.M.; Serrano, M. and Valero, D. (2012) Alginate coatings preserve fruit quality and bioactive compounds during storage of sweet cherry fruit. *Food Bioprocess Technol.* **5**(8), 2990–2997 (doi: 10.1007/s11947-011-0599-2).
Donovan J.; Meyer A.S. and Waterhouse, A.L. (1998) Phenolic composition and antioxidant activity of prunes and prune juice (*Prunus domestica*). *J. Agric. Food Chem.* **46**, 1247–1252.

Dragovic-Uzelac, V.; Levaj, B.; Mrkic, V. *et al.* (2007) The content of polyphenols and carotenoids in three apricot cultivars depending on stage of maturity and geographical region. *Food Chem.* **102**, 966–975.

Drogoudi, P.D.; Vemmos, S.; Pantelidis, G. *et al.* (2008) Physical characters and antioxidant, sugar, and mineral nutrient contents in fruit from 29 apricot (*Prunus armeniaca* L.) cultivars and hybrids. *J. Agric. Food Chem.* **56**, 10754–10760.

Egea, M.I.; Murcia, M.A.; Sanchez-Bel, P. *et al.* (2006) Influence of different storage treatments on antioxidant systems of apricot var. 'Bulida'. *Acta Hort.* **717**, 337–341.

Egea, M.I.; Sanchez-Bel, P.; Martınez-Madrid, C. *et al.* (2007) The effect of beta ionization on the antioxidant potential of 'Búlida' apricot and its relationship with quality. *Postharvest Biol. Technol.* **46**, 63–70.

Enomoto, S.; Yanaoka, K.; Utsunomiya, H. *et al.* (2010). Inhibitory effects of Japanese apricot (*Prunus mume* Siebold et Zucc.; Ume) on *Helicobacter pylori*-related chronic gastritis. *Eur. J. Clin. Nutr.* **64**, 714–719.

Erdogan, S. and Erdemoglu, S. (2011) Evaluation of polyphenol contents in differently processed apricots using accelerated solvent extraction followed by high-performance liquid chromatography–diode array detector. *Int. J. Food Sci. Nut.* **62**, 729–739.

Esmaillzadeh, A.; Tabas, F.; Gaieni, I. *et al.* (2004) Concentrate pomegranate juice improves lipid profiles in diabetic patients with hyperlipidemia. *J. Med. Food*, **7**, 305–308.

Esti, M.; Cinquanta, L.; Sinesio, F. *et al.* (2002) Physicochemical and sensory fruit characteristics of two sweet cherry cultivars after cool storage. *Food Chem.* **76**, 399–405.

Fernández-Panchón, M.S.; Villano, D.; Troncoso, A.M. and García-Parrilla, M.C. (2008) Antioxidant activity of phenolic compounds: from in vitro results to in vivo evidence. *Crit. Rev. Food Sci. Nutr.* **48**, 649–671.

Ferreti, G.; Bachetti, T.; Belleggia, A. and Neri, D. (2010) Cherry antioxidants: from farm to table. *Molecules*. **15**, 6993–7005.

Fraser, P.D. and Bramley, P.M. (2004) The biosynthesis and nutritional uses of carotenoids. *Progr. Lipid Res.* **43**, 228–265.

Freedman, N.D.; Park, Y.; Subar, A.F. *et al.* (2008) Fruit and vegetable intake and head and neck cancer risk in a large United States prospective cohort study. *Int. J. Cancer*, **122**, 2330–2336.

Fuhrman, B. and Aviram, M. (2006) Protection against cardiovascular diseases. In: *Pomegranates: Ancient Roots to Modern Medicine* (eds Seeram, N.P.; Schulman, R.; Heber, D.). CRC Taylor and Francis, Boca Raton, FL, pp. 63–89.

Fujii, T.; Ikami, T.; Xu, J.-W. and Ikeda, K. (2006) Prune extract (*Prunus domestica* L.) suppresses the proliferation and induces the apoptosis of human colon carcinoma caco-2. *J. Nut. Sci. Vitamin.* **52**, 389–391.

Gabrielian, E. and Zohary, D. (2004) Wild relatives of food crops native to Armenia and Nakhichevan. *Flora Mediterranea*, **14**, 5–80.

Gao, L. and Mazza, G. (1995) Characterization, quantification, and distribution of anthocyanins and colorless phenolic in sweet cherries. *J. Agric. Food Chem.* **43**, 343–346.

García-Parra, J.; González-Cebrino, F.; Delgado, J. *et al.* (2011) Effect of thermal processing and high-pressure processing on the nutritional value and quality attributes of a nectarine puree with industrial origin during the refrigerated storage. *J. Food Sci.* **76**, C618–C625.

Gil, M.I.; García-Viguera, C.; Artés, F. and Tomás-Barberán, F.A. (1995a) Changes in pomegranate juice pigmentation during ripening. *J. Sci. Food Agric.*, **68**, 77–81.

Gil, M.I.; Sherif, J.; Ayed, N. *et al.* (1995b) Influence of cultivar, maturity stage and geographical location on the juice pigmentation of Tunisian pomegranates. *Z. Lebensm. Unters. Forsch.* **201**, 361–364.

Gil, M.I.; Sanchez, R.; Marin, J.G. and Artes, F. (1996a) Quality changes in pomegranates during ripening and cold storage. *Z. Lebensm. Unters. Forsch.* **202**, 481–485.

Gil, M.I.; Artés, F. and Tomás-Barberán, F.A. (1996b) Minimal processing and modified atmosphere packaging effects on the pigmentation of pomegranate seeds. *J. Food Sci.* **61**, 161–164.

Gil, M.I.; Tomás-Barberán, F.A., Hess-Pierce, B. *et al.* (2000) Antioxidant activity of pomegranate juice and its relationship with phenolic composition and processing, *J. Agric. Food Chem.* **48**, 4581–4589

Gil, M.I.; Tomás-Barberán, F.A.; Hess-Pierce, B. and Kader, A.A. (2002) Antioxidant capacity, phenolic compounds, carotenoids and vitamin C of nectarine, peach and plum cultivars from California. *J. Agric. Food Chem.* **50**, 4976–4982.

Gonçalves B.; Landbo A.K.; Knudse D. *et al.* (2004) Effect of ripeness and postharvest storage on the phenolic profiles of cherries (*Prunus avium* L.). *J. Agric. Food Chem.* **52**, 523–530.

González-Flores, D.; Velardo, B.; Garrido, M. *et al.* (2011) Ingestion of Japanese plums (*Prunus salicina* Lindl. cv. Crimsonglobe) increases the urinary 6-sulfatoxymelatonin and total antioxidant capacity levels in young, middle-aged and elderly humans: Nutritional and functional characterization of their content. *J. Food Nut. Res.* **50**, 229–236.

González-Sarrías, A.; Azorín-Ortuño, M.; Yañez-Gascón, M.J. *et al.* (2009a) Dissimilar in vitro and in vivo effects of ellagic acid and its microbiota-derived metabolites, urolithins, on the cytochrome P450 1 A 1. *J. Agric. Food Chem.* **57**, 5623–5632.

González-Sarrías A, Espín J-C, Tomás-Barberán F-A. and García-Conesa M.T. (2009b) Gene expression, cell cycle arrest and MAPK signaling regulation in Caco-2 cells exposed to ellagic acid and its metabolites, urolithins. *Mol. Nut. Food Res.* **53**, 686–698.

González-Sarrías, A.; Larrosa, M.; Tomás-Barberán, F.A. *et al.* (2010a) NF-kB-dependent anti-inflammatory activity of urolithins, gut microbiota ellagic acid-derived metabolites, in human colonic fibroblasts. *British J. Nutr.* **104**, 503–512.

González-Sarrías, A.; Giménez-Bastida, J.; García-Conesa; M.T. *et al.* (2010b) Occurrence of urolithins, gut microbiota ellagic acid metabolites, and proliferation markers expression response in human prostate gland upon consumption of walnuts and pomegranate juice. *Mol. Nut. Food Res.* **54**, 311–322.

Gurrieri, F.; Audergon, J.M.; Albagnac, G. and Reich, M. (2001) Soluble sugars and carboxylic acids in ripe apricot fruit as parameters for distinguishing different cultivars. *Euphytica* **117**, 183–189.

Haciseferogullari, H.; Gezer, I.; Özcan, M.M. and Muratasma, B. (2007) Postharvest chemical and physical-mechanical properties of some apricot varieties cultivated in Turkey. *J. Food Eng.* **79**, 364– 373.

Hegedüs, A.; Engel, R.; Abrankó, L. *et al.* (2010) Antioxidant and antiradical capacities in apricot (*Prunus armeniaca* L.) fruits: variations from genotypes, years, and analytical methods. *J. Food Sci.* **75**, 722–730.

Heinonen, M.; Meyer, A.S. and Frankel, E.N. (1998) Antioxidant activity of berry phenolics on human low-density lipoprotein and liposome oxidation. *J. Agric. Food Chem.* **46**, 4107–4111.

Hernandez, F.; Melgarejo, P.; Tomas-Barberan, F.A. and Artes, F. (1999) Evolution of juice anthocyanins during ripening of new selected pomegranate (*Punica granatum*) clones. *Eur. Food Res. Technol.* **210**, 39–42.

Holcroft, D.M.; Gil, M.I. and Kader, A.A. (1998) Effect of carbon dioxide on anthocyanins, phenylalanine ammonia lyase and glucosyl transferase in the arils of stored pomegranates. *J. Am. Soc. Hort. Sci.* **123**, 136–140.

Hong, Y.J.; Barrett, D.M. and Mitchell, A.E. (2004) Liquid chromatography/mass spectrometry investigation of the impact of thermal processing and storage on peach procyanidins. *J. Agric. Food Chem.* **52**. 2366–2371.

Hort, A. (1980) *Theophrastus Enquiry into Plants*, Vol. II. Harvard University Press, Cambridge (also available at http://www.biodiversitylibrary.org/item/66171, last accessed 28 February 2013).

Howard, L.A.; Wong, A.D.; Perry, A.K. and Klein, B.P. (1999) Beta-carotene and ascorbic acid retention in fresh and processed vegetables. *J. Food Sci.* **64**, 929–936.

Igual, M.; García-Martínez, E.; Martín-Esparza, M.E. and Martínez-Navarrete, N. (2012) Effect of processing on the drying kinetics and functional value of dried apricot. *Food Res. Int.* **47**, 284–290.

Jacob, R.A.; Spinozzi, G.M.; Simon, V.A. *et al.* (2003) Consumption of cherries lowers plasma urate in healthy women. *J. Nutr.* **133**, 1826–1829.

Jeong, J.T.; Moon, J.H.; Park, K.H. and Shin, C.S. (2006) Isolation and characterization of a new compound from *Prunus mume* fruit that inhibits cancer cells. *J. Agric. Food Chem.* **54**, 2123–2128

Jiménez, A. M.; Martínez-Tomé, M.; Egea, I. *et al.* (2008) Effect of industrial processing and storage on antioxidant activity of apricot. *Eur. Food Res. Technol.* **227**, 125–134.

Jones, R.B. (2007) The effects of postharvest handling conditions and cooking on anthocyanin, lycopene and glucosinolate content and bioavailability in fruits and vegetables. New Zealand. *J. Crop Hort. Sci.* **35**, 219–227.

Kahlon, T.S. and Smith, G.E. (2007) In vitro binding of bile acids by bananas, peaches, pineapple, grapes, pears, apricots and nectarines. *Food Chem.* **101**, 1046–1051.

Kamisli, F. and Karatas, F. (2009) Effects of sulphurisation on vitamins (A, C and E) and malondialdehyde in apricots. *Int. J. Food Sci.Technol.* **44**, 987–993.

Kan, T. and; Bostan, S.Z. (2010) Changes of contents of polyphenols and vitamin a of organic and conventional fresh and dried apricot cultivars (*Prunus armeniaca* L.). *World J. Agric. Sci.* **6**, 120–126.

Kan, J.; Che, J.; Xie, H.Y. and Jin, C.H. (2011) Effect of 1-methylcyclopropene on postharvest physiological changes of 'Zaohong' plum. *Act. Physiol. Plant.* **33**, 1669–1677.

Karabulut, I.; Topcu, A.; Duran, A. *et al.* (2007) Effect of hot air drying and sun drying on color values and b-carotene content of apricot (*Prunus armeniaca* L.). *Lebensm. Wiss. Technol.* **40**, 753–758.

Karatas, F. and Kamisli, F. (2007) Variations of vitamins (A, C and E) and MDA in apricots dried in IR and microwave. *J. Food Engineer.* **78**, 662–668.

Katayama, S.; Ogawa, H. and Nakamura, S. (2011) Apricot carotenoids possess potent anti-amyloidogenic activity in vitro. *J. Agric. Food Chem.* **59**, 12691–12696

Kawaii, S. and Lnaski, E.P. (2004) Differentiation-promoting activity of pomegranate (*Punica granatum*) fruit extracts in HL-60 human promyelocytic leukemia cells. *J. Med. Food* **7**, 13–18.

Khan, A. S.; Singh, Z. and Swinny, E. E. (2009) Postharvest application of 1-Methylcyclopropene modulates fruit ripening, storage life and quality of 'Tegan Blue' Japanese plum kept in ambient and cold storage. *Int. J. Food Sci. Technol.* **44**, 1272–1280.

Kim, N.D.; Mehta, R.; Yu, W.P. *et al.* (2002) Chemotherapeutic adjuvant therapeutic potential of pomegranate (*Punica granatum*) for human breast cancer. *Breast Cancer Res. Treat.* **71**, 203–217.

Kim, D.-O.; Jeong, S.W. and Lee, C.Y. (2003) Antioxidant capacity of phenolic phytochemicals from various cultivars of plums. *Food Chem.* **81**, 321–326.

Kim, D.O.; Heo, H.J.; Kim, Y.J. *et al.* (2005) Sweet and sour cherries phenolics and their protective effects on neuronal cells. *J. Agric. Food Chem.* **53**, 9921–9927.

Konopacka, D.; Jesionkowska, K.; Kruczynska, D. *et al.* (2010) Apple and peach consumption habits across European countries. *Appetite* **55**, 478–483.

Korekar, G.; Stobdan, T.; Arora, R. *et al.* (2011) Antioxidant capacity and phenolics content of apricot (*Prunus armeniaca* L.) kernel as a function of genotype. *Plant Foods Hum. Nutr.* **66**, 376–383.

Krinsky, N.I. and Johnson, E.J. (2005) Carotenoid actions and their relation to health and disease. *Mol. Asp. Med.* **26**, 459–516.

Kurus, M.; Ugras, M.; Ates, M. and Otlu, A. (2009) Apricot ameliorates alcohol induced testicular damage in rat model. *Food Chem. Toxicol.* **47**, 2666–2672.

Kutlu, T.; Durmaz, G.; Ates, B. and Erdogan, A. (2009). Protective effect of dietary apricot kernel oil supplementation on cholesterol levels and antioxidant status of liver in hypercholesteremic rats. *J. Food Agric. Environm.* **7**, 61–65.

Lansky, E.P.; Jiang, W.G.; Mo, H.B. *et al.* (2005) Possible synergistic prostate cancer suppression by anatomically discrete pomegranate fractions. *Invest. New Drugs* **23**, 11–20.

Larrosa, M.; González-Sarrías, A.; Yáñez-Gascón, M.J. *et al.* (2010a) Antiinflammatory properties of a pomegranate extract and its metabolite urolithin A in a colitis rat model and the effect of colon inflammation on the phenolic metabolism. *J. Nutr. Biochem.* **21**, 717–725.

Larrosa, M.; García-Conesa, M.T.; Espín, J.C. and Tomás-Barberán, F.A. (2010b) Ellagitannins, Ellagic acid and Vascular Health. *Mol. Asp. Med.* **31**, 513–539.

Lea, M.A.; Ibeh, C.; DesBordes, C. *et al.* (2008). Inhibition of growth and induction of differentiation of colon cancer cells by peach and plum phenolic compounds. *Anticancer Res.* **28**, 2067–2076.

Leccese, A.; Bureau, S.; Reich, M. *et al.* (2010). Pomological and nutraceutical properties in apricot fruit: cultivation systems and cold storage fruit management. *Plant Foods Hum. Nutr.* **65**, 112–120.

Ledbetter, C.; Peterson, S. and Jenner, J. (2006) Modification of sugar profiles in California adapted apricots (*Prunus armeniaca* L.) through breeding with Central Asian germplasm. *Euphytica* **148**, 251–259.

Lee, S.-H.; Lillehoj, H.S.; Cho, S.-M. *et al.* (2009) Immunostimulatory effects of oriental plum (*Prunus salicina* Lindl.). *Comp. Immunol. Microbiol. Infect. Dis.* **32**, 407–417.

Leong, S.Y. and Oey, I. (2012) Effects of processing on anthocyanins, carotenoids and vitamin C in summer fruits and vegetables. *Food Chem.* **133**, 1577–1587.

Lewicki, P. (2006). Design of hot air drying for better foods. *Trends Food Sci. Technol.* **17**, 153–163.

Liverani, A.; Giovannini, D.; Versari, N. *et al.* (2010) Japanese and European plum cultivar evaluation in the Po Valley of Italy: Yield and climate influence. *Acta Hort.* **874**, 327–336.

Madrau, M.A.; Piscopo, A.; Sanguinetti, A.M. *et al.* (2009) Effect of drying temperature on polyphenolic content and antioxidant activity of apricots. *Eur. Food Res. Technol.*, **228**, 441–448.

Malik, A.; Afaq, F.; Sarfaraz, S. *et al.* (2005) Pomegranate fruit juice for chemoprevention and chemotherapy of prostate cancer. *Proc. Natl. Acad. Sci. USA* **102**, 14813–14818.

Marlett, J. A. and Vollendorf, N.W. (1994) Dietary fiber content and composition of different forms of fruits. *Food Chem.* **51**, 39–44.

Martinez-Romero, D.; Dupille, E.; Guillen, F. *et al.* (2003a) 1-methylcyclopropene increases storability and shelf life in climacteric and nonclimacteric plums. *J. Agric. Food Chem.* **51**, 4680–4686.

Martinez-Romero, D.; Serrano, M.; Guillen, F. and Valero, D. (2003b) 1-methylcyclopropene increased storability in plum (*Prunus salicina* Lindl. cv. Golden Japan). *Acta Hort.* **599**, 71–77.

McCune, L. M.; Kubota, C.; Stendell-Hollins, N. R. and Thomson, C. A. (2011) Cherries and health: A review. *Crit. Rev. Food Sci. Nutr.* **51**, 1–12.

Mehlenbacher, S.A.; Cociu, V. and Hough, L.F. (1991) Apricots (*Prunus*). In: *Genetic resources of temperate fruit and nut crops* (eds Moore, J.N. and Ballington, J.R.). International Society for Horticultural Science, Wageningen, The Netherlands, pp. 65–107.

Menniti, A.M.; Gregori, R. and Donati, I. (2004) 1-Methylcyclopropene retards postharvest softening of plums. *Postharvest Biol. Technol.* **31**, 269–275.

Menniti, A.M.; Donati, I. and Gregori, R. (2006) Responses of 1-MCP application in plums stored under air and controlled atmospheres. *Postharvest Biol. Technol.* **39**, 243–246.

Metha, R. and Lansky, E.P. (2004) Breast cancer chemopreventive properties of pomegranate (*Punica granatum*) fruit extracts in a mouse mammary organ culture. *Eur. J. Cancer Prev.* **13**, 345–348.

Miller, A. (2004) *Punica protopunica.* IUCN Red List of Threatened Species. Version 2012.2. http://www.iucnredlist.org/details/30404/0 (last accessed 20 March 2013).

Miranda, G.; Berna, A.; Salazar, D. and Mulet, A. (2009) Sulphur dioxide evolution during dried apricot storage. *Lebensm. Wiss. Technol.* **42**, 531–533.

Moldenke, H. and Moldenke, A. (1952) Plants of the Bible. *Chronica Botanica* **28**, 1–328.

Moller, B. and Herrmann, K. (1983) Quinic acid esters of hydroxycinnamic acids in stone and pome fruits. *Phytochemistry* **22**, 477–481.

Mori, S.; Sawada, T.; Okada, T. *et al.* (2007) New antiproliferative agent, MK615, from Japanese apricot 'Prunus mume' induces striking autophagy in colon cancer cells in vitro. *World J. Gastroenterol.* **13**, 6512–6517.

Morimoto, Y.; Kawahara, K.I.; Kikuchi, K. *et al.* (2009) MK615 attenuates *Porphyromonas gingivalis* lipopolysaccharide-induced pro-inflammatory cytokine release via MAPK inactivation in murine macrophage-like RAW264.7 cell. *Biochem. Biophys. Res. Com.* **389**, 90–94.

Morimoto-Yamashita, Y.; Matsuo, M.; Komatsuzawa, H. *et al.* (2011) MK615: A new therapeutic approach for the treatment of oral disease. *Medical Hypotheses* **77**, 258–260

Mozetiĉ B.; Trebŝe P. and Hribar J. (2002) Determination and quantisation of anthocyanins and hydroxycinnamic acids in different cultivars of sweet cherries (*Prunus avium* L.) from Nova Gorica Region (Slovenia). *Food Technol. Biotechnol.* **40**, 207–212.

Mozetiĉ B.; Simĉiĉ M. and Trebŝe P. (2006) Anthocyanins and hidroxycinnamic acids of Lambert Compact cherries (*Prunus avium* L.) after cold storage and 1-methylcyclopropene treatments. *Food Chem.* **97**, 302–309.

Munzuroglu, O.; Karatas, F. and Geckil, H. (2003) The vitamin and selenium contents of apricot fruit of different varieties cultivated in different geographical regions. *Food Chem.* **83**, 205–212.

Nakagawa, A.; Sawada, T.; Okada, T. *et al.* (2007) New antineoplastic agent, MK615, from UME (a variety of) Japanese apricot inhibits growth of breast cancer cells in vitro. *Breast J.* **13**, 44–49.

National Cancer Institute (2004) Glycemic Index database based on CS-FII 96 data. DHQ Nutrient Database. National Cancer Institute, Bethesda, MD.

Negishi, H.; Onobayashi, Y.; Xu, J. *et al.* (2007) Effects of prune extract on blood pressure elevation in stroke-prone spontaneously hypertensive rats. In: *Clinical and Experimental Pharmacology and Physiology.* Proceedings of the 2nd International Symposia on 'Lifestyle-related Disease – Perspective for Primary Prevention and Treatment in Animal Models and Humans'. 34 Supplement **1**:47–48.

Noratto, G.; Porter, W.; Byrne, D. and Cisneros-Zevallos, L. (2009) Phenolics in peaches and plums preferentially suppress the growth of estrogen-independent MDA-MB-435 breast cancer cells. *Acta Hort.* **841**, 567–570.

Okada, T.; Sawada, T.; Osawa, T. *et al.* (2008) MK615 inhibits pancreatic cancer cell growth by dual inhibition of Aurora A and B kinases. *World J. Gastroenterol.* **14**, 1378–1382

Okie, W. (2006) Introgression of *Prunus* species in Plum. *New York Fruit Quarterly* **14**(1), 29–37.

Omoni, A.O. and Aluko, R.E. (2005) The anti-carcinogenic and anti-atherogenic effects of lycopene: A review. *Trends Food Sci. Technol.* **16**, 344–350.

Ozsahin, A.D. and Yilmaz, O. (2010) *Prunus armeniaca* L. cv. Hacihaliloglu fruits extracts prevent lipid peroxidation and protect the unsaturated fatty acids in the fenton reagent. *Environment. Asian J. Chem.* **22**, 8022–8032.

Park, C.; Jin, C.Y.; Kim, G.Y. *et al.* (2011) Induction of apoptosis by ethanol extract of *Prunus mume* in U937 human leukemia cells through activation of caspases. *Oncol Rep.* **26**, 987–93.

Parlakpinar, H.; Olmez, E.; Acet, A. *et al.* (2009) Beneficial effects of apricot-feeding on myocardial ischemia-reperfusion injury in rats. *Food Chem.Toxicol.* **47**, 802–808.

Pérez, R.; Gómez, M. and Morales, R. (2008) Agromorphological characterization of traditional Spanish sweet cherry (*Prunus avium* L.), sour cherry (*Prunus cerasus* L.) and duke cherry (*Prunus x gondouinii* Rehd.) cultivars. *Spanish J. Agric. Res.* **6**, 42–55.

Pliakoni, E.D.; Nanos, G.D. and Gil, M.I. (2010) Two-season study of the influence of regulated deficit irrigation and reflective mulch on individual and total phenolic compounds of necatrines at harvest and during storage. *J. Agric. Food Chem.* **58**, 11783–11789.

Plumb, G.W., Pascual-Teresa, S.; Santos-Buelga, C. *et al.* (2002) Antioxidant properties of gallocatechin and prodelphinidins from pomegranate peel. *Redox. Rep.* **7**, 41–46.

Poiana, M.A.; Moigradean, D.; Dogaru, D. *et al.* (2011) Processing and storage impact on the antioxidant properties and color quality of some low sugar fruit jams. *Romanian Biotechnol. Let.* **16**, 6504–6512.

Prior, R.L.; Gu, L.; Wu, X. *et al.* (2007) Plasma antioxidant capacity changes following a meal as a measure of the ability of a food to alter in vivo antioxidant status. *J. Am. Coll. Nutr.* **26**, 170–181.

Puerta-Gómez, A.F. and Cisneros-Zevallos, L. (2011) Postharvest studies beyond fresh market eating quality: Phytochemical antioxidant changes in peach and plum fruit during ripening and advanced senescence. *Postharvest Biol. Technol.* **60**, 220–224.

Radi, M.; Mahrouz, M.; Jaouad, A. *et al.* (1997) Phenolic composition, browning susceptibility, and carotenoid content of several apricot cultivars at maturity. *HortScience* **32**, 1087–1091.

Radi, M.; Mahrouz, M.; Jaouad, A. and Amioti, M. J. (2003) Influence of mineral fertilization (NPK) on the quality of apricot fruit (cv. Canino). The effect of the mode of nitrogen supply. *Agronomie* **23**, 737–745.

Ramadan, M.F.; Zayed, R.; Abozid, M. and Asker, M.M.S. (2011) Apricot and pumpkin oils reduce plasma cholesterol and triacylglycerol concentrations in rats fed a high-fat diet. *Grasas y Aceites*, **62**, 443–452.

Rendina E.; Lim YF.; Marlow D. *et al.* (2012) Dietary supplementation with dried plum prevents ovariectomy-induced bone loss while modulating the immune response in C57BL/6J mice. *J. Nut. Biochem.* **23**, 60–68.

Rickman, J.C.; Barret, D.M. and Bruhn, C.M. (2007) Nutritional comparison of fresh, frozen and canned fruits and vegetables. Part 1. Vitamins C and B and phenolic compounds. *J. Sci. Food Agric.* **87**, 930–944.

Rivera, D.; Matilla, G.; Obón, C. and Alcaraz, F. (2012) *Plants and Humans in the Near East and The Caucasus*. Editum – Plants and Humans, Murcia, Spain.

Romano, G.S.; Cittadini, E.D.; Pugh, B. and Schouten, R. (2006) Sweet cherry quality in the horticultural production chain. *Stewart Postharvest Review*, **1**:1 (doi: 10.2212/spr.2006.6.2).

Rossello, C.; Canellas, J.; Santiesteban, I. and Mulet, A. (1993) Simulation of the absorption process of sulphur dioxide in apricots. *Lebensm. Wiss. Technol.* **26**, 322–328.

Roussos, P.A.; Sefferou, V.; Denaxa, N.K. *et al.* (2011) Apricot (*Prunus armeniaca* L.) fruit quality attributes and phytochemicals under different crop load. *Sci. Hortic.* **129**, 472–478.

Ruiz, D. and Egea, J. (2008) Phenotypic diversity and relationships of fruit quality traits in apricot (*Prunus armeniaca* L.) germplasm. *Euphytica* **163**, 143–158.

Ruiz, D.; Egea, J.; Tomás-Barberán, F.A. and Gil, M.I. (2005a). Characterization and quantitation of phenolic compounds in new apricot (*Prunus armeniaca* L.) varieties. *J. Agric. Food Chem.* **53**, 9544−9552.

Ruiz, D.; Egea, J.; Tomás-Barberán, F.A. and Gil, M.I. (2005b) Carotenoids from new apricot (*Prunus armeniaca* L.) varieties and their relationship with flesh and skin color. *J. Agric. Food Chem.* **53**, 6368−6374.

Ryley, J. and Kayda, P. (1993) Vitamins in thermal processing. *Food Chem.* **49**, 119–129.

Sass-Kiss, A.; Kiss, J.; Milotay, P. *et al.* (2005) Differences in anthocyanin and carotenoid content of fruits and vegetables. *Food Res. Internat.* **38**, 1023–1029.

Sadilova, E.; Carle, R. and Stintzing, F.C. (2007) Thermal degradation of anthocyanins and its impact on color and in vitro antioxidant capacity. *Mol. Nutr. Food Res.* **51**, 1461–1471.

Sarić, A.; Sobocanec, S.; Balog, T. *et al.* (2009) Improved antioxidant and anti-inflammatory potential in mice consuming sour cherry juice (*Prunus cerasus* cv. Maraska). *Plant Foods Hum. Nutr.* **64**, 231–237.

Schmitz-Eiberger, M.A. and Blanke, M.M. (2012) Bioactive components in forced sweet cherry fruit (Prunus avium L.), antioxidative capacity and allergenic potential as dependent on cultivation cover. *LWT – Food Sci. Technol.* **46**, 388–392.

Schmitzer, V.; Slatnar, A.; Mikulic-Petkovsek, M. *et al.* (2011) Comparative study of primary and secondary metabolism in apricot (*Prunus armeniaca* L.) cultivars. *J. Sci. Food Agric.* **91**, 860–868.

Schreiner, M. and Huyskens-Keil, S. (2006) Phytochemicals in fruit and vegetables: health promotion and postharvest elicitors. *Crit. Rev. Plant Sci.* **25**, 267–278.

Seeram, N.P.; Zhang, Y. and Nair, M.G. (2003) Inhibition of proliferation of human cancer cells and cyclooxygenase enzymes by anthocyanidins and catechins. *Nutr. Cancer.* **46**, 101–106.

Seeram, N.P.; Adams, L.S.; Henning, S.M. *et al.* (2005) In vitro antiproliferative, apoptotic and antioxidant activities of punicalagin, ellagic acid and a total pomegranate tannin extract are enhanced in combination with other polyphenols as found in pomegranate juice. *J. Nutr. Biochem.* **16**, 360–367.

Seeram, N.P.; Zhang, Y.; Errd, J.D. *et al.* (2006) Pomegranate phytochemicals. In: *Pomegranates: Ancient Roots to Modern Medicine* (eds Seeram, N.P.; Schulman, R.; Heber, D.) CRC Taylor and Francis, Boca Raton, FL, pp. 3–29.

Serrano, M.; Guillén, F.; Martínez-Romero, D. *et al.* (2005) Chemical constituents and antioxidant activity of sweet cherry at different ripening stages. *J. Agric. Food Chem.* **53**, 2741–2745.

Serrano, M.; Díaz-Mula, H.; Zapata, P. J. *et al.* (2009) Maturity stage at harvest determines the fruit quality and antioxidant potential after storage of sweet cherry cultivars. *J. Agric. Food Chem.* **57**, 3240–3246.

Serrano, M.; Díaz-Mula, H.M. and Valero D. (2011) Antioxidant compounds in fruits and vegetables and changes during postharvest storage and processing. *Stewart Postharvest Review*, **1**:1 (doi: 10.2212/spr.2011.1.1).

Shishodia, S.; Adams, L.; Bhatt, I.D. and Aggarwal, B.B. (2006) Anticancer potential of pomegranate. In: *Pomegranates: Ancient Roots to Modern Medicine* (eds Seeram, N.P.; Schulman, R.; Heber, D.) CRC Taylor and Francis, Boca Raton, FL, pp. 107–116.

Sizova, N.V. and Andreeva, N.Y. (2007) Estimation of vitamin E in plant oils by microcalorimetry. *Pharm. Chem. J.* **41**, 337–340.

Skog, L.J.; Schaefer, B.H. and Smith, P.G. (2003) Effect of ripeness at harvest on response of plum to treatment with 1-methylcyclopropene. *Acta Hort.* **599**, 49–52.

Smith B.J. (2009) Dried plum improves bone density and microarchitecture by up-regulating osteoblast activity and down-regulating osteoclast differentiation. *Acta Hort.* **841**, 441–448.

Sochor, J.; Zitka, O.; Skutkova, H. *et al.* (2010) Content of phenolic compounds and antioxidant capacity in fruits of apricot genotypes. *Molecules* **15**, 6285–6305.

Sochor, J.; Skutkova, H.; Babula, P. *et al.* (2011) Mathematical evaluation of the amino acid and polyphenol content and antioxidant activities of fruits from different apricot cultivars. *Molecules* **16**, 7428–7457.

Soni, M.; Mohanty, P.K. and Jaliwala, Y.A. (2003) Hepatoprotective activity of fruits of "*Prunus domestica*". *Int. J. Pharmacol. Biol. Sci.* **2**, 439–453.

Souty, M.; Audergon, J.M. and Chambroy, Y. (1990) Abricot: les critères de qualité. *L'Arboriculture Fruitière* **430**, 16–24.

Stacewicz-Sapuntzakis, M.; Bowen, P.E.; Hussain, E.A. *et al.* (2001). Chemical composition and potential health effects of prunes: A functional food? *Crit. Rev. Food Sci. Nut.* **41**, 251–286.

Steiner, A.; Abreu, M.; Correia, L. *et al.* (2006) Metabolic response to mild heat pre-treatments and modified atmosphere packaging of fresh-cut peaches. *Eur. Food Res. Technol.* **222**, 217–222.

Tavarini, S.; Gil, M.I.; Tomas-Barberan, F.A. *et al.* (2011) Effects of water stress and rootstocks on fruit phenolic composition and physical/chemical quality in Suncrest peach. *Ann. Appl. Biol.* **158**, 226–233.

Tinker, L.F.; Schneeman, B.O., Davis, P.A. *et al.* (1991) Consumption of prunes as a source of dietary fiber in men with mild hypercholesterolemia. *Am. J. Clin. Nut.* **53**, 1259–1265.

Tomás-Barberán F.A. and Espín J.C. (2001) Phenolic compounds and related enzymes as determinants of quality in fruits and vegetables. *J. Sci. Food Agric.* **81**, 853–876.

Tomás-Barberán, F.A.; Gil, M.I.; Cremin, P. *et al.* (2001) HPLC-DAD-ESIMS analysis of phenolic compounds in nectarines, peaches and plums. *J. Agric. Food Chem.* **49**, 4748–4760.

Turan, S.; Topcu, A.; Karabulut, I. *et al.* (2007) Fatty acid, triacylglycerol, phytosterol, and tocopherol variations in kernel oil of Malatya apricots from Turkey. *J. Agric. Food Chem.* **55**, 10787–10794.

Ugras, M.; Kurus, M.; Ates, B. *et al.* (2010) *Prunus armeniaca* L. (apricot) protects rat testes from detrimental effects of low-dose x-rays. *Nutr. Res.* **30**, 200–208.

USDA (2012) USDA, ARS, National Genetic Resources Program. Germplasm Resources Information Network – (GRIN) (Online Database). National Germplasm Resources Laboratory, Beltsville, Maryland. (http://www.ars-grin.gov/cgi-bin/npgs/html/taxgenform.pl?language=en, last accessed 28 February 2013).

USDA Carotenoid Database. http://www.nal.usda.gov/fnic/foodcomp/Data/Carot/file1.dat (last accessed 28 February 2013).

USDA Food Composition. http://fnic.nal.usda.gov/food-composition (last accessed 28 February 2013).

Usenik, V.; Kastelec, D.; Veberic, R. and Stampar, F. (2008a) Quality changes during ripening of plums (*Prunus domestica* L.). *Food Chem.* **111**, 830–836.

Usenik V.; Fabčič J. and Štampar F. (2008b) Sugars, organic acids, phenolic composition and antioxidant activity of sweet cherry (*Prunus avium* L.). *Food Chem.* **107**, 185–192.

Valero, D. and Serrano, M. (2010) *Postharvest Biology and Technology for Preserving Fruit Quality*. CRC Press-Taylor and Francis, Boca Raton, FL.

Valero, D.; Martinez-Romero, D.; Valverde, J.M. *et al.* (2003) Quality improvement and extension of shelf life by 1-methylcyclopropene in plum as affected by ripening stage at harvest. *Innov. Food Sci. Emerg. Technol.* **4**, 339–348.

Valero, D.; Díaz-Mula, H.M.; Zapata, P.J. *et al.* (2011) Postharvest treatments with salicylic acid, acetylsalicylic acid or oxalic acid delayed ripening and enhanced bioactive compounds and antioxidant capacity in sweet cherry. *J. Agric. Food Chem.* **59**, 5483–5489.

Verma, N.; Mohanty, A. and Lal, A. (2010) Pomegranate Genetic Resources and Germplasm Conservation: A Review. *Fruit, Vegetable and Cereal Sci. Biotechnol.* **4**, 120–125.

Vizzotto, M.; Cisneros-Zevallos, L.; Byrne, D.H. *et al.* (2007) Large variation found in the phytochemical and antioxidant activity of peach and plum germplasm. *J. Amer. Soc. Hort. Sci.* **132**, 334–340.

Voi, A.L.; Impembo, M.; Fasanro, G. and Castaldo, D. (1995) Chemical characterization of apricot puree. *J. Food Comp. Anal.* **8**, 78–85.

Voutilainen, S.; Nurmi, T.; Mursu, J. and Rissanen, T.H. (2006) Carotenoids and cardiovascular health. *Amer. J. Clin. Nutr.* **83**, 1265–1271.

Wang, S.Y. (2006) Pre-harvest conditions and antioxidant capacity in fruits. *Acta Hort.* **712**, 1–10.

Wang, L.S. and Stoner, G.D. (2008) Anthocyanins and their role in cancer prevention. *Cancer Lett.* **269**, 281–290.

Wang, L. and Vestrheim, S. (2003) Controlled atmosphere storage of Norwegian grown plums (*Prunus domestica* L.). *Acta Agriculturae Scandinavica Section B – Soil and Plant Science*, **53**, 33–37.

Wang, S.; Melnyk, J.P.; Tsao, R. and Marcone, M.F. (2011) How natural dietary antioxidants in fruits, vegetables and legumes promote vascular health. *Food Res. Int.* **44**, 14–22.

Wikipedia (2012) Prunus. http://en.wikipedia.org/wiki/Prunus (last accessed 28 February 2013).

Wilms, L.C.; Hollman, P.C.; Boots, A.W. and Kleinjans, J.C. (2005) Protection by quercetin and quercetin-rich fruit juice against induction of oxidative DNA damage and formation of BPDE-DNA adducts in human lymphocytes. *Mutat. Res.* **582**, 155–162.

Wu, S.F.; Chen, Y.S.; Lin, C.L. *et al.* (2011) Qualitative and quantitative analyses of the anti-allergic constituent of commercial Prunus mume products in Taiwan. *J. Food Drug Analysis* **19**, 66–72.

Yu, M.H.; Gwon Im, H.; Gyu Lee, S. *et al.* (2009) Inhibitory effect of immature plum on PMA-induced MMP-9 expression in human hepatocellular carcinoma. *Nat. Prod. Res.* **23**, 704–718.

Yuan, Z.; Yin, Y.; Qu, J. *et al.* (2007) Population genetic diversity in Chinese pomegranate (*Punica granatum* L.) cultivars revealed by fluorescent-AFLP markers. *J. Genet. Genom.* **34**, 1061–1071.

Zhang, J.; Gu, H.D.; Zhang, L. *et al.* (2011) Protective effects of apricot kernel oil on myocardium against ischemia–reperfusion injury in rats. *Food Chem. Toxicol.* **49**, 3136–3141.

Zohary, D. and Hopf M. (1994) *Domestication of Plants in the Old World*. Oxford University Press, Oxford.

Zhivondov, A. and Uzundzhalieva, K. 2012. Taxonomic classification of plum-apricot hybrids. *Acta Hort.* **966**, 211–217.

8 The Potential Health Benefits of the Subtropical Fruits Kiwifruit, Feijoa and Tamarillo

Stephen J.M. Skinner, Denise Hunter, Suengmok Cho and Margot Skinner

INTRODUCTION

Kiwifruit (*Actinidia* spp. Lindl), a native of China and adjoining countries, and feijoa (*Feijoa sellowiana* Berg. (syn *Acca sellowiana*)) and tamarillo (*Solanum betaceum* (syn. *Cyphomandra betacea*)), indigenous to subtropical regions of South America, are cultivated as commercial fruiting vines and trees in regions of the world that have the appropriate climate. Tamarillo was the first of these fruits to be introduced to New Zealand, being introduced in 1891 with small-scale production occurring from the 1920s (Scarrow, 2009). Feijoa and kiwifruit were both introduced to New Zealand in the early 1900s, and the commercial production of these fruits in New Zealand is largely attributable to pioneer plant breeder, Hayward Wright (Scarrow, 2009; Campbell and Haggerty, 2009). The popularity of kiwifruit in New Zealand grew during World War II when restrictions of fruit imports were in place, although even by 1948 the size and value of the kiwifruit crop was only a little more than one-tenth that of tamarillos (Ferguson and Bollard, 1990). In the present day, the commercial value of kiwifruit to New Zealand has far exceeded those of tamarillo and feijoa, contributing $NZ962.6 million in exports, compared with $0.4 and $0.2 million, respectively (Aitken and Hewett, 2011). Kiwifruit, feijoa and tamarillo contain vitamins, minerals, dietary fibre and an array of phytochemicals that may give them some unique health benefits. In this chapter, the supporting data for the health benefits of kiwifruit and various extracts made from the fruit and other parts of the plant are reviewed and the status of intervention trials that begin to validate these health benefits summarized. Investigations into the health benefits of feijoa and tamarillo are in their infancy but an understanding of the presence and concentration of major phytochemicals and dietary fibre in these fruits may lead to research on potential health benefit effects in the future.

KIWIFRUIT

The kiwifruit originated in southwestern China and associated regions. It was known in China as 'mihoutao' (monkey peach), and was recommended in the early Chinese pharmacopoeia for a variety of symptoms. They were used to aid digestion, reduce irritability, relieve rheumatism, prevent kidney or urinary tract stones and cure hemorrhoids, dyspepsia and vomiting (Ferguson, 1990). They were also used for the treatment of many types of cancer,

Bioactives in Fruit: Health Benefits and Functional Foods, First Edition. Edited by Margot Skinner and Denise Hunter.

such as breast cancer and cancers of the digestive system (Ai *et al.*, 1982; Motohashi *et al.*, 2002). The bioactive constituents in kiwifruit that are recognized in traditional Chinese medicine include polysaccharides (Yan *et al.*, 1995; Zhu *et al.*, 1996), alkaloids (Sakan *et al.*, 1959; Tang *et al.*, 1997), saponins (Qing *et al.*, 1999) and organic acids (Heatherbell, 1975; Lu *et al.*, 2007b).

When kiwifruit were first introduced and grown in New Zealand, they were known as 'Chinese gooseberries'. As the popularity of the fruit increased, the growers and marketers considered exporting their produce and the name 'kiwifruit' was introduced and became widely accepted in most of the countries where it was commercially grown. The most commonly consumed kiwifruit are green, predominantly *Actinidia deliciosa* 'Hayward' (A. Chev.) C.F. Liang et A.R. Ferguson or gold, *A. chinensis* 'Hort16A' Planch. cultivars, although some others are now being produced, for example the kiwiberry (*A. arguta* Sieb. et Zucc. Planch. ex Miq) and other variants of green, gold and red kiwifruit. More than 90% of the kiwifruit in international trade are of a single cultivar, and the most commonly traded yellow-fleshed kiwifruit is 'Hort16A' (ZESPRI® GOLD Kiwifruit), subsequently referred to as gold kiwifruit. Smaller amounts of kiwiberry, also called 'hardy kiwifruit' or 'baby kiwi' (*A. arguta*), are produced and green kiwifruit with a red centre (e.g. *A. chinensis* var. *rufopulpa* 'Hongyang') have begun commercial production.

On average, green kiwifruit contain approximately 85 mg of ascorbate (vitamin C) per 100 g fresh weight (FW), with gold kiwifruit often containing up to 20% more (Ferguson and Ferguson, 2002). Kiwifruit contain up to 2 or 3% of dietary fibre, and consequently one kiwifruit can contribute a significant proportion of recommended daily dietary fibre intake (Ferguson and Ferguson, 2002). Kiwifruit also contain vitamins E and K, folate, the minerals potassium, magnesium and copper, as well as phytochemicals, including carotenoids and polyphenols. Typically, green-fleshed kiwifruit (e.g. *A. deliciosa*) contain chlorophylls *a* and *b*, and carotenoids such as β-carotene, lutein, violaxanthin and 9'-*cis*-neoxanthin; yellow-fleshed kiwifruit (e.g. *A. chinensis* or *A. macrosperma*) also contain carotenoids but little or no chlorophyll (McGhie *et al.*, 2002); and red-fleshed genotypes (e.g. *A. chinensis* 'Hongyang') contain anthocyanins (Montefiori *et al.*, 2005). In addition, 'Hayward' leaf tissue and juice prepared from the fruit have been shown to contain polyphenolics, specifically flavonols (Dawes and Keene, 1999; Webby, 1990). Phytochemicals are well recognized for their antioxidant and cell protection activities, and they are becoming increasingly recognized for other properties that may be important for health. However, the phytochemical composition of kiwifruit can vary greatly according to the particular species or genotype. Many experimental studies give inadequate details of the species or cultivar of kiwifruit used. Most studies, especially those from outside China, have probably used fruit of only *A. deliciosa* 'Hayward' but, even then, the composition of 'Hayward' fruit could be affected by varying growing conditions, the maturity or ripeness of the fruit or different periods in cold storage. Some authors do not specify the plant part analysed.

Benefits from vitamin C

Kiwifruit contain relatively high concentrations of vitamin C. Humans must acquire vitamin C from their diet and recommendations for intake have traditionally been intended to prevent scurvy. However, evidence from metabolic, pharmacokinetic and epidemiological evidence strongly suggests that vitamin C intake above the current recommended daily intake (RDI) might contribute to prevention of chronic diseases, particularly cardiovascular diseases, certain cancers and common infections associated with the upper respiratory tract such as colds and influenza (Frei *et al.*, 2012; Hunter *et al.*, 2012). The recent identification of an extended role for ascorbate in regulating the hydroxylation of hypoxia-inducible factor (HIF) and histone demethylases highlights the necessity for the maintenance of adequate ascorbate in all body tissues to control cell metabolism and gene expression. The Gulo mouse provides a useful model of vitamin C deficiency, developing scurvy-like symptoms after ascorbate removal and showing changes in collagen synthesis, aortic wall structure, atherosclerotic plaque formation and inflammatory processes when maintained on suboptimal ascorbate supplements. Both green and gold kiwifruit have been shown to be much better vehicles for ascorbate delivery to tissues than supplementation in the Gulo mouse model, suggesting some synergistic effect from other components of the fruit (Vissers *et al.*, 2011). It will be interesting to determine whether this effect is kiwifruit specific and whether similar effects occur in humans.

Protection from oxidative stress, mutagenicity and cancer

Cells are exposed to reactive oxygen species (ROS) during normal metabolism but at a much higher rate during acute and chronic disease, smoking, periods of excessive alcohol consumption, poor diet and environmental pollution (Hunter *et al.*, 2012). ROS are mostly produced by intracellular mitochondria and at high concentrations cause damage to DNA, proteins and lipids. Dietary antioxidants have great potential in protecting against ROS damage that is a key feature of ageing and degenerative diseases (Ames *et al.*, 1993). However, there is some evidence that ageing can also be mediated by ROS-independent mechanisms in mitochondria (Sanz *et al.*, 2012). The antioxidants in fruits and vegetables are vitamins C and E, carotenoids and polyphenolic compounds (Dawes and Keene, 1999). Diseases that may benefit from these vitamins and compounds are cardiovascular diseases (Joshipura *et al.*, 2001) and cancers (Block *et al.*, 1992), brain disease (Hosoya *et al.*, 2004) and diabetes (Hosoya *et al.*, 2004).

Chemical measurements of antioxidant capacity are the Trolox™ equivalent antioxidant capacity (TEAC) assay, the ferric reducing antioxidant power (FRAP) assay, the oxygen radical absorption capacity (ORAC) assay, the DPPH (2,2′-diphenyl-1-picrylhydrazyl) free radical scavenging potential and the total radical absorption potential (TRAP) assay. Fruits and their extracts need to be tested using at least two of these assays, since different fruit components are not equivalent in any one of the assays (Ozgen *et al.*, 2006). By the ORAC assay, kiwifruit ranked fifth out of thirteen fruits on a fresh weight basis (Wang *et al.*, 1996). Similar results were obtained using a modified TEAC assay (Chun *et al.*, 2005). The antioxidant capacity of 'Hayward' kiwifruit consistently ranges from 6.0 to 9.2 μmol Trolox equivalent (TE)/g fresh weight, when assayed by the ORAC method (Wang *et al.*, 1996; Wu *et al.*, 2004a, 2004b); the antioxidant activity of 'Hort16A' kiwifruit, is reportedly 12.1 μmol TE/g fresh weight (USDA, 2007). However, other non-commercial genotypes of kiwifruit may have a higher antioxidant capacity, with *A. eriantha* and *A. latifolia* rating highest in a study of eight genotypes, when analysed using several assays (Du *et al.*, 2009). The antioxidant capacity of different kiwifruit cultivars correlates well with their total polyphenol content and vitamin C content (Hunter *et al.*, 2008; Du *et al.*, 2009). The antioxidant capacity of kiwifruit flesh is only average compared with that of other fruits, while the kiwifruit skin (undefined species and cultivar) has three times the activity of the flesh. However, only in the kiwiberry is the skin commonly eaten, and since this is a commercially minor player, the kiwifruit skin-derived antioxidants may be of little significance. Another factor that requires attention is the relevance of chemical measurements of antioxidants in biological systems.

It would be far more appropriate to test antioxidant activity in cellular or *in vivo* assays; of these, cell-based assays may be the most practical alternative (Hunter *et al.*, 2008). It should also be noted that antioxidants are very different in their physical chemistry and this affects their uptake into the cells. Vitamin C is very water soluble and is only taken into cells by transporter proteins (SVCT 1 and 2). At the blood–brain barrier and blood–retina barrier, it is predominantly transported as the oxidized form (dehydroascorbate) and accumulates in the brain and retina as ascorbate after reduction (Hosoya *et al.*, 2004). The SVCT proteins are present in many other tissues and organs, such as the human placenta (Choi *et al.*, 2012) and, in the context of Types 1 and 2 diabetes, are strongly inhibited by moderately high concentrations of glucose, allowing ROS to rise unchecked and cause damage in Types 1 and 2 diabetics where glucose regulation has been neglected (Hosoya *et al.*, 2004). Once in the cells, ascorbate increases the concentration of the antioxidant glutathione. kiwifruit polyphenols are also potentially able to give antioxidant support in diabetes and pre-diabetes (Margina *et al.*, 2012).

Many cancers common in the Western world, such as colon, prostate, cervix and breast cancers, have been related to dietary and environmental mutagens, including compounds in cooked meat, N-nitroso compounds and fungal toxins, and to dietary habits, including high meat and saturated fat consumption (Ferguson *et al.*, 2004). Alcohol and tobacco may also contribute to the risk. Dietary antimutagens may slow the progression towards cancer and there are numerous mechanisms by which they might act (Ferguson and Philpott, 2008). They could inhibit mutagen uptake by adsorbing mutagens and promoting faecal bulking, thereby reducing the transit time and contact time of the mutagen in the gut (Ferguson *et al.*, 2004). 'Hayward' kiwifruit contain relatively high amounts of lignin, which contributes to their insoluble dietary fibre content (Bunzel and Ralph, 2006). Lignin is hydrophobic and higher lignin content increases adsorption of heterocyclic aromatic amines, potential mutagenic compounds and potential carcinogens following absorption and metabolic destruction by the liver (Funk *et al.*, 2007).

Dietary antimutagens can also inhibit the endogenous formation of mutagens. Vitamins C and E, and plant polyphenols inhibit endogenous production of N-nitroso compounds from amino precursors (Bartsch *et al.*, 1988; Ferguson *et al.*, 2004; Ferguson and Philpott, 2008). Ethanolic extracts of kiwifruit also prevent the mutagenic activity of N-nitrosoamines *in vitro* (Ikken *et al.*, 1999). The prebiotic effect of dietary fibre may modulate colonic bacteria populations and this could decrease the production of mutagens following fermentation of dietary chemicals by gut bacteria (Ferguson *et al.*, 2004). Dietary antioxidants, including vitamins C and E, and polyphenols, might also protect DNA, either directly by scavenging radical species, or indirectly through promoting endogenous antioxidant activity (Ferguson *et al.*, 2004). Juice from green kiwifruit inhibited nitrosation reactions by scavenging nitrite, reportedly because of the vitamin C content and the potential breakdown product of vitamin C, 3-hydroxy-3 pyranone (Normington *et al.*, 1986).

Increased susceptibility to mutagenicity is positively correlated with the incidence of cancer. DNA repair has been used as a marker for susceptibility to mutagenicity (Collins and Gaivao, 2007). Lymphocyte DNA repair was increased in healthy volunteers supplemented with β-carotene or lycopene, but this might have been due to a direct antioxidant effect, rather than to any antimutagenic effect (Torbergsen and Collins, 2000). Kiwifruit contain β-carotene, and similar results have been achieved by feeding subjects with kiwifruit. In a pilot trial, two groups of six healthy subjects were given dietary and physical activity advice, and one group was given one kiwifruit daily ('Hayward'; L.R. Ferguson, personal communication) for every 30 kg of body weight (Rush *et al.*, 2002). Leukocytes from subjects eating kiwifruit daily were better able to repair DNA after a peroxide challenge *ex vivo*. However, this was not specifically shown to be due to carotenoid content, and the vitamin C, folate and fibre contents of the fruit may also have contributed to the improvement. In a randomized crossover trial, Collins and colleagues fed 14 subjects one, two or three kiwifruit (species and cultivar undefined) for three weeks (Collins *et al.*, 2003). Kiwifruit consumption was associated with a decrease in endogenous oxidation of DNA pyrimidines and purines, suggesting that a small intake of kiwifruit, as is manageable in a normal diet, may provide protection against DNA damage that might otherwise lead to mutations and cancer (Collins *et al.*, 2003).

Toxicity to cancer cells provides a further mechanism by which kiwifruit might assist prevention or treatment of cancer. Such cytotoxicity is usually determined *in vitro*. For example, selected fractions from the peel of green kiwifruit were shown to be cytotoxic to two human oral tumour cell lines (HSC-2 and HSG), but not to a 'normal' oral cell line (human gingival fibroblast) (Motohashi *et al.*, 2001). These findings indicate that extracts from *Actinidia* species may be toxic to cancer cells, but it should be noted that these effects have only been demonstrated *in vitro*, not in a clinical setting.

Many chemotherapeutic agents cause serious myelotoxic side effects, such as anemia, leucopenia, and thrombocytopenia (Gentile and Epremian, 1987). Various growth factors and some traditional medicines have been used to stimulate bone marrow proliferation in cancer patients (Danova and Aglietta, 1997; Hisha *et al.*, 1997). Methanolic extracts from kiwiberry stems stimulated both proliferation and colony formation of bone marrow (a measure of the potential of progenitor cells to develop into myeloid cells) in isolated bone marrow cells from mouse femurs (Takano *et al.*, 2003). The stimulation was largely due to (+)–catechin and (–)–epicatechin. These findings were confirmed *in vivo* using a mouse model of reduced bone marrow cellularity. Catechin, found in kiwiberry stems (Takano *et al.*, 2003) and also in small quantities in juice of green kiwifruit (Dawes and Keene, 1999), stimulated bone marrow proliferation and could, therefore, be effective in reducing the toxicity of chemotherapeutic agents.

Gut health and digestion

Traditionally, kiwifruit were used to aid digestion. Extracts of the fruit were found to contain a thiol protease, which was named actinidin (Boland and Hardman, 1972). Thiol proteases are found in other fruits, such as figs, papaya and pineapple. Recent work has used *in vitro* models to determine whether actinidin from green kiwifruit influences the digestion of proteins in the stomach and small intestine. The addition of an actinidin extract enhanced the digestion of alpha-, beta- and kappa-caseins in sodium caseinate by 37, 33 and 48%, respectively, under simulated gastric conditions, but did not enhance digestion of all food proteins more than that observed with pepsin alone (Kaur *et al.*, 2010a). Enhanced digestion of whey protein isolate, zein, gluten and gliadin was observed in the presence of a kiwifruit extract containing actinidin in the small intestine model, and different peptide patterns from those seen after digestion with pepsin and pancreatin alone were observed (Kaur *et al.*, 2010b). Using an *in vivo* model, Rutherford *et al.* (2011) demonstrated that actinidin markedly increased the gastric digestion of the main components of soy protein isolate, beef muscle protein, gelatine and gluten, and ileal degradability of some proteins in whey protein isolate, beef muscle and gluten. Enhanced digestion of proteins may be of benefit because some proteins are resistant to either gastric digestion or small intestinal digestion, or both, and poorly digested proteins may lead to reduced stomach emptying and a feeling of over-fullness (Rutherford *et al.*

2011). Furthermore, the proteins and peptides released following digestion may confer health benefits themselves, including having immunomodulatory, antihypertensive, osteoprotective and antilipemic activity (Möller *et al.*, 2008).

Kiwifruit are now widely recognized for their strong laxative properties, and these have been confirmed by controlled clinical studies. Healthy elderly subjects fed 'Hayward' kiwifruit daily for three weeks, one fruit per 30 kg of body weight, reported an improvement in laxation parameters, such as frequency and ease of defecation, stool bulk and stool softness (Rush *et al.*, 2002). It was concluded that kiwifruit, eaten in these realistic quantities, could be useful in maintaining regularity in elderly people who otherwise have no major bowel problems. A subsequent trial indicated that 'Hayward' kiwifruit could relieve chronic constipation (Chan *et al.*, 2007). Importantly, there appeared to be no adverse effects, with no participants reporting diarrhoea. Irritable bowel syndrome (IBS) is a common functional disorder of the gastrointestinal system; is characterized by abdominal pain, diarrhoea (IBS/D), constipation (IBS/C) and alternating diarrhoea and constipation (IBSC/A). After a four-week intervention with two kiwifruit per day, weekly defecation frequency significantly increased and colon transit time significantly decreased in the IBS/C group of participants who consumed kiwifruit. These findings suggest that kiwifruit consumption could improve bowel function and help adults diagnosed with IBS/C (Chang *et al.*, 2010).

The laxative effects of kiwifruit have been ascribed to their content of dietary fibre (Rush *et al.*, 2002; Chan *et al.*, 2007), approximately 2–3% fresh weight (Fourie and Hansman, 1992; Rupérez *et al.*, 1995) and about 10% of recommended daily intake (Ferguson and Ferguson, 2002). The cell walls of kiwifruit are unusual in that during ripening they swell much more than those of other fruits (Hallett *et al.*, 1992), and this may reflect an exceptionally high water-holding capacity, important for faecal bulking and enhancement of laxation (Rush *et al.*, 2002). Other kiwifruit components suggested as having laxative properties include the proteolytic enzyme actinidin and non-digestible oligosaccharides. Further studies are required to elucidate their role in the promotion of laxation by kiwifruit.

Gut health and immune function are strongly influenced by microbial colonization of the gastrointestinal tract. Gut microflora act as an effective barrier against opportunistic and pathogenic microorganisms (Cummings *et al.*, 2004). Stimulating proliferation of beneficial bacteria with prebiotic components in the diet (dietary fibre) could reduce colonization by harmful bacteria. Dietary fibre is the '... non-digestible food ingredient that beneficially affects the host by selectively stimulating the growth and/or activity of one or a limited number of bacteria in the colon' (Gibson and Roberfroid, 1995). Kiwifruit are a reasonable source of dietary fibre (Lund *et al.*, 1983), mostly cellulose, pectin polysaccharides and hemicelluloses, such as glucuronoxylans and xyloglucans (Martin-Cabrejas *et al.*, 1995). Kiwifruit polysaccharides have the potential to prevent adhesion of enteropathogens and enhance adhesion of probiotic bacteria to Caco-2 (colon epithelial-derived) cells (Ying *et al.*, 2007). Aqueous extracts and extracts from the edible flesh of 'Hayward' and 'Hort16A' also promoted the growth of lactic acid bacteria and reduced the growth of *Escherichia coli* using a faecal batch fermentation model (Molan *et al.*, 2007). Furthermore, using an *in vitro* batch model of gastric-ileal digestion and colonic fermentation, whole kiwifruit samples increased proliferation of *Bifidobacterium* spp. and the *Bacteroides-Prevotella-Porphyromonas* group, as well as enhanced adhesion of bifidobacteria to Caco-2 cells (Parkar *et al.*, 2012). With the proliferation of beneficial gut microbiota, such as bifidobacteria, suppression or competitive exclusion of pathogens may be achieved (Parkar *et al.*, 2012) and changes to gut flora populations may also modulate immune function, influencing immunoglobulin (Ig) concentrations and cytokine production (Lim *et al.*, 2005). The contribution of such modulation to clinical and health outcomes remains uncertain, but the dietary fibre in kiwifruit may contribute to the immunomodulatory activity found in various kiwifruit cultivars.

Immune health and protection from infection

The maintenance of well-regulated immune responses is essential for a healthy life, particularly when seasonal infections seriously affect well-being. The surveillance and response to these infections is provided by two independent but linked systems: innate and adaptive immunity (Skinner *et al.*, 2011). Innate immunity acts as a local or first line of defence and is non-specific. Adaptive immunity is highly specific and provides systemic as well as local defence. Adaptive immunity retains a specific 'memory' of previous infections and so retains the ability to respond rapidly to future attacks from the same or similar infectious agents. These two immune arms have specific roles but they link, in that the innate immune cells initiate and regulate the response of the cells of the adaptive system and together they provide defence against viruses, bacteria, fungi, protozoa and helminths, killing the infectious agent or the cells infected with them.

The importance of nutrition in maintaining immunity is recognized and accepted. Deficiencies in certain nutrients such as zinc, copper, iron, selenium and the vitamins A, C, E, B9, B12 and folate can cause impaired immune capability that may lead to greater susceptibility to infection and chronic diseases (Calder and Kew, 2002). Supplementation with one, or more commonly all, of the recognized nutrients may restore immune function. However, supplementation in adequately nourished persons may have little effect on the immune system in general and the balance between the adaptive and innate systems. The potential beneficial effects of phytochemicals in fruits and vegetables, such as carotenoids and polyphenols, on immune function have only recently been investigated (Murakami and Ohigashi, 2007; Mainardi *et al.*, 2009; Tibe *et al.*, 2012). It has also been demonstrated that consumption of fruit products can decrease the incidence of upper respiratory tract (URT) infections in at-risk populations, such as the elderly (Meydani *et al.*, 2004) and athletes undergoing intensive training (Nieman *et al.*, 2007). In addition to these small molecules, the fibres of cell wall carbohydrates, which escape digestion, can be substrates for microbial fermentation by colonic microbes to produce short chain fatty acids and thus provide direct immunomodulatory activity (Vinolo *et al.*, 2011) or through prebiotic effects on immune function (Macfarlane and Macfarlane, 2012). Kiwifruit contain a range of these small compounds and cell wall macromolecules, as well as vitamins and minerals that may promote a healthy immune system.

A first step by which kiwifruit may contribute to innate immune defences is by direct antimicrobial action. The antimicrobial activity of kiwifruit may protect the plant against infection, and this may confer protection to animals that eat the fruit or other parts of kiwifruit plants. For example, antifungal proteins isolated from green and gold kiwifruit are active against the plant fungal pathogens *Botrytis cinerea* and *Fusarium oxysporum*, respectively (Wang and Ng, 2002; Xia and Ng, 2004). Antimicrobial activity against human microbial pathogens has also been demonstrated. For example, extracts from the skin, pulp and seeds of *A. chinensis* demonstrated bacteriostatic activity against *Staphylococcus aureus* and *Streptococcus pyogenes* (Basile *et al.*, 1997). These *in vitro* studies suggest that ingestion of kiwifruit may provide a line of defence against infection in humans if antimicrobial activity in kiwifruit can influence pathogenic bacteria infections *in vivo*. Furthermore, the use of kiwifruit in the development of antimicrobial products may slow the progression of bacterial and fungal resistance to traditional drugs across populations.

Direct mucosal antimicrobial activity is provided by the host in the form of antimicrobial peptides, such as β-defensins. Recently, colonic epithelial cell studies *in vitro* have shown that human faecal microbial fermentation products from *in vitro* digested and fermented green kiwifruit increased the production of the gastrointestinal defensins, human beta defensin-1 and 2 (HBD-1, HBD-2) (Bentley-Hewitt *et al.*, 2012). The first line of defence in the gastrointestinal tract provided by the epithelial cells not only involves the production of antimicrobial peptides but also of a physical barrier underneath the mucus barrier, to prevent pathogenic invasion of the underlying tissues (Ashida *et al.*, 2012). Epithelial barrier integrity is maintained by tight junctions between the cells. Again using colonic epithelial cell cultures, the products of fermented green and gold kiwifruit were shown to contribute towards the maintenance of barrier integrity after an inflammatory insult (Skinner *et al.*, 2013). Other recent *in vitro* studies used human cells to demonstrate that kiwifruit extracts can modulate innate immune cell function. A water extract of a pasteurized puree from gold kiwifruit was shown to enhance natural killer cell activity in peripheral blood mononuclear cells isolated from human blood donors (Skinner, 2012).

Other *in vitro* work has focused on the effects of gold kiwifruit on adaptive immune responses using human blood cells. An initial screening of a selection of processed products prepared from gold kiwifruit for their ability to support measures of both innate and adaptive immune responses *in vitro* led to the selection of a pasteurized puree to test further (Skinner *et al.*, 2011). Blood from human volunteers previously vaccinated against tetanus was used to determine effects of a water extract of the pasteurized puree on an *in vitro* recall response to the tetanus vaccination. As expected, the addition of tetanus toxoid to the blood cell cultures increased IFNγ secretion compared with the control to which no antigen had been added. Co-incubation of tetanus toxoid with the kiwifruit sample resulted in a significant additional increase in IFNγ secretion compared with tetanus toxoid stimulation alone. These results suggested that the puree would support an appropriate Th1-type response, as there was no effect on IL-4 secretion (Skinner *et al.*, 2011).

The fruit of *A. polygama* have been used to treat inflammatory diseases such as rheumatoid arthritis. Alpha-linolenic acid isolated from *A. polygama* was shown to down-regulate the inflammatory mediators cycloxygenase-2 (COX-2) and inducible nitric oxide synthase (iNOS), as well as TNF-α in lipopolysaccharide (LPS)-activated mouse macrophage cell lines (Kim *et al.*, 2003). Dietary α-linolenic acid is a rare omega-3 fatty acid and was found to inhibit the breakdown of inhibitory factor κBα, thus blocking transcription factor nuclear factor (NF)-κB activation. NFκB activation leads to the transcription of inflammatory genes, including those that led to increased concentrations of TNF-α, IL-1β, COX-2 and IL-6 (Calder, 2002; Ren and Chung, 2007). The pro-inflammatory cytokines TNF-α and IL-1β were shown to be reduced by aqueous extracts of both green and gold kiwifruit in LPS-stimulated human monocytes and whole blood (Farr *et al.*, 2007), supporting the role of kiwifruit in decreasing the expression of pro-inflammatory cytokines in human cells in an inflammatory situation. Both aqueous and ethyl acetate extracts from gold and green kiwifruit were shown to exhibit anti-inflammatory effects in LPS-stimulated macrophages and in large intestinal epithelial cells isolated from $IL10^{-/-}$ (B6.129P2-$Il10^{tm1Cgn}$) mice that are used as a mouse model for a Crohn's disease-like inflammatory colitis (Edmunds *et al.*, 2011). The authors suggested that kiwifruit mediated its effects via inhibition of toll-like receptor 4 (TLR-4) signalling. Interestingly, IL-6, which is primarily produced during inflammation (Neurath and Finotto, 2011), was significantly reduced in large intestinal epithelial cells from wild-type mice in the absence of LPS simulation, indicating an alternative mechanism by which kiwifruit might act as an anti-inflammatory mediator (Edmunds *et al.*, 2011).

Research using cell lines clearly identifies the phenotype of the cell whose activity is being modulated. Some studies have used mixtures of cells, for example whole blood, peripheral blood mononuclear cells and cells isolated from the epithelium of the large intestine. Consequently, although the effects of kiwifruit have been demonstrated on cells belonging to both the innate and adaptive arms of the immune system, it is not yet clear which cell type directly mediates the effect, or by which mechanism. This will only be clarified when direct effects on purified cell types are shown, and the ways in which specific signalling pathways are modified have been established.

Encouraging results from *in vitro* studies provide preliminary data that require validation *in vivo* using animal models and human intervention trials. Some models have been used to demonstrate the positive immunomodulatory effects of feeding green and gold kiwifruit juice to untreated mice (Iwasawa *et al.*, 2010) or by giving aqueous extracts of *A. macrosperma* to mice with transplanted with tumours (Lu *et al.*, 2007a). Two different models have been used to demonstrate that consumption of kiwifruit can enhance adaptive immunity in a positive way. In the work of Shu and colleagues, a mouse cholera vaccination model (of 11 days' duration) and a diphtheria/tetanus toxoid vaccination model (of 29 days' duration) were used (Shu *et al.*, 2008). Mice were given either test diets (standard diet incorporated with either green or gold kiwifruit extract) or control diets *ad libitum* throughout the respective experimental periods. The kiwifruit extracts significantly enhanced specific intestinal mucosal and serum antibody responses to the vaccines and promoted IFN-γ and natural killer cell activity, with no improvement in proliferative cell response, phagocytic activity or IL-4 production. The overall results of the study demonstrated the ability of kiwifruit extracts to enhance markers of both innate and acquired immunity in the test model. In the studies of Hunter and colleagues, the same gold kiwifruit pasteurized puree mentioned above in the '*in vitro*' section was tested for its ability to modulate mouse immune responses to a model protein, ovalbumin (OVA), given with a suboptimal dose of a mucosal adjuvant (cholera toxin), to determine effects on a weak gut-associated immune response. When fed to mice at a dose of 200 μl per day for 20 days, the kiwifruit puree enhanced the production of OVA-specific antibody (IgG) in the serum, and the ability of lymphoid cells from the mesenteric lymph nodes, but not the spleen, to proliferate on re-exposure *in vitro* to the immunizing antigen, OVA (Hunter *et al.*, 2008). In a further feeding trial it was established that OVA-specific antibody isotypes representative of both a Th1 (IgG2b and 2c) and a Th2 (IgG1) immune response were produced. In addition, IL-5, a hallmark cytokine of Th2 responses, was produced by mesenteric lymph node cells from the mice when they were re-exposed to OVA *in vitro* (Skinner *et al.*, 2007). These results suggested that the gold kiwifruit puree enhanced both Th1 and Th2-type immune responses. Interestingly, a concentrated juice (40 °Brix) made from gold kiwifruit had no effect on this weak immune response when fed to the mice at a similar dose. The two processed kiwifruit products contained similar amounts of vitamins and minerals, but the puree contained a higher concentration of carotenoids and possibly a higher concentration of soluble polysaccharides (Hunter *et al.*, 2008). Together with some of the *in vitro* data, these results collectively provided supporting data to proceed to three human intervention trials.

Upper respiratory tract infections typified by symptoms of colds and influenza (flu) are some of the most common illnesses, with a US study showing adults suffer 1–3 incidents per year and children 4–6 per year (Monto and Sullivan, 1993). Unfortunately, there is no effective antiviral medication for these infections. Nutritional status is important for maintaining optimal immune function and recent evidence suggests that fruit, vegetable (Li and Werler, 2010; Roll *et al.*, 2011) and tea consumption (Rowe *et al.*, 2007) can reduce the symptoms and incidence of colds and flu. There are segments of the population that are particularly susceptible to these upper respiratory tract infections, for example pregnant women, athletes, students before examinations, people with chronic health conditions and also the young and the aged. Two randomized crossover studies demonstrated that consumption of gold kiwifruit reduced symptoms of colds and flu and their severity in two at-risk populations, adults aged 65 years or over (Hunter *et al.*, 2012) and pre-school children aged 2–5 years (Skinner, 2012).

In the adult study, 32 healthy adults aged $\geq$65 years consumed either two gold kiwifruit plus two freeze-dried kiwifruit or two freeze-dried bananas per day for four weeks with a two-week washout period. The primary outcome variables were the duration and severity of symptoms of colds and flu recorded by subjects using the Wisconsin Upper Respiratory Symptom Survey (WURSS21). Plasma vitamin C, carotenoid content and lipid peroxidation and red cell folate were measured. The rationale for providing banana as the 'placebo' was to provide an alternative fruit that had relatively similar energy content but lower nutritional value than gold kiwifruit with respect to most minerals, vitamins and carotenoids. In the first study with older adults, there were two symptoms, sore throat and head congestion, which people experienced for significantly fewer days when they ate kiwifruit rather than banana. They experienced a sore throat for a mean of 5.4 days when they ate banana compared with only 2.0 days when they ate kiwifruit ($P = 0.02$). For the head congestion this was reduced from 4.7 days with banana to less than 1 day (0.9 days) with kiwifruit ($P = 0.03$). The severity of head congestion was also significantly reduced ($P = 0.01$). Consumption of gold kiwifruit compared with banana led to significantly higher concentrations of plasma vitamin C (72.9 μM vs 53.7 μM, $P < 0.0001$), α-tocopherol (35.6 μM vs 32.5 μM, $P < 0.001$) and lutein/zeaxanthin (601 nM vs 498 nM, $P < 0.05$), and enhanced the concentration of red cell folate (611 nmol/l vs 559 nmol/l, $P = 0.03$) (Hunter *et al.*, 2012).

In the second study with 66 preschool children aged 2–5 years, the overall incidence of a cold or flu-like illness was considerably reduced when children ate kiwifruit compared with banana. The odds ratio of having a cold or flu-like illness was 0.55 when kiwifruit was consumed compared with banana, which means that the children had almost a 50% reduction in the chance of having a cold or flu when they ate kiwifruit. When individual symptoms were analysed, there were significant improvements in a number of symptoms. Children had a better appetite ($P = 0.01$), did not feel so unwell ($P = 0.04$), had more energy ($P = 0.01$), cried less ($P = 0.04$), and they suffered less from headaches ($P = 0.02$) and sore throats ($P = 0.04$) (Skinner, 2012). The reduction in these last two symptoms is in agreement with the reduced head congestion and sore throat symptoms observed in the study with the older adults (Hunter *et al.*, 2012).

Relative to many other fruits, there is a large amount of evidence available demonstrating a modulatory effect of kiwifruit on immune function. It appears that kiwifruit may be a valuable nutritional tool to limit the incidence and severity of common infections such as colds and flu, as well as to support appropriate immune responses to specific antigens and inflammation, when included as part of a healthy diet.

Cardiovascular disease

Modification of the diet by increased consumption of fruits and vegetables has been associated with reduced risk of cardiovascular disease (CVD) (Miura *et al.*, 2004; Lauretani *et al.*, 2008). The result is decreased mortality from cardiovascular disease and also reduced risk of ischemic stroke (Miura *et al.*, 2004). This has been ascribed to phytochemicals in fruits and vegetables, such as carotenoids, flavonoids, polyphenols and other antioxidants (Emsaillzadeh and Azadbakht, 2008; Lauretani *et al.*, 2008). Both green and gold kiwifruit have been shown to be effective in cardiovascular studies (Skinner *et al.*, 2011) and kiwifruit have been demonstrated to decrease platelet aggregation and lower blood pressure 24 hours after ingestion (Rabasseda, 2012), even in smokers (Karlsen *et al.*, 2012).

Crude ethanolic and aqueous extracts of kiwifruit (probably 'Hayward') have been shown to have antioxidant, antihypertensive, hypocholesterolemic and fibrinolytic activity using *in vitro* assays (Jung *et al.*, 2005). Because of the strong antioxidant activity of kiwifruit, it has been suggested that consumption of kiwifruit could reduce oxidation of cholesterol, thereby lowering the formation of atherosclerotic lesions. Antihypertensive activity was assessed by inhibition of angiotensin I-converting enzyme (ACE) which is important in the regulation of blood pressure (Jung *et al.*, 2005). There is increasing interest in the possibility of reducing hypertension by ACE inhibitors present in food, rather than increasing the use of drugs. Hypocholesterolemic activity was assayed by inhibition of 3-hydroxy-3-methyl-glutaryl coenzyme A (HMG-CoA), a key enzyme in the biosynthesis of cholesterol. Kiwifruit extracts were only weak inhibitors (Jung *et al.*, 2005).

Kiwifruit extracts also mildly stimulated fibrinolytic activity. Fibrinolysis is the process by which fibrin in blood clots is dissolved, and increased fibrinolytic activity is associated with a reduced risk of thromboembolic and cardiovascular disease. These results indicated that kiwifruit have potential as cardiovascular protectants through various mechanisms and although some effects observed were relatively minor, cumulatively they may represent a significant reduction in overall risk. Even so, it is important to acknowledge the limitation of *in vitro* studies, and the *in vivo* effects of kiwifruit consumption may well depend on the metabolism of kiwifruit compounds by gut microorganisms and the absorption of bioactive parent and metabolite compounds in the gut before uptake by the 'target' cells.

Adhesion and aggregation of platelets at the site of injury in atherosclerotic vessel walls are very important in the pathogenesis of CVD (Duttaroy, 2007). Drugs such as aspirin are commonly used to inhibit platelet aggregation, and treatment with aspirin reduces myocardial infarction, stroke and death. However, aspirin has only relatively weak antiplatelet activity and, moreover, can cause serious side effects in some patients. *In vitro* data suggest that natural products, including fruits, may also inhibit platelet activity by a different mechanism than aspirin, with tomatoes and kiwifruit having particularly high inhibitory activity (Duttaroy, 2007). This has been confirmed *in vivo* and the consumption of two or three kiwifruit per day by healthy volunteers was shown to inhibit platelet aggregation and also to raise plasma antioxidant concentrations and lower triglyceride concentrations (Duttaroy and Jorgensen, 2004). Regular consumption of kiwifruit could, therefore, be of great benefit in preventing and halting the processes that lead to CVD.

Obesity is a strong risk factor for diseases such as diabetes and CVD. Inhibitors of pancreatic lipase, a key enzyme for lipid absorption, could be useful for treatment of obesity. Ursolic acid and a coumaryl triterpene isolated from *A. arguta* roots are useful inhibitors of pancreatic lipase activity *in vitro* (Jang *et al.*, 2008). Although much less potent than Orlistat, the widely used anti-obesity drug, they warrant further study. Ursolic acid has also been identified in the peel of *A. deliciosa* 'Hayward' (Fiorentino *et al.* 2009), which may be more relevant in terms of consumption of fruit, for those who consume the peel, or for potential bioactive compounds that may be recovered from the side stream from kiwifruit processing.

Sleep disorders and the sleep-inducing effects of kiwifruit

Sleep is vital to maintain health and well-being because of its primary function of providing rest and restoring the body's energy (Krueger *et al.*, 2008). Sleep sustains physical and cognitive performance, the immune system, stable mood, productivity and quality of life

(Krueger *et al.*, 2008; Imeri and Opp, 2009). Disorders and deprivation of sleep impair cognitive and psychological functioning and worsen physical health (Brand and Kirov, 2011). Obesity and cardiovascular disease are deeply related to sleep disorders (Miller and Cappuccio, 2007). Although sleep is fundamental to maintain health, insomnia is currently a widespread health complaint, and has become a prevalent and disruptive problem in modern society (Doghramji, 2006; Erman, 2008). According to the numerous surveys conducted worldwide, approximately 10–15% of the adult population suffers from chronic insomnia, while an additional 25–35% has transient or occasional insomnia (National Institutes of Health Consensus Development Program, 2005). According to a recent survey conducted by the National Sleep Foundation in 2011, the majority (87%) of adults in the USA have at least one sleep problem for at least a few days a week (Swanson *et al.* 2011).

Natural sleep aids, which contain specific constituents or extracts of foods and medicinal plants, have recently become popular as alternative medications to prescription sleep drugs to improve sleep quality and avoid side effects (Meletis and Zabriskie, 2008). According to the 2002 National Health Interview Survey, more than 1.6 million American adults use alternative medicines to treat insomnia (Pearson *et al.*, 2006). Many insomnia patients prefer natural sleep aids because they think that they have fewer adverse effects and interactions and they do not require a medical prescription (Fernández-San-Martín *et al.*, 2011). Sleep-inducing effects of sedative plants, such as valerian (*Valeriana officinalis*), St. John's wort (*Hypericum perforatum*), passion flower (*Passiflora incarnata*), hops (*Humulus lupulus*) and kava-kava (*Piper methysticum*), have been widely investigated. Recently, several studies on sleep-inducing effects of kiwifruit have been reported.

Lin *et al.* (2011) at the Taipei Medical University reported that administration of two green kiwifruit (*Actinidia deliciosa*, ZESPRI® GREEN Kiwifruit, 100 g each), one hour before bedtime for four weeks, resulted in improved sleep onset, duration and efficiency in human adults. This clinical result has been derived from the Chinese version of Pittsburgh Sleep Quality Index (CPSQI), sleep diary and actigraphy for 24 subjects (two males and 22 females, between 20 and 55 years of age). According to the results, total sleep time (13.4%) and sleep efficiency (5.4%) were significantly increased. Significant decreases in waking time after sleep onset (28.9%) and sleep latency (35.4%) were also observed. They have suggested three possible mechanisms explaining the sleep-inducing effects of kiwifruit. One possible mechanism is the abundance of antioxidants (flavonoids, anthocyanins and carotenoids) in kiwifruit that may prevent damage from ROS, which reportedly increase in concentration with sleep disorders and emotional stress (Tsaluehidu *et al.*, 2008). The second is a high serotonin content in kiwifruit (pulp edge: 6.8 μg/g; pulp center: 3.0 μg/g, as per 30 mg of kiwifruit) (Feldman and Lee, 1985). Deficiency of serotonin may cause insomnia (Wyatt *et al.*, 1970). Finally, the folate intake from the raw kiwifruit was suggested as a beneficial mechanism for improving sleep quality. Folate deficiency has been reported to induce insomnia and restless leg syndrome (Kelly, 1998). Generally, the kiwifruit contains 0.23 μg/g of folate (Wyatt *et al.*, 1970). In the study of Lin *et al.* (2011), experimental evidence for these possible mechanisms of sleep-inducing effects of kiwifruit was not provided.

Sleep-inducing effects of green (*Actinidia deliciosa*) and gold (*Actinidia chinensis*) kiwifruit peels have been reported (Yang *et al.*, 2013). The sleep-inducing effects of green (GRPE) and gold (GOPE) kiwifruit peel ethanol extracts were evaluated using a pentobarbital-induced sleep test in mice. The oral administration (125, 250, 500, and 1000 mg/kg) of GRPE and GOPE produced a dose-dependent decrease in sleep latency and an increase in sleep duration. The hypnotic effects of GRPE and GOPE (1000 mg/kg, p.o.)

were significantly inhibited by flumazenil (8 mg/kg, i.p.), which is a specific GABAA-BZD receptor antagonist. Among solvent fractions (ethyl acetate, butanol and H_2O) of GRPE and GOPE, only their ethyl acetate fractions (250 mg/kg, p.o.) with the highest flavonoid content showed significant sleep-inducing activity. In addition, their hypnotic effects were also blocked by flumazenil like GRPE and GOPE. These results suggest that the sleep-inducing effects of GRPE and GOPE may be attributed to the positive allosteric modulation of their flavonoids. The peel of kiwifruit is a particularly good source of flavonoids (Fiorentino *et al.*, 2009). Flavonoids, which modulate the γ-aminobutyric acid type A-benzodiazepine (GABAA-BZD) receptor, have been known as one of the major sedative-hypnotic agents of medicinal plants (Johnston, 2005). The kiwifruit peel contains various flavonoids, such as naringenin, quercetin, tricin, kaempferol, epicatechin, catechin, gallo-catechin and rutin (Fiorentino *et al.*, 2009). The flavonoid composition of green kiwifruit peel is very similar to that of the well-known sedative herb, hops (*Humulus lupulus*) (Zanoli and Zavatti, 2008). According to the results of Fernández *et al.* (2006), rutin potentiated pentobarbital-induced sleep in mice. Kaempferol (Grundmann *et al.*, 2009) and (−)–epigallocatechin gallate (Vignes *et al.*, 2006) also produced anxiolytic properties through the positive allosteric modulation of GABAA receptors. One million metric tonnes of kiwifruit are processed annually worldwide, producing about 30% (wet weight) of by-product, such as pulp and peel (Djilas *et al.*, 2009). Therefore, the peel of kiwifruit may have potential as a good resource for development of natural sleep aids.

In the studies of Lin *et al.* (2011) and Lin *et al.* Yang *et al.* (2013), raw kiwifruit and its peel showed sleep-inducing effects in a human clinical trial and an animal model, respectively. Based on these results, kiwifruit is a potential resource for development of natural sleep aids. Further studies are needed to evaluate the effects on sleep architecture and precise mechanisms of flavonoid constituents or other active compounds in kiwifruits.

Other health benefits from kiwifruit

In addition to the 'mainstream' health targets commonly used to define the benefits from functional foods and fruits, a small number of studies demonstrate more novel bioactive effects of kiwifruit. These often arise from the use of less common *Actinidia* species or of fractions from commercial kiwifruit cultivars. For example, polysaccharide extracts from *A. chinensis* stimulated proliferation of keratinocyte and fibroblast cells *in vitro*, as well as stimulating collagen production from stable dermal equivalents (Deters *et al.*, 2005); this finding led to the conclusion that kiwifruit could have potential as pharmacologically active agents in dermatology.

Less convincing is the suggestion that kiwifruit juice (species and cultivar not specified, but probably 'Hayward') might be useful in dislodging meat bolus obstructions (Thomas *et al.*, 2004), which had previously been suggested to be a result of the proteolytic enzyme (actinidin) content (Karanjia and Rees, 1993). Whether kiwifruit juice could actually be used clinically for removal of bolus obstructions does not seem to have been tested.

Some of the proteins and polypeptides in kiwifruit may have specific biological effects and could have pharmacological benefits. A good example is kissper, a small, cysteine-rich peptide, present in large amounts in ripe kiwifruit (probably 'Hayward') (Ciardiello *et al.*, 2008). When ingested, kissper may affect gastrointestinal physiology and, it is conjectured, might have potential for treatment of disorders such as cystic fibrosis, in which ion transport mechanisms are disturbed. Further research and clinical trials are required to determine whether the many bioactive compounds in kiwifruit have any real effects when the fruit are consumed.

Kiwifruit allergies and other detrimental health effects

As kiwifruit consumption has become more common around the world, so too has the incidence of kiwifruit allergy. Allergic responses to kiwifruit range from localized oral allergy syndrome to life-threatening anaphylaxis (Lucas *et al.*, 2003), and several clinical subgroups have been established (Alemán *et al.*, 2004). The allergenicity of kiwifruit is not in doubt but it is currently ill-defined and poorly understood (Lucas *et al.*, 2003; Lucas and Atkinson, 2008). Three of the allergens identified, actinidin, thaumatin and kiwellin, are proteins often present in kiwifruit in large amounts and their identification as allergens may simply reflect this abundance, rather than their allergenicity *per se* (Lucas and Atkinson, 2008). Actinidin content and protease activity vary in different kiwifruit. Fruit of some *A. arguta* cultivars contain more actinidin than 'Hayward', whereas actinidin was not detectable or was barely detectable in two *A. rufa* selections (Nishiyama *et al.*, 2004) and some cultivars of *A. chinensis*, including 'Hort16A' (Bublin *et al.*, 2004). Further work on kiwifruit allergens is required, as future commercial kiwifruit cultivars should preferably be less allergenic than the cultivars now available.

kiwifruit contain small amounts of oxalate. This is commonly associated with the 'catch' in the throat sometimes experienced when fresh kiwifruit and, especially, processed kiwifruit products are eaten (Perera *et al.*, 1990; Walker and Prescott, 2003). Oxalate is present in cultivars of *A. deliciosa* (e.g. 'Hayward') as insoluble, fine needle-like, calcium oxalate raphide crystals (Perera *et al.*, 1990), but in other *Actinidia* species the crystal shape varies (Watanabe and Takahashi, 1998). The fruit of *A. chinensis* contain 18–45 mg/100 g fresh weight oxalate, similar to amounts in the mature fruit of *A. deliciosa* and *A. eriantha* (Rassam and Laign, 2005). High dietary intake of oxalate may be a major risk factor for calcium oxalate kidney stone formation (Siener *et al.*, 2006) and oxalate consumption also has an antinutritional effect, reducing the bioavailability of Ca^{2+}, Mg^{2+} and Fe^{2+} (Massey, 2003). Normal dietary intake of oxalate is estimated to be 50–200 mg per day and other foods, such as spinach, have considerably higher amounts of total oxalate (1959 mg/100 g fresh weight) than kiwifruit (Siener *et al.*, 2006). Although kiwifruit contain oxalate raphide crystals, the amount is probably not of concern to healthy individuals who maintain a well-balanced diet; the effect of raphides on kiwifruit palatability is of more significance.

FEIJOA

Feijoa sellowiana Berg. (syn *Acca sellowiana*; feijoa) fruit have a green skin with a creamy coloured flesh and translucent seed pulp; they are highly flavored (Thorp, 2008). Only the flesh and seed pulp are typically consumed. Other common names for feijoa include 'goiabeira serrana' (Brazil), 'guayabo del pais' (Uruguay) and pineapple guava (mainland USA) (Schotsmans *et al.*, 2011). Originating from Brazil and Uruguay, feijoa were introduced to Europe around the start of the twentieth century (Thorp and Bieleski, 2002), were distributed widely and are now primarily cultivated commercially in New Zealand and California (Manabe and Isobe, 2005). In addition to consumption *in natura,* feijoa have traditionally been used for a number of purposes, including for medicinal purposes (particularly for the digestive system), in candy and confections, alcoholic beverages and fresh juice, for flower consumption, as firewood, for use in tools, poles and handicrafts, as well as bait for the insect *Anastrepha fraterculus* (dos Santos *et al.*, 2009). Because of its unique flavour and aroma, the chemical composition and health properties of feijoa have received some attention.

As is commonly observed, the nutritional content of feijoa fruit varies depending on the geographical origin (Romero-Rodriguez *et al.*, 1994) and different genotypes (Beyhan *et al.*, 2011). Nevertheless, nutritional analyses of feijoa indicate that the fruit may be a useful source of fibre (approximately 5%), vitamin C (approximately 16–25 mg/100 g FW), potassium (approximately 90–130 mg/100 g FW) and organic acids (Visser and Burrows, 1983; Romero-Rodriguez *et al.*, 1992, 1994). Feijoa also contain phosphorus, calcium, magnesium, iron, manganese, zinc and copper (Visser and Burrows, 1983; Romero-Rodriguez *et al.*, 1994; Beyhan *et al.*, 2011). The seeds of feijoa contain a number of fatty acids, particularly mono- and polyunsaturated fatty acids, with linoleic acid being the predominant fatty acid (Andrade *et al.*, 2012). It was suggested that fatty acids, such as those present in feijoa seeds, might have anti-inflammatory, antioxidant and antidiabetic effects (Andrade *et al.*, 2012); however, it is not known whether the fatty acids would be available upon consumption of the seeds.

Feijoa fruit contain a large number of phytochemicals; many are the volatile compounds that give feijoa its unique aroma and flavour. In total, 85 volatile compounds have been identified in feijoa fruit, including 29 hydrocarbons, 20 alcohols, 18 esters, eight aldehydes, seven ketones, linalool oxide, mint sulfide, and flavone (Binder and Flath, 1989). However, the major volatile constituents contributing to the aroma of feijoa, confirmed by GC/MS headspace analysis, are methyl benzoate, ethyl benzoate and ethyl butanoate (Shaw *et al.*, 1983). Although the leaves of the feijoa have been analysed for polyphenolic content (Ruberto and Tringali, 2004; Lapcik *et al.*, 2005; El-Shenawy *et al.*, 2008), the polyphenolic and carotenoid content of feijoa fruit, particularly the flesh, is less defined. Feijoa fruit are purported to be high in phenolic content (Weston, 2010). An ethanolic extract from fresh feijoa fruit was reported to have 17.68 μg/g gallic acid equivalent of total polyphenols (Beyhan *et al.*, 2010), although others have reported much greater polyphenolic content (e.g. 59 mg/100 g; Isobe *et al.*, 2003, in Weston, 2010). The polyphenolic content of feijoa fruit is known to include catechins, leucoanthocyanins, flavonols, proanthocyanidins, naphthoquinones and tannins (Salvo *et al.*, 1987, in Rossi *et al.*, 2007). Furthermore, an active component in the acetonic extract from feijoa fruit has been shown to be flavone (Bontempo *et al.*, 2007). Although the peel is seldom eaten, its content has been analysed in detail; it contains many of the same constituents as the fruit flesh, but also is rich in terpenes (Shaw *et al.*, 1989; Fernandez *et al.*, 2004), including β-caryophylline, ledene, α-humulene, β-elemene and δ-cadinene. As with the incomplete information available about the phytochemical composition of feijoa, research into the health properties of feijoa fruit is in its infancy compared with that of more common fruits, such as kiwifruit. To date, indications of health benefits of feijoa have largely been provided using *in vitro* cell-based assays, although a small number of animal studies are emerging. No human intervention trials using feijoa as a functional component have been reported. A review of the potential health benefits of feijoa fruit is provided here.

Potential protection from infection

Initial studies into feijoa examined the efficacy of extracts on inhibiting the proliferation of a range of bacteria. A whole-fruit extract inhibited proliferation of a range of Gram-negative and Gram-positive bacteria *in vitro*, but was found to be only bactericidal against Gram-negative bacteria (Vuotto *et al.*, 2000). The thick peptidoglycan layer present in Gram-positive bacteria may provide some resistance against the antibacterial activity of feijoa extracts. Extracts from the skin, pulp and seeds of the fruit demonstrated bacteriostatic activity, halting growth and replication across eight bacterial strains, and having greater efficacy than *A. chinensis* extracts (Basile *et al.*, 1997). Subsequently, fractionation of the skin extract demonstrated antibacterial activity against *Staphylococcus epidermis*, *Escherichia coli* and *Pseudomonas aeruginosa*, as well as antifungal activity against *Candida albicans* and *C. glabrata*, and weak anti-HIV activity from one fraction (Motohashi *et al.*, 2000).

Activity of feijoa extracts in immune responses to infection has gained attention, at least using *in vitro* model systems. Respiratory burst is part of the immune system's 'first line of defence', targeting reactive oxygen species (ROS) to invading pathogens, thereby causing oxidative damage and inactivating or destroying the pathogen. However, loss of regulation of this process can damage tissues and cause inflammation, through the action of excessive ROS production. Feijoa fruit extract has been shown to provide antioxidant activity in an *in vitro* model of oxidative burst, using both whole blood (Ielpo *et al.* 2000) and isolated polymorphonuclear leukocyte preparations (Vuotto *et al.*, 2000). The action of the feijoa extract was suggested to be through direct scavenging of ROS or functional changes that cells undergo in the presence of the extract (Ielpo *et al.*, 2000). Whilst encouraging, these studies were carried out using *in vitro* systems and the efficacy of feijoa in reducing or preventing infection or reducing tissue damage as a result of infection remains to be determined *in vivo*.

Potential antioxidant activity

In addition to the antioxidant activity described in response to respiratory burst, further evidence is available that suggests feijoa may provide protection from oxidative stress using a number of models. In the ferric reducing antioxidant power assay, methanolic extracts of dried and fresh feijoa fruit exhibited reducing power, demonstrating the electron donor properties of the extracts, which would neutralize ROS by forming stable products (Beyhan *et al.*, 2010). The same extracts were also shown to chelate ferrous ions and scavenge ROS in the DPPH free radical scavenging assay, both actions that may be physiologically important for reducing lipid peroxidation (Beyhan *et al.*, 2010), if the active components of the extract were stable through digestion and were bioavailable.

Some evidence is available to suggest that feijoa may enhance antioxidant capacity *in vivo*. For example, 3,4-methylenedioxymethamphetamine (MDMA/ecstasy) induces formation of ROS, oxidative stress and hepatotoxicity, but in mice administered a methanolic extract of feijoa peel, the effects of MDMA were reduced (Karami *et al.*, 2011). Specifically, the concentration of two enzymes used as markers for hepatotoxicity (aspartate aminotransferase and alanine aminotransferase) were reduced, and the concentration of glutathione (GSH, an endogenous antioxidant compound) increased in a dose-dependent manner, compared with that in the positive control mice. Subsequently, Keles *et al.* (2012) demonstrated enhanced antioxidant defence in rats fed feijoa fruit extracts (50 mg/kg) daily for four weeks. A number of extracts were used in this study, including methanol, *n*-hexane, chloroform, ethyl acetate, *n*-butanol and an aqueous extract, and their effects on blood and tissue concentrations of several markers of antioxidant capacity and oxidative stress were examined. The ethyl acetate extract was most efficacious, significantly decreasing blood malondialdehyde concentrations (a marker for lipid peroxidation), and significantly increasing GSH and plasma antioxidant activity compared with the positive controls and the other extracts. In addition, histopathological examination of the vital organs showed that the extracts did not induce changes or damage (Keles *et al.*, 2012). Furthermore, antioxidant activity from a methanolic feijoa fruit extract has been suggested to provide some protection to the gastric mucosa (Leuzzi *et al.*, 2009). Using a rat model, the extract (2 g/kg) prevented the development of ethanol-induced gastric ulcers, with the gastric protection purported to arise from a synergistic effect of enhanced antioxidant activity, and stimulation of prostaglandin E_2 associated with an increase in mucus secretion. This study may suggest that the antioxidant effect of feijoa fruit would be useful in protecting against oxidative stress in the gastrointestinal tract, and thereby confer protection irrespective of bioavailability of active compounds in blood and tissues. Nevertheless, enhancement of antioxidant defence in plasma and tissues, as demonstrated by Keles *et al.* (2012), does suggest that bioactive compounds present in feijoa have an effect *in vivo* past the gut, but this requires confirmation in a human intervention trial.

Potential anti-inflammatory activity

Inflammation plays an important role in the immune response, to inactivate or destroy invading organisms, remove irritants and set the stage for tissue repair (Guzik *et al.*, 2003). However, prolonged inflammation can contribute to ageing and chronic degenerative diseases (Salminen *et al.*, 2008). Therefore, functional or plant-based foods with anti-inflammatory properties could be useful in the treatment of low-grade inflammation as an alternative to traditional non-steroidal anti-inflammatory drugs, which often have serious side effects when taken long term. Rossi *et al.* (2007) found that an acetonic extract of feijoa fruit displayed potential anti-inflammatory activity, by inhibiting nitrite production and expression of inducible nitric oxide synthase in a monocyte/macrophage cell line, J774. The mechanism of action was through attenuation of nuclear factor κB and/or mitogen-activated protein kinases, signal molecules that stimulate an inflammatory response. Upon fractionation of the acetonic extract, two fractions exhibited activity and their main components were flavone and stearic acid (Rossi *et al.*, 2007). Although these results are promising, particularly given that the mechanism of action was defined, the potential anti-inflammatory activity of feijoa remains to be proven *in vivo*.

Potential anticancer effect

To date, the evidence to suggest that feijoa fruit could be beneficial in preventing or slowing the development of cancer is weak because its efficacy has only been demonstrated *in vitro*. A crude acetonic feijoa fruit extract inhibited proliferation of selected cancer cell lines, including HeLa, MCF-7, SKBR-3, MDA-MB231 and NB4 cells, which was accompanied by cell cycle and apoptosis modulation (Bontempo *et al.*, 2007). The LnCap prostate cancer cell line was minimally sensitive to the feijoa extract, suggesting that the extract was not displaying general toxicity but rather cancer cell selectivity. Interestingly, Bontempo *et al.* (2007) identified flavone as the compound responsible for the apoptotic action, but although purified flavone also demonstrated apoptotic action, it did not exert the same effect on cell cycle and differentiation, suggesting other components in the crude extract accounted for these activities. However, whilst interesting, this work is preliminary and much more evidence regarding the anticancer potential of feijoa, using animal models, epidemiological data and human intervention trials, is required before any firm conclusions can be drawn.

The pool of evidence that feijoa has a number of potential health benefits continues to grow. Interest in feijoa is probably due to its unusual flavour and aroma, suggesting a unique complement of phytochemical bioactive compounds, although the purported health benefits to date are similar to those identified for many other, more common fruits. However, further research into the bioactivity of feijoa may uncover new health benefits and, at least, will add to the evidence gathered so far. In particular, the use of feijoa in human studies is urgently required before any firm claims can be made on its benefit to human health.

TAMARILLO

Tamarillo (*Cyphomandra betacea* Sendt.), also known as tree tomato, is a subtropical fruit thought to be native to the Andes region of Peru, Chile, Ecuador and Bolivia, and is cultivated in a number of South American countries including Argentina, Brazil, Colombia and Venezuela, as well as New Zealand (de Rosso and Mercadante, 2007). Tamarillo belongs to the Solanaceae family, producing red, yellow/orange or purple fruit (Boyes and Strubi, 1997) that are ovoid, smooth skinned and have small seeds (Romero-Rodriguez *et al.*, 1994). The flavour of tamarillo flesh is said to be reminiscent of tomato (Rodriguez-Amaya *et al.*, 1983) and can be astringent, although astringency may be dependent on variety (Boyes and Strubi, 1997). The fruit are eaten fresh, blended with milk or water, cooked in stews and sauces, and incorporated into desserts and salads (Vasco *et al.*, 2009). Traditionally, the tamarillo fruit has been used in Ecuador as a therapeutic treatment for tonsillitis, high cholesterol and stomach pain (Tene *et al.*, 2007). However, to date, few publications are available that consider the potential health benefits of tamarillo, either *in vitro* or *in vivo*. Nevertheless, tamarillo has an interesting phytochemical content that may indicate areas of health research in the future; the composition of tamarillo and the few papers addressing health aspects of tamarillo are reviewed here.

Tamarillo has an interesting flavour and its nutritional profile includes fibre, reportedly 129 g/kg (Schmeda-Hirschmann *et al.*, 2005), or approximately 3–6% (Romero-Rodriguez *et al.*, 1994), vitamin A (approximately 2475 IU/100 g edible portion) (Rodriguez-Amaya *et al.*, 1983), vitamin C (approx. 20–35 mg/100 g FW) (Romero-Rodriguez *et al.*, 1994) and minerals such as sodium, calcium, iron, copper, zinc, magnesium, manganese and phosphate (Romero-Rodriguez *et al.*, 1994; Schmeda-Hirschmann *et al.*, 2005; Vasco *et al.*, 2009). Most noteworthy is its potassium content, which reportedly ranges from approximately 290–440 mg/100 g FW (Romero-Rodriguez *et al.*, 1994; Vasco *et al.*, 2009) to 800 mg/100 g FW (Schmeda-Hirschmann *et al.*, 2005), similar to or exceeding that of banana (USDA, 2011). Only minor differences in the macro- and micronutrient contents were observed between red and yellow tamarillo (Romero-Rodriguez *et al.*, 1994; Vasco *et al.*, 2009).

The estimated total soluble phenolic contents of whole tamarillo were 125 and 187 mg gallic acid equivalents/100 g FW for a golden and a purple-red variety, respectively (Vasco *et al.*, 2009). Partitioning between the peel, pulp and seed-jelly revealed that the greatest contributor to the phenolic content by weight is the peel, followed by the seed-jelly and pulp. Analysis indicated that five hydroxycinnamic acid derivatives and three flavonol glycosides were present in both golden/yellow and purple-red tamarillos. However, anthocyanins were only detected in the purple-red variety (Vasco *et al.*, 2009). Although there is some discrepancy in the reported anthocyanin species present in tamarillo, the most sensitive analysis used to date reported three anthocyanins were present in a whole-fruit extract: delphinidin 3-rutinoside, pelargonidin 3-glucoside-5-rhamnoside and cyanidin

3-rutinoside, representing 62.0, 31.5 and 6.5% of the total anthocyanin content, respectively (de Rosso and Mercadante, 2007).

Tamarillos also contain carotenoids, and studies indicate that they are present in both golden/yellow and purple-red tamarillos (Vasco *et al.*, 2009). The total carotenoid content in tamarillo is between 3.1 and 5.9 mg/100 g (Rodriguez-Amaya *et al.*, 1983; de Rosso and Mercadante, 2007), and similar content is reported in the skin and pulp (Rodriguez-Amaya *et al.*, 1983). Detailed analysis indicates 20–26 carotenoids are present in tamarillo (de Rosso and Mercadante, 2007; Mertz *et al.*, 2010), with discrepancies probably caused by different provenance and/or variety of fruit, as well as analytical differences. However, the major carotenoids present include β-cryptoxanthin (45.3%), β-carotene (26.1%), zeaxanthin (5.1%) and antheraxanthin (4.0%) (de Rosso and Mercadante, 2007).

Tentative health benefits of tamarillo

Much less research is available reporting the potential health benefits of tamarillo than for kiwifruit or even feijoa. The antioxidant activity of tamarillo preparations have been determined using chemical-based methods (Vasco *et al.*, 2009; Ordonez *et al.*, 2010), but are considered low to moderate antioxidant activity compared with those of other fruits (Vasco *et al.*, 2009). Perhaps of slightly more physiological relevance, inhibition of LDL oxidation by a tamarillo extract was demonstrated *in vitro* (Kou *et al.*, 2009) and efficacy of tamarillo preparations in protecting against oxidative stress using cell models (PC12 and HepG2 cells) has also been demonstrated (Kou *et al.*, 2009; Ordonez *et al.*, 2010). Interestingly, proteins isolated from tamarillo have also been shown to have *in vitro* antioxidant, antimutagenic and antimicrobial activity (Ordonez *et al.*, 2006, 2011), but their description is beyond the scope of this chapter. Information is lacking regarding the efficacy of tamarillo extracts *in vivo*, using either animal or human subjects; therefore, any firm conclusions on the potential health benefit of tamarillo cannot be made. However, at this stage its use for protection from oxidative stress may warrant further investigation.

CONCLUSION

Health professionals, including nutritionists and dieticians have the responsibility for providing consumers with scientifically supported knowledge to help them make informed dietary decisions. The scientific information supporting the unique health benefits of kiwifruit, feijoa and tamarillo is growing rapidly, but evidence of health benefits must be validated by human intervention trials.

While in some instances it may be claimed that these subtropical fruits, particularly kiwifruit, and their processed products, are of value in preventing or moderating certain diseases, it is still early in the investigative process and several other potential benefits may be substantiated in the future. It seems likely that regular consumption may be a nutritious, pleasant and economic way of improving health and well-being. To obtain maximum benefit, consumers will need much more guidance on which varieties, what quantities and the state of ripeness required of these fruits, to mention a few key parameters that may well influence their impact on health. From a scientific viewpoint it is fascinating to see yet another example of where the barriers between nutrition and medicine are being dismantled, or at least reconsidered, with many insights available from historical documents.

REFERENCES

Ai X-P, Li Y-G, Zhang M-X. (1982) Chinese medicines against cancers. Heilong-jiang Science Technology Publishers, Heilong-Jiang Province.

Aitken AG, Hewett EW. 2011. FreshFacts: New Zealand Horticulture. http://www.freshfacts.co.nz/file/fresh-facts-(2011)pdf (last accessed 3 March 2013).

Alemán A, Sastre J, Quirce S, *et al.* (2004) Allergy to kiwi: A double-blind, placebo-controlled food challenge study in patients from a birch-free area. *Journal of Allergy and Clinical Immunology* **113**: 543–550.

Ames BN, Shigenaga MK, Hagen TM. (1993) Oxidants, antioxidants, and the degenerative diseases of aging. *Proceedings of the National Academy of Sciences USA* **90**: 7915–7922.

Andrade JMdM, Marin R, Apel MA, *et al.* (2012) Comparison of the fatty acid profiles of edible native fruit seeds from southern Brazil. *International Journal of Food Properties* **15**(4): 815–822.

Ashida H, Ogawa M, Kim M, *et al.* (2012) Bacteria and host interactions in the gut epithelial barrier. *Nature Chemical Biology* **8**, 36–45.

Bartsch H, Ohshima H, Pignatelli B. (1988) Inhibitors of endogenous nitrosation mechanisms and implications in human cancer prevention. *Mutation Research/Fundamental and Molecular Mechanisms of Mutagenesis* **202**: 307–324.

Basile A, Vuotto ML, Violante U, *et al.* (1997) Antibacterial activity in *Actinidia chinensis*, *Feijoa sellowiana* and *Aberia caffra*. *International Journal of Antimicrobial Agents* **8**: 199–203.

Bentley-Hewitt KL, Blatchford PA, Parkar SG, *et al.* (2012) Digested and fermented green kiwifruit increases human β-Defensin 1 and 2 production *in vitro*. *Plant Foods for Human Nutrition* **67**(3): 208–214.

Beyhan O, Elmastas M, Gedikli F. (2010) Total phenolic compounds and antioxidant capacity of leaf, dry fruit and fresh fruit of feijoa (Acca sellowiana, Myrtaceae). *Journal of Medicinal Plants Research* **4**(11): 1065–1072.

Beyhan O, Bozkurt MA, Boysal SC. (2011) Determination of macri-micro nutrient contents in dried fruit and leaves and some pomological characteristics of selected feijoa genotypes (*Feijoa sellowianai* Berg.) from Sakarya provinces in Turkey. *Journal of Animal and Plant Sciences* **21**(2): 251–255.

Binder RG, Flath RA. 1989. Volatile Components of Pineapple Guava. *Journal of Agricultural and Food Chemistry* **37**(3): 734–736.

Block G, Patterson B, Subar A. (1992) Fruit, vegetables, and cancer prevention: a review of the epidemiological evidence. *Nutrition and Cancer* **18**: 1–29.

Brand S, Kirov R. (2011) Sleep and its importance in adolescence and in common adolescent somatic and psychiatric conditions. *International Journal of General Medicine* **4**: 425–442.

Boland MJ, Hardman MJ. (1972) Kinetic studies on the thiol protease from *Actinidia chinensis*. *FEBS Letters* **27**: 282–4.

Bontempo P, Mita L, Miceli M, *et al.* (2007) Feijoa sellowiana derived natural Flavone exerts anti-cancer action displaying HDAC inhibitory activities. *International Journal of Biochemistry and Cell Biology* **39**: 1902–1914.

Boyes S, Strubi P. (1997) Organic acid and sugar composition of three New Zealand grown tamarillo varieties (Solanum betaceum (Cav)). *New Zealand Journal of Crop and Horticultural Science* **25**(1): 79–83.

Bublin M, Mari A, Ebner C, *et al.* (2004) IgE sensitization profiles toward green and gold kiwifruits differ among patients allergic to kiwifruit from 3 European countries. *Journal of Allergy and Clinical Immunology* **114**: 1169–1175.

Bunzel M, Ralph J. (2006) NMR characterization of lignins isolated from fruit and vegetable insoluble dietary fibre. *Journal of Agricultural and Food Chemistry* **54**: 8352–8361.

Calder PC. 2002. Dietary modification of inflammation with lipids. *Proceeding of the Nutrition Society* **61**: 345–358

Calder PC, Kew S. 2002. The immune system: a target for functional foods? *British Journal of Nutrition* **88** (Suppl 2): S165–S177.

Campbell H, Haggerty J. (2009) Kiwifruit – The hairy berry. Te Ara – the Encyclopedia of New Zealand, updated 13-Jul-12. http://www.teara.govt.nz/en/kiwifruit/page-1 (last accessed 20 March 2013).

Chan AO, Leung G, Tong T, Wong NY. (2007) Increasing dietary fibre intake in terms of kiwifruit improves constipation in Chinese patients. *World Journal of Gastroenterology* **13**: 4771–4775.

Chang CC, Lin YT, Lu YT, *et al.* (2010) Kiwifruit improves bowel function in patients with irritable bowel syndrome with constipation. *Asia Pacific Journal of Clinical Nutrition* **19**: 451–457.

Choi DY, Lee YJ, Hong JT, Lee HJ. (2012) Antioxidant properties of natural polyphenols and their therapeutic potentials for Alzheimer's disease. *Brain Research Bulletin* **87**(2–3): 144–153.

Chun OK, Kim D-O, Smith N, *et al.* 2005. Daily consumption of phenolics and total antioxidant capacity from fruit and vegetables in the American diet. *Journal of the Science of Food and Agriculture* **85**: 1715–1724.

Ciardiello MA, Meleleo D, Saviano G, *et al.* 2008. Kissper, a kiwifruit peptide with channel-like activity: Structural and functional features. *Journal of Peptide Science* **14**: 742–754.

Collins AR, Gaivao I. (2007) DNA base excision repair as a biomarker in molecular epidemiology studies. *Molecular Aspects of Medicine* **28**: 307–322.

Collins AR, Harrington V, Drew J, Melvin R. 2003. Nutritional modulation of DNA repair in a human intervention study. *Carcinogenesis* **24**: 511–515.

Cummings JH, Antoine J-M, Azpiroz F, *et al.* (2004) PASSCLAIM – Gut health and immunity. *European Journal of Nutrition* **43**(Suppl 2): II/118–II/173.

Danova M, Aglietta M. (1997) Cytokine receptors, growth factors and cell cycle in human bone marrow and peripheral blood hematopoietic progenitors. *Haematologica* **82**: 622–629.

Dawes HM, Keene JB. (1999) Phenolic composition of kiwifruit juice. *Journal of Agricultural and Food Chemistry* **47**: 2398–2403.

de Rosso VV, Mercadante AZ. (2007) HPLC-PDA-MS/MS of anthocyanins and carotenoids from dovyalis and tamarillo fruits. *Journal of Agricultural and Food Chemistry* **55**(22): 9135–9141.

Deters AM, Schröder KR, Hensel A. (2005) Kiwifruit (*Actinidia chinensis* L.) polysaccharides exert stimulating effects on cell proliferation via enhanced growth factor receptors, energy production, and collagen synthesis of human keratinocytes, fibroblasts, and skin equivalents. *Journal of Cellular Physiology* **202**: 717–722.

Djilas S, Canadanovic-Brunet J, Cetkovic G. (2009) By-products of fruits processing as a source of phytochemicals. *Chemical Industry and Chemical Engineering Quarterly* **15**: 191–202.

Doghramji K. (2006) The epidemiology and diagnosis of insomnia. *American Journal of Managed Care* **12**: 214–220.

dos Santos KL, Peroni N, Guries RP, Nodari RO. (2009) Traditional Knowledge and Management of Feijoa (Acca sellowiana) in Southern Brazil. *Economic Botany* **63**(2): 204–214.

Du G, Li M, Ma F, Liang D. (2009) Antioxidant capacity and the relationship with polyphenol and vitamin C in *Actinidia* fruits. *Food Chemistry* **113**: 557–562.

Duttaroy AK. (2007) Kiwifruits and cardiovascular health. *Acta Horticulturae* **753**: 819–824.

Duttaroy AK, Jørgensen A. (2004) Effects of kiwifruit consumption on platelet aggregation and plasma lipids in healthy human volunteers. *Platelets* **15**: 287–292.

Edmunds, S.J., Roy, N.C., Love, D.R., Laing, W.A., (2011) Kiwifruit extracts inhibit cytokine production by lipopolysaccharide-activated macrophages, and intestinal epithelial cells isolated from IL10 gene deficient mice. *Cellular Immunolology* **270**, 70–79.

El-Shenawy S, Marzouk MS, El-Dib RA, *et al.* (2008) Polyphenols and biological activities of *Feijoa sellowiana* leave and twigs. *Revista Latinoamericana de Quimica* **36**(3): 103–120.

Emsaillzadeh A, Azadbakht L. (2008) Dietary flavonoid intake and cardiovascular mortality. *British Journal of Nutrition* **100**: 695–697.

Erman MK. (2008) New perspectives in the diagnosis and management of insomnia. *CNS Spectrums* **13**(Suppl 17): 3.

Farr J, Hurst S, Skinner M, (2007) Anti-inflammatory effects of kiwifruit. *Asia Pacific Journal of Clinical Nutrition* **16**, s73.

Feldman JM, Lee EM. (1985) Serotonin content of foods: effect on urinary excretion of 5-hydroxyindoleaceti acid. *American Journal of Clinical Nutrition* **42**: 639–643.

Ferguson AR. (1990) The kiwifruit in China. In: *Kiwifruit: science and management* (Warrington IJ and Weston GC, eds), pp. 155–164, Ray Richards Publisher/New Zealand Society for Horticultural Science, Auckland.

Ferguson AR, Bollard EG. (1990) Domestication of the kiwifruit. In: *Kiwifruit: science and management* (Warrington IJ and Weston GC, eds.), pp. 165–246, Ray Richards Publisher/New Zealand Society for Horticultural Science, Auckland.

Ferguson AR, Ferguson LR. (2002) Are kiwifruit really good for you? *Acta Horticulturae* **610**: 131–138.

Ferguson LR, Philpott M. (2008) Nutrition and mutagenesis. *Annual Review in Nutrition* **28**: 313–329.

Ferguson LR, Philpott M, Karunasinghe N. (2004) Dietary cancer and prevention using antimutagens. *Toxicology* **198**: 147–159.

Fernandez X, Loiseau AM, Poulain S, *et al.* (2004) Chemical composition of the essential oil from feijoa (Feijoa sellowiana berg.) peel. *Journal of Essential Oil Research* **16**(3): 274–275.

Fernández SP, Wasowski C, Loscalzo LM, *et al.* (2006) Central nervous system depressant action of flavonoid glycosides. *European Journal of Pharmacology* **539**: 168–176.

Fernández-San-Martín MI, Masa-Font R, Palacios-Soler L, *et al.* (2011) Effectiveness of valerian on insomnia: a meta-analysis of randomized placebo-controlled trials. *Sleep Medicine* **11**: 505–511.

Fiorentino A, D'Abrosca B, Pacifico S, *et al.* (2009) Identification and assessment of antioxidant capacity of phytochemicals from Kiwifruits. *Journal of Agricultural and Food Chemistry* **57**: 4148–4155.

Fourie PC, Hansmann CF. (1992) Fruit composition of four South African-grown kiwifruit cultivars. *New Zealand Journal of Crop and Horticultural Science* **20**: 449–452.

Frei B, Birlouez-Aragon I, Lykkesfeldt J. (2012) Authors' perspective: what is the optimum intake of vitamin C in humans? *Critical Reviews in Food Science and Nutrition* **52**: 815–829.

Funk C, Braune A, Grabber J.H, *et al.* (2007) Model studies of lignified fibre fermentation by human fecal microbiota and its impact on heterocyclic aromatic amine adsorption. *Mutation Research/Fundamental and Molecular Mechanisms of Mutagenesis* **624**: 41–48.

Gentile P, Epremian BE. (1987) Approaches to ablating the myelotoxicity of chemotherapy. *Critical Reviews in Oncolology and Hematology* **7**(1): 71–87.

Gibson GR, Roberfroid MB. (1995) Dietary modulation of the human colonic microbiota: introducing the concept of prebiotics. *Journal of Nutrition* **125**: 1401–1412.

Grundmann O, Nakajima J, Kamata K, *et al.* (2009) Kaempferol from the leaves of Apocynum venetum possesses anxiolytic activities in the elevated plus maze test in mice. *Phytomedicine* **16**: 295–302.

Guzik TJ, Korbut R, Adamek-Guzik T. (2003) Nitric oxide and superoxide in inflammation and immune regulation. *Journal of Physiology and Pharmacology* **54**(4): 469–487.

Hallett IC, MacRae EA, Wegrzyn TF. (1992) Changes in kiwifruit cell wall ultrastructure and cell packing during postharvest ripening. *International Journal of Plant Science* **153**: 49–60.

Heatherbell DA. 1975. Identification and quantitative analysis of sugars and non-volatile organic acids in Chinese gooseberry fruit (*Actinidia chinensis* Planch.). *Journal of the Science of Food and Agriculture* **26**: 815–20.

Hisha H, Yamada H, Sakurai MH, *et al.* (1997) Isolation and identification of hematopoietic stem cell-stimulating substances from kampo (Japanese herbal) medicine, Juzen-Taiho-To. *Blood* **90**: 1022–1030.

Hosoya K, Minamizono A, Katayama K, *et al.* (2004) Vitamin C transport in oxidized form across the rat blood–retinal barrier. *Investigative Ophthalmology and Visual Science* **45**: 1232–1239.

Hunter DC, Denis M, Parlane NA, *et al.* (2008) Feeding ZESPRI™ GOLD Kiwifruit puree to mice enhances serum immunoglobulins specific for ovalbumin and stimulates ovalbumin-specific mesenteric lymph node cell proliferation in response to orally administered ovalbumin. *Nutrition Research* **28**: 251–257.

Hunter DC, Skinner MA, Wolber FM, *et al.* (2012) Consumption of gold kiwifruit reduces severity and duration of selected upper respiratory tract infection symptoms and increases plasma vitamin C concentration in healthy older adults. *British Journal of Nutrition* **108**: 1235–1245.

Ielpo MTL, Basile A, Miranda R, *et al.* 2000. Immunopharmacological properties of flavonoids. *Fitoterapia* **71**: S101–S109.

Ikken Y, Morales P, Martinez A, *et al.* 1999. Antimutagenic effect of fruit and vegetable ethanolic extracts against N-nitrosamines evaluated by the Ames test. *Journal of Agricultural and Food Chemistry* **47**: 3257–3264.

Imeri L, Opp MR. (2009) How (and why) the immune system makes us sleep. *Nature Reviews Neuroscience* **10**: 199–210.

Isobe Y, Kase Y, Narita M, Komiya T. (2003) Antioxidative activity of tropical fruit, *Feijoa sellowiana* Berg. *Nippon Kasei Gakkaishi* **54**(11): 945–949.

Iwasawa H, Morita E, Ueda H, Yamazaki M. (2010) Influence of kiwifruit on immunity and its anti-oxidant effects in mice. *Food Science and Technology Research* **16**, 135–142.

Jang DS, Lee GY, Kim J, *et al.* (2008) A new pancreatic lipase inhibitor isolated from the roots of *Actinidia arguta*. *Archives of Pharmacal Research* **31**: 666–670.

Johnston GAR. (2005) GABAA receptor channel pharmacology. *Current Pharmaceutical Design* **11**: 1867–1885.

Joshipura KJ, Hu FB, Manson JE, *et al.* 2001. The effect of fruit and vegetable intake on risk for coronary heart disease. *Annals of Internal Medicine* **134**: 1106–1114.

Jung K-A, Song T-C, Han D-S, *et al.* (2005) Cardiovascular protective properties of kiwifruit extracts *in vitro*. *Biological and Pharmaceutical Bulletin* **28**: 1782–1785.

Karami M, Nokabadi F, Ebrahimzadeh MA. (2011) Protective effects of feijoa extract on MDMA treated mice. *Pharmacologyonline* **3**: 1032–1038.

Karanjia ND, Rees M. (1993) The use of Coca-Cola in the management of bolus obstruction in benign oesophageal stricture. *Annals of the Royal College of Surgeons of England* **75**: 94–95.

Karlsen A, Svendsen M, Seljeflot I, *et al.* (2012) Kiwifruit decreases blood pressure and whole-blood platelet aggregation in male smokers. *Journal of Human Hypertension* 19 January. doi: 10.1038/jhh.(2011)116 (epub).

Kaur L, Rutherford SM, Moughan PJ, *et al.* 2010a. Actinidin enhances gastric protein digestion as assessed using an *in vitro* gastric digestion model. *Journal of Agricultural and Food Chemistry* **58**: 5068–5073.

Kaur L, Rutherford SM, Moughan PJ, *et al.* 2010b. Actinidin enhances protein digestion in the small intestine as assessed using an *in vitro* digestion model. *Journal of Agricultural and Food Chemistry* **58**: 5074–5080.

Keles H, Ince S, Kucukkurt I, *et al.* (2012) The effects of *Feijoa sellowiana* fruits on the antioxidant defense system, lipid peroxidation, and tissue morphology in rats. *Pharmaceutical Biology* **50**(3): 318–325.

Kelly GS. 1998. Folates: supplemental forms and therapeutic applications. *Alternative Medicine Review* **3**: 208–220.

Kim YK, Kang HJ, Lee KT, *et al.* (2003) Anti-inflammation activity of *Actinidia polygama*. *Archives of Pharmacal Research* **26**(12):1061–6.

Kou M-C, Yen J-H, Hong J-T, *et al.* (2009) Cyphomandra betacea Sendt. phenolics protect LDL from oxidation and PC12 cells from oxidative stress. *Lwt-Food Science and Technology* **42**(2): 458–463.

Krueger JM, Rector DM, Roy S, *et al.* (2008) Sleep as a fundamental property of neuronal assemblies. *Nature Reviews Neuroscience* **9**: 910–919.

Lapcik O, Klejdus B, Kokoska L, *et al.* (2005) Identification of isoflavones in *Acca sellowiana* and two *Psidium* species (Myrtaceae). *Biochemical Systematics and Ecology* **33**(10): 983–992.

Lauretani F, Semba RD, Dayhoff-Brannigan M, *et al.* (2008) Low total plasma carotenoids are independent predictors of mortality among older persons. The InCHIANTI study. *European Journal of Nutrition* **47**: 335–340.

Leuzzi A, Galati EM, Mondello MR, Monforte MT. (2009) Antiulcer activity of *Feijoa sellowiana* L. (Mirtaceae): morphological study. *Planta Medica* **75**(9): 1055–1055.

Li L, Werler MM. 2010. Fruit and vegetable intake and risk of upper respiratory tract infection in pregnant women. *Public Health Nutrition* **13**(2): 276–282.

Lim CC, Ferguson LR, Tannock GW. (2005) Dietary fibres as "prebiotics": Implications for colorectal cancer. *Molecular Nutrition and Food Research* **49**: 609–619.

Lin HH, Tsai PS, Fang SC, Liu JF. (2011) Effect of kiwifruit consumption on sleep quality in adults with sleep problems. *Asia Pacific Journal of Clinical Nutrition* **20**: 169–174.

Lu Y, Fan J, Zhao YP, *et al.* (2007a) Immunomodulatory activity of aqueous extract of Actinidia macrosperma. *Asia Pacific Journal of Clinical Nutrition* **16**: 261–265.

Lu Y, Zha, YP, Wang ZC, *et al.* (2007b) Composition and antimicrobial activity of the essential oil of Actinidia macrosperma from China. *Natural Product Research* **21**: 227–233.

Lucas JSA, Atkinson RG. (2008) What is a food allergen? *Clinical and Experimental Allergy* **38**: 1095–1099.

Lucas JSA, Lewis SA, Hourihane JOB. (2003) Kiwifruit allergy: A review. *Pediatric Allergy and Immunology* **14**: 420–428.

Lund ED, Smoot JM, Hall NT. (1983) Dietary fibre content of eleven tropical fruits and vegetables. *Journal of Agricultural and Food Chemistry* **31**: 1013–1016.

Macfarlane GT, Macfarlane S. (2012) Bacteria, colonic fermentation, and gastrointestinal health. *Journal of AOAC International* **95**: 50–60.

McGhie TK, Ainge GD. (2002) Color in fruit of the genus Actinidia: carotenoid and chlorophyll compositions. *Journal of Agricultural and Food Chemistry* **50**: 117–121.

Mainardi T, Kapoor S, Bielory L. (2009) Complementary and alternative medicine: herbs, phytochemicals and vitamins and their immunologic effects. *Journal of Allergy and Clinical Immunology* **123**: 283–294.

Manabe M, Isobe Y. (2005) Suppressing effects of *Feijoa sellowiana* Berg (Feijoa) on cytokine secretion by intestinal epithelium. *Food Science and Technology Research* **11**(1): 71–76.

Margina D, Ilie M, Gradinaru D. (2012) Quercetin and epigallocatechin gallate induce *in vitro* a dose-dependent stiffening and hyperpolarizing effect on the cell membrane of human mononuclear blood cells. International *Journal of Molecular Science* **13**: 4839–4859.

Martín-Cabrejas MA, Esteban RM, López-Andreu FJ, *et al.* (1995) Dietary fibre content of pear and kiwi pomaces. *Journal of Agricultural and Food Chemistry* **43**: 662–666.

Massey LK. (2003) Dietary influences on urinary oxalate and risk of kidney stones. *Frontiers of BioScience* **8**: s584–s594.

Meletis CD, Zabriskie N. (2008) Natural approaches for optimal sleep. *Alternative and Complementary Therapies* **14**: 181–188.
Mertz C, Brat P, Caris-Veyrat C, Gunata Z. (2010) Characterization and thermal lability of carotenoids and vitamin C of tamarillo fruit (*Solanum betaceum* Cav.). *Food Chemistry* **119**: 653–659.
Meydani SN, Leka LS, Fine BC, *et al.* (2004) Vitamin E and respiratory tract infections in elderly nursing home residents – A randomized controlled trial. *Journal of the American Medical Association* **292**: 828–836.
Miller MA, Cappuccio FP. (2007) Inflammation, sleep, obesity and cardiovascular disease. *Current Vascular Pharmacology* **5**: 93–102.
Miura K, Greenland P, Stamler J, *et al.* (2004) Relation of vegetable, fruit, and meat intake to 7-year blood pressure change in middle-aged men: The Chicago Western Electric Study. *American Journal of Epidemiology* **159**: 572–580.
Molan AL, Kruger MC, Drummond LN. (2007) The ability of kiwifruit to positively modulate key markers of gastrointestinal function. *Proceedings of the Nutrition Society of New Zealand* **32**: 66–71.
Möller NP, Scholz-Ahrens KE, Roos N, Schrezenmeir J. (2008) Bioactive peptides and proteins from foods: indication for health effects. *European Journal of Nutrition* **47**: 171–182.
Montefiori M, McGhie TK, Costa G, Ferguson AR. (2005) Pigments in the fruit of red-fleshed kiwifruit (*Actinidia chinensis* and *Actinidia deliciosa*). *Journal of Agricultural and Food Chemistry* **53**: 9526–9530.
Monto AS, Sullivan KM. (1993) Acute respiratory illness in the community: frequency of illness and the agents involved. *Epidemiology and Infection* **110**(1): 145–160.
Motohashi N, Kawase M, Shirataki Y, *et al.* (2000) Biological activity of Feijoa peel extracts. *Anticancer Research* **20**(6B): 4323–4329.
Motohashi N, Shirataki Y, Kawase M, *et al.* (2001) Biological activity of kiwifruit peel extracts. *Phytotherapy Research* **15**: 337–343.
Motohashi N, Shirataki Y, Kawase M, *et al.* (2002) Cancer prevention and therapy with kiwifruit in Chinese folklore medicine: a study of kiwifruit extracts. *Journal of Ethnopharmacology* **81**: 357–364.
Murakami A, Ohigashi H. (2007) Targeting NOX, INOS and COX-2 in inflammatory cells: chemoprevention using food phytochemicals. *International Journal of Cancer* **121**: 2357–2363.
National Institutes of Health (NIH) Consensus Development Program (2005) NIH State-of-the-Science Conference Statement on Manifestations and Management of Chronic Insomnia in Adults. http://consensus.nih.gov/2005/insomnia.htm (last accessed 4 March 2013).
Neurath MF, Finotto S. (2011) IL-6 signaling in autoimmunity, chronic inflammation and inflammation-associated cancer. *Cytokine and Growth Factor Reviews* **22**(2): 83–89.
Nieman DC, Henson DA, Gross SJ, *et al.* (2007) Quercetin reduces illness but not immune perturbations after intensive exercise. *Medicine and Science in Sports and Exercise* **39**: 1561–1569.
Nishiyama I, Fukuda T, Oota T. (2004) Varietal differences in actinidin concentration and protease activity in the fruit juice of *Actinidia arguta* and *Actinidia rufa*. *Journal of the Japanese Society for Horticultural Science* **73**: 157–162.
Normington KW, Baker I, Molina M, *et al.* (1986) Characterization of a nitrite scavenger, 3-hydroxy-2-pyranone, from Chinese wild plum juice. *Journal of Agricultural and Food Chemistry* **34**: 215–217.
Ordonez RM, Ordonez AAL, Sayago JE, *et al.* (2006) Antimicrobial activity of glycosidase inhibitory protein isolated from Cyphomandra betacea Sendt. fruit. *Peptides* **27**(6): 1187–1191.
Ordonez RM, Luz Cardozo M, Catiana Zampini I, Ines Isla M. (2010) Evaluation of antioxidant activity and genotoxicity of alcoholic and aqueous beverages and pomace derived from ripe fruits of Cyphomandra betacea Sendt. *Journal of Agricultural and Food Chemistry* **58**(1): 331–337.
Ordonez RM, Zampini IC, Rodriguez F, *et al.* (2011) Radical scavenging capacity and antimutagenic properties of purified proteins from Solanum betaceum Fruits and Solanum tuberosum tubers. *Journal of Agricultural and Food Chemistry* **59**(16): 8655–8660.
Ozgen M, Reese RN, Tulio AZJ, *et al.* (2006) Modified 2,2-azino-bis-3-ethylbenzothiazoline-6-sulfonic acid (ABTS) method to measure antioxidant capacity of selected small fruits and comparison to ferric reducing antioxidant power (FRAP) and 2,2'-diphenyl-1-picrylhydrazyl (DPPH) methods. *Journal of Agricultural and Food Chemistry* **54**: 1151–1157.
Parkar SG, Rosendale D, Paturi G, *et al.* (2012) *In vitro* utilization of gold and green kiwifruit oligosaccharides by human gut microbial populations. *Plant Foods for Human Nutrition* **67**: 200–207.
Pearson NJ, Johnson LL, Nahin RL. (2006) Insomnia, trouble sleeping, and complementary and alternative medicine: Analysis of the 2002 national health interview survey data. *Archives of Internal Medicine* **166**: 1775–1782.

Perera CO, Hallett IC, Nguyen TT, Charles JC. (1990) Calcium oxalate crystals: the irritant factor in kiwifruit. *Journal of Food Science* **55**: 1066–1069.

Qing YM, Chen XH, Cai MT, *et al.* (1999) Studies on the triterpenoids constituents of indochina Actinidia root (*Actinidia indochinensis*). *Journal of Chinese Medicinal Materials* **30**: 323–326.

Rabasseda X. (2012) A report from the American Heart Association Scientific Sessions 2011 (November 12–16, 2011, Orlando, Florida, USA). *Drugs Today (Barc).* **48**: 79–94.

Rassam M, Laing WA. (2005) Variation in ascorbic acid and oxalate levels in the fruit of *Actinidia chinensis* tissues and genotypes. *Journal of Agricultural and Food Chemistry* **53**: 2322–2326.

Ren J, Chung SH. (2007) Anti-inflammatory effect of α-linolenic acid and its mode of action through the inhibition of nitric oxide production and inducible nitric oxide synthase gene expression via NFκB and mitogen activated protein kinase pathways. *Journal of Agricultural and Food Chemistry* **55**: 5073–5080.

Rodriguez-Amaya DB, Bobbio PA, Bobbio FO. (1983) Carotenoid Composition and Vitamin-a Value of the Brasilian Fruit Cyphomandra-Betacea. *Food Chemistry* **12**(1): 61–65.

Roll S, Nocon M, Willich SN. (2011) Reduction of common cold symptoms by encapsulated juice powder concentrate of fruits and vegetables: a randomised, double-blind, placebo-controlled trial. *British Journal of Nutrition* **105**(1): 118–122.

Romero-Rodriguez MA, Vazquez Oderiz ML, Lopez Hernandez J, Simal Lozano J (1992) Determination of vitamin C and organic acids in various fruits by HPLC. *Journal of Chromatographic Science* **30**(11): 433–442.

Romero-Rodriguez MA, Vazquez-Oderiz ML, Lopez Hernandez J, Simal Lozano J. (1994) Composition of babaco, feijoa, passion-fruit and tamarillo produced in Galicia (NW Spain). *Food Chemistry* **49**: 251–255.

Rossi A, Rigano D, Pergola C, *et al.* (2007) Inhibition of inducible nitric oxide synthase expression by an acetonic extract from Feijoa sellowiana berg. fruits. *Journal of Agricultural and Food Chemistry* **55**(13): 5053–5061.

Rowe CA, Nantz MP, Bukowski JF, Percival SS. (2007) Specific formulation of *Camellia sinensis* prevents cold and flu symptoms and enhances gamma delta T cell function: a randomized, double-blind, placebo-controlled study. *Journal of the American College of Nutrition* **26**: 445–452.

Ruberto G, Tringali C (2004) Secondary metabolites from the leaves of *Feijoa sellowiana* Berg. *Phytochemistry* **65**(21): 2947–2951.

Rupérez P, Bartolomé AP, Fernández-Serrano MI. (1995) Dietary fibre in Spanish kiwifruit. *European Journal of Clinical Nutrition* **49**(Suppl 3): S274–S276.

Rush EC, Patel M, Plank LD, Ferguson LR. (2002) Kiwifruit promotes laxation in the elderly. *Asia Pacific Journal of Clinical Nutrition* **11**: 164–168.

Rutherford SM, Montoya CA, Zou ML, *et al.* (2011) Effect of actinidin from kiwifruit (*Actinidia deliciosa* cv. Hayward) on the digestion of food proteins determined in the growing rat. *Food Chemistry* **129**: 1681–1689.

Sakan T, Fujino A, Murai F, *et al.* (1959) On the structure of actinidine and matatabilactone, the effective components of *Actinidia polygama*. *Bulletin of the Chemical Society of Japan* **32**: 315.

Salminen A, Huuskonen J, Ojala J, *et al.* (2008) Activation of innate immunity system during aging: NF-kappa B signaling is the molecular culprit of inflamm-aging. *Ageing Research Reviews* **7**(2): 83–105.

Salvo F, Toscano MA, Dugo G. (1987) Chemical composition of *Feijoa sellowiana* fruit. *Rivista della Societa Italiana di Scienza dell'Alimentazione* **16**: 471–474.

Sanz A, Scialo F, Mallikarjun V, Stefanatos R. (2012) Regulation of lifespan by the mitochondrial electron transport chain: ROS-dependent and ROS-independent mechanisms. *Antioxidants and Redox Signaling,* 3 September. [Epub ahead of print].

Scarrow S. (2009) Citrus, berries, exotic fruit and nuts – Tamarillos, passionfruit and feijoas. Te Ara – the Encyclopedia of New Zealand (updated 13-Jul-12). http://www.TeAra.govt.nz/en/citrus-berries-exotic-fruit-and-nuts/3 (last accessed 4 March 2013).

Schmeda-Hirschmann G, Feresin G, Tapia A, *et al.* (2005) Proximate composition and free radical scavenging activity of edible fruits from the Argentinian Yungas. *Journal of the Science of Food and Agriculture* **85**: 1357–1364.

Schotsmans WC, East A, Thorp G, Woolf AB. (2011) Feijoa (*Acca sellowiana* Berg Burret). In: *Postharvest biology and technology of tropical and subtropical fruits. Volume 3: cocona to mango* (Yahia, EM. ed.), pp. 115–133, Woodhead Publishing Ltd, Cambridge, UK.

Shaw GJ, Ellingham PJ, Birch EJ. (1983) Volatile constituents of feijoa – headspace analysis of intact fruit. *Journal of the Science of Food and Agriculture* **34**(7): 743–747.

Shaw GJ, Allen JM, Yates MK. (1989) Volatile flavor constituents in the skin oil from Feijoa-Sellowiana. *Phytochemistry* **28**(5): 1529–1530.

Shu Q, De Silva UM, Chen S, *et al.* (2008) Kiwifruit extract enhances markers of innate and acquired immunity in a murine model. *Food and Agricultural Immunology* **19**: 149–161.

Siener R, Hönow R, Seidler A, *et al.* (2006) Oxalate contents of species of the Polygonaceae, Amaranthaceae and Chenopodiaceae families. *Food Chemistry* **98**: 220–224.

Skinner MA. 2012. Wellness foods based on the health benefits of fruit: Gold Kiwifruit for immune support and reducing symptoms of colds and influenza. *Journal of Food and Drug Analysis* **20**(S1): 261–264.

Skinner MA, Bentley-Hewitt K, Rosendale D, *et al.* (2013) Effects of kiwifruit on innate and adaptive immunity and symptoms of upper respiratory tract infections. *Advances in Food and Nutrition Research* **68**: 301–320.

Skinner MA, Hunter DC, Denis M, *et al.* (2007) Health benefits of ZESPRI^{TM} GOLD Kiwifruit: effects on muscle performance, muscle fatigue and immune responses. *Proceedings of the Nutrition Society of New Zealand* **32**; 49–59.

Skinner MA, Loh JM, Hunter DC, Zhang J. (2011) Gold kiwifruit (*Actinidia chinensis* 'Hort16A') for immune support. *Proceedings of the Nutrition Society* **70**: 276–280.

Swanson LM, Arnedt JT, Rosekind MR, *et al.* (2011) Sleep disorders and work performance: findings from the 2008 National Sleep Foundation Sleep in America poll. *Journal of Sleep Research* **20**(3): 487–494.

Takano F, Tanaka T, Tsukamoto E, *et al.* (2003) Isolation of (+)-catechin and (-)-epicatechin from *Actinidia arguta* as bone marrow cell proliferation promoting compounds. *Planta Medica* **69**: 321–326.

Tang SH, Zhang KM. (1997) *Actinidia chinensis* Planch actinidine composition resistance cancer. *Journal of Jishou University* **18**: 69–71.

Tene V, Malagon O, Finzi PV, *et al.* (2007) An ethnobotanical survey of medicinal plants used in Loja and Zamora-Chinchipe, Ecuador. *Journal of Ethnopharmacology* **111**(1): 63–81.

Thomas L, Low C, Webb C, *et al.* (2004) Naturally occurring fruit juices dislodge meat bolus obstruction *in vitro*. *Clinical Otolaryngology* **29**: 694–697.

Thorp G. (2008) Myrtaceae: *Acca sellowiana* Feijoa. In: *Encyclopedia of Fruits and Nuts* (Janick J and Paull RE eds), CAPI, Wallingford, Oxfordshire, UK.

Thorp G, Bieleski R. (2002) *Feijoa: origins, cultivation and uses*. David Bateman Ltd, Auckland, New Zealand.

Tibe O, Pernthaner A, Sutherland I, *et al.* (2012) Condensed tannins from Botswanan forage plants are effective priming agents of $\gamma\delta$ T cells in ruminants. *Veterinary Immunology and Immunopathology* **146**: 237–244.

Torbergsen AC, Collins AR. (2000) Recovery of human lymphocytes from oxidative DNA damage; the apparent enhancement of DNA repair by carotenoids is probably simply an antioxidant effect. *European Journal of Nutrition* **39**: 80–85.

Tsaluehidu S, Cocchi M, Tonollo L, Puri BK. (2008) Fatty acids and oxidative stress in psychiatric disorders. *BMC Psychiatry* **8**(Suppl 1): 5.

USDA (2007) Oxygen radical absorbance capacity (ORAC) of selected foods – 2007. US Department of Agriculture, Agricultural Research Service, pp. 1–34.

USDA (2011) USDA National Nutrient Database for Standard Reference, Release 24. US Department of Agriculture, Agricultural Research Service (Nutrient Data Laboratory Home Page: http://www.ars.usda.gov/ba/bhnrc/ndl; last accessed 4 March 2013).

Vasco C, Avila J, Ruales J, *et al.* (2009) Physical and chemical characteristics of golden-yellow and purple-red varieties of tamarillo fruit (*Solanum betaceum* Cav.). *International Journal of Foods Sciences and Nutrition* **60**(S7): 278–288.

Vignes M, Maurice T, Lanté F, *et al.* (2006) Anxiolytic properties of green tea polyphenol (-)-epigallocatechin gallate (EGCG). *Brain Research* **1110**: 102–115.

Vinolo MAR, Rodrigues HG, Hatanaka E, *et al.* (2011) Suppressive effect of short-chain fatty acids on production of proinflammatory mediators by neutrophils. *Journal of Nutritional Biochemistry* **22**: 849–855.

Visser FR, Burrows JK. (1983) Composition of New Zealand foods. I: Characteristic fruits and vegetables. Bulletin 235, New Zealand Department of Scientific and Industrial Research.

Vissers MC, Bozonet SM, Pearson JF, Braithwaite LJ. (2011) Dietary ascorbate intake affects steady state tissue concentrations in vitamin C-deficient mice: tissue deficiency after suboptimal intake and superior bioavailability from a food source (kiwifruit). *American Journal of Clinical Nutrition* **93**: 292–301.

Vuotto ML, Basile A, Moscatiello V, *et al.* (2000) Antimicrobial and antioxidant activities of Feijoa sellowiana fruit. *International Journal of Antimicrobial Agents* **13**(3): 197–201.

Walker S, Prescott J. (2003) Psychophysical properties of mechanical oral irritation. *Journal of Sensory Studies* **18**: 325–346.

Wang H, Ng TB. (2002) Isolation of an antifungal thaumatin-like protein from kiwifruits. *Phytochemistry* **61**: 1–6.

Wang H, Cao G-H, Prior RL. (1996) Total antioxidant capacity of fruits. *Journal of Agricultural and Food Chemistry* **44**: 701–705.

Webby RF. (1990) Flavonoid complement of cultivars of *Actinidia deliciosa* var. *deliciosa*, kiwifruit. *New Zealand Journal of Crop and Horticultural Science* **18**: 1–4.

Weston RJ. (2010) Bioactive products from fruit of the feijoa (*Feijoa sellowiana*, Myrtaceae): A review. *Food Chemistry* **121**(4): 923–926.

Wu X-L, Beecher GR, Holden JM, *et al.* (2004a) Lipophilic and hydrophilic antioxidant capacities of common foods in the United States. *Journal of Agricultural and Food Chemistry* **52**: 4026–4037.

Wu X-L, Gu L-W, Holden J, *et al.* (2004b) Development of a database for total antioxidant capacity in foods: a preliminary study. *Journal of Food Composition and Analysis* **17**: 407–422.

Wyatt RJ, Engelman K, Kupfer DJ, *et al.* (1970) Effects of L-tryptophan (a natural sedative) on human sleep. *Lancet* **2**: 842–846.

Xia L, Ng TB. (2004) Actinchinin, a novel antifungal protein from the gold kiwifruit. *Peptides* **25**: 1093–1098.

Yan JQ, Wang J Y, Zhao M. (1995) *Actinidia chinensis* Planch. polysaccharide superoxide free radical hydroxyl free radical electron spin resonance. *Chinese Journal of Biochemical Pharmaceutics* **16**: 12–14.

Yang H, Lee YC, Han KS, Singh H, Yoon M, Park JH, Cho CW, Cho S. (2013) Green and gold kiwifruit peel ethanol extracts potentiate pentobarbital-induced sleep in mice via a GABAergic mechanism. *Food Chemistry* **136**(1): 160–163.

Ying DY, Parkar S, Luo XX, *et al.* (2007) Microencapsulation of probiotics using kiwifruit polysaccharide and alginate chitosan. *Acta Horticulturae* **753**: 801–808.

Zanoli P, Zavatti M. (2008) Pharmacognostic and pharmacological profile of *Humulus lupulus* L. *Journal of Ethnopharmacology* **116**:383–96.

Zhu CC, Xu GJ. (1996) *Actinidia rubricaulis* var. coriacea polysaccharides Arps-2 Arps-3. *Journal of Chinese Medicinal Materials* **19**: 623–625.

9 Effect of Health-promoting Properties of Grapes, Including Resveratrol

Susan S. Percival and Robin L. West

INTRODUCTION

Although there are 40–60 known grape species, only three are relevant to human consumption: *Vitis vinifera,* the common table grape; *Vitis labrusca*, the Concord or purple grape; and *Vitis rotundifolia,* also known as the muscadine grape or scuppernong. The Nagano Purple grape is a cross between *V. vinifera* and *V. labrusca* and is found mostly in Asia. All varieties are consumed as fresh fruit, used for wine production or dried to make raisins.

Postharvest processing of grapes results in multiple forms for consumption. Considering just wines, a multitude of cultivars and blends exist. One online database (Mercer-Hursh, 2009) claimed 9235 identifiable wines from 50 different countries using 11 different species. California is the top grape producer in the United States. California's total utilized grape production was 6.678×10^6 tons of grapes in 2012; over half of it was processed into wine, 15% was consumed fresh and 30% was dried into raisins. In 2012, grape growing in the United States produced 7.337×10^6 tons of fresh equivalents valued at US$ 4911 million, the highest of all non-citrus fruits. The second highest production of a non-citrus fruit was apple at 4.498×10^6 tons valued at US$ 3069 million (USDA, 2013). Grapes are a valuable crop produced for many forms of consumption.

Knowledge of the composition of grapes (Tables 9.1–9.3) is important in understanding the mechanisms of their health benefits, the primary one being their antioxidant activity. For example, Resveratrol, a stilbene with antioxidant and histone deacetylase activity, is a compound found in the skin of grapes that provides health-promoting antioxidant benefits. Grape juice is one of the top 50 foods for antioxidant content, whether expressed per gram or per serving (Halvorsen *et al.*, 2006). Halvorsen *et al.* analysed several brand names of grape juice and found that the brand primarily composed of Concord grapes had more activity than the other brands, although the grape species was not noted for the other brands. However, variation in grape composition exists due to harvest times, soil and climate conditions of that particular harvest, differences in processing and other reasons. Table 9.1 summarizes some of the literature on composition pertaining to the flavonoid family of compounds. Concord grapes have the highest level of anthocyanins, which give them their deep purple colour. Conversely, white or green table grapes have no detectable anthocyanins. The flavan-3-ols have been widely studied for their health benefits; this family of compounds is found in red

Bioactives in Fruit: Health Benefits and Functional Foods, First Edition. Edited by Margot Skinner and Denise Hunter.

Table 9.1 Composition of fresh grape phytochemical families (flavonoids, mg/100 g, edible portion).[1]

Class	Flavonoid	Grapes, red	Grapes, white or green	Grapes, Concord
Anthocyanidins	Cyanidin	0.95	0.00	13.16
	Delphinidin	2.10	0.00	39.58
	Malvidin	36.20	0.00	3.47
	Pelargonidin	0.02	0.00	0.00
	Peonidin	2.90	0.00	2.62
	Petunidin	1.80	0.00	8.78
Flavan-3-ols	(−)–Epicatechin	0.96	1.70	2.14
	(−)–Epicatechin-3-gallate	0.17	0.25	0.00
	(−)–Epigallocatechin	0.08	0.02	0.00
	(−)–Epigallocatechin-3-gallate	0.00	0.00	0.00
	(+)–Catechin	0.82	3.73	0.00
	(+)–Gallocatechin	0.00	0.01	0.00
Flavones	Apigenin	0.00	0.00	0.00
	Luteolin	1.30	0.00	0.00
Flavonols	Kaempferol	0.00	0.04	0.00
	Myricetin	0.01	0.22	0.00
	Quercetin	1.04	1.12	3.11

[1]US Department of Agriculture, Agricultural Research Service; 2011. USDA Database for Flavonoid Content of Selected Foods, Release 3. Nutrient Data Laboratory Home Page: http://www.ars.usda.gov/nutrientdata.

and green grapes. Benefits attributed to grapes is a result of their composition; no one group of compounds is responsible.

The purpose of this chapter is to review recent studies investigating health benefits of grapes or grape products in humans. The focus is on studies that investigated the effects of consuming whole grape products, but the effects of resveratrol on human subjects are also described.

BIOAVAILABILITY AND METABOLIC FATE OF COMPOUNDS

The bioavailability of compounds from fruits and vegetables in humans is controversial. Detection methods, the food matrix given with the compound of interest, the conjugation form of the compound and the fact that multiple compounds are derived from the parent compound during absorption and first pass through the liver affects how bioavailability is calculated. Although estimates of the bioavailability of compounds are usually low, if all compounds that are derived from the parent cannot be identified then bioavailability will be underestimated. Stable isotope labelling of the parent compound would be a way to determine true bioavailability, but synthesis of labelled compounds are expensive. Furthermore, fermentation of unabsorbed compounds may occur in the gut. These fermented compounds may be absorbed and provide benefit or could benefit the host indirectly by supporting beneficial bacteria. Therefore, benefits can be derived from grape consumption even if they are not absorbed into the blood stream. Interactions with the human gut immune cells and fermentation in the large intestine may both provide benefit to the body.

Early studies on bioavailability of compounds from grapes focused on the anthocyanin family of compounds. Malvidin-3-glucoside bioavailability from red wine, dealcoholized red wine or red grape juice was studied after human consumption. No aglycone, only malvidin-3-glucoside, was found in urine and plasma. The different beverages did not differ in absorptive capacity; malvidin-3-glucoside was bioavailable from all. In one study, approximately 120 mg of malvidin-3-glucoside was consumed and 27 μg excreted (Bub *et al.*, 2001), roughly 0.2% of the dose. Frank *et al.* focused on the bioavailability of anthocyanins from red grape juice compared with red wine (Frank *et al.*, 2003). Nine people were given equivalent volumes and content of anthocyanin-3-glucosides from each beverage. The investigators looked only for the glucosides and no other conjugates in the blood and urine. They found nanogram quantities in the blood and urine from milligrams of quantity consumed. Another study by the same group (Bitsch *et al.*, 2004) found total anthocyanins in the urine was greater for red grape juice compared with red wine, although still less than 1%. Total plasma antioxidant activity was also higher after juice ingestion in this study. Concord grape juice contains delphinidin-3-*O*-glucoside and cyanidin-3-*O*-glucoside as the main anthocyanins (Stalmach *et al.*, 2011). The Nagano Purple grape's main anthocyanin is cyanidin-3-glucoside (Kamiyama *et al.*, 2009). The bioavailability of anthocyanins appears to be low.

The food form or the conjugated form of the grape compounds influences its bioavailability. In one study, human subjects were given either the single compounds quercetin or resveratrol as aglycones, or grape juice, which contain the conjugates of these compounds. After consuming the pure aglycones, metabolites of the parent compounds were detected in the urine as conjugates. After grape juice consumption, these conjugated metabolites were also detected in the urine, but the authors suggested that there was lower bioavailability of the conjugated compounds from grapes compared to the single aglycone compounds (Meng *et al.*, 2004). Quercetin bioavailability and tissue distribution has been studied in several *in vivo* models (Olthof *et al.*, 2000; Bieger *et al.*, 2008). Quercetin glycosides are bioavailable from grape juice after a single dose (Davalos *et al.*, 2006). Conjugation reduces bioavailability but aglycones are converted once absorbed.

Polymerization of flavan-3-ols creates proanthocyanidins, which also have health benefits. Bioavailability of the proanthocyanidins is associated with length of the polymer; monomers are better absorbed than the dimers and trimers (Serra *et al.*, 2010) (Table 9.2).

One grape compound that has received much scientific and commercial attention is resveratrol (Table 9.3). Resveratrol-3-*O*-sulfate, resveratrol-4′-*O*-glucuronide and resveratrol-3-*O*-glucuronide are major plasma metabolites in humans after consumption of unconjugated resveratrol. The concentration of metabolites in plasma was greater than the unconjugated resveratrol in this pharmacokinetic study that used up to 5 g per day (Brown *et al.*, 2010).

Table 9.2 Proanthocyanidin content of grapes and grape products (mg/100 g, edible portion).[1]

Compound	Grapes, red	Grapes, green	Raisins	Grape juice, purple	Grape juice, white
Monomers	1.36	0.96	Undetectable	1.69	0.16
Dimers	2.38	2.33	Undetectable	3.18	0.24
Trimers	1.01	1.88	Undetectable	1.78	0.00
4–6 mers	6.07	8.35	Undetectable	7.49	0.00
7–10 mers	6.23	9.15	Undetectable	6.46	0.00
Polymers	44.56	58.87	Undetectable	28.37	0.00

[1]US Department of Agriculture, Agricultural Research Service; 2011. USDA Database for the Proanthocyanidin Content of Selected Foods, Release 3. Nutrient Data Laboratory Home Page: http://www.ars.usda.gov/nutrientdata.

Table 9.3 Resveratrol content of grapes and grape products.

Grape species	Cultivar	Form studied	Resveratrol	Study
Vitis labrusca		Fresh fruit	10.9 mg/kg	(Lago-Vanzela *et al.*, 2011b)
Vitis vinifera	Morena Clara	Fresh fruit Not available	3.91 mg/kg Not detected	(Lago-Vanzela *et al.*, 2011a)
Vitis vinifera	various	Wine	16.1–31.2 molar %	(Perez-Trujillo *et al.*, 2011)
Vitis vinifera	Malbec	Fresh fruit	0.8–2.8 ppm	(Berli *et al.*, 2008)
Vitis rotundifolia	Noble Carlos	Hot-press juice Not available	0.1 mg/l <0.1 mg/l	(Leblanc *et al.*, 2008)
Vitis rotundifolia		Fresh fruit	1.0 mg/kg	(Pastrana-Bonilla *et al.*, 2003)

The tissues take up resveratrol, as a study by Patel *et al.* (2010) indicates. Twenty patients with histologically confirmed colorectal cancer consumed eight daily doses of resveratrol at 0.5 g or 1.0 g before surgical resection. Normal and malignant biopsy tissue samples were obtained before and after dosing. Parent compound plus its metabolites, resveratrol-3-*O*-glucuronide, resveratrol-4′-*O*-glucuronide, resveratrol-3-*O*-sulfate, resveratrol-4′-*O*-sulfate, resveratrol sulfate glucuronide and resveratrol disulfate were identified in the normal and malignant tissues (Patel *et al.*, 2010). Although this study specifically measured tissue uptake, absorption by the body and distribution of phytochemicals, in general, is still not well understood.

Metabolomics is proving to be a useful approach to examine the metabolic fate of polyphenols derived from grapes. Otherwise healthy adults who had mild hypertension were randomized in a double-blind, crossover, nutrition intervention where urinary metabolites were determined. A capsule of red grape juice extract was compared to placebo in one arm and a mix of dried red wine extract plus dried red grape juice extract was compared to placebo in the other. The metabolites derived from the grape products were significantly different from those in the placebo. Each grape mix was slightly different from the other in the urinary phenolic profiles, indicating detectable shifts due to diet and polyphenol intake (van Dorsten *et al.*, 2010). Stalmach *et al.* recently reported plasma and urinary polyphenolic compounds derived after consumption of Concord grape juice. After acute consumption of 350 ml by eight subjects, 41 metabolites were identified in plasma and urine (Stalmach *et al.*, 2011). Further identification of metabolites from grapes will help our understanding of how polyphenols are handled by the body.

Determining the bioavailability of compounds from grapes is controversial. Detecting the presence and amounts of metabolites formed during absorption and metabolism is not precise. However, there are notable health benefits associated with consumption of grapes and grape products that promote their use in a healthy diet, even if the exact compounds are not known.

HEART HEALTH BENEFITS AND POTENTIAL MECHANISMS OF ACTION

The benefits of grape and grape products in reducing risk of heart disease has been studied with a variety of end points. If oxidation of low density lipoprotein (LDL) cholesterol molecules is crucial to the progression of plaque formation, then consumption of any antioxidant-containing food has potential benefit. Several studies show that grapes and grape products are beneficial in reducing risk of heart disease by a number of mechanisms. A recent observational study examined the flavonoid consumption of individuals in the CPS II Nutrition Cohort. The subjects were excluded if they had any heart problems, stroke or diabetes or were pre- or perimenopausal. Almost 100 000 men and women's diets were analysed for flavonoid consumption. A total of 2771 deaths due to cardiovascular disease (CVD) occurred during follow-up. Inverse associations were observed for anthocyanidins, flavan-3-ols, flavones, flavonols and proanthocyanidins in relation to heart disease (McCullough *et al.*, 2012). Grapes and grape products contribute considerably to the dietary intake of flavonoids. Other studies are summarized in the following sections.

Platelet aggregation

Studies show that consumption of grape products alters platelet aggregation, blood pressure, blood lipids, lipoproteins and endothelial function. In a study by Keevil *et al.*, ten subjects consumed the juice of grapefruits, oranges or Concord grapes for 7–10 days. After consumption of Concord grape juice, platelet aggregation was reduced 77%, while consumption of orange juice and grapefruit juice had no effect. The authors concluded that these results were due to anthocyanins or other polyphenols that are unique to grape juice, as well as the total amount of polyphenols in grape juice, which was three times more than in the citrus juices (Keevil *et al.*, 2000). Pignatelli *et al.* also demonstrated reduced platelet aggregation after acute consumption of either red or white wine (Pignatelli *et al.*, 2002). In a longer study, twenty subjects consumed Concord grape juice for 14 days, which resulted in a reduction in platelet aggregation, regardless of whether it was induced by phorbol 12-myristate 13-acetate (PMA), ADP or collagen agonists. Consumption of Concord grape juice led to an increase in platelet-derived nitric oxide (NO) release and significantly decreased platelet superoxide production, as well as an increase in antioxidant capacity in protein-free plasma (Freedman *et al.*, 2001). Albers *et al.* also found a decrease in platelet superoxide production (Albers *et al.*, 2004). In hypercholesterolemic rabbits, platelet aggregation was inhibited by consuming red wine (with or without alcohol). *In vitro*, adding Concord grape juice to platelets increased NO release and decreased platelet aggregation (Wang *et al.*, 2002). Reducing platelet aggregation is but one mechanism by which grapes reduced heart disease risk.

In another study, twenty healthy volunteers drank red or white wine at 300 ml/day for 15 days. Nitrite plus nitrate concentrations roughly doubled in the supernatant of arachidonic acid-stimulated platelets in both wine drinking groups; there was no difference between red or white (Gresele *et al.*, 2008). This study also quantitated resveratrol in the plasma. Plasma resveratrol increased from 0.7 to 1.3 μmol/l in white wine drinkers and 0.7 to 1.7 μmol/l in red wine drinkers. Gresele *et al.* (2008) extended these studies *in vitro*. Resveratrol was added to cells in culture at levels that were determined from the plasma levels measured after 15 days of wine consumption. At these levels, resveratrol significantly increased (i) production of NO from stimulated platelets, (ii) activity of platelet nitric oxide synthase (pNOS), (iii) phosphorylation of protein kinase B, an activator of the endothelial NOS (eNOS) and (iv) phosphorylation of vasodilator-activated protein, an expression of the biologic activity of NO in platelets. Simultaneously, treatment with resveratrol decreased phosphorylation of P38 mitogen-activated protein kinase (p38MAPK), a proinflammatory pathway in human platelets. Nicotinamide adenine dinucleotide phosphate (NADPH) oxidase and reactive oxygen species (ROS) were reduced. Resveratrol, at concentrations attainable after moderate wine intake, activates platelet eNOS and in this way blunts the proinflammatory pathway linked to p38MAPK, inhibiting ROS production and ultimately, platelet function.

Rathel and others also suggested resveratrol is a promising candidate for increasing eNOS by altering transcription rate of the enzyme. However, they found it was not responsible for the entire effect. In an *in vitro* study, significant differences in eNOS transcription were found when testing red wines, and red wine had a significantly greater effect than white wine, but there was no consistent pattern regarding geographical area or cultivar (Rathel *et al.*, 2007).

Not all data on platelet aggregation show an effect. Albers (Albers *et al.*, 2004) and Coimbra (Coimbra *et al.*, 2005) found no effect on platelet aggregation in subjects having either stable coronary artery disease or having hypercholesterolemia, respectively. Variation in grape products, amounts consumed or number of days consumed may contribute to the variability seen among studies.

Blood pressure

Reducing high blood pressure is a goal in reducing the risk of heart disease. Data on the effect of consuming grapes or grape products in reducing blood pressure are mixed. Park *et al.* examined borderline hypertensives who were not on medication and found consumption of Concord grape juice reduced both systolic and diastolic blood pressure. Individuals on the placebo also had reduced blood pressure, but not as much as the subjects consuming Concord grape juice (Park *et al.*, 2004). Hansen *et al.* found no effect on the blood pressure of healthy adults consuming red wine or red grape juice (Hansen *et al.*, 2005), and Dohadwala *et al.* observed no effect of Concord grape juice on the blood pressure of prehypertensive and stage 1 hypertensive individuals (Dohadwala *et al.*, 2010). No general conclusions about the effect of grape products on blood pressure can be made.

Blood lipids and inflammatory factors

Another risk-altering characteristic of grapes and grape products that has been investigated is related to blood lipids. Concord grape juice consumption was compared to vitamin E supplementation (RRR-alpha-tocopherol) in a two-week study. In healthy adults, Concord grape juice increased serum antioxidant capacity and protected LDL cholesterol against oxidation to an extent similar to that obtained with vitamin E. Plasma protein oxidation was significantly reduced by Concord grape juice relative to vitamin E. The difference between the polyphenols of grape juice and the fat-soluble vitamin may be related to their hydrophobicity (O'Byrne *et al.*, 2002).

One of the earliest studies of the benefits of grape juice consumption was a short-term crossover study of Concord grape juice with and without additional vitamin C supplementation. The susceptibility of LDL cholesterol lipoproteins to oxidation was improved by Concord grape juice, regardless of vitamin C intake, suggesting protection of the lipoproteins could be achieved from the grape juice phytochemicals (Vinson *et al.*, 2000).

Albers studied the effects of Concord grape juice in subjects with stable coronary artery disease for 14 days. There was a significant reduction in sCD40L and platelet-dependent superoxide production and a significant increase in high density lipoprotein (HDL) cholesterol. There were no effects on glucose, total cholesterol, insulin, platelet aggregation, NO, P-selectin, thromboxane B2, interleukin (IL)-8 or C-reactive protein (CRP) (Albers *et al.*, 2004).

Healthy adults (N = 69) were studied in a four-armed (15–19 per group), placebo-controlled, four-week intervention. All subjects were placed on a restricted diet with no red/blue/purple fruits, juices or vegetables. The treatment groups consisted of red wine, red grape extract with polyphenol content equivalent to the wine, a half dose of the red grape extract or a placebo capsule. Red wine increased HDL cholesterol production but there were no other changes in blood lipids. Fibrinogen levels were variable but a modest decrease was seen after red wine consumption. The authors estimated the benefits of red wine to lower cardiovascular disease risk by 10–25% (Hansen *et al.*, 2005).

A dietary recall study partially agreed with results of Hansen *et al.* (2005). Three thousand two hundred men who participated in the British Regional Heart Study gave a seven-day recall of alcohol intake and a fasting serum sample. Alcohol of any sort was positively associated with HDL cholesterol production. Other positive associations included increased factor IX, tissue plasminogen activator, hematocrit and blood viscosity. They found a negative association with triglycerides, white cell count, fibrinogen and von Willebrand factor (vWF). Among beverages, wine gave most marked reduction in fibrinogen and vWF and was the only beverage associated with lower CRP. Red wine was more effective than white wine (Wannamethee *et al.*, 2003).

A study on heart disease performed with 24 premenopausal and 20 postmenopausal women was conducted as a crossover, single-blind, four-week intervention with a three-week washout between periods. Lyophilized grape powder prepared by the California Table Grape Commission was compared to a fructose:dextrose placebo. Seven-day food records were collected and the subjects were told to avoid tea, grape products, fruit juice, citrus, berries, onions, apples and broccoli. The authors reported that grape treatment lowered triglycerides and LDL cholesterol in pre- and postmenopausal women. Lecithin cholesterol acyltransferase activity and urinary isoprostane concentration was reduced, although modestly, by grape consumption (Zern *et al.*, 2005).

Pignatelli *et al.* examined urinary biomarkers of inflammation before and after wine consumption (Pignatelli *et al.*, 2006). Twenty healthy volunteers drank red or white wine or placebo for 15 days. Urinary PGF2α-III was significantly reduced by consumption of white wine (23%) and to a greater extent by red wine (38%). Levels of urinary PGF2α-III in individuals that did not drink throughout the study remained the same as at baseline. The decrease in an individual's urinary PGF2α-III was significantly and positively correlated with their serum polyphenol concentrations.

Research on the Nagano Purple grape is limited, although its polyphenol content is able to inhibit *ex vivo* oxidation of LDL cholesterol after grape consumption compared with fasting. The main anthocyanin found in the blood was Cy-3-glc (Kamiyama *et al.*, 2009).

A potential mechanism of lipoprotein protection was suggested by the work of Blanco-Colío *et al.* (Blanco-Colio *et al.*, 2000). Simultaneous intake of red wine during consumption of a high-fat meal prevented nuclear factor-kappaB activation in peripheral blood mononuclear cells of healthy volunteers during postprandial lipemia. Inhibiting nuclear factor-kappaB prevents transcription of inflammatory mediators that often accompany postprandial lipemia.

Forty healthy men in a randomized crossover intervention drank red wine or gin for 28 days. Wine compared to gin reduced oxidative stress biomarkers in serum and cells (Estruch *et al.*, 2011). In a subsequent, randomized, crossover trial in men at high risk for metabolic disease, red wine was compared to dealcoholized red wine or gin and serum concentrations of adhesion molecules and cytokines determined. These biomarkers were favourably altered after consumption of red wine or dealcoholized red wine compared to gin.

Furthermore, the results indicated that both ethanol and non-alcoholic compounds may be responsible for the anti-inflammatory effects (Chiva-Blanch *et al.*, 2012).

In stressed individuals

Haemodialysis (HD) patients often have oxidative stress and increased cardiovascular risks. Their LDL cholesterol is more oxidized than that of healthy individuals. Castilla *et al.* studied 38 clinically stable subjects who had been on HD for three months or more. Red grape juice was given for 14 days to the subjects and to 15 healthy controls. They found no changes to glucose, uric acid, plasma protein or albumin by red grape juice in either HD or healthy subjects. However, both healthy and HD subjects showed lower total cholesterol, lower LDL cholesterol, higher HDL cholesterol, higher ApoA-I and lower ApoB after consuming red grape juice. Higher plasma antioxidant capacity and lower oxidized LDL cholesterol was exhibited in both groups. No change in γ-tocopherol, lycopene, carotenoids or ascorbic acid was observed due to consumption of red grape juice, but α-tocopherol was increased. MCP-1 was reduced in plasma in both groups but there were no changes to adhesion molecules (VCAM-1, ICAM-1), CRP or C3 (complement) in either group (Castilla *et al.*, 2006).

A follow-up study was performed with red grape juice, with and without supplementation of vitamin E. Clinically stable HD subjects ($N = 32$) were given red grape juice for two weeks. Red grape juice lowered total cholesterol, LDL cholesterol and ApoB, while increasing HDL cholesterol, HDL phospholipids and ApoA-1. Vitamin E alone did nothing to serum lipids, but the combination of E and red grape juice was as effective as the juice alone. Vitamin E reduced levels of ICAM-1 adhesion molecule and red grape juice reduced MCP-1, a cardiovascular disease marker. Neutrophil NADPH oxidase activity (superoxide production) was reduced by the treatments, which benefits HD patients because neutrophils are activated by the haemodialysis procedure (Castilla *et al.*, 2008). The conclusion from these two studies is that concentrated red grape juice exerts antioxidant, hypolipidemic and anti-inflammatory effects in haemodialysis patients who are at risk for oxidative stress.

Endothelial cell function

The function of the endothelium is often assessed as an index of cardiovascular disease risk. The endothelium must filter fluids, maintain blood vessel tone and prevent clot formation. Endothelial dysfunction is a key event in the development of cardiovascular disease. Endothelial function can be evaluated by measuring flow-mediated vasodilation (FMD) of the brachial artery using B-mode ultrasound. Chou *et al.* studied 22 adults with angiographically documented coronary artery disease. The 56-day study concluded that FMD was improved by consumption of Concord grape juice (Chou *et al.*, 2001). Lekakis *et al.* showed similar improvement in FMD after consumption of red grapes (Lekakis *et al.*, 2005).

Coimbra *et al.* (2005) studied FMD and other parameters in hypercholesterolemic subjects. In a crossover design study, sixteen hypercholesterolemic and 24 normal subjects were randomized to consume either the Concord grape juice or red wine (Pinot Noir) for 14 days, wash out for 14 days and then consume the other beverage for 14 days. In hypercholesterolemic patients, body mass index, heart rate, blood pressure, total LDL and HDL cholesterol, triglycerides, glucose and VCAM did not change after grape juice or red wine consumption. Serum ICAM levels significantly decreased after consumption of grape juice but not red wine. Consumption of red wine and grape juice both improved endothelial-dependent flow-mediated dilation, while only the red wine improved endothelial-independent dilation.

Spaak *et al.* sought to show whether any effect on FMD measurement was due to alcohol consumption. Thirteen volunteers (24–47 years; 7 men, 6 women) drank wine, ethanol and water in a randomized, single-blind trial on three occasions, two weeks apart. Blood pressure was not affected, and the effect on FMD was dependent on the dose consumed (Spaak *et al.*, 2008).

Huang *et al.* extended these heart-healthy findings to the endothelial progenitor cell and showed red wine consumption for three weeks improved FMD, plasma NO and circulating numbers of endothelial progenitor cells (Huang *et al.*, 2010).

Resveratrol also has effects on the endothelium. *In vitro*, resveratrol modified the endothelium-dependent and NO-mediated vasodilation in the human internal mammary artery but only partially in saphenous vein rings, although it improved their endothelial reactivity (Rakici *et al.*, 2005).

IMMUNITY

The natural immune response, when left unchecked, creates free radicals and potentially damages the host. Inflammatory biomarkers, which are derived from immune cells, are peptide and lipid mediators that create problems when unregulated. These mediators are critical signalling mechanisms during an immune response but their chronic synthesis is associated with risk of disease. Antioxidant compounds in grapes facilitate not only risk reduction during an uncontrolled immune response but also immune-modulating activity, which results in a better resolution of inflammation and more efficient production of mediators.

In a study by Watzl *et al.*, no effect of an acute consumption of alcohol or wine (24 hours) was shown on phagocytic activity of neutrophils or monocytes, the production of TNF-α, IL-2 or IL-4, lymphocyte proliferation and lytic activity of natural killer cells in six healthy men (Watzl *et al.*, 2002). The short nature of this study most likely precluded the ability to measure effects. Another study by this group examined daily, moderate consumption of wine for two weeks. Twenty-four individuals drank red wine at 12% ethanol, dealcoholized red wine, red grape juice or 12% ethanol. Production of cytokines by activated peripheral blood mononuclear cells, phagocytosis of granulocytes and monocytes, apoptosis of lymphocytes, lymphocyte proliferation and natural killer cell lysis activity were not different among the groups (Watzl *et al.*, 2004). The authors concluded that the treatment did not suppress the immune response, but neither did it show benefit.

We performed a similar experiment in mice. Animals were placed in groups that drank from their drinking bottles either cabernet sauvignon wine (*V. vinifera*), a wine derived from muscadine grapes (*V. rotundifolia*), water or ethanol equivalent to the ethanol content of the wines (12%). Using animals allowed us to test not only surveillance immunity but also the immune response after lipopolysaccharide (LPS) injection. Ethanol caused some reduction in baseline immunity and in the immune response to LPS, but results from mice drinking the wines did not differ from those of mice drinking water, suggesting a protective effect of the red wine polyphenols against ethanol damage. We also demonstrated changes in P450 enzymes and conjugating enzymes that were very different among the groups. These results suggested that the metabolism of different wines differed among the groups, thus altering the time of exposure to alcohol and/or limiting free radicals formed from ethanol's metabolism (Percival and Sims, 2000).

We also examined immune function in a group of 85 men and women (mean age, 58 y). Individuals consuming Concord grape juice had significantly greater numbers of circulating

$\gamma\delta$T cells and higher serum vitamin C levels than those in the placebo group. Individuals consuming the placebo had lower serum antioxidant activity, less $\gamma\delta$T-cell proliferation and increased DNA strand breaks when challenged with H_2O_2. Analysis of the data by structural equation modelling confirmed that 61% of the variance in biological functions (i.e. T-cell proliferation and antioxidant values) at nine weeks was due to grape juice consumption (Rowe *et al.*, 2011).

CANCER

Cancer prevention is difficult to study in humans, although epidemiological data clearly indicate that consumption of fruits and vegetables is associated with a reduced risk of certain cancers. For example, in a Korean case-control study, a high grape intake was associated with a reduced risk of breast cancer in postmenopausal women (Do *et al.*, 2007). Although this chapter is focused on data acquired from human studies, data from animal studies are important to examine because they are derived from methods that cannot be used in human studies to help determine cancer risk reduction.

Cancer was induced with 7,12-dimethylbenz[a]anthracene (DMBA) in rats that were given two levels of Concord grape juice or water. Concord grape juice significantly inhibited mammary adenocarcinoma multiplicity and final tumour mass in both groups, compared with control rats. Furthermore, consumption of grape juice phenolics also significantly inhibited *in vivo* mammary DMBA-DNA adduct formation. In this study, the mechanism suggested was increased activity of phase II metabolizing enzymes (Singletary *et al.*, 2003), which we also showed in mice (Percival and Sims, 2000). This study and another by Jung *et al.* (2006) showed that specific constituents or combinations of phytochemicals that are found in purple grape juice blocked the initiation stage of DMBA-induced rat mammary tumorigenesis. The tumour inhibitory effect associated with the suppression of mammary DMBA-DNA adduct formation was explained by Jung *et al.* in part by increased liver activity of the phase II metabolizing enzyme glutathione S-transferase.

Numerous *in vitro* studies of the effect of polyphenols on cancer cell growth have been performed. We showed a synergy among grape compounds quercetin, ellagic acid and resveratrol in regulating the cell cycle of cultured cancer cells (Mertens-Talcott *et al.*, 2003; Mertens-Talcott and Percival, 2005).

In a study by Patel *et al.* twenty patients with histologically confirmed colorectal cancer consumed resveratrol at 0.5 or 1.0 g for eight days before surgical resection. Normal and malignant biopsy tissue samples were obtained after dosing and analysed for resveratrol and its metabolites. Both tumour and normal tissue accumulated resveratrol at concentrations much greater than were found in the plasma. They estimated cell proliferation in the biopsies with Ki-67 staining. Compared with pre-dose biopsies, Ki-67 staining was reduced by approximately 5%. The authors acknowledged that the biological effect of such a small change is difficult to interpret but pointed out that resveratrol has a pharmacological effect in the human colon (Patel *et al.*, 2010).

In another study, 42 healthy volunteers underwent baseline assessment of cytochrome P450 (CYP) and phase II detoxification enzymes. After the baseline evaluation, study participants took 1 g of resveratrol once daily for four weeks. Results indicated that resveratrol can modulate enzyme systems involved in carcinogen activation and detoxification, which may be one mechanism by which resveratrol inhibits carcinogenesis. However, pharmacologic doses of resveratrol may lead to increased adverse drug reactions or altered drug efficacy due to inhibition or induction of certain CYPs (Chow *et al.*, 2010).

Brown *et al.* analysed the effect of resveratrol on cancer biomarkers. Compared with predosing values, the ingestion of resveratrol caused a decrease in circulating IGF-I and IGFBP-3 in volunteers. The decrease was most marked at the 2.5 g dose level. The results suggest that repeated administration of high doses of resveratrol generates micromolar concentrations of parent and much higher levels of glucuronide and sulfate conjugates in the plasma. The observed decrease in circulating IGF-I and IGFBP-3 was suggested to contribute to cancer chemopreventive activity (Brown *et al.*, 2010).

Ingersoll and others suggested a potential use for Concord grape juice in reducing side effects of chemotherapy. Nausea and vomiting frequency, duration and distress were lower among experimental group members who consumed Concord grape juice, although a high attrition rate (50%) resulted in insufficient power to demonstrate significance. Nonetheless, the research demonstrates a beneficial physiological effect of consuming grape juice (Ingersoll *et al.*, 2010).

Overall, there appears to be a role for grapes and grape products in reducing the risk of cancer although much is left to study. Whether they benefit a specific cancer or are part of the larger effect of fruits and vegetables in general remains to be investigated.

OBESITY, METABOLIC SYNDROME AND DIABETES

Studies on grapes and grape products have been conducted to investigate their effects on the related disorders of metabolic syndrome, obesity, and diabetes. Obesity – in combination with other disorders such as hypertension and dyslipidemia – contributes to metabolic syndrome, a condition that greatly increases risk for heart disease and diabetes. Obesity, in and of itself, is considered to be a case of chronic, low grade inflammation. Continual assault of inflammatory mediators is associated with tissue damage that leads to more severe symptoms and risks of disease.

Hashemi *et al.* analysed the effects of Concord grape juice on thirty adolescents with metabolic syndrome (mean age, 13.4 y). Subjects were given a rather large amount of juice (18 ml/kg) to consume daily for one month. After consumption of the juice, E-selectin and IL-6 were lower than at baseline, but glucose, total cholesterol, LDL and HDL cholesterol, triglycerides and sICAM-1 did not change. After drinking the juice, flow-mediated dilation was significantly improved at four hours and at one month (Hashemi *et al.*, 2010).

Type 2 diabetes is associated with oxidative stress, which leads to further disease consequences. In a study that is one of the few that used the *V. rotundifolia* grape to examine physiological changes, three groups of type 2 diabetics drank either muscadine juice, muscadine wine or dealcoholized muscadine wine. Researchers included a non-diabetic control that did not take a supplement ($n = 15$) and a non-diabetic control that drank muscadine juice ($n = 8$). Although serum glucose and insulin levels were reduced in members of all muscadine groups, significance was achieved only in the dealcoholized muscadine wine for lower insulin levels. Researchers found no effect on blood pressure. The muscadine beverages were associated with moderate changes in blood lipids, liver enzymes and erythrocyte membrane fatty acids, but results suggest better glycemic control (Banini *et al.*, 2006).

Hollis *et al.* investigated the effects of consumption of Concord grape juice compared with a grape-flavoured product on weight gain. Daily consumption of the Concord grape juice for 12 weeks did not lead to significant weight gain but consumption of the grape-flavoured, sugar-sweetened, polyphenol-free placebo beverage did. Energy and macronutrient intake were not different among the groups. Individuals in the placebo group tended to have higher hunger and lower fullness rankings compared with individuals in the Concord grape juice group. There were no differences among the groups in plasma lipids, fasting glucose, insulin AUC, oral glucose tolerance and plasma oxygen radical absorbance capacity. The authors speculated that the lack of weight gain in the Concord grape juice group may have been because of dietary compensation and perhaps the polyphenol effect on thermogenesis or substrate oxidation (Hollis *et al.*, 2009).

In an animal study, Dal-Pan *et al.* investigated the metabolic effect of resveratrol supplementation on a non-human primate, the grey mouse lemur (*Microcebus murinus*). Resveratrol was supplemented for five weeks during the animals' seasonal weight gain and was found to reduce the gain. The animals on resveratrol had reduced food intake and an increase in resting metabolic rate but no alteration in plasma glucagon-like peptide 1, pancreatic polypeptide, peptide YY and glucose-dependent insulinotropic polypeptide, anorectic gut hormones known to influence food intake and metabolic rate, suggesting that the hormones were not the reason for the reduction in weight gain (Dal-Pan *et al.*, 2010). After one year of supplementation, animals on resveratrol still showed an increase in resting metabolic rate and daily energy expenditure and a decrease in food intake. The effect on body mass was apparent during their seasonal weight gain but was no different from the controls during the long day when they are more active (Dal-Pan *et al.*, 2011b).

Brasnyo *et al.* studied the effect of resveratrol in type 2 diabetic subjects. Resveratrol, at a moderate 10 mg/day dose for four weeks, increased the time for glucose to reach maximum after a test meal. Glucose was significantly lower 30 minutes after the test meal. No effect was seen on insulin, HOMAβ, amylin, glucagon-like peptide 1 or glucose-dependent insulinotropic peptide. The Akt signalling pathway is involved in insulin sensitivity and resveratrol consumption resulted in a significantly greater phosphorylated Akt in platelets. Oxidative stress was reduced by resveratrol, as evidenced by reduced *o*-tyrosine excretion. Insulin sensitivity (HOMAir) was significantly reduced compared with the placebo, which did not change over the four weeks (Brasnyo *et al.*, 2011).

Consumption of grapes and grape products, including the single compound, resveratrol, is associated with better glycemic control and reduced free radical damage, suggesting value as part of a healthy diet in combatting disease risk associated with obesity.

COGNITION

Memory and cognition are complex processes that may be influenced by the foods we eat. Memory is typically described as being composed of three types of information storage and three types of information processing. Information is stored in the Sensory Information Store (SIS), the Short-Term Store (STS), and the Long-Term Store (LTS). Information processing involves encoding (putting information into a store), maintenance (keeping it 'alive') and retrieval (finding encoded information). Cognition refers to the mental processes involved in the application of stored information, such as paying attention, remembering, producing and understanding language, solving problems and making decisions.

Research has examined the effect of grapes or grape products in memory and cognition. Chemical components of the grape may aid in prevention of age-related memory decline. As with cancer prevention studies, data from animal studies may be as important as those generated in human studies.

Older rats, aged 19–21 months, received 10 or 50% Concord grape juice versus a control beverage that was matched for calories, sugars and acidity. The group given 50% grape juice supplementation showed improved performance on motor tests, which rely on balance, coordination and strength. The group given 10% juice was not significantly different than placebo, except in the Morris water maze test of spatial memory. Dopamine release was significantly greater after consuming 10% grape juice, suggesting an improved sensitivity of the muscarinic receptor in the aging rat (Shukitt-Hale *et al.*, 2006).

In another animal model, the Tg2576 transgenic mouse model of Alzheimer's disease, *V. rotundifolia* (muscadine) grape juice and *V. vinifera* (cabernet sauvignon) grape juice improved spatial memory, as determined by the Morris water maze test. The animals also had decreased amyloid plaque in the cortex and hippocampus relative to controls. *In vitro* studies suggested that the *V. rotundifolia* improved spatial memory by a different mechanism than the *V. vinifera* (Ho *et al.*, 2009). In other studies by the same group, *V. vinifera* reduced the generation of Alzheimer's-type amyloid-beta peptides, while *V. rotundifolia* treatment interfered with the oligomerization of amyloid-beta peptides. *V. rotundifolia* prevented formation of soluble, high-molecular-weight, amyloid-beta oligomer species that are responsible for a cascade of cellular events that result in cognitive decline. *V. vinifera* aids in processing of amyloid precursor protein, which ultimately prevents the formation of amyloid-beta peptides (Wang *et al.*, 2006).

We conducted a study in which healthy subjects (aged 55–70 y) underwent a battery of cognitive tests before, during and after 10 weeks of Concord grape juice consumption (Rowe *et al.*, 2011). Animal studies focused on spatial memory, but in this study of the impact of grape juice on human cognition, a wide range of assessments were tested.

Healthy community-dwelling participants (N = 79; mean age = 58.1 y, SD = 6.3; mean education = 15.9 years SD = 3.2) were randomly assigned to receive either Concord grape juice (N = 40) or a calorie and sugar matched (52% fructose/48% glucose) placebo drink (N = 39), both of which were provided by Welch's, Concord, MA. Participants and experimenters were blinded to condition assignment. Each participant was instructed to drink 12 ounces of liquid throughout the day, every day for nine weeks. Blinding and compliance were assessed and found adequate.

The cognitive measures included the Brief Test of Adult Cognition by Telephone (Tun and Lachman 2006) which was administered at the beginning and end by telephone. In addition, participants completed six comparable self-testing packets at regular intervals across the 10-week period. These measures are listed in Table 9.4.

Traditional analyses of variance were conducted to identify potential changes in performance at 10 weeks (for the BTACT, using week 1 as a covariate) and across the six self-testing packets. None of these analyses showed any significant changes in performance as a function of grape juice consumption, although both groups gained from practice, as expected, across the testing interval. To consider potential *a priori* differences between the groups, and to minimize the impact of error in measurement, structural equation models were also employed to evaluate the impact of age, self-rated health, years of education, gender and grape juice condition on latent factors established for verbal memory, speed of processing and inductive reasoning. Even considering the sample size, none of the models were supported by the data, and none revealed any significant effects of grape juice consumption on cognition that were large enough to surpass the observed practice effects across both groups.

The lack of significant findings is likely due to the relatively young and healthy participants used in the study, or to a dose-response issue, that is, insufficient exposure to polyphenols with daily 12-ounce consumption. It may also be that a food's effect on mental health must

Table 9.4 Cognitive tests used in the study of Concord grape juice.

Brief Test of Adult Cognition by Telephone (BTACT)	Episodic verbal memory (immediate–delayed list recall, 15 items) Working memory (backwards digit span; sequences of 2–8 digits) Verbal fluency (category fluency, 1 min) Inductive reasoning (number series, 5 items) Speed of processing (backward counting from 100, 30 s) Reaction time (task switching, 52 trials)
Self-testing packets	Episodic memory, complex stimulus (immediate–delayed story recall, 15 sentences, 2-min study) Episodic verbal memory (immediate-delayed list recall, 15 items, 1 min study) Inductive reasoning (letter series, 30 problems, 4 min) Speed of processing (number comparison of 3–13 digits, 48 pairs to compare, 90 s)

be studied in a group already experiencing some cognitive decline, where the rate of decline might be slowed.

Krikorian *et al.* studied a group of older adults who displayed memory decline but who had not yet advanced to dementia. They enrolled twelve older adults with memory decline in a randomized, placebo-controlled, double-blind trial with Concord grape juice supplementation for 12 weeks. Significant improvement was found in verbal learning, as was a non-significant trend for enhanced verbal and spatial recall (Krikorian *et al.*, 2010).

Resveratrol as a single compound has been studied for its effect on cognition and mental health status. In a non-human primate model using young and old animals, resveratrol supplementation lowered body temperature and decreased locomotor activity (Pifferi *et al.*, 2011). Resveratrol was dosed at 200 mg/kg, which is roughly equivalent to 65 mg/kg for a human, as determined by body surface area. For a 70-kg man, this is 3.8 g per day, which would be considered a high dose relative to what can be consumed in supplement form. This same group studied the effect of resveratrol (200 mg/kg) on caloric restriction (70% of control *ad lib* calories) compared with control *ad lib* fed in the same primate model of adult male grey mouse lemurs (Dal-Pan *et al.*, 2011b). Resveratrol improved spatial memory performance and resulted in fewer errors in the circular platform task.

The same group conducted an 18-month treatment in which the effects of a moderate, chronic caloric restriction or an oral supplementation with resveratrol on cognitive and motor performances were studied in non-human primates, grey mouse lemurs. Compared with control animals, neither resveratrol nor caloric restriction changed motor performances, evaluated by rotarod and jump tests, but they did increase spontaneous locomotor activity. Working memory was improved by both treatments in the spontaneous alternation task. Resveratrol supplementation increased spatial memory performances in the circular platform task, although there was only a trend for caloric restriction. The data provided support for the growing evidence that nutritional interventions can have beneficial effects on brain functions even in adults (Dal-Pan *et al.*, 2011a).

Finally, in a human crossover study, resveratrol supplementation was studied in relation to cognitive performance and localized cerebral blood flow in healthy human adults. Forty-five minutes after a single dose, the presence of resveratrol and its conjugates was demonstrated in blood, and an increase in cerebral blood flow and enhanced oxygen extraction during a task were found (Kennedy *et al.*, 2010).

The data on the effect of grapes and grape products on human cognition are mixed, most likely because of the difficulties in interpreting data from the human models. These studies may not be sufficiently powered or may not be using the appropriate end points. More studies in this area are needed.

CONCLUSION

In general, studies on the effects of grapes and grape products are conducted using wine or juice, most likely because of the convenience and ease in distributing these products to subjects; beverages also have the ability to be concentrated for research purposes, although many studies use the actual products found in the aisles of grocery stores. Differentiating between products is important to get an accurate message to the consumer. Each of the studies reviewed in this chapter examined the effects of grape and grape products on different biomarkers and some of the findings differ. However, the evidence overwhelmingly indicates that consumption of grapes and grape products beneficially influence the human body in many ways and are a part of a healthful diet.

ACKNOWLEDGEMENTS

The authors would like to thank the graduate students, research participants, staff and colleagues who participated in the studies on grape products performed at the University of Florida. The support of IFAS Agricultural Experiment Station and the ICBR Flow Cytometry Core Facility is gratefully acknowledged. Welch Foods, Inc. and Florida Grape Growers Association also provided funding in support of the research program of SSP.

REFERENCES

Albers, A.R., Varghese, S., Vitseva, O. *et al.* (2004) The antiinflammatory effects of purple grape juice consumption in subjects with stable coronary artery disease. *Arterioscler.Thromb.Vasc.Biol.*, **24**, (11) e179–e180.

Banini, A.E., Boyd, L.C., Allen, J.C. *et al.* (2006) Muscadine grape products intake, diet and blood constituents of non-diabetic and type 2 diabetic subjects. *Nutrition.*, **22**, (11–12) 1137–1145.

Berli, F., D'Angelo, J., Cavagnaro, B. *et al.* (2008) Phenolic composition in grape (Vitis vinifera L. cv. Malbec) ripened with different solar UV-B radiation levels by capillary zone electrophoresis. *J Agric.Food Chem.*, **56**, (9) 2892–2898.

Bieger, J., Cermak, R., Blank, R. *et al.* (2008) Tissue distribution of quercetin in pigs after long-term dietary supplementation. *J Nutr.*, **138**, (8) 1417–1420.

Bitsch, R., Netzel, M., Frank, T. *et al.* (2004) Bioavailability and biokinetics of anthocyanins from red grape juice and red wine. *J Biomed.Biotechnol.*, **2004**, (5) 293–298.

Blanco-Colio, L.M., Valderrama, M., Alvarez-Sala, L.A. *et al.* (2000) Red wine intake prevents nuclear factor-kappaB activation in peripheral blood mononuclear cells of healthy volunteers during postprandial lipemia. *Circulation*, **102**, (9) 1020–1026.

Brasnyo, P., Molnar, G.A., Mohas, M. *et al.* (2011) Resveratrol improves insulin sensitivity, reduces oxidative stress and activates the Akt pathway in type 2 diabetic patients. *Br. J Nutr.*, **106**, (3) 383–389.

Brown, V.A., Patel, K.R., Viskaduraki, M. *et al.* (2010) Repeat dose study of the cancer chemopreventive agent resveratrol in healthy volunteers: safety, pharmacokinetics, and effect on the insulin-like growth factor axis. *Cancer Res.*, **70**, (22) 9003–9011.

Bub, A., Watzl, B., Heeb, D. *et al.* (2001) Malvidin-3-glucoside bioavailability in humans after ingestion of red wine, dealcoholized red wine and red grape juice. *Eur. J Nutr.*, **40**, (3) 113–120.

Castilla, P., Davalos, A., Teruel, J.L. *et al.* (2008) Comparative effects of dietary supplementation with red grape juice and vitamin E on production of superoxide by circulating neutrophil NADPH oxidase in hemodialysis patients. *Am. J Clin. Nutr.*, **87**, (4) 1053–1061.

Castilla, P., Echarri, R., Davalos, A. *et al.* (2006) Concentrated red grape juice exerts antioxidant, hypolipidemic, and antiinflammatory effects in both hemodialysis patients and healthy subjects. *Am. J Clin. Nutr.*, **84**, (1) 252–262.

Chiva-Blanch, G., Urpi-Sarda, M., Llorach, R. *et al.* (2012) Differential effects of polyphenols and alcohol of red wine on the expression of adhesion molecules and inflammatory cytokines related to atherosclerosis: a randomized clinical trial. *Am. J. Clin. Nutr.*, **95**, (2) 326–334.

Chou, E.J., Keevil, J.G., Aeschlimann, S. *et al.* (2001) Effect of ingestion of purple grape juice on endothelial function in patients with coronary heart disease. *Am. J. Cardiol.*, **88**, (5) 553–555.

Chow, H.H., Garland, L.L., Hsu, C.H. *et al.* (2010) Resveratrol modulates drug- and carcinogen-metabolizing enzymes in a healthy volunteer study. *Cancer Prev. Res. (Phila).*, **3**, (9) 1168–1175.

Coimbra, S.R., Lage, S.H., Brandizzi, L. *et al.* (2005) The action of red wine and purple grape juice on vascular reactivity is independent of plasma lipids in hypercholesterolemic patients. *Braz. J. Med. Biol. Res.*, **38**, (9) 1339–1347.

Dal-Pan, A., Blanc, S., and Aujard, F. (2010) Resveratrol suppresses body mass gain in a seasonal non-human primate model of obesity. *BMC Physiol.*, **10**:11. doi: 10.1186/1472-6793-10-11.

Dal-Pan, A., Pifferi, F., Marchal, J. *et al.* (2011a) Cognitive performances are selectively enhanced during chronic caloric restriction or resveratrol supplementation in a primate. *PLoSOne*, **6**, (1) e16581.

Dal-Pan, A., Terrien, J., Pifferi, F. *et al.* (2011b) Caloric restriction or resveratrol supplementation and ageing in a non-human primate: first-year outcome of the RESTRIKAL study in Microcebus murinus. *Age (Dordr.).*, **33**, (1) 15–31.

Davalos, A., Castilla, P., Gomez-Cordoves, C., and Bartolome, B. (2006) Quercetin is bioavailable from a single ingestion of grape juice. *Int.J Food Sci.Nutr.*, **57**, (5–6) 391–398.

Do, M.H., Lee, S.S., Kim, J.Y. *et al.* (2007) Fruits, vegetables, soy foods and breast cancer in pre- and postmenopausal Korean women: a case-control study. *Int. J. Vitam. Nutr. Res.*, **77**, (2) 130–141.

Dohadwala, M.M., Hamburg, N.M., Holbrook, M. *et al.* (2010) Effects of Concord grape juice on ambulatory blood pressure in prehypertension and stage 1 hypertension. *Am. J. Clin. Nutr.*, **92**, (5) 1052–1059.

Estruch, R., Sacanella, E., Mota, F. *et al.* (2011) Moderate consumption of red wine, but not gin, decreases erythrocyte superoxide dismutase activity: a randomised cross-over trial. *Nutr. Metab. Cardiovasc. Dis.*, **21**, (1) 46–53.

Frank, T., Netzel, M., Strass, G. *et al.* (2003) Bioavailability of anthocyanidin-3-glucosides following consumption of red wine and red grape juice. *Can. J. Physiol. Pharmacol.*, **81**, (5) 423–435.

Freedman, J.E., Parker, C., III, Li, L. *et al.* (2001) Select flavonoids and whole juice from purple grapes inhibit platelet function and enhance nitric oxide release. *Circulation*, **103**, (23) 2792–2798.

Gresele, P., Pignatelli, P., Guglielmini, G. *et al.* (2008) Resveratrol, at concentrations attainable with moderate wine consumption, stimulates human platelet nitric oxide production. *J. Nutr.*, **138**, (9) 1602–1608.

Halvorsen, B.L., Carlsen, M.H., Phillips, K.M. *et al.* (2006) Content of redox-active compounds (ie, antioxidants) in foods consumed in the United States. *Am. J. Clin. Nutr.*, **84**, (1) 95–135.

Hansen, A.S., Marckmann, P., Dragsted, L.O. *et al.* (2005) Effect of red wine and red grape extract on blood lipids, haemostatic factors, and other risk factors for cardiovascular disease. *Eur. J. Clin. Nutr.*, **59**, (3) 449–455.

Hashemi, M., Kelishadi, R., Hashemipour, M. *et al.* (2010) Acute and long-term effects of grape and pomegranate juice consumption on vascular reactivity in paediatric metabolic syndrome. *Cardiol. Young.*, **20**, (1) 73–77.

Ho, L., Chen, L.H., Wang, J. *et al.* (2009) Heterogeneity in red wine polyphenolic contents differentially influences Alzheimer's disease-type neuropathology and cognitive deterioration. *J. Alzheimers. Dis.*, **16**, (1) 59–72.

Hollis, J.H., Houchins, J.A., Blumberg, J.B., and Mattes, R.D. (2009) Effects of concord grape juice on appetite, diet, body weight, lipid profile, and antioxidant status of adults. *J. Am. Coll. Nutr.*, **28**, (5) 574–582.

Huang, P.H., Chen, Y.H., Tsai, H.Y. *et al.* (2010) Intake of red wine increases the number and functional capacity of circulating endothelial progenitor cells by enhancing nitric oxide bioavailability. *Arterioscler. Thromb. Vasc. Biol.*, **30**, (4) 869–877.

Ingersoll, G.L., Wasilewski, A., Haller, M. *et al.* (2010) Effect of concord grape juice on chemotherapy-induced nausea and vomiting: results of a pilot study. *Oncol. Nurs. Forum.*, **37**, (2) 213–221.

Jung, K.J., Wallig, M.A. and Singletary, K.W. (2006) Purple grape juice inhibits 7,12-dimethylbenz[a]anthracene (DMBA)-induced rat mammary tumorigenesis and in vivo DMBA-DNA adduct formation. *Cancer Lett.*, **233**, (2) 279–288.

Kamiyama, M., Kishimoto, Y., Tani, M. *et al.* (2009) Inhibition of low-density lipoprotein oxidation by Nagano purple grape (Vitis viniferaxVitis labrusca). *J. Nutr. Sci. Vitaminol. (Tokyo).*, **55**, (6) 471–478.

Keevil, J.G., Osman, H.E., Reed, J.D., and Folts, J.D. (2000) Grape juice, but not orange juice or grapefruit juice, inhibits human platelet aggregation. *J. Nutr.*, **130**, (1) 53–56.

Kennedy, D.O., Wightman, E.L., Reay, J.L. *et al.* (2010) Effects of resveratrol on cerebral blood flow variables and cognitive performance in humans: a double-blind, placebo-controlled, crossover investigation. *Am. J. Clin. Nutr.*, **91**, (6) 1590–1597.

Krikorian, R., Nash, T.A., Shidler, M.D. *et al.* (2010) Concord grape juice supplementation improves memory function in older adults with mild cognitive impairment. *Br. J. Nutr.*, **103**, (5) 730–734.

Lago-Vanzela, E.S., Da-Silva, R., Gomes, E. *et al.* (2011a) Phenolic composition of the Brazilian seedless table grape varieties BRS Clara and BRS Morena. *J. Agric. Food Chem.*, **59**, (15) 8314–8323.

Lago-Vanzela, E.S., Da-Silva, R., Gomes, E. *et al.* (2011b) Phenolic composition of the edible parts (flesh and skin) of Bordo Grape (Vitis labrusca) using HPLC-DAD-ESI-MS/MS. *J. Agric. Food Chem.*, **59**, (24) 13136–13146.

Leblanc, M.R., Johnson, C.E. and Wilson, P.W. (2008) Influence of pressing method on juice stilbene content in muscadine and bunch grapes. *J. Food Sci.*, **73**, (4) H58–H62.

Lekakis, J., Rallidis, L.S., Andreadou, I. *et al.* (2005) Polyphenolic compounds from red grapes acutely improve endothelial function in patients with coronary heart disease. *Eur. J. Cardiovasc. Prev. Rehabil.*, **12**, (6) 596–600.

McCullough, M.L., Peterson, J.J., Patel, R. *et al.* (2012) Flavonoid intake and cardiovascular disease mortality in a prospective cohort of US adults. *Am. J. Clin. Nutr.*, **95**, (2) 454–464.

Meng, X., Maliakal, P., Lu, H. *et al.* (2004) Urinary and plasma levels of resveratrol and quercetin in humans, mice, and rats after ingestion of pure compounds and grape juice. *J. Agric. Food Chem.*, **52**, (4) 935–942.

Mercer-Hursh, T. (2009) Grape Varieties Database. http://a-muse-in-the-cellar.com/content/grape-varieties-database (last accessed 4 March 2013).

Mertens-Talcott, S.U. and Percival, S.S. (2005) Ellagic acid and quercetin interact synergistically with resveratrol in the induction of apoptosis and cause transient cell cycle arrest in human leukemia cells. *Cancer Letters*, **218**, (2) 141–151.

Mertens-Talcott, S.U., Talcott, S.T., and Percival, S.S. (2003) Low concentrations of quercetin and ellagic acid synergistically influence proliferation, cytotoxicity and apoptosis in MOLT-4 human leukemia cells. *J. Nutr.*, **133**, (8) 2669–2674.

O'Byrne, D.J., Devaraj, S., Grundy, S.M., and Jialal, I. (2002) Comparison of the antioxidant effects of Concord grape juice flavonoids alpha-tocopherol on markers of oxidative stress in healthy adults. *Am. J. Clin. Nutr.*, **76**, (6) 1367–1374.

Olthof, M.R., Hollman, P.C., Vree, T.B., and Katan, M.B. (2000) Bioavailabilities of quercetin-3-glucoside and quercetin-4′-glucoside do not differ in humans. *J. Nutr.*, **130**, (5) 1200–1203.

Park, Y.K., Kim, J.S. and Kang, M.H. (2004) Concord grape juice supplementation reduces blood pressure in Korean hypertensive men: double-blind, placebo controlled intervention trial. *Biofactors*, **22**, (1–4) 145–147.

Pastrana-Bonilla, E., Akoh, C.C., Sellappan, S., and Krewer, G. (2003) Phenolic content and antioxidant capacity of muscadine grapes. *J. Agric. Food Chem.*, **51**, (18) 5497–5503.

Patel, K.R., Brown, V.A., Jones, D.J. *et al.* (2010) Clinical pharmacology of resveratrol and its metabolites in colorectal cancer patients. *Cancer Res.*, **70**, (19) 7392–7399.

Percival, S.S. and Sims, C.A. (2000) Wine modifies the effects of alcohol on immune cells of mice. *J. Nutr.*, **130**, (5) 1091–1094.

Perez-Trujillo, J.P., Hernandez, Z., Lopez-Bellido, F.J., and Hermosin-Gutierrez, I. (2011) Characteristic phenolic composition of single-cultivar red wines of the Canary Islands (Spain). *J. Agric. Food Chem.*, **59**, (11) 6150–6164.

Pifferi, F., Dal-Pan, A., Menaker, M., and Aujard, F. (2011) Resveratrol dietary supplementation shortens the free-running circadian period and decreases body temperature in a prosimian primate. *J. Biol. Rhythms.*, **26**, (3) 271–275

Pignatelli, P., Lenti, L., Pulcinelli, F.M. *et al.* (2002) Red and white wine differently affect collagen-induced platelet aggregation. *Pathophysiol. Haemost. Thromb.*, **32**, (5–6) 356–358.

Pignatelli, P., Ghiselli, A., Buchetti, B. *et al.* (2006) Polyphenols synergistically inhibit oxidative stress in subjects given red and white wine. *Atherosclerosis*, **188**, (1) 77–83.

Rakici, O., Kiziltepe, U., Coskun, B. *et al.* (2005) Effects of resveratrol on vascular tone and endothelial function of human saphenous vein and internal mammary artery. *Int. J. Cardiol.*, **105**, (2) 209–215.

Rathel, T.R., Samtleben, R., Vollmar, A.M., and Dirsch, V.M. (2007) Activation of endothelial nitric oxide synthase by red wine polyphenols: impact of grape cultivars, growing area and the vinification process. *J. Hypertens.*, **25**, (3) 541–549.

Rowe, C.A., Nantz, M.P., Nieves, C., Jr *et al.* (2011) Regular consumption of concord grape juice benefits human immunity. *J. Med. Food.*, **14**, (1–2) 69–78.

Serra, A., Macia, A., Romero, M.P. *et al.* (2010) Bioavailability of procyanidin dimers and trimers and matrix food effects in in vitro and in vivo models. *Br. J. Nutr.*, **103**, (7) 944–952.

Shukitt-Hale, B., Carey, A., Simon, L. *et al.* (2006) Effects of Concord grape juice on cognitive and motor deficits in aging. *Nutrition*, **22**, (3) 295–302.

Singletary, K.W., Stansbury, M.J., Giusti, M. *et al.* (2003) Inhibition of rat mammary tumorigenesis by concord grape juice constituents. *J. Agric. Food Chem.*, **51**, (25) 7280–7286.

Spaak, J., Merlocco, A.C., Soleas, G.J. *et al.* (2008) Dose-related effects of red wine and alcohol on hemodynamics, sympathetic nerve activity, and arterial diameter. *Am. J. Physiol. Heart Circ. Physiol.*, **294**, (2) H605–H612.

Stalmach, A., Edwards, C.A., Wightman, J.D. and Crozier, A. (2011) Identification of (poly)phenolic compounds in concord grape juice and their metabolites in human plasma and urine after juice consumption. *J. Agric. Food Chem.*, **59**, (17) 9512–9522.

Tun, P.A. and Lachman, M.E. (2006) Telephone assessment of cognitive function in adulthood: the Brief Test of Adult Cognition by Telephone. *Age Ageing*, **35**, (6) 629–632.

USDA (United States Department of Agriculture) (2013) Noncitrus Fruits and Nuts, 2012 Preliminary Summary. http://www.usda.gov/nass/PUBS/TODAYRPT/ncit0113.pdf (last accessed 20 March 2013)

van Dorsten, F.A., Grun, C.H., van Velzen, E.J. *et al.* (2010) The metabolic fate of red wine and grape juice polyphenols in humans assessed by metabolomics. *Mol. Nutr. Food Res.*, **54**, (7) 897–908.

Vinson, J.A., Yang, J., Proch, J. and Liang, X. (2000) Grape juice, but not orange juice, has *in vitro*, *ex vivo*, and *in vivo* antioxidant properties. *J. Med. Food.*, **3**, (4) 167–171.

Wang, J., Ho, L., Zhao, Z. *et al.* (2006) Moderate consumption of Cabernet Sauvignon attenuates Abeta neuropathology in a mouse model of Alzheimer's disease. *FASEB J.*, **20**, (13) 2313–2320.

Wang, Z., Huang, Y., Zou, J. *et al.* (2002) Effects of red wine and wine polyphenol resveratrol on platelet aggregation in vivo and in vitro. *Int. J. Mol. Med.*, **9**, (1) 77–79.

Wannamethee, S.G., Lowe, G.D., Shaper, G. *et al.* (2003) The effects of different alcoholic drinks on lipids, insulin and haemostatic and inflammatory markers in older men. *Thromb. Haemost.*, **90**, (6) 1080–1087.

Watzl, B., Bub, A., Briviba, K. and Rechkemmer, G. (2002) Acute intake of moderate amounts of red wine or alcohol has no effect on the immune system of healthy men. *Eur. J. Nutr.*, **41**, (6) 264–270.

Watzl, B., Bub, A., Pretzer, G. *et al.* (2004) Daily moderate amounts of red wine or alcohol have no effect on the immune system of healthy men. *Eur. J. Clin. Nutr.*, **58**, (1) 40–45.

Zern, T.L., Wood, R.J., Greene, C. *et al.* (2005) Grape polyphenols exert a cardioprotective effect in pre- and postmenopausal women by lowering plasma lipids and reducing oxidative stress. *J. Nutr.*, **135**, (8) 1911–1917.

10 Potential Health Benefits of Blackcurrants

Rosalind Miller, Sophie Putnam, Michael Edwards, Gary Woodward and Colin Kay

INTRODUCTION

Berry fruits are reportedly consumed in relatively low quantities in the UK diet, in comparison to other fruits such as apples and bananas (Henderson *et al.*, 2002). However, anthocyanins from berries were reported to make up 10.1% of the overall anthocyanin intake in Northern European subjects, which is only marginally lower than the amount of anthocyanins coming from other more commonly consumed fruits, such as grapes, apples and pears (approximately 12.7% in Northern Europe) (Zamora-Ros *et al.*, 2011). Despite the reported low rate of consumption of anthocyanins from berries, the small soft-fleshed vibrantly coloured fruits have been assigned numerous health benefits (Seeram, 2008). Although the fibre, vitamin and mineral content of these fruits are significant, the health benefits are often associated with the high levels of polyphenols present in the fruit. The presence and concentration of specific polyphenols varies from berry to berry and, as such, differing health promoting activities may be evident. The current chapter focuses on the activities of blackcurrants, blackcurrant extracts and anthocyanins (either extracted and purified from blackcurrants or synthetically derived) that are present in significant quantities in blackcurrants. Numerous other berry species have been explored for their health effects, including blueberries, blackberries, elderberries, chokeberries, strawberries and raspberries, to name a few (Seeram, 2008). Below a few commonly consumed species are highlighted. Other chapters provide more detailed reviews of the health benefits of other fruits.

Strawberries, *Fragari* × *ananassa,* are an aggregate accessory fruit first cultivated in France and are well known for their red flesh and sweet aromatic flavour. There is limited research on the health benefits of strawberries. However, strawberry consumption has been associated with reduced risk of cardiovascular disease (CVD) (Mink *et al.*, 2007; Sesso *et al.*, 2007; Huntley, 2009). Human intervention studies have also shown decreases in oxidative stress and lipid peroxidation (Jenkins *et al.*, 2008; Cao *et al.*, 1998; Basu *et al.*, 2009; Burton-Freeman *et al.*, 2010; Tulipani *et al.*, 2011), hypolipidaemic effects (Erlund *et al.*, 2008; Basu *et al.*, 2009; Burton-Freeman *et al.*, 2010), anti-inflammatory (Edirisinghe *et al.*, 2011a; Ellis *et al.*, 2011) and antidiabetic effects (Edirisinghe *et al.*, 2011a). Cancer chemopreventative effects have also been demonstrated in cell (Xue *et al.*, 2001, Wang *et al.*, 2005) and animal studies (Carlton *et al.*, 2001), but further evidence is required to elucidate the health benefits of strawberries.

Bioactives in Fruit: Health Benefits and Functional Foods, First Edition. Edited by Margot Skinner and Denise Hunter.

The red raspberry, *Rubus idaeus,* is also an aggregate fruit native to Europe and northern Asia and is commonly consumed across the world. The black raspberry, *Rubus occidentalis,* is less commonly grown and consumed, but interestingly is more frequently studied due to its high polyphenol content. Human intervention studies are again limited, but possible anticarcinogenic effects of raspberries, in particular black raspberries, are beginning to emerge (Kresty *et al.*, 2006; Weisel *et al.*, 2006; Stoner *et al.*, 2007; Mallary *et al.*, 2008; Wang L *et al.*, 2011a).

The blackcurrant, *Ribes nigrum*, is a small perennial shrub native to Europe and northern Asia which produces small, dark purple edible berries. The berries and leaves have been used in traditional herbal medicine in both Europe and Asia for a variety of health problems and over the past 25 years scientific interest in these berries has been accumulating. The multitude of nutrients within the fruit includes structural and non-structural carbohydrates, polyunsaturated fatty acids, ascorbic acid, potassium, calcium, magnesium and iron (Gopalan *et al.*, 2012). These nutrients may contribute to the perceived health effects of blackcurrants but it is the tannins, stilbenoids and flavonoids which have elicited the most interest. One particular group of flavonoids, the anthocyanins, which are up to fourfold more abundant in the blackcurrant compared to other common fruits, have been suggested to be at least partly responsible for the proposed health benefits. The anthocyanins found in abundance in blackcurrants are the main focus of this chapter (delphinidin-3-*O*-glucoside, delphinidin-3-*O*-rutinoside, cyanidin-3-*O*-glucoside and cyanidin-3-*O*-rutinoside) (Gopalan *et al.*, 2012). However, it should be noted that anthocyanins are not limited to blackcurrants and are also commonly found in other dark red and purple fruits and vegetables. Therefore, any health benefits attributed to anthocyanins are relevant, but may not be unique to blackcurrants.

Occurrence of anthocyanins in the diet

Anthocyanins are commonly occurring plant pigments found in abundance in flowers, fruits and vegetables, where they provide red, blue and purple colouration (Kong *et al.*, 2003; McGhie and Walton, 2007; Crozier *et al.*, 2009). In nature, anthocyanins occur exclusively as glycoside (bound to a sugar moiety) or acylglycoside conjugates of anthocyanidins (which represent the aglycone form) (Wu *et al.*, 2006; Crozier *et al.*, 2009). Approximately 17 anthocyanidins occur in plants, of which six (cyanidin, delphinidin, malvidin, pelargonidin, peonidin and petunidin) are distributed ubiquitously (Wu *et al.*, 2006; Crozier *et al.*, 2009). Berry fruits contain the highest concentrations of anthocyanins when compared with most other dietary sources (Mazza and Miniati, 1993), with cyanidin displaying the widest distribution in foods (Table 10.1), although with considerable variation in concentration (Manach *et al.*, 2005; Wu *et al.*, 2006; Zamora-Ros *et al.*, 2011). Of the many anthocyanin species commonly present in berry fruits, cyanidin, pelargonidin and peonidin are the most prevalent, with cyanidin-3-*O*-glucoside, cyanidin-3-*O*-rutinoside, cyanidin-3-*O*-galactoside, cyanidin-3-*O*-arabinoside and peonidin-3-*O*-glucoside being the most abundant (Maatta-Riihinen *et al.*, 2004; Wu *et al.*, 2004; Tian *et al.*, 2005; Wu and Prior, 2005; Netzel *et al.*, 2006; Seeram *et al.*, 2006; Lin and Harnly, 2007). A detailed catalogue of anthocyanin-containing foods and their respective content can be found in the USDA (USDA, 2011) and Phenol-Explorer (Neveu *et al.*, 2010) databases. Although the anthocyanin content of berries varies depending on a number of factors (including geographical location, climate and method of cultivation amongst others), estimated anthocyanin content has been reported for the majority of berry fruits.

Table 10.1 Anthocyanin composition of commonly consumed dietary sources.[1]

Food	Anthocyanidin aglycones (milligrams) per 100 grams fresh weight[2]						
	Cy	***Dp***	***Mv***	***Pg***	***Pn***	***Pt***	***Total***
Blackberries, raw	90.31	—	—	0.15	—	—	90.46
Blueberries, wild raw	42.47	92.71	103.80	—	23.49	58.23	320.70
Cranberries, raw	41.81	7.66	0.310	—	42.10	—	91.88
Currants, European black, raw	85.63	181.11	—	1.17	0.66	3.87	272.44
Eggplant (aubergine), raw	0.02	13.76	—	0.02	—	—	13.80
Elderberries, raw	758.48	—	61.35	1.13	—	—	820.96
Grapes, red, raw	1.46	3.67	34.71	0.02	2.89	2.11	44.86
Pistachio nuts, raw	6.06	—	—	—	—	—	6.06
Radishes, raw	—	—	—	25.66	—	—	25.66
Raspberries, black	323.47	—	—	0.15	0.55	—	324.17
Raspberries, red, raw	35.84	0.29	0.70	1.85	—	—	38.68
Strawberries, raw	1.96	0.32	—	31.27	—	0.08	33.63

[1]Prominent food sources of anthocyanins and their amounts are derived from the food composition Appendix table (containing 1877 food items) provided by Zamora-Ros *et al.* (2011), based upon dietary recall data from 36 037 participants in the European Prospective Investigation into Cancer and Nutrition (EPIC) study (covering 27 centres in ten European countries).
[2]*Anthocyanins* calculated as sum of available forms (glycosides and aglycones) in selected raw food sources (Zamora-Ros *et al.*, 2011).
Abbreviations: Cy, cyanidin; Dp, delphinidin; Mv, malvidin; Pg, pelargonidin; Pn, peonidin; Pt, petunidin.

The anthocyanin content of raspberries is approximately 1037 mg/100 g fresh weight, occurring mainly as cyanidin-3-*O*-rutinoside and cyanidin-3-*O*-sophoroside, with trace amounts of cyanidin-3-*O*-glucoside and pelargonidin-3-*O*-rutinsoide. The anthocyanins in blackcurrants are predominantly delphinidin-3-*O*-glucoside, delphinidin-3-*O*-rutinoside, cyanidin-3-*O*-glucoside, and cyanidin-3-*O*-rutinoside. With regard to relative hierarchical content of anthocyanins in berries, chokeberries have one of the highest known anthocyanin concentrations (1478.1 mg/100 g), which is derived almost exclusively from cyanidin glycosides (Chandra *et al.*, 2001; Kay *et al.*, 2004; Maatta-Riihinen *et al.*, 2004; Nakajima *et al.*, 2004; Wu *et al.*, 2004), followed by elderberries (1370.1 mg/100 g; primarily cyanidin glycosides) (Chandra *et al.*, 2001; Nakajima *et al.*, 2004; Wu *et al.*, 2004; Neveu *et al.*, 2010), blackcurrants (592.23 mg/100 g; primarily delphinidin-3-*O*-rutinoside), lowbush blueberries (187.23 mg/100 g; primarily malvidin-3-*O*-glucoside), strawberries (73.00 mg/100 g; primarily pelargonidin-3-*O*-glucoside) and black grapes (72.10 mg/100 g; primarily malvidin-3-*O*-glucoside) (Neveu *et al.*, 2010).

Anthocyanin intake in the United States of America has been estimated at 12.5 mg/day/person (Wu *et al.*, 2006), although dietary intakes vary within and between regions. The average daily intake in Germany was reported as 2.7 mg/person in 2002 (Wu *et al.*, 2006), whereas a mean anthocyanidin intake of 18.88 mg/day was described for a Spanish adult population (Zamora-Ros *et al.*, 2010), and mean dietary intake in Finland has been estimated at 82 mg/day (Manach *et al.*, 2005). Whilst mean intakes of anthocyanins vary significantly between populations, intake levels also range significantly within populations, with ranges reported between 4.8 mg/day and 19.4 mg/day in US cohort studies (range between quintile 1 and 5) (Cassidy *et al.*, 2012; McCullough *et al.*, 2012). The high content of anthocyanins found in berries indicates that a substantial daily intake may be achieved with regular consumption of berries and berry-derived beverages (Manach *et al.*, 2005; Wu *et al.*, 2006), such that doses of 400–500 mg of anthocyanins could be consumed in a single serving (Kay *et al.*, 2009). This is of particular relevance for berry and grape juices, which could therefore serve as highly concentrated sources of dietary anthocyanins. As a large proportion of berry fruits are consumed as commercially processed products, the preservation of anthocyanins during industrial processing and storage remains a focus in the food industry, both in terms of preserving colour and potential biological activity of processed food products (Woodward *et al.*, 2011a). For example, the processing of fresh berries to juice concentrates is a multistage process, including milling, pressing, pasteurisation, filtration, clarification and concentration. The accumulated data from several studies indicate that anthocyanins remain relatively stable during fruit juice processing and are found in high concentrations in the final processed juice (Skrede *et al.*, 2000; Lee *et al.*, 2002; Fang *et al.*, 2005; Woodward *et al.*, 2011a). However, studies report a loss of anthocyanins during prolonged storage (Hollands *et al.*, 2008); this does not appear to be the result of anthocyanin degradation (Woodward *et al.*, 2011a) but is likely the result of adduct formation or polymerisation in concentrated juices (Rein *et al.*, 2005; Castañeda-Ovando *et al.*, 2009). Identifying these anthocyanin products in stored juices remains an active area of research as it may impact (either positively or negatively) the quality and health-related benefits of commercial berry products.

Anthocyanin classification and structural properties

Anthocyanins are derivatives of flavylium (2-phenylbenzopyrylium) salts characterized by their glycoside and polyhydroxy/methoxy moieties (Kong *et al.*, 2003). In common with other flavonoids, the anthocyanin molecule contains two benzoyl rings (A and B), linked by

(a)

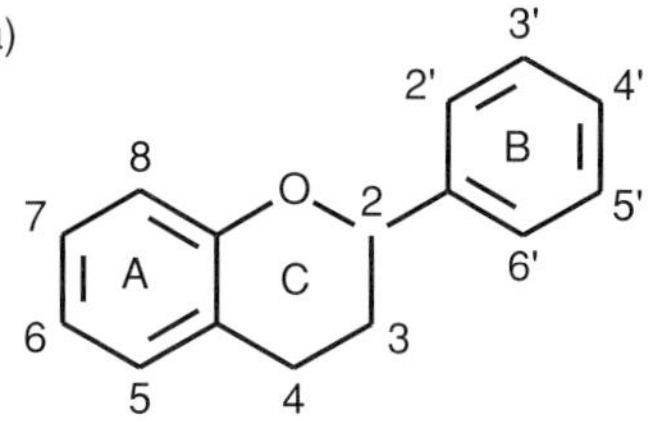

(b)

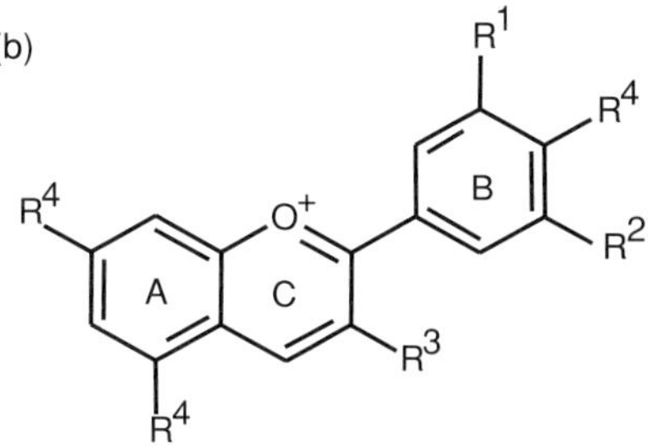

Anthocyanidin	Substituent	
	R^1	R^2
Pelargonidin	–H	–H
Cyanidin	–OH	–H
Peonidin	$-OCH_3$	–H
Delphinidin	–OH	–OH
Petunidin	$-OCH_3$	–OH
Malvidin	$-OCH_3$	$-OCH_3$

Fig. 10.1 Chemical structure of flavonoid 'backbone' (a) and an anthocyanin molecule (b). (a) Basic flavonoid structure depicting two aromatic rings (A-ring and B-ring) linked by a heterocyclic C-ring [adapted from (Erdman *et al.*, 2007)]. (b) Flavylium cation form of anthocyanin and attached substituent groups. R^1, R^2, hydrogen (H), hydroxyl (OH), methoxy (OCH_3); R^3, R^4, hydrogen (H) or sugar (adapted from Kong *et al.*, 2003).

a heterocyclic oxygen ring, giving the characteristic C_6–C_3–C_6 structure (Figure 10.1) (Kong *et al.*, 2003; McGhie and Walton, 2007; Wallace, 2011). Commonly occurring glycosides are 3-monosides, 3-biosides (disaccharides), 3-triosides and 3,5-diglycosides; associated with the sugars glucose, galactose, rhamnose, arabinose and xylose (Kong *et al.*, 2003; McGhie and Walton, 2007).

Anthocyanin stability

Anthocyanins consist of a chromophore formed by eight conjugated double bonds and a positively charged oxygen in the heterocyclic C-ring, forming the flavylium cation (Figure 10.2) (Wallace, 2011). However, the stability of this flavylium cation is significantly influenced by pH, and within aqueous solutions several molecular forms of anthocyanins exist in dynamic equilibrium (Figure 10.2) (McGhie and Walton, 2007). At pH less than 2, the flavylium cation is the predominant form present, giving rise to a characteristic red colour (McGhie and Walton, 2007; Wallace, 2011). However, rapid de-protonation occurs with increasing pH, to generate a quinonoidal structure (pH 6–8) and resulting blue colouration (Fleschhut *et al.*, 2006; McGhie and Walton, 2007). A concurrent but substantially slower hydration of the flavylium cation also ensues to generate a colourless hemiketal form (pH 4–5), which then tautomerizes through opening of the C-ring to yield colourless *cis* and *trans* chalcone forms (pH 7–8) (Fleschhut *et al.*, 2006; McGhie and Walton, 2007).

Anthocyanidins, the aglycone form of anthocyanins, are formed by hydrolysis of the glycoside linkage. Anthocyanidins are not present in nature but may be generated following biological, enzymatic or acid hydrolysis (Keppler and Humpf, 2005; Fleschhut *et al.*,

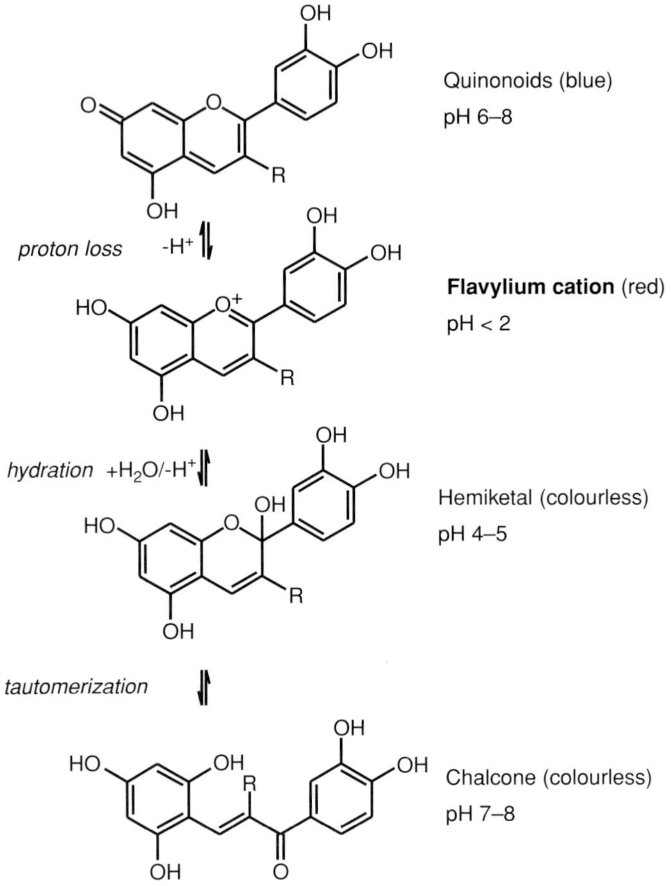

Fig. 10.2 Anthocyanin molecular structures generated under varying pH conditions (adapted from Fleschhut *et al.*, 2006; McGhie and Walton, 2007; Del Rio *et al.*, 2010).

2006). Anthocyanidins form highly reactive α-diketone structures at pH > 3, which are susceptible to decomposition into monomeric phenolic acid and aldehyde compounds (Figure 10.3) (Keppler and Humpf, 2005). The presence of the sugar moiety confers a degree of stability upon anthocyanins, as compared with the aglycone form (Fleschhut *et al.*, 2006), but degradation of glycosides also occurs at a more alkaline pH (Woodward *et al.*, 2009; Kay *et al.*, 2009).

Stability under physiological conditions

On entry into the body, anthocyanins may undergo significant structural modification. Regions of the gastrointestinal tract are characterized by their differential pH, enzymatic and microbial characteristics, each having differing effects on the molecular structure of anthocyanins. Many bioavailability studies have consistently shown that less than 1% of their ingested intact/parent structure is recovered in the systemic circulation (Manach *et al.*, 2005). Currently, no conclusive evidence has yet accounted for their disappearance following ingestion; however, their extensive degradation and subsequent metabolism has been proposed as the cause of their low recoveries (McGhie and Walton, 2007; Williamson and Clifford, 2010).

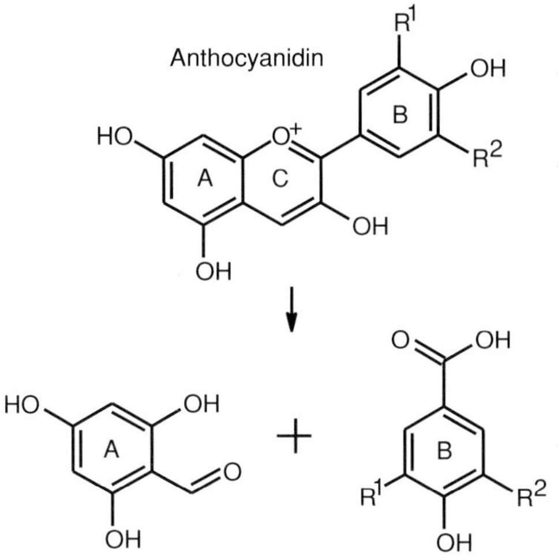

Phloroglucinol aldehyde Phenolic acid

Anthocyanidin		B-ring degradation product (phenolic acid)
Cyanidin	→	Protocatechuic acid
Delphinidin	→	Gallic aid
Malvidin	→	Syringic acid
Pelargonidin	→	4-hydroxybenzoic acid
Peonidin	→	Vanillic acid
Petunidin	→	3-methylgallic acid

Fig. 10.3 Monomeric phenolic acids and aldehyde formed by anthocyanidin degradation (adapted from Keppler and Humpf, 2005; Fleschhut *et al.*, 2006; Forester and Waterhouse, 2008; Williamson and Clifford, 2010).

It has long been known that anthocyanins are prone to degradation, forming smaller phenolic constituents (Markakis, 1974; Keppler and Humpf, 2005) and the extent to which this occurs is influenced by their structure. Owing to the steric demand of the glycoside conjugates, a distinction between anthocyanin and anthocyanidin stability is evident, where anthocyanins (glycosides) show greater pH stability than their respective aglycones (Fleschhut *et al.*, 2006). Evidence suggests that anthocyanin glycosides are most stable at pH 2, with moderate instability between pH 3.3 and 3.8 (Nielsen *et al.*, 2003b), and increasing instability above pH 3.8–4.5 and beyond (Nielsen *et al.*, 2003b). At pH 7.4 and 37°C, the anthocyanidins cyanidin, delphinidin, peonidin and malvidin have been shown to degrade completely within 60 minutes, with delphinidin characterized as the least stable, followed by malvidin, cyanidin, peonidin and pelargonidin, respectively (Fleschhut *et al.*, 2006; Kern *et al.*, 2007). A recent study demonstrated that in physiological buffer (at pH 7.4), cyanidin, delphinidin and pelargonidin underwent rapid degradation with the generation of phenolic acids and a common aldehyde (Woodward *et al.*, 2009). Decomposition of the respective 3-glucoside conjugates of these anthocyanins accounted for total losses of 20%, 40% and 100% over 12 hours (for pelargonidin-3-*O*-glucoside, cyanidin-3-*O*-glucoside and delphinidin-3-*O*-glucoside, respectively), with increased anthocyanin B-ring hydroxylation associated with a decrease in stability (Woodward *et al.*, 2009). Moreover, Kay *et al.* (2009) examined the stability of cyanidin and cyanidin-3-*O*-glucoside during incubation in phosphate buffer and cell culture media (Kay *et al.*, 2009) and reported losses of 96% and 57% for cyanidin and its 3-glucoside respectively, after a four-hour incubation in cell-free Dulbecco's Modified Eagle Medium (pH 7). Losses mirrored those observed in cultured caco-2 (intestinal epithelial) cells, suggesting anthocyanin degradation was spontaneous; degradation products were identified as protocatechuic acid and phloroglucinol aldehyde (Kay *et al.*, 2009).

Using *in vitro* models of gastric and pancreatic digestion (McDougall *et al.*, 2005a, 2005b), it was shown that the recovery of pure phenolic acids following peptic digestion was near 100% (McDougall *et al.*, 2005a), while the recovery of anthocyanins was between 70 and 90%. Furthermore, poor anthocyanin recovery has been reported following *in vitro* pancreatic and bile salt digestion (pH 6–7), representing intestinal processing, with a less than 40% recovery of the anthocyanins reported (Perez-Vicente *et al.*, 2002; McDougall *et al.*, 2005a, 2005b; Woodward *et al.*, 2011b). This evidence suggests that anthocyanins remain relatively stable during peptic digestion (representing digestion in the stomach at pH 2) and are likely passed intact to the upper intestinal tract, where they rapidly degrade in the higher pH environment.

Anthocyanin metabolism and bioavailability

Anthocyanidins are predominantly more hydrophobic than their glycoside counterparts and, although unstable, theoretically could be absorbed via passive diffusion. In contrast, anthocyanins (glycoside conjugates) are more hydrophilic and likely require either an active transport system or hydrolysis prior to absorption (Kay, 2006). Nevertheless, the exact mechanism of anthocyanin absorption has yet to be sufficiently resolved but previous *in vitro* evidence has been presented suggesting mechanisms involving both the sodium-dependent glucose cotransporter (SGLT) (Yi *et al.*, 2006) or extracellular hydrolysis via lactase phloridzin hydrolase (LPH) followed by passive diffusion (Day *et al.*, 2000). However, studies by Hollman *et al.* (1999) and Arts *et al.* (2004) demonstrated that inhibition of both SGLT and LPH did not significantly reduce the absorption of cyanidin-3-*O*-glucoside by jejunal tissue (Hollman *et al.*, 1999; Arts *et al.*, 2004). Alternatively, anthocyanins, or conjugated metabolites, may also serve as substrates for the family of intestinal ATP-binding cassette active transporter proteins, mediating efflux into the intestinal lumen or across enterocyte basolateral membranes (Brand *et al.*, 2006). Despite evidence of these possible routes of absorption, recent studies indicating the rapid degradation of anthocyanins to phenolic intermediates under physiological conditions suggests lower molecular weight phenolic degradants may be the primary absorbed form of anthocyanins; although this has yet to be established in human feeding studies.

In relation to other flavonoids, anthocyanins have invariably been considered to be poorly bioavailable (McGhie and Walton, 2007). Manach *et al.* (2005) and Kay (2006) reviewed feeding studies which generally describe rapid excretion of anthocyanins (6–8 hours), with little of the ingested dose appearing in urine (range 0.003–5%; <1% on average) (Manach *et al.*, 2005; Kay, 2006; McGhie and Walton, 2007). Moreover, maximal plasma concentrations are generally reached in a relatively short period (range 0.3–4.0 hours; 1.5 hours on average) at levels in the low nanomolar range (range 0.0024–592 nM; ~90 nM on average)(Miyazawa *et al.*, 1999; Bub *et al.*, 2001; Cao *et al.*, 2001; Matsumoto *et al.*, 2001; Mazza *et al.*, 2002; Wu *et al.*, 2002; Frank *et al.*, 2003; Nielsen *et al.*, 2003a; Bitsch *et al.*, 2004a; Kay *et al.*, 2004; Harada *et al.*, 2004; Frank *et al.*, 2005; Kay *et al.*, 2005; Kurilich *et al.*, 2005; Netzel *et al.*, 2005; Stoner *et al.*, 2005; Mullen *et al.*, 2008a; Charron *et al.*, 2009; Garcia-Alonso *et al.*, 2009; Milbury *et al.*, 2010; Wiczkowski *et al.*, 2010; Giordano *et al.*, 2011). Although these levels are reported for the parent flavylium structure and its methyl, glucuronide and sulfate derivatives, few studies have identified lower molecular weight degradation products or their conjugates in biological fluids. In addition, measurement of anthocyanins in blood or excreta is usually based upon regeneration of the flavylium cation, which may exclude modified structures unable to regenerate this form (McGhie and Walton, 2007; Williamson and Clifford, 2010).

Postabsorption, elimination of anthocyanins and/or metabolites may occur by urinary or faecal excretion (McGhie and Walton, 2007), and potentially through the lungs [as reported for radiolabelled quercetin (Walle *et al.*, 2001)]. Urinary excretion of anthocyanins has been reported by numerous human intervention studies and, as noted above, low recovery of an ingested dose is documented (McGhie and Walton, 2007). Any compounds which are eliminated by biliary excretion may be subject to enterohepatic recycling and/or further metabolism by intestinal microflora and subsequent reabsorption, as described for quercetin (Grassi *et al.*, 2009; Wang and Ho, 2009); however, there is little human evidence for enterohepatic re-circulation of anthocyanins at this time.

Many studies over the past decade have detailed the excretion of ingested anthocyanins as intact glycosylated or glucuronidated compounds (Miyazawa *et al.*, 1999; Bub *et al.*, 2001; Cao *et al.*, 2001; Matsumoto *et al.*, 2001; Netzel *et al.*, 2001; Mazza *et al.*, 2002; Mulleder *et al.*, 2002; Wu *et al.*, 2002; Rechner *et al.*, 2002b; Felgines *et al.*, 2003; Nielsen *et al.*, 2003a; Bitsch *et al.*, 2004b; Kay *et al.*, 2004, 2005; Manach *et al.*, 2005; Tian *et al.*, 2006; Charron *et al.*, 2007; Vitaglione *et al.*, 2007). Following ingestion of anthocyanins, parent compounds are generally perceived as representing around 32% of the total anthocyanins detected in serum, with 68% identified as conjugated metabolites of the parent flavylium structure (Kay *et al.*, 2004, 2005). Similar to serum availability, only 32% of anthocyanins excreted in urine are reported as parent structures, where 67% are shown to be conjugated metabolites. Of these metabolites, glucuronidation is generally reported as the major metabolic pathway for anthocyanin conjugation (representing 59% and 57% of total anthocyanins detected in the blood and urine, respectively), followed by methylation (43% and 51% of total anthocyanins detected in the serum and urine, respectively) and sulfation (2.4% of the total metabolites) (Felgines *et al.*, 2003). It should be noted that these percentages are relative to the total identified anthocyanins and not the recovery relative to the initial dose.

Current pharmacokinetic data suggest that around 99% of the ingested anthocyanin dose is undetectable in urine or plasma, with no conclusive explanation for the disappearance of the parent anthocyanin. It remains to be shown whether this disappearance is due to degradation of anthocyanins to constituent phenolic acids and aldehydes and subsequent conjugation (liver or microbial), or poor absorption. As the appearance of anthocyanins in the serum is rapid (Hollman and Katan, 1998), it is unlikely that their disappearance reflects poor absorption. Furthermore, it is probable that previously unidentified metabolites comprise undetected colonic metabolites. Whilst uptake of nutrients is typically associated with the small intestine, the colon is also capable of absorption, and is likely to be of particular relevance to anthocyanins (McGhie and Walton, 2007; Williamson and Clifford, 2010). Studies involving individuals who have undergone an ileostomy (where the colon has been excised and the ileum is drained into an affixed exterior pouch) suggest that a high proportion of ingested anthocyanins transit the small intestine unchanged, and therefore reach the colon (Williamson and Clifford, 2010). Moreover, the colon contains an extensive microbial population (Williamson and Clifford, 2010) and unabsorbed anthocyanins could be degraded by colonic microflora to generate smaller aromatic molecules, which are then absorbed (Williamson and Clifford, 2010). There is now significant evidence of this occurrence for other flavonoids such as quercetin and catechin (Li *et al.*, 2000; Rechner *et al.*, 2002a; Rios *et al.*, 2003; Mullen *et al.*, 2008b; Roowi *et al.*, 2010) and more recently in a human study using a berry extract rich in anthocyanins (Nurmi *et al.*, 2009). However, confirming colonic metabolite absorption requires feeding of labelled anthocyanins to establish the origin of the phenolic (or their lower molecular weight) intermediates. Following the ingestion of anthocyanins (cyanidin and pelargonidin) by rats, the phenolic acids protocatechuic acid and 4-hydroxybenzoic acid have been detected (Tsuda *et al.*, 1999; Ichiyanagi *et al.*, 2004). In humans, it has been suggested that the low systemic presence of anthocyanins may be accounted for by the formation of protocatechuic acid (Vitaglione *et al.*, 2007). Indeed, current opinion suggests that anthocyanins are likely metabolized to phenolic acid constituents (Tsuda *et al.*, 1999; Kay *et al.*, 2005, 2009; Vitaglione *et al.*, 2007; Han *et al.*, 2009). In a recent *in vitro* simulated gastrointestinal digestion and microsomal conjugation study, the degradation products of cyanidin (protocatechuic acid) and pelargonidin (4-hydroxybenzoic acid) were detected as mono-glucuronidated conjugates (Woodward *et al.*, 2011b). Similarly, formation of sulfated and glucuronidated conjugates of protocatechuic acid and phloroglucinol aldehyde, following incubation of caco-2 cells with cyanidin-3-*O*-glucoside, has been described (Kay *et al.*, 2009). Finally, *in vitro* synthesis of glutathione (GSH) conjugates of cyanidin-3-*O*-glucoside and delphinidin-3-*O*-glucoside by rat liver microsomes was detailed by Fernandes *et al.* (2009). Although these conjugates appear biologically plausible, they have yet to be sufficiently described in human feeding studies, despite evidence from animal studies.

Of the limited human data available suggesting absorption of colonic metabolites, one study is of particular importance. Nurmi *et al.* (2009) conducted a human feeding study involving bilberry/lingonberry (as a purée) and reported an extensive increase of phenolic acids (with 18 identified in total) in biofluids 4–6 hours post consumption (Nurmi *et al.*, 2009). Homovanillic and vanillic acids were identified as the most abundant of the proposed anthocyanin metabolites. Although excretion of these phenolic acids increased substantially following the intervention, it was not possible to determine whether they were derived exclusively from anthocyanins or from the ingestion and metabolism of other phenolic substrates in the diet. In a ^{14}C-labelled anthocyanin feeding study in mice conducted by Felgines *et al.* (2010), a total of 92% of administered radioactivity was ultimately recovered (Felgines *et al.*, 2010). Felgines *et al.* also conducted a parallel feeding study in order to derive structural information regarding the kinetics of anthocyanins and their metabolites in the blood (through HPLC analysis). The reported peak concentration of anthocyanins did not match the peak recovery of radioactivity in the blood, and the authors concluded that the peak radioactivity most likely represented metabolites associated with anthocyanin degradation, as a higher proportion (approximately sixfold) of labelled metabolites lacking the flavylium skeleton were detected. These two reports suggest that colonic metabolites of anthocyanins may account for previously unrecovered anthocyanin metabolites post consumption; however, labelling studies in humans are necessary to prove this with certainty.

BIOACTIVITY OF ANTHOCYANINS AND METABOLITES

Metabolic syndrome

Metabolic syndrome describes a condition characterized by the presence of coexisting traditional risk factors for CVD, such as hypertension, dyslipidaemia, glucose intolerance and obesity, in addition to non-traditional CVD risk factors, such as inflammation and abnormalities of blood coagulation. Although the specific aetiology for metabolic syndrome is not known, insulin resistance, a clinical state in which a normal or elevated insulin concentration reflects an impaired biological response, is present and is considered a key pathophysiologic abnormality. As such, metabolic syndrome can be considered to be a pre-diabetic state (Cefalu *et al.*, 2008).

Glucoregulation

Regular consumption of diets with a high glycaemic impact may increase the risk of obesity, type 2 diabetes and CVD via various mechanisms (Ludwig, 2002). Such meals can induce early postprandial hyperglycaemia, which stimulates insulin secretion resulting in hyperinsulinaemia. Repeated postprandial hyperglycaemia and hyperinsulinaemia may cause insulin resistance, beta-cell dysfunction and inflammation (Matsui *et al.*, 2002; Hanamura *et al.*, 2006). An exaggerated postprandial insulin response may cause transient late postprandial hypoglycaemia and may elicit the secretion of stress hormones (Ludwig, 2002). Berry mixtures containing blackcurrants and certain anthocyanins abundant in blackcurrants have been reported to positively mediate carbohydrate metabolism and postprandial glycaemia in humans, and thus are purported to present potential adjunct therapies for individuals with diabetes or pre-diabetes. However, the currently available evidence to support such an effect is minimal and further research in this area is needed to ascertain the extent of the activity of blackcurrants and anthocyanins in ameliorating the diabetic condition.

Two intervention studies have been carried out in healthy adults assessing the effect of berry purees on glucose metabolism, insulin secretion/resistance and other mediators of the glycaemic response. Törrönen *et al.* investigated the effects on postprandial glycaemic response of 150 g berry puree (composed of 37.5 g of each of the following berries: bilberries, blackcurrants, cranberries and strawberries) sweetened with 35 g sucrose versus a carbohydrate-matched load. It was estimated from previous data that the amount of polyphenols present in the berry meal was 800 mg, with approximately 300 mg as anthocyanins. Plasma glucose concentrations at 15 and 30 minutes post-ingestion of the berry meal were significantly lower than after ingestion of the control meal and the glucose concentration peaked significantly later after consumption of the berry meal as compared to the control meal (Törrönen *et al.*, 2010).

In a second study by Törrönen *et al.*, in which the same intervention was administered, the glycaemic profile for sucrose consumed with and without berries was investigated, along with the effect of the berry mixture on postprandial regulation of glucose homeostasis, including insulin and glucagon-like peptide 1 (GLP-1) responses (Törrönen *et al.*, 2012). Consistently with Törrönen's 2010 study, the berry meal induced a delayed and attenuated glucose response. Additionally, the berry meal stimulated a lower and prolonged insulin response compared with the control meal, and no signs of exaggerated insulin response were seen (Törrönen *et al.*, 2012). No significant effect was observed on GLP-1, the main action of which is to stimulate glucose-dependent insulin secretion when glucose is absorbed through the gastrointestinal tract and the blood glucose concentration is above the normal physiological range (D'Alessio and Vahl, 2004; Holst, 2007).

Although these two human intervention studies provide promising reports of an effect of berries on postprandial hyperglycaemia, the interpretation of such results must be carried out with the following factor in mind. The control meal was not fibre-matched for the berry meal; it has been proposed that soluble fibre can help attenuate the postprandial glucose response (Wood, 2007) and although the soluble fibre content within the berry meal was low (1.5 g), it cannot be discounted as a confounding factor.

Various mechanisms of action have been proposed to account for the observed advantageous effects of berries and their anthocyanins on postprandial glycaemic control, predominantly proposing an effect of these compounds on digestive enzyme inhibition, competitive binding with glucose transporters or modulation of signalling pathways involved in insulin resistance.

In particular, polyphenol-rich extracts of blueberries, blackcurrants, strawberries and raspberries have been shown to inhibit α-glucosidase activity *in vitro* (McDougall *et al.*, 2005c; da Silva Pinto *et al.*, 2008; Adiisakwattana *et al.*, 2009, 2011), an enzyme responsible for the digestion of sucrose to glucose in the intestinal epithelium. In particular, Adiisakwattana *et al.* showed that a synthetically-derived anthocyanin, cyanidin-3-*O*-rutinoside, found in blackcurrants and other berries, was a potent α-glucosidase inhibitor in both *in vitro* and animal studies (Adiisakwattana *et al.*, 2011). Inhibition of α-glucosidase is one of the therapeutic targets for the control of hyperglycaemia and type 2 diabetes. It is proposed that α-glucosidase inhibition delays digestion of disaccharides to absorbable monosaccharides, thus leading to a reduction in glucose absorption and subsequently an attenuation of postprandial hyperglycaemia (McDougall *et al.*, 2005c).

Anthocyanins may also influence postprandial glycaemia by affecting adipocyte function, which plays an important role in the development of insulin resistance (Sasaki *et al.*, 2007). By synthesizing and secreting adipocytokines (Scherer, 2006), such as adiponectin, adipocytes are able to improve the insulin sensitivity of target cells (Yamauchi *et al.*, 2001). Serum levels of adiponectin protein correlate with systemic insulin sensitivity (Berg *et al.*, 2002) and are decreased in insulin resistant, diabetic and obese subjects (Lindsay *et al.*, 2002). In an obese or diabetic state, the expression of glucose transporter 4 (GLUT4) is reduced in adipocytes; the reduction is accompanied by an increase in RBP4 expression and secretion into the blood. The increase causes impairment of insulin signalling in skeletal muscle and stimulates glucose production in the liver. These changes lead to high glucose concentrations in the blood. Therefore, dysregulation of the adipocyte GLUT4-RBP4 system is strongly associated with type 2 diabetes, and lowering RBP4 is a new potentially important target for prevention and management of type 2 diabetes. Sasaki *et al.* found that cyanidin-3-*O*-glucoside, reduced blood glucose levels and enhanced insulin sensitivity in type 2 diabetic mice. It was also found that cyanidin-3-*O*-glucoside regulated the GLUT4-RBP4 system and related inflammatory adipocytokines (Sasaki *et al.*, 2007).

More recently, a study by Scazzochio *et al.* demonstrated that cyandin-3-*O*-glucoside, and its metabolite protocatachuic acid, enhanced glucose uptake and GLUT4 translocation in both insulin-stimulated human omental adipocytes and an adipocytic cell line (3T3-L1). Cyanidin-3-*O*-glucoside elicited the same response in the absence of insulin, therefore indicating an insulin-like activity; specifically, up-regulating PPARγ activity and the expression and secretion of its target gene adiponectin, which is responsible for improving the insulin sensitivity of target cells (Scazzochio *et al.*, 2011).

A further potential mechanism of action by which anthocyanins could modulate postprandial hyperglycaemia has been put forward by Guo *et al.* who investigated the influence of cyanidin-3-*O*-glucoside on insulin resistance of 3T3-L1 adipocytes, induced by exposure to H_2O_2 or tumour necrosis factor alpha (TNF-α) (Guo *et al.* 2008), both of which are potent inducers of insulin resistance (Stephens *et al.*, 1997; Rudich *et al.*, 1998) and have physiological relevance *in vivo* (Wellen and Hotamisligil, 2005). Furthermore, elevated levels of reactive oxygen species (ROS) or TNF-α (or both) have been associated with insulin-resistant states such as steatohepatitis, obesity and metabolic syndrome (Furukawa *et al.*, 2004; Svegliati-Baroni *et al.*, 2006). Guo *et al.* reported that cyanidin-3-*O*-glucoside affords a significant protection against H_2O_2- or TNF-α-induced insulin resistance in 3T3-L1 adipocytes by blocking JNK-mediated Ser307 phosphorylation of IRS1 (Guo *et al.*, 2008).

The evidence provided herein supports a positive effect of anthocyanins on glucose modulation and insulin resistance, associated with metabolic syndrome; however, this area of research is in its infancy and further robust studies in humans are required to validate the two existing human studies and elucidate the mechanism(s) of action responsible for the observed effects.

Lipid regulation

It is well established that the development of obesity is usually accompanied by dyslipidaemia and hypertriglyceridaemia (Flier, 2004). Triglyceride (TG) metabolism is regulated by several factors such as TG intake from food, synthesis and oxidation in various tissues (Mu and Porsgaard, 2005), with the hydrolysis of TG by lipoprotein lipase (LPL) to free fatty acids an important determinant in TG metabolism (Mead *et al.*, 2002). LPL therefore regulates fat accumulation in adipose tissue and fat oxidation in skeletal muscle tissue (Cryer, 1981; Mead *et al.*, 2002), and thus is a potential target in anti-obesity strategies.

Cyanidin-3-*O*-glucoside has been shown to exert beneficial effects on lipid metabolism, although the data in this area are limited to *in vitro* and animal studies. A recent study by Wei *et al.* reported that cyanidin-3-*O*-glucoside ingestion (1 g/kg body weight) in a mouse model of type 2 diabetes was able to mitigate hypertriglyceridaemia and decrease adipose storage in visceral adipose tissue, which ultimately led to a significant loss of body weight in comparison to the control group. It was proposed that increased lipoprotein lipase (LPL) activity in the plasma and skeletal muscle in conjunction with decreased LPL expression in the visceral adipose tissue (via adenosine monophosphate-activated protein kinase phosphorylation [pAMK] activation, upstream of LPL activation) was the likely mechanism by which hypertriglyceridaemia and visceral adipose storage was mediated (Kim *et al.*, 2007; Ohira *et al.*, 2009; Wei *et al.*, 2011).

Several studies have looked at the effects of cyanidin and cyanidin-3-*O*-glucoside treatment on adipocyte gene expression *in vitro*. Adipocytes store energy as fat and accumulate TG during nutritional excess. However, their role is not limited to storage and, indeed, they are implicated in multiple physiological processes via the production of adipokines; these are proteins involved in energy balance, insulin sensitivity and glucose and lipid metabolism, amongst many other functions (Wood *et al.*, 2009). Tsuda *et al.* reported that cyanidin-3-*O*-glucoside- and cyanidin-treated rat adipocytes secreted more adiponectin and leptin than control cells (Tsuda *et al.*, 2004); leptin is known to reduce food intake and increase energy expenditure *in vivo* (Friedman and Halaas, 1998), whilst plasma adiponectin concentrations and mRNA expression levels are decreased in the obese and insulin resistant state (Maeda *et al.*, 1996; Arita *et al.*, 1999). Adiponectin mRNA levels were also shown to increase in mice fed cyanidin-3-*O*-glucoside for 12 weeks (Tsuda *et al.*, 2004). In two follow up studies by the same group, rat adipocytes were treated with 100 μM cyandin-3-*O*-glucoside or cyanidin for 24 hours. Multiple genes were up-regulated by the anthocyanin treatments, of particular note were those associated with lipid metabolism (un-coupling protein 2, acyl-CoA oxidase 1 and perilipin). Furthermore, up-regulation of hormone sensitive lipase and enhanced lipolytic activity were observed upon treatment with the two anthocyanins (Tsuda *et al.*, 2005, 2006).

A further manifestation of metabolic syndrome is non-alcoholic fatty liver disease (NAFLD), which can significantly increase the risk of liver cancer or cirrhosis (Flegal *et al.*, 2010) and it is thought that the dysregulation of TG metabolism may be the primary contributing factor to this disease (Gaemers and Groen, 2006). It has been reported that cyanidin-3-*O*-glucoside ingestion (100 mg/kg body weight) in a type 2 diabetic mouse model resulted in fewer or smaller TG droplets in the livers of treatment mice, along with a decreased overall TG liver mass. In the same study by Guo *et al.*, it was shown that cyanidin-3-*O*-glucoside (1, 10 and 100 μM) was able to abolish high glucose-induced mitochondrial GPAT-1 activation in human liver cells (HepG2) via a mechanism involving protein kinase-c activation (Guo *et al.*, 2011). GPAT-1 is a lipogenic enzyme that plays a critical part in the development of fatty liver disease (Gonzalez-Baro *et al.*, 2007). A further study investigating the effects of cyanidin-3-*O*-glucoside on hepatic lipid homeostasis *in vitro* showed that fatty acid oxidation was significantly increased in the treatment versus the control group and that AMP-activated protein kinase (AMPK) plays a central role in control of hepatic lipid metabolism via regulation of certain downstream pathways, such as acetyl CoA carboxylase (ACC) and carnitine palmitoyl transferase 1 (CPT-1) (Guo *et al.*, 2012).

Cardiovascular disease

Cardiovascular disease (CVD) arises from disorders of the heart and vasculature, as manifested in pathophysiologies such as peripheral arterial disease, cerebrovascular disease (stroke) and coronary heart disease (CHD) (de Pascual-Teresa *et al.*, 2010; WHO, 2011). It is responsible for almost half of all deaths in Europe annually (Chong *et al.*, 2010), and it is estimated that approximately 23.6 million people will die owing to CVD by 2030 (WHO, 2011). Epidemiological data have associated consumption of flavonoid-rich foodstuffs with decreased risk of CVD (Vita, 2005; Hooper *et al.*, 2008; Wallace, 2011) and anthocyanin intake has been inversely related to mortality owing to CHD and CVD (Mink *et al.*, 2007; McCullough *et al.*, 2012) and relative risk of hypertension (Cassidy *et al.*, 2010). Data from over 30 000 postmenopausal women enrolled in the Iowa Women's Health Study suggested a significant inverse relationship between dietary anthocyanin consumption (median intake 0.2 mg/day versus no intake) and mortality due to CHD and CVD over a 16 year follow-up period (Mink *et al.*, 2007; Wallace, 2011). Furthermore, anthocyanin intake (highest versus lowest consumers) has been associated with a reduced relative risk of hypertension in a prospective study (14 year follow-up) of participants enrolled from the Nurse's Health Study I and II (46 672 and 87 242 women, respectively) and Health Professionals Follow-Up Study (23 043 men) (Cassidy *et al.*, 2010). More recently, anthocyanin intake (top versus bottom quintile of intake) was associated with a lower risk of fatal CVD for subjects enrolled in the Cancer Prevention Study II Nutrition Cohort (38 180 men and 60 289 women) during seven years of follow up (McCullough *et al.*, 2012).

A recent review of human dietary intervention studies indicated that fruits such as berries and purple grapes (which have a relatively high content of anthocyanins, flavonols and procyanidins) appeared more effective at reducing CVD risk, as compared to other fruits studied; particularly with regard to antihypertensive activity, elevation of endothelium-dependent vasodilatation and inhibition of platelet aggregation (Chong *et al.*, 2010). Furthermore, a randomized, double-blind, placebo-controlled trial involving 120 dyslipidaemic participants consuming a supplement of 320 mg/day anthocyanins (consisting of 17 anthocyanins purified from blackcurrant and bilberry) for 12 weeks, reported increased circulating high density lipoprotein (HDL) cholesterol and reduced low density lipoprotein (LDL) cholesterol concentrations (Qin *et al.*, 2009). Likewise, ingestion of a daily supplement of 320 mg anthocyanins by 75 hypercholesterolaemic subjects for 12 weeks resulted in significant increases in HDL cholesterol and endothelial function (as assessed by flow-mediated dilatation) and a significant reduction in LDL cholesterol, as compared with 75 hypercholesterolaemic subjects receiving a placebo, in a double-blind parallel group intervention (Zhu *et al.*, 2011). However, not all human intervention studies support this protective association (Karlsen *et al.*, 2007; Erlund *et al.*, 2008; Jin *et al.*, 2011; Hassellund *et al.*, 2012). For example, Hassellund *et al.*

found an increase in HDL cholesterol but no significant difference in total cholesterol and LDL-cholesterol after a four-week daily intake of 640 mg anthocyanins extracted from bilberries and blackcurrants (Hassellund *et al.*, 2012).

Rodent studies have also reported reductions in total cholesterol with 8 week and 12 week cyanidin-3-*O*-glucoside supplementation (Tamura *et al.*, 2004; Wang D *et al.*, 2011b, 2012a; Wang Y *et al.*, 2012c) and total HDL- and LDL-cholesterol with four-week protocatechuic acid supplementation (Tamura *et al.*, 2004). However, again, findings have been inconsistent, with significant increases in plasma cholesterol and LDL-cholesterol reported in a study carried out in hyperlipidaemic rabbits fed a purified anthocyanin fraction from blackcurrants for 16 weeks. The study also found no significant effect on total plasma cholesterol but a reduction in very low density lipoprotein (VLDL) cholesterol with blackcurrant juice consumption (Finné Nielsen *et al.*, 2005). In addition, Frank *et al.* found no significant effect on plasma cholesterol with four weeks cyanidin-3-*O*-glucoside or blackcurrant concentrate supplementation in rodents (Frank *et al.*, 2002). *In vitro* studies have found reductions in cholesterol concentrations in isolated erythrocytes after incubation with relatively high concentrations (10 μmol/l) of cyanidin-3-*O*-glucoside (Duchnowicz *et al.*, 2012). Anthocyanins have also been shown to induce cholesterol efflux in macrophages and macrophage-derived foam cells, potentially through the peroxisome proliferator receptor-γ-liver X receptor α-ABCA1 signalling pathway (Xia *et al.*, 2005). A similar pathway has also been demonstrated with protocatechuic acid but not cyanidin-3-*O*-glucoside (Wang D *et al.*, 2012b).

Data from animal and *in vitro* studies have demonstrated a variety of potential cardiovascular activities of anthocyanins (Domitrovic, 2011), including vasodilatory and vasoprotective effects (Bell and Gochenaur, 2006; Ziberna *et al.*, 2011), antithrombotic activity (Rechner and Kroner, 2005; de Pascual-Teresa *et al.*, 2010; Yang *et al.*, 2011), activation and/or upregulation of endothelial nitric oxide synthase (eNOS) (Xu *et al.*, 2004a, 2004b; Lazze *et al.*, 2006; Chalopin *et al.*, 2010; Edirisinghe *et al.*, 2011b; Simoncini *et al.*, 2011), decreased production of the vasoconstrictor endothelin-1 (ET-1) (Lazze *et al.*, 2006), and inhibition of angiotensin-converting enzyme (Persson *et al.*, 2009; Hidalgo *et al.*, 2012). Thus the beneficial effects of dietary anthocyanins on cardiovascular health may be elicited by various mechanisms (Basu *et al.*, 2010; Tsuda, 2011; Wallace, 2011; Landberg *et al.*, 2012); however, proof of these activities in humans is not consistent across clinical studies.

The role of ROS generation in the pathogenesis of atherosclerosis is well documented (Schissel *et al.*, 1996; Tabas, 1999; Skalen *et al.*, 2002; Medeiros *et al.*, 2004; Stocker and Keaney, 2004; Williams and Fisher, 2005) and anthocyanins have consistently been shown to possess novel and potent antioxidant properties *in vitro* (Igarashi *et al.*, 1989; Costanino *et al.*, 1992; Tamura and Yamagami, 1994; Tsuda *et al.*, 1994). A number of studies have reported the antioxidant activity of berry extracts and pure anthocyanins by several methods (including the total antioxidant activity, Trolox equivalent antioxidant activity, oxygen radical absorbance capacity, and 2,2-diphenyl-1-picrylhydrazyl/DPPH radical scavenging capacity assays) (Bors *et al.*, 1990; Rice-Evans *et al.*, 1996). However, owing to the low reported serum and tissue levels of anthocyanins, and the controversy regarding the failure of some antioxidant supplementation trials (Fruebis *et al.*, 1994; Witting *et al.*, 1999; Stocker and Keaney, 2004), the antioxidant activity of anthocyanins *in vivo* has been brought into question. Although the redox status of cellular environments is of direct importance to disease pathologies such as CVD, the redox status is primarily maintained by endogenous antioxidants such as GSH (Betteridge, 2000; Sies, 2007; Leopold and Loscalzo, 2009). GSH, in addition to other endogenous radical scavengers, exists at concentrations many orders of magnitude above those reported for phytochemicals such as anthocyanins and other flavonoids. Therefore, the likelihood of dietary phytochemicals dramatically altering redox status through direct scavenging mechanisms is questionable. More recent research suggests a mechanism of action that is not directly related to radical scavenging but associated with the regulation of key proteins involved in cellular protection/defence. These activities are still a matter of debate, and are described here in relation to vascular disease pathologies.

Vascular function

Dysfunction of the vascular endothelium is the initial process in the development of atherosclerosis (Landmesser *et al.*, 2004; Higashi *et al.*, 2009); in particular, reduced bioavailability of nitric oxide (NO) and impairment of endothelium-dependent relaxation by various mechanisms (Hobbs *et al.*, 1999; Kawashima and Yokoyama, 2004; Michel and Vanhoutte, 2010). NO also inhibits expression of the cell surface adhesion molecules, vascular cell adhesion molecule 1 (VCAM-1), intercellular adhesion molecule 1 (ICAM-1), P-selectin and monocyte chemoattractant protein-1, thereby reducing leucocyte recruitment, adhesion and extravasation (Ross, 1999; Granger *et al.*, 2004; Naseem, 2005; Michel and Vanhoutte, 2010). Furthermore, NO acts to inhibit platelet activation and aggregation, proliferation of vascular smooth muscle cells and oxidation of LDL, all of which are important processes in atherogenesis (Kawashima and Yokoyama, 2004; Naseem, 2005; Michel and Vanhoutte, 2010). Therefore, any decrease in NO bioavailability not only limits endothelium-mediated vasodilatation but can also enhance inflammatory processes underlying atherosclerosis (Landmesser *et al.*, 2004; Michel and Vanhoutte, 2010). This may arise from decreased production of NO by eNOS (whether by down-regulation of enzyme expression or loss of activity) or increased scavenging of NO by ROS, thus leading to loss of NO signalling (Landmesser *et al.*, 2004; Naseem, 2005; Higashi *et al.*, 2009).

Up-regulation of eNOS by cyanidin-3-*O*-glucoside, found in blackcurrants, in bovine artery endothelial cells (BAECs) has been described (Xu *et al.*, 2004a, 2004b). Incubation of BAECs with cyanidin-3-*O*-glucoside at 1–100 nM elicited a dose-dependent up-regulation in eNOS expression, (Xu *et al.*, 2004b). Concerning anthocyanin bioactivity on human cells, Lazzé *et al.* have described the effects of cyanidin and delphinidin on the expression of eNOS and ET-1 in cultured human umbilical vein endothelial cells (HUVECs) (Lazze *et al.*, 2006). Significant decreases in ET-1 secretion and increases in eNOS protein level were measured following 24-hour incubations with 100 μM cyanidin, and 50 or 100 μM delphinidin (Lazze *et al.*, 2006). However, the instability of anthocyanidins at physiological pH must be noted (Woodward *et al.*, 2009), in conjunction with the high levels of anthocyanidins to which cells were exposed (Crozier *et al.*, 2009). Therefore, anthocyanins, or their degradation products and/or metabolites *in vivo*, could increase both eNOS expression and activity, with enhanced endothelial function resulting from increased NO bioavailability (Wallace, 2011).

The superoxide-generating NADPH oxidase (NOX) enzymes constitute a major source of ROS in the vasculature (Zalba *et al.*, 2000; Touyz, 2004; Bedard and Krause, 2007; Bonomini *et al.*, 2008), and inhibition of NOX activity, with decreased scavenging of NO by generated superoxide, has been proposed as a 'short-term' mechanism by which certain flavonoids may enhance NO bioavailability and, thereby, improve vascular function (Schewe *et al.*, 2008). Therefore, it is possible that anthocyanins and other flavonoids could act similarly to increase NO bioavailability *in vivo* by both direct (upregulation/activation of eNOS) and indirect [NOX inhibition with reduced scavenging of NO by superoxide, and possibly decreased oxidative degradation of eNOS cofactor tetrahydrobiopterin] mechanisms, thus improving vascular endothelial function (Kawashima and Yokoyama, 2004)].

Anthocyanins may modulate the expression of the haem oxygenase (HO) enzyme, which catalyses the oxidative catabolism of haem in an oxygen- and NADPH-dependent reaction (Ryter *et al.*, 2002, 2006). Haem oxygenase is expressed as inducible (HO-1) and constitutive (HO-2) isoenzymes (Abraham and Kappas, 2008), and HO-1 demonstrates protective activity against cellular oxidative stress (Siow *et al.*, 1999; He *et al.*, 2010), which may be mediated in part by the activity of generated carbon monoxide (Ryter *et al.*, 2002). Cyanidin-3-*O*-glucoside has been reported to up-regulate expression of HO-1 in human iliac artery endothelial cells (Sorrenti *et al.*, 2007); therefore, up-regulation of HO-1 by anthocyanins could enhance protection against oxidative stress in endothelial cells.

Expression and/or activation of adhesion molecules on endothelial cells and circulating inflammatory cells is a key process in the recruitment of leucocytes to sites of inflammation (Granger *et al.*, 2004) and of particular relevance to CVD, which is increasingly viewed as a chronic inflammatory process (Ross, 1999; Libby *et al.*, 2002; Granger *et al.*, 2004; Bonomini *et al.*, 2008). Anthocyanins may act to modulate expression of adhesion molecules by endothelial cells in response to pro-inflammatory stimuli (Chen *et al.*, 2011; Wallace, 2011). For example, Wang *et al.* (2010) investigated the effects of protocatechuic acid (a potential cyanidin-3-*O*-glucoside metabolite) on monocyte adhesion in a cell model of atherosclerotic development (Wang *et al.*, 2010). Pretreatment of mouse aortic endothelial cells (MAECs) with 20 or 40 μM protocatechuic acid for 24 hours elicited a significant reduction in human monocyte adhesion to TNF-α- activated MAECs. In addition, protocatechuic acid pretreatment partially inhibited TNF-α-stimulated expression of VCAM-1 and ICAM-1 expression by MAECs, and decreased TNF-α-induced activation of the transcription factor nuclear factor kappa B (NF-κB), which can up-regulate transcription of both these adhesion molecules (Wang *et al.*, 2010). The effect of protocatechuic acid *in vivo* was also examined in an established animal model of atherosclerosis, namely apolipoprotein E (ApoE)-deficient mice (Wang *et al.*, 2010). ApoE is present on HDL (Naseem, 2005) and in the absence of ApoE, lipoprotein remnants are not transported to the liver for metabolism, resulting in hypercholesterolaemia and development of atherosclerotic lesions (Ross, 1999). Supplementation of the diet of ApoE-deficient mice with 0.03 g/kg protocatechuic acid for 20 weeks resulted in reduced aortic expression of VCAM-1 and ICAM-1, in association with reduced serum levels of these proteins and NF-κB activation. Moreover, protocatechuic acid delayed the development of atherosclerosis in this mouse model, as assessed by measurement of lesions in the aortic sinus (Wang *et al.*, 2010). Effects on platelet reactivity have also been investigated using anthocyanins and potential colonic metabolites; 1 μM delphinidin-3-*O*-rutinoside and 10 μM dihydroferulic acid demonstrated an inhibitory effect on platelet aggregation induced by the agonist thrombin receptor activating protein (Rechner and Kroner, 2005).

In summary, the cardioprotective effects of anthocyanins observed in epidemiological studies may ultimately result from the ability of anthocyanins to prevent endothelial insult and subsequent dysfunction. Whilst anthocyanins do possess significant radical scavenging capacities, the available evidence suggests that their bioactivity results from mechanisms involving the modulation of key vascular and inflammatory mediators and enzymes (such as NO, eNOS, superoxide, NOX, ET-1, VCAM-1, TNF-α and HO-1) as opposed to direct antioxidant effects. Despite a significant amount of *in vitro* and cell culture evidence detailing the mechanisms of action of anthocyanins, in particular cyanidin-3-*O*-glucoside and its metabolite protocatechuic acid, still very little is known of their metabolism *in vivo* and the bioactivity of their metabolites. Elucidating the exact mechanisms by which anthocyanins and their metabolites modulate vascular function represents a major focus for future anthocyanin research.

Immunoregulatory

Inflammation is an important mechanism to activate tissue repair. However, continuing chronic inflammation can have detrimental effects and result in chronic inflammatory diseases. Flavonoids have been shown *in vitro* to have immunomodulatory properties (Middleton, 1998). Anthocyanins in particular have been reported to have a number of anti-inflammatory effects. Bilberry and blackcurrant anthocyanins have been shown to suppress LPS-induced activation of NF- κB, which controls the expression of genes involved in the inflammatory response, in cultured monocytes. Significant decreases in NF-κB inflammatory mediators including interleukin-8 and interferon-α were also found after a three-week supplementation with the bilberry and blackcurrant anthocyanins in 59 healthy subjects compared to controls (Karlsen *et al.*, 2007). Other cellular models have also demonstrated inhibition of pro-inflammatory cytokines after anthocyanin and/or berry concentrate exposure (Herath *et al.*, 2003; Cimino *et al.*, 2006; Lyall *et al.*, 2009). In animal asthma models, anthocyanins have been shown to inhibit airway inflammation (Rossi *et al.*, 2003; Park *et al.*, 2007). However, Hurst *et al.* found that proanthocyanin-enriched but not anthocyanin-enriched blackcurrant extract suppressed interleukin-4 and interleukin-13-stimulated eotaxin-3 secretion in human alveolar epithelial cells, a major factor in airway inflammation and a symptom of atopic asthma (Hurst *et al.*, 2010).

There is limited information available on the immunomodulatory effects of blackcurrants and its constituents. However, a placebo-controlled blackcurrant seed oil study demonstrated improved delayed-type hypersensitivity skin response and an increase in peripheral blood mononuclear cell proliferation after two months of supplementation in older adults (Wu *et al.*, 1999). Anthocyanins at 250–500 μM and anthocyanin-rich extracts, including blackcurrant extract at 500 μg/ml, have been shown to significantly induce TNF-α production (regulatory cytokine) and reduce NO production (reducing oxidative stress) in activated macrophages (Wang and Mazza, 2002). Bub *et al.* found increases in immune system markers after a two-week consumption of an anthocyanin-rich juice containing aronia, blueberries and boysenberries in healthy men. Lymphocyte proliferative responsiveness, interleukin-2 secretion by activated lymphocytes and the lytic activity of natural killer cells were increased compared to baseline (Bub *et al.*, 2003). However, it is worth noting the study did not include a placebo control and similar increases were found with the comparative flavanol-rich juice containing green tea, apricot and lime. More research is needed to confirm these findings *in vivo* and to determine the mechanism(s) of action.

Antimicrobial

To date, the majority of evidence for antibacterial or antiviral effects of blackcurrant extracts and/or their constituent anthocyanins has been generated from *in vitro* studies. Indeed, a recent search of the literature yielded no published human intervention studies in this area.

Two studies have investigated the antiviral activities of the whole crude extract of *Ribes nigrum* L (Chinese blackcurrant, known as Kurokarin), which has been used medicinally in China for over 1000 years (Knox *et al.* 2003). Suzutani *et al.* investigated the antiherpes virus activity of an extract of *Ribes nigrum* L *in vitro* and found that the extract inhibited herpes simplex virus type 1 attachment on the cell membrane completely at a 100-fold dilution (Suzutani *et al.*, 2003). It also inhibited the plaque formation of herpes simplex virus type 1 and 2 and varicella-zoster virus by 50% at a 400-fold dilution (thus inhibiting virus replication). The mechanism of action is thought to be due to inhibition of protein synthesis in infected cells by a yet to be defined component of the blackcurrant extract.

Similarly, the antiviral activities of *Ribes nigrum* L were shown against influenza virus A (IVA) and B (IVB) with an IC50 value of 3.2 μg/ml and at 10 μg/ml, both IVA and IVB were almost entirely inactivated (up to 99%). At higher concentrations of blackcurrant extract (100 μg/ml), virus titres of IVA in culture fluids were completely suppressed six hours post-infection, indicating the extract inhibited virus release from the infected cells. This discovery was not replicated for IVB however. Inhibition of growth of both IVA and IVB in cells treated with the extract pre-infection was less effective than those treated post-infection, indicating a viral-suppressant effect of blackcurrant extract during the late stage of virus growth (Knox *et al.*, 2003).

A previous study carried out by the same group (Knox *et al.*, 2001) investigated the effect of the anythocyanin component of blackcurrant extract on IVA and IVB. Fractionation of the anthocyanins yielded four fractions with potent antiviral activity against IVA and IVB. One of the fractions contained the known anthocyanins 3-*O*-alpha-L-rhamnopyranosyl-beta-D-glucopyranosyl-cyanidin and 3-*O*-beta-D-glucopyranosyl-cyanidin, which did not inactivate the flu virus but inhibited virus adsorption to cells and also virus release from infected cells.

A further two *in vitro* studies have investigated the effects of blackcurrant extracts and cordials on bacterial activity. A commercial blackcurrant cordial (100% fruit) as well as a fresh blackcurrant extract were shown to inhibit the growth of 12 bacteria, along with the fungus *Candida albicans*, at a 1:5 concentration. At a lower 1:10 concentration, all bacteria were still inhibited but no inhibitory effect was seen against *C. albicans* (Cavanagh *et al.*, 2003).

Finally, Puupponen-Pimia *et al.* investigated the ability of certain whole berry extracts and their phenolic fractions to inhibit the growth of intestinal pathogens. Although the most active antibacterial extracts in this case were those of raspberry and cloudberry, blackcurrant extract also exerted a significant antibacterial effect on selected pathogens. At 2 and 10 mg/ml, a whole blackcurrant extract significantly inhibited the growth of *Salmonella enterica* and *Staphylococcus aureus*; however, the phenolic fraction was only effective against *S. aureus* (Puupponen-Pimia *et al.*, 2005).

The antibacterial properties of blackcurrants and anthocyanins make them useful targets for further investigation. However, it remains to be determined whether the concentrations of blackcurrant extract and anthocyanins used in the studies identified herein are physiologically relevant.

Antineurodegenerative

Chronic neurodegenerative diseases become more prevalent with increased age. However, the causes of many such diseases are yet to be fully elucidated. The neuropathology of many neurodegenerative diseases has been linked to increases in brain oxidative stress and, historically, many efforts have been directed at exploring antioxidant effects on neuronal damage (Spencer, 2010). However, more recently it has been proposed that, apart from their proposed antioxidant effect, polyphenols may have other effects, especially in the brain where they are found at low concentrations and may directly or indirectly influence the brain's architecture for memory (Williams *et al.*, 2008; Spencer, 2008, 2009). It has been proposed that their effects in the brain are mediated by multiple mechanisms including, but not limited to, an interaction with neuronal signalling and synaptic function, influence on blood flow and neurogenesis and inhibition of neurodegeneration and neuroinflammation (Spencer, 2010).

The majority of studies in this area are carried out using synthetic anthocyanins in cellular models of oxidative stress. To date, no studies have been carried out investigating the effects of a whole blackcurrant extract on cognitive function or neurodegeneration either *in vitro* or *in vivo*. However, it should be noted that clinical trials have investigated the effects of an anthocyanin-rich blueberry extract on cognitive function in humans with encouraging results (Chapter 11).

There is growing interest in therapeutic strategies with dietary antioxidants aimed at counteracting oxidative stress-induced damage associated with neurodegenerative diseases, including Alzheimer's disease and Parkinson's disease (Lau *et al.*, 2005). Oxidative stress, involving ROS and other oxidizing agents, is one of the major factors responsible for the dysfunction or death of neuronal cells that contributes to the initiation and progression of these neurological diseases (Barnham *et al.*, 2004). Five studies to date have investigated the effects of certain anthocyanins, found in blackcurrants, on oxidative stress-induced neurodegeneration.

Tarozzi *et al.* investigated the effects of cyanidin-3-*O*-glucopyranoside, known to have the highest antioxidant capacity of any anthocyanin (Amorini *et al.*, 2001) and its aglycone (cyanidin) and *in vivo* metabolite, protocatachuic acid, against H_2O_2-induced oxidative stress in a human neuronal cell line (SH-SY5Y). It was found that pretreatment of SH-SY5Y cells with cyanidin-3-*O*-glucopyranoside (at 50–100 μM), cyanidin (at 25–100 μM) and protocatechuic acid (50–100 μM) inhibited H_2O_2-induced ROS formation at different cellular levels: cyanidin-3-*O*-glucopyranoside at the membrane level, protocatechuic acid at the cytosolic level and cyanidin at both the membrane and cytosolic levels. In addition, cyanidin showed a higher antioxidant activity at membrane and cytosolic level than cyanidin-3-*O*-glucopyranoside and protocatechuic acid, respectively (Tarozzi *et al.*, 2007). These results suggest that cyanidin and protocatechuic acid may be considered as neuroprotective and may play an important role in the promotion of brain health. Further to this, Andres-Lacueva *et al.* reported that cyanidin-3-*O*-glucopyranoside is able to cross the blood–brain barrier and localize in various brain regions important for learning and memory (Andres-Lacueva *et al.*, 2005).

More recently, Kelsey *et al.* assessed the effects of synthetic cyanidin-3-*O*-glucoside (an anthocyanin found in blackcurrants at about 10.3% of total anthocyanin content) and mitochondrial oxidative stress (MOS) induced apoptosis *in vitro* in primary cultures of rat cerebellar granule neurons, a model of neuronal injury. It was reported that cyanidin-3-*O*-glucoside at both concentrations significantly decreased the percentage of cell apoptosis; this was due to its ability to preserve mitochondrial GSH, an endogenous antioxidant, and thus to prevent cardiolipin oxidation and mitochondrial fragmentation (Kelsey *et al.*, 2011). Therefore, it is likely that anthocyanins protect neurons from MOS by acting locally at the level of the mitochondria to enhance the resistance of these organelles to oxidative damage.

One further study carried out by Dreiseitel *et al.* used an *in vitro* assay in which the affinity of anthocyanins, particularly synthetic cyanidin-3-*O*-glucoside, for monoamine oxidase (MAO) isoforms (A and B) was investigated. Monoamine oxidases (MAO) are mitochondrial enzymes that catalyse the oxidation of monoamines in multiple tissues, including the brain. Elevated MAO activity has long been implicated in the aetiology of depression, anxiety and neurodegenerative disease. It was found that multiple anthocyanidins (such as cyanidin and delphinidin) and anthocyanidin-3-*O*-glycosides (such as delphinidin-3-*O*-glucoside and cyanidin-3-*O*-rutinoside), commonly found in blackcurrants, achieved MAO A and B inhibition *in vitro* in a concentration-dependent manner with IC_{50} values in the low micromolar range. However, these anthocyanins did not reach the levels of inhibition observed for known MAO A or MAO B inhibitors (Dreiseitel *et al.*, 2009).

Although the data reported herein indicate a positive effect of anthocyanins on the prevention of neurodegeneration, these findings must be interpreted in light of the anthocyanin concentrations used in these *in vitro* studies. Indeed in both the Tarozzi *et al.* study and that of Kelsey *et al.* (Tarozzi *et al.*, 2007; Kelsey *et al.*, 2011) the anthocyanin concentrations (25–200 μM) used were significantly higher than plasma concentrations that are achieved *in vivo* (<1 μM) following an anthocyanin-rich diet (Prior and Wu, 2006).

Although the mechanisms of action supporting the pharmacological effects of anthocyanins are unclear, Dreiseitel *et al.* (2008) hypothesised that the proteasome may be one of the key cellular targets of anthocyanins (Chen *et al.*, 2004). Proteasome activity controls the degradation of cellular proteins and is closely implicated in signal transduction, development, cell cycle progression (Naujokat *et al.*, 2006), antigen processing, immune response (Kloetzel, 2004) and inflammation (Visekruna *et al.*, 2006) plus protection from oxidative stress (Yamamoto *et al.*, 2007) relevant to neurodegenerative diseases (Williams *et al.*, 2006; Nencioni *et al.*, 2007). Dreiseitel *et al.* investigated this hypothesis *in vitro* by treating HL-60 cells (a human cell line) with synthetic anthocyanins, anthocyanidins and proanthocyanidins and measuring the impact on proteasome activity. Phytochemicals found in blackcurrant, such as cyanidin-3,5-*O*-diglucoside, cyanidin-3-*O*-glucoside, delphinidin-3-*O*-glucoside, cyanidin-3-*O*-rutinoside, cyanidin and delphinidin, all showed concentration-dependent inhibition of the proteasome with IC_{50} values ranging from 11 to 32 μM (Dreiseitel *et al.*, 2008).

Only one study to date has reported a protective effect of a commercially prepared anthocyanin supplement *in vivo (*Medox). Medox contains a combination of anthocyanins from blackcurrant and bilberry. Shin *et al.* investigated the effect of anthocyanins in middle cerebral artery occlusion and a reperfusion model of cerebral ischemia in rats. Ischemic stroke results from a transient or permanent reduction in cerebral blood flow to the major brain arteries. The major pathobiological mechanisms of ischemia/reperfusion injury include excitotoxicity, oxidative stress, inflammation and apoptosis. It was reported that pretreatment of rats with anthocyanins (300 mg/kg) significantly reduced the brain infarct volume and the number of terminal deoxynucleotidyl transferase-mediated deoxyuridine triphosphate (UTP)-biotin nick end-labelling (TUNEL) positive cells caused by middle cerebral artery occlusion and reperfusion. In immunohistochemical analyses, anthocyanins markedly reduced the number of cells showing up-regulation of markers associated with neuronal cell death. It was suggested that anthocyanins reduced neuronal damage induced by focal cerebral ischemia through blocking the JNK and p53 signalling pathways that are associated with cell death (Shin *et al.*, 2006).

Taken together, the current body of literature supports a potential protective effect of anthocyanins and their constituent metabolites against neurodegeneration. However, specific clinical trials with blackcurrant extract versus isolated synthetic anthocyanins are lacking.

Anticarcinogenic

Epidemiological evidence has indicated a negative correlation between consumption of fruits and vegetables, and more specifically phenolics, with the incidence of various cancers (Decker *et al.*, 1995; Steinmetz and Potter, 1996; Kris-Etherton *et al.*, 2002; Riboli and

Norat, 2003). Although research in this area is limited, that which does exist indicates an anticarcinogenic effect of certain polyphenols (Stoner *et al.*, 2007, 2008; Stoner, 2009; Seeram *et al.*, 2010). *In vitro* analysis of 0.25 and 0.5% blackcurrant extract demonstrated inhibition of colon cancer and breast carcinoma cell proliferation (Olsson *et al.*, 2004). These findings have also been repeated in similar studies using blackcurrant and blackcurrant skin extracts on colon and liver cancer cells (Wu *et al.*, 2007; Bishayee *et al.*, 2010). Blackcurrant juice at a concentration of 10–50 μl/ml was also found to possess the second highest inhibitory effect on cancer cell proliferation from a range of 13 berry fruits tested on stomach, prostate, colorectal and breast cancer cells (Boivin *et al.*, 2007). Other similar studies using extracts of berries, including blackcurrant, have also shown cell growth inhibition in colon and breast cancer cells (Olsson *et al.*, 2004; Wu *et al.*, 2007). In a rodent model, blackcurrant skin extract was found to dose-dependently decrease the incidence and proliferation of chemically-induced preneoplastic liver nodules (Bishayee *et al.*, 2011).

Investigations on specific components of blackcurrants, such as anthocyanins and the polysaccharide-rich substance, cassis polysaccharide, have reported anticarcinogenic properties (Duthie *et al.*, 2004; Cooke *et al.*, 2005; Ding *et al.*, 2006; Takata *et al.*, 2007; Wang and Stoner, 2008). Indeed, anthocyanins have been shown to significantly inhibit the growth of tumour cells to a larger extent than some other flavonoid groups (Kamei *et al.*, 1995). Cyanidin-3-*O*-rutinoside and delphinidin were reported to induce apoptosis in leukemic cells and hepatocellular carcinoma cells, respectively (Feng *et al.*, 2007, 2010). In addition, cyanidin-3-*O*-rutinoside and cyanidin-3-*O*-glucoside demonstrably inhibited the growth of oesophageal cancer cells and the migration and invasion of lung cancer cells (Chen *et al.*, 2006; Zikri *et al.*, 2009). It remains to be determined whether these findings can be extrapolated to humans; the levels and duration of exposure of blackcurrant extracts and blackcurrant components used *in vitro* is often deemed unachievable *in vivo* due to low serum bioavailability. However, considerable levels of ingested polyphenols are thought to reach the colon, and thus levels used in colon cancer cell screens could be deemed physiologically relevant. Nevertheless, whether these polyphenols remain intact within the colon environment is still to be fully determined.

Blackcurrant anthocyanins may act as direct antimutagenic agents or inhibit the activation of carcinogens (Kay *et al.*, 2005; Duthie, 2007). Other possible mechanisms include inhibition of growth factors, receptors and enzymes, thereby modulating cellular processes such as proliferation and apoptosis (Kay *et al.*, 2005, Aiyer *et al.*, 2012). For example, Bishayee *et al.* demonstrated anthocyanin-rich blackcurrant skin extract induced apoptosis in liver cancer cells through translational level effects on the apoptopic (cellular death) regulating proteins, Bcl-2 and Bax (Bishayee *et al.*, 2010). An increase in the expression of p21WAF1, a cell cycle regulator and a member of the cyclin kinase inhibitors, was also reported with exposure of 40 mg/ml blackcurrant extract to colon cancer cells for 24 hours. However, the same study failed to show significant differences in Bcl-2 and Bax expression (Wu *et al.*, 2007). With regards to specific anthocyanins, delphinidin at a concentration of $\geq$5 μM has been reported to act as an inhibitor of the receptor tyrosine kinase (RTK) superfamily (Teller *et al.*, 2009). These receptors are known to have an important role in tumour growth and metastasis, and in many cancer cell types they have been found to be mutated or stimulated. Inhibition of one particular RTK, epidermal growth factor receptor, has been repeatedly found in studies using cyanidin and delphinidin with simultaneous reduction in cell growth (Meiers *et al.*, 2001; Marko *et al.*, 2004; Teller *et al.*, 2009). The inflammatory-mediating cyclooxygenase-2 (COX-2) enzyme has been shown to be inhibited by cyanidin-3-*O*-glucoside by blocking the activation of the Fyn signalling pathway (Lim *et al.*, 2011). COX-2 is thought to play an important role in epithelial cancer development, by promoting cell proliferation and inhibiting apoptosis.

The anthocyanin metabolite, protocatechuic acid, has also been shown to possess pro-apotopic and antiproliferative effects *in vitro* (Tseng *et al.*, 2000; Anter *et al.*, 2011). The pro-apoptopic activity of protocatechuic acid has been proposed to be due to effects on regulatory proteins and signalling pathways. For example, protocatechuic acid at 2 mM has been shown to reduce the phosphorylation of retinoblastoma protein, reduce Bcl-2 expression and increase Bax expression, thereby increasing the likelihood of cell apoptosis (Tseng *et al.*, 2000). Although *in vitro* investigations are very useful for the elucidation of potential mechanisms, it is worth noting that the concentrations of specific blackcurrant constituents tested may not be physiologically relevant. Further work is needed in order to draw any definitive conclusions.

FUTURE DIRECTION OF RESEARCH

Whilst a significant amount of scientific literature is available detailing the biological activities of anthocyanins, there are still considerable gaps in the current knowledge base; particularly in relation to dietary availability, bioavailability, metabolism, dose response, and the biological activity of anthocyanin metabolites. Future *in vitro* research into the health benefits of anthocyanins must utilise not only parent glycosides, but also identified *in vivo* degradants and/or metabolites, and all species should be assayed at concentrations deemed to be physiologically relevant following human consumption. In addition, further human intervention studies are required to fully elucidate the health benefits of blackcurrants.

LIST OF ABBREVIATIONS

ABCA1	ATP-binding cassette transporter A1
ApoE	Apolipoprotein E
ATP	Adenosine triphosphate
BAEC	Bovine artery endothelial cell
CHD	Coronary heart disease
COX-2	cyclooxygenase 2
CVD	Cardiovascular disease
Cy	Cyanidin
Dp	Delphinidin
eNOS	Endothelial nitric oxide synthase
ET-1	Endothelin-1
GLP1	glucagon-like peptide 1
GLUT4	Glucose transporter 4
GSH	Glutathione
HDL	High density lipoprotein
HO	Haem oxygenase
HPLC	High performance liquid chromatography
HUVEC	Human umbilical vein endothelial cell
ICAM-1	Intercellular adhesion molecule 1

IVA	Influenza Virus A
IVB	Influenza Virus B
LDL	Low-density lipoprotein
LPH	Lactase phloridzin hydrolase
LPL	lipoprotein lipase
MAEC	Mouse aortic endothelial cell
MAO	Monoamine oxidases
MOS	Mitochondrial oxidative stress
Mv	Malvidin
NADPH	Nicotinamide adenine dinucleotide phosphate (reduced form)
NF-κB	Nuclear factor kappa B
NO	Nitric oxide
NOX	NADPH oxidase
Pn	Peonidin
Pg	Pelargonidin
Pt	Petunidin
ROS	Reactive oxygen species
RTK	Receptor tyrosine kinase
SGLT	Sodium-dependent glucose co-transporter
TG	Triglyceride
TNF-α	Tumour necrosis factor
USDA	United States Department of Agriculture
VCAM-1	Vascular cell adhesion molecule 1

REFERENCES

Abraham, N.G. and Kappas, A. 2008. Pharmacological and clinical aspects of heme oxygenase. *Pharmacol Rev*, **60**, 79–127.

Adisakwattana, S., Charoenlertkul, P. and Yibchok-Anun, S. 2009. Alpha-glucosidase inhibitory activity of cyanidin-3-galactoside and synergistic effect with acarbose. *J Enzyme Inhib Med Chem,* **24**(1), 65–69.

Adisakwattana, S., Yibchok-Anun, S., Charoenlertkul, P. and Wongsasiripat, N. 2011. Cyanidin-3-rutinoside alleviates postprandial hyperglycemia and its synergism with acarbose by inhibition of intestinal α-glucosidase. *J Clin Biochem Nutr,* **49**(1), 36–41.

Aiyer, H.S., Warri, A.M., Woode, D.R. *et al.* 2012. Influence of berry polyphenols on receptor signalling and cell-death pathways: Implications for breast cancer prevention. *J Agric Food Chem*, **60**(23), 5693–5708.

Amorini, A.M., Fazzina, G., Lazzarino, G. *et al.* 2001. Activity and mechanism of the antioxidant properties of cyanidin-3-*O*-beta-glucopyranoside. *Free Radic Res*, **35**, 953–966.

Andres-Lacueva, C., Shukitt-Hale, B., Galli, R.L. *et al.* 2005. Anthocyanins in aged blueberry-fed rats are found centrally and may enhance memory. *Nutr Neurosci*, **8**, 111–120.

Anter, J., Romero-Jiménez, M., Fernández-Bedmar, Z. *et al.* 2011. Antigenotoxicity, cytotoxicity, and apoptosis induction by apigenin, bisabolol, and protocatechuic acid. *J Med Food*, **14**(3), 276–283.

Arita, Y., Kihara, S., Ouchi, N. *et al.* 1999. Paradoxical decrease of an adipose-specific protein, adiponectin, in obesity. *Biochem Biophys Res Commun*, **257**, 79–83.

Arts, I.C., Sesink, A.L., Faassen-Peters, M. and Hollman, P.C. 2004. The type of sugar moiety is a major determinant of the small intestinal uptake and subsequent biliary excretion of dietary quercetin glycosides. *Br J Nutr*, **91**, 841–847.

Barnham, K.J., Masters, C.L. and Bush, A.I. 2004. Neurodegenerative diseases and oxidative stress. *Nat Rev Drug Discov*, **3**, 205–214.

Basu, A., Wilkinson, M., Penugonda, K. *et al.* 2009. Freeze-dried strawberry powder improves lipid profile and lipid peroxidation in women with metabolic syndrome: baseline and post intervention effects. *Nutr J*, **8**, 43.

Basu, A., Rhone, M. and Lyons, T.J. 2010. Berries: emerging impact on cardiovascular health. *Nutr Rev*, **68**, 168–177.

Bedard, K. and Krause, K.H. 2007. The NOX family of ROS-generating NADPH oxidases: physiology and pathophysiology. *Physiol Rev*, **87**, 245–313.

Bell, D.R. and Gochenaur, K. 2006. Direct vasoactive and vasoprotective properties of anthocyanin-rich extracts. *J Appl Physiol*, **100**, 1164–70.

Berg, A.H., Combs, T.P. and Scherer, P.E. 2002. ACRP30/adiponectin: an adipokine regulating glucose and lipid metabolism. *Trends Endocrinol Metab*, **13**, 84–89.

Betteridge, D.J. 2000. What is oxidative stress? *Metabolism*, **49**, 3–8.

Bishaysee, A., Haznagy-Radnai, E., Mbimba, T. *et al.* 2010. Anthocyanin-rich blackcurrant extract suppresses the growth of human hepatocellular carcinoma cells. *Nat Prod Commun*, **5** (10), 1613–8.

Bishaysee, A., Mbimba, T., Thoppil, R.J. *et al.* 2011. Anthocyanin-rich black currant (*Ribes nigrum* L.) extract affords chemoprevention against diethylnitrosamine-induced hepatocellular carcinogenesis in rats. *J Nutr Biochem*, **22**, 1035–1046.

Bitsch, R., Netzel, M., Frank, T. *et al.* 2004a. Bioavailability and biokinetics of anthocyanins from red grape juice and red wine. *J Biomed Biotechnol*, **2004**(5), 293–298.

Bitsch, R., Netzel, M., Sonntag, S. *et al.* 2004b. Urinary excretion of cyanidin glucosides and glucuronides in healthy humans after elderberry juice ingestion. *J Biomed Biotechnol*, **2004**(5), 343–345.

Boivin, D., Blanchette, M., Barrette, S. *et al.* 2007. Inhibition of cancer cell proliferation and suppression of TNF-induced activation of NFκB by edible berry juice. *Anticancer Res*, **27**, 937–948.

Bonomini, F., Tengattini, S., Fabiano, A. *et al.* 2008. Atherosclerosis and oxidative stress. *Histol Histopathol*, **23**, 381–90.

Bors, W., Heller, W., Michel, C. and Saran, M. 1990. Flavonoids as antioxidants: determination of radical-scavenging efficiencies. *Methods Enzymol*, **186**, 343–355.

Brand, W., Schutte, M.E., Williamson, G. *et al.* 2006. Flavonoid-mediated inhibition of intestinal ABC transporters may affect the oral bioavailability of drugs, food-borne toxic compounds and bioactive ingredients. *Biomed Pharmacother*, **60**, 508–519.

Bub, A., Watzl, B., Heeb, D. *et al.* 2001. Malvidin-3-glucoside bioavailability in humans after ingestion of red wine, dealcoholized red wine and red grape juice. *Eur J Nutr*, **40**, 113–20.

Bub, A., Watzl, B., Blockhaus, M. *et al.* 2003. Fruit juice consumption modulates antioxidative status, immune status and DNA damage. *J Nutr Biochem*, **14** (2), 90–98.

Burton-Freeman, B., Linares, A., Hyson, D. and Kappagoda, T. 2010. Strawberry to highfat meal in overweight hyperlipidemic men and women. *J Am Coll Nutr*, **29**, 46–54.

Cao G., Russell R.M., Lischner N. and Prior R.L. 1998. Serum antioxidant capacity is increased by consumption of strawberries, spinach, red wine or vitamin C in elderly women. *J Nutr*, **128**, 2383–2390.

Cao, G., Muccitelli, H.U., Sanchez-Moreno, C. and Prior, R.L. 2001. Anthocyanins are absorbed in glycated forms in elderly women: a pharmacokinetic study. *Am J Clin Nutr*, **73**, 920–926.

Carlton, P.S., Kresty, L.A., Siglin, J.C. *et al.* 2001. Inhibition of N-nitrosomethylbenzylamine-induced tumorigenesis in the rat esophagus by dietary freeze-dried strawberries. *Carcinogenesis*, **22**, 441–446.

Cassidy, A., O'Reilly, E.J., Kay, C. *et al.* 2010. Habitual intake of flavonoid subclasses and incident hypertension in adults. *Am J Clin Nutr*, **93**, 338–347.

Cassidy, A., Rimm, E.B., O'Reilly, É.J. *et al.* 2012. Dietary flavonoids and risk of stroke in women. *Stroke*, **43**(4), 946–951.

Castañeda-Ovando, A., Pacheco-Hernández, M.D.L. and Páez-Hernández, M.E. *et al.* 2009. Chemical studies of anthocyanins: A review. *Food Chem*, **113**, 859–871.

Cavanagh, H., Hipwell, M. and Wilkinson, J. 2003. Antibacterial activity of berry fruits used for culinary purposes. *J Medicin Food*, **6** (1).

Cefalu, W.T., Ye, J., Zuberi, A. *et al.* 2008. Botanicals and the metabolic syndrome. *Am J Clin Nutr*, **87**(2), 481S–487S.

Chalopin, M., Tesse, A., Martinez, M.C. *et al.* 2010. Estrogen receptor alpha as a key target of red wine polyphenols action on the endothelium. *PLoS One*, **5**, e8554.

Chandra, A., Rana, J. and Li, Y. 2001. Separation, identification, quantification, and method validation of anthocyanins in botanical supplement raw materials by HPLC and HPLC-MS. *J Agric Food Chem*, **49**, 3515–3521.

Charron, C.S., Clevidence, B.A., Britz, S.J. and Novotny, J.A. 2007. Effect of dose size on bioavailability of acylated and nonacylated anthocyanins from red cabbage (*Brassica oleracea* L. Var. capitata). *J Agric Food Chem*, **55**, 5354–5362.

Charron, C.S., Kurilich, A.C., Clevidence, B.A. *et al.* 2009. Bioavailability of anthocyanins from purple carrot juice: effects of acylation and plant matrix. *J Agric Food Chem*, **57**, 1226–1230.

Chen, M.S., Chen, D. and Dou, Q.P. 2004. Inhibition of proteasome activity by various fruits and vegetables is associated with cancer cell death. *In Vivo*, **18**, 73–80.

Chen, P.N., Chu, S.C., Chiou, H.L. *et al.* 2006. Mulberry anthocyanins, cyanidin 3-rutinoside and cyanidin 3-glucoside, exhibited an inhibitory effect on the migration and invasion of a human lung cancer cell line. *Cancer Lett*, **235**(2), 248–259.

Chen, C.-Y., Yi, L., Jin, X. *et al.* 2011. Inhibitory effect of delphinidin on monocyte–endothelial cell adhesion induced by oxidized low-density lipoprotein via ROS/p38MAPK/NF-κB pathway. *Cell Biochem Biophys*, **61**(2), 337–348.

Chong, M. F.-F., Macdonald, R. and Lovegrove, J.A. 2010. Fruit polyphenols and CVD risk: a review of human intervention studies. *Br J Nutr*, **104**, S28–S39.

Cimino, F., Ambra, R., Canali, R. *et al.* 2006. Effect of cyanidin-3-O-glucoside on UVB-induced response in human keratinocytes. *J Agric Food Chem*, **54**(11), 4041–4047.

Cooke, D., Steward, W.P., Gescher, A.J. and Marczylo, T. 2005. Anthocyans from fruits and vegetables–does bright colour signal cancer chemopreventive activity? *Eur J Cancer*, **41**(13), 1931–1940.

Costanino, L., Albasini, A., Rastelli, G. and Benvenuti, S. 1992. Activity of polyphenolic crude extracts as scavengers of superoxide radicals and inhibitors of xanthine oxidase. *Plant Med*, **58**, 342–344.

Crozier, A., Jaganath, I.B. and Clifford, M.N. 2009. Dietary phenolics: chemistry, bioavailability and effects on health. *Nat Prod Rep*, **26**, 1001–1043.

Cryer, A. 1981. Tissue lipoprotein lipase activity and its action in lipoprotein metabolism. *Int J Biochem*, **13**, 525–541.

D'Alessio, D.A., and Vahl, T.P. 2004. Glucagon-like peptide 1: evolution of an incretin into a treatment for diabetes. *Am J Physiol Endocrinol Metab*, **286**, E882–E890.

Da Silva Pinto, M., Kwon, Y.I., Apostolidis, E. *et al.* 2008. Functionality of bioactive compounds in Brazilian strawberry (Fragaria x ananassa Duch.) cultivars: evaluation of hyperglycemia and hypertension potential using in vitro models. *J Agric Food Chem*, **56**, 4386–4392.

Day, A.J., Cañada, F.J., Díaz, J.C. *et al.* 2000. Dietary flavonoid and isoflavone glycosides are hydrolysed by the lactase site of lactase phlorizin hydrolase. *FEBS Lett*, **468**, 166–170.

Decker, E.A. 1995. The role of phenolics, conjugated linoleic acid, carnosine, and pyrroloquinoline quinone as nonessential dietary antioxidants. *Nutr Rev*, **53**(3), 49–58.

De Pascual-Teresa, S., Moreno, D.A. and Garcia-Viguera, C. 2010. Flavanols and anthocyanins in cardiovascular health: a review of current evidence. *Int J Mol Sci*, **11**, 1679–703.

Del Rio, D., Borges, G. and Crozier, A. 2010. Berry flavonoids and phenolics: bioavailability and evidence of protective effects. *Br J Nutr*, **104**, S67–S90.

Ding, M., Feng, R., Wang, S.Y. *et al.* 2006. Cyanidin-3-glucoside, a natural product derived from blackberry, exhibits chemopreventive and chemotherapeutic activity. *J Biol Chem.* **281**(25), 17359–68.

Domitrovic, R. 2011. The molecular basis for the pharmacological activity of anthocyans. *Curr Med Chem*, **18**, 4454–69.

Dreiseitel, A., Schreier, P., Oehme, A. *et al.* 2008. Inhibition of proteasome activity by anthocyanins and anthocyanidins. *Biochem Biophys Res Commun*, **372**, 57–61.

Dreiseitel, A., Korte, G., Schreier, P. *et al.* 2009. Berry anthocyanins and their aglycons inhibit monoamine oxidases A and B. *Pharmacol Res*, **59**, 306–311.

Duchnowicz, P., Broncel, M., Podsędek, A. and Koter-Michalak, M. 2012. Hypolipidemic and antioxidant effects of hydroxycinnamic acids, quercetin, and cyanidin 3-glucoside in hypercholesterolemic erythrocytes (in vitro study). *Eur J Nutr*, **51**(4), 435–443.

Duthie, S.J. 2007. Berry phytochemicals, genomic stability and cancer: evidence for chemoprotection at several stages in the carcinogenic process. *Mol Nutr Food Res*, **51**(6), 665–674.

Duthie, S.J., Gardner, P.T., Morrice, P.C. *et al.* 2004. DNA stability and lipid peroxidation in vitamin E-deficient rats in vivo and colon cells in vitro: modulation by the dietary anthocyanin, cyanidin-3-glycoside. *Eur J Nutr*, **44**, 195–203.

Edirisinghe, I., Banaszewski, K., Cappozzo, J. *et al.* 2011a. Strawberry anthocyanin and its association with postprandial inflammation and insulin. *Br J Nutr*, **106**(6), 913–922.

Edirisinghe, I., Banaszewski, K., Cappozzo, J. *et al.* 2011b. Effect of blackcurrant anthocyanins on the activation of eNOS in-vitro in human endothelial cells. *J Agric Food Chem*, **59**(16), 8616–8624.

Ellis, C.L., Edirisinghe, I., Kappagoda, T. and Burton-Freeman, B.J. 2011. Attenuation of meal-induced inflammatory and thrombotic responses in overweight men and women after 6-week daily strawberry (Fragaria) intake. A randomized placebo-controlled trial. *Atheroscler Thromb*, **18**(4), 318–327.

Erdman, J.W., Jr., Balentine, D., Arab, L. *et al.* 2007. Flavonoids and heart health. Proceedings of the ILSI North America Flavonoids Workshop, May 31-June 1, 2005, Washington, DC. *J Nutr*, **137**, 718S–737S.

Erlund, I., Koli, R., Alfthan, G. *et al.* 2008. Favorable effects of berry consumption on platelet function, blood pressure, and HDL cholesterol. *Am J Clin Nutr*, **87**, 323–331.

Fang, Z., Zhang, M., Sun, Y. and Sun, J. 2005. How to improve bayberry (Myrica rubra Sieb. et Zucc.) juice color quality: effect of juice processing on bayberry anthocyanins and polyphenolics. *J Agric Food Chem*, **54**, 99–106.

Felgines, C., Krisa, S., Mauray, A. *et al.* 2010. Radiolabelled cyanidin 3-O-glucoside is poorly absorbed in the mouse. *Br J Nutr*, **103**, 1738–1745.

Felgines, C., Talavera, S., Gonthier, M.P. *et al.* 2003. Strawberry anthocyanins are recovered in urine as glucuro- and sulfoconjugates in humans. *J Nutr*, **133**, 1296–1301.

Feng, R., Ni, H.M., Wang, S.Y. *et al.* 2007. Cyanidin-3-rutinoside, a natural polyphenol antioxidant, selectively kills leukemic cells by induction of oxidative stress. *J Biol Chem*, **282**(18), 13468–13476.

Feng, R., Wang, S.Y., Shi, Y.H. *et al.* 2010. Delphinidin induces necrosis in hepatocellular carcinoma cells in the presence of 3-methyladenine, an autophagy inhibitor. *J Agric Food Chem*, **58**(7), 3957–3964.

Fernandes, I., Azevedo, J., Faria, A. *et al.* 2009. Enzymatic hemisynthesis of metabolites and conjugates of anthocyanins. *J Agric Food Chem*, **57**, 735–475.

Finné Nielsen, I.L., Elbøl Rasmussen, S., Mortensen, A. *et al.* 2005. Anthocyanins increase low-density lipoprotein and plasma cholesterol and do not reduce atherosclerosis in Watanabe Heritable Hyperlipidemic rabbits. *Mol Nutr Food Res* **49**(4), 301–308.

Flegal, K.M., Carroll, M.D., Ogden, C. and Curtin, L. 2010. Prevalence and trends in obesity among US adults. *JAMA*, **303**, 235–241.

Fleschhut, J., Kratzer, F., Rechkemmer, G. and Kulling, S.E. 2006. Stability and biotransformation of various dietary anthocyanins in vitro. *Eur J Nutr*, **45**, 7–18.

Flier, J.S. 2004. Obesity wars: molecular progress confronts an expanding epidemic. *Cell*, **116**, 337–350.

Forester, S.C. and Waterhouse, A.L. 2008. Identification of Cabernet Sauvignon anthocyanin gut microflora metabolites. *J Agric Food Chem*, **56**, 9299–9304.

Frank, J., Kamal-Eldin, A., Lundh, T. *et al.* 2002. Effects of dietary anthocyanins on tocopherols and lipids in rats. *J Agric Food Chem*, **50**(25), 7226–7230.

Frank, T., Netzel, M., Strass, G. *et al.* 2003. Bioavailability of anthocyanidin-3-glucosides following consumption of red wine and red grape juice. *Can J Physiol Pharm*, **81**, 423–435.

Frank, T., Janßen, M., Netzel, M. *et al.* 2005. Pharmacokinetics of anthocyanidin-3-glycosides following consumption of Hibiscus sabdariffa L. extract. *J Clin Pharm*, **45**, 203–210.

Friedman, J. and Halaas, J. 1998. Leptin and the regulation of body weight in mammals. *Nature*, **395**, 763–770.

Fruebis, J., Steinberg, D., Dresel, H.A. and Carew, T.E. 1994. A comparison of the antiatherogenic effects of probucol and of a structural analogue of probucol in low density lipoprotein receptor-deficient rabbits. *J Clin Invest*, **94**, 392–398.

Furukawa, S., Fujita, T., Shimabukuro, M. *et al.* 2004. Increased oxidative stress in obesity and its impact on metabolic syndrome. *J Clin Invest*, **114**, 1752–1761.

Gaemers, I. and Groen, A. 2006. New insights in the pathogenesis of non-alcoholic fatty liver disease. *Curr Opin Lipidol*, **17**, 268–273.

Garcia-Alonso, M., Minihane, A.-M., Rimbach, G. *et al.* 2009. Red wine anthocyanins are rapidly absorbed in humans and affect monocyte chemoattractant protein 1 levels and antioxidant capacity of plasma. *J Nutr Biochem*, **20**, 521–529.

Gonzalez-Baró, M.R., Lewin, T.M. and Coleman, R.A. 2009. Regulation of triglyceride metabolism. II. Function of mitochondrial GPAT1 in the regulation of triacylglycerol biosynthesis and insulin action. *Am J Physiol Gastrointest Liver Physiol*, **292**(5), G1195–G1199.

Guo, H., Ling, W., Wang, Q. *et al.* 2008. Cyanidin 3-glucoside protects 3T3-L1 adipocytes against H2O2- or TNF-a-induced insulin resistance by inhibiting c-Jun NH2-terminal kinase activation. *Biochem Pharmacol*, **75**, 1393–1401.

Guo, H., Li, D., Ling, W. *et al.* 2011. Anthocyanin inhibits high glucose-induced hepatic mtGPAT1 activation and prevents fatty acid synthesis through PKC. *J Lipid Res*, **52**, 908–922.

Guo, H., Liu, G., Zhong, R. *et al.* 2012. Cyanidin-3-O-b-glucoside regulates fatty acid metabolism via an AMP-activated protein kinase dependent signalling pathway in human HepG2 cells. *Lipids Health Dis*, **11**, 10.

Giordano, L., Coletta, W., Tamburrelli, C. *et al.* 2011. Four-week ingestion of blood orange juice results in measurable anthocyanin urinary levels but does not affect cellular markers related to cardiovascular risk: a randomized cross-over study in healthy volunteers. *Eur J Nutr*, **51**, 541–548.

Granger, D.N., Vowinkel, T. and Petnehazy, T. 2004. Modulation of the inflammatory response in cardiovascular disease. *Hypertension*, **43**, 924–931.

Grassi, D., Desideri, G., Croce, G. *et al.* 2009. Flavonoids, vascular function and cardiovascular protection. *Curr Pharm Des*, **15**, 1072–1084.

Gopalan, A., Reuben, S.C., Ahmed, S. *et al.* 2012. The health benefits of blackcurrants. *Food Funct*, **3**, 795–809.

Han, S.J., Ryu, S.N., Trinh, H.T. *et al.* 2009. Metabolism of cyanidin-3-O-beta-D-glucoside isolated from black colored rice and its antiscratching behavioral effect in mice. *J Food Sci*, **74**, H253-H258.

Hanamura, T., Mayama, C., Aoki, H. *et al.* 2006. Antihyperglycemic effect of polyphenols from acerola (Malpighia emarginata DC.) fruit. *Biosci Biotechnol Biochem*, **70**, 1813–1820.

Harada, K., Kano, M., TakayanagI, T. *et al.* 2004. Absorption of acylated anthocyanins in rats and humans after ingesting an extract of *Ipomoea batatas* purple sweet potato tuber. *Biosci Biotech Biochem*, **68**, 1500–1507.

Hassellund, S.S., Flaa, A., Sandvik, L. *et al.* 2012. Effects of anthocyanins on blood pressure and stress reactivity: a double-blind randomized placebo-controlled crossover study. *J Hum Hypertens*, **26**(6):396–404.

He, M., Siow, R.C., Sugden, D. *et al.* 2010. Induction of HO-1 and redox signaling in endothelial cells by advanced glycation end products: A role for Nrf2 in vascular protection in diabetes. *Nutr Metab Cardiovasc Dis*, **21**, 277–285.

Henderson, L., Gregory, J. and Swan, G. 2002. *National Diet And Nutrition Survey Adults Aged 19 To 64 Years*. Volume **1**. UK Department of Health and Food Standards Agency, London.

Herath, H.M., Takano-Ishikawa, Y. and Yamaki, K. 2003. Inhibitory effect of some flavonoids on tumor necrosis factor-alpha production in lipopolysaccharide-stimulated mouse macrophage cell line J774.1. *J Med Food*, **6**(4), 365–370.

Hidalgo, M., Martin-Santamaria, S., Recio, I. *et al.* 2012. Potential anti-inflammatory, anti-adhesive, anti/estrogenic, and angiotensin-converting enzyme inhibitory activities of anthocyanins and their gut metabolites. *Genes Nutr*, **7**, 295–306.

Higashi, Y., Noma, K., Yoshizumi, M. and Kihara, Y. 2009. Endothelial function and oxidative stress in cardiovascular diseases. *Circ J*, **73**, 411–418.

Hobbs, A.J., Higgs, A. and Moncada, S. 1999. Inhibition of nitric oxide synthase as a potential therapeutic target. *Annu Rev Pharmacol Toxicol*, **39**, 191–220.

Hollands, W., Brett, G.M., Radreau, P. *et al.* 2008. Processing blackcurrants dramatically reduces the content and does not enhance the urinary yield of anthocyanins in human subjects. *Food Chem*, **108**, 869–878.

Hollman, P.C. and Katan, M.B. 1998. Absorption, metabolism and bioavailability of flavonoids. In: *Flavonoids in health and disease* (eds Rice-Evans, C. and Packer, L.). CRC Press, p. 495.

Hollman, P.C., Bijsman, M.N., Van Gameren, Y. *et al.* 1999. The sugar moiety is a major determinant of the absorption of dietary flavonoid glycosides in man. *Free Radic Res*, **31**, 569–573.

Holst, J.J. 2007. The physiology of glucagon-like peptide 1. *Physiol Rev*, **87**, 1409–1439.

Hooper, L., Kroon, P.A., Rimm, E.B. *et al.* 2008. Flavonoids, flavonoid-rich foods, and cardiovascular risk: a meta-analysis of randomized controlled trials. *Am J Clin Nutr*, **88**, 38–50.

Hurst, S.M., Mcghie, T.K., Cooney, J.M. *et al.* 2010. Blackcurrant proanthocyanidins augment IFN-gamma-induced suppression of IL-4 stimulated CCL26 secretion in alveolar epithelial cells. *Mol Nutr Food Res*, **54**, S159–S170.

Huntley, A.L. 2009. The health benefits of berry flavonoids for menopausal women: cardiovascular disease, cancer and cognition. *Maturitas*, **63**, 297–301.

Ichiyanagi, T., Rahman, M.M., Kashiwada, Y. *et al.* 2004. Absorption and metabolism of delphinidin 3-O-beta-D-glucopyranoside in rats. *Free Radic Biol Med*, **36**, 930–937.

Igarashi, K., Takanashi, K., Makino, M. and Yasui, T. 1989. Antioxidative activity of major anthocyanin isolated from wild grapes. *Kogyo Gakkaishi*, **36**, 852–856.

Jenkins, D.J.A., Nguyena, T.H., Kendalla, C.W.C. *et al.* 2008. The effect of strawberries in a cholesterol-lowering dietary portfolio. *Metab Clin Exp*, **57**, 1636–1644.

Jin, Y., Alimbetov, D., George, T. *et al.* 2011. A randomised trial to investigate the effects of acute consumption of a blackcurrant juice drink on markers of vascular reactivity and bioavailability of anthocyanins in human subjects. *Eur J Clin Nutr*, **65**, 849–856.

Kamei, H., Kojima, T., Hasegawa, M. *et al.* 1995. Suppression of tumor cell growth by anthocyanins in vitro. *Cancer Invest*, **13**(6), 590–594.

Karlsen, A., Retterstøl, L., Laake, P. *et al.* 2007. Anthocyanins inhibit nuclear factor-kappaB activation in monocytes and reduce plasma concentrations of pro-inflammatory mediators in healthy adults. *J Nutr*, **137**(8), 1951–1954.

Kawashima, S. and Yokoyama, M. 2004. Dysfunction of endothelial nitric oxide synthase and atherosclerosis. *Arterioscler Thromb Vasc Biol*, **24**, 998–1005.

Kay, C.D. 2006. Aspects of anthocyanin absorption, metabolism and pharmacokinetics in humans. *Nutr Res Rev*, **19**, 137–146.

Kay, C.D., Mazza, G., Holub, B.J. and Wang, J. 2004. Anthocyanin metabolites in human urine and serum. *Br J Nutr*, **91**, 933–942.

Kay, C.D., Mazza, G.J. and Holub, B.J. 2005. Anthocyanins exist in the circulation primarily as metabolites in adult men. *J Nutr*, **135**, 2582–2588.

Kay, C.D., Kroon, P.A. and Cassidy, A. 2009. The bioactivity of dietary anthocyanins is likely to be mediated by their degradation products. *Mol Nutr Food Res*, **53**(Suppl 1), S92–101.

Kelsey, N., Hulick, W., Winter, A. *et al.* 2011. Neuroprotective effects of anthocyanins on apoptosis induced by mitochondrial oxidative stress. *Nutr Neurosci*, **14**(6), 249.

Keppler, K. and Humpf, H.U. 2005. Metabolism of anthocyanins and their phenolic degradation products by the intestinal microflora. *Bioorg Med Chem*, **13**, 5195–205.

Kern, M., Fridrich, D., Reichert, J. *et al.* 2007. Limited stability in cell culture medium and hydrogen peroxide formation affect the growth inhibitory properties of delphinidin and its degradation product gallic acid. *Molecular Nutrition and Food Research*, **51**, 1163–1172.

Kloetzel, P.M. (2004) The proteasome and MHC class I antigen processing. *Biochim Biophys Acta*, **1695**, 217–225.

Kim, S.J., Nian, C. and Mcintosh, C.H. 2007. Activation of lipoprotein lipase by glucose-dependent insulinotropic polypeptide in adipocytes: a role for a protein kinase B, LKB1, and AMP-activated protein kinase cascade. *J Biol Chem*, **282**, 8557–8567.

Knox, Y.M., Hayashi, K., Suzutani, T. *et al.* 2001. Activity of anthocyanins from fruit extract of Ribes nigrum L. against influenza A and B viruses. *Acta Virol*, **45**(4), 209–215.

Knox, Y., Suzutani, T., Yosida, I. and Azuma, M. 2003. Anti-influenza virus activity of crude extract of Ribes nigrum L. *Phytotherapy Res*, **17**, 120–122.

Kong, J.M., Chia, L.S., Goh, N.K. *et al.* 2003. Analysis and biological activities of anthocyanins. *Phytochem*, **64**, 923–933.

Kresty, L.A., Frankel, W.L., Hammond, C.D. *et al.* 2006. Transitioning from preclinincal to clinical chemo-preventive assessments of lyophilized black raspberries: interim results show berries modulate markers of oxidative stress in Barrett's esophagus patients. *Nutr Cancer*, **54**, 148–156.

Kris-Etherton, P.M., Hecker, K.D., Bonanome, A. *et al.* 2002. Bioactive compounds in foods: their role in the prevention of cardiovascular disease and cancer. *Am J Med*, **113** 9B, 71S–88S.

Kurilich, A.C., Clevidence, B.A., Britz, S.J. *et al.* 2005. Plasma and urine responses are lower for acylated vs nonacylated anthocyanins from raw and cooked purple carrots. *J Agric Food Chem*, **53**, 6537–6542.

Landberg, R., Naidoo, N. and Van Dam, R.M. 2012. Diet and endothelial function: from individual components to dietary patterns. *Curr Opin Lipidol*, **23**, 147–155.

Landmesser, U., Hornig, B. and Drexler, H. 2004. Endothelial function: a critical determinant in atherosclerosis? *Circulation*, **109**, 1127–1133.

Lau, F.C., Shukitt-Hale, B. and Joseph, J.A. 2005. The beneficial effects of fruit polyphenols on brain aging. *Neurobiol Aging*, **26**, 128–132.

Lazze, M.C., Pizzala, R., Perucca, P. *et al.* 2006. Anthocyanidins decrease endothelin-1 production and increase endothelial nitric oxide synthase in human endothelial cells. *Mol Nutr Food Res*, **50**, 44–51.

Lee, J., Durst, R.W. and Wrolstad, R.E. 2002. Impact of juice processing on blueberry anthocyanins and polyphenolics: comparison of two pretreatments. *J Food Sci*, **67**, 1660–1667.

Leopold, J.A. and Loscalzo, J. 2009. Oxidative risk for atherothrombotic cardiovascular disease. *Free Radic Biol Med*, **47**, 1673–1706.

Li, C., Lee, M.J., Sheng, S. *et al.* 2000. Structural identification of two metabolites of catechins and their kinetics in human urine and blood after tea ingestion. *Chem Res Toxicol*, **13**, 177–184.

Libby, P., Ridker, P.M. and Maseri, A. 2002. Inflammation and atherosclerosis. *Circulation*, **105**, 1135–1143.

Lim, T.G, Kwon, J.Y., Kim, J. *et al.* 2011. Cyanindin-3-glucoside suppresses b[a]PDE-induced cyclooxygenase-2 expression by directly inhibiting Fyn kinase activity. *Biochem Pharmacol*, **82**, 167–174.

Lin, L.Z. and Harnly, J.M. 2007. A screening method for the identification of glycosylated flavonoids and other phenolic compounds using a standard analytical approach for all plant materials. *J Agric Food Chem*, **55**, 1084–1096.

Lindsay, R.S., Funahashi, T., Hanson, R.L. *et al.* 2002. Adiponectin and development of type 2 diabetes in the Pima Indian population. *Lancet*, **360**, 57–58.

Ludwig, D.S. 2002. The glycemic index. Physiological mechanisms relating to obesity, diabetes, and cardiovascular disease. *JAMA*, **287**, 2414–2423.

Lyall, K.A., Hurst, S.M., Cooney, J. *et al.* 2009. Short-term blackcurrant extract consumption modulates exercise-induced oxidative stress and lipopolysaccharide-stimulated inflammatory responses. *Am J Physiol Regul Integr Comp Physiol*, **297**(1), R70–81.

Maatta-Riihinen, K.R., Kamal-Eldin, A., Mattila, P.H. *et al.* 2004. Distribution and Contents of Phenolic Compounds in Eighteen Scandinavian Berry Species. *J Agric Food Chem*, **52**, 4477–4486.

Maeda, K., Okubo, K., Shimomura, I. *et al.* 1996. cDNA cloning and expression of a novel adipose specific collagen-like factor, apM1 (AdiPose Most abundant Gene transcript 1). *Biochem Biophys Res Commun*, **221**, 16286–16289.

Mallery, S.R., Zwick, J.C., Pei, P. *et al.* 2008. Topical application of a bioadhesive black raspberry gel modulates gene expression and reduces cyclooxygenase 2 protein in human premalignant oral lesions. *Cancer Res*, **68**, 4945–4957.

Manach, C., Williamson, G., Morand, C. *et al.* 2005. Bioavailability and bioefficacy of polyphenols in humans. I. Review of 97 bioavailability studies. *Am J Clin Nutr*, **81**, 230S–242S.

Markakis, P. 1974. Anthocyanins and their stability in foods. *CRC Crit Rev Food Technol*, **4**, 437–456.

Marko, D., Puppel, N., Tjaden, Z. *et al.* 2004. The substitution pattern of anthocyanidins affects different cellular signaling cascades regulating cell proliferation. *Mol Nutr Food Res*, **48**(4), 318–325.

Matsui, T., Ebuchi, S., Kobayashi, M. *et al.* 2002. Antihyperglycemic effect of diacylated anthocyanin derived from Ipomoea batatas cultivar Ayamurasaki can be achieved through the a-glucosidase inhibitory action. *J Agric Food Chem*, **50**, 7244–7248.

Matsumoto, H., Inaba, H., Kishi, M. *et al.* 2001. Orally administered delphinidin 3-rutinoside and cyanidin 3-rutinoside are directly absorbed in rats and humans and appear in the blood as the intact forms. *J Agric Food Chem*, **49**, 154615–51.

Mazza, G., Kay, C.D., Cottrell, T. and Holub, B.J. 2002. Absorption of anthocyanins from blueberries and serum antioxidant status in human subjects. *J Agric Food Chem*, **50**, 7731–7737.

Mazza, G. and Miniati, E. 1993. *Anthocyanins in fruits, vegetables and grains*. CRC Press Inc., Boca Raton, FL.

McCullough, M.L., Peterson, J.J., Patel, R. *et al.* 2012. Flavonoid intake and cardiovascular disease mortality in a prospective cohort of US adults. *Am J Clin Nutr*, **95**(2):454–464.

McDougall, G.J., Dobson, P., Smith, P. *et al.* 2005a. Assessing potential bioavailability of raspberry anthocyanins using an *in vitro* digestion system. *J Agric Food Chem*, **53**, 5896–5904.

McDougall, G.J., Fyffe, S., Dobson, P. and Stewart, D. 2005b. Anthocyanins from red wine – their stability under simulated gastrointestinal digestion. *Phytochemistry*, **66**, 2540–2548.

McDougall, G.J., Shpiro, F., Dobson, P. *et al.* 2005c Different polyphenolic components of soft fruits inhibit a-amylase and a-glucosidase. *J Agric Food Chem*, **53**, 2760–2766.

McGhie, T.K. and Walton, M.C. 2007. The bioavailability and absorption of anthocyanins: towards a better understanding. *Mol Nutr Food Res*, **51**, 702–713.

Mead, J., Irvine, S. and Ramji, D. 2002. Lipoprotein lipase: structure, function, regulation, and role in disease. *J Mol Med*, **80**, 753–769.

Medeiros, L.A., Khan, T., El Khoury, J.B. *et al.* 2004. Fibrillar amyloid protein present in atheroma activates CD36 signal transduction. *J Biol Chem*, **279**, 10643–10648.

Meiers, S., Kemény, M., Weyand, U. *et al.* 2001. The anthocyanidins cyanidin and delphinidin are potent inhibitors of the epidermal growth-factor receptor. *J Agric Food Chem*, **49**(2), 958–962.

Michel, T. and Vanhoutte, P.M. 2010. Cellular signaling and NO production. *Pflugers Arch*, **459**, 807–816.

Middleton, E. Jr. 1998. Effect of plant flavonoids on immune and inflammatory cell function. *Adv Exp Med Biol*, **439**, 175–182.

Milbury, P.E., Vita, J.A. and Blumberg, J.B. 2010. Anthocyanins are bioavailable in humans following an acute dose of cranberry juice. *J Nutr*, **140**, 1099–1104.

Mink, P.J., Scrafford, C.G., Barraj, L.M. *et al.* 2007. Flavonoid intake and cardiovascular disease mortality: a prospective study in postmenopausal women. *Am J Clin Nutr*, **85**, 895–909.

Miyazawa, T., Nakagawa, K., Kudo, M. *et al.* 1999. Direct intestinal absorption of red fruit anthocyanins, cyanidin-3-glucoside and cyanidin-3,5-diglucoside, into rats and humans. *J Agric Food Chem*, **47**, 1083–1091.

Mu, H. and Porsgaard, T. 2005. The metabolism of structured triacylglycerols. *Prog Lipid Res*, **44**, 430–448.

Mulleder, U., Murkovic, M. and Pfannhauser, W. 2002. Urinary excretion of cyanidin glycosides. *J Biochem Biophys Methods*, **53**, 61–66.

Mullen, W., Edwards, C.A., Serafini, M. and Crozier, A. 2008a. Bioavailability of pelargonidin-3-O-glucoside and its metabolites in humans following the ingestion of strawberries with and without cream. *J Agric Food Chem*, **56**, 713–719.

Mullen, W., Rouanet, J.M., Auger, C. *et al.* 2008b. Bioavailability of [2-(14)C]quercetin-4′-glucoside in rats. *J Agric Food Chem*, **56**, 12127–12137.

Nakajima, J.I., Tanaka, I., Seo, S. *et al.* 2004. LC/PDA/ESI-MS profiling and radical scavenging activity of anthocyanins in various berries. *J Biomed Biotechnol*, **2004**(5), 241–247.

Naseem, K.M. 2005. The role of nitric oxide in cardiovascular diseases. *Mol Aspects Med*, **26**, 33–65.

Naujokat, C., Berges, C., Höh, A. *et al.* 2006. Proteasomal chymotrypsin-like peptidase activityis required for essential functions of human monocyte-derived dendritic cells. *Immunology*, **120**, 120–132.

Nencioni, A., Grünebach, F., Patrone, F. *et al.* 2007. Proteasomeinhibitors: antitumor effects and beyond. *Leukemia*, **21**, 30–36.

Netzel, M., Strass, G., Janssen, M. *et al.* 2001. Bioactive anthocyanins detected in human urine after ingestion of blackcurrant juice. *J Environ Pathol Toxicol Oncol*, **20**, 89–95.

Netzel, M., Strass, G., Herbst, M. *et al.* 2005. The excretion and biological antioxidant activity of elderberry antioxidants in healthy humans. *Food Res Int*, **38**, 905–910.

Netzel, M., Netzel, G., Tian, Q. *et al.* 2006. Sources of antioxidant activity in Australian native fruits. identification and quantification of anthocyanins. *J Agric Food Chem*, **54**, 9820–9826.

Neveu, V., Perez-Jiménez, J., Vos, F. *et al.* 2010. Phenol-Explorer: an online comprehensive database on polyphenol contents in foods. http://www.phenol-explorer.eu/ (last accessed 22 March 2013).

Nielsen, I.L., Dragsted, L.O., Ravn-Haren, G. *et al.* 2003a. Absorption and excretion of black currant anthocyanins in humans and watanabe heritable hyperlipidemic rabbits. *J Agric Food Chem*, **51**, 2813–2820.

Nielsen, I.L., Haren, G.R., Magnussen, E.L. *et al.* 2003b. Quantification of anthocyanins in commercial black currant juices by simple high-performance liquid chromatography. Investigation of their pH stability and antioxidative potency. *J Agric Food Chem*, **51**, 5861–5866.

Nurmi, T., Mursu, J., Heinonen, M. *et al.* 2009. Metabolism of Berry Anthocyanins to Phenolic Acids in Humans. *J Agric Food Chem*, **57**, 2274–2281.

Ohira, M., Miyashita, Y., Murano, T. *et al.* 2009. Metformin promotes induction of lipoprotein lipase in skeletalmuscle through activation of adenosine monophosphate-activated protein kinase. *Metabolism*, **58**, 1408–1414.

Olsson, M.E., Gustavsson, K.E., Andersson, S. *et al.* 2004. Inhibition of cancer cell proliferation in vitro by fruit and berry extracts and correlations with antioxidant levels. *J Agric Food Chem*, **52**(24), 7264–7271.

Park, S.J., Shin, W.H., Seo, J. and Kim, E.J. 2007. Anthocyanins inhibit airway inflammation and hyperresponsiveness in a murine asthma model. *Food Chem Toxicol*, **45**(8), 1459–1467.

Perez-Vicente, A., Gil-Izquierdo, A. and Garcia-Viguera, C. 2002. In vitro gastrointestinal digestion study of pomegranate juice phenolic compounds, anthocyanins, and vitamin C. *J Agric Food Chem*, **50**, 2308–2312.

Persson, I.A.L., Persson, K. and Andersson, R.G.G. 2009. Effect of vaccinium myrtillus and its polyphenols on angiotensin-converting enzyme activity in human endothelial cells. *J Agric Food Chem*, **57**, 4626–4629.

Prior, RL. and Wu, X. 2006. Anthocyanins: structural characteristics that result in unique metabolic patterns and biological activities. *Free Radic Res* **40**, 1014–1028.

Puupponen-Pimia, R., Nohynek, L., Hartmann-Schmidlin, S. *et al.* 2005. Berry phenolics selectively inhibit the growth of intestinal pathogens. *J Appl Microbiol*, **98**, 991–1000.

Qin, Y., Xia, M., Ma, J. *et al.* 2009. Anthocyanin supplementation improves serum LDL- and HDL-cholesterol concentrations associated with the inhibition of cholesteryl ester transfer protein in dyslipidemic subjects. *Am J Clin Nutr*, **90**, 485–492.

Rechner, A.R. and Kroner, C. 2005. Anthocyanins and colonic metabolites of dietary polyphenols inhibit platelet function. *Thromb Res*, **116**, 327–334.

Rechner, A.R., Kuhnle, G., Bremner, P. *et al.* 2002a. The metabolic fate of dietary polyphenols in humans. *Free Radic Biol Med*, **33**, 220–235.

Rechner, A.R., Kuhnle, G., Hu, H. *et al.* 2002b. The metabolism of dietary polyphenols and the relevance to circulating levels of conjugated metabolites. *Free Radical Res*, **36**, 1229–1241.

Rein, M.J., Ollilainen, V., Vahermo, M. *et al.* 2005. Identification of novel pyranoanthocyanins in berry juices. *Eur Food Res Techn*, **220**, 239–244.

Riboli, E. and Norat, T. 2003. Epidemiologic evidence of the protective effect of fruit and vegetables on cancer risk. *Am J Clin Nutr*, **78**, 559S–569S.

Rice-Evans, C.A., Miller, N.J. and Paganga, G. 1996. Structure-antioxidant activity relationships of flavonoids and phenolic acids. *Free Radic Biol Med*, **20**, 933–956.

Rios, L.Y., Gonthier, M.-P., Rémésy, C. *et al.* 2003. Chocolate intake increases urinary excretion of polyphenol-derived phenolic acids in healthy human subjects. *Am J Clin Nutr*, **77**, 912–918.

Roowi, S., Stalmach, A., Mullen, W. *et al.* 2010. Green tea flavan-3-ols: colonic degradation and urinary excretion of catabolites by humans. *J Agric Food Chem*, **58**, 1296–1304.

Ross, R. 1999. Atherosclerosis – an inflammatory disease. *N Engl J Med*, **340**, 115–126.

Rossi, A., Serraino, I., Dugo, P. *et al.* 2003. Protective effects of anthocyanins from blackberry in a rat model of acute lung inflammation. *Free Radic Res*, **37**(8), 891–900.

Rudich, A., Tirosh, A., Potashnik, R. *et al.* 1998. Prolonged oxidative stress impairs insulin-induced GLUT4 translocation in 3T3-L1 adipocytes. *Diabetes*, **47**, 1562–1569.

Ryter, S.W., Otterbein, L.E., Morse, D. and Choi, A.M. 2002. Heme oxygenase/carbon monoxide signaling pathways: regulation and functional significance. *Mol Cell Biochem*, **234–235**, 249–263.

Ryter, S.W., Alam, J. and Choi, A.M. 2006. Heme oxygenase-1/carbon monoxide: from basic science to therapeutic applications. *Physiol Rev*, **86**, 583–650.

Sasaki, R., Nishimura, N., Hoshino, H. *et al.* 2007. Cyanidin 3-glucoside ameliorates hyperglycemia and insulin sensitivity due to downregulation of retinol binding protein 4 expression in diabetic mice. *Biochem Pharmacol*, **74**, 1619–1627.

Scherer, P.E. 2006. Adipose tissue: from lipid storage compartment to endocrine organ. *Diabetes* **55**, 1537–1545.

Schewe, T., Steffen, Y. and Sies, H. 2008. How do dietary flavanols improve vascular function? A position paper. *Arch Biochem Biophys*, **476**, 102–106.

Schissel, S.L., Tweedie-Hardman, J., Rapp, J.H. *et al.* 1996. Rabbit aorta and human atherosclerotic lesions hydrolyze the sphingomyelin of retained low-density lipoprotein. Proposed role for arterial-wall sphingomyelinase in subendothelial retention and aggregation of atherogenic lipoproteins. *J Clin Invest*, **98**, 1455–1464.

Seeram, N.P. 2008. Berry fruits: compositional elements, biochemical activities and impact of their intake on human health. *J Agric Food Chem*, **56**, 627–629.

Seeram, N.P. 2010. Recent trends and advances in berry health benefits research. *J Agric Food Chem*, **58**(7), 3869–7380.

Seeram, N.P., Adams, L.S., Zhang, Y. *et al.* 2006. Blackberry, black raspberry, blueberry, cranberry, red raspberry, and strawberry extracts inhibit growth and stimulate apoptosis of human cancer cells in vitro. *J Agric Food Chem*, **54**, 9329–9339.

Sesso, H.D., Gaziano, J.M., Jenkins, D.J. and Buring, J.E. 2007. Strawberry intake, lipids, C-reactive protein, and the risk of cardiovascular disease in women. *J Am Coll Nutr*, **26**, 303–310.

Shin, W.H., Park, S.J. and Kim, E.J. 2006. Protective effect of anthocyanins in middle cerebral artery occlusion and reperfusion model of cerebral ischemia in rats. *Life Sci*, **79**(2), 130–137.

Sies, H. 2007. Total antioxidant capacity: appraisal of a concept. *J Nutr*, **137**, 1493–1495.

Simoncini, T., Lenzi, E., Zochling, A. *et al.* 2011. Estrogen-like effects of wine extracts on nitric oxide synthesis in human endothelial cells. *Maturitas*, **70**, 169–175.

Siow, R.C., Sato, H. and Mann, G.E. 1999. Heme oxygenase-carbon monoxide signalling pathway in atherosclerosis: anti-atherogenic actions of bilirubin and carbon monoxide? *Cardiovasc Res*, **41**, 385–394.

Skalen, K., Gustafsson, M., Rydberg, E.K. *et al.* 2002. Subendothelial retention of atherogenic lipoproteins in early atherosclerosis. *Nature*, **417**, 750–754.

Skrede, G., Wrolstad, R.E. and Durst, R.W. 2000. Changes in anthocyanins and polyphenolics during juice processing of highbush blueberries (*Vaccinium corymbosum* L.). *J Food Sci*, **65**, 357–364.

Sorrenti, V., Mazza, F., Campisi, A. *et al.* 2007. Heme oxygenase induction by cyanidin-3-O-beta-glucoside in cultured human endothelial cells. *Mol Nutr Food Res*, **51**, 580–586.

Spencer, J.P.E. 2008. Flavonoids: modulators of brain function? *Br J Nutr*, **99E**, ES60–ES77.

Spencer, J.P.E. 2009. Flavonoids and brain health: multiple effects underpinned by common mechanisms. *Genes Nutr*, **4**, 243–250.

Spencer, J.P.E. 2010. The impact of fruit flavonoids on memory and cognition. *Br J Nutr*, **104**, S40–S47.

Steinmetz, K.A. and Potter, J.D. 1996. Vegetables, fruit, and cancer prevention: a review. *J Am Diet Assoc*, **96**(10), 1027–1039.

Stephens, J.M., Lee, J. and Pilch, P.F. 1997. Tumor necrosis factor-alpha-induced insulin resistance in 3T3-L1 adipocytes is accompanied by a loss of insulin receptor substrate-1 and GLUT4 expression without a loss of insulin receptormediated signal transduction. *J Biol Chem*, **272**, 971–976.

Stocker, R. and Keaney, J.F., Jr. 2004. Role of oxidative modifications in atherosclerosis. *Physiol Rev*, **84**, 1381–1478.

Stoner, G.D. 2009. Foodstuffs for preventing cancer: the preclinical and clinical development of berries. *Cancer Prev Res*, **2** (3), 187–94.

Stoner, G.D., Sardo, C., Apseloff, G. *et al.* 2005. Pharmacokinetics of anthocyanins and ellagic acid in healthy volunteers fed freeze-dried black raspberries daily for 7 days. *J Clin Pharmacol*, **45**, 1153–1164.

Stoner, G.D., Wang, L.S., Zikri, N. *et al.* 2007. Cancer prevention with freeze-dried berries and berry components. *Semin Cancer Biol*, **17** (5), 403–410.

Stoner, G.D., Wang, L.S. and Casto, B.C. 2008. Laboratory and clinical studies of cancer chemoprevention by antioxidants in berries. *Carcinogenesis*, **29**(9), 1665–1674.

Suzutani, T., Ogasawara, M., Yoshida, I. *et al.* 2003. Anti-herpes virus activity of an extract of Ribes nigrum L. *Phytotherapy Res*, **17**, 609–613.

Svegliati-Baroni, G., Candelaresi, C., Saccomanno, S. *et al.* 2006. A model of insulin resistance and nonalcoholic steatohepatitis in rats: role of peroxisome proliferator-activated receptor-alpha and n-3 polyunsaturated fatty acid treatment on liver injury. *Am J Pathol*, **169**, 846–860.

Tabas, I. 1999. Nonoxidative modifications of lipoproteins in atherogenesis. *Annu Rev Nutr*, **19**, 123–139.

Takata, R., Yanai, T., Yamamoto, R. and Konno, T. 2007. Improvement of the antitumor activity of black currant polysaccharide by an enzymatic treatment. *Biosci Biotechnol Biochem*, **71**(5), 1342–1344.

Tamura, H. and Yamagami, A. 1994. Antioxidative activity of monoacylated anthocyanins isolated from Muscat Bailey A grape. *J Agric Food Chem*, **42**, 1612–1615.

Tamura, A., Fukushima, M., Shimada, K. *et al.* 2004. Cholesterol metabolism in rat is affected by protocatechuic acid. *J Nutr Sci Vitaminol*, **50**(1), 13–18.

Tarozzi, A., Morroni, F., Hrelia, S. *et al.* 2007. Neuroprotective effects of anthocyanins and their in vivo metabolites in SH-SY5Y cells. *Neurosci Let*, **424**, 36–40.

Teller, N., Thiele, W., Boettler, U. *et al.* 2009. Delphinidin inhibits a broad spectrum of receptor tyrosine kinases of the ErbB and VEGFR family. *Mol Nutr Food Res*, **53**(9), 1075–1083.

Tian, Q., Giusti, M.M., Stoner, G.D. and Schwartz, S.J. 2005. Screening for anthocyanins using high-performance liquid chromatography coupled to electrospray ionization tandem mass spectrometry with precursor-ion analysis, product-ion analysis, common-neutral-loss analysis, and selected reaction monitoring. *J Chromatogr A*, **1091**, 72.

Tian, Q., Giusti, M.M., Stoner, G.D. and Schwartz, S.J. 2006. Urinary excretion of black raspberry (Rubus occidentalis) anthocyanins and their metabolites. *J Agric Food Chem*, **54**, 1467–1472.

Torronen, R., Sarkkinen, E., Tapola, N. *et al.* 2010. Berries modify the postprandial plasma glucose response to sucrose in healthy subjects. *Br J Nutr*, **103**, 1094–1097.

Torronen, R., Sarkkinen, E., Niskanen, T. *et al.* 2012. Postprandial glucose, insulin and glucagon-like peptide 1 responses to sucrose ingested with berries in healthy subjects. *Br J Nutr*, **107**, 1445–1451.

Touyz, R.M. 2004. Reactive oxygen species, vascular oxidative stress, and redox signaling in hypertension: what is the clinical significance? *Hypertension*, **44**, 248–52.

Tseng, T.H., Kao, T.W., Chu, C.Y. *et al.* 2000. Induction of apoptosis by hibiscus protocatechuic acid in human leukemia cells via reduction of retinoblastoma (RB) phosphorylation and Bcl-2 expression. *Biochem Pharmacol*, **60**(3), 307–315.

Tsuda, T. 2011. Dietary anthocyanin-rich plants: Biochemical basis and recent progress in health benefits studies. *Mol Nutr Food Res*. **56**, 159–170.

Tsuda, T., Watanabe, M., Ohshima, K. *et al.* 1994. Antioxidative activity of the anthocyanin pigments cyanidin 3-O-beta-D-glucoside and cyanidin. *J Agric Food Chem*, **42**, 2407–2410.

Tsuda, T., Horio, F. and Osawa, T. 1999. Absorption and metabolism of cyanidin 3-O-beta-D-glucoside in rats. *FEBS Lett*, **449**, 179–182.

Tsuda, T., Ueno, Y., Aoki, H. *et al.* 2004. Anthocyanin enhances adipocytokine secretion and adipocyte-specific gene expression in isolated rat adipocytes. *Biochem Biophys Res Commun*, **316**, 149–157.

Tsuda, T., Ueno, Y., Kojo, H. *et al.* 2005. Gene expression profile of isolated rat adipocytes treated with anthocyanins. *Biochimica et Biophysica Acta*, **1733**, 137–147.

Tsuda, T., Ueno, Y., Yoshikawa, T. *et al.* 2006. Microarray profiling of gene expression in human adipocytes in response to anthocyanins. *Biochem pharmacol*, **71**, 1184–1197.

Tulipani, S., Alvarez-Suarez, J.M., Busco, F. *et al.* 2011. Strawberries improve plasma antioxidant status and erythrocyte resistance to oxidative hemolysis. *Food Chem*, **128**, 180–186.

USDA 2011. USDA Database for the Flavonoid Content of Selected Foods (Release 3.0). US Department of Agriculture, Agricultural Research Service (http://www.ars.usda.gov/SP2UserFiles/Place/12354500/Data/Flav/Flav_R03.pdf, last accessed 22 March 2013).

Visekruna, A., Joeris, T., Seidel, D. *et al.* 2006. Proteasome-mediated degradation of IjBa and processing of p105 in Crohn disease and ulcerative colitis. *J Clin Invest*, **116**, 3195–3203.

Vita, J.A. 2005. Polyphenols and cardiovascular disease: effects on endothelial and platelet function. *Am J Clin Nutr*, **81**, 292S–297S.

Vitaglione, P., Donnarumma, G., Napolitano, A. *et al.* 2007. Protocatechuic acid is the major human metabolite of cyanidin-glucosides. *J Nutr*, **137**, 2043–2048.

Wallace, T.C. 2011. Anthocyanins in Cardiovascular Disease. *Advances in Nutrition*, **2**, 1–7.

Walle, T., Walle, U.K. and Halushka, P.V. 2001. Carbon dioxide is the major metabolite of quercetin in humans. *J Nutr*, **131**, 2648–52.

Wang, Y. and Ho, C.T. 2009. Metabolism of flavonoids. *Forum Nutr*, **61**, 64–74.

Wang, J. and Mazza, G.J. 2002. Effects of anthocyanins and other phenolic compounds on the production of tumor necrosis factor alpha in LPS/IFN-gamma-activated RAW 264.7 macrophages. *Agric Food Chem*, **50**(15), 4183–4189.

Wang L.S. and Stoner G.D. 2008. Anthocyanins and their role in cancer prevention. *Cancer Lett*, **269**(2), 281–290.

Wang, S.Y., Feng, R.T., Lu, Y.J. *et al.* 2005. Inhibitory effect of activator protein1,nuclear factor-kappaB,andcell transformationbyextracts of strawberries (*Fragaria ananassa* Duch). *J Agric Food Chem*, **53**, 4187–4193.

Wang, D., Wei, X., Yan, X. *et al.* 2010. Protocatechuic acid, a metabolite of anthocyanins, inhibits monocyte adhesion and reduces atherosclerosis in apolipoprotein E-deficient mice. *J Agric Food Chem*, **58**, 12722–12728.

Wang, L.S., Arnold, M., Huang, Y.M. *et al.* 2011a. Modulation of genetic and epigenetic biomarkers of colorectal cancer in humans by black raspberries: a phase I pilot study. *Clin Cancer Res*, **17**, 598–610.

Wang, D., Zou, T., Yang, Y. *et al.* 2011b. Cyanidin-3-O-beta-glucoside with the aid of its metabolite protocatechuic acid, reduces monocyte infiltration in apolipoprotein E-deficient mice. *Biochem Pharmacol*, **82**, 713–719.

Wang, D., Xia, M., Gao, S. *et al.* 2012a. Cyanidin-3-O-β-glucoside upregulates hepatic cholesterol 7α-hydroxylase expression and reduces hypercholesterolemia in mice. *Mol Nutr Food Res*, **56**(4), 610–621.

Wang, D., Xia, M., Yan, X. *et al.* 2012b. Gut Microbiota Metabolism of Anthocyanin Promotes Reverse Cholesterol Transport in Mice Via Repressing miRNA-10b. *Circ Res*. **111**(8), 967–981.

Wang, Y., Zhang, Y., Wang, X. *et al.* 2012c. Supplementation with cyanidin-3-O-β-glucoside protects against hypercholesterolemia-mediated endothelial dysfunction and attenuates atherosclerosis in apolipoprotein E-deficient mice. *J Nutr*, **142**(6), 1033–1037.

Wei, X., Wang, D., Yang, Y. *et al.* 2011. Cyanidin-3-O-β-glucoside improves obesity and triglyceridemetabolism in KK-Ay mice by regulating lipoprotein lipase activity. *Sci Food Agric*, **91**, 1006–1013.

Weisel, T., Baum, M., Eisenbrand, G. *et al.* 2006. An anthocyanin/polyphenolic-rich fruit juice reduces oxidative DNA damage and increases glutathione level in healthy probands. *Biotechnol J.*, **1**, 388–397.

Wellen, K.E. and Hotamisligil, G.S. 2005. Inflammation, stress, and diabetes. *J Clin Invest*, **115**, 1111–1119.

WHO 2011. *Global status report on noncommunicable diseases 2010*. World Health Organization, Geneva, Switzerland (ISBN 978 92 4 156422 9).

Wiczkowski, W., Romaszko, E. and Piskula, M.K. 2010. Bioavailability of cyanidin glycosides from natural chokeberry (Aronia melanocarpa) juice with dietary-relevant dose of anthocyanins in humans. *J Agric Food Chem*, **58** (23), 12130–12136.

Williams, K.J. and Fisher, E.A. 2005. Oxidation, lipoproteins, and atherosclerosis: which is wrong, the antioxidants or the theory? *Curr Opin Clin Nutr Metab Care*, **8**, 139–146.

Williams, A.J., Dave, J.R. and Tortella, F.C. 2006. Neuroprotection with the proteasome inhibitor MLN519 in focal ischemic brain injury: relation to nuclear factor kappaB (NF-kappaB), inflammatory gene expression, and leukocyte infiltration. *Neurochem Int*, **49**, 106–112.

Williams, C.M., El Mohsen, M.A. and Vazour, D. 2008. Blueberry-induced changes in spatial working memory correlate with changes in hippocampal CREB phosphorylation and brain-derived neurotrophic factor (BDNF) levels. *Free Radic Biol Med*, **45**, 295–305.

Williamson, G. and Clifford, M.N. 2010. Colonic metabolites of berry polyphenols: the missing link to biological activity? *Br J Nutr*, **104**, S48-S66.

Witting, P., Pettersson, K., Ostlund-Lindqvist, A.M. *et al.* 1999. Dissociation of atherogenesis from aortic accumulation of lipid hydro(pero)xides in Watanabe heritable hyperlipidemic rabbits. *J Clin Invest*, **104**, 213–220.

Wood, P.J. (2007) Cereal b-glucans in diet and health. *J Cereal Sci*, **46**, 230–238.

Wood, S., Perez De Heredia, F., Wang, B. and Trayhurn, P. 2009. Cellular hypoxia and adipose tissue dysfunction in obesity. *Proc Nutr Soc*, **68**, 370–377.

Woodward, G., Kroon, P., Cassidy, A. and Kay, C. 2009. Anthocyanin stability and recovery: implications for the analysis of clinical and experimental samples. *J Agric Food Chem*, **57**, 5271–5278.

Woodward, G.M., Mccarthy, D., Pham-Thanh, D. and Kay, C.D. 2011a. Anthocyanins remain stable during commercial blackcurrant juice processing. *J Food Sci*, **76**, S408–414.

Woodward, G.M., Needs, P.W. and Kay, C.D. 2011b. Anthocyanin-derived phenolic acids form glucuronides following simulated gastrointestinal digestion and microsomal glucuronidation. *Mol Nutr Food Res*, **55**, 378–386.

Wu, X. and Prior, R.L. 2005. Systematic Identification and Characterization of Anthocyanins by HPLC-ESI-MS/MS in Common Foods in the United States: Fruits and Berries. *J Agric Food Chem*, **53**, 2589–2599.

Wu, D., Meydani, M., Leka, L.S. *et al.* 1999. Effect of dietary supplementation with black currant seed oil on the immune response of healthy elderly subjects. *Am J Clin Nutr*, **70**(4), 536–543.

Wu, X., Cao, G. and Prior, R.L. 2002. Absorption and metabolism of anthocyanins in elderly women after consumption of elderberry or blueberry. *J Nutr*, **132**, 1865–1871.

Wu, X., Gu, L., Prior, R.L. and Mckay, S. 2004. Characterization of anthocyanins and proanthocyanidins in some cultivars of Ribes, Aronia, and Sambucus and their antioxidant capacity. *J Agric Food Chem*, **52**, 7846–7856.

Wu, X., Beecher, G.R., Holden, J.M. *et al.* 2006. Concentrations of anthocyanins in common foods in the United States and estimation of normal consumption. *J Agric Food Chem*, **54**, 4069–4075.

Wu, Q.K., Koponen, J.M., Mykkänen, H.M. and Törrönen, A.R. 2007. Berry phenolic extracts modulate the expression of p21(WAF1) and Bax but not Bcl-2 in HT-29 colon cancer cells. *J Agric Food Chem*, **55**(4), 1156–1163.

Xu, J.W., Ikeda, K. and Yamori, Y. 2004a. Cyanidin-3-glucoside regulates phosphorylation of endothelial nitric oxide synthase. *FEBS Lett*, **574**, 176–180.

Xu, J.W., Ikeda, K. and Yamori, Y. 2004b. Upregulation of endothelial nitric oxide synthase by cyanidin-3-glucoside, a typical anthocyanin pigment. *Hypertension*, **44**, 217–222.

Xue, H., Aziz, R.M., Sun, N. *et al.* 2001. Inhibition of cellular transformation by berry extracts. *Carcinogenesis*, **22**, 351–356.

Yamamoto, N., Sawada, H., Izumi, Y. *et al.* 2007. Proteasomeinhibition induces glutathione synthesis and protects cells from oxidative stress – relevance to Parkinson disease. *J Biol Chem*, **282**, 4364–4372.

Yamauchi, T., Kamon, J., Waki, H. *et al.* 2001. The fat-derived hormone adiponectin reverses insulin resistance associated with both lipoatrophy and obesity. *Nat Med*, **7**, 941–946.

Yang, Y., Andrews, M.C., Hu, Y. *et al.* 2011. Anthocyanin extract from black rice significantly ameliorates platelet hyperactivity and hypertriglyceridemia in dyslipidemic rats induced by high fat diets. *J Agric Food Chem*, **59**, 6759–6764.

Yi, W., Akoh, C.C., Fischer, J. and Krewer, G. 2006. Absorption of Anthocyanins from Blueberry Extracts by Caco-2 Human Intestinal Cell Monolayers. *J. Agric. Food Chem.*, **54**, 5651–5658.

Zalba, G., Beaumont, F.J., San Jose, G. *et al.* 2000. Vascular NADH/NADPH oxidase is involved in enhanced superoxide production in spontaneously hypertensive rats. *Hypertension*, **35**, 1055–1061.

Zamora-Ros, R., Andres-Lacueva, C., Lamuela-Raventos, R.M. *et al.* 2010. Estimation of dietary sources and flavonoid intake in a Spanish adult population (EPIC-Spain). *J Am Diet Assoc*, **110**, 390–398.
Zamora-Ros, R., Knaze, V., Lujan-Barroso, L. *et al.* 2011. Estimation of the intake of anthocyanidins and their food sources in the European Prospective Investigation into Cancer and Nutrition (EPIC) study. *Br J Nutr*, **106**, 1090–1099.
Zhu, Y., Xia, M., Yang, Y. *et al.* 2011. Purified anthocyanin supplementation improves endothelial function via NO-cGMP activation in hypercholesterolemic individuals. *Clin Chem*. **57**, 1524–1533.
Ziberna, L., Lunder, M., Tramer, F. *et al.* 2011. The endothelial plasma membrane transporter bilitranslocase mediates rat aortic vasodilation induced by anthocyanins. *Nutr Metab Cardiovasc Dis*, **23**(1):68–74.
Zikri, N.N., Riedl, K.M., Wang, L.S. *et al.* 2009. Black raspberry components inhibit proliferation, induce apoptosis, and modulate gene expression in rat esophageal epithelial cells. *Nutr Cancer*, **61**(6), 816–826.

11 Overview of the Health Properties of Blueberries

Carrie M. Elks, Joseph Francis, April J. Stull, William T. Cefalu, Barbara Shukitt-Hale and Donald K. Ingram

INTRODUCTION

Although they are found worldwide, blueberries (*Vaccinium* species) are one of the few fruit crops native to North America. The *Vaccinium* genus is comprised of over 400 species, including cranberry, bilberry (wild European blueberry, *V. myrtillis*), lingonberry and huckleberry (Prodorutti *et al.*, 2007). North American blueberry species include highbush blueberry (*V. corymbosum*) and lowbush blueberry (*V. angustifolium*), among others. Highbush blueberries are cultivated, while lowbush blueberries grow in the wild.

Blueberries were not truly domesticated until the twentieth century by US Department of Agriculture botanist, Frederick Coville (Kaplan, 2011). In the early 1900s, Coville demonstrated that blueberries required moist, acidic soil for successful cultivation and also needed cross-pollenation for crop propagation. In 1911, Coville developed the first successful crosses of blueberry plants from one highbush plant and one lowbush plant (Kaplan, 2011). Coville died in 1937, but his blueberry varieties remain a part of the pedigrees of many varieties grown today.

In the 1920s, European highbush blueberry cultivation began in The Netherlands and spread soon thereafter to Poland and Germany, where the first blueberry breeding products (crosses of North American varieties) were introduced. As of 2007, the production of blueberries in Europe is concentrated in Germany, Poland, France, Lithuania, Romania, Italy, Spain and The Netherlands (Prodorutti *et al.*, 2007). Chile and Argentina are the major South American blueberry producers; most of their product is exported to the United States and Europe (Prodorutti *et al.*, 2007). Blueberry production in the United States increased by 660% from 1970–2008 (USDA, 2010). Likewise, world production of blueberries increased by 640% in the same time frame (USDA, 2010). This increased interest is likely due to the reported health benefits of blueberries, which are highlighted in this chapter.

BIOACTIVE COMPONENTS OF BLUEBERRIES

Bioactive compounds within fruits and vegetables are critical in protecting the plants from damage. The blueberry's nutritional value has recently been emphasized, as its polyphenolic compounds can exert benefits to human health. Values for selected macronutrients and

Bioactives in Fruit: Health Benefits and Functional Foods, First Edition. Edited by Margot Skinner and Denise Hunter.

Table 11.1 Selected nutrients in one cup (148 g edible portion) of raw blueberries (USDA, 2012).

Proximates	
Water (g)	124.63
Energy (kcal)	84
Protein (g)	1.1
Fat (g)	0.49
Carbohydrate (g)	21.45
Fibre (g)	3.6
Sugars (g)	14.74
Vitamins and Minerals	
Vitamin C (mg)	14.4
Folate (mcg DFE)	9
Vitamin A (mcg RAE)	4
Vitamin K (mcg)	28.6
Calcium (mg)	9
Magnesium (mg)	9
Phosphorus (mg)	18
Potassium (mg)	114
Sodium (mg)	1

Abbreviations: g = grams, kcal = kilocalories, mg = milligrams, mcg = micrograms, DFE = dietary folate equivalents, RAE = retinol activity equivalents.

micronutrients found in raw blueberries appear in Table 11.1, which was adapted from the USDA Nutrient Database for Standard Reference (USDA, 2012). The antioxidant capacity of blueberries is among the highest of fruits and vegetables (Ehlenfeldt and Prior, 2001; Halvorsen *et al.*, 2002). Although blueberries are rich in vitamins and minerals (Bushway *et al.*, 1983), the contribution of these micronutrients to their antioxidant capacity is minimal (Prior *et al.*, 1998). Blueberries contain high levels of flavonoid polyphenols (Prior *et al.*, 1998) and are especially high in anthocyanins. Six anthocyanidins are commonly found in nature; blueberries contain five of these anthocyanidins (Gao and Mazza, 1994). Malvidin, delphinidin, cyanidin, petunidin and peonidin are found in all *Vaccinium* species; they are usually found in the form of anthocyanidin glycosides, which are referred to as anthocyanins (from the Greek words 'anthos', meaning flower and 'kyanos', meaning blue) and are responsible for giving the blueberry its unique colouring (Gao and Mazza, 1994). Anthocyanins play a vital role in the blueberry's skin by protecting the fruit's flesh from overexposure to ultraviolet light.

Significant differences exist in both the types of anthocyanins and total anthocyanin content among *Vaccinium* subspecies (Kalt *et al.*, 2001). North American wild lowbush blueberries have higher total anthocyanin levels per 100 g fresh weight than do highbush varieties; this is because the majority of anthocyanins are present in the skin of the berry and lowbush blueberries have a greater amount of surface area per volume of fruit (Kalt *et al.*, 2001). Highbush blueberries are more readily available as whole, relatively unprocessed fruit (either fresh or frozen) in grocery stores, while lowbush varieties are used predominantly in processed foods and in flavoured yogurt (Kalt *et al.*, 1999, 2000) and within supermarkets as frozen whole berries. Processing significantly affects the antioxidant capacity of blueberries and blueberry products (Kalt *et al.*, 1999, 2000). Although freezing is a form of processing and does result in a slight loss of antioxidant capacity, frozen blueberries and fresh blueberry purees retained higher levels of anthocyanins than all other forms of processed blueberries and blueberry products measured in a study by Kalt and colleagues (Kalt *et al.*, 2000).

Increased susceptibility to the long-term effects of oxidative stress and inflammation is involved in the majority of debilitating, degenerative conditions. The impact caused by these diseases in terms of quality of life and lifespan, as well as their economic burden on society, has led to the increased search for preventative, mitigative and curative strategies. Research suggests that polyphenols in foods such as blueberries can extend a multitude of health benefits to humans, mainly through a combination of additive and/or synergistic effects derived from their antioxidant and anti-inflammatory capabilities. Much of this research is discussed in this chapter.

OBESITY AND INSULIN RESISTANCE

Metabolic syndrome is a pre-diabetic state characterized by phenotypes including insulin resistance, hyperlipidemia, inflammation and obesity, all of which are also risk factors for cardiovascular disease. If left unmanaged, metabolic syndrome can progress to type 2 diabetes mellitus (T2DM) and lead to cardiovascular disease, both of which remain major public health concerns. Emerging preclinical and clinical evidence suggests that blueberries can be effective in mitigating the alterations in blood pressure, lipid levels and insulin response seen in metabolic syndrome (Basu and Lyons, 2012).

Obesity

In a preclinical study, mice were fed high-fat diets with or without whole blueberry or strawberry powder or purified anthocyanins (Prior *et al.*, 2008). Results indicated that mice fed the blueberry-supplemented high-fat diet actually had significantly greater body weight gains, fat mass and epididymal fat pad weights than high-fat-fed controls. However, when purified blueberry anthocyanins were administered in drinking water to high-fat-fed mice, body weight gains and fat mass were significantly lower than in the high-fat-fed controls (Prior *et al.*, 2008). A later study by the same group determined that purified blueberry anthocyanins, but not whole blueberries in the diet, decreased cholesterol, triglycerides and leptin in high-fat-fed mice, thereby suggesting that the form in which bioactives are delivered may be critical in determining their effectiveness (Prior *et al.*, 2009). In another study, male C57BL/6 mice were fed a high-fat diet supplemented with blueberry juice or purified blueberry anthocyanins in drinking water (Prior *et al.*, 2010). In non-supplemented high-fat-fed mice, body weight and fat mass were higher and lean mass was lower when compared to low-fat-fed mice. However, in high-fat-fed mice supplemented with blueberry juice or the low dose of anthocyanins, fat mass did not differ from control fed mice. The purified anthocyanins decreased fasting serum glucose to levels similar to those of low-fat fed mice, while the juice did not (Prior *et al.*, 2010). Both treatments decreased serum leptin levels in high-fat-fed mice.

Pre-diabetes

A study by DeFuria and colleagues (DeFuria *et al.*, 2009) found that in the same mouse model used by Prior *et al.*, addition of highbush blueberry powder to a high-fat diet protected the rodents from insulin resistance; there was also a trend in reduction of hyperglycemia. These effects were attributed to reductions in adipocyte death. There were no effects on energy intake, metabolic rate, body weight or adiposity when blueberries were fed to animals for eight weeks (DeFuria *et al.*, 2009). Further, supplementation seemed to attenuate the up-regulation of several inflammatory genes, including tumour necrosis factor-α, interleukin-6, monocyte chemoattractant protein 1 and inducible nitric oxide synthase, in adipose tissue (DeFuria *et al.*, 2009). Another preclinical study conducted by Seymour and colleagues (Seymour *et al.*, 2011) reported similar effects in the Zucker fatty rat model of obesity and insulin resistance. Zucker fatty or Zucker lean rats were fed either a high-fat diet or a control diet; both diets were supplemented with highbush blueberry powder (Seymour *et al.*, 2011). In high-fat-fed fatty rats, the addition of blueberries led to reductions in triglycerides, fasting insulin, homeostasis model index and glucose area under the curve. Reductions in abdominal fat and increases in peroxisome proliferator-activated receptor activity were also seen in these rats, suggesting a partial resolution of these metabolic syndrome components. Interestingly, the addition of blueberries to the low-fat diet led to significant reductions in fat mass, liver weight and body weight in Zucker fatty rats (Seymour *et al.*, 2011). These results add to the body of evidence that suggests that blueberries can exert their effects on the phenotypes of metabolic syndrome across several species.

In a clinical study from our Center for the Study of Botanicals and Metabolic Syndrome, obese, insulin-resistant participants were asked to consume blueberry smoothies containing highbush blueberry powder or placebo twice daily for six weeks (Stull *et al.*, 2010). Inflammatory biomarkers, insulin sensitivity and adiposity were assessed at baseline and at study end. Increases in insulin sensitivity were significantly greater in the blueberry-supplemented group than in the placebo group (Stull *et al.*, 2010). However, there were no significant alterations in energy intake, adiposity or inflammatory biomarkers. This is the first clinical study to demonstrate that blueberries can improve insulin sensitivity and cardiovascular risk factors in obese, insulin-resistant subjects. Follow-up studies are now underway in other susceptible populations to determine the clinical effectiveness of blueberry supplementation.

TYPE 2 DIABETES MELLITUS

Unmanaged pre-diabetes often leads to the development of T2DM, the incidence of which continues to rapidly increase in developed countries. Recent epidemiological data suggest that consumption of two or more servings of blueberries per week is associated with lower risk of T2DM, and suggest that the high levels of flavanoids, particularly anthocyanins, in blueberries are responsible for their positive effects (Wedick *et al.*, 2012). Research from as early as the 1920s has suggested that blueberries may have beneficial effects on diabetes. In his 1928 paper, E.M. Watson describes the effectiveness of blueberry leaf extract as an adjunct therapy to insulin in the treatment of diabetes in elderly patients, stating that '. . . blueberry leaf extract has a stabilizing influence upon the carbohydrate tolerance of certain cases of diabetes' (Watson, 1928). Lowbush blueberry (*V. angustifolium*) leaf extract is still recommended in Canada as a natural treatment for T2DM. A 2006 study from the Montreal Diabetes Research Center examined the effects of ethanol extracts of *V. angustifolium* root, stem, fruit and leaf on pancreatic beta cells and cells from skeletal muscle and adipose tissues (Martineau *et al.*, 2006). Root, stem and leaf extracts significantly enhanced glucose transport in muscle cells in both the presence and absence of insulin. In adipose cells, root and stem extracts enhanced glucose uptake in the absence of insulin. The fruit extract was found to increase thymidine incorporation in replicating pancreatic beta cells, while stem, leaf and fruit extracts reduced apoptosis in cells exposed to elevated glucose concentrations for several days (Martineau *et al.*, 2006). The protection afforded by blueberry extracts against glucose toxicity and ability to enhance beta cell proliferation demonstrate that wild blueberries do indeed have antidiabetic properties.

Fermentation of lowbush blueberry juice with bacteria (*Serratia vaccinii)* isolated from blueberry flora increases its phenolic content and antioxidant activity and can also modify its biological activity, suggesting that the fermentation process may offer additional benefits to blueberry juices (Martin and Matar, 2005; Vuong *et al.*, 2006). Fermented lowbush blueberry juice was shown to affect glucose uptake, adipogenesis and glucose transport in skeletal muscle cells and adipocytes (Vuong *et al.*, 2007). Glucose uptake was significantly increased in myotubes and in adipocytes with a short-term treatment of fermented blueberry juice, whereas non-fermented juice had no effect on transport. The increase in glucose uptake seen with the fermented blueberry juice could be due to the activation of a cellular signalling protein in the insulin-independent pathway involving AMP-activated protein kinase phosphorylation in myotubes and adipocytes (Vuong *et al.*, 2007). Triglyceride content was also markedly reduced in fermented juice-treated adipocytes.

An *in vivo* study by the same group examined the effects of fermented blueberry juice in the KKA^y mouse model of T2DM (Vuong *et al.*, 2009). Acute and chronic studies were conducted. In the initial anti-obesity study, young mice consumed fermented blueberry juice during weeks 4–7 of life, which are considered critical periods for developing obesity and diabetes. The fermented blueberry juice reduced the body weight gain and glucose area under the curve when compared to the control group. In the acute study, older mice were divided into three groups and daily gavages of water, fermented blueberry juice or the antidiabetic medication, metformin (positive control), were administered and blood glucose measured at varying time points. Blood glucose levels in the blueberry juice group were significantly lower when compared to levels of controls. In the chronic study, mice were divided into six groups. The groups received water, normal blueberry juice or fermented blueberry juice for three weeks (Vuong *et al.*, 2009). Each mouse in the fermented blueberry juice group was matched to a mouse in a separate group that received sugar water (to control for the sugar intake associated with blueberry juice). Mice receiving sugar water were also pair-fed the amount of food consumed by their blueberry juice mate the previous day. Chronic administration of fermented blueberry juice was found to reduce food intake and body weight in addition to increasing adiponectin levels. The blood glucose levels of the blueberry-treated mice decreased by the third day of the study, an effect which persisted for the remainder of the study. The blueberry-treated mice also exhibited increased glucose tolerance and insulin sensitivity. Results from this study are twofold: they demonstrate that fermented blueberry juice can not only decrease glycemia in overtly diabetic mice but that it can also prevent young pre-diabetic mice from developing obesity and diabetes.

Overall, preclinical and clinical research to date has revealed that blueberry bioactives in many forms can exert their insulin-sensitizing and glucose-lowering effects across several species. It is worth noting that despite the abundance of evidence demonstrating the positive effects of blueberries on obesity and metabolic syndrome in mature animals and humans, there are no data on how blueberries would affect obesity in young subjects. Since obesity is becoming increasingly prevalent in children and adolescents, there is a great need for this type of research to be conducted.

CARDIOVASCULAR DISEASES

Inflammation and oxidative stress are large contributors to the development and progression of cardiovascular diseases including hypertension, atherosclerosis, and heart failure. Atherosclerosis is a condition affecting the coronary arteries, where the gradual uptake of oxidized low density lipoprotein (LDL) by the endothelium and the resulting inflammatory response leads to plaque deposition in arterial walls. Over time, the plaques can accumulate and lead to restriction of blood flow, which can exacerbate hypertension and cause myocardial infarction and, eventually, heart failure (McMurray and Stewart, 2000).

Endothelial dysfunction

Endothelial dysfunction is a well-established risk factor for cardiovascular disease, as endothelial cells are constantly exposed to reactive oxygen species (ROS) from monocytes and neutrophils (Neto, 2007). In response to insult, endothelial cells produce proinflammatory cytokines and adhesion molecules, events which play a key role in early atherogenesis and eventual plaque rupture (Ross, 1993). One of the earliest studies of the response of endothelial cells to blueberry supplementation was conducted *in vitro* by Youdim and colleagues (Youdim *et al.*, 2002). The potential antioxidant and anti-inflammatory effects of anthocyanins and hydroxycinnamic acids from blueberry and cranberry against damage induced by treatment with hydrogen peroxide and tumour necrosis factor in human microvascular endothelial cells were assessed. Polyphenolic fractions from both berries protected membrane lipids from oxidation. The polyphenols also reduced the up-regulation of inflammatory mediators involved in the recruitment of leukocytes to sites of endothelial inflammation or damage (Neto, 2007; Youdim *et al.*, 2002). These results were some of the first to suggest that blueberry bioactives may have protective, beneficial actions against oxidative and inflammatory stress in the vascular system.

Blueberry bioactives can alter the composition of glycosaminoglycans (GAG) produced by vascular endothelial and smooth muscle cells. These molecules are important structural components of the arterial wall and participate in organization of the extracellular matrix; they can also participate in blood coagulation and lipoprotein metabolism via interactions with other molecules (Kalea *et al.*, 2006; Neto, 2007). In a study with Sprague-Dawley rats, the effect of a diet enriched with wild blueberries on aortic GAG composition and abundance was examined. Results indicated a significant increase in the total amount of aortic GAG in blueberry-fed rats when compared to control rats, with decreased GAG disaccharide sulfation also seen in these animals. It was concluded that these alterations may affect endothelial and vascular smooth muscle signal transduction as well as lipoprotein lipase activity, both of which can impact cardiovascular disease risk (Kalea *et al.*, 2006). Another study by the same group examined the effect of seven weeks' feeding of the same diet on vascular reactivity in weanling Sprague-Dawley rats (Kalea *et al.*, 2009). Consumption of the blueberry-supplemented diet was found to suppress phenylephrine-induced vasoconstriction in aortic rings, and this depressed response was attenuated with an inhibitor of nitric oxide. Acetylcholine-induced vasorelaxation was also enhanced in blueberry-fed rats, and this effect was also attenuated by inhibition of nitric oxide (Kalea *et al.*, 2009). These results suggest that blueberries improve vascular reactivity, possibly through their antioxidant properties and effects on nitric oxide signalling.

Lipids

Interestingly, blueberry bioactives can also alter plasma lipid concentrations in pigs and rodents. Pigs are a good model for evaluating dietary effects on plasma lipids, as they are omnivorous and will consume many foods common to the human diet (Kalt *et al.*, 2008). In a study by Kalt and colleagues, two feeding trials were conducted with pigs to determine the effects of blueberry supplementation on plasma lipid levels. The diets in the first trial contained high levels of plant-based components (70%), and supplementation with varied amounts of Jersey blueberry powder resulted in decreases in total, LDL- and high density lipoprotein (HDL)-cholesterol, with the greatest reduction observed in the pigs fed a 2% blueberry diet (Kalt *et al.*, 2008). The diets in the second trial contained only 20% plant-based ingredients, and with these diets the lipid-modulating effect of blueberries was attenuated, unless the diets were supplemented with cholesterol. This result suggests that blueberry bioactives do influence lipid profiles but that the magnitude of their effects may depend on the basal diet used.

The Syrian Golden hamster is another commonly used model for cholesterol metabolism, as it shares similarity with humans in its sensitivity to development of hypercholesterolemia from dietary cholesterol intake (Suckling and Jackson, 1993; Kim *et al.*, 2010). In a recent study, hamsters were fed high-fat diets supplemented with Tifblue blueberry pomace, blueberry pomace ethanol extract or blueberry residue from ethanol extract for three weeks (Kim *et al.*, 2010). The ethanol blueberry extract resulted in a significant decrease in serum LDL, while all blueberry supplements resulted in decreased total cholesterol levels when compared to non-supplemented animals. Hepatic gene expression profiles of all blueberry-supplemented hamsters revealed decreases in genes associated with endogenous cholesterol synthesis, which is a likely contributor to the cholesterol-lowering effects seen. Up-regulation of genes involved in bile acid synthesis was also seen in blueberry-supplemented hamsters (Kim *et al.*, 2010).

The apolipoprotein E null (apo$E^{-/-}$) mouse has been widely used in cardiovascular research as it develops atherosclerotic lesions resembling those in humans. Apo$E^{-/-}$ mice were fed a rodent diet alone (control) or supplemented with freeze-dried whole wild blueberries for 20 weeks (Wu *et al.*, 2010). At study completion, the mean atherosclerotic lesion area for apo$E^{-/-}$ mice fed blueberries was reduced both in the descending aorta and in the aortic sinus when compared with control mice, and F_2-isoprostanes (a marker of lipid peroxidation and oxidative stress) were significantly lower in the livers of blueberry-fed mice. Additionally, expression of major antioxidant enzymes was up-regulated in the aortae of blueberry-fed mice, with concomitant increases in the activities of superoxide dismutase and glutathione reductase in liver and/or serum. The results from these two rodent studies suggest that blueberries, as a whole food and in extract form, are protective against atherosclerosis, and that this protection may be provided via a reduction in oxidative stress and enhancement of endogenous antioxidant defenses.

Blood pressure

Hypertension remains a leading cause of heart and kidney diseases in developed countries. In many cases, hypertension can be controlled with pharmacological treatment. However, antihypertensive drugs can also have undesirable side effects. Only a handful of studies have examined the effects of blueberry consumption on hypertension. The first was conducted by Shaughnessy and colleagues in the stroke-prone spontaneously hypertensive rat (SHR) (Shaughnessy *et al.*, 2009). Stroke-prone SHR and normotensive Wistar control rats were fed a control diet or a diet supplemented with freeze-dried blueberries for eight weeks. Systolic blood pressure was lowered after four weeks in SHR fed the blueberry diet, and decreased even further after eight weeks of feeding. Proteinuria and kidney nitrites were also lower in these animals, suggesting that blueberries may prevent oxidative damage in the kidney and thereby mitigate the severity of hypertension (Shaughnessy *et al.*, 2009).

In a follow-up study using an identical blueberry diet, the same group sought to determine whether the beneficial effects of blueberries on blood pressure are due to the inhibition of angiotensin-converting enzyme (ACE), which serves to convert the peptide, angiotensin I, to the potent vasoconstrictor, angiotensin II (Wiseman *et al.*, 2011). Stroke-prone SHR and their normotensive controls were fed control or blueberry diets. After two weeks on a control diet, the hypertensive rats exhibited significantly higher blood ACE activity levels than did normotensive rats. Conversely, the blueberry diet lowered ACE activity in SHR but not in control rats. Blueberry feeding inhibited increases in plasma ACE in SHR after six weeks of feeding (Wiseman *et al.*, 2011). Interestingly, the blueberry diet had no significant effects on ACE activity in other rat tissues. These results suggest that blueberries may be effective in mitigating early hypertension, and that this may be due to the inhibition of soluble ACE activity.

We recently conducted a study in the SHR model that examined effects of chronic and acute feeding of blueberries on blood pressure and renal outcomes (Elks *et al.*, 2011). Male stroke-prone SHR received a blueberry-enriched diet (highbush blueberry powder) or an isocaloric control diet for six or twelve weeks or two days. Rats fed the blueberry-enriched diet for six or twelve weeks exhibited significantly lower systolic blood pressure, improved renal haemodynamics, and decreased renovascular resistance compared to control SHR. Significant decreases in total ROS, peroxynitrite and superoxide production rates in kidney tissues were also seen in rats on long-term dietary treatment, consistent with reduced pathology and improved function. In blueberry-fed rats, renal glutathione and catalase activities increased markedly when compared to control animals (Elks *et al.*, 2011). However, similar measurements made in SHR fed the same diet for only two days yielded evidence of increased oxidative stress. These results were evidence of hormesis during brief exposure, which dissipated with time. Our conclusions from this study were that long-term feeding of a blueberry-enriched diet lowered blood pressure, preserved renal haemodynamics and improved redox status in kidneys of hypertensive rats, and concomitantly demonstrated the potential to delay or attenuate development of hypertension-induced renal injury; these effects appeared to be mediated by a short-term hormetic response manifested as a transient, mild oxidative stress (Elks *et al.*, 2011).

In contrast with the aforementioned preclinical results, a previous clinical study examining the effects of consumption of fresh blueberries thrice weekly for three weeks on blood pressure, ACE activity and oxidative stress in chronic smokers showed no significant effect of blueberry on ACE activity or blood pressure in these subjects (McAnulty *et al.*, 2005). However, lipid hydroperoxides (a measure of oxidative stress) were significantly decreased after blueberry consumption. These differing results may be due to the use of rodents versus humans and dosing or duration of treatments. The smoking status of the subjects in this clinical study also may have affected outcomes.

In a later clinical study by Basu and colleagues, blueberries were found to decrease several features of metabolic syndrome in obese men and women. In this study, subjects consumed a beverage containing highbush blueberries twice per day for eight weeks (Basu *et al.*,

2010). Anthropometric and blood pressure measurements, along with serum analyses, were conducted at the beginning, midpoint and conclusion of the study. Blood pressure, oxidized LDL and markers of oxidative stress, including malondialdehyde and hydroxynonenal levels, were all significantly decreased in the subjects consuming blueberries when compared to the control subjects. However, serum glucose, insulin and lipids were not significantly affected. There were also no significant effects on body weight or waist circumference. Notably, these subjects all had normal levels of serum glucose, insulin and lipids prior to the intervention, which may explain the null response of these parameters to the blueberry supplementation (Basu *et al.*, 2010). Overall, it is clear from the results of all of these studies that blueberries can have positive effects on hypertension, and that these effects are likely a result of decreased oxidative stress.

Cardiac function

Myocardial ischemia and hypertension can ultimately result in heart failure, in which cardiac output is insufficient to meet the needs of the lungs and the body. A 2009 study by Ahmet and colleagues examined the effects of blueberry consumption on ischemic injury in the heart (Ahmet *et al.*, 2009b). Fischer-344 (F344) rats were fed the same diets used by Elks and colleagues for three months. Mitochondrial permeability transition of cardiomyocytes was assessed in rats from each group and those animals fed a blueberry-enriched diet demonstrated an increase in mitochondrial permeability transition threshold, indicating an improvement in mitochondrial integrity with blueberry feeding. Animals from each group were also subjected to myocardial ischemia by left descending coronary artery ligation and resulting infarct size was significantly lower in rats fed the blueberry-enriched diet, suggesting an overall cardioprotective effect of blueberries (Ahmet *et al.*, 2009b). In another study by the same group, male Wistar rats were subjected to coronary artery ligation; two weeks following ligation surgery, rats were divided into two groups of similar infarct size as measured by echocardiography (Ahmet *et al.*, 2009a). A sham group was also included, in which rats underwent the same surgery but coronary arteries were not ligated. The rats were then assigned to consume for the next year a control diet or the identical blueberry diet used previously (this lengthy period was chosen to allow rats to develop congestive heart failure). Mortality over the year period was significantly decreased in heart failure rats consuming the blueberry diet when compared to those on the control diet (Ahmet *et al.*, 2009a). Significant attenuation of several echocardiographic indices of heart failure was also seen in the blueberry-fed rats. These two studies suggest that blueberry-enriched diets positively affect the course of heart failure by reducing morbidity and mortality, and that these effects may occur through reducing infarct size and preserving cardiac function.

CANCER

The putative *in vitro* anticancer properties of blueberries have been studied since the 1990s, with the earliest report appearing as a result of an examination of ethyl acetate extracts of several *Vaccinium* species, including blueberries (Bomser *et al.*, 1996; Neto, 2007). All extracts were found to inhibit ornithine decarboxylase (an enzyme involved in polyamine synthesis that is up-regulated in cancer) and to induce the phase II xenobiotic detoxification enzyme, quinone reductase (QR) (Bomser *et al.*, 1996). A crude acetone extract of lowbush blueberries was later found to induce QR in a dose-dependent manner in chemically-induced carcinogenesis in Hepa cells (Smith *et al.*, 2000). More recently, the focus of *in vitro* and *in vivo* studies has shifted to determining which phytochemicals in blueberries are responsible for their anticancer effects. Some of these studies are highlighted here.

In a 2005 study, the effects of three lowbush blueberry (*V. angustifolium*) extracts on human prostate cancer cells were evaluated. Cells were treated with varied doses of crude anthocyanin- or proanthocyanidin-enriched blueberry methanol extracts (Matchett *et al.*, 2005). All extracts decreased the activity of matrix metalloproteinases (MMP) 2 and 9 (essential in regulating extracellular matrix proteolysis, a critical step in metastasis) and increased the activity of tissue inhibitor of metalloproteinases 1, but the degrees of the cellular responses varied according to the fraction applied. Notably, the most robust MMP inhibition was seen with the proanthocyanidin fraction (Matchett *et al.*, 2005). Protein kinase C and mitogen-activated protein kinase inhibitors affected blueberry-induced MMP down-regulation, suggesting that these pathways are involved in the regulation of MMP activity by blueberries (Matchett *et al.*, 2006). Similar results have been seen with other proanthocyanidin-rich berries, including the cranberry.

The putative antiproliferative role for blueberries may be due, in part, to apoptosis induction in tumour cells by blueberry bioactives. An evaluation of bioactive compounds from different cultivars of rabbiteye blueberries was conducted using two colon cancer cell lines (Yi *et al.*, 2005). Polyphenol extracts from the cultivars were freeze dried and fractionated into phenolic acids, tannins, flavonols and anthocyanins. The fractions were applied to cells to assess antiproliferation activity and induction of apoptosis. All fractions were deemed effective, but the greatest effect was seen from the anthocyanin fraction, with a decrease in cell growth of over 50% (Yi *et al.*, 2005). The anthocyanin fraction also resulted in significant increases in DNA fragmentation, which, along with other cellular processes, can indicate apoptosis induction. In a similar study, anthocyanin fractions from several highbush blueberry cultivars were found to increase apoptosis in colon cancer cells as determined by uniform DNA fragmentation and caspase-3 activity (Srivastava *et al.*, 2007). Prominent DNA fragmentation was seen with fractions from all cultivars tested. A significant dose-dependent difference in caspase-3 activity was found between the control cells and anthocyanin-treated cells (Srivastava *et al.*, 2007). However, dose-dependent decreases in QR and glutathione-S-transferase activities were found in all anthocyanin-treated cells when compared to control cells. These results suggest that blueberry anthocyanins can induce apoptosis of cancer cells at varying doses, but that some doses also negatively affect the activities of some protective phase II enzymes. In a study that examined stomach, prostate, intestine and breast cancer cell lines, lowbush blueberry juice (50 μl/ml culture medium) exhibited antiproliferative properties (Boivin *et al.*, 2007). The antiproliferative effects were not found to involve apoptosis but instead involved cell-cycle arrest, down-regulation of TNFα-induced cyclooxygenase activation and decreased activity of the transcription factor, nuclear factor kappa B (NF-κB), which suggests a potential anti-inflammatory effect.

The triple-negative breast cancer phenotype lacks expression of progesterone receptor, estrogen receptor and the HER-2 epidermal growth factor receptor, hence the 'triple-negative' title (Reis-Filho and Tutt, 2008; Adams *et al.*, 2010). A 2010 study by Adams and colleagues investigated the chemopreventive activity of highbush blueberry extract in triple-negative breast cancer cell lines both *in vitro* and *in vivo* (Adams *et al.*, 2010). Blueberry extract decreased cell proliferation and cell motility in all cell lines. The extract also decreased the activity of MMP-9 while increasing tissue inhibitor of metalloproteinase-1 and plasminogen activator inhibitor-1 secretion in conditioned medium (Adams *et al.*, 2010). Treatment with blueberries decreased phosphatidylinositol 3-kinase/Akt and NF-κB activation in cells, while the protein kinase C and extracellular signal-related kinase pathways were not affected. *In vivo*, a xenograft model was used to evaluate the effects of blueberry treatment on tumour weight and proliferation in female nude mice. Tumour weight and proliferation were decreased in mice gavaged with blueberry extract each day for seven weeks, and apoptosis was increased. In tumours from blueberry-treated mice, decreased activation of AKT and the p65 subunit of NF-κB were detected. Taken together, these data suggest that the inhibitory effect of blueberry bioactives on the growth and metastatic potential of triple-negative breast cancer cells may occur, in part, through modulation of the PI3K/AKT/NF-κB pathway.

Blueberry bioactives may also be effective in the context of estrogen-induced breast cancers. In a study using August Copenhagen Irish rats, the effects of a diet containing whole blueberry powder on estrogen-mediated mammary tumourigenesis were assessed. Animals received implants of 17β-estradiol to induce tumor formation, and were euthanized after six months. At study end, although no significant differences were found in tumour incidence, the tumour volume in blueberry-treated rats was significantly decreased in comparison with the untreated controls (Aiyer *et al.*, 2008). Contrasted to the positive results above, Stoner and colleagues have demonstrated that diets containing whole blueberry powder were unable to inhibit N-nitrosomethylbenzylamine–induced esophageal tumours in rats (Stoner *et al.*, 2008). Simmen and colleagues examined the efficacy of a blueberry diet on azoxymethane-induced formation of aberrant crypt foci in the colons of male and female Sprague-Dawley rats, and surprisingly found increases in formation in female blueberry-supplemented rats (Simmen *et al.*, 2009).

Although they seem quite promising, results from the *in vitro* studies mentioned above should be interpreted with great caution for two reasons: (i) the cell lines used did not exhibit inhibition of apoptosis (which generally occurs with cancer) and (ii) the behaviour of cancer cells in culture is often not indicative of their behaviour *in vivo*. The results of *in vivo* studies to date suggest that consumption of blueberries may be chemopreventive in some cases, *but the benefits seem to vary in magnitude depending on the type and the cause of the tumour, and should be investigated further.* Further, the variability in the blueberry bioactives used was a likely contributor to the variable effects seen.

LONGEVITY

The nematode *Caenorhabditis elegans* is often used as an aging model, due to its short lifespan and since many aspects of aging are similar between nematodes and humans. The effects of blueberry polyphenols on lifespan and aging of *C. elegans* were assessed by Wilson and colleagues (Wilson *et al.*, 2006). Wild blueberry (*V. angustifolium*) polyphenols or proanthocyanadins were found to increase lifespan and slow aging-related functional declines in *C. elegans*. Interestingly, blueberry treatment increased survival during acute heat stress but did not protect against paraquat-induced acute oxidative stress, suggesting that some of the beneficial effects of blueberry supplementation were not antioxidant in nature (Wilson *et al.*, 2006). Also, the effects of blueberry polyphenols on lifespan were found to be genetically based and to require the presence of a calcium/calmodulin-dependent protein kinase pathway that mediates osmotic stress resistance (Wilson *et al.*, 2006). This study was the first to demonstrate that blueberry bioactives had antioxidant-independent benefits during aging (Wilson *et al.*, 2006).

A recent investigation of the effects of addition of blueberry extract into diets on longevity in fruit flies (*Drosophila melanogaster*) revealed a significant increase in mean lifespan of blueberry-supplemented flies (Peng *et al.*, 2012). This extension was accompanied by increased expression of several antioxidant genes. Blueberry extracts were also found to significantly increase survival rate and partially restore climbing ability in flies subjected to paraquat-induced oxidative stress (Peng *et al.*, 2012). Similar to the study in *C. elegans*, these effects were also genotype-dependent (Peng *et al.*, 2012). Taken together, the results from these studies support a possible role for blueberry bioactives in increasing longevity across several species.

Dietary restriction extends lifespan in multiple species via several mechanisms. The most commonly used dietary restriction paradigms include 40% restriction or intermittent fasting (food is withheld every other day). Aires and colleagues recently conducted a longevity study in C57BL/6 mice, where mice were placed on one of three high fat diet feeding schedules: *ad libitum*, intermittent or intermittent with a diet supplemented with blueberry extract, epigallocatechin gallate and pomegranate powder as sources of polyphenol antioxidants (Aires *et al.*, 2012). As expected, the intermittently fed mice lived significantly longer than the control mice; interestingly, the intermittently fed mice given the antioxidants outlived the intermittently fed mice given the control diet. Body weights were lower in the intermittently fed mice when compared to the control mice, but there was no significant difference in body weights between the mice fed control and antioxidant diets intermittently (Vuong *et al.*, 2006). Addition of antioxidants to the intermittent feeding regimen led to the down-regulation of a number of gene pathways, many of which were related to inflammation and immunity, thereby supporting the suggestion that polyphenol antioxidants can have anti-inflammatory effects (Vuong *et al.*, 2006); this is one possible mechanism by which polyphenol antioxidants can extend lifespan.

NEUROPROTECTION AND AGING

Cognitive function

The first studies examining the nutritional neuroprotection afforded by blueberries were conducted by Dr James A. Joseph and colleagues; Dr Joseph remained a leader in the field of berry research until his death in 2010. One of the first studies Joseph *et al.* conducted compared the effects of a control diet and long-term diets supplemented with strawberry or spinach extracts on cognitive and motor function and neuronal signal transduction in aging rats (Joseph *et al.*, 1998). Six-month-old F344 rats were fed various supplemented diets for eight months prior to being subjected to a battery of psychomotor tests (rod walking, wire suspension, plank walking, inclined screen and accelerating rotarod) and the working memory version of the Morris water maze as a cognitive test (Joseph *et al.*, 1998). The Morris water maze is a spatial learning and memory paradigm that requires rats to find the location of a hidden platform just below the surface of a pool of water, with the reward for accurate navigation being escape from the water (Shukitt-Hale *et al.*, 2007). In the rats fed supplemented diets, performance on age-sensitive cognitive tests and in the Morris water maze was significantly improved when compared to the performance of the control group (Joseph *et al.*, 1998). In a subsequent experiment, a group of 19-month-old F344 rats was also fed the same supplemented diets, with the addition of a blueberry-enriched diet, for eight weeks (Joseph *et al.*, 1999). Cognitive and motor decline was reversed in these animals, with the most robust effects seen in the animals fed the blueberry-enriched diets (Joseph *et al.*, 1999). Specifically, the blueberry-treated group had improved performance on rod walking and increased latency in falling from the accelerating rotarod, suggesting improvements in balance and coordination (Joseph *et al.*, 2005). None of the other supplemented groups exhibited these differences when compared to the control group. These were the first studies to demonstrate the reversal of age-related cognitive and motor decline by dietary antioxidant supplementation and, more specifically, by blueberries.

In addition to improving brain function in healthy aging animals, blueberries have also been shown to positively impact neuronal function and memory in rodent models of accelerated aging (Shukitt-Hale *et al.*, 2007). Animal models of accelerated aging exhibit enhanced oxidative stress and inflammation, and disruption of the dopaminergic system, making them very similar to healthy aging rodent models. One method used to produce profound deficits in behaviour and neuronal function that resemble those seen with aging, is the exposure of young rats to particles of high energy and charge (Shukitt-Hale *et al.*, 2007). Young male Sprague-Dawley rats were fed blueberry or strawberry diets for two months prior to irradiation with iron-56 particles. When they reached four months of age, half of the animals from each diet group was irradiated, while the other half served as the non-irradiated controls. One month following radiation exposure, rats were subjected to Morris water maze testing. After sacrifice, dopamine release from striatal slices was measured. Results indicated that both diets were beneficial; however, the strawberry diet was more protective against spatial deficits, while the blueberry diet seemed to improve reversal learning (Shukitt-Hale *et al.*, 2007).

Positive results have also emerged from the only human clinical trial assessing the effects of blueberry supplementation on memory in older adults conducted at the University of Cincinnati, under the supervision of Krikorian and colleagues (Krikorian *et al.*, 2010). In that study, nine adults with mild cognitive decline, as determined by the Clinical Dementia Rating, consumed wild blueberry juice daily for 12 weeks (Krikorian *et al.*, 2010). Measures of memory function were performed at baseline and during the final week of the study; these tests relied on hippocampal processing. The subjects' moods were also assessed. Results of this study indicated significant performance improvements on both memory tests, as well as decreases in symptoms of depression, in the group that consumed the blueberry juice (Krikorian *et al.*, 2010).

A recent epidemiologic study examined the effects of berry intake on cognitive decline in a large, prospective cohort of women from the Nurses' Health Study (Devore *et al.*, 2012). Results from that analysis indicated that increased intakes of blueberries and strawberries, as well as increased intakes of anthocyanidins and total flavanoids, were associated with slowing the rate of cognitive decline by up to 2.5 years (Devore *et al.*, 2012).

Cellular signalling in the brain

Blueberry supplementation also seems to increase protective cellular signals and inactivate stress signals in the brain, as also evidenced by results from studies with both young and aged animals (Goyarzu *et al.*, 2004; Shukitt-Hale *et al.*, 2008; Williams *et al.*, 2008). Aged male F344 rats fed a blueberry-supplemented diet for four months exhibited an attenuation of the age-associated increases normally seen in NF-κB expression in the striatum, hippocampus and frontal cortex of the brain (indicative of inflammatory processes) when compared to those of unsupplemented controls (Goyarzu *et al.*, 2004). In young rats given hippocampal injections of kainic acid (an excitotoxin that produces neuronal lesions), similar decreases in NF-κB were seen with blueberry supplementation (Shukitt-Hale *et al.*, 2008). Hippocampal expression of the proinflammatory cytokines IL-1β and TNFα was also reduced, while IGF-1 expression was increased. In 2005, it was found that the anthocyanins in blueberries can indeed cross the blood–brain barrier and accumulate in brain areas important for learning and memory, and may therefore produce their effects by acting directly in the central nervous system (Andres-Lacueva *et al.*, 2005). These direct actions are a likely cause of the beneficial effects noted in the studies above.

The same blueberry-enriched diet, when fed to aged rodents for 12 weeks, has also been shown to regulate acquisition and spatial working memory, as well as improving spatial learning in the X-maze (Williams *et al.*, 2008). The animals' memory performance in this study was correlated with hippocampal cAMP-response element-binding protein (CREB) activation and increased brain-derived neurotrophic factor (BDNF) in the hippocampus. These changes in blueberry-supplemented animals were accompanied by increased extracellular signal-related kinase (ERK1/2) phosphorylation, as well as increases in Akt and other signals associated with *de novo* protein synthesis. Indeed, activation of these signals would be necessary for memory formation to occur (Williams *et al.*, 2008; Joseph *et al.*, 2009). These results suggest that improvements in spatial working memory in blueberry-supplemented animals may be facilitated by increased activation of the ERK-CREB-BDNF pathway in the brain by blueberry bioactives.

Evidence for blueberry-induced alterations in molecular signalling has also been seen in a double transgenic (amyloid precursor protein/presenilin 1; APP/PS1) mouse model, which possesses two genetic mutations that promote amyloid beta (Aβ) production and the development of Alzheimer's disease (AD)-like plaques in several regions of the brain. Mice were fed blueberry-supplemented diets from four months of age to twelve months of age and Y-maze performance evaluated. The blueberry-treated APP/PS1 mice demonstrated maze performance similar to that of non-transgenic mice, which was significantly improved when compared to control APP/PS1 mice (Joseph *et al.*, 2003, 2005). Interestingly, there were no significant differences between blueberry-treated and non-treated APP/PS1 mice in the number of brain plaques as a pathological marker of AD. However, the blueberry-supplemented APP/PS1 mice had higher hippocampal ERK and striatal and hippocampal protein kinase C concentrations (both important contributors in converting short-term memory to long-term memory) than non-treated mice (Joseph *et al.*, 2003, 2005). Blueberry-induced enhancement of memory-associated neuronal signalling was thought to contribute to the positive effects seen in the APP/PS1 mice. These data were the first to suggest that genetic predisposition to AD could be affected by dietary interventions (Joseph *et al.*, 2003). A later study in murine primary culture microglia demonstrated that blueberry treatment suppresses the p42/44 mitogen-activated protein kinase (MAPK) pathway, thereby inhibiting Aβ aggregation, increasing microglial clearance of Aβ and suppressing microglial activation (Zhu *et al.*, 2008). These results provided additional mechanistic insight to support the behavioural recovery seen with blueberry supplementation in APP/PS1 mice.

Cerebral ischemia

In addition to being beneficial in models of cognitive decline, blueberries have also been shown to protect the brain from ischemic insults, including stroke. In a 2002 study by Sweeney and colleagues, rats were fed a blueberry diet for six weeks prior to induction of stroke by ligation of the left common carotid artery (Sweeney *et al.*, 2002). In non-supplemented rats, hypoxia-ischemia resulted in a significant loss of hippocampal neurons in the left cerebral hemisphere, while blueberry-supplemented rats lost significantly fewer neurons in the ischemic hippocampus (Sweeney *et al.*, 2002). In a later study, Sprague-Dawley rats were fed control, blueberry, spinach or spirulina diets for four weeks prior to induction of stroke (Wang *et al.*, 2005). In the supplemented animals, infarct size was significantly lower and poststroke locomotor activity was increased. Less apoptosis was also evident in the ischemic hemispheres of supplemented animals, suggesting that the supplements provided have protective anti-apoptotic effects in addition to their antioxidant effects (Wang *et al.*, 2005).

Parkinson's disease and Huntington's disease

Neural tissue transplantation has been considered as a potential therapy for age-related neurodegenerative diseases, including Parkinson's disease and Huntington's disease, but a major drawback to this method is the poor survival of transplanted tissues. Transplantation-induced host immune response, inflammation and subsequent oxidative stress are likely contributors to cell death in these procedures (McGuire *et al.*, 2006). In a 2005 study by Granholm and colleagues, dietary blueberry supplementation was found to significantly improve the growth of intraocular hippocampal grafts in middle-aged F344 host animals, with the cellular organization of the grafts being comparable to that of young host animals (Willis *et al.*, 2005). In a follow-up study, the same group found that the blueberry diet did not affect vascularization of neural transplants (Willis *et al.*, 2008). It was later determined that the positive effects of the blueberry-enriched diet on intraocular hippocampal grafts were due to long-term attenuation of microglial activation and astrogliosis in the hosts (Willis *et al.*, 2010).

The traditional model for induction of Parkinson's disease in rodents is the 6-hydroxydopamine (6-OHDA) model. In this model, 6-OHDA, a neurotoxin, is injected unilaterally into specific sites in the rat brain, thereby producing a major loss of dopamine neurons on the injected side while sparing contralateral dopamine neurons. McGuire and colleagues examined the effects of a blueberry-enriched diet on survival of embryonic dopamine neurons transplanted into 6-OHDA lesioned F-344 rat brains (McGuire *et al.*, 2006). Here, 6-OHDA was injected into both the right median forebrain bundle and the right substantia nigra to induce a near total destruction of the right nigrostriatal pathway (McGuire *et al.*, 2006). Animals were then placed on a highbush blueberry diet or a corn control diet for six weeks postlesion (6 weeks prior to transplantation) through eight weeks post-transplantation. Six weeks following 6-OHDA lesioning, embryonic dopamine neurons were injected into the denervated striatum and rotational behaviour (dependent on intact striatal activity) was assessed at every two weeks post-transplantation. After eight weeks, the blueberry-enriched diet was found to improve graft survival, increase graft size and increase innervated area when compared to the control diet. These results provide further support of the neuroprotective properties of blueberry bioactives in the context of neural transplantation paradigms (McGuire *et al.*, 2006).

The neuroprotective and nootropic properties of blueberry bioactives are well exemplified in the studies referenced above. Notably, in addition to acting through the antioxidant mechanisms for which they are known, these bioactives also seem to be involved in regulating gene expression, signalling pathways and plasticity. Although the evidence is overwhelmingly positive in animal models, more evidence of the efficacy of blueberries in humans should be obtained before decisions are made regarding any clinical therapeutic recommendations for use in neurodegenerative disorders. Such studies are currently underway in the Kirkorian laboratory.

GASTROINTESTINAL DISEASES

In addition to their effects on the diseases and conditions mentioned above, blueberry bioactives have been shown to exert other beneficial health effects. The anti-inflammatory properties of blueberries and probiotics were recently studied in the dextran sulfate sodium (DSS)-induced rat model of ulcerative colitis (Osman *et al.*, 2008). Addition of DSS to drinking water for several days will induce acute colitis (inflammatory infiltration of the colon, bloody diarrhoea and ulcerations) in mice or rats. In this study, a disease activity index was used to assess colitis severity (Osman *et al.*, 2008). Whole blueberries were added to normal rat chow and fed to animals for the duration of the study. On days 9 and 10 post-DSS administration, the disease severity was significantly decreased in all blueberry and probiotic groups compared to the colitis control. Additionally, when given alone or with various probiotic species, blueberries reduced both bacterial translocation and inflammation; these mechanisms were considered responsible for the beneficial effects of blueberries on this colitis model.

Similar results were found in a recent study using the multidrug resistance gene deficient mouse, an inflammatory bowel disease model, to assess the effects of blueberries or broccoli on colon morphology and cecal microbiota (Paturi *et al.*, 2012). Mice were fed diets supplemented with freeze-dried New Zealand highbush blueberries or broccoli for 21 weeks. In comparison to mice fed the control diet, both blueberry and broccoli supplementation altered cecum microbiota in similar fashion. Blueberry- and broccoli-supplemented diets increased colon crypt size and the number of goblet cells per crypt (Paturi *et al.*, 2012). However, only the broccoli diet was able to reduce colonic inflammation in this model. Results from these two studies warrant further investigation to establish whether blueberry consumption can have similar effects on colon health in humans.

VISION

Rhodopsin is the photoreceptor that plays a major role in low light vision. Using frog rod outer membrane segments, Matsumoto *et al.* have shown *in vitro* that black currant anthocyanins, and cyanidin-3-glucoside in particular, stimulated rhodopsin regeneration (Matsumoto *et al.*, 2003). Cyanidin 3-glucoside is also found in blueberries (Zheng and Wang, 2002) and may also be responsible for the possible effects of blueberry bioactives on vision. Another *in vitro* study was conducted to investigate the cytoprotective role of blueberries on human retinal cells (Dutot *et al.*, 2008). Blueberry extract was incubated with human retinal cells. Following incubation with blueberry extract, cells were incubated with tert-butylhydroperoxide to induce oxidative stress. Intracellular metabolism, reactive oxygen species, superoxide anion and mitochondrial apoptosis were evaluated. Blueberry extract was found to protect cells against *tert*-butylhydroperoxide-induced cytotoxicity. Addition of the extract led to increased retinal cell viability and decreased oxidative stress (Dutot *et al.*, 2008). In another study, human retinal pigment epithelium cells were incubated with an anthocyanin extract made from whole Chinese blueberries, to assess the effects of the anthocyanins on senescence and photodamage (Liu *et al.*, 2011). A replicative senescence method was used in the cells to mimic the effects of aging; this was done by repeatedly passaging the cells from the fourth passage to the tenth passage. Senescence was determined by assessment of cell proliferation and percentage of cells positive for beta-galactosidase (increased in senescence). Photodamage to the cells was caused by exposure to white light irradiation. The anthocyanins were found to have dose-dependent effects on maintenance of cell viability and protection against light-induced damage (Liu *et al.*, 2011).

An *in vivo* study by the same group has demonstrated that blueberries have positive effects on vision (Liu *et al.*, 2010). Rabbits were fed wild Chinese blueberries for four weeks prior to induction of light-induced retinal damage. Photodamage to rabbit eye retinas was found to be significantly lower in blueberry-treated rabbits, suggesting that blueberries may also protect vision in mammals (Liu *et al.*, 2010). More recently, a study focused on examining the effect of bilberry (wild European blueberry; *Vaccinium myrtillus* L.) on visual function was conducted (Miyake *et al.*, 2012). Researchers generated a mouse model of endotoxin-induced uveitis by injecting lipopolysaccharide. Some mice were pretreated with bilberry extract and retinal effects were examined. The bilberry extract prevented functional impairment of photoreceptor cell function as measured by electroretinogram. Further, uveitis-associated rhodopsin decline and outer segment shortening were attenuated in bilberry-treated animals. The extract also prevented activation of signal transducer and activator of transcription 3 (STAT3), which induces inflammation-related rhodopsin decrease, and increased interleukin-6 expression, which activates STAT3. The bilberry extract, in addition to its anti-inflammatory effects, also attenuated reactive oxygen species in the inflamed retina (Miyake *et al.*, 2012). The authors concluded that bilberry extract has a protective effect on vision in response to retinal inflammation.

Although the limited *in vitro* research and animal studies seem promising, there is still a great deal of controversy surrounding the benefits of blueberry bioactives on vision in humans. As reviewed by Kalt and Dufour, European research conducted with bilberry from the 1960s to the 1980s suggested that anthocyanins from the bilberry can improve night vision and adaptation to low light conditions (Kalt and Dufour, 1997). In 2004, Canter and Ernst conducted meta-analysis of placebo-controlled clinical trials of bilberry effects on night vision and found that evidence from these trials did not support a beneficial role for these bioactives in normal night vision. Although eight of the thirty clinical trials demonstrated positive results on night vision, only one of those eight was a randomized-controlled trial, while the other seven used a pre-post design. Negative outcomes in the clinical trials reported were associated with more rigorous methodology, lower dose levels, and with extracts from various sources that may differ in bioactive composition. However, there was no rigorous research into the effects of bilberry extracts on those exhibiting impaired night vision due to pathological eye conditions (Canter and Ernst, 2004). Further, the behaviour of retinal cells *in vitro* is not necessarily indicative of *in vivo* function, so results of cell culture studies should be interpreted cautiously.

CONCLUSION

Blueberries have emerged as a rich source of bioactive components that offer therapeutic potential for attenuating the progression of several diseases, in addition to having protective effects in the gastrointestinal tract and eye. The principal mechanisms behind these protective effects include potential antioxidant activity and possible attenuation of inflammatory gene expression and oxidative stress. Although it is clear from current evidence that blueberries can exert health benefits in a variety of conditions, further studies are needed. Specifically, prospective, well-designed clinical trials should be conducted to compare the efficacy of whole berries versus fractionated bioactives and to examine dose response, longevity of any health benefits and any possible effects – positive or negative – that result from interactions between blueberry bioactives and other foods and drugs *in vivo*.

REFERENCES

Adams, L.S., Phung, S., Yee, N. *et al.* (2010) Blueberry phytochemicals inhibit growth and metastatic potential of MDA-MB-231 breast cancer cells through modulation of the phosphatidylinositol 3-kinase pathway. *Cancer Research,* **70**, 3594–3605.

Ahmet, I., Spangler, E., Shukitt-Hale, B. *et al.* (2009a) Survival and cardioprotective benefits of long-term blueberry enriched diet in dilated cardiomyopathy following myocardial infarction in rats. *PLoS ONE,* **4**, e7975.

Ahmet, I., Spangler, E., Shukitt-Hale, B. *et al.* (2009b) Blueberry-enriched diet protects rat heart from ischemic damage. *PLoS ONE,* **4**, e5954.

Aires, D.J., Rockwell, G., Wang, T. *et al.* (2012) Potentiation of dietary restriction-induced lifespan extension by polyphenols. *Biochimica et Biophysica Acta (BBA) – Molecular Basis of Disease,* **1822**, 522–526.

Aiyer, H.S., Srinivasan, C. and Gupta, R.C. (2008) Dietary berries and ellagic acid diminish estrogen-mediated mammary tumorigenesis andisin; ACI rats. *Nutrition and Cancer*, **60**, 227–234.

Andres-Lacueva, C., Shukitt-Hale, B., Galli, R. *et al.* (2005) Anthocyanins in aged blueberry-fed rats are found centrally and may enhance memory. *Nutritional Neuroscience,* **8**, 111–120.

Basu, A., Du, M., Leyva, M.J. *et al.* (2010) Blueberries decrease cardiovascular risk factors in obese men and women with metabolic syndrome. *The Journal of Nutrition,* **140**, 1582–1587.

Basu, A. and Lyons, T.J. (2012) Strawberries, blueberries, and cranberries in the metabolic syndrome: clinical perspectives. *Journal of Agricultural and Food Chemistry*, **60**(23), 5687–5692

Boivin, D., Blanchette, M., Barrette, S. *et al.* (2007) Inhibition of cancer cell proliferation and suppression of TNF-induced Activation of NFκB by edible berry juice. *Anticancer Research,* **27**, 937–948.

Bomser, J., Madhavi, D.L., Singletary, K. and Smith, M.A.L. (1996) In vitro anticancer activity of fruit extracts from vaccinium species. *Planta Med*, **62**, 212–216.

Bushway, R.J., Gann, D.F.M., Cook, W.P. and Bushway, A.A. (1983) Mineral and vitamin content of lowbush blueberries (Vaccinium angustifolium Ait.). *Journal of Food Science,* **48**, 1878–1878.

Canter, P.H. and Ernst, E. 2004. Anthocyanosides of Vaccinium myrtillus (Bilberry) for night vision—a systematic review of placebo-controlled trials. *Survey of Ophthalmology*, **49**, 38–50.

DeFuria, J., Bennett, G., Strissel, K.J. *et al.* (2009) Dietary blueberry attenuates whole-body insulin resistance in high fat-fed mice by reducing adipocyte death and its inflammatory sequelae. *The Journal of Nutrition,* **139**, 1510–1516.

Devore, E.E., Kang, J.H., Breteler, M.M.B. and Grodstein, F. (2012) Dietary intakes of berries and flavonoids in relation to cognitive decline. *Annals of Neurology*, doi: 10.1002/ana.23594.

Dutot, M., Rambaux, L., Warnet, J.M. and Rat, P. (2008) Oxidative stress modulation using polyphenol-rich blueberries: application on a human retinal cell model. *Journal français d'ophtalmologie*, **31**, 975–980.

Ehlenfeldt, M.K. and Prior, R.L. (2001) Oxygen radical absorbance capacity (ORAC) and phenolic and anthocyanin concentrations in fruit and leaf tissues of highbush blueberry. *Journal of Agricultural and Food Chemistry*, **49**, 2222–2227.

Elks, C.M., Reed, S.D., Mariappan, N. *et al* (2011) A blueberry-enriched diet attenuates nephropathy in a rat model of hypertension via reduction in oxidative stress. *PLoS ONE*, **6**, e24028.

Gao, L. and Mazza, G. (1994) Quantitation and distribution of simple and acylated anthocyanins and other phenolics in blueberries. *Journal of Food Science*, **59**, 1057–1059.

Goyarzu, P., Malin, D.H., Lau, F.C. *et al.* 2004. Blueberry supplemented diet: effects on object recognition memory and Nuclear Factor-kappa B levels in aged rats. *Nutritional Neuroscience*, **7**, 75–83.

Halvorsen, B.L., Holte, K., Myhrstad, M.C.W. *et al.* 2002. A systematic screening of total antioxidants in dietary plants. *The Journal of Nutrition*, **132**, 461–471.

Joseph, J.A., Shukitt-Hale, B., Denisova, N.A. *et al.* (1998) Long-term dietary strawberry, spinach, or vitamin e supplementation retards the onset of age-related neuronal signal-transduction and cognitive behavioral deficits. *The Journal of Neuroscience,* **18**, 8047–8055.

Joseph, J.A., Shukitt-Hale, B., Denisova, N.A. *et al.* (1999) Reversals of age-related declines in neuronal signal transduction, cognitive, and motor behavioral deficits with blueberry, spinach, or strawberry dietary supplementation. *The Journal of Neuroscience,* **19**, 8114–8121.

Joseph, J.A., Arendash, G., Gordon, M. *et al.* (2003) Blueberry supplementation enhances signaling and prevents behavioral deficits in an Alzheimer disease model. *Nutritional Neuroscience,* **6**, 153–162.

Joseph, J.A., Shukitt-Hale, B. and Casadesus, G. (2005) Reversing the deleterious effects of aging on neuronal communication and behavior: beneficial properties of fruit polyphenolic compounds. *The American Journal of Clinical Nutrition*, **81**, 313S–316S.

Joseph, J.A., Shukitt-Hale,B. and Willis, L.M. (2009) Grape juice, berries, and walnuts affect brain aging and behavior. *The Journal of Nutrition*, **139**, 1813S–1817S.

Kalea, A.Z., Clark, K., Schuschke, D.A. and Klimis-Zacas, D.J. (2009) Vascular reactivity is affected by dietary consumption of wild blueberries in the Sprague-Dawley rat. *Journal of Medicinal Food*, **12**, 21–28.

Kalea, A.Z., Lamari, F.N., Theocharis, A.D. *et al.* (2006) Wild blueberry (Vaccinium angustifolium) consumption affects the composition and structure of glycosaminoglycans in Sprague-Dawley rat aorta. *The Journal of Nutritional Biochemistry,* **17**, 109–116.

Kalt, W. and Dufour, D. (1997) Health functionality of blueberries. *HortTechnology*, **7**, 216–221.

Kalt, W., Forney, C.F., Martin, A. and Prior, R.L. (1999) Antioxidant capacity, vitamin C, phenolics, and anthocyanins after fresh storage of small fruits. *Journal of Agricultural and Food Chemistry*, **47**, 4638–4644.

Kalt, W., McDonald, J.E. and Donner, H. (2000) Anthocyanins, phenolics, and antioxidant capacity of processed lowbush blueberry products. *Journal of Food Science*, **65**, 390–393.

Kalt, W., Ryan, D. A. J., Duy, J. C. *et al.* (2001) Interspecific variation in anthocyanins, phenolics, and antioxidant capacity among genotypes of highbush and lowbush blueberries (Vaccinium Section cyanococcus spp.). *Journal of Agricultural and Food Chemistry,* **49**, 4761–4767.

Kalt, W., Foote, K., Fillmore, S.A.E. *et al.* (2008) Effect of blueberry feeding on plasma lipids in pigs. *British Journal of Nutrition,* **100**, 70–78.

Kaplan, J.K. (2011) Blueberry growing comes to the National Agricultural Library. *Agricultural Research.* US Department of Agriculture, Beltsville, MD.

Kim, H., Bartley, G.E., Rimando, A.M. and Yokoyama, W. (2010) Hepatic gene expression related to lower plasma cholesterol in hamsters fed high-fat diets supplemented with blueberry peels and peel extract. *Journal of Agricultural and Food Chemistry*, **58**, 3984–3991.

Krikorian, R., Shidler, M.D., Nash, T.A. *et al.* (2010) Blueberry supplementation improves memory in older adults. *Journal of Agricultural and Food Chemistry,* **58**, 3996–4000.

Liu, Y., Song, X., Han, Y. *et al.* (2010) Identification of anthocyanin components of wild chinese blueberries and amelioration of light-induced retinal damage in pigmented rabbit using whole berries. *Journal of Agricultural and Food Chemistry,* **59**, 356–363.

Liu, Y., Song, X., Zhang, D. *et al.* (2011) Blueberry anthocyanins: protection against ageing and light-induced damage in retinal pigment epithelial cells. *British Journal of Nutrition*, **108**, 16–27.

Martin, L.J. and Matar, C. (2005) Increase of antioxidant capacity of the lowbush blueberry (Vaccinium angustifolium) during fermentation by a novel bacterium from the fruit microflora. *Journal of the Science of Food and Agriculture*, **85**, 1477–1484.

Martineau, L.C., Couture, A., Spoor, D. *et al.* (2006) Anti-diabetic properties of the Canadian lowbush blueberry Vaccinium angustifolium Ait. *Phytomedicine,* **13**, 612–623.

Matchett, M.D., MacKinnon, S.L., Sweeney, M.I. *et al.* R. (2005) Blueberry flavonoids inhibit matrix metalloproteinase activity in DU145 human prostate cancer cells. *Biochemistry and Cell Biology,* **83**, 637–643.

Matchett, M.D., MacKinnon, S.L., Sweeney, M.I. *et al.* (2006) Inhibition of matrix metalloproteinase activity in DU145 human prostate cancer cells by flavonoids from lowbush blueberry (Vaccinium angustifolium): possible roles for protein kinase C and mitogen-activated protein-kinase-mediated events. *The Journal of Nutritional Biochemistry,* **17**, 117–125.

Matsumoto, H., Nakamura, Y., Tachibanaki, S. *et al.* (2003) Stimulatory effect of cyanidin 3-glycosides on the regeneration of Rhodopsin. *Journal of Agricultural and Food Chemistry,* **51**, 3560–3563.

McAnulty, S.R., McAnulty, L.S., Morrow, J.D. *et al.* (2005) Effect of daily fruit ingestion on angiotensin converting enzyme activity, blood pressure, and oxidative stress in chronic smokers. *Free Radical Research,* **39**, 1241–1248.

McGuire, S., Sortwell, C., Shukitt-Hale, B. *et al.* (2006) Dietary supplementation with blueberry extract improves survival of transplanted dopamine neurons. *Nutritional Neuroscience,* **9**, 251–258.

McMurray, J. J. and Stewart, S. (2000) Epidemiology, aetiology, and prognosis of heart failure. *Heart*, **83**, 596–602.
Miyake, S., Takahashi, N., Sasaki, M. *et al.* (2012) Vision preservation during retinal inflammation by anthocyanin-rich bilberry extract: cellular and molecular mechanism. *Laboratory Investigation*, **92**, 102–109.
Neto, C.C. (2007) Cranberry and blueberry: Evidence for protective effects against cancer and vascular diseases. *Molecular Nutrition and Food Research*, **51**, 652–664.
Osman, N., Adawi, D., Ahrne, S. *et al.* (2008) Probiotics and blueberry attenuate the severity of dextran sulfate sodium (DSS)-induced colitis. *Digestive Diseases and Sciences*, **53**, 2464–73.
Paturi, G., Mandimika, T., Butts, C.A. *et al.* (2012) Influence of dietary blueberry and broccoli on cecal microbiota activity and colon morphology in mdr1a−/− mice, a model of inflammatory bowel diseases. *Nutrition*, **28**, 324–330.
Peng, C., Zuo, Y., Kwan, K.M. *et al.* (2012) Blueberry extract prolongs lifespan of Drosophila melanogaster. *Experimental Gerontology*, **47**, 170–178.
Prior, R.L., Cao, G., Martin, A. *et al.* (1998) Antioxidant capacity as influenced by total phenolic and anthocyanin content, maturity, and variety of Vaccinium species. *Journal of Agricultural and Food Chemistry*, **46**, 2686–2693.
Prior, R.L., Wilkes, S.E., Rogers, T.R. *et al.* (2010) Purified blueberry anthocyanins and blueberry juice alter development of obesity in mice fed an obesogenic high-fat diet. *Journal of Agricultural and Food Chemistry*, **58**, 3970–3976.
Prior, R. L., Wu, X., Gu, L. *et al.* (2008) Whole Berries versus Berry Anthocyanins: Interactions with Dietary Fat Levels in the C57BL/6J Mouse Model of Obesity. *Journal of Agricultural and Food Chemistry*, **56**, 647–653.
Prior, R.L., Wu, X., Gu, L. *et al.* (2009) Purified berry anthocyanins but not whole berries normalize lipid parameters in mice fed an obesogenic high fat diet. *Molecular Nutrition and Food Research*, **53**, 1406–1418.
Prodorutti, D., Pertot, I., Giongo, L. and Gessler, C. (2007) Highbush blueberry: cultivation, protection, breeding and biotechnology. *European Journal of Plant Science and Biotechnology*, **1**, 44–56.
Reis-Filho, J.S. and Tutt, A.N.J. (2008) Triple negative tumours: a critical review. *Histopathology*, **52**, 108–118.
Ross, R. 1993. The pathogenesis of atherosclerosis: a perspective for the 1990s. *Nature*, **362**, 801–809.
Seymour, E.M., Tanone, I.I., Urcuyo-Llanes, D.E. *et al.* (2011) Blueberry intake alters skeletal muscle and adipose tissue peroxisome proliferator-activated receptor activity and reduces insulin resistance in obese rats. *Journal of Medicinal Food*, **14**, 1511–1518.
Shaughnessy, K.S., Boswall, I.A., Scanlan, A.P. *et al.* (2009) Diets containing blueberry extract lower blood pressure in spontaneously hypertensive stroke-prone rats. *Nutrition Research*, **29**, 130–138.
Shukitt-Hale, B., Carey, A.N., Jenkins, D. *et al.* (2007) Beneficial effects of fruit extracts on neuronal function and behavior in a rodent model of accelerated aging. *Neurobiology of Aging*, **28**, 1187–1194.
Shukitt-Hale, B., Lau, F., Carey, A. *et al.* (2008) Blueberry polyphenols attenuate kainic acid-induced decrements in cognition and alter inflammatory gene expression in rat hippocampus. *Nutritional Neuroscience*, **11**, 172–182.
Simmen, F.A., Frank, J.A., Wu, X. *et al.* (2009) Lack of efficacy of blueberry in nutritional prevention of azoxymethane-initiated cancers of rat small intestine and colon. *BMC Gastroenterology*, **9**, 67.
Smith, M.A.L., Marley, K.A., Seigler, D. *et al.* 2000. Bioactive properties of wild blueberry fruits. *Journal of Food Science*, **65**, 352–356.
Srivastava, A., Akoh, C.C., Fischer, J. and Krewer, G. (2007) Effect of anthocyanin fractions from selected cultivars of Georgia-grown blueberries on apoptosis and phase II enzymes. *Journal of Agricultural and Food Chemistry*, **55**, 3180–3185.
Stoner, G.D., Wang, L.-S. and Casto, B.C. (2008) Laboratory and clinical studies of cancer chemoprevention by antioxidants in berries. *Carcinogenesis*, **29**, 1665–1674.
Stull, A.J., Cash, K.C., Johnson, W.D. *et al.* (2010) Bioactives in blueberries improve insulin sensitivity in obese, insulin-resistant men and women. *The Journal of Nutrition*, **140**, 1764–1768.
Suckling, K.E. and Jackson, B. 1993. Animal models of human lipid metabolism. *Prog Lipid Res*, **32**, 1–24.
Sweeney, M.I., Kalt, W., MacKinnon, S.L. *et al.* 2002. Feeding rats diets enriched in lowbush blueberries for six weeks decreases ischemia-induced brain damage. *Nutritional Neuroscience*, **5**, 427–431.
USDA (United States Department of Agriculture) (2010) *U.S. Blueberry Industry* [Online]. USDA ERS. http://usda.mannlib.cornell.edu/MannUsda/viewDocumentInfo.do?documentID=1765 (last accessed 7 March 2013).

USDA (United States Department of Agriculture) (2012) USDA National Nutrient Database for Standard Reference, Release 25. USDA Agricultural Research Service .http://ndb.nal.usda.gov/ (last accessed 20 March 2013).

Vuong, T.R.I., Martin, L.U.C. and Matar, C. (2006) Antioxidant activity of fermented berry juices and their effects on nitric oxide and tumor necrosis factor-alpha production in macrophages 264.7 gamma NO(–) cell line. *Journal of Food Biochemistry*, **30**, 249–268.

Vuong, T., Martineau, L.C., Ramassamy, C. *et al.* (2007) Fermented Canadian lowbush blueberry juice stimulates glucose uptake and AMP-activated protein kinase in insulin-sensitive cultured muscle cells and adipocytes. *Canadian Journal of Physiology and Pharmacology,* **85**, 956–965.

Vuong, T., Benhaddou-Andaloussi, A., Brault, A. *et al.* (2009) Antiobesity and antidiabetic effects of biotransformed blueberry juice in KKAy mice. *International Journal of Obesity,* **33**, 1166–1173.

Wang, Y., Chang, C.-F., Chou, J. *et al.* (2005) Dietary supplementation with blueberries, spinach, or spirulina reduces ischemic brain damage. *Experimental Neurology,* **193**, 75–84.

Watson, E.M. (1928) Some observations on the effect of blueberry leaf extract in diabetes mellitus. *Canadian Medical Association Journal*, **19**, 166–171.

Wedick, N.M., Pan, A., Cassidy, A. *et al.* (2012) Dietary flavonoid intakes and risk of type 2 diabetes in US men and women. *The American Journal of Clinical Nutrition,* **95**, 925–933.

Williams, C.M., El Mohsen, M.A., Vauzour, D. *et al.* (2008) Blueberry-induced changes in spatial working memory correlate with changes in hippocampal CREB phosphorylation and brain-derived neurotrophic factor (BDNF) levels. *Free Radical Biology and Medicine,* **45**, 295–305.

Willis, L., Bickford, P., Zaman, V. *et al.* (2005) Blueberry extract enhances survival of intraocular hippocampal transplants. *Cell Transplantation*, **14**, 213–223.

Willis, L.M., Small, B.J., Bickford, P.C. *et al.* (2008) Dietary blueberry supplementation affects growth but not vascularization of neural transplants. *Journal of Cerebral Blood Flow and Metabolism*, **28**, 1150–1164.

Willis, L.M., Freeman, L., Bickford, P.C. *et al.* (2010) Blueberry supplementation attenuates microglial activation in hippocampal intraocular grafts to aged hosts. *Glia,* **58**, 679–690.

Wilson, M.A., Shukitt-Hale, B., Kalt, W. *et al.* (2006) Blueberry polyphenols increase lifespan and thermotolerance in *Caenorhabditis elegans*. *Aging Cell,* **5**, 59–68.

Wiseman, W., Egan, J.M., Slemmer, J.E. *et al.* (2011) Feeding blueberry diets inhibits angiotensin II-converting enzyme (ACE) activity in spontaneously hypertensive stroke-prone rats. *Canadian Journal of Physiology and Pharmacology,* **89**, 67–71.

Wu, X., Kang, J., Xie, C. *et al.* (2010) Dietary blueberries attenuate atherosclerosis in apolipoprotein e-deficient mice by upregulating antioxidant enzyme expression. *The Journal of Nutrition,* **140**, 1628–1632.

Yi, W., Fischer, J., Krewer, G. and Akoh, C.C. (2005) Phenolic compounds from blueberries can inhibit colon cancer cell proliferation and induce apoptosis. *Journal of Agricultural and Food Chemistry*, **53**, 7320–7329.

Youdim, K.A., McDonald, J., Kalt, W. and Joseph, J.A. (2002) Potential role of dietary flavonoids in reducing microvascular endothelium vulnerability to oxidative and inflammatory insults. *The Journal of Nutritional Biochemistry*, **13**, 282–288.

Zheng, W. and Wang, S.Y. (2002) Oxygen radical absorbing capacity of phenolics in blueberries, cranberries, chokeberries, and lingonberries. *Journal of Agricultural and Food Chemistry*, **51**, 502–509.

Zhu, Y., Bickford, P., Sanberg, P. *et al.* (2008) Blueberry opposes β-amyloid peptide-induced microglial activation via inhibition of p44/42 mitogen-activation protein kinase. *Rejuvenation Research,* **11**, 891–901.

12 Cranberry Polyphenols in the Promotion of Urinary Tract, Cardiovascular and Emerging Health Areas

Kerrie L. Kaspar and Christina Khoo

INTRODUCTION

The American cranberry, a member of the Ericaceae family, is native to the cool, temperate, acidic soils and peat wetlands of the Northeastern United States and southern Canada. Cranberries were first used by Native Americans, who recognized the versatility of the fruit as a food, fabric dye and healing agent. The name 'cranberry' derives from the Pilgrim name for the fruit, 'craneberry', because the small, pink blossoms that appear in the spring resemble the head and bill of a Sandhill crane. European settlers adopted the Native American uses for the fruit and found the berry to be a valuable bartering tool. The first commercial endeavour to produce cranberry as a crop was in Dennis, MA, in 1816 by Captain Henry Hall (Winston *et al.*, 2002).

There are two major species of cranberry: the American cranberry (*Vaccinium macrocarpon*) and the European cranberry (*V. oxycoccos*). The United States and Canada together account for more than 90% of the world's production but New Zealand, countries in Europe and in South America (Chile) have also been growing cranberries (Neto and Vinson, 2011). In North America, cranberries are harvested in the fall, from mid-September to November. Farmers flood the fields for wet harvesting, while a small percentage of bogs are dry harvested for fresh fruit. Today, cranberries are commercially grown throughout the northern part of the United States and are available in both fresh and processed forms. Cranberry products such as concentrate, dried cranberries, juice and sauce are growing in popularity as consumer products and ingredients and use 95% of cranberry produced. In 2011, production was 7.71 million barrels with Wisconsin leading all US states in the production of cranberries, at 4.41 million barrels, followed by Massachusetts with 2.32 million. New Jersey, Oregon and Washington also had substantial cranberry production (NASS, 2011; Neto and Vinson, 2011).

Natural food products, especially those rich in polyphenols, are gaining increasing popularity for helping to maintain health while alleviating oxidative stress and inflammatory symptoms (Scalbert *et al.*, 2005b). Polyphenols are common constituents of foods plant origin and also major antioxidants in our diet. The main dietary sources of polyphenols are fruits and beverages. A recent publication on polyphenols showed that beverages (wine, tea), fruits (berries, apples, plums) and vegetables contain a large diversity of polyphenol compounds with 105, 112 and 92 types of polyphenols, respectively (Perez-Jimenez *et al.*, 2010). Fruits

Bioactives in Fruit: Health Benefits and Functional Foods, First Edition. Edited by Margot Skinner and Denise Hunter.

like cranberries, apples, grapes, pears, cherries and various berries contain up to 200–300 mg polyphenols per 100 g fresh weight. Typically, a glass of red wine, a glass of cranberry juice cocktail or a cup of tea or coffee contains about 100 mg polyphenols. Total polyphenol intake in different countries is quite variable with crude estimates of consumption in Western countries ranging between 65 and 250 mg/day, primarily coming from tea, red wine, fruits, chocolate, vegetables and legumes (Williamson and Manach, 2005). Globally, the estimated dietary intake ranges from 100–200 mg/d to 1 g/d, which is more than 10 times higher than other known dietary antioxidants, including vitamin C, vitamin E and carotenoids (Mennen *et al.*, 2006; Jennings *et al.*, 2012; Knaze *et al.*, 2012).

Due to its rich content of polyphenols, specifically flavonoids, the cranberry is linked to different bioactivities, including inhibition of bacterial adhesion, bacteriostatic, anti-inflammatory activities and modulation of signal transduction, which may lead to improved urinary tract, cardiovascular, oral and gastrointestinal health. The beautiful, rich red colour of the berry and associated health properties have increased the popularity of this small berry and its products for use as both an ingredient and as a functional food.

BIOACTIVES IN CRANBERRY

Cranberries contain recognized nutrients, such as vitamin C (13.3 mg), lutein and zeaxanthin (91 μg) and fibre (4.6 g), along with minor components, such as potassium (85 mg), zinc (100 μg) and manganese (360 μg) per 100 g fresh weight (USDA, 2011). American whalers and mariners carried cranberries on their voyages to prevent scurvy. Compared to other fresh fruits, cranberries have a unique profile of carbohydrates with a low fructose to glucose ratio (1:5) and a low total sugar content of only 4%. This is less than half the amount found in other commonly consumed fruits (9–16%); therefore, cranberry products need to be sweetened for palatability. Benzoic and other phenolic acids, including hydroxyl-cinnamic acids, represent 0.57% of the weight of fresh cranberries (Zuo *et al.*, 2002). Benzoic acid forms 80% of the total phenolic acids contained in cranberry juice. There exist 14 other phenolic acids, as well as organic acids such as citric, quinic and malic acids, which have bioactivity but are outside the scope of this discussion. By far the most well-known nutrients in cranberries are the phytonutrients called polyphenols, specifically flavonoids, and phenolic acids (McKay and Blumberg, 2007).

The flavonoid class of polyphenols consists of flavonols, flavones, catechins, flavanones and anthocyanins. The catechins are monomeric flavan-3-ol units that may combine to form oligomers, and polymers, known as proanthocyanidins. Flavonoids are plant phenolics with a similar core structure, but have different heterocyclic rings, leading to several types of compounds. Therefore, flavonoids are not one compound or structure, but a family of similar, yet different compounds. They are synthesized as the secondary metabolites of a complex common pathway in plants and many of these secondary metabolites exhibit health benefits. In addition, flavonoids are attached to other molecules, such as carbohydrates (simple sugars and fibre), organic acids and even to each other (Reed *et al.*, 2005). All of these factors influence how the body metabolizes the flavonoid and contribute to bioavailability and activity of these complex families of compounds. The complexity and the wide range in the diet have made research in this area both challenging and intriguing.

Cranberries contain one of the highest levels of flavonoids of most commonly consumed fruits. The major flavonoid classes, which have been widely screened for bioactivity, are anthocyanins, flavonols and flavan-3-ols, particularly proanthocyanidins (Figure 12.1). These

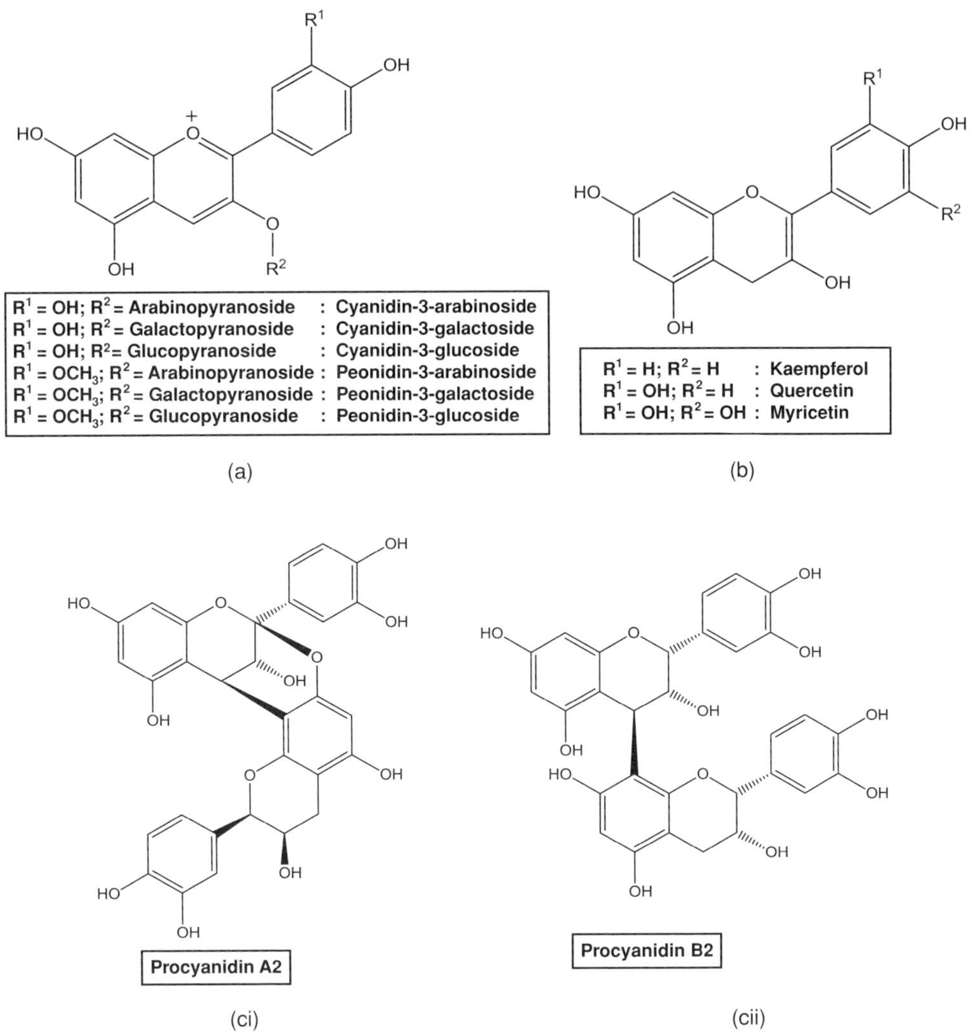

Fig. 12.1 Examples of some key flavonoids found in cranberry: (a) anthocyanins; (b) flavonols; (c) (i) procyanidin A2, (ii) procyanidin B2.

activities range from antimicrobial to preventing bacterial adhesion, reducing oxidative stress, protecting cell health and promoting vascular health (Kondo *et al.*, 2003; Neto, 2007; Reed and Howell, 2009). The fruit is typically consumed in the form of juice, dried cranberries and sauce. The polyphenol content of cranberry products is found in various databases globally, ranging from 100–300 mg per 100 g, which can contribute significantly to daily flavonoid intake (Table 12.1) (Neveu *et al.*, 2010; Bhagwat *et al.*, 2011).

Flavonols

Flavonols, with a characteristic yellow colour, have been studied for decades for their health properties and are found quite abundantly in fruits and vegetables such as broccoli, onions

Table 12.1 Range of flavonoids content in cranberry products.[1]

Flavonoid class	Products	Content per 100g (mg)	Serving size	Content recommended serving size (as is unless stated otherwise) (mg)
Proanthocyanidins[2]	Fresh cranberries (100 g FW)	420[a]	28 g	117.3
	Cranberry juice drink/cocktail	21.6[a], 35[b]	8 oz or 240 ml	50, 130
	Sweetened dried cranberries	61–82[c]	40 g	24–32
Anthocyanins[3]	Fresh cranberries	49.9–85.2	28g	14–23.8
	Cranberry juice drink/cocktail	0.2–3.2	8 oz or 240 ml	1.0–7.6
	Sweetened dried cranberries	8.2–10.3	40 g	0.4–3.3
Flavonols (sum of)	Fresh cranberries	21.6–28.1	28 g	6.1–7.8
	Cranberry juice drink/cocktail	2.4–5.0	8 oz or 240 ml	6–12.1
	Sweetened dried cranberries	6.95	40 g	2.8

[1]Data obtained from USDA Flavonoids database where available and supplemented with information from Cunningham *et al.* (2004), Phenol Explorer database Neveu *et al.* (2010), Grace *et al.* (2012) and Ocean Spray Cranberries, Inc., internal database and as analysed by an independent third party laboratory.
[2]Methodologies used as described in (a) USDA database (normal phase HPLC with catechin standard), (b) Grace *et al.* (2012) (DMAC method, with PAC A2 standard) and (c) Cunningham *et al.* (2004).
[3]Reverse phase HPLC with diode array detector with cyanidin-3-galactoside as a standard.

and apples (Graf *et al.*, 2005). The total amount of flavonols in cranberry ranges from 20 to 40 mg/100 g, comprising 20 different glycosides (Zheng and Wang, 2003; Cunningham *et al.*, 2004; Vvedenskaya *et al.*, 2004).

The flavonol content of cranberry is almost twice as high as 12 other commonly consumed fruit juices, including pomegranate and grape (Aherne and O'Brien, 2002; Mullen *et al.*, 2007). In a review by Erdman *et al.*, intake of flavonols in the United States was estimated at 9.4 mg/day, while the rest of the world ranged between 21 and 64 mg/day (Erdman *et al.*, 2007). Cranberry contains one of the highest levels of quercetin, a biologically active flavonol known for its anti-inflammatory properties. Quercetin is the most abundant flavonol in cranberry; it varies from 11 to 25 mg/100 g, primarily as the 3-*o*-galactoside. Myricetin is the second most abundant, followed by kaempferol (Manach *et al.*, 2004; Vvedenskaya *et al.*, 2004).

Proanthocyanidins and anthocyanins

Two other important flavonoid compounds in cranberries are proanthocyanidins and anthocyanins, which have been linked to urinary tract health and cardiovascular protection (Foo *et al.*, 2000a; Howell, 2002; Caton *et al.*, 2010). Cranberries not only contain one of the highest levels of proanthocyanidins (PACs) among commonly consumed fruits and berries, they also contain a mixture of A type (double) and B type (single) interflavan bonds (Foo *et al.*, 2000a) (Figure 12.1c). Cranberries are uniquely rich in the A type oligomers and polymers that are thought to promote the anti-adhesion activity of cranberry by inhibiting attachment of certain pathogenic bacteria such as uropathogenic *E. coli* and the ulcer-causing *H. pylori* (Howell, 2007; Cote *et al.*, 2010). Cranberries are unique because they have more type A PACs than other commonly consumed fruits, such as blueberries, strawberries, apples or grapes (Bhagwat *et al.*, 2011).

Anthocyanins impart the characteristic red colour of cranberries. The anthocyanins in cranberry, other berries and wine are associated with modulation of inflammatory signalling molecules and signalling pathways, cytokines, adhesion molecules and NADPH oxidase. There is a strong association between anthocyanin intake and cardiovascular protection (Wallace, 2011). Acute bioavailability and pharmacokinetic studies with cranberry beverages found anthocyanins in plasma and their metabolites in urine shortly after consumption, demonstrating bioavailability (Zhang and Zuo, 2004; Singh *et al.*, 2008; Milbury *et al.*, 2010).

Dietary intake of flavonoids

The data on consumption of flavonoids are quite variable due to insufficient information on the content of polyphenols in food composition databases, accurate analytical methods and inadequate food content information. However, due to growing evidence regarding the important health benefits of flavonoids, there are many ongoing activities to address these issues. The most comprehensive databases to date are the series of USDA databases containing flavonoids and proanthocyanidins (Bhagwat *et al.*, 2011; USDA, 2004). Total intake of flavonoids of the US population was estimated to be 189.7 mg/day (Chun *et al.*, 2007), although in the highest quintile it can reach up to 1 g/day. If one serving of a cranberry product can contribute anywhere between 50 and 200 mg flavonoids, then the amount of flavonoid intake in Western diets should be increased by at least two times if not more than the current intake levels (Chun *et al.*, 2007, 2010; Cassidy *et al.*, 2011; McCullough *et al.*, 2012).

As reviewed by Del Rio *et al.* consumption of berry flavonoids are linked to health benefits. Different flavonoids are found in plasma and tissues in varying amounts, either through direct absorption or through metabolism by gut flora (Del Rio *et al.*, 2010). As absorption is low, increasing the dietary intake may help promote health. New studies continue to emerge confirming the importance of flavonoids in maintaining health (McCullough *et al.*, 2012).

As discussed in the following sections, the evidence for urinary tract, cardiovascular and oral/gastrointestinal health is linked to the flavonoid and proanthocyanidin content of the cranberry.

HEALTH BENEFITS OF CRANBERRY

Urinary tract health

In the twentieth century, US folklore made cranberry well-known for its beneficial effect of preventing urinary tract infections (UTIs) (Henig and Leahy, 2000). The purported benefits of cranberries for urinary tract health have now been supported by more than 50 years of scientific research, including epidemiological, mechanistic and clinical trials. There is ongoing research on proposed mechanisms of action as discussed in this chapter, but current literature from interventional studies supports that cranberry may help prevent UTIs, particularly in women with recurrent UTIs (Jepson and Craig, 2008; Wang *et al.*, 2012).

In the United States, UTIs are the most common bacterial infection in the ambulatory care setting (Schappert and Rechtsteiner, 2011). Although males and females can develop a UTI, infections occur more frequently in women (Foxman and Brown, 2003). It is estimated that more than 50% of women will experience at least one UTI in their lifetime (Griebling, 2005), and 20–30% of women who experience a UTI will have two or more recurrent episodes (Foxman, 1990). Other populations at risk of developing UTIs include children, pregnant women, the elderly, patients with spinal cord injuries, catheterized patients and those with chronic and/or immune-compromising diseases such as diabetes and HIV/AIDS (Foxman, 2002). The prevalence of UTIs results in a high economic cost, with estimates in the billions of dollars. Due to the frequency of infection, high recurrence rates and antibiotic treatment prescribed for clearance, there is concern over the issue of antibiotic resistance. Increasingly, uropathogenic bacterial isolates have been identified as being antibiotic resistant (Zhanel *et al.*, 2006). To minimize antibiotic resistance, cranberry may be recommended as a natural, preventative alternative for UTIs.

Etiology and pathogenesis of UTIs

In the literature, there is no universal definition for UTIs. Clinicians typically use a combination of the following criteria to diagnose a UTI: the presence of uropathogenic bacteria in the urine ranging from 10^2 to 10^5 CFUs/ml urine, pyuria and associated symptoms (Wang *et al.*, 2012). Patients typically present with symptoms that can include: dysuria with or without frequency, suprapubic pain, hematuria, fever, chills, flank pain, costovertaebral-angle tenderness, nausea and vomiting (Hooton, 2012). The pathogenesis of UTIs involves bacterial colonization and subsequent adhesion to mucosal uroepithelial cells that line the urinary tract. Although *Escherichia coli* (*E. coli*) accounts for about 80% of infections, other uropathogenic species such as *Staphylococcus saprophyticus*, *Klebsiella* spp. and *Enterococcus faecalis* have also been identified as causative agents in non-complicated UTIs (Ronald, 2003).

Cystitis and pyelonephritis are the two common types of infection caused by uropathogenic bacteria in the urinary tract. Cystitis is an infection of the bladder, whereas pyelonephritis is an infection of the kidney due to the ascension of bacteria from the bladder. *E. coli* adhesion organelles, or fimbriae, are often associated with infection type: type 1 fimbriated *E. coli* are more prevalent with cystitis and p-fimbriated *E. coli* with pyelonephritis; however, both fimbriae types can be found in either infection and can also be co-expressed (Domingue *et al.*, 1985; Johnson *et al.*, 2005).

Uropathogenic *E. coli* has been shown to reside in rectal, vaginal and periurethral areas (Gruneberg, 1969). The bacterial strains found at these sites originate from the gastrointestinal tract but once in the bladder may express virulence factors associated with increased pathogenicity (Moreno *et al.*, 2006). In women, vaginal intercourse is considered to be the main risk factor that increases exposure of uropathogenic bacteria to the urinary tract (Foxman and Brown, 2003).

Clinical trials

Cranberry products have been investigated in clinical trials for prophylaxis and treatment of UTIs. Many clinical trials have demonstrated that cranberry juice or cranberry powders are effective in preventing recurrent urinary tract infections (Jepson and Craig, 2008, Wang *et al.*, 2012), but limited evidence is available that demonstrates cranberry can cure or treat a UTI.

The first randomized, double-blind, placebo controlled trial conducted with the intent to demonstrate UTI prophylaxis was conducted in a nursing home population. This study found that consumption of low calorie cranberry juice cocktail (27% cranberry juice, sweetened with saccharin) reduced the incidence of bacteriuria (10^5 CFUs/ml urine) with pyuria in a nursing home population (Avorn *et al.*, 1994). Since then, many trials using cranberry products have been conducted to confirm these initial results. While some studies have demonstrated cranberry consumption to be efficacious, others have shown no effect. Potential reasons for discrepancies between results may include the chosen study population, sample size, primary objective definition and study treatment compliance. Additionally, cranberry treatments used in trials can differ considerably: cranberry capsules or tablets made from powders, extracts and juices with different formulations and/or percentage cranberry juice, or juice concentrates have all been administered. Historically, cranberry treatments have not been standardized and clinical trials have not consistently demonstrated an efficacious dose.

Herein, clinical trials based on the four populations generally studied are reviewed, with the intent to demonstrate UTI prophylaxis: women with recurrent UTIs, pregnant women, children and patients with neurogenic bladder.

Women with recurrent UTIs

Cranberry products have been evaluated in five clinical trials to determine their efficacy in preventing recurrent UTIs in women. Three of five clinical trials had a successful primary outcome regarding cranberry consumption and the prevention of recurrent UTIs (Walker *et al.*, 1997; Kontiokari *et al.*, 2001; Stothers, 2002), while the other studies did not demonstrate an effect (Barbosa-Cesnik *et al.*, 2011; Stapleton *et al.*, 2012). Only three trials were randomized, double-blind, placebo controlled (Walker *et al.*, 1997; Barbosa-Cesnik *et al.*, 2011; Stapleton *et al.*, 2012), none of which were adequately powered.

All trials recruited healthy women, ages 18–72 years, with a history of at least one UTI within the previous year. The conditions for study entry differed among the trials. Walker *et al.* began intervention prior to antibiotic completion for the presenting UTI at time of enrolment while Barbosa-Cesnik *et al.* randomized women after presenting with a UTI, with no mention of the antibiotic treatment regimen. The remaining studies enrolled women who did not have a UTI at the time of enrolment. The enrolment conditions (after a treated UTI or not) did not appear to affect the outcome because both positive and negative results for cranberry prophylaxis was observed in both recruiting styles.

Cranberry treatments and dosing varied greatly among these studies. Walker *et al.* provided participants with 400 mg of encapsulated cranberry solids taken once per day for three months. Kontiokari *et al.* used cranberry–lingonberry juice made from concentrates, primarily containing cranberry: 7.5 g cranberry concentrate and 1.7 g lingonberry concentrate diluted in 50 ml water. This beverage was consumed once per day for six months. Stothers *et al.* had two cranberry treatment arms, administered as a juice or tablet. Participants in the juice arm consumed 240 ml of 'pure, unsweetened' cranberry juice three times per day and the tablet arm received a 1:30 parts concentrated cranberry juice tablet twice per day for 12 months. Both Barbosa-Cesnik *et al.* and Stapleton *et al.* used the same low calorie cranberry juice cocktail beverage (27% cranberry juice, sweetened with sucralose) in their studies, administered by the National Center for Complementary and Alternative Medicine (NCCAM). Participants in Barbosa-Cesnik *et al.* study consumed 2–240 ml cranberry beverages per day for six months, with a total average of 224 mg PACs/d (analysed by DMAC method). Stapleton *et al.* participants consumed the same juice beverage, assigned to either 120 ml/d (56 mg PACs) or 240 ml/d (112 mg PACs) for six months. These beverages are the only noted standardized beverages, based on PAC content (Barbosa-Cesnik *et al.*, 2011; Stapleton *et al.*, 2012).

The two trials that did not demonstrate an effect were both underpowered studies: Barbosa-Cesnick *et al.* and Stapleton *et al.* Barbosa-Cesnick had low recurrence rates for infection: 14% in the cranberry group and 20% in the placebo group, which are considerably lower than expected (Foxman *et al.*, 2000a), and would require an increased sample size for adequate statistical power (Eells *et al.*, 2011). Stapleton *et al.* had recruitment issues, therefore could not enrol the number of subjects initially anticipated to enter the trial. Although Stapleton *et al.* did not show a positive outcome for preventing recurrent UTIs, they did show that women who consumed cranberry juice had significantly less episodes of bacteriuria or UTI by p-fimbriated *E. coli* compared to the control group. This is a positive finding for cranberry since the focus of *in vitro* work has been on the anti-adhesion effects of cranberry using p-fimbriated *E. coli*. These findings demonstrate that cranberry may have a preventative effect on p-fimbriated *E. coli* and not type 1 fimbriated *E. coli*. This may explain why not all clinical trials have been successful only 27–42% of UTIs diagnosed as cystitis are caused by p-fimbriated *E. coli*. (Dowling *et al.*, 1987; Johnson *et al.*, 2005).

There is still a lack of randomized, double-blind, placebo controlled clinical trials that are statically powered to detect UTI recurrence differences between cranberry and placebo. Additionally, for clinical trial outcome comparison purposes, analytical composition of cranberry treatments should be published in order to determine efficacious doses.

Pregnant women

Asymptomatic bacteriuria (ASB) is of particular concern in pregnant women due to their association with pre-term delivery and low birth weight (Romero *et al.*, 1989; Sheiner *et al.*, 2009). ASB is defined as $> 10^5$ CFUs/ml of uropathogenic bacteria in the urine, without the traditional symptoms associated with UTIs. Untreated ASB in pregnancy may progress to a UTI, particularly pyelonephritis (Kass, 1960) which is associated with increased risk to both the neonate and pregnant woman (Farkash *et al.*, 2012).

Considering the substantial risks of developing ASB or UTI during pregnancy, there is increased interest to use cranberry to prevent ASB in pregnant women. The first study published to investigate the effect of cranberry on ASB/UTI in pregnant women did not find any statistical difference in ASB, UTI or neonatal outcomes among participants who were compliant with zero, one or two 8 oz (240 ml) servings of cranberry per day. Initially, all participants were asked to consume three 8 oz beverages per day (placebo, cranberry or combination of placebo and cranberry), but due to lack of compliance and a high withdraw rate of 39%, beverage treatments were reduced to two servings per day (Wing *et al.*, 2008). Although this study was underpowered, women compliant with two 8 oz servings of cranberry per day saw a 57% reduction in ASB and 41% reduction in UTIs, indicating that cranberry may be efficacious in preventing ASB and UTIs in pregnant women. Further studies would help solidify this area of importance for pregnant women and UTI prevention.

Children

UTIs are a common infection in children, especially those diagnosed with vesicoureteral reflux (VUR) (Foxman, 2002). VUR is a condition in which there is abnormal urine flow from the bladder to the upper urinary tract. In the trials discussed here, it should be noted that primary VUR was not an exclusion criteria, but $\leq$4% of enrolled subjects were diagnosed with this condition.

Paediatric trials that have included only relatively healthy children, including males and females, have demonstrated a benefit from cranberry consumption. The three trials described here used a variety of cranberry products: 7.5 g cranberry concentrate, plus 1.7 g of lingonberry concentrate diluted in 50 ml of water per day for six months (Ferrara *et al.*, 2009), commercially available cranberry juice containing 8.2 g of cranberry concentrate per 200 ml water administered at 5 ml /kg body weight per day for six months (Salo *et al.*, 2012) and cranberry juice containing 37% PACs (PAC analytical method not specified) administered at 2 ml/kg body weight for one year (Afshar *et al.*, 2012). Primary outcomes analysed demonstrated that cranberry treatment was efficacious in reducing UTI risk by 65% (Afshar *et al.*, 2012) and preventing UTI recurrences (Ferrara *et al.*, 2009). The primary outcome of reducing the number of children who experienced a recurrent UTI was not statistically significant in the Salo *et al.* trial; however, cranberry treatment did reduce the number of recurrent UTIs and the number of days on antibiotics.

Mechanism of action

The mechanism by which cranberries are efficacious for the prevention of UTIs has yielded many hypotheses since the 1920s. To date, researchers have not yet elucidated an exact *in vivo* mechanism of action, although there is substantial *in vitro* and *ex* vivo evidence indicating that cranberry and cranberry type A PACs stimulate an activity that results in inhibiting bacteria, particularly *E. coli*, from adhering to the mucosal lining of the urinary tract (Sobota, 1984; Zafriri *et al.*, 1989; Howell *et al.*, 1998; Gupta *et al.*, 2007). Cranberries contain a rich and diverse mixture of polyphenolic compounds, in addition to other potentially bioactive compounds. More research is needed to determine how these compounds contribute to the anti-adhesion effect, or modulate other systems, such as the immune system or GI tract, in order to help prevent UTIs.

Cranberry was originally thought to prevent UTIs by lowering urinary pH, providing a bacteriostatic effect in the urinary tract. It was hypothesized that urinary hippuric acid concentrations increased after cranberry juice consumption, thereby lowering the urinary pH enough to prevent infection. Research has since indicated that consuming cranberry juice does not reduce urinary pH (Bodel *et al.*, 1959; Avorn *et al.*, 1994), nor does it provide a bacteriostatic effect independent of pH (Tong *et al.*, 2006). Current research focuses on anti-adhesion, in addition to the reduction of uropathogenic bacteria strains in the GI tract and suppression or inhibition of biofilm formation.

Anti-adhesion

In order for bacteria to cause infection in the urinary tract, they must gain entry to uroepithelial cells by binding their fimbriae to certain cell receptors. There are two major *E. coli* fimbrial types implicated in UTIs, type 1 and p-fimbriae. Type 1 fimbriae primarily bind a cell receptor called uroplankin Ia, which is made up of glycoproteins with terminally exposed mannosides (Xie *et al.*, 2006). P-fimbriae specifically bind to uroepithelail cell glycolipids with α-D-Gal(1$\rightarrow$4)-β-D-Gal terminal end (Kallenius *et al.*, 1981).

The anti-adhesion activity of cranberry was first recognized by Sobota *et al.* (1984). A series of experiments using cranberry juice and uropathogenic strains of *E. coli* demonstrated that cranberry juice contains a compound or compounds that inhibit bacterial adherence to uroepitheial cells. A similar effect was found in both murine and human urine after cranberry consumption, indicating that the active compound or compounds are present in the urine after metabolism.

Further research suggested that there were two different cranberry compounds responsible for the anti-adhesion effect, fructose and a non-dialyzable fraction that contained high molecular weight compounds, later identified as proanthocyanins (Zafriri *et al.*, 1989; Howell *et al.*, 1998). Binding studies demonstrate that fructose has a high affinity to the type 1 FimH adhesin, although 15-fold less than its natural mannose ligand (Bouckaert *et al.*, 2005). The metabolic pathway of fructose can potentially lead to its urinary excretion (Kawasaki *et al.*, 2002; Tasevska *et al.*, 2005), which could result in an anti-adhesion effect, although this has not been studied in detail in human populations.

Research has focused on the PAC fraction of cranberry as type A PACs have been identified as the key compounds responsible for anti-adhesion (Foo *et al.*, 2000a, 2000b). The hypothesis that type A PACs are responsible for anti-adhesion activity by preventing fimbrial adhesion to urepithelial cells is widely accepted by researchers, although it is not clear if this mechanism is due to PACs binding to fimbriae, thereby preventing their adhesion, or compressing fimbriae in a manner such that they cannot bind cells (Liu *et al.*, 2008), *etc*.

However, questions remain regarding the *in vivo* anti-adhesion activity due to lack of both bioavailability and published *ex vivo* anti-adhesion data from clinical trials. Recently, type A PACs have been shown to be bioavailable in small concentrations (Zampariello *et al.*, 2012). Maximal urinary concentrations of A2 PACs averaged 24.4 ng/mg creatinine at 11 hours post cranberry consumption. This finding was helpful in order to establish the missing link between the presence of type A PACs in urine and anti-adhesion, but more research is needed. *In vitro*, milligram concentrations of type A PACs are needed in order to promote anti-adhesion activity (Foo *et al.*, 2000b). The discrepancy between the small concentrations of bioavailable PACs (nanogram) and the large concentrations needed to prevent bacterial adhesion *in vitro* (milligram) indicates that other polyphenols, in addition to PACs, may be contributing to the observed anti-adhesion effect.

Biofilms

Biofilms are sessile communities of bacteria that can bind to organic and inorganic matter and each other, form an extracellular matrix for functional and structural integrity, exhibit an altered phenotype regarding growth rate and gene transcription, and are associated with long-term persistence (Dolan *et al.*, 2010). Biofilms are typically associated with catheter-associated UTIs, but recently biofilms have been implicated in the etiology of recurrent UTIs (Anderson *et al.*, 2004).

The risk of a recurrent UTI is 24% within six months of a first time UTI, even with successful clearance of infection with antibiotic use (Foxman *et al.*, 2000b). These recurring infections have been shown to be caused by the same uropathogenic strain of bacteria from the initial infection (Ikaheimo *et al.*, 1996; Stapleton and Stamm, 1997). One way bacteria may be able to persist in the bladder long enough to cause a subsequent UTI is by colonizing the bladder epithelium cells in order to remain undetected by the immune system. These biofilms, or intracellular bacterial colonies (IBCs), have demonstrated their ability to evade the immune system by remaining dormant, or 'quiescent', within the uroepithelial cells (Mulvey, 2002). They can remain quiescent for periods of up to several weeks without detection from the immune system and remain unaffected by antibiotic treatment even though urinary bacterial loads are reduced (Hooton and Stamm, 1997; Hvidberg *et al.*, 2000).

The role of cranberry bioactives in biofilm research has mostly been direct applications, such as oral care or *in vitro* studies. Cranberry juice (27% cocktail) and extracts containing flavonols, anthocyanins and PACs have demonstrated the ability to inhibit biofilm formation in gram negative and gram positive bacteria (Pinzon-Arango *et al.*, 2011; Laplante *et al.*,

2012), prevent bacterial adhesion to plastic material surfaces (Eydelant and Tufenkji, 2008), and interfere with biofilm-producing mechanisms such as acidogenicity and glucan-mediated processes (Duarte *et al.*, 2006; Koo *et al.*, 2006). *Ex vivo* biofilm studies have demonstrated that urine from cranberry juice consumers can prevent biofilm formation in uropathogenic *E. coli* (Di Martino *et al.*, 2005; Camesano *et al.*, 2011), implying that active metabolites of cranberry are present in the urine post consumption. These studies indicate that cranberries may have a role in *in vivo* biofilm inhibition and can be used in direct applications such as oral health or used as a coating on inert surfaces to inhibit bacterial adhesion and subsequent biofilm formation.

Cardiovascular health

In a comparative analyses of commonly consumed polyphenol-rich beverages in the United States, blueberry juice and cranberry juice were listed among the top ten beverages of high antioxidant potency measured by the sum of four assays [1,1-Diphenyl-2-picryl hydrazyl radical (DPPH), oxygen radical absorbance capacity (ORAC), Trolox equivalent antioxidant capacity(TEAC) and Ferric reducing ability of plasma (FRAP)] (Seeram *et al.*, 2008). These antioxidant measurements, although indicative of the antioxidant potential of the beverages, are no longer thought to be the primary mechanism of action for polyphenols. Current evidence suggests that many of the biological effects of flavonoids are related to their ability to modulate cell-signalling pathways and decrease oxidative stress through inhibiting endogenous enzymes that generate free radicals or through activating antioxidant enzymes like superoxide dismutase that can reduce free radicals (Stevenson and Hurst, 2007; Basu and Lyons, 2012). Flavonoids, including those in cranberries and blueberries, have been shown to exhibit anti-atherogenic, apoptotic and anti-inflammatory activities resulting in increased HDL-C, lower blood pressure, decreased LDL oxidation and improved endothelial function in *in vitro*, animal and human studies (Neto, 2007; McCullough *et al.*, 2012).

Human intervention studies using chokeberries, cranberries, blueberries and strawberries with high polyphenol content showed significant improvements in LDL oxidation, lipid peroxidation, total plasma antioxidant capacity, dyslipidemia and glucose metabolism, supporting berries as an essential fruit group in a cardiovascular or heart-healthy diet (Basu *et al.*, 2010). As reviewed by McKay and Blumberg, researchers have been studying the effects of cranberry on heart health for the last decade (McKay and Blumberg, 2007).

In particular, cranberry consumption has been shown in *in vitro* and human clinical studies to significantly reduce LDL oxidation (Wilson *et al.*, 1998; Ruel *et al.*, 2005, 2008). Oxidation of LDL particles in the arterial wall is considered a key event in atherosclerosis. Some mechanisms thought to prevent LDL oxidation include antioxidative mechanisms (scavenging of free radicals, chelation of metals, sparing of vitamin E and carotenoids in the LDL particle); down-regulation of oxidation pathways in macrophage, endothelial cells and smooth muscle cells; inhibition of oxidative enzymes; and changes in LDL oxidation scavenging and elimination (Aviram and Fuhrman, 1998; Reed and Howell, 2009). In a randomized, double-blind, placebo controlled trial in women with metabolic syndrome, consumption of two servings of a low calorie cranberry juice cocktail per day containing 460 mg of flavonoids resulted in improved LDL oxidation (Basu *et al.*, 2011). Cranberry juice also significantly increased plasma antioxidant capacity (1.5 +/− 0.6 to 2.2 +/− 0.4 μmol; $P < 0.05$) and decreased malondialdehyde (3.4 +/− 1.1 to 1.7 +/− 0.7 μmol/l, $P < 0.05$) at eight weeks versus placebo.

Another benefit of a polyphenol-rich diet is improvement of vascular health. In a study looking at flavonoid intake in a cross-sectional analysis of 1898 women in the TwinsUK registry, the authors reported that higher anthocyanin intake was inversely associated with arterial stiffness. Arterial stiffness is increasingly becoming a recognized predictor of future cardiovascular events (Collaboration, 2010) and an intake difference of approximately 44 mg/day can result in that change (Jennings *et al.*, 2012). An emerging body of evidence also supports the same benefit with cranberries. Subjects with stable coronary artery disease consuming a serving of a low calorie cranberry beverage (480 ml/day) containing 835 mg of total polyphenols and 94 mg anthocyanins, had significant reduction in arterial stiffness compared to subjects on placebo (Dohadwala *et al.*, 2011). A pilot study conducted earlier by the same authors showed that an acute reduction in flow mediated dilation two and four hours after consumption was also shown with the same dose of the beverage (Dohadwala *et al.*, 2011) . This is consistent with other findings to support the role of polyphenols for vascular health (Goyarzu *et al.*, 2004; Scalbert *et al.*, 2005a) and evidence for the contribution of cranberry polyphenols (Caton *et al.*, 2010; Dohadwala *et al.*, 2011; Flammer *et al.*, 2013). Flavonoids are also linked to reducing hypertension in adults by increasing endothelial derived nitric oxide (NO), free radical induced conversion of NO by NADPH oxidase and modulation of e-NOS activity and expression (Cassidy *et al.*, 2011). Cranberry procyanidins have been shown to reduce expression of enothelin-1, a vasoconstrictor, in cultured endothelial cells (Caton *et al.*, 2010). In support of mechanistic studies, intervention trials with cranberries showed that consumption of a low calorie cranberry beverage may reduce blood pressure (Caton *et al.*, 2010).

Although limited and heterogenous, evidence that cranberry consumption may improve HDL cholesterol is emerging. HDL cholesterol was increased in several intervention studies after consumption of one to two servings of a low calorie cranberry beverage (Ruel *et al.*, 2006; Dohadwala *et al.*, 2011) and unchanged or not measured in others (Ruel *et al.*, 2005; Duthie *et al.*, 2006; Lee *et al.*, 2008; Basu *et al.*, 2011). There may be a correlation between duration of study and positive effect – the positive studies were generally conducted for 12 weeks or longer and the unchanged studies were generally shorter. In a study where the HDL cholesterol carrier protein, ApoA-1, was measured, the level was reported to be increased (Ruel *et al.*, 2006). Emerging science based on interventional studies have also suggested that berries and berry products may improve insulin resistance and glycaemic response, as well as reduce risk factors associated with metabolic syndrome (Basu and Lyons, 2012). In a crossover study with 12 subjects with type 2 diabetes, subjects on a low calorie cranberry juice beverage containing 27% juice and had a significantly lower glycaemic response than those on a conventionally sweetened beverage (Wilson *et al.*, 2008). This study suggests that a low calorie cranberry juice provides polyphenols important for managing glycaemic response. *In vitro* and mechanistic studies have also been conducted with cranberry flavonoids showing they have anti-inflammatory and antiglycation activities (Liu *et al.*, 2011).

Oral and gastrointestinal health

The anti-adhesion properties of cranberries, established through urinary tract health research, have generated new hypotheses for research on other conditions linked to bacterial adhesion such as oral and gastrointestinal diseases.

Interest in using cranberry for oral health applications originated after dental research identified the role of biofilm formation in dental caries and other oral diseases. Cranberry bioactives studied in the oral health field have mostly involved non-dialyzable material (NDM) and type A PACs, but flavonols, anthocyanins and phenolic fractions have also been studied for the prevention of caries and periodontal diseases, all of which involve biofilm formation in disease pathogenesis (Bodet *et al.*, 2008).

Dental biofilms are formed from bacterial adherence to surfaces of the oral cavity. The predominant bacteria found in biofilms associated with cariogenesis are the serotype mutans streptococci. The first step in this process involves bacterial adhesion to the surface of the tooth. Studies have demonstrated that cranberry bioactives can prevent bacterial adhesion to the oral cavity by inhibiting a sucrose-dependent mechanism in which bacteria synthesize fructans and glucans via glucosyltransferase and fructosyltransferase to provide themselves with binding sites required for hard-surface adhesion to teeth (Steinberg *et al.*, 2004; Duarte *et al.*, 2006). Additionally, cranberry can inhibit hydrophobic interactions involved with adhesion, independent of the sucrose-mediated pathway (Yamanaka *et al.*, 2004; Steinberg *et al.*, 2005). Oral biofilm bacteria create an acidic environment as a result of their metabolic activity, leading to promotion of tooth decay. In order to maintain a more alkaline cellular pH, protons are transported from the bacteria to the extracellular matrix by the F-ATPase proton pump (Sturr and Marquis, 1992). Cranberry PAC and flavonol fractions inhibited F-ATPase activity, resulting in a higher pH (4.7–4.9); however, it should be noted that this range is below the critical enamel dissolution pH (5.0–5.5) (Duarte *et al.*, 2006). Individual flavonols and proanthocyanidins have also demonstrated inhibition of F-ATPases, but enhanced inhibition was observed when these compounds were combined, suggesting a synergistic effect (Gregoire *et al.*, 2007). A preliminary trial with healthy volunteers demonstrated that a cranberry mouthwash made with NDM reduced the total oral bacterial count compared to the control mouthwash (Weiss *et al.*, 2004), indicating that cranberry polyphenols may aid in preventing dental caries.

Periodontal disease can affect the alveolar bone, periodontal ligament, cementum or gingiva. Initially, biofilm formation is necessary for disease pathogenesis, which includes the production of proteolytic enzymes that destroy host tissue and promotes inflammation. The host cytokine response to bacterial lipopolysaccharide (LPS) produced by immune cells is a major contributor to disease progression. Cranberry polyphenol fractions inhibited biofilm formation and proteolytic enzymes from periodontopathogens (Bodet *et al.*, 2007b; Yamanaka *et al.*, 2007).The NDM fraction also demonstrated anti-inflammatory activity by reducing pro-inflammatory mediators such as IL-1, IL-6, IL-8, TNF-α and prostaglandin E2 in immune cells (Bodet *et al.*, 2006). These activities were later found to be modulated by the transcription factor activator protein-1 (AP-1) in gingival fibroblasts (Bodet *et al.*, 2007a). Type A PACs have also demonstrated similar activity by inhibiting biofilm formation and reducing cytokine production through the NF-κB p65 pathway (La *et al.*, 2010). These studies demonstrate that cranberry compounds can inhibit different transcription pathways that are involved in oral disease pathogenesis.

Another area of emerging interest is the effect of cranberries on controlling *H.pylori* prevalence in the stomach. A multicentric, randomized, controlled, double-blind trial was carried out in 295 asymptomatic children (6–16 y of age) who tested positive for *H. pylori* by 13C-urea breath test (UBT). The study showed that daily consumption of 80 ml of cranberry juice for three weeks resulted in a significantly higher eradication rate compared to placebo (Gotteland *et al.*, 2008). Two other trials with adults also showed that daily consumption of one to two servings of cranberry beverages compared to placebo were shown to reduce the prevalence of *H. pylori* or reduced the use of antibiotics (Zhang *et al.*, 2005; Shmuely *et al.*, 2007). In mechanistic studies, cranberry proanthocyanidins and other polyphenols were able to inhibit adhesion of the pathogen to gastric mucus, provide bacteriostatic compounds and anti-inflammatory activities to control the infection (Burger *et al.*, 2002; Chatterjee *et al.*, 2004; Maatta-Riihinen *et al.*, 2005). A recent mechanistic study suggested that cranberry polyphenols induced a morphological change, inducing *H. pylori* to develop a coccoid form, and inhibiting its growth bacteriostatically, while polyphenol-poor juices had no effect (Matsushima *et al.*, 2008). *H.pylori* infection is highly prevalent in developing countries, sometimes as high as 80%. Gastric colonization by *H. pylori* results in a chronic gastritis which will remain asymptomatic in most subjects, but it is an etiologic factor for peptic ulcer and a risk factor for gastric cancer. Current standard treatment for eradication includes triple antibiotic therapy and a proton pump inhibitor, which could potentially lead to increased antibiotic resistance.

Cranberry bioactives have anti-adhesion, antibiofilm and anti-inflammatory properties related to oral and gastrointestinal diseases. There is also active ongoing research on the role of cranberry phenolic acids and metabolites of cranberry polyphenols on gut health, but this outside the scope of this discussion (Puupponen-Pimia *et al.*, 2005). As the oral cavity and the stomach have direct contact with cranberry compounds, including Type A proanthocyanidins shown to inhibit adhesion of certain pathogenic bacteria, and no circulatory uptake is required, cranberry makes an ideal alternative prophylactic for maintaining health.

CONCLUSION

Cranberries have been long known for their urinary tract health promoting properties. Research from the past 50 years has given us much insight to how cranberry may promote urinary tract health. Type A PACs have been identified as compounds that are primarily responsible for anti-adhesion activity *in vitro*, but more research on bioavailability and clinical intervention is needed to confirm the mechanism of action. Emerging hypotheses suggest, in unison with anti-adhesion, other mechanisms such as biofilm formation, gastrointestinal health and immunity may play a role in UTI prevention. Clinical trials from the last 20 years have identified populations for which cranberry seems to be most efficacious for preventing UTIs, such as women with recurrent urinary tract infections and children. More clinical trials are needed to determine optimal efficacious doses of cranberry juice and extracts. An emerging body of evidence also supports cranberry consumption for cardiovascular health, consistent with the growing literature on flavonoids, due to the rich content of flavonoids in cranberry. Due to their low bioavailability, the general scientific opinion is that flavonoids are acting as signal modulators that require lower levels than needed for free radical scavenging activities. Moreover, flavonoids are metabolized by the gut flora to metabolites that are likely biologically active. This remains an active area of investigation. There is little doubt that consuming fruits and vegetables contributes to health. As the wealth of evidence suggests, consuming a diet rich in polyphenols provides a wide range of health benefits linked to these bioactive compounds.

REFERENCES

Afshar, K., Stothers, L., Scott, H. and Macneily, A.E. (2012) Cranberry juice for the prevention of pediatric urinary tract infection: a randomized controlled trial. *Journal of Urology*, **188**, 1584–1587.

Aherne, S.A. and O'Brien, N.M. (2002) Dietary flavonols: chemistry, food content, and metabolism. *Nutrition*, **18**, 75–81.

Anderson, G.G., Dodson, K.W., Hooton, T.M. and Hultgren, S.J. (2004) Intracellular bacterial communities of uropathogenic *Escherichia coli* in urinary tract pathogenesis. *Trends in Microbiology,* **12**, 424–430.

Aviram, M. and Fuhrman, B. (1998) Polyphenolic flavonoids inhibit macrophage-mediated oxidation of LDL and attenuate atherogenesis. *Atherosclerosis*, **137**, S45–S50.

Avorn, J., Monane, M., Gurwitz, J.H. *et al.* (1994) Reduction of bacteriuria and pyuria after ingestion of cranberry juice. *Journal of the American Medical Association,* **271**, 751–754.

Barbosa-Cesnik, C., Brown, M.B., Buxton, M. *et al.* (2011) Cranberry juice fails to prevent recurrent urinary tract infection: results from a randomized placebo-controlled trial. *Clinical Infectious Diseases,* **52**, 23–30.

Basu, A. and Lyons, T.J. (2012) Strawberries, blueberries, and cranberries in the metabolic syndrome: clinical perspectives. *Journal of Agricultural and Food Chemistry*, **60**(23), 5687–5692.

Basu, A., Rhone, M. and Lyons, T.J. (2010) Berries: emerging impact on cardiovascular health. *Nutrition Reviews*, **68**, 168–177.

Basu, A., Betts, N.M., Ortiz, J. *et al.* (2011) Low-energy cranberry juice decreases lipid oxidation and increases plasma antioxidant capacity in women with metabolic syndrome. *Nutrition Research*, **31**, 190–196.

Bhagwat, S.A., Haytowitz, D.B. and Holden, J.M. (2011) USDA Database for the Flavonoid Content of Selected Foods, Release 3.0, 2011. US Department of Agriculture, Agricultural Research Service. Nutrient Data Laboratory Home Page: http://www.ars.usda.gov/Services/docs.htm?docid=6231 (last accessed 7 March 2013).

Bodel, P.T., Cotran, R. and Kass, E.H. (1959) Cranberry juice and the antibacterial action of hippuric acid. *Journal of Laboratory and Clinical Medicine*, **56**, 881–888.

Bodet, C., Chandad, F. and Grenier, D. (2006) Anti-inflammatory activity of a high-molecular-weight cranberry fraction on macrophages stimulated by lipopolysaccharides from periodontopathogens. *Journal of Dental Research*, **85**, 235–239.

Bodet, C., Chandad, F. and Grenier, D. (2007a) Cranberry components inhibit interleukin-6, interleukin-9 and prostaglandin E2 production by lipopolysaccharide-activated gingival fibroblasts. *European Journal of Oral Science*, **115**, 64–70.

Bodet, C., Chandad, F. and Grenier, D. (2007b) Inhibition of host extracellular matrix destructive enzyme production and activity by a high-molecular-weight cranberry fraction. *Journal of Periodontal Research*, **42**, 159–168.

Bodet, C., Grenier, D., Chandad, F. *et al.* (2008) Potential oral health benefits of cranberry. *Critical Reviews in Food Science and Nutrition,* **48**, 672–680.

Bouckaert, J., Berglund, J., Schembri, M. *et al.* (2005) Receptor binding studies disclose a novel class of high-affinity inhibitors of the *Escherchia coli* FimH adhesion. *Molecular Microbiology,* **55**, 441–455.

Burger, O., Weiss, E., Sharon, N. *et al.* (2002) Inhibition of *Helicobacter pylori* adhesion to human gastric mucus by a high-molecular-weight constituent of cranberry juice. *Critical Reviews in Food Science and Nutrition,* **42**, 279–284.

Camesano, T.A., Zhang, Y., Tao, Y. *et al.* (2011) Prevention of biofilm formation and bacterial adhesion after consumption of cranberry juice cocktail. 242nd National Meeting of the American Chemical Society (ACS), Denver, CO.

Cassidy, A., O'Reilly, E.J., Kay, C. *et al.* (2011) Habitual intake of flavonoid subclasses and incident hypertension in adults. *American Journal of Clinical Nutrition,* **93**, 338–347.

Caton, P.W., Pothecary, M.R., Lees, D.M. *et al.* (2010) Regulation of vascular endothelial function by procyanidin-rich foods and beverages (dagger). *Journal of Agricultural and Food Chemistry*, **58**, 4008–4013.

Chatterjee, A., Yasmin, T., Bagchi, D. and Stohs, S.J. (2004) Inhibition of *Helicobacter pylori* in vitro by various berry extracts, with enhanced susceptibility to clarithromycin. *Molecular and Cellular Biochemistry,* **265**, 19–26.

Chun, O.K., Chung, S.J. and Song, W.O. (2007) Estimated dietary flavonoid intake and major food sources of U.S. adults. *Journal of Nutrition,* **137**, 1244–1252.

Chun, O.K., Floegel, A., Chung, S.-J. *et al.* (2010) Estimation of antioxidant intakes from diet and supplements in U.S. adults. *Journal of Nutrition,* **140**, 317–324.

Collaboration, A.S. (2010) Determinants of pulse wave velocity in healthy people and in the presence of cardiovascular risk factors: 'establishing normal and reference values'. *European Heart Journal*, **31**, 2338–2350.

Cote, J., Caillet, S., Doyon, G. *et al.* (2010) Bioactive compounds in cranberries and their biological properties. *Critical Reviews in Food Science and Nutrition,* **50**, 666–679.

Cunningham, D.G., Vannozzi, S.A., Turk, R., *et al.* (2004) Cranberry phytochemicals and their health benefits. *Nutraceutical Beverages: chemistry, nutrition and health effects.* American Chemical Society, Washington, DC.

Del Rio, D., Costa, L.G., Lean, M.E. and Crozier, A. (2010) Polyphenols and health: what compounds are involved? *Nutrition, Metabolism and Cardiovascular Diseases,* **20**, 1–6.

Di Martino, P., Agniel, R., Gaillard, J. and Denys, L.P. (2005) Effects of cranberry juice on uropathogenic *Escherichia coli in vitro* biofilm formation. *Journal of Chemotherapy,* **1**, 63–65.

Dohadwala, M.M., Holbrook, M., Hamburg, N.M. *et al.* (2011) Effects of cranberry juice consumption on vascular function in patients with coronary artery disease. *American Journal of Clinical Nutrition,* **93**, 934–940.

Dolan, L.C., Potter, S.M. and Burdock, G.A. (2010) Evidence-based review on the effect of normal dietary consumption of fructose on blood lipids and body weight of overweight and obese individuals. *Critlcal Reviews in Food Science and Nutrition,* **50**, 889–918.

Domingue, G.J., Roberts, J.A., Laucirica, R. *et al.* (1985) Pathogenic significance of P-fimbriated *Escherichia coli* in urinary tract infections. *Journal of Urology,* **133**, 983–989.

Dowling, K.J., Roberts, J.A. and Kaack, M.B. (1987) P-fimbriated *Escherichia coli* urinary tract infection: a clinical correlation. *Southern Medical Journal,* **80**, 1533–1536.

Duarte, S., Gregoire, S., Singh, A.P. *et al.* (2006) Inhibitory effects of cranberry polyphenols on formation and acidogenicity of *Streptococcus mutans* biofilms. *FEMS Microbiology Letters,* **257**, 50–56.

Duthie, S.J., Jenkinson, A.M., Crozier, A. *et al.* (2006) The effects of cranberry juice consumption on antioxidant status and biomarkers relating to heart disease and cancer in healthy human volunteers. *European Journal of Nutrition,* **45**, 113–122.

Eells, S.J., McKinnell, J.A. and Miller, L.G. (2011) Daily cranberry prophylaxis to prevent recurrent urinary tract infections may be beneficial in some populations of women. *Clinical Infectious Disesaes,* **52**, 1393–1394; author reply 1394–1395.

Erdman, J.W., Balentine, D., Arab, L. *et al.* (2007) Flavonoids and Heart Health: Proceedings of the ILSI North America Flavonoids Workshop, 31 May–1 June, Washington, DC. *Journal of Nutrition,* **137**, S718–S737.

Eydelant, I.A. and Tufenkji, N. (2008) Cranberry derived proanthocyanidins reduce bacterial adhesion to selected biomaterials. *Langmuir,* **24**, 10273–10281.

Farkash, E., Weintraub, A.Y., Sergienko, R. *et al.* (2012) Acute antepartum pyelonephritis in pregnancy: a critical analysis of risk factors and outcomes. *European Journal of Obstetrics and Gynecology and Reproductive Biology,* **162**, 24–27.

Ferrara, P., Romaniello, L., Vitelli, O. *et al.* (2009) Cranberry juice for the prevention of recurrent urinary tract infections: A randomized controlled trial in children. *Scandinavian Journal of Urology and Nephrology,* **43**, 369–372

Flammer, A.J., Martin, E.A., Gossl, M. *et al.* (2013) Polyphenol-rich cranberry juice has a neutral effect on endothelial function but decreases the fraction of osteocalcin-expressing endothelial progenitor cells. *European Journal of Nutrition,* **52**(1), 289–296.

Foo, L.Y., Lu, Y., Howell, A.B. and Vorsa, N. (2000a) The structure of cranberry proanthocyanidins which inhibit adherence of uropathogenic P-fimbriated *Escherichia coli* in vitro. *Phytochemistry,* **54**, 173–181.

Foo, L.Y., Lu, Y., Howell, A.B. and Vorsa, N. (2000b) A-Type proanthocyanidin trimers from cranberry that inhibit adherence of uropathogenic P-Fimbriated *escherichia coli. Journal of Natural Products,* **63**, 1225–1228.

Foxman, B. (1990) Recurring urinary tract infection: incidence and risk factors. *American Journal of Public Health,* **80**, 331–333.

Foxman, B. (2002) Epidemiology of urinary tract infections: incidence, morbidity, and economic costs. *American Journal of Medicine,* **113** (Suppl 1A), 5S–13S.

Foxman, B. and Brown, P. (2003) Epidemiology of urinary tract infections. Transmission and risk factors, incidence and costs. *Infectious Disease Clinics of North America,* **17**, 227–241.

Foxman, B., Barlow, R., D'Arcy, H. *et al.* (2000a) Urinary tract infection: self-reported incidence and associated costs. *Annals of Epidemiology,* **10**, 509–515.

Foxman, B., Gillespie, B., Koopman, J. *et al.* (2000b) Risk factors for second urinary tract infection among college women. *American Journal of Epidemiology,* **151**, 1194–1205.

Gotteland, M., Andrews, M., Toledo, M. *et al.* (2008) Modulation of *Helicobacter pylori* colonization with cranberry juice and *Lactobacillus johnsonii* La1 in chidren. *Nutrition,* **24**, 421–426.

Goyarzu, P., Malin, D., Lau, F.C. *et al.* (2004) Blueberry supplemented diet: effects on object recognition memory and nuclear factor-kappa B levels in aged rats. *Nutritional Neuroscience,* **7**, 75–83.

Grace, M.H., Massey, A.R., Mbeunkui, F. *et al.* (2012) Comparison of health-relevant flavonoids in commonly consumed cranberry products. *Journal of Food Science*, **77**, H176–183.

Graf, B.A., Milbury, P.E. and Blumberg, J.B. (2005) Flavonols, flavones, flavanones, and human health: epidemiological evidence. *Journal of Medicinal Food*, **8**, 281–290.

Gregoire, S., Singh, A.P., Vorsa, N. and Koo, H. (2007) Influence of cranberry phenolics on glucan synthesis by glucosyltransferases and streptococcus mutans acidogenicity. *Journal of Applied Microbiology*, **103**, 1960–1968.

Griebling, T.L. (2005) Urologic diseases in america project: trends in resource use for urinary tract infections in women. *Journal of Urology*, **173**, 1281–1287.

Gruneberg, R.N. (1969) Relationship of infecting urinary organism to the faecal flora in patients with symptomatic urinary infection. *Lancet*, **2**, 766–768.

Gupta, K., Chou, M.Y., Howell, A. *et al.* (2007) Cranberry products inhibit adherence of p-fimbriated escherichia coli to primary cultured bladder and vaginal epithelial cells. *Journal of Urology,* **177**, 2357–2360.

Henig, Y.S. and Leahy, M.M. (2000) Cranberry juice and urinary-tract health: Science supports folklore. *Nutrition*, **16**, 684–687.

Hooton, T.M. (2012) Clinical practice. Uncomplicated urinary tract infection. *New England Journal of Medicine*, **366**, 1028–1037.

Hooton, T.M. and Stamm, W.E. (1997) Diagnosis and treatment of uncomplicated urinary tract infection. *Infectious Disease Clinics of North America*, **11**, 551–581.

Howell, A.B. (2002) Cranberry proanthocyanidins and the maintenance of urinary tract health. *Critical Reviews in Food Science and Nutrition*, **42**, 273–278.

Howell, A.B. (2007) Bioactive compounds in cranberries and their role in prevention of urinary tract infections. *Molecular Nutrition and Food Research*, **51**, 732–737.

Howell, A.B., Vorsa, N., Marderosian, A.D. and Foo, L.Y. (1998) Inhibition of the adherence of p-fimbriated *Escherichia coli* to uroepithelial-cell surfaces by proanthocyanidin extracts from cranberries. *New England Journal of Medicine,* **339**, 1085–1086.

Hvidberg, H., Struve, C., Krogfelt, K.A. *et al.* (2000) Development of a long-term ascending urinary tract infection mouse model for antibiotic treatment studies. *Antimicrobial Agents and Chemotherapy,* **44**, 156–163.

Ikaheimo, R., Siitonen, A., Heiskanen, T. *et al.* (1996) Recurrence of urinary tract infection in a primary care setting: analysis of a 1-year follow-up of 179 women. *Clinical Infectious Diseases,* **22**, 91–99.

Jennings, A., Welch, A.A., Fairweather-Tait, S.J. *et al.* (2012) Higher anthocyanin intake is associated with lower arterial stiffness and central blood pressure in women. *American Journal of Clinical Nutrition*, **96**, 781-788.

Jepson, R.G. and Craig, J.C. (2008) Cranberries for preventing urinary tract infections. *Cochrane Database of Systematic Reviews* 1 (Art. No.: CD001321). doi: 10.1002/14651858.CD001321.pub4.

Johnson, J.R., Murray, A.C., Kuskowski, M.A. *et al.* (2005) Distribution and characteristics of *Escherichia coli* Clonal Group A. *Emerging Infectious Diseases,* **11**, 141–145.

Kallenius, G., Svenson, S., Mollby, R. *et al.* (1981) Structure of carbohydrate part of receptor on human uroepithelial cells for pyelonephritogenic *Escherichia coli. Lancet,* **2**, 604–606.

Kass, E.H. (1960) Bacteriuria and pyelonephritis of pregnancy. *Archives of Internal Medicine*, **105**, 194–198.

Kawasaki, T., Akanuma, H. and Yamanouchi, T. (2002) Increased fructose concentrations in blood and urine in patients with diabetes. *Diabetes Care*, **25**, 353–357.

Knaze, V., Zamora-Ros, R., Lujan-Barroso, L. *et al.* (2012) Intake estimation of total and individual flavan-3-ols, proanthocyanidins and theaflavins, their food sources and determinants in the European Prospective Investigation into Cancer and Nutrition (EPIC) study. *British Journal of Nutrition*, **108**, 1095-1108.

Kondo, M., Lamoureaux, T.L., Neto, C.C. and Sweeney-Nixon, M. (2003) Flavonoid and proanthocyanidin-rich extracts from cranberry fruits: neuroprotective effects, antitumor activity and antifungal activity *in vitro*. 226th American Chemical Society National Meeting, New York, NY.

Kontiokari, T., Sundqvist, K., Nuutinen, M. *et al.* (2001) Randomised trial of cranberry-lingonberry juice and *Lacrobacillus* GG drink for the prevention of urinary tract infections in women. *British Medical Journal,* **322**, 1571–1575.

Koo, H., Nino de Guzman, P., Schobel, B.D. *et al.* (2006) Influence of cranberry juice on glucan-mediated processes involved in *Streptococcus mutans* biofilm development. *Caries Research,* **40**, 20–27.

La, V.D., Howell, A.B. and Grenier, D. (2010) Anti-Porphyromonas gingivalis and Anti-inflammatory Activities of A-Type Cranberry Proanthocyanidins. *Antimicrobial Agents and Chemotherapy*, **54**, 1778–1784.

Laplante, K.L., Sarkisian, S.A., Woodmansee, S. *et al.* (2012) Effects of cranberry extracts on growth and biofilm production of *Escherichia coli* and *Staphylococcus* species. *Phytotherapy Research,* **26**, 1371-1374.

Lee, I.T., Chan, Y.C., Lin, C.W. *et al.* (2008) Effect of cranberry extracts on lipid profiles in subjects with type 2 diabetes. *Diabetic Medicine,* **25**, 1473–1477.

Liu, H., Liu, H., Wang, W. *et al.* (2011) Cranberry phytochemicals inhibit glycation of human hemoglobin and serum albumin by scavenging reactive carbonyls. *Food and Function,* **2**, 475–482.

Liu, Y., Gallardo-Moreno, A.M., Pinzon-Arango, P.A. *et al.* (2008) Cranberry changes the physicochemical surface properties of *E.coli* and adhesion with uroepithelial cells. *Colloids and Surfaces B: Biointerfaces,* **65**, 35–42.

Maatta-Riihinen, K.R., Kahkonen, M.P., Torronen, A.R. and Heinonen, I.M. (2005) Catechins and procyanidins in berries of vaccinium species and their antioxidant activity. *Journal of Agricultural and Food Chemistry*, **53**, 8485–8491.

Manach, C., Scalbert, A., Morand, C. *et al.* (2004) Polyphenols: food sources and bioavailability. *American Journal of Clinical Nutrition,* **79**, 727–747.

Matsushima, M., Suzuki, T., Masui, A. *et al.* (2008) Growth inhibitory action of cranberry on *Helicobacter pylori*. *Journal of Gastroenterology and Hepatology,* **23**, S175–S180.

McCullough, M.L., Peterson, J.J., Patel, R. *et al.* (2012) Flavonoid intake and cardiovascular disease mortality in a prospective cohort of US adults. *American Journal of Clinical Nutrition,* **95**, 454–464.

McKay, D.L. and Blumberg, J.B. (2007) Cranberries (*Vaccinium macrocarpon*) and cardiovascular disease risk factors. *Nutrition Reviews,* **65**, 490–502.

Mennen, L.I., Sapinho, D., Ito, H. *et al.* (2006) Urinary flavonoids and phenolic acids as biomarkers of intake for polyphenol-rich foods. *British Journal of Nutrition*, **96**, 191–198.

Milbury, P.E., Vita, J.A. and Blumberg, J.B. (2010) Anthocyanins are Bioavailable in Humans following an Acute Dose of Cranberry Juice. *Journal of Nutrition*, **140**, 1099–1104.

Moreno, E., Andreu, A., Perez, T. *et al.* (2006) Relationship between *Escherichia coli* strains causing urinary tract infection in women and the dominant faecal flora of the same hosts. *Epidemiology and Infection,* **134**, 1015–1023.

Mullen, W., Marks, S.C. and Crozier, A. (2007) Evaluation of phenolic compounds in commercial fruit juices and fruit drinks. *Journal of Agricultural and Food Chemistry*, **55**, 3148–3157.

Mulvey, M.A. (2002) Adhesion and entry of uropathogenic *Escherichia coli*. *Cellular Microbiology,* **4**, 257–271.

NASS (2011) Agricultural Statistics. USDA National Agricultural Statistics Service (http://www.nass.usda.gov/Publications/Ag_Statistics/ last accessed 7 March 2013).

Neto, C.C. (2007) Cranberry and blueberry: evidence for protective effects against cancer and vascular diseases. *Molecular Nutrition and Food Research*, **51**, 652–644.

Neto, C.C. and Vinson, J.A. (2011) Cranberry. In: Benzie, I. F. F. and Wachtel-Galor, S. (eds) *Herbal Medicine: Biomolecular and Clinical Aspects*, 2nd edn. CRC Press, Boca Raton, FL.

Neveu, V., Perez-Jimenez, J., Vos, F. *et al.* (2010) Phenol-Explorer: an online comprehensive database on polyphenol contents in foods. http://www.phenol-explorer.eu/ (last accessed 20 March 2013).

Perez-Jimenez, J., Neveu, V., Vos, F. and Scalbert, A. (2010) Systematic analysis of the content of 502 polyphenols in 452 foods and beverages: an application of the phenol-explorer database. *Journal of Agriculture and Food Chemistry*, **58**, 4959–4969.

Pinzon-Arango, P.A., Holguin, K. and Camesano, T.A. (2011) Impact of cranberry juice and proanthocyanidins on the ability of *Escherichia coli* to form biofilms. *Food Science and Biotechnology,* **20**, 1315–1321.

Puupponen-Pimia, R., Nohynek, L., Alakomi, H.L. and Oksman-Caldentey, M. (2005) Bioactive berry compounds – novel tools against human pathogens. *Applied Microbiology and Biotechnology*, **67**, 8–18.

Reed, J.D. and Howell, A.B. (2009) Biological activity of cranberry proanthocyanidins: effects on oxidation, microbial adhesion, inflammation, and health. In: Cooper, R. And Kronenberg, F. (eds) *Botanical Medicine*. Mary Ann Liebert, Inc., New Rochelle, NY.

Reed, J.D., Krueger, C.G. and Vestling, M.M. (2005) MALDI-TOF mass spectrometry of oligomeric food polyphenols. *Phytochemistry*, **66**, 2248–2263.

Romero, R., Oyarzun, E., Mazor, M. *et al.* (1989) Meta-analysis of the relationship between asymptomatic bacteriuria and preterm delivery/low birth weight. *Obstetrics and Gynecology,* **73**, 576–582.

Ronald, A. (2003) The etiology of urinary tract infection: traditional and emerging pathogens. *Disease-a-Month*, **49**, 71–82.

Ruel, G., Pomerleau, S., Couture, P. *et al.* (2005) Changes in plasma antioxidant capacity and oxidized low-density lipoprotein levels in men after short-term cranberry juice consumption. *Metabolism,* **54**, 856–861.

Ruel, G., Pomerleau, S., Couture, P. *et al.* (2006) Favourable impact of low-calorie cranberry juice consumption on plasma HDL-cholesterol concentrations in men. *British Journal of Nutrition,* **96**, 357–364.

Ruel, G., Pomerleau, S., Couture, P. *et al.* (2008) Low-calorie cranberry juice supplementation reduces plasma oxidized LDL and cell adhesion molecule concentrations in men. *British Journal of Nutrition,* **99**, 352–359.

Salo, J., Uhari, M., Helminen, M. *et al.* (2012) Cranberry juice for the prevention of recurrences of urinary tract infections in children: a randomized placebo-controlled trial. *Clinical Infectious Diseases,* **54**, 340–346.

Scalbert, A., Johnson, I.T. and Saltmarsh, M. (2005a) Polyphenols: antioxidants and beyond. *American Journal of Clinical Nutrition,* **81**(suppl), 215S–217S.

Scalbert, A., Manach, C., Morand, C. *et al.* (2005b) Dietary polyphenols and the prevention of diseases. *Critical Reviews in Food Science and Nutrition,* **45**, 287–306.

Schappert, S.M. and Rechtsteiner, E.A. (2011) Ambulatory medical care utilization estimates for 2007. *Vital and Health Statistics*, **169**, 1–38.

Seeram, N.P., Aviram, M., Zhang, Y. *et al.* (2008) Comparison of antioxidant potency of commonly consumed polyphenol-rich beverages in the United States. *Journal of Agricultural and Food Chemistry,* **56**, 1415–1422.

Sheiner, E., Mazor-Drey, E. and Levy, A. (2009) Asymptomatic bacteriuria during pregnancy. *Journal of Maternal-Fetal and Neonatal Medicine*, **22**, 423–427.

Shmuely, H., Yahav, J., Samra, Z. *et al.* (2007) Effect of cranberry juice on eradication of helicobacter pylori in patients treated with antibiotics and a proton pump inhibitor. *Molecular Nutrition and Food Research*, **51**, 746–751.

Singh, M., Arseneault, M., Sanderson, T. *et al.* (2008) Challenges for research on polyphenols from foods in Alzheimer's disease: bioavailability, metabolism, and cellular and molecular mechanisms. *Journal of Agricultural and Food Chemistry*, **56**, 4855–4873.

Sobota, A.E. (1984) Inhibition of bacterial adherence by cranberry juice: potential use for the treatment of urinary tract infections. *Journal of Urology*, **131**, 1013–1016.

Stapleton, A. and Stamm, W.E. (1997) Prevention of urinary tract infection. *Infectious Disease Clinics of North America*, **11**, 719–733.

Stapleton, A.E., Dziura, J., Hooton, T.M. *et al.* (2012) Recurrent urinary tract infection and urinary *Escherichia coli* in women ingesting cranberry juice daily: a randomized controlled trial. *Mayo Clinic Proceedings*, **87**, 143–150.

Steinberg, D., Feldman, M., Ofek, I. and Weiss, E.I. (2004) Effect of a high-molecular-weight component of cranberry on constituents of dental biofilm. *Journal of Antimicrobial Chemotherapy*, **54**, 86–89.

Steinberg, D., Feldman, M., Ofek, I. and Weiss, E.I. (2005) Cranberry high molecular weight constituents promote *Streptococcus sobrinus* desorption from artificial biofilm. *International Journal of Antimicrobial Agents,* **25**, 247–251.

Stevenson, D.E. and Hurst, R.D. (2007) Polyphenolic phytochemicals – just antioxidants or much more? *Cellular and Molecular Life Sciences*, **64**, 2900–2916.

Stothers, L. (2002) A randomized trial to evaluate effectiveness and cost effectiveness of naturopathic cranberry products as prophylaxis against urinary tract infection in women. *Canadian Journal of Urology*, **9**, 1558–1562.

Sturr, M.G. and Marquis, R.E. (1992) Comparative acid tolerances and inhibitor sensitivities of isolated F-ATPases of oral lactic acid bacteria. *Applied Environmental Microbiology*, **58**, 2287–2291.

Tasevska, N., Runswick, S.A., McTaggart, A. and Bingham, S.A. (2005) Urinary sucrose and fructose as biomarkers for sugar consumption. *Cancer Epidemiology, Biomarkers and Prevention*, **14**, 1287–1294.

Tong, H., Heong, S. and Chang, S. (2006) Effect of ingesting cranberry juice on bacterial growth in urine. *American Journal of Health System Pharmacy*, **63**, 1417–1419.

USDA (2004) USDA Database for the Proanthocyanidin Content of Selected Food. US Department of Agriculture, Agricultural Research Service. Nutrient Data Laboratory Home Page: http://www.ars.usda.gov/Services/docs.htm?docid=5843 (last accessed 7 March 2013). (This database was created through a collaborative effort between the Nutrient Data Laboratory and Food Composition Laboratory, USDA-ARS; The Arkansas Children's Nutrition Center, USDA-ARS; Mars, Inc.; and Ocean Spray Cranberries, Inc.)

USDA (2011) USDA National Nutrient Database for Standard Reference, Release 24. US Department of Agriculture, Agricultural Research Service. 2011. Nutrient Data Laboratory Home Page: http://www.ars.usda.gov/ba/bhnrc/ndl (last accessed 7 March 2013).

Vvedenskaya, I.O., Rosen, R.T., Guido, J.E. *et al.* (2004) Characterization of flavonols in cranberry (*Vaccinium macrocarpon*) powder. *Journal of Agricultural and Food Chemistry,* **52**, 188–195.

Walker, E.B., Barney, D.P., Mickerlsen, J.N. *et al.* (1997) Cranberry concentrate: UTI prophylaxis. *Journal of Family Practice,* **45**, 167–168.

Wallace, T.C. (2011) Anthocyanins in cardiovascular disease. *Advances in Nutrition*, **2**, 1–7.

Wang, C.-H., Fang, C.-C., Chen, N.-C. *et al.* (2012) Cranberry-containing products for prevention of urinary tract infections in susceptible populations: a systematic review and meta-analysis of randomized controlled trials. *Archives of Internal Medicine,* **172**, 988–996.

Weiss, E.I., Kozlovsky, A., Steinberg, D. *et al.* (2004) A high molecular mass cranberry constituent reduces mutans streptococci level in saliva and inhibits in vitro adhesion to hydroxyapatite. *FEMS Microbiology Letters,* **232**, 89–92.

Williamson, G. and Manach, C. (2005) Bioavailability and bioefficacy of polyphenols in humans. II. Review of 93 intervention studies. *American Journal of Clinical Nutrition*, **81**(suppl), 243S–255S.

Wilson, T., Meyers, S.L., Singh, A.P. and Vorsa, N. (2008) Favorable glycemic response of type 2 diabetics to low-calorie cranberry juice. *Journal of Food Science*, **73**, H241–H245.

Wilson, T., Porcari, J.P. and Harbin, D. (1998) Cranberry extract inhibits low density lipoprotein oxidation. *Life Sciences*, **62**, 381–386.

Wing, D., Rumney, P.J., Preslicka, C. and Chung, J. (2008) Daily cranberry juice for the prevention of asymptomatic bacteriuria in pregnancy. *Journal of Urology*, **180**, 1367–1372.

Winston, D., Graff, A., Brinckmann, J. *et al.* (2002) Cranberry fruit: *Vaccinium macrocarpon* Aiton. American Herbal Pharmacopoeia, Scotts Valley, CA.

Xie, B., Zhou, G., Chan, S.-Y. *et al.* (2006) Distinct glycan structures of uroplakins Ia and Ib: structural basis for the selective binding of FimH adhesin to uroplakin Ia. *Journal of Biological Chemistry*, **281**, 14644–14653.

Yamanaka, A., Kimizuka, R., Kato, T. and Okuda, K. (2004) Inhibitory effects of cranberry juice on attachment of oral streptococci and biofilm formation. *Oral Microbiology and Immunology (Copenhagen)*, **19**, 150–154.

Yamanaka, A., Kouchi, T., Kasai, K. *et al.* (2007) Inhibitory effect of cranberry polyphenol on biofilm formation and cystein proteases of *Porphyromonas gingivalis*. *Journal of Periodontal Research,* **42**, 589–592.

Zafriri, D., Ofek, I., Adar, R. *et al.* (1989) Inhibitory activity of cranberry juice on adherence of type 1 and type P fimbriated *Escherichia coli* to eucaryotic cells. *Antimicrobial Agents and Chemotherapy,* **33**, 92–98.

Zampariello, C.A., McKay, D.L., Dolnikowski, G. *et al.* (2012) Determination of cranberry proanthocyanidin A2 in human plasma and urine using LC-MS/MS. *the Journal of the Federation of American Societies for Experimental Biology*, **26**, 124–128.

Zhanel, G.G., Hisanaga, T.L., Laing, N.M. *et al.* (2006) Antibiotic resistance in Escherichia coli outpatient urinary isolates: final results from the North American Urinary Tract Infection Collaborative Alliance (NAUTICA). *International Journal of Antimicrobial Agents,* **27**, 468–475.

Zhang, K. and Zuo, Y. (2004) GC-MS determination of flavonoids and phenolic and benzoic acids in human plasma after consumption of cranberry juice. *Journal of Agricultural and Food Chemistry*, **52**, 222–227.

Zhang, L., Ma, J., Pan, K. *et al.* (2005) Efficacy of cranberry juice on *Helicobacter pylori* infection: a double-lind, randomized placebo-controlled trial. *Helicobacter,* **10**, 139–145.

Zheng, W. and Wang, S.Y. (2003) Oxygen radical absorbing capacity of phenolics in blueberries, crannberries, chokeberries, and lingonberries. *Journal of Agricultural and Food Chemistry*, **51**, 502–509.

Zuo, Y., Wang, C. and Zhan, J. (2002) Separation and characterization of benzoic and phenolic antioxidants in american cranberry fruit by HPLC and GC–MS. 223rd American Chemical Society National Meeting, Orlando, FL.

FURTHER READING/RESOURCES

www.cranberryinstitute.org
www.uscranberries.com

13 Vegetable Fruits: A Cornucopia of Health Benefits

Carolyn E. Lister

INTRODUCTION

Is it a fruit or a vegetable? The term 'vegetable' is largely based on culinary and cultural tradition and has no botanical definition, although it typically means an edible plant or part of a plant. However, it usually excludes seeds and most sweet fruit, so typically means the leaf, stem or root of a plant. The application of the word is somewhat arbitrary and subjective and can depend on cultural usage. The term 'fruit' is used both as a culinary term and has a botanical definition, and these two usages can be quite different. In botanical terms fruits are 'a part that developed from the ovary of a flowering plant'. Botanically, pineapples, oranges and apples are 'fruits', but so are 'vegetables' like tomatoes and cucumbers. In culinary usage 'fruit' is only applied to those botanical fruits which are edible and considered to be a sweet or dessert food. There are a number of edible botanical fruits (e.g. tomato, eggplant, capsicum – Table 13.1 shows a more detailed list) that are not considered to be a sweet or dessert food but instead are almost always used as part of a savoury dish. This is the reason that they are labelled as 'vegetables' or vegetable fruits. Thus, a plant part may scientifically be referred to as a 'fruit', even though it is used in cooking or food preparation as a vegetable. Languages other than English often have categories that can be identified with the common English meanings of 'fruit' and 'vegetable', but their precise meaning often depends on local culinary traditions. Cultural usage can mean in one country what is regarded as a vegetable may be a fruit in another. For example, in Brazil the avocado is traditionally consumed with sugar as a dessert or in milkshakes, and hence it is regarded as a culinary fruit, whereas in other countries (including Mexico and the United States) it is used in salads and dips, and hence considered to be a vegetable. Likewise, pumpkin is mainly consumed as a vegetable in most cultures but can also made into a dessert such as pumpkin pie.

This chapter largely focuses on fruits in two families, the Solanaceae (tomato and *Capsicum* spp.) and Cucurbitaceae (pumpkin, squash and vegetable melons/gourds) because of the amount of study they have received. Other key vegetable fruits are avocado and olive, and these are covered in other chapters in this book. Most other vegetable fruits have received little scientific study but are summarized here (Table 13.2).

Bioactives in Fruit: Health Benefits and Functional Foods, First Edition. Edited by Margot Skinner and Denise Hunter.

Table 13.1 Details of vegetable fruits.

Common Name	Alternative names	Scientific name	Family
Ackee	achee, akee, akee apple, vegetable brain	*Blighia sapida* K.D. Koenig (*Cupania sapida*)	Sapindaceae
Avocado	alligator pear	*Persea americana* Mill. (*P. gratissima*)	Lauraceae
Bitter melon	balsam pear, bitter cucumber, bitter gourd, bitter squash, wild cucumber	*Momordica charantia* L.	Cucurbitaceae
Bottle gourd	calabash, calabaza, dudhi, ghia, lauki, long melon, opo squash	*Lagenaria siceraria* (Molina) Standl. (*L. vulgaris, Cucurbita lagenaria*)	Cucurbitaceae
Breadfruit		*Artocarpus altilis* (Parkinson ex F.A. Zorn) Fosberg (*A. communis*)	Moraceae
Caigua	lady's slipper, slipper gourd, sparrow gourd, stuffing cucumber, wild cucumber	*Cyclanthera pedata* (L.) Schrad. (*Momordica pedata*)	Cucurbitaceae
Capsicum	bell pepper, chilli pepper, paprika, pepper, sweet pepper	*Capsicum annum* L. along with other species (see text)	Solanaceae
Chayote	choko, cho-cho, christophene, mango squash, mirliton, pear squash, vegetable pear	*Sechium edule* (Jacq.) Sw. (*Chayota edulis, Sicyos edulis*)	Cucurbitaceae
Cucumber		*Cucumis sativus* L.	Cucurbitaceae
Eggplant	aubergine, brinjal, garden egg, guinea squash, melanzana, melongene, patlican	*Solanum melongena* L. (*S. ovigerum*, *S. trongum*)	Solanaceae
Ivy gourd	baby watermelon, gentleman's toes, kundru, little gourd	*Coccinia grandis* (L.) Voigt (*C. indica, Cephalandra indica*)	Cucurbitaceae
Luffa	angled luffa, bath sponge, dish cloth gourd loofa, ridge gourd, ridged luffa, silk squash, sponge gourd, vegetable gourd, vegetable sponge	*Luffa acutangula,* (L.) Roxb. and *L. cylindrica* (L.) M. Roem. (*L. aegyptiaca*)	Cucurbitaceae
Okra	gumbo, lady finger	*Abelmoschus esculentus* (L.) Moench (*Hibiscus esculentus*)	Malvaceae
Olive		*Olea europaea* L.	Oleaceae
Pumpkin/squash	Summer squash, winter squash	*Cucurbita maxima* Duch., *C. pepo* L., *C. moschata* Lam., *C. ficifolia* Bouche	Cucurbitaceae
Snake gourd	chichinga, padwal, serpent gourd, snake tomato	*Trichosanthes cucumerina* L., *T. anguina* L.	Cucurbitaceae
Tomatillo	fresadilla, ground cherry, husk tomato, jamberry	*Physalis philadelphica* Lam. (*P. ixocarpa*)	Solanaceae
Tomato	love apple, wolf peach	*Lycopersicon esculentum* Mill. (*L. lycopersicum, Solanum lycopersicon*)	Solanaceae
Wax gourd	ash gourd, ash pumpkin, Chinese watermelon, gourd melon, kundur, tallow gourd, white gourd, white pumpkin, winter melon	*Benincasa hispida* (Thunb.) Cogn.	Cucurbitaceae

Table 13.2 Phytochemical composition and health benefits of vegetable fruits.

Common Name	Key phytochemicals	Health benefits/activities	Reference(s)
Ackee	Alkaloids: hypoglycin A and B Phenolics: quinines Saponins: blighosides A, B, and C Terpenes: tri- and sesquiterpenes	Traditional usage: conjunctivitis, fever, parasites, skin conditions Scientific study: • *in vitro* – antioxidant, hypoglycaemic	Parkinson, 2007
Avocado	See chapter 14		
Bitter melon	Alkaloids: momordicine Carotenoids: cryptoxanthin, beta-carotene, zeaxanthin Peptides (insulin-like): gumarin Phenolics: catechin and epicatechin; gallic, caffeic, coumaric, gentisic and chlorogenic acids Saponins: goyasaponins Triterpenoids: cucurbitane-type glycosides (momordin, cucurbitacin, charantin) and oleanane-type glycosides	Traditional usage: diabetes and associated complications, fevers, gastrointestinal complaints, painful menstruation, skin conditions, wound healing Scientific study: • human – diabetes (low quality) • animal – cancer, cardioprotective, immunomodulatory, ulcers • *in vitro* – antibacterial, antiviral including HIV	Gurbuz *et al.*, 2000; Grover and Yadav, 2004; Puri *et al.*, 2009; Leung *et al.*, 2009; Nerurkar and Ray, 2010; Gadang *et al.*, 2011; Fang and Ng, 2011; Chaturvedi, 2012
Bottle gourd	Cardolenides: periplogenin glycosides Carotenoids: beta-carotene Phenolics: (E)-4-hydroxymethyl-phenyl-6-O-caffeoyl-b-D-glucopyranoside, 1-(2-hydroxy-4-hydroxymethyl)-phenyl-6-O-caffeoyl-b-Dgluco-pyranoside, protocatechuic acid gallic acid, caffeic acid and 3,4-dimethoxy cinnamic acid flavone-C glycosides, ellagitannins Polysaccharides: galacturonates Triterpenes: cucurbitane and friedooleanane-type Sterols: fucosterol, campesterol	Traditional usage: coughs, diabetes, diuretic, general tonic, heart conditions, insomnia, rheumatism Scientific study: • animal – analgesic, antioxidant, cancer, cardioprotective, cholesterol lowering, CNS depressant, diabetes, diuretic, hepatoprotective, hypertension, immunomodulatory • *in vitro* – antioxidant, anti-inflammatory	Panda and Kar, 2011; Tyagi *et al.*, 2012

(continued)

Table 13.2 (*Continued*)

Common Name	Key phytochemicals	Health benefits/activities	Reference(s)
Breadfruit	Carotenoids (yellow-fleshed varieties) Phenolics: geranyl flavanoid derivatves Stilbenes Sterols: cycloartenol Triterpenes	Traditional usage: liver tonic Scientific study: • *in vitro* – antioxidant, anti-inflammatory, chemopreventive	Hsu *et al.*, 2011; Lin *et al.*, 2011, 2012
Caigua	Phenolics: flavone glycosides Saponins Triterpenes: cucurbitacins Sterols: sitosterol	Traditional usage: diabetes, gastrointestinal disorders, high blood pressure, inflammation, tonsilitis Scientific study: • human – cholesterol lowering (low quality) • *in vitro* – angiotensin-converting enzyme (ACE) inhibitory, antioxidant	Gonzales *et al.*, 1995; Ranilla *et al.*, 2010
Capsicum	Capsaicinoids: capsaicin, dihydrocapsaicin, capsiate, dihydrocapsiate Carotenoids: capsanthin, beta-carotene, beta-cryptoxanthin, capsorubin, antheraxanthin Phenolics: apigenin glycosides, hydroxycinnamic acids (coumaric and caffeic acid derivatives), luteolin glycosides, quercetin glycosides, lignans	Traditional usage: asthma, blood circulation, cancer, neuralgia, pneumonia, rheumatism, sores, stomach problems Scientific study: • human – cancer, diabetes, dyspepsia, migraine, pain relief, pneumonia, weight management (energy balance/obesity/satiety) • animal – allergic airway inflammation, arthritis, CVD, cognitive function, gastroprotective, neuroprotective, thermoregulatory • *in vitro* – anti-inflammatory, antimicroial, antioxidant, antiplatelet, inhibition of alpha-glucosidase and ACE	See Table 13.7
Chayote	Alkaloids Phenolics: vicenin-2 and vitexin Saponins Sterols	Traditional usage: anti-inflammatory, diuretic, heart conditions Scientific study: • animal – diabetes, hypertension • *in vitro* – antibacterial, antioxidant, antiproliferative, cytotoxic (cancer cells)	Gordon, 2000; Tiwari *et al.*, 2011

Cucumber	Carotenoids (mainly in skin): beta-carotene Lignans: lariciresinol, pinoresinol, secoisolariciresinol Phenolics: caffeic and chlorogenic acids, isoquercetrin, apigenin Triterpenes: cucurbitacins	Traditional usage: constipation, digestive aid, diuretic, fever, skin problems Scientific study: • animal – anti-inflammatory, cholesterol lowering, diabetes • *in vitro* – antimicrobial, antimutagenic, antioxidant	Karthiyayini *et al.*, 2009; Dixit and Kar, 2010
Eggplant	Glycoalkaloids: solasodine, solamargine and solasonine Phenolics: chlorogenic and caffeic acid, delphinidin glycosides (nasunin), coumarins, tyramine	Traditional usage: coughs, digestive complaints, diuretic, heart disease, high cholesterol, obesity, rheumatism, warts Scientific study: • animal – cholesterol-lowering, diabetes • *in vitro* – antiangiogenic, anticancer, anti-inflammatory, antioxidant, cardioprotective, hepatoprotective, inhibition of alpha-glucosidase and ACE	Sudheesh *et al.*, 1997; Noda *et al.*, 2000; Derivi *et al.*, 2002; Maiani *et al.*, 2005; Kwon *et al.*, 2008; Akanitapichat *et al.*, 2010
Ivy gourd	Alkaloids Carotenoids: beta-carotene, cryptoxanthin Phenolics: ellagic acid Saponins Steroids	Traditional usage: asthma, diabetes, eczema, fever, jaundice, leprosy, liver conditions, sores Scientific study: • animal – allergic reactions, diabetes, cough suppressant, diabetes, hepatoprotective, mast cell stabilizing • *in vitro* – anthelmintic, antimicrobial, antioxidant, stimulation glucose transporter (GLUT) 2	Yadav *et al.*, 2010; Taur and Patil, 2011
Luffa	Peptides: luffin Phenolics: apigenin, diosmetin glucuronide Saponins: lucyin, luc(y)osides Triterpenes: cucurbitacins, oleanolic acid	Traditional usage: constipation, demulcent, diuretic, jaundice Scientific study: • animal – allergic reactions, analgesic, asthma, anti-inflammatory, cholesterol lowering, cough suppressant, cardioprotective, diabetes, expectorant, immunostimulatory, sedative • *in vitro* – antiangiogenic, antimicrobial, antioxidant, antiproliferative	Reddy *et al.*, 2009, 2010, 2011; Tiwari *et al.*, 2011; Velmurugan *et al.*, 2011; Sangh *et al.*, 2012

(continued)

Table 13.2 (*Continued*)

Common Name	Key phytochemicals	Health benefits/activities	Reference(s)
Okra	Glycoproteins Phenolics: catechin, procyanidin B2, epicatechin, quercetin glycosides (rutin), hydroxycinnamic acids Polysaccharides: pectic rhamnogalacturonan Saponins Triterpenes	Traditional usage: anemia, arthritis, demulcent, diabetes, digestive complaints (including constipation and dysentery), fever, general debility, headache, skin itchiness, sore throat associated with coughing, urinary problems Scientific study: • human – cholesterol lowering • animal – diabetes, cholesterol lowering, cytoprotection, diuretic, gastroprotective, hepatoprotective, neuroprotective • *in vitro* – anticomplementary, antioxidant, antiadhesive (protection against *H. pylori*), antiproliferative, immunomodulatory, propapoptotic	Sabitha *et al.*, 2011; Lengsfeld *et al.*, 2004; Khomsug *et al.*, 2010; Danamma *et al.*, 2011; Joshi *et al.*, 2011
Olive	See chapter 15		
Pumpkin/ squash	Carotenoids: beta-carotene, alpha-carotene, cryptoxanthin, cucurbitaxanthin, lutein, violaxanthin GABA Glycolipids: glyceroglycolipids, galactolipids Phenolics: chlorogenic, caffeic, syringic, vanillic, *p*-coumaric and sinapic acids, luteolin Polsaccharides: D-*chiro*-inositol Polyamines Proteins: mosch(at)ins, cucurmosin Sterols: beta-sitosterol Triterpenoids: cucurbitane type	Traditional usage: blood purifier, boils, cooling, diabetes, diuretic, fever, haemorrhoids, inflammation, leprosy, prostate and bladder health, wounds, worms and other parasites Scientific study: • human – diabetes (low quality) • animal – cancer, cholesterol lowering, cough suppressant, gastroduodenal protective (including ulcers), obesity, worms • *in vitro* – antimutagenic, antioxidant, antimicrobial, antitumour, immunomodulatory, inhibition of ACE, alpha-amylase and alpha glucosidase	Fu *et al.*, 2006; Choi *et al.*, 2007; Sarkar and Guha, 2008; Nosál'ová *et al.*, 2011; Adams *et al.*, 2011; Badr *et al.*, 2011; Dhiman *et al.*, 2012

Snake gourd	Carotenoids: beta-carotene, lycopene, lutein Phenolics: flavonoids, tannins Saponins Sterols: beta-sitosterol Triterpenes: cucurbitacins	Traditional usage: antibiotic, cough, fever, heartconditions, liver disorders, purgative, vermifuge Scientific study: • animal studies – gastroprotective, anti-inflammatory, hepatoprotective • *in vitro* – anticancer, antioxidant	Kongtun *et al.*, 2009; Kumar *et al.*, 2009; Arawwawala *et al.*, 2010a, 2010b
Tomatillo	Alkaloids Phenolics: anthocyanins in purple varieties, quercetin glycosides Steroidal lactones: withanolides (ixocarpalactone A, withaphysacarpin), physalins	Traditional usage: respiratory conditions, gastrointestinal disorders Scientific study: • *in vitro* – antimicrobial, antioxidant, antiproliferative, apoptotic, phase 2 enzyme induction	Su *et al.*, 2004; Choi *et al.*, 2006; Gonzalez-Mendoza *et al.*, 2010
Tomato	Carotenoids: lycopene, beta-carotene, phytoene, phytofluene GABA Glycoalkaloids: tomatine Phenolics: chlorogenic acid, naringenin/naringenin chalcone, querectin glycosides	Traditional usage: bleeding, fever, food poisoning and other digestive problems, headache, high blood pressure, insect bites, kidney and liver conditions Scientific study: • human – asthma, bone health, brain health, cancer, CVD, eye health, gingivitis, infertility, liver injury, osteoporosis, pneumonia, UV light-induced erythema, stroke • animal – anticoagulant, cataracts, diabetes, immune function, obesity, pancreatitis, Parkinson's disease • *in vitro* – antioxidant, anti-inflammatory, antiproliferative	See text and Tables 13.4 and 13.5; Ozkan *et al.*, 2012
Wax gourd	Carotenoids: carotenes Peptides: benincasins, hispin Phenolics Sterols: lupeol, lupeol acetate, beta-sitosterol Triterpenes: alnusenol, multiflorenol, isomultiflorenol Uronic acid	Traditional usage: diabetes, diuretic, epilepsy, peptic ulcer Scientific study: • animal – Alzheimer's, bronchospasm, compulsive disorders, diarrhoea, morphine withdrawal, obesity, renoprotective • *in vitro* – antioxidant and ACE inhibitor	Zaini *et al.*, 2011; Dhiman *et al.*, 2012

TOMATOES

The tomato (*Lycopersicon esculentum*) is a member of the Solanaceae and is an important agricultural commodity worldwide. It is one of the most widely used and versatile vegetable fruits, being consumed raw, juiced, cooked and as a component of other dishes. Originating in South America, the tomato was spread around the world following the Spanish colonization of the Americas. Most cultivars produce red fruit, but yellow, orange, pink, purple, green, black or white varieties are also available. Tomatoes have a large body of evidence for health benefits and have had considerable study in humans.

Phytochemical composition of tomatoes

Nutritionally tomatoes are a good source of antioxidant vitamins, including vitamin C, vitamin A (in the form of β-carotene) and vitamin E. Tomatoes also contain a significant number of phytochemicals in reasonable quantities (Table 13.3). Of these, lycopene is the one that has received particular interest.

Lycopene and other carotenoids

Lycopene is only present in a few foods, the most common being tomatoes, with watermelon, pink grapefruit, Chilean guava, red papaya and rosehips being other sources. The lycopene content of fresh tomatoes can vary from virtually none up to over 20 mg/100 g fresh weight (FW) in high pigment varieties (Ilahy *et al.*, 2011), but most values for typical, red tomatoes are between 5 and 8 mg/100 g FW (Dumas *et al.*, 2003). The carotenoid composition of

Table 13.3 Concentrations of the main phytochemicals reported in fresh, ripe red tomatoes.

Component	Concentration[a] (mg/100 g FW)
Ascorbic acid	2–30
Carotenoids (total)	4–24
β-carotene	0.4–1.0
lycopene	2.5–25
phytoene	1–3
phytofluene	~1
Total phenolics	22–82
Phenolic acids	16–29
caffeic acid	0.2–10
chlorogenic acid	0.6–20
coumaric acid	0.1–1.6
ferulic acid	0.1–0.7
Flavonoids	
naringenin/chalconaringenin	0.1–23
quercetin glycosides (primarily rutin)	0.3–4.3
kaempferol glycosides	0–0.1
total flavonols	0.1–7.9
Glycoalkaloids	0.5–15
GABA	22–190

*Data from: Gross, 1991; Friedman, 2002; Dumas *et al.*, 2003; Saito *et al.*, 2008; USDA, 2011; Lister, personal data.

tomatoes has been reviewed by Gross (1991). Carotenes, mainly lycopene, make up 90–95% of the carotenoids in tomatoes. In addition to lycopene, tomatoes contain beta-carotene, zeta-carotene, gamma-carotene and neurosporene along with the colourless carotenoid precursors phytoene and phytofluene. Yellow tomatoes and orange tomatoes generally lack lycopene, or compounds other than lycopene are present at higher concentrations. In some cultivars beta-carotene predominates and in others delta-carotene. The predominant lycopene isomer present in tangerine tomatoes is prolycopene (*tetra-cis*-lycopene), a geometric isomer of *all-trans*-lycopene, which gives this fruit a characteristic orange color (Unlu *et al.*, 2007).

The concentrations of lycopene may vary according to such factors as cultivar, growing conditions, method of ripening, processing and storage conditions (Abushita *et al.*, 2000; Dumas *et al.*, 2003). The amount of lycopene present in processed foods is often much higher than that found in fresh foods given that processing often involves concentration via water loss. For example per 100 g: tomato juice ~10 mg, tomato puree/sauce/ketchup ~15 mg and tomato paste up to 40 mg. Box 13.1 provides more information on dietary sources and suggested intake of lycopene.

Box 13.1 Dietary intake of lycopene and suggested beneficial intake

Tomatoes are the predominant dietary source of lycopene at around 80%, and this is derived mainly from processed products such as ketchup, juice, spaghetti sauce and pizza sauce. Various estimates have been made of lycopene intake in different populations, with huge variation both between countries and studies (mainly due to method of estimation). Estimates of typical daily intake range from 0.7 mg in Finland, 1.1 mg in the United Kingdom, 4.2 mg in New Zealand, 5.5 mg in the USA (although one study notes that about 50% of the population consume less than 2 mg), 7.6 mg in Italy and Spain, to 11 mg in Greece. Although the beneficial effects of lycopene have been well documented it is not yet recognized as an essential nutrient. As a result there is no official recommended intake level set by health professionals and government regulatory agencies. Researchers vary in what is considered an efficacious dose, with most suggestions ranging from 5 to 35 mg per day. Some studies suggest for a healthy individual daily intake of 5–7 mg may maintain circulating lycopene at a level sufficient to combat oxidative stress and prevent chronic diseases (Rao and Shen, 2002). These levels could be achieved by consuming at least one or two servings of tomatoes/tomato products every day. Under the condition of disease (e.g. cancer and cardiovascular disease) higher levels, ranging from 35 to 75 mg per day, may be required (Heath *et al.*, 2006).

An important consideration with regards to the health benefits of any compound is its bioavailability. Lycopene bioavailability has been well studied in humans and can be affected by a number of factors, including food processing and dietary composition (Story *et al.*, 2010). Lycopene can occur in several forms in fresh plant foods, including carotenoid–protein complexes in chloroplasts or in crystalline form inside chromoplasts. Numerous studies have demonstrated that for thermally processed tomato products lycopene is more bioavailable than fresh tomatoes (Story *et al.*, 2010). Thermal processing disrupts cellular membranes, which allows lycopene to be released from the tissue matrix, and in addition protein–carotenoid complexes may be weakened. Composition of the diet also impacts on lycopene bioavailability. Most importantly, because lycopene is a lipid-soluble compound, consuming it with fat increases its bioavailability. Competition by other carotenoids or cholesterol may also influence lycopene absorption along with other factors, such as probiotics. Human organs store lycopene to varying degrees with the highest concentrations in the liver, testes, adrenal glands and adipose tissues with lower levels in the kidney, ovary, lung and prostate (Kun *et al.*, 2006).

Phenolics

In most cases the individual phenolics in tomato are not present in large amounts (Table 13.3). The key phenolics are naringenin and its isomer chalconaringenin (although much of the literature treats these two compounds together as naringenin), chlorogenic acid and rutin (quercetin 3-rutinoside). Cultivar, effects of environment and agronomic practices plus fruit maturity are all important factors influencing the composition and content of phenolics in tomato fruit (Slimestad and Verheul, 2009).

With the interest in anthocyanins for their health benefits there has been research into the development of purple, anthocyanin-rich tomatoes. Purple coloured fruits exist in some wild tomato species but generally they are limited to the skin and the fruit are not commercially acceptable. In some heirloom varieties the black and purple colours are due to mutations affecting chlorophyll breakdown and carotenoid biosynthesis and are not related to anthocyanin production (Mes *et al.*, 2008). A number of research groups have used genetic modification to achieve purple tomatoes. Although many have had limited or no success, Butelli *et al.* (2008) achieved high concentrations of anthocyanins in both the peel and the flesh of tomato fruit (total anthocyanins averaged 283 mg/100 g fresh weight). A mouse feeding trial with the genetically modified fruit found they prolonged the life of cancer-susceptible mice compared with a standard diet or with the inclusion of standard red tomatoes. Because there is consumer resistance to genetically modified food there is still a focus on the development of purple tomatoes by traditional breeding methods. However, concentrations of anthocyanins are relatively low; a purple tomato achieved through traditional cross-breeding had an anthocyanin concentration of 72 mg per 100 g dry weight (Li *et al.*, 2011). There are also reported differences in the anthocyanin composition of transgenic and non-transgenic tomatoes. In the transgenic fruit the two main anthocyanins were delphinidin 3-(trans-coumaroyl)-rutinoside-5-glucoside and petunidin 3-(trans-coumaroyl)-rutinoside-5-glucoside (Butelli *et al.*, 2008), whereas in non-transgenic fruit the main anthocyanin was the petunidin glycoside with no delphinidin glycoside reported (Li *et al.*, 2011).

Glycoalkaloids

Many members of the Solanaceae contain glycoalkaloids and have attracted scientific interest because they are reported to be involved in host-plant resistance and also may have a variety of health effects in animals and humans. Tomato glycoalkaloids and their role in the diet have been reviewed by Friedman (2002). The glycoalkaloid known as tomatine was first isolated in 1948 and subsequently found to be a mixture of two glycoalkaloids, α-tomatine and dehydrotomatine. Both compounds are present in all parts of the tomato plant, including the fruit. Immature green tomatoes contain up to 50 mg of α-tomatine/100 g but it is largely degraded as the tomato ripens, so in mature red tomatoes the level is only 0.5 mg/100 g. Consumers of green tomatoes, high-tomatine red tomatoes and tomato products such as pickled green and green fried tomatoes consume significant amounts of tomatine. However, compared with eating high glycoalkaloid potatoes, consumption of glycoalkaloid tomatoes does not appear to present health problems. It is thought that the absence of a 5,6-double bond in the B-ring of tomatidine results in a much less toxic molecule than the potato glycoalkaloids, α-chaconine and α-solanine, which contain the double bond.

Other

In addition to carotenoids and phenolics, several other compounds have been highlighted in studies of the health benefits of tomatoes. One of these is γ-aminobutyric acid (GABA), a ubiquitous four carbon non-protein amino acid present in a wide range of organisms including plants, bacteria and vertebrates. Screening of various plants showed that tomato contains one of the highest concentrations of GABA of all vegetable crops (Saito *et al.*, 2008). Within tomato there is considerable variation in GABA concentrations, with cultivation under saline or drought stress promoting its accumulation and postharvest handling practices also having an impact (Saito *et al.*, 2008; Mae *et al.*, 2012). Cystine-knot miniproteins (named TCMP-1 and TCMP-2) have been identified from tomato that show potential health benefits (Cavallini *et al.*, 2011). In addition, there is a group of 37 constituents in lycopene-free and water-soluble tomato concentrate (WSTC) that has been used in the development of tomato products and is the basis for a health claim in Europe (EFSA, 2009). There is limited characterization of the components published but they are reported to contain nucleosides and derivatives along with phenolic compounds and Maillard products (O'Kennedy *et al.*, 2006).

Health benefits of tomatoes

Hundreds of papers have been published on the potential health benefits of tomatoes and lycopene. Evidence comes from a range of types of studies with much of the evidence supporting the role of tomatoes/lycopene in reducing the risk of chronic diseases being based on epidemiological studies. More recently, clinical and intervention studies have been undertaken. In addition to human studies, animal studies and *in vitro* investigations are contributing a better understanding of the health effects and mechanisms by which lycopene exerts its effects. Two health areas of greatest focus have been cancer and cardiovascular disease (CVD) but some other areas are emerging where tomatoes may have beneficial effects.

Cancer

There have been a number of key reviews examining the relationships between tomato and/or lycopene consumption and the risk of various cancers (Table 13.4). Probably the most significant review, from the perspective of possible claims on tomato products, has been the 'The US food and drug administration's evidence-based review for qualified health claims: Tomatoes, lycopene, and cancer' (Kavanaugh *et al.*, 2007). However, the limits for the review were very narrow, being restricted to a subset of clinical intervention trials and observational studies. Another key document is the World Cancer Research Fund report 'Food, Nutrition, Physical Activity, and the Prevention of Cancer' (2007). This systematic literature review took a similar, although slightly different, approach to the FDA. It was decided to gather evidence from all relevant epidemiological and experimental studies, together with biological findings. The third key collection of research findings is a database that has been collated by the Tomato Products Wellness Council (2012). In addition, there are numerous other

Table 13.4 Summary of the findings from key reviews of human studies with tomatoes and reduction of cancer risk.

Cancer Site	Summary of Findings	Reference(s)
Bladder	Data of too low quality, too inconsistent, or the number of studies too few to allow conclusions to be reached.	Kavanaugh *et al.*, 2007
	No studies found statistically significant associations, although tendencies for inverse associations noted. RE: N = 4	Giovannucci, 1999
Brain	Not reported in key reviews but a recent paper indicates a potential benefit in treatment of high-grade gliomas.	Puri *et al.*, 2010
Breast	Neutral relationship with intake of processed tomatoes. RE: (–) = 2, N = 5	TPWC, 2012
	No credible evidence to support tomatoes/tomato-based foods. RE: N = 2	Kavanaugh *et al.*, 2007
	Suggestion of a benefit, although data are very limited.	Giovannucci, 1999
Cervical	Possible protective relationship, however studies are of limited quality. RE: (–) = 3	TPWC, 2012
	No credible evidence to support a relationship (all studies identified eliminated due to scientific deficiencies).	Kavanaugh *et al.*, 2007
	Suggestion of a benefit.	Giovannucc,i 1999
Colorectal	Relationship with tomato/tomato-based foods inconclusive. RE: (–) = 1, N = 1	TPWC, 2012
	No credible evidence supporting an association. RE: N = 2	Kavanaugh *et al.*, 2007
	Suggestion of a benefit.	Giovannucci, 1999
Endometrial	No credible evidence available to support a relationship (no quality studies identified).	Kavanaugh *et al.*, 2007
Gastric/oral (upper GI)	Protective effect. RE: (–) = 7	TPWC, 2012
	Unlikely that tomatoes reduce the risk. RE: (–) = 3, N = 4	Kavanaugh *et al.*, 2007
	No prospective studies identified but consistent inverse association observed in diverse populations strongly suggests a protective effect.	Giovannucci, 1999
Head and neck	Neutral relationship with serum lycopene concentrations. RE: (–) = 1, N = 2	TPWC, 2012
Liver	Review of *in vitro*, animal and clinical studies indicate lycopene may possibly prevent the development of hepatocellular carcinoma.	Seren *et al.*, 2008b

Lung	Protective relationship. RE: (–) = 6	TPWC, 2012
	No credible evidence for an association. RE: N = 3	Kavanaugh *et al.*, 2007
	Protective effect (10 of 14 studies showed either a statistically significant or a suggestive inverse association).	Giovannucci, 1999
Ovarian	Neutral relationship. RE: (–) = 1, N = 1	TPWC, 2012
	Highly uncertain. One study suggests that consumption of tomato sauce two times per week may reduce the risk while this same study shows that consumption of tomatoes or tomato juice had no effect on risk.	Kavanaugh *et al.*, 2007
	No association. RE: N = 1	Giovannucci, 1999
Pancreatic	Potential protection, although data limited. RE: (–) = 2	TPWC, 2012
	Highly unlikely that tomatoes reduce the risk. RE: (–) = 1, N = 1	Kavanaugh *et al.*, 2007
	Inverse association. RE: (–) = 4	Giovannucci, 1999
Prostate	Protective relationship (processed tomato intake may be the most beneficial). RE: (–) = 24, N = 9, (+) = 2	TPWC, 2012
	Very limited and preliminary scientific research suggests that eating one-half to one cup of tomatoes and/ or tomato sauce a week may reduce the risk. However, there is little scientific evidence supporting this claim. RE: (–) = 6, N = 7	Kavanaugh *et al.*, 2007
	Foods containing lycopene probably protective. Substantial amount of consistent evidence, in particular on tomato products, from both cohort and case-control studies. There is evidence for plausible mechanisms.	WCRFR, 2007
	Of all the cancers examined the evidence for a benefit was strongest.	Giovannucci, 1999
Renal	Data limited. RE: (–) = 1	TPWC, 2012
Skin	Insufficient data to allow conclusions to be reached.	WCRFR, 2007

TPWC = Tomato Products Wellness Council; WCRFR = World Cancer Research Fund Report; RE = Risk Estimate, (–) = negative association, N = neutral association, (+) = positive association.

literature reviews (e.g. Hwang and Bowen, 2002; Seren *et al.*, 2008a; van Breemen and Pajkovic, 2008), although most are focused on prostate cancer (e.g. Haseen *et al.*, 2009; Tan *et al.*, 2010; Ilic *et al.*, 2011).

There is variation in the conclusions drawn by the various key reviews. Most notably the recommendations arrived at by the FDA (limited evidence for tomatoes and prostate cancer and no evidence to support relationships between tomatoes and risk reduction of any other cancers) are not always in agreement with conclusions drawn by others. There are a number of reasons for this and Coates (2007) made some very pertinent comments. Very few studies met the inclusion criteria for the FDA review, with many studies excluded for various reasons. The literature relating tomatoes and/or lycopene to the incidence of cancer is heterogeneous and there are few clinical trials of any great size. Instead, clues about whether lycopene or tomatoes might be associated with a reduced risk of cancer have come from preclinical and observational studies and from small trials. If the totality of evidence is included it is suggested that tomatoes/lycopene are associated with a reduced risk of prostate cancer and, perhaps, of other forms of cancer.

Although a range of cancers have been investigated the largest body of research on the health benefits of tomatoes/lycopene is for prostate cancer. In addition to those reports identified in Table 13.4, there have been a number of other reviews (Van Patten *et al.*, 2008; Colli and Amling 2009; Ellinger *et al.*, 2009). Overall, prospective and retrospective epidemiological studies indicating an inverse relationship between lycopene intake and prostate cancer risk have been supported by *in vitro* and *in vivo* experiments showing that oral lycopene is bioavailable and accumulates in prostate tissue. However, evidence is still somewhat debated and it cannot be concluded whether the benefits of tomato/tomato-based foods are due to lycopene alone or if other compounds play a role as well. Further large-scale trials are required to investigate the effects of potential foods or bioactive components of foods for longer durations and in more people to understand the potential of their action more fully. In addition to the prevention of prostate cancer, there has been considerable interest in the development of treatments for prostate cancer (Van Patten *et al.*, 2008; Ellinger *et al.*, 2009; Haseen *et al.*, 2009). Conclusions have been varied and in most cases there is insufficient evidence to draw a firm conclusion but some suggest lycopene ingestion might be effective in prostate cancer therapy. Well-designed trials are warranted to expand knowledge, replicate findings and further assess the impact of diet and dietary supplement interventions on recurrence and treatment-associated morbidities.

Other cancers have been less researched with regards to tomato/lycopene consumption. The cancers where tomatoes/lycopene show the most promise include gastric, lung and pancreatic (Table 13.4). For some cancer types there are complications in interpretation of potential effects due to the nature of the cancer type, human genotype and other factors. For example, with regards to ovarian cancer menopausal status may be an important factor for determining if lycopene/lycopene-rich foods have an impact. One study suggested the benefit of dietary lycopene was specific to premenopausal women (Cramer *et al.*, 2001).

Heart disease

There is an accumulating body of evidence that regular intake of tomatoes and tomato products may have benefits for heart health. Over 100 studies have been published investigating the association between tomato/tomato-based foods and/or lycopene and CVD risk. The role of lycopene for heart health has recently been reviewed by Boehm (2012), the Tomato Products Wellness Council (2012) and Mordente *et al.* (2011). In addition, Ried and Fakler

(2011) summarized current evidence on the effect of lycopene on serum lipid concentrations and blood pressure (key factors in heart health). The Tomato Products Wellness Council (2012) drew the following conclusions: 'the association between tomato/tomato-based foods remains relatively strong for a protective relationship between consumption of these foods and lower risk of cardiovascular disease, although there is variation depending on the endpoints measured'. Both Boehm (2012) and Mordente *et al.* (2011) concluded that epidemiological studies support a role for lycopene in the prevention of CVD including studies showing a dose-response relationship between lycopene and CVD. They also note that *in vitro* data and results from animal experiments partly show promising preventive mechanisms of lycopene. However, results from human intervention studies mostly fail to show any CVD prevention, although there is evidence of mechanisms of action that indicate potential protection. Ried and Fakler (2011) conducted a meta-analysis which suggested that lycopene (taken in doses $\geq$25 mg daily) is effective in reducing low density lipoprotein (LDL) cholesterol by about 10% in patients with slightly elevated cholesterol. More research is needed to confirm suggested beneficial effects on total serum cholesterol and systolic blood pressure.

There may be an explanation why controlled clinical trials and dietary intervention studies have not provided clear evidence that lycopene prevents cardiovascular diseases yet epidemiological evidence points to a benefit of tomato/tomato-based foods. Clinical trials have focussed on lycopene yet, at least in part, other compounds in tomatoes besides lycopene may exert protective effects. There are a number of studies that indicate this, including those with GABA and a range of water-soluble tomato components (WSTC I and II) mentioned above. In recent years GABA has attracted attention for its potential antihypertensive benefits in humans (Shimada *et al.*, 2009) and an animal study demonstrated an antihypertensive effect of a GABA-rich tomato cultivar (Yoshimura *et al.*, 2010). An European Food Safety Authority (EFSA) panel evaluating health claims considered that WSTC I and II are sufficiently characterised to support a health claim around platelet aggregation (EFSA, 2009).

Other diseases

Numerous other diseases have also been investigated in relation to tomatoes/lycopene (Table 13.5). For many diseases there is limited evidence but there are several areas showing considerable promise. Evidence from a number of studies points to a beneficial effect of tomato/tomato-based foods intake on protection from UV light-induced erythema. However, much of the work (5 out of 7 studies) supporting this conclusion is from a single laboratory. A positive influence of lycopene on bone health *in vitro* has been supported with limited clinical data although more studies are required to fully substantiate a benefit. Limited data support a modest inverse relationship between plasma/serum lycopene concentrations and cognitive function. There is preliminary evidence from *in vitro* studies and animal models to support a role of tomatoes/lycopene in brain health but research on cognitive performance endpoints in humans is required. Reducing the oxidative and inflammatory burden in an oxidative-inflammatory susceptible system (aged brain) is purported to be a possible mechanism for tomato/lycopene benefit. Population-based studies have indicated lycopene may have a benefit for eye health, as do other carotenoids (although unlike some of those carotenoids, such as beta-carotene, lycopene cannot be converted to vitamin A by the body). This potential benefit is supported by animal studies. Lycopene had an inhibitory effect on cataract development, offered protection against galactose induced cataract changes in lens tissue and selenite-induced stress in rat models (Pollack *et al.*, 1996; Trivedi *et al.*, 2001a, 2001b; Gupta *et al.*, 2003).

Table 13.5 Summary of the potential benefits of tomatoes/lycopene for other health conditions in addition to cancer and CVD.

Disease	Type of study and outcome	Reference(s)
Allergic rhinitis	Small-scale human clinical showed regular intake of tomato extract improved the nasal symptoms of perennial allergic rhinitis.	Yoshimura *et al.*, 2007
Asthma	Two small human clinical trials indicate potential benefit of tomatoes/lycopene including reducing the inflammation.	Riccioni *et al.*, 2007; Wood *et al.*, 2008
Bone health	Limited tomato-specific study but human studies (9) suggest a beneficial relationship between dietary sources of lycopene and elevated plasma/serum concentrations of lycopene and bone health as measured by bone mass and markers of bone metabolism.	TPWC, 2012
Brain health	Review of human studies (8) supports an inverse relationship between plasma/serum lycopene concentrations and cognitive function but there is insufficient data on the effect of tomato-specific intake on cognitive performance endpoints in humans.	TPWC, 2012
Diabetes	Review of human studies (34) show variable results depending on treatment and further study required to clarify potential benefits.	TPWC, 2012
Eye health	Population-based, case-controlled study reported individuals with low serum levels of lycopene were twice as likely to have age-related macular degeneration as those with higher serum levels.	Mares-Perlman *et al.*, 1995
Gingivitis	Small human randomized, placebo-controlled study with lycopene showed significant reductions in gingivitis and bleeding index.	Chandra *et al.*, 2007
Infertility	Two small human clinical trials showed lycopene significantly improved quality of sperm and seminal fluid – increased sperm density, motility and morphology.	Mohanty *et al.*, 2001; Gupta and Kumar, 2002
Liver conditions	Review of *in vitro*, animal and clinical studies suggested that lycopene may attenuate liver injury.	Seren *et al.*, 2008b
Pneumonia	Human trial (infants and children with pneumonia) showed decreased respiratory rate and significantly reduced IgE.	Mohamed *et al.*, 2008
Skin health	Review of human studies (7) suggests a beneficial effect of tomato (lycopene) intake in reducing UV light-induced erythema.	TPWC, 2012
Stroke	Prospective, nested case-control human study demonstrated baseline plasma concentration of lycopene tended to be inversely related to risk of ischemic stroke.	Hak *et al.*, 2004

A number of studies also suggest tomatoes may have a beneficial effect for risk factor management in people with diabetes (e.g. impact on glycation variables, lipids, oxidative stress and insulin). Both feeding tomato/tomato products and lycopene supplementation decreased oxidative stress and improved lipid status in people with diabetes (Tomato Products Wellness Council, 2012). Observational data on tomato intake are mixed with neutral, negative and positive relationships with diabetes/risk factors reported. The reason for mixed effects like other studies may reflect the types of tomato products consumed, as it may be processed tomato products that have potential benefits. These findings require additional research for clarity.

Proposed mechanisms of action of lycopene and other tomato phytochemicals

There have been numerous reviews of both the antioxidant mechanisms and other molecular mechanisms of lycopene (Erdman *et al.*, 2009; Wertz 2009). Lycopene attracted attention in the late 1980s when it was found that of all the carotenoids, including the better known beta-carotene, it was the most potent quencher of highly reactive singlet oxygen. It is also a potent scavenger of peroxyl radicals and nitrogen dioxide. It was thought that by increasing lycopene concentrations in the body, oxidative stress was reduced and antioxidant potential increased, and thus protection against various diseases conferred. However, there is now considerable debate if this is the primary mechanism for health effects and a recent review concluded that 'there is limited experimental support for the "antioxidant hypothesis" as a major mechanism of lycopene's *in vivo* action' (Erdman *et al.*, 2009). As an alternative hypothesis, the authors suggest that the lycopenoids, the metabolic products of lycopene, may be more relevant to health outcomes. The lycopenoids would not be expected to have direct antioxidant activity, due to an insufficient number of conjugated double bonds, but are more likely to act by altering gene expression. In addition lycopene, and its breakdown products, may regulate gene functions, improve intercell communication, moderate hormone function and immune response or regulate metabolism, induce apoptosis and antiproliferation in cancer cells, induce antimetastatic activity and up-regulate the antioxidant response element leading to the synthesis of cytoprotective enzymes. It is also possible that some of these mechanisms are interrelated and operate simultaneously. There may also be other modes of action that have not yet been uncovered.

Because prevention of prostate cancer has been a primary focus of studies there been considerable focus of the mechanisms by which this may occur. Herzog and Wertz (2005) summarized that in several preclinical studies lycopene improved oxidative stress defence, activated phase II drug metabolism and improved intercellular communication by enhancing gap junctional communication. Lycopene also modulated intracellular signalling pathways, for example insulin-like growth factor (IGF)-1 and androgen signalling, leading to reduced proliferation of normal and cancerous prostate epithelial cells. Animal trials have also shown down-regulation of inflammatory markers in prostate tissue and serum, which may also contribute to a cancer-preventive effect (Herzog *et al.*, 2005; Pannellini *et al.*, 2010).

With regards to cardiovascular protection it has often been thought that the primary mechanism of action of lycopene is as an antioxidant and thus to protect LDL from oxidation (Rao, 2002). More recent research has highlighted other mechanisms of action that may be beneficial. These include prevention of endothelial injury, modulation of lipid metabolism through a control of cholesterol synthesis and oxysterol toxic activities, reduction of inflammatory response through changes in cytokine production, and inhibition of smooth muscle cell proliferation through regulation of molecular pathways involved in cell proliferation and apoptosis (Palozza *et al.*, 2010).

Substances other than lycopene may also be responsible for the heart protective effects of tomatoes. Inhibition of platelet aggregation is a target mechanism for cardiovascular health due to aggregation of platelets triggering the cascade of reactions leading to blood clot formation (thrombosis). A number of studies have shown that water- and methanol-soluble compounds in tomatoes, rather than lycopene, inhibit platelet aggregation (Palomo *et al.*, 2012). In addition, an animal study demonstrated an antihypertensive effect of a GABA-rich tomato cultivar (Yoshimura *et al.*, 2010).

Other tomato components, such as the phenolics, glycoalkaloids and cystine-knot miniproteins, may also have mechanisms of action that explain the health benefits. The glycoalkaloid tomatine is a biologically active molecule and its potential effects on enzymes, cells, tissues and animals have been summarized by Friedman (2002). Tomatine may induce lowering of plasma cholesterol and triglycerides as well as enhancing the immune system via contribution to the 'oxidative burst'. The main health effect attributed to tomato phenolics is their contribution to antioxidant activity (Toor *et al.*, 2005). Although this may be true *in vitro* the significance of this *in vivo* is under debate and the benefits of the phenolics may be due to other potential mechanisms of action (Stevenson and Hurst, 2007; Crozier *et al.*, 2009). Some mechanisms are in common with lycopene and it is likely that carotenoids and phenolics may act together to exert benefits in tomato. In one human study on tomato flavonoids, naringenin absorption from cooked tomato paste was reported (Bugianesi *et al.*, 2002). They stated that their results support the hypothesis that tomato benefits could be attributed to a positive synergistic action *in vivo* between lycopene and other tomato components, such as naringenin, rather than lycopene alone. Cystine-knot miniproteins (named TCMP-1 and TCMP-2) identified from tomato have shown antiangiogenic effects *in vitro* (Cavallini *et al.*, 2011), although the significance *in vivo* is unknown at this point.

CAPSICUM

Capsicum is the name of a group of annual plants also in the Solanaceae family; it includes the sweet/bell peppers as well as hot peppers (e.g. cayenne, jalapeno) and paprika. The nomenclature can be confusing as they have a variety of names in different countries: chilli pepper, red or green pepper in North America, sweet pepper in Britain and typically 'capsicum' in Australia and New Zealand. The large mild form is called bell pepper in the United States and Canada. They are called paprika in some other countries (although paprika can also refer to the powdered spice made from various *Capsicum* fruit). The name 'pepper' came into use because of their similar flavour to the condiment black pepper (*Piper nigrum*) but there is no botanical relationship with this plant, or with Sichuan pepper. *Capsicums* are native to Mexico and Central America, where they have been cultivated for thousands of years. Although botanically a fruit they are used as vegetables, spices and medicines. There are about 25 species of *Capsicum* but cultivated varieties belong to one of the five major groups: *Capsicum annuum* (sweet/bell and Jalapeno), *C. frutescans* (Tabasco), *C. chinense* (Habanero), *C. baccatum* (Aji) and *C. pubescans* (Manzano). These five species are inter-crossable and can produce fertile and heterogeneous hybrids.

Phytochemical composition of *Capsicum*

Capsicums are an exceptional source of vitamin C, with much more than the traditional vitamin C source, oranges. They are also a rich source of several classes of phytochemicals including carotenoids, flavonoids and capsaicinoids (Table 13.6). Of these, capsaicin (a capsaicinoid) is the most studied. Fruits of different pepper varieties show variation in the composition and level of phytochemicals and they are also influenced by genotype, developmental and environmental factors (Wahyuni *et al.*, 2011).

Capsaicinoids

Capsaicinoids comprise a series of homologous branched and straight-chain alkyl vanillylamides that are classified in general as alkaloids. Although crude extracts of *Capsicum* fruits contain at least 100 different volatile chemical constituents, the main active compound in pungent fruit is capsaicin or, chemically, 8-methyl-N-vanillyl-6-nonenamide (Cordell and Araujo, 1993). Capsaicin is present in the greatest amount in placenta and seeds as compared to the pericarp. In addition to capsaicin, the next most common capsaicinoids are dihydrocapsaicin, nordihydrocapsaicin, homodihydrocapsaicin and homocapsaicin. Capsaicin and dihydrocapsaicin are the most pungent and make up about 90% of the total capsaicinoids in chilli peppers. They accumulate during ripening reaching a maximum in mature fruit, after which the content typically decreases. However, the capsaicinoids can continue to accumulate during postharvest ripening under continuous light (Macheix *et al.*, 1990).

Table 13.6 Concentrations of the main phytochemicals reported in *Capsicum*.

Component	Concentration (mg/100 g FW)
Ascorbic acid	20–393
Carotenoids (total)	0.9–141
Provitamin A carotenoids	Trace–18
ß-carotene	0–7
Capsanthin	0–28
β-cryptoxanthin	0–9
Capsorubin	0–1
Zeaxanthin	0–3
Antheraxanthin	0–1
Lutein + zeaxanthin	0.3–0.8
Total phenolics	70–543
Flavonoids	Trace–85
quercetin glycosides	0.3–51
luteolin glycosides	Trace–10.4
apigenin glycosides	0–1.4
Capsaicinoids	0–1120
capsaicin	0–660
dihydrocapsaicin	0–448
capsiate	1–37
dihydrocapsiate	2–9

Data from: Yazawa *et al.*, 1989; Lee *et al.*, 1995; Hornero-Mendez *et al.*, 2000; Garces-Claver *et al.*, 2006; Antonious *et al.*, 2009; Singh *et al.*, 2009a; Tanaka *et al.*, 2009; Cheema and Pant, 2011; USDA, 2011; Wahyuni *et al.*, 2011.

Many of the pharmacological studies of *Capsicum* have focused on capsaicin but there is growing evidence that other compounds may contribute to the medicinal properties. Considerable amounts of capsaicin-like compounds have been reported in sweet peppers (Yazawa *et al.*, 1989) but until recently the lack of pungency of these capsaicin analogues has long made them unattractive targets for research. Capsiate, the non-pungent ester isoster of capsaicin, and its dihydroderivatives are the major capsaicinoids of sweet peppers (Kobata *et al.*, 1999). The difference between the sensory properties of these plants is due to the way the vanillyl and the acyl moieties of this basic structural motif are linked: via amide bond in hot pepper (capsaicin) and via ester bond in sweet pepper (capsiate).

Carotenoids

The variation in colour of *Capsicum* fruits is largely due to the composition and concentration of a range of carotenoids, although for some it may also be influenced by comparative concentrations of chlorophyll and anthocyanins. In general, the order of carotenoid concentration is red > orange > yellow > green >>> white. In red cultivars the main carotenoid is the red capsanthin, which is largely present in the form of esters. Another red carotenoid, capsorubin, is present along with beta-cryptoxanthin, beta-carotene, zeaxanthin, antheraxanthin, vioaxanthin and neoxanthin. Yellow cultivars lack the red carotenoids; the yellow carotenoids violaxanthin and lutein usually predominate, with neoxanthin, antheraxanthin, zeaxanthin, alpha- and beta-carotene, alpha- and beta-cryptoxanthin also present. Orange cultivars, as expected, contain a mix or red and yellow carotenoids: violaxanthin, zeaxanthin, alpha-carotene, capsanthin, casporubin, beta-cryptoxanthin and antheraxanthin. Green fruit contain a narrower range of carotenoids, with beta-carotene, alpha-carotene, cryptoxanthin and lutein usually present. Brown fruit accumulate the same carotenoid pigments found in red-fruited genotypes but in combination with green chlorophyll. Shades ranging from violet to black pigmentation are attributed to anthocyanin accumulation. Chlorophyll concentration is higher in black fruit than in violet fruit. Carotenoids present in purple/violet fruit include beta-carotene, lutein, violaxanthin and neoxanthin, with concentrations higher in black fruit. During ripening of all colours total carotenoids increase and many of the hydroxylated carotenoids are progressively esterified with fatty acids (Hornero-Mendez and Minguez-Mosquera, 2000).

Flavonoids and other phenolics

There has been limited study of the phenolic composition of *Capsicum*. The major phenolics reported are phenolic acids: gallic, benzoic, 3,4-dihydroxybenzoic acid and salicylic acids. Flavonoids present include quercetin, luteolin and apigenin glycosides plus catechin. It is reported that quercetin glycosides are only present in the form of *O*-glycosides, with quercetin 3-*O*-rhamnoside and quercetin 3-*O*-rhamnoside-7-*O*-glucoside the most abundant in a range of genotypes (Wahyuni *et al.*, 2011). Apigenin glycosides are present as both *C*-glycosides and *O*-glycosides while luteolin is present in only *C*-glycosylated forms. Total flavonoid concentrations can vary hugely between cultivars, commonly up to 2000-fold. Some researchers have reported correlations between flavonoid composition and *Capsicum* type or fruit colour but others have not noted such correlations (Wahyuni *et al.*, 2011).

Studies report that phenolic acid profiles of peppers are related to their maturity stage and colour. It is reported that flavonoids are highest in immature fruit and decrease as the fruits ripen (Sukrasno and Yeoman, 1993). Another study showed that although flavonoid content decreased during maturation other phenolic acids increased (Howard *et al.*, 2000). In violet to black varieties anthocyanins are present and these are solely delphinidin glycosides, mainly delphinidin-3-*p*-coumaroylrutinoside-5-glucoside (Lightbourne *et al.*, 2008). Of note is that black varieties are also high chlorophyll which modifies the appearance.

Potential health benefits of *Capsicum*

Capsicums and their phytochemicals have received considerable scientific study including numerous human trials (Tables 13.2 and 13.7). In particular, they are regarded as having potential for treatment/prevention of obesity and diabetes. Capsaicinoids in particular are found to exert multiple pharmacological and physiological effects, including for pain relief (Box 13.2), cancer prevention and weight loss, plus benefits for the cardiovascular and gastrointestinal systems. Meta-analysis of energy balance studies concluded that 'purposeful inclusion of these compounds in the diet may aid weight management, albeit modestly' (Ludy *et al.*, 2012).

Box 13.2 Capsaicin for pain relief

Natural occurring substances have been used for hundreds of years to treat pain, one of these is capsaicin. Capsaicin has been approved by the US Food and Drug Administration (FDA) for use on the skin. It has been included in topical treatments aimed at relief of different pain conditions, although it can produce skin irritation. Capsaicin has been used in topical creams and patches to treat chronic pain syndromes, such as post-herpetic neuralgia, musculoskeletal pain, diabetic neuropathy, osteoarthritis and rheumatoid arthritis. It has also been applied to treat pain from rashes, psoriasis, mastectomy and bladder disorders.

Capsaicin has been studied quite extensively with regards to its potential health benefits (Reyes-Escogido *et al.*, 2011). Research around 20 years ago identified that capsaicin reduced adipose tissue in rodents by enhancing energy and lipid metabolism. This has resulted in numerous studies into its possible therapeutic effects in obesity treatment (Diepvens *et al.*, 2007; Leung, 2008). Despite a range of potential benefits there are some areas of debate as to the health effects. With regards to cancer studies it has been documented to have both positive and negative effects and so no firm conclusions have been able to be drawn. Research into the possible protective role of capsaicin in the gastrointestinal tract has produced controversial results, although this may be the result of dose effects. For example, at low doses capsaicin is reported to protect the gastric mucous against ulceration but at higher concentrations it is reported to worsen damage to the gastric mucous caused by ethanol or aspirin.

Various health benefits of capsaicin have been established for some time but there is now a growing body of evidence for the health benefits of the capsinoids: capsiate and dihydrocapsiate. Watanabe *et al.* (2011) reviewed the physiological and biological effects of capsinoids along with their metabolism and toxicology. Like capsaicin, capsiate may be important in energy metabolism and hence treatment/prevention of obesity. Capsiate was shown to increase adrenalin consumption, body temperature and oxygen consumption in mice. The increase in body temperature and oxygen consumption was also observed in humans. Another human study showed CH-19 Sweet pepper (containing capsiate) activated the sympathetic nervous system and enhanced thermogenesis as effectively as hot red pepper.

Table 13.7 Summary of the potential health benefits of *Capsicum*.

Disease/activity	Extract/phytochemical	Type(s) of study	Outcome(s)	Reference(s)
Allergic airway inflammation	*C. annuum*	Animal	Reduced the pathophysiological signs of allergic airway disease.	Jang *et al.*, 2011
Antimicrobial	Chilli peppers, capsaicin, dihydrocapsaicin, phenolic acids	*In vitro*	Effective against a range of bacteria including foodborne pathogens and *Helicobacter pylori*.	Qureshi *et al.*, 2008; Simonne *et al.*, 2011
Arthritis	Capsaicin	Animal	Ameliorated the progression of arthritis.	Hood *et al.*, 2001
Cancer	Capsaicin and red pepper	Human, animal, *in vitro*	Variable results (some positive and some negative) and no firm conclusions can be drawn.	Simonne *et al.*, 2011; Singletary, 2011
Cardiovascular disease	*Capsicum* and various phytochemicals	Animal, *in vitro*	Although there is no direct evidence for a benefit the modes of action of *Capsicum* phytochemicals indicate potential effects.	Simonne *et al.*, 2011
Cognitive function	Capsaicin	Animal	Stimulated sensory neurons and improved cognitive function as well as enhanced angiogenesis and neurogenesis.	Harada *et al.*, 2009
Diabetes	*C. frutescens*	Human	Lowering of plasma glucose concentrations and elevation of plasma insulin levels.	Chaiyasit *et al.*, 2009
Dyspepsia	Red pepper powder (capsaicin)	Human	Decreased abdominal pain and dyspepsia.	Bortolotti *et al.*, 2002
Energy balance	Capsaicin and Capsiate	Human	Meta-analyses indicate that capsaicin and capsiate both augment energy expenditure and enhance fat oxidation, especially at high doses. They also suppress orexigenic sensations. The magnitude of these effects is small.	Ludy *et al.*, 2012

Exercise endurance	Capsiate	Animal	Increased endurance swimming capacity.	Haramizu *et al.*, 2006
Gastroprotective (including ulcer)	Red pepper and capsaicin	Animal	Protected against induced oxidative injury (effects included increased antioxidant enzymes and mucosal glycoproteins).	Satyanarayana, 2006; Prakash and Srinivasan, 2010
Increased body temperature	CH-19 Sweet, a non-pungent cultivar of red pepper (high capsiate)	Human	Increased thermogenesis and energy consumption.	Ohnuki *et al.*, 2001
Migraine	Capsaicin	Cuman	Intranasal application lessened migraine attacks in chronic migraine patients.	Meents *et al.*, 2010
Neuroprotective	Capsaicin	Animal	Improved brain and liver dysfunction in a mouse model of hepatic failure.	Avraham *et al.*, 2009
Obesity	Capsaicin	Epidemiological evidence	Association between consumption of capsaicin and lower incidence of obesity.	Wahlqvist and Wattanapenpaiboon, 2001
Pain relief (pharmacological use)	Capsaicin (topical application)	Human	Strong evidence that effective in treating some pain conditions.	Singletary, 2011
Pneumonia	Capsiates	Cuman	Useful for triggering swallowing reflex in elderly patients with aspiration pneumonia.	Yamasaki *et al.*, 2010
Satiety	Red pepper, capsaicin	Cuman	Some studies have demonstrated increased satiety (which may be a mediator for the effect on energy balance) but results are not consistent.	Singletary, 2011; Ludy *et al.*, 2012
Thermoregulatory	Capsaicin	Animal	Effects in both cold and warm environments.	Szikszay *et al.*, 1982

These changes may be significant in terms of obesity and, in fact, it was demonstrated that capsiate promoted energy metabolism and suppressed body fat accumulation in mice. Results of a human trial indicate capsinoid intake would be able to enhance energy expenditure and fat burning in humans, particularly those with a high body mass index (BMI) (Inoue *et al.*, 2007). Since capsinoids are less toxic than capsaicinoids, they may have the advantages over capsaicinoids in clinical applications such as cancer prevention and weight loss.

In addition to the capsaicinoids other phytochemicals may confer health benefits, in particular the carotenoids. There has been limited study of capsanthin, which is found exclusively in red *Capsicum*. Unlike some carotenoids, such as beta-carotene, it has no vitamin A activity but may have activities in common with other carotenoids. A review of capsanthin notes that epidemiological studies indicate it may have an inhibitory effect on colon carcinogenesis (Kim *et al.*, 2009). It has also been shown to be protective against chemically induced hepatotoxicity and inhibit lipid peroxidation (reducing production of malondialdehyde). Capsanthin is bioavailable but its clearance from the human body is faster than that of lycopene.

Proposed mechanisms of action of capsaicin and other *Capsicum* phytochemicals

Many of the phytochemicals in *Caspicum* are regarded as antioxidants that exert their biological effects through free-radical scavenging, protein binding and interaction with human signal transduction pathways (Edge *et al.*, 1997; Padayatty *et al.*, 2003). Most early studies on the mechanism of action of capsaicin centred on its effects on the nervous system. Capsaicin, and other pungent vanilloids, activate a cell surface receptor called transient receptor potential vanilloid 1 (TRPV1; previously known as the vanilloid receptor, VR), which is expressed mainly in nociceptive neurons. TRPV1 is a cation channel which is activated by pungent vanilloid compounds, extracellular protons or noxious heat and plays a central role in neurogenic inflammation (Szallasi, 2001). Capsiate has been shown to activate TRPV1 in cells with a similar potency to capsaicin and nordihydrocapsiate is also an activator (Watanabe *et al.*, 2011). Capsaicin's effects in the nervous system are not exclusively analgesic, with studies showing that it participates in release of somatostatine, calcitonin gene related peptide (CGRP) and endotheline (Reyes-Escogido *et al.*, 2011).

Capsaicin and other vanilloids have a diversity of biological activities, some of which are mediated by the activation of cellular targets different from TRPV1 (Szallasi and Blumberg, 1999). These mechanisms include inhibition of plasma membrane NADH oxidase and induction of apoptosis in tumour cells and in activated T cells in a TRPV1-independent way (Macho *et al.*, 2000). Oyagbemi *et al.* (2010) reviewed the underlying molecular mechanisms of action of capsaicin with regards to cancer. These include blocking signalling pathways that are required for carcinogenesis, inhibition of inducible cyclooxygenase -2 (COX-2) mRNA expression resulting in anti-inflammatory activity, and inhibition of metabolic activation of xenobiotics by cytochrome P450. Capsaicin also generates reactive oxygen species in cells resulting in induction of apoptosis and cell cycle arrest.

Capsiates also target a variety of pathways involved in cancer development and inflammation. In a mouse model capsiates induced apoptosis that was preceded by an increase in the production of reactive oxygen species and by a subsequent loss of mitochondria transmembrane potential (Macho *et al.*, 2003). Anti-inflammatory activity has been shown in an *in vivo* mouse model (Sancho *et al.*, 2002). Capsiate also suppressed UVB-induced inflammatory and angiogenic activity via inhibition of extracellular signal-regulated kinase and nuclear factor-kappaB activation in skin (Lee *et al.*, 2008). Potent activity against vascular endothelial growth factor-induced angiogenesis and vascular permeability have also been reported (Pyun *et al.*, 2008). These all point to the chemopreventative potential for capsiate and dihydrocapsiate, as is reported for capsaicin.

Uncoupling proteins (UCPs) and thyroid hormones play important roles in energy expenditure, the maintenance of body weight, and thermoregulation. Masuda *et al.*, (2003) demonstrated that capsiate up-regulated UCP1 and UCP2 but had no effect on thyroid hormones. In addition, there may be other related benefits to capsiate impacts on UCPs, including increased physical performance. It was demonstrated that capsiate increased endurance swimming capacity of mice (Haramizu *et al.*, 2006). This was reportedly due to enhanced lipolysis and spared carbohydrate use via stimulation of the vanilloid receptors. In a further study capsiate administration resulted in an UCP-3 down-regulation, an enhanced muscle oxidative capacity and a decreased abdominal fat content *in vivo* (Faraut *et al.*, 2009). Research by Iwai *et al.* (2003) supports that capsiate is well absorbed into blood and may influence lipid catabolism through secretion of adrenaline via activation of the adrenal sympathetic nerve.

OTHER SOLANACEOUS VEGETABLE FRUITS

There are a two other key members of the Solanaceae that fall into the vegetable fruits category (eggplant and tomatillo) but they have received limited scientific study, especially at a human level.

Eggplant

Eggplant/aubergine (*Solanum melongena*) is botanically classified as a berry although is used as a vegetable. The most widely cultivated varieties are oval and have dark purple/black skin, although a much wider range of shapes, sizes and colours is available. Colours include lavender, pink, jade green, orange, and yellow-white and there is also variation in size and shape, from that of a small tomato to a large zucchini. Although it has a long and rich food history, at one point eggplant had a reputation of being able to cause insanity, leprosy and cancer, partly because of its slightly bitter taste. Eggplant contains nicotine and, in fact, has the highest amount of nicotine of any other edible plant. However, you would have to consume about 20 eggplants in order to get the same amount of nicotine as one cigarette.

Eggplant may have a range of potential health benefits (Table 13.2). It contains a variety of phytochemicals, such as phenolics and flavonoids, which may be responsible for these benefits. Hydroxycinamic acid derivatives such as chlorogenic acid are the most common phenolics (Singh *et al.*, 2009b). Eggplant has attracted attention because it has been reported in the top 10 vegetables in terms of antioxidant capacity (Cao *et al.*, 1996). However, as mentioned earlier, the relevance of *in vitro* antioxidant activity to *in vivo* activity is now under scrutiny. One of the phytochemicals from eggplant that has attracted study is an acylated anthocyanin, nasunin (delphinidin-3-(*p*-coumaroylrutinoside)-5-glucoside) which is present in *cis* and *trans* isomers (Ichiyanagi *et al.*, 2005). It is thought acylated anthocyanins are relatively more stable than their non-acylated relatives and may have some unique physiological functions. Nasunin is an antioxidant *in vitro* and has been shown to protect against lipid peroxidation brain homogenates (Noda *et al.*, 2000) and is bioavailable in rats (Ichiyanagi *et al.*, 2006). Although nasunin shows promise, further research is required to determine if there are any health benefits *in vivo*.

One of the few areas receiving any *in vivo* study has been the effect of eggplant on cholesterol. Studies have shown that eggplant extracts have hypolipidemic activity in rats fed normal as well as high fat diets (Sudheesh *et al.*, 1997). Eggplant capsules have been used in various South American countries to reduce cholesterol and there have been some human studies with mixed results. The results of these studies have not been published in English and the quality of studies is unclear. It has also been suggested that some of the eggplant phytochemicals interfere with analytical results, casting doubt on the outcomes (Brietzig *et al.*, 2005). Further research on the potential health benefits of eggplant is clearly warranted.

Like other members of the Solanaceae eggplant contains glycoalkaloids (α-solasonine and α-solamargine), although the total glycoalkaloid content is relatively low (Milner *et al.*, 2011). The mesocarp with seeds was found to contain 7–38 mg glycoalkaloids per 100 g, mesocarp without seeds 1–4 mg/100 g while the peel possessed negligible quantities. When compared with potato glycoalkaloids, eggplant glycoalkaloids are relatively non-toxic and there have been no reports of the ill effects. They may, in fact, have some health benefits. α-Solamargine has been shown to inhibit growth in a number of cancer cell lines and is regarded as apoptotic. Other activities also point to a role for eggplant phytochemicals in cancer prevention, although studies are *in vitro* only at this point. Extracts have been shown to suppress the formation of blood vessels required for tumour growth and metastasis (Matsubara *et al.*, 2005) and inhibit inflammation (Han *et al.*, 2003).

Tomatillo

Tomatillos (*Physalis philadelphica*) are a staple in Mexican cuisine and are related to the Cape Gooseberry (*Physalis peruviana*). They produce small, spherical and green fruit, although yellow, red or purple variants do occur. Tomatillos are most commonly used as a vegetable (e.g. in green sauces) but some varieties, mainly purple and red which are sweeter, are used as a fruit (e.g. in jams and preserves). Even though tomatillos are sometimes called 'green tomatoes', they should not be confused with them, as although they are in the same family they are a different genus.

The tomatillo has attracted little scientific study on its health benefits and what there has been is *in vitro* with no human or animal studies (Table 13.2). The fruit contains several compounds that have anticancer properties *in vitro*. A number of withanolides, including ixocarpalactone A, prevent proliferation of human colon cancer cells and induce apoptosis (Choi *et al.*, 2006). The withanolides also induce quinone reductase (a phase II enzymes responsible for detoxification of carcinogens) activity *in vitro*. In Guatemala, tomatillo is use as a folk remedy for respiratory infections and the withanolides and physalins do possess antibacterial activity *in vitro* against bacteria that cause these infections (Silva *et al.*, 2005; Gibson *et al.*, 2012). Other species of Physalis also contain withanolides, and other phytochemicals in common with tomatillo, and have demonstrated medicinal properties in animal and *in vitro* studies. These properties include antiasthmatic, anti-inflammatory, antiulcer, cholesterol lowering and hepatoprotective activities (Ramandan, 2011).

CUCURBITACEAE

The Cucurbitaceae is an important family consisting of around 120 genera and over 800 species. There are a number of commonly used vegetable fruits in this family, including cucumbers, pumpkins and squashes. The family also contains the gourds and vegetable melons, which are widely used in Eastern medicine. The Cucurbits come in many colours, including orange, yellow, red, green, white and blue. The family is predominantly distributed around the tropics, where those with edible fruits were amongst the earliest cultivated plants in both the Old and New Worlds. Pumpkin/squash are comprised of *Cucurbita moschata*, *C. pepo*, *C. maxima*, *C. mixta*, *C. ficifolia* and *Telfairia occidentalis*. Three of these, *C. pepo*, *C. maxima* and *C. moschata*, represent economically important species cultivated worldwide and have high production. The names pumpkin and squash are used inconsistently and in some cases it can be difficult to be sure what variety is being referred to in the literature. Typically *C. maxima* includes hubbard squash, buttercup squash and some types of large pumpkin, while *C. moschata* includes butternut squash/pumpkin, *C. pepo* acorn squash, summer squash (zucchini/courgette) and some pumpkins. In North America, squash is loosely grouped into summer squash or winter squash, depending on whether they are harvested as immature fruit (summer squash) or mature fruit (autumn squash or winter squash). The term 'pumpkin' is used for the types of squash that are carved to make Jack O' Lanterns at Halloween or used in pumpkin pie. In many other parts of the world the name 'pumpkin' generally refers to the broader category called winter squash in North America. In addition to the fruit pulp, the seeds and seed oil of many Cucurbits have shown considerable health benefits.

Vegetable melons and gourds are an extremely interesting group and are often used widely in Eastern medicine. *Momordica charantia* is commonly known as bitter gourd or bitter melon in English. All parts of the plant, including the fruit, have a bitter taste. The fruit is oblong and resembles a small cucumber; young fruit is emerald green that turns to orange-yellow when ripe. It is cultivated throughout the tropics particularly India, China, East Africa and South America. Other melons and gourds are also widely used and have many common properties.

Phytochemical composition of the Cucurbitaceae

Carotenoids

Although carotenoids are widespread in fruit and vegetables they are particularly abundant in some members of the Cucurbitaceae. There is considerable variation in the literature both with regards the quantity and composition of the carotenoids (Table 13.8). Total carotenoids can vary from trace amounts in white fleshed fruit up to around 40 mg/100 g in orange varieties. Some of the variation in concentrations is due to methodology used but typically total carotenoids in orange pumpkin would appear to be between 5 and 10 mg per 100 g. Many pumpkins are regarded as an excellent source of provitamin A carotenoids (e.g. beta-carotene) because they can provide more than 100% of the recommended daily intake (RDI).

Within the Cucurbitaceae that fall into the vegetable fruit category, beta-carotene is probably the most predominant carotenoid but alpha-carotene, beta-cryptoxanthin and lutein/zeaxanthin are also present in various species in significant amounts. Minor carotenoids present include zeta-carotene, violaxanthin, beta-carotene 5,6-epoxide, taraxanthin, luteoxanthin, antheraxanthin, auroxanthin, phytofluene, neurosporene and neoxanthin. In early examination of squash it was thought two of the carotenoids were distinct from others that

Table 13.8 The level of total phenolics, total flavonoids and carotenoids in some members of the Cucurbitaceae (analysed raw).

Vegetable	Total phenolics (mg GAE/100 g)	Total flavonoids (mg/100 g)	Carotenoids (μg/100 g)			
			Beta-carotene	Alpha-carotene	Beta-cryptoxanthin	Lutein + zeaxanthin
Bitter melon	—	—	190	185	—	170
Cucumber	10–29	0.2	45	11	26	23
Pumpkin	16–157	0.8–1.6	3100	515	2145	1500
Squash, summer, all varieties	12–26	0.4–0.7	120	0	0	2125
Squash, summer, crookneck and straightneck	—	—	90	0	0	290
Squash, summer, scallop	—	—	130	0	0	2308
Squash, winter, all varieties	—	—	820	0	0	38
Squash, winter, acorn	—	—	220	0	0	38
Squash, winter, butternut	—	—	4226	834	3471	0
Squash, winter, hubbard	—	—	820	0	0	0
Squash, winter, spaghetti	—	—	64	16	0	0

Data from: USDA, 2011; Lister, personal data.

had been identified; these were given the names cucurbitene and cucurbitaxanthin. However, later it was found that cucurbitene was a mixture of lutein and zeaxathin. In many summer squash the carotenoids are almost exclusively concentrated in the skin whereas with winter squash and pumpkins the pulp has high carotenoids, although concentrations are still higher in the skin. The carotenoid composition of pumpkins and squashes has been summarized elsewhere (Gross, 1991).

The exact composition of carotenoids in pumpkin/squash varies among studies and often relates to the particular type of squash analysed. In some studies β-carotene is the predominant carotenoid while in others lutein is predominant. The carotenoid composition of squashes and pumpkins from Brazil was determined by HPLC-DAD, complemented by HPLC-MS for identification (Azevedo-Meleiro and Rodriguez-Amaya, 2007). *Cucurbita moschata* 'Menina Brasileira' and *C. moschata* 'Goianinha' had similar profiles, with β-carotene and α-carotene as the major carotenoids. The hybrid 'Tetsukabuto' resembled the *Cucurbita pepo* 'Mogango', lutein and β-carotene being the principal carotenoids. *Cucurbita maxima* 'Exposicao' had a different profile, with the predominance of violaxanthin, followed by β-carotene and lutein.

Cucurbitacins

Cucurbitacins are found in a wide variety of foods, including the brassica vegetables, some mushrooms and even some mollusks. However, they are named for the Cucurbitaceae family due to their initial discovery in its members. Cucurbitacins constitute a group of diverse tetracyclic triterpenoid substances that are highly oxygenated, containing a cucurbitane skeleton characterized as 19-(10→9β)-abeo-10α-lanost-5-ene (also known as 9β-methyl-19-nor lanosta-5-ene). They are extremely bitter tasting to animals as well as humans, and they are considered to be part of the plants' natural defence mechanisms. The same properties that make cucurbitacins potentially toxic to some animals and microorganisms also make them effective as antiviral, antibacterial and anti-inflammatory substances. Che *et al.* (1985) classified the naturally occurring cucurbitacins into cucurbitacin A, B, C, D, E, F, I, L, 23, 24-dihydrocucurbitacin F and hexanorcucurbitacin F, as well as the three acetylated derivatives. They differ from each other by hydroxylation, the presence of ketone function, presence of a double bond and acetylation of a hydroxyl group. The cucurbitacins are also present in glycosidic forms, for example cucurbitacin B glucoside. The pattern of cucurbitacins varies between species of this family.

Other

The study of other phytochemicals in pumpkins/squash is limited. Polysaccharides, including protein-bound polysaccharides, are often considered the key bioactive components in pumpkin, particularly with regards their benefits for diabetes. However, there is limited understanding of structure–function relationships and this is an area requiring further research. Alkaloids have been reported in *Momordica*, and saponins have been found in *Cucurbita, Citrullus, Lagenaria* and *Momordica* species. Proteins with ribosome-inhibiting properties have been isolated from several cucurbit species. These proteins include momorcharin (from *Momordica charantia*), luffaculin (from *Luffa operculata*) and trichosanthins (from *Trichosanthes spp.*) and vary in their activities and effectiveness. Like tomatoes pumpkin contains GABA; other phytochemical components noted in the literature are polyamines and galactolipids.

Little is known about the phenolic compounds of pumpkin and squash. Small amounts of vanillic acid, *p*-coumaric acid and sinapic acid were found in pumpkin skin (Schmidtlein and Herrmann, 1975). The flavone luteolin was found in pumpkin puree at the concentration of 16.3 mg/kg (Lugasi and Hovari, 2002). A more recent study found the major phenolic compounds in squash (*C. maxima* cv. 'Turkinja') to be chlorogenic and syringic acids with smaller amounts of caffeic (Dragovic-Uzelac *et al.*, 2005). The concentrations of these phenolic acids were very low though at less than 5 mg/100 g FW. The total phenolic content of pumpkin and squash is typically between 10 and 200 mg/100 g (Table 13.8). There are some reports of higher concentrations with Jacobo-Valenzuela *et al.* (2011) noting a value of 477 mg/100 g in *C. moschata*. The amount of phenolics in the skin alone can be higher. A recent report examining the antioxidant activity of the pericarp of *C. maxima* found total polyphenolic, flavonoid and flavonol contents of methanol extracts to be 1200 mg/100 g (gallic acid equivalent), 380 mg/100 g and 80 mg/100 g (rutin equivalent), respectively (Attarde *et al.*, 2010).

Potential health benefits of the Cucurbitaceae

There has been relatively little study of the most common members of the Cucurbitaceae in the Western world. Naturally, with the high concentrations of carotenoids present in the pumpkins, these may be expected to confer benefits, although they have not been specifically the subject of study. The wider health benefits of the carotenoids have been extensively reviewed (Higdon, 2007; Maiani *et al.*, 2009). Of the Cucurbita species the only member subject to human study is the fig leaf gourd (*Cucurbita ficifolia*). This species has a history of usage in the treatment of diabetes and a single human study supports this benefit (Acosta-Patino *et al.*, 2001), although this study is not of high quality (only 10 patients). Animal and *in vitro* studies suggest pumpkins and squashes may have a diversity of health benefits (Table 13.2).

Other Cucurbitaceae have received much greater attention, in particular bitter melon (*Momordica charantia*), which is one of the traditional medicines most commonly used for diabetes. The fruit is most often eaten green or as it is beginning to turn yellow. Preclinical studies with bitter melon have documented the antidiabetic and hypoglycaemic effects but clinical trial data with human subjects are limited and flawed by poor study design and low statistical power (Leung *et al.*, 2009). Also of note are anticancer activities that have been documented in both *in vitro* and animal models (Nerurkar and Ray, 2010) but many other activities are also reported (Table 13.2). Leung *et al.* (2009) also noted toxicity and side effects, although in most cases relatively minor.

The bottle gourd (*Lagenaria sicereria*) is a vine grown for its fruit, which can either be harvested young and used as a vegetable or harvested mature, dried and used as a bottle, utensil or pipe. It has been used as a medicine in many countries including India, China, parts of Europe, Brazil and Hawaii. Although there have been no human clinical trials with bottle gourd there have been numerous small animal trials investigating a diversity of pharmacological activities (Table 13.2). A review by Tyagi *et al.* (2012) highlighted both the phytochemical composition and health benefits of bottle gourd which include benefits for diabetes and heart health. Despite these health benefits caution needs to be taken with consumption of bottle gourd, particularly if it is bitter tasting. Like other members of the Cucurbitaceae family, they contain cucurbitacins that are known to be cytotoxic and excessive consumption of bottle gourd juice causes symptoms including gut discomfort, nausea, vomiting, diarrhoea and gastrointestinal bleeding (Sharma *et al.*, 2012). Other Lagenaria species have received less study but in most cases tend to show similar properties, although some additional potential benefits have been reported. Studies with *L. breviflora* suggest it may have potential as an antifertility drug, as it had impacts on sperm count, and it has anticonvulsant, analgesic, antinociceptive and anti-inflammatory properties (Adesina and Akinwusi, 1984; Saba *et al.*, 2009; Onasanwo *et al.*, 2011). *L. vulgaris* fruit had significant anti-ulcer activity in rats (Patil and Garad, 2005). In all Laneraria species, human studies are required to support the evidence from animal and *in vitro* studies

Another member of the Cucurbitaceae that has attracted research interest is caigua. A human trial conducted in Peru has reported that caigua can lower cholesterol levels in humans (Gonzales *et al.*, 1995). Other human trials have also been conducted but have not been published in English, so it is difficult to ascertain details and the quality of these studies. Other members of the Cucurbitaceae have received less study but where reported these tend to have similar biological activities and health properties probably due to similarities in their phytochemical composition (e.g. presence of cucurbitacins, saponins). Cleary, the Cucrbitaceae warrant further study, particularly human clinical trials with regards diabetes.

Proposed mechanisms of action of Cucurbitaceae phytochemicals

A large number of biological activities have been attributed to cucurbitacins and to their glycosylated derivatives. Anti-inflammatory activities of some cucurbitacins are linked with inhibition of cyclooxygenase enzymes (Dhiman *et al.*, 2012). They have been studied as inducers of apoptosis in various cancer cell lines (Rios *et al.*, 2012). The most significant mechanisms with regard to the apoptotic effects of cucurbitacins are their ability to modify transcriptional activities via nuclear factors or genes and their capability to activate or inhibit pro- or anti-apoptotic proteins. Cucurbitacins are also selective inhibitors of the Janus kinase (JAK) and signal transducer and activator of transcription (STAT) pathways, but they may also affect the mitogen-activated protein kinase (MAPK) pathways (known to be important for cancer cell proliferation and survival), poly (ADP-ribose) polymerase (PARP) cleavage, expression of active caspase-3, decreased pSTAT3 and JAK3 concentrations, as well as decreases in various downstream STAT3 targets which are implicated in apoptosis and the cell cycle.

Saponins may be responsible, at least in part, for the health effects of many members of the Cucrbitaceae. Saponins are well known bioactive phytochemicals and have been investigated for a multitude of activities, including antimicrobial, cytotoxic, anti-inflammatory and immunostimulatory (Francis *et al.*, 2002). Although saponins are known to have hypoglycaemic activity, the cellular and molecular mechanisms of action are only beginning to be explored, and may be varied. Recent *in vitro* studies have shown that saponins from *M. charantia* may lower blood glucose by promoting insulin secretion (Keller *et al.*, 2011). However, saponins have limited bioavailability, so further work is required to understand what may happen *in vivo*.

With regards to bitter melon there has been considerable research on various potential mechanisms using *in vitro* assays and animal trials. Chaturvedi (2012) has reviewed the mechanisms behind the hypoglycaemic and hypolipidemic effects of bitter melon. Extracts inhibited glucose uptake in the intestine, suppressed the activity of intestinal maltase, sucrase and pancreatic lipase. With regards to the potential chemoprevention and anti-tumour activities of bitter melon there are a number of mechanisms of action that have been identified.

These include modulation of cell cycle regulation, inducing apoptosis, PARP cleavage and caspase activation, and decreasing cell proliferation (Nerurkar and Ray, 2010).

OTHER VEGETABLE FRUITS

Although members of the Solanaceae and Cucurbitaceae are the most common and abundant vegetable fruits, and have attracted the most scientific study, there are a few other vegetable fruits of interest.

Ackee

Ackee fruit (*Blighia sapida*) is a tropical to subtropical plant in the soapberry family (Sapindaceae) and indigenous to equatorial Africa. It is cultivated in the West Indies, Central and South America, and Florida for its edible yellow fruit arils. Ackee is the primary constituent of the national dish of Jamaica (ackee and salted cod fish) and has been consumed by the locals for centuries. Capsules of the fruits contain saponins, which lather in water and are used for washing. In many Africa countries ackee is used for making soap for treatment of scabies, tinea, dermatosis and burns as well as a general antiseptic and for softening the skin (Ekue *et al.*, 2010). Numerous other traditional medicinal uses of ackee have been reported, including dental decay, fever, malaria, internal haemorrhage, dysentery, burns, eyes inflammation, yellow fever, constipation, cutaneous infections, whitlow and head lice. All parts (bark, fruit, seeds, roots, leaves) of the plant are used although bark and leaves are the most common. Despite the extensive traditional usage there has been little scientific study with only antioxidant activity reported in the available literature (Table 13.2) and caution may need to be exercised when consuming ackee (Box 13.3).

Box 13.3 Caution needed with ackee intake

There has been a lot of controversy surrounding the inclusion of ackee in the diet due to the presence of toxic cyclopropyl peptides hypoglycin A and B (HGA and HGB). The consumption of immature ackee possessing high concentrations of HGA has been shown to be the cause of an acute condition termed Jamaican Vomiting Sickness or Toxic Hypoglycaemic Syndrome (Bowen-Forbes and Minnott, 2009). Children and individuals with compromised nutritional status are more commonly and severely affected, with symptoms ranging from prostration, vomiting, hypoglycaemia and, in rare cases, coma and death.

Breadfuit

Breadfruit (*Artocarpus altilis*) is a member of the Moraceae (fig) family. Baked or roasted in a fire, the fruit has a starchy texture and fragrance that is reminiscent of fresh baked bread. Breadfruit has been an important staple crop in the Pacific for more than 3000 years. This 'tree of bread' is regarded as having potential to play a significant role in alleviating hunger in the tropics but also has medical usage. Hawaiians used the bark, buds, fruit and roots of the breadfruit tree medicinally but the milky sap appears to be the part of the tree most frequently employed in medicinal preparations. Breadfruit tree preparation is believed to lower hypertension and has also been studied to treat taeniasis (a digestive tract infection caused by tapeworms belonging to the genus Taenia), diabetes, sore eyes, sciatica, enlarged spleen, skin infections, boils, burns, fractures, gout, oliguria and rheumatism, and as an anodyne. Leaf extracts of the breadfruit tree have been used to treat toothaches and diarrhoea. Fruit has been used less extensively and has had very little scientific study (Table 13.2).

Okra

Okra (*Abelmoschus esculentus*) is cultivated for its fibrous fruits or pods containing round, white seeds. The fruits are harvested when immature and eaten as a vegetable. Okra has a long history of medicinal usage in many cultures, including the Egyptians who believed that eating okra prevented the development of urinary calculi. Another common usage was for gastrointestinal complaints, as the mucilage okra contains is also used as soothing emollient medicine in the treatment of diarrhoea, dysentery and gastric ulcers. There is some scientific evidence to support this usage, with a gastroprotective effect of the methanolic extract of okra in ethanol-induced gastric ulcer in rats (Gurbuz *et al.*, 2003). A range of other activities has also been reported in animal and *in vitro* studies (Table 13.2). One abstract was found reporting cholesterol-lowering effects of okra in Senegalese adult men (Bangana *et al.*, 2007). However, no further details could be found.

It is recognized that okra are rich in oxalates and it is suggested people with kidney stones may benefit from restricting such foods. The formation of kidney stones containing oxalate is an area of controversy in clinical nutrition with respect to dietary restriction of oxalate. It is important to note that the ability of various oxalate-rich foods to augment urinary oxalate excretion depends not only on oxalate content but on the bioavailability. One study showed that although okra has a moderate oxalate content it had a negligible bioavailable oxalate (Brinkley *et al.*, 1990).

CONCLUSIONS

Vegetable fruits offer great diversity not only in their appearance and taste but also their phytochemical composition and health benefits. There is a large range of phytochemicals present in the key species, with carotenoids and phenolics (including flavonoids) common across all families. Most vegetable fruits have significant concentrations of carotenoids and, in some cases, they contain specific compounds with fairly narrow distribution in foods. For example, tomatoes contain lycopene and *Capsicum* species capsorubin. Some classes of phytochemicals are distinctive of each family group: Solanaceae contain glycoalkaloids, *Capsicum* species capsaicinoids and Cucurbitaceae cucurbitacins. These phytochemicals may be responsible for at least some of the health benefits but in some cases they may also be detrimental to health.

Of the vegetable fruits tomatoes have the largest body of scientific evidence for health benefits, with a link between tomato/lycopene consumption and prevention of prostate cancer well documented from *in vitro* to human studies. However, the evidence is not sufficient to make definitive health claims and more extensive human clinical trials are required. Tomatoes may also have benefits for heart health and there is growing evidence to support a role in skin, bone and eye health. *Capsicum* species have also received considerable attention with health effects from anticancer to weight management. Many of the health benefits of these fruits are attributed to capsaicin and capsinoids. Most common members of the Cucurbitaceae consumed in the Western world have had relatively little study but fruits used in Eastern medicine, such as bitter melon, have shown potential benefits, in particular for diabetes. The weight of scientific evidence supports the regular consumption of vegetable fruits such as tomatoes and *Capsicum* and their use in the development of functional foods. Other species require more study, particular in humans, to ascertain their health benefits.

REFERENCES

Abushita, A.A., Daood, H.G. and Biacs, P. (2000) Change in carotenoids and antioxidant vitamins in tomato as a function of varietal and technological factors. *Journal of Agricultural and Food Chemistry*, **48**(6), 2075–2081.

Acosta-Patino, J.L., Jimenez-Balderas, E., Juarez-Oropeza, M.A. and Diaz-Zagoya, J.C. (2001) Hypoglycemic action of *Cucurbita ficifolia* on Type 2 diabetic patients with moderately high blood glucose levels. *Journal of Ethnopharmacology* **77**(1), 99–101.

Adams, G.G., Imran, S., Wang, S. *et al.* (2011) The hypoglycaemic effect of pumpkins as anti-diabetic and functional medicines. *Food Research International* **44**(4), 862–867.

Adesina, S.K. and Akinwusi, D.D. (1984) Biological effects of *Calliandra portoricensis* and *Lagenaria breviflora* extracts. *Fitoterapia* **55**(6), 339–342.

Akanitapichat, P., Phraibung, K., Nuchklang, K. and Prompitakkul, S. (2010) Antioxidant and hepatoprotective activities of five eggplant varieties. *Food and Chemical Toxicology* **48**(10), 3017–3021.

Antonious, G.F., Lobel, L., Kochhar, T. *et al.* (2009) Antioxidants in *Capsicum chinense*: Variation among countries of origin. *Journal of Environmental Science and Health Part B – Pesticides Food Contaminants and Agricultural Wastes* **44**, 621–626.

Arawwawala, L.D.A.M., Thabrew, M.I. and Arambewela, L.S.R. (2010a) Gastroprotective activity of *Trichosanthes cucumerina* in rats. *Journal of Ethnopharmacology* **127**(3), 750–754.

Arawwawala, M., Thabrew, I., Arambewela, L. and Handunnetti, S. (2010b) Anti-inflammatory activity of *Trichosanthes cucumerina* Linn. in rats. *Journal of Ethnopharmacology* **131**(3), 538–543.

Attarde, D.L., Kadu, S.S., Chaudhari, B.J. *et al.* (2010) *In vitro* antioxidant activity of pericarp of *Cucurbita maxima* Duch. ex Lam. *International Journal of PharmTech Research* **2**(2), 1533–1538.

Avraham, Y., Grigoriadis, N.C., Magen, I. *et al.* (2009) Capsaicin affects brain function in a model of hepatic encephalopathy associated with fulminant hepatic failure in mice. *British Journal of Pharmacology* **158**(3), 896–906.

Azevedo-Meleiro, C.H. and Rodriguez-Amaya, D.B. (2007) Qualitative and quantitative differences in carotenoid composition among *Cucurbita moschata*, *Cucurbita maxima*, and *Cucurbita pepo*. *Journal of Agricultural and Food Chemistry* **55**(10), 4027–4033.

Badr, S.E.A., Shaaban, M., Elkholy, Y.M. *et al.* (2011) Chemical composition and biological activity of ripe pumpkin fruits (*Cucurbita pepo* L.) cultivated in Egyptian habitats. *Natural Product Research* **25**(16), 1524–1539.

Bangana, A., Dossou, N., Wade, S. *et al.* (2007) Cholesterol lowering effects of okra (*Hibiscus esculentus*) in Senegalese adults men. *Annals of Nutrition and Metabolism* **51**, 283–283.

Boehm, V. (2012) Lycopene and heart health. *Molecular Nutrition and Food Research* **56**(2), 296–303.

Bortolotti, M., Coccia, G., Grossi, G. and Miglioli, M. (2002) The treatment of functional dyspepsia with red pepper. *Alimentary Pharmacology and Therapeutics* **16**(6), 1075–1082.

Bowen-Forbes, C.S. and Minott, D.A. (2009) Structural characterization of hypoglycin B, a diastereomeric dipeptide from the ackee fruit (*Blighia sapida* Koenig) by NMR experiments. *Magnetic Resonance in Chemistry* **47**(11), 1004–1006.

Brietzig, E.G., Falkenberg, M.B. and Freitas, S.F.T. (2005) Evaluation of the in vitro interference from dry eggplant extract (*Solanum melongena* L.) on laboratory tests. *Acta Bioquimica Clinica Latinoamericana* **39**(4), 493–501.

Brinkley, L.J., Gregory, J. and Pak, C.Y.C. (1990) A further study of oxalate bioavailability in foods. *Journal of Urology* **144**(1), 94–96.

Bugianesi, R., Catasta, G., Spigno, P. *et al.* (2002) Naringenin from cooked tomato paste is bioavailable in men. *Journal of Nutrition* **132**(11), 3349–3352.

Butelli, E., Titta, L., Giorgio, M. *et al.* (2008) Enrichment of tomato fruit with health-promoting anthocyanins by expression of select transcription factors. *Nature Biotechnology* **26**(11), 1301–1308.

Cao, G.H., Sofic, E. and Prior, R.L. (1996) Antioxidant capacity of tea and common vegetables. *Journal of Agricultural and Food Chemistry* **44**(11), 3426–3431.

Cavallini, C., Trettene, M., Degan, M. *et al.* (2011) Anti-angiogenic effects of two cystine-knot miniproteins from tomato fruit. *British Journal of Pharmacology* **162**(6), 1261–1273.

Chaiyasit, K., Khovidhunkit, W. and Wittayalertpanya, S. (2009) Pharmacokinetic and the effect of capsaicin in *Capsicum frutescens* on decreasing plasma glucose level. *Journal of the Medical Association of Thailand* **92**(1), 108–113.

Chandra, R.V., Prabhuji, M.L.V., Roopa, D.A. *et al.* (2007) Efficacy of lycopene in the treatment of gingivitis: a randomised, placebo-controlled clinical trial. *Oral Health and Preventive Dentistry* **5**(4), 327–336.

Chaturvedi, P. (2012) Antidiabetic potentials of *Momordica charantia*: multiple mechanisms behind the effects. *Journal of Medicinal Food* **15**(2), 101–107.

Che, C.T., Fang, X.D., Phoebe, C.H. *et al.* (1985) High-field H-1-NMR spectral-analysis of some cucurbitacins. *Journal of Natural Products* **48**(3), 429–434.

Cheema, S.K. and Pant, M.R. (2011) Estimation of capsaicin in seven cultivated varieties of *Capsicum annuum* L. *Research Journal of Pharmaceutical, Biological and Chemical Sciences* **2**, 701–706.

Choi, H., Eo, H., Park, K. *et al.* (2007) A water-soluble extract from *Cucurbita moschata* shows anti-obesity effects by controlling lipid metabolism in a high fat diet-induced obesity mouse model. *Biochemical and Biophysical Research Communications* **359**(3), 419–425.

Choi, J.K., Murillo, G., Su, B.-N. *et al.* (2006) Ixocarpalactone A isolated from the Mexican tomatillo shows potent antiproliferative and apoptotic activity in colon cancer cells. *Febs Journal* **273**(24), 5714–5723.

Coates P.M. (2007) Evidence-based reviews in support of health policy decisions. *Journal of the National Cancer Institute* **99**, 1059–1059.

Colli J.L. and Amling C.L. 2009. Chemoprevention of prostate cancer: what can be recommended to patients? *Current Urology Reports* **10**(3), 165–171.

Cordell, G.A. and Araujo, O.E. (1993) Capsaicin - identification, nomenclature, and pharmacotherapy. *Annals of Pharmacotherapy* **27**(3), 330–336.

Cramer, D.W., Kuper, H., Harlow, B.L. and Titus-Ernstoff L. (2001) Carotenoids, antioxidants and ovarian cancer risk in pre- and postmenopausal women. *International Journal of Cancer* **94**, 128–134.

Crozier, A., Jaganath, I.B. and Clifford, M.N. (2009) Dietary phenolics: chemistry, bioavailability and effects on health. *Natural Product Reports* **26**(8), 1001–1043.

Danamma, B., Kumari, K.A., Goud, B.J. and Basha, S.N. (2011) Diuretic activity and study of biochemical parameters in the methanol extract of *Hibiscus esculentus* (okra) fresh fruits. *International Journal of Pharmacy and Biological Sciences* **1**(3), 160–169.

Derivi, S.C.N., Mendez, M.H.M., Francisconi, A.D. *et al.* (2002) Hypoglycemic effect of eggplant (*Solanum melongena*, L.) in rats. *Ciencia e Tecnologia de Alimentos* **22**(2), 164–169.

Dhiman, K., Gupta, A., Sharma, D.K. *et al.* (2012) A review on the medicinally important plants of the family Cucurbitaceae. *Asian Journal of Clinical Nutrition* **4**(1), 16–26.

Diepvens, K., Westerterp, K.R. and Westerterp-Plantenga, M.S. (2007) Obesity and thermogenesis related to the consumption of caffeine, ephedrine, capsaicin, and green tea. *American Journal of Physiology-Regulatory Integrative and Comparative Physiology* **292**, R77–R85.

Dixit, Y. and Kar, A. (2010) Protective role of three vegetable peels in alloxan induced diabetes mellitus in male mice. *Plant Foods for Human Nutrition* **65**(3), 284–289.

Dragovic-Uzelac, V., Delonga, K., Levaj, B. *et al.* (2005) Phenolic profiles of raw apricots, pumpkins, and their purees in the evaluation of apricot nectar and jam authenticity. *Journal of Agricultural and Food Chemistry* **53**(12), 4836–4842.

Dumas, Y., Dadomo, M., Di Lucca G. and Grolier P. (2003) Effects of environmental factors and agricultural techniques on antioxidant content of tomatoes. *Journal of the Science of Food and Agriculture* **83**(5), 369–382.

Edge, R., McGarvey, D.J. and Truscott, T.G. (1997) The carotenoids as anti-oxidants – a review. *Journal of Photochemistry and Photobiology B – Biology* **41**(3), 189–200.

EFSA (European Food Safety Authority) (2009) Water-soluble tomato concentrate (WSTC I and II) and platelet aggregation. *EFSA Journal* **7**(5), article 1101 (http://www.efsa.europa.eu/en/efsajournal/doc/1101.pdf, last accessed 10 March 2013).

Ekue, M.R.M., Sinsin, B., Eyog-Matig, O. and Finkeldey, R. (2010) Uses, traditional management, perception of variation and preferences in ackee (*Blighia sapida* KD Koenig) fruit traits in Benin: implications for domestication and conservation. *Journal of Ethnobiology and Ethnomedicine* **6**:12. doi: 10.1186/1746-4269-6-12.

Ellinger, S., Ellinger, J., Muller, S.C. and Stehle P. (2009). Tomatoes and lycopene in prevention and therapy – is there an evidence for prostate diseases? *Aktuelle Urologie* **40**(1), 37–43.

Erdman, J.W. Jr., Ford, N.A. and Lindshield B.L. (2009) Are the health attributes of lycopene related to its antioxidant function? *Archives of Biochemistry and Biophysics* **483**(2), 229–235.

Fang, E.F. and Ng, T.B. (2011) Bitter Gourd (*Momordica charantia*) is a cornucopia of health: a review of its credited antidiabetic, anti-HIV, and antitumor properties. *Current Molecular Medicine* **11**(5), 417-436.

Faraut, B., Giannesini, B., Matarazzo, V. *et al.* (2009) Capsiate administration results in an uncoupling protein-3 downregulation, an enhanced muscle oxidative capacity and a decreased abdominal fat content in vivo. *International Journal of Obesity* **33**(12), 1348–1355.

Francis, G., Kerem, Z., Makkar, H.P.S. and Becker, K. (2002) The biological action of saponins in animal systems: a review. *British Journal of Nutrition* **88**(6), 587–605.

Friedman, M. (2002) Tomato glycoalkaloids: role in the plant and in the diet. *Journal of Agricultural and Food Chemistry* **50**(21), 5751–5780.

Fu, C., Shi, H. and Li, Q. (2006) A review on pharmacological activities and utilization technologies of pumpkin. *Plant Foods for Human Nutrition* **61**(2), 73–80.

Gadang, V., Gilbert, W., Hettiararchchy, N. *et al.* (2011) Dietary bitter melon seed increases peroxisome proliferator-activated receptor-gamma gene expression in adipose tissue, down-regulates the nuclear factor-kappa B expression, and alleviates the symptoms associated with metabolic syndrome. *Journal of Medicinal Food* **14**(1–2), 86–93.

Garces-Claver, A., Arnedo-Andres, M.S., Abadia, J. *et al.* (2006) Determination of capsaicin and dihydrocapsaicin in *Capsicum* fruits by liquid chromatography-electrospray/time-of-flight mass spectrometry. *Journal of Agricultural and Food Chemistry* **54**, 9303–9311.

Gibson, K.A., Reese, R.N., Halaweish, F.T. and Ren, Y. (2012) Isolation and characterization of a bactericidal withanolide from *Physalis virginiana*. *Pharmacognosy Magazine* **8**(29), 22–28.

Giovannucci, E. (1999) Tomatoes, tomato-based products, lycopene, and cancer: review of the epidemiologic literature. *Journal of the National Cancer Institute* **91**, 317–331.

Gonzales, G.F., Gonez, C. and Villena, A. (1995) Serum-lipid and lipoprotein levels in postmenopausal women - short-course effect of caigua. *Menopause* **2**(4), 225–234.

Gonzalez-Mendoza, D., Grimaldo-Juarez, O., Soto-Ortiz, R. *et al.* (2010) Evaluation of total phenolics, anthocyanins and antioxidant capacity in purple tomatillo (*Physalis ixocarpa*) genotypes. *African Journal of Biotechnology* **9**(32), 5173–5176.

Gordon, E.A., Guppy, L.J. and Nelson, M. (2000) The antihypertensive effects of the Jamaican Cho-Cho (*Sechium edule*). *West Indian Medical Journal* **49**(1), 27–31.

Gross J. (1991) *Pigments in Vegetables: Chlorophylls and Carotenoids*. Van Nostrand Reinhold, New York.

Grover, J.K. and Yadav, S.P. (2004) Pharmacological actions and potential uses of *Momordica charantia*: a review. *Journal of Ethnopharmacology* **93**(1), 123–132.

Gupta, N.P. and Kumar, R. (2002) Lycopene therapy in idiopathic male infertility – a preliminary report. *International Urology and Nephrology* **34**(3), 369–372.

Gupta, S.K., Trivedi, D., Srivastava, S. *et al.* (2003) Lycopene attenuates oxidative stress induced experimental cataract development: an in vitro and in vivo study. *Nutrition* **19**(9), 794–799.

Gurbuz, I., Akyuz, C., Yesilada, E. and Sener, B. (2000) Anti-ulcerogenic effect of *Momordica charantia* L. fruits on various ulcer models in rats. *Journal of Ethnopharmacology* **71**(1–2), 77–82.

Gurbuz, I., Ustun, O., Yesilada, E. *et al.* (2003) Anti-ulcerogenic activity of some plants used as folk remedy in Turkey. *Journal of Ethnopharmacology* **88**, 93–97.

Hak, A.E., Ma, J., Powell, C.B. *et al.* (2004) Prospective study of plasma carotenoids and tocopherols in relation to risk of ischemic stroke. *Stroke* **35**(7), 1584–1588.

Han, S.W., Tae, J., Kim, J.A. *et al.* (2003) The aqueous extract of *Solanum melongena* inhibits PAR2 agonist-induced inflammation. *Clinica Chimica Acta* **328**(1-2), 39–44.

Harada, N., Narimatsu, N., Kurihara, H. *et al.* (2009) Stimulation of sensory neurons improves cognitive function by promoting the hippocampal production of insulin-like growth factor-I in mice. *Translational Research* **154**(2), 90–102.

Haramizu, S., Mizunoya, W., Masuda, Y. *et al.* (2006) Capsiate, a nonpungent capsaicin analog, increases endurance swimming capacity of mice by stimulation of vanilloid receptors. *Bioscience Biotechnology and Biochemistry* **70**(4), 774–781.

Haseen, F., Cantwell, M.M., O'Sullivan, J.M. and Murray, L.J. (2009) Is there a benefit from lycopene supplementation in men with prostate cancer? A systematic review. *Prostate Cancer and Prostatic Diseases* **12**(4), 325–332.

Heath, E., Sahin, K., Seren, S. and Kucuk, O. (2006) Lycopene: prospects for chemoprevention and treatmentof prostate cancer. *CAB Reviews: Perspectives in Agriculture, Veterinary Science, Nutrition and Natural Resources* **1**(064): 9 pp.

Herzog, A. and Wertz, K. (2005) Lycopene effects for improved prostate health. *Agro Food Industry Hi-Tech* **16**(1), 13–15.

Herzog, A., Siler, U., Spitzer, V. *et al.* (2005) Lycopene reduced gene expression of steroid targets and inflammatory markers in normal rat prostate. *FASEB Journal* **19**(2), 272–274.

Higdon J. (2007) *An Evidence-Based Approach to Dietary Phytochemicals*. Thieme Medical Publishers, New York.

Hood, V.C., Cruwys, S.C., Urban, L. and Kidd, B.L. (2001) The neurogenic contribution to synovial leucocyte infiltration and other outcome measures in a guinea pig model of arthritis. *Neuroscience Letters* **299**(3), 201–204.

Hornero-Mendez, D. and Minguez-Mosquera, M.I. (2000) Xanthophyll esterification accompanying carotenoid overaccumulation in chromoplast of *Capsicum annuum* ripening fruits is a constitutive process and useful for ripeness index. *Journal of Agricultural and Food Chemistry* **48**(5), 1617–1622.

Howard, L.R., Talcott, S.T., Brenes, C.H. and Villalon, B. (2000) Changes in phytochemical and antioxidant activity of selected pepper cultivars (*Capsicum* species) as influenced by maturity. *Journal of Agricultural and Food Chemistry* **48**(5), 1713–1720.

Hsu, C.-L., Shyu, M.-H., Lin, J.-A. *et al.* (2011) Cytotoxic effects of geranyl flavonoid derivatives from the fruit of *Artocarpus communis* in SK-Hep-1 human hepatocellular carcinoma cells. *Food Chemistry* **127**(1), 127–134.

Hwang, E.-S. and Bowen, P.E. (2002) Can the consumption of tomatoes or lycopene reduce cancer risk? *Integrative Cancer Therapies* **1**, 121–132.

Ichiyanagi, T., Kashiwada, Y., Shida, Y. *et al.* (2005) Nasunin from eggplant consists of cis-trans isomers of delphinidin 3- 4-(p-coumaroyl)-L-rhamnosyl (1 -> 6)glucopyranoside -5-glucopyranoside. *Journal of Agricultural and Food Chemistry* **53**(24), 9472–9477.

Ichiyanagi, T., Terahara, N., Rahman, M.M. and Konishi, T. (2006) Gastrointestinal uptake of nasunin, acylated anthocyanin in eggplant. *Journal of Agricultural and Food Chemistry* **54**(15), 5306-5312.

Ilahy, R., Hdider, C., Lenucci, M.S. *et al.* (2011) Phytochemical composition and antioxidant activity of high-lycopene tomato (*Solanum lycopersicum* L.) cultivars grown in Southern Italy. *Scientia Horticulturae* **127**, 255–261.

Ilic, D., Forbes, K.M. and Hassed, C. (2011) Lycopene for the prevention of prostate cancer. *Cochrane Database of Systematic Reviews* **11** (Art. No.: CD008007). doi: 10.1002/14651858.CD008007.pub2.

Inoue, N., Matsunaga, Y., Satoh, H. and Takahashi, M. (2007) Enhanced energy expenditure and fat oxidation in humans with high BMI scores by the ingestion of novel and non-pungent capsaicin analogues (capsinoids). *Bioscience Biotechnology and Biochemistry* **71**(2), 380–389.

Iwai, K., Yazawa, A. and Watanabe, T. (2003) Roles as metabolic regulators of the non nutrients, capsaicin and capsiate, supplemented to diets. *Proceedings of the Japan Academy Series B – Physical and Biological Sciences* **79**(7), 207–212.

Jacobo-Valenzuela, N., Marostica-Junior M.R., Zazueta-Morales, J. de J., and Gallegos-Infante, J.A. (2011) Physicochemical, technological properties, and health-benefits of *Cucurbita moschata* Duchense vs. Cehualca a review. *Food Research International* **44**(9), 2587–2593.

Jang, H.Y., Kim, S.M., Yuk, J.E. *et al.* (2011) *Capsicum annuum* L. methanolic extract inhibits ovalbumin-induced airway inflammation and oxidative stress in a mouse model of asthma. *Journal of Medicinal Food* **14**(10), 1144–1151.

Joshi, S.V., Kedar, K.A., Markana, U.V. *et al.* (2011) Alteration of gastric mucus secretion in rats treated with *Abelmoschus esculentus* seed mucilage. *Der Pharmacia Lettre* **3**(5), 183–188.

Karthiyayini, T., Rajesh, K., Kumar, K.L.S. *et al.* (2009) Evaluation of antidiabetic and hypolipidemic effect of Cucumis sativus fruit in streptozotocin-induced-diabetic rats. *Biomedical and Pharmacology Journal* **2**(2), 351–355.

Kavanaugh, C.J., Trumbo, P.R. and Ellwood, K.C. (2007) The US food and drug administration's evidence-based review for qualified health claims: Tomatoes, lycopene, and cancer. *Journal of the National Cancer Institute* **99**(14), 1074–1085.

Keller, A.C., Ma, J., Kavalier, A. *et al.* (2011) Saponins from the traditional medicinal plant *Momordica charantia* stimulate insulin secretion in vitro. *Phytomedicine* **19**(1), 32–37.

Khomsug, P., Thongjaroenbuangam, W., Pakdeenarong, N. *et al.* (2010) Antioxidative activities and phenolic content of extracts from okra (*Abelmoschus esculentus* L.). *Research Journal of Biological Sciences* **5**(4), 310–313.

Kim, S., Ha, T.Y. and Hwang, I.K. (2009) Analysis, bioavailability, and potential healthy effects of capsanthin, natural red pigment from *Capsicum* spp. *Food Reviews International* **25**(3), 198–213.

Kobata, K., Sutoh, K., Todo, T. *et al.* (1999) Nordihydrocapsiate, a new capsinoid from the fruits of a nonpungent pepper, *Capsicum annuum*. *Journal of Natural Products* **62**(2), 335–336.

Kongtun, S., Jiratchariyakul, W., Kummalue, T. *et al.* (2009) Cytotoxic properties of root extract and fruit juice of *Trichosanthes cucumerina*. *Planta Medica* **75**(8), 839–842.

Kumar, S.S., Kumar, B.R. and Mohan, G.K. (2009) Hepatoprotective effect of *Trichosanthes cucumerina* Var cucumerina L. on carbon tetrachloride induced liver damage in rats. *Journal of Ethnopharmacology* **123**(2), 347–350.

Kun, Y., Lule, U.S. and Xiao-Lin, D. (2006) Lycopene: Its properties and relationship to human health. *Food Reviews International* **22**(4), 309–333.

Kwon, Y.I., Apostolidis, E. and Shetty, K. (2008) *In vitro* studies of eggplant (*Solanum melongena*) phenolics as inhibitors of key enzymes relevant for type 2 diabetes and hypertension. *Bioresource Technology* **99**(8), 2981–2988.

Lee, Y., Howard, L.R. and Villalon, B. (1995) Flavonoids and antioxidant activity of fresh pepper (*Capsicum annum*) cultivars. *Journal of Food Science* **60**, 473–476.

Lee, E., Lee, H. and Kim, T. (2008) Capsiate suppresses UVB-incluced inflammatory and angiogenic activity via inhibition of ERK and NF-kB activation in skin. *Journal of Investigative Dermatology* **128**, S16–S16.

Lengsfeld, C., Titgemeyer, F., Faller, G. and Hensel, A. (2004) Glycosylated compounds from okra inhibit adhesion of *Helicobacter pylori* to human gastric mucosa. *Journal of Agricultural and Food Chemistry* **52**(6), 1495–1503.

Leung F.W. (2008) Capsaicin-sensitive intestinal mucosal afferent mechanism and body fat distribution. *Life Sciences* **83**, 1–5.

Leung, L., Birtwhistle, R., Kotecha, J. *et al.* (2009) Anti-diabetic and hypoglycaemic effects of *Momordica charantia* (bitter melon): a mini review. *British Journal of Nutrition* **102**(12), 1703–1708.

Li, H., Deng, Z., Liu, R. *et al.* (2011) Characterization of phytochemicals and antioxidant activities of a purple tomato (*Solanum lycopersicum* L.). *Journal of Agricultural and Food Chemistry* **59**(21), 11803–11811.

Lightbourn, G.J., Griesbach, R.J., Novotny, J.A. *et al.* (2008) Effects of anthocyanin and carotenoid combinations on foliage and immature fruit color of *Capsicum annuum* L. *Journal of Heredity* **99**(2), 105–111.

Lin, J.-A., Fang, S.-C., Wu, C.-H. *et al.* (2011) Anti-inflammatory effect of the 5,7,4'-trihydroxy-6-geranylflavanone isolated from the fruit of *Artocarpus communis* in S100B-induced human monocytes. *Journal of Agricultural and Food Chemistry* **59**(1), 105–111.

Lin, J.-A., Wu, C.-H., Fang, S.-C. and Yen, G.-C. (2012) Combining the observation of cell morphology with the evaluation of key inflammatory mediators to assess the anti-inflammatory effects of geranyl flavonoid derivatives in breadfruit. *Food Chemistry* **132**(4), 2118–2125.

Ludy, M.J., Moore, G.E. and Mattes, R.D. (2012) The effects of capsaicin and capsiate on energy balance: critical review and meta-analyses of studies in humans. *Chemical Senses* **37**(2), 103–121.

Lugasi, A. and Hovari, J. (2002) Flavonoid aglycons in foods of plant origin II. Fresh and dried fruits. *Acta Alimentaria* **31**, 63–71.

Macheix, J.J., Fleuriet, A. and Billot, J. (1990) *Fruit Phenolics*. CRC Press, Boca Raton, FL.

Macho, A., Lucena, C., Calzado, M. A. *et al.* (2000) Phorboid 20-homovanillates induce apoptosis through a VR1-independent mechanism. *Chemistry and Biology* **7**(7), 483–492.

Macho, A., Sancho, R., Minassi, A. *et al.* (2003) Involvement of reactive oxygen species in capsaicinoid-induced apoptosis in transformed cells. *Free Radical Research* **37**(6), 611–619.

Mae, N., Makino, Y., Oshita, S. *et al.* (2012) Accumulation mechanism of gamma-aminobutyric acid in tomatoes (*Solanum lycopersicum* L.) under low O_2 with and without CO_2. *Journal of Agricultural and Food Chemistry* **60**(4), 1013–1019.

Maiani, G., Caston, M.J.P., Catasta, G. *et al.* (2005) Antiangiogenic activity of nasunin, an antioxidant anthocyanin, in eggplant peels. *Journal of Agricultural and Food Chemistry* **53**(16), 6272–6275.

Maiani, G., Caston, M.J.P., Catasta, G. *et al.* (2009) Carotenoids: Actual knowledge on food sources, intakes, stability and bioavailability and their protective role in humans. *Molecular Nutrition & Food Research* **53**, S194–S218.

Mares-Perlman, J.A., Brady, W.E., Klein, R. *et al.* (1995) Serum antioxidants and age-related macular degeneration in a population-based case-control study. *Archives of Ophthalmology* **113**(12), 1518–1523.
Masuda, Y., Haramizu, S., Oki, K. *et al.* (2003) Upregulation of uncoupling proteins by oral administration of capsiate, a nonpungent capsaicin analog. *Journal of Applied Physiology* **95**(6), 2408–2415.
Matsubara, K., Kaneyuki, T., Miyake, T. and Mori, M. (2005) Antiangiogenic activity of nasunin, an antioxidant anthocyanin, in eggplant peels. *Journal of Agricultural and Food Chemistry* **53**(16), 6272–6275.
Meents, J.E., Neeb, L. and Reuter, U. (2010) TRPV1 in migraine pathophysiology. *Trends in Molecular Medicine* **16**(4), 153–159.
Mes, P.J., Boches, P., Myers, J.R. and Durst, R. (2008) Characterization of tomatoes expressing anthocyanin in the fruit. *Journal of the American Society for Horticultural Science* **133**(2), 262–269.
Milner, S.E., Brunton, N.P., Jones, P.W. *et al.* (2011) Bioactivities of glycoalkaloids and their aglycones from *Solanum* species. *Journal of Agricultural and Food Chemistry* **59**(8), 3454–3484.
Mohamed, M.S., El-Mougi, M.T., Mansour, E.H. and Saad, H.H. (2008) Administration of lycopene and beta-carotene decreased risks of pneumonia among children. *Pakistan Journal of Nutrition* **7**(2), 273–277.
Mohanty, N.K., Kumar, R. and Gupta, N.P. (2001) Lycopene therapy in the management of idiopathic oligoasthenospermia. *Indian Journal of Urology* **56**, 102–103.
Mordente, A., Guantario, B., Meucci, E. *et al.* (2011) Lycopene and cardiovascular diseases: an update. *Current Medicinal Chemistry* **18**(8), 1146–1163.
Nerurkar, P. and Ray, R.B. (2010) Bitter melon: antagonist to cancer. *Pharmaceutical Research* **27**(6), 1049–1053.
Noda, Y., Kneyuki, T., Igarashi, K. *et al.* (2000) Antioxidant activity of nasunin, an anthocyanin in eggplant peels. *Toxicology* **148**(2–3), 119–123.
Nosál'ová, G., Prisenžňáková, L.U., Košťálová, Z. *et al.* (2011) Suppressive effect of pectic polysaccharides from *Cucurbita pepo* L. var. Styriaca on citric acid-induced cough reflex in guinea pigs. *Fitoterapia* **82**(3), 357–364.
Ohnuki, K., Niwa, S., Maeda, S. *et al.* (2001) CH-19 Sweet, a non-pungent cultivar of red pepper, increased body temperature and oxygen consumption in humans. *Bioscience Biotechnology and Biochemistry* **65**(9), 2033–2036.
O'Kennedy, N., Crosbie, L., van Lieshout, M. *et al.* (2006) Effects of antiplatelet components of tomato extract on platelet function in vitro and ex vivo: a time-course cannulation study in healthy humans. *American Journal of Clinical Nutrition* **84**(3), 570–579.
Onasanwo, S.A., Saba, A.B., Oridupa, O.A. *et al.* (2011) Anti-nociceptive and anti-inflammatory properties of the ethanolic extract of *Lagenaria breviflora* whole fruit in rat and mice. *Nigerian Journal of Physiological Sciences* **26**(1), 71–76.
Oyagbemi, A.A., Saba, A.B. and Azeez, O.I. (2010) Capsaicin: A novel chemopreventive molecule and its underlying molecular mechanisms of action. *Indian Journal of Cancer* **47**(1), 53–58.
Ozkan, E., Akyuz, C., Dulundu, E. *et al.* (2012) Protective effects of lycopene on cerulein-induced experimental acute pancreatitis in rats. *Journal of Surgical Research* **176**(1), 232–238.
Padayatty, S.J., Katz, A., Wang, Y.H. *et al.* (2003) Vitamin C as an antioxidant: evaluation of its role in disease prevention. *Journal of the American College of Nutrition* **22**(1), 18–35.
Palomo, I., Fuentes, E., Padro, T. and Badimon, L. (2012) Platelets and atherogenesis: Platelet anti-aggregation activity and endothelial protection from tomatoes (*Solanum lycopersicum* L.). *Experimental and Therapeutic Medicine* **3**(4), 577–584.
Palozza, P., Parrone, N., Simone, R.E. and Catalano, A. (2010) Lycopene in atherosclerosis prevention: An integrated scheme of the potential mechanisms of action from cell culture studies. *Archives of Biochemistry and Biophysics* **504**, 26–33.
Panda, S. and Kar, A. (2011) Periplogenin, isolated from *Lagenaria siceraria*, ameliorates L-T(4)-induced hyperthyroidism and associated cardiovascular problems. *Hormone and Metabolic Research* **43**(3), 188–193.
Pannellini, T., Iezzi, M., Liberatore, M. *et al.* (2010) A dietary tomato supplement prevents prostate cancer in TRAMP mice. *Cancer Prevention Research* **3**(10), 1284–1291.
Parkinson, A.A. (2007) *Phytochemical analysis of ackee (Blighia sapida) pods.* City University of New York, NY.
Patil, K.S. and Garad, S.V. (2005) Anti-ulcer activity of fruit of *Lagenaria vulgaris. Nigerian Journal of Natural Products and Medicine* **9**, 26–28.
Pollack, A., Madar, Z., Eisner, Z. *et al.* (1996) Inhibitory effect of lycopene on cataract development in galactosemic rats. *Metabolic Pediatric and Systemic Ophthalmology* **19-20**, 31–36.

Prakash, U.N.S. and Srinivasan, K. (2010) Gastrointestinal protective effect of dietary spices during ethanol-induced oxidant stress in experimental rats. *Applied Physiology Nutrition and Metabolism* **35**(2), 134-141.

Puri, M., Kaur, I., Kanwar, R.K. *et al.* (2009) Ribosome inactivating proteins (RIPs) from *Momordica charantia* for anti-viral therapy. *Current Molecular Medicine* **9**(9), 1080–1094.

Puri, T., Goyal, S., Julka, P.K. *et al.* (2010) Lycopene in treatment of high-grade gliomas: A pilot study. *Neurology India* **58**, 20–23

Pyun, B.-J., Choi, S., Lee, Y. *et al.* (2008) Capsiate, a nonpungent capsaicin-like compound, inhibits angiogenesis and vascular permeability via a direct inhibition of Src kinase activity. *Cancer Research* **68**(1), 227–235.

Qureshi, A.A., Omer, S., Sanghai, D.B. *et al.* (2008) Capsaicin: preclinical and clinical studies. *Plant Archives* **8**(1), 7–11.

Ramadan, M.F. (2011) Bioactive phytochemicals, nutritional value, and functional properties of cape gooseberry (*Physalis peruviana*): an overview. *Food Research International* **44**(7), 1830–1836.

Ranilla, L.G., Kwon, Y.-I., Apostolidis, E. and Shetty, K. (2010) Phenolic compounds, antioxidant activity and in vitro inhibitory potential against key enzymes relevant for hyperglycemia and hypertension of commonly used medicinal plants, herbs and spices in Latin America. *Bioresource Technology* **101**(12), 4676–4689.

Rao, A.V. (2002) Lycopene, tomatoes, and the prevention of coronary heart disease. *Experimental Biology and Medicine (Maywood)* **227**(10), 908–913.

Rao, A.V. and Shen, H.L. (2002) Effect of low dose lycopene intake on lycopene bioavailability and oxidative stress. *Nutrition Research* **22**, 1125–1131.

Reddy, B.P., Goud, R.K., Mohan, S.V. and Sarma, P.N. (2009) Antiproliferative and antiangiogenic effects of partially purified *Luffa acutangula* fruit extracts on human lung adenocarcinoma epithelial cell line (A-549). *Current Trends in Biotechnology and Pharmacy* **3**(4), 396–404.

Reddy, B.P., Mohan, S.V. and Sarma, P.N. (2010) Antioxidant and antimicrobial activity of *Luffa acutangula* fruit extract based on total phenolic content as index. *Journal of Chemical and Pharmaceutical Sciences* **3**(4), 218–223.

Reddy, B.P., Mohan, S.V. and Sarma, P.N. (2011) The protective role of *Luffa acutangula* fruit methanolic fraction against t-BHP induced oxidative damage in human erythrocytes. *Current Trends in Biotechnology and Pharmacy* **5**(1), 1073–1082.

Reyes-Escogido, M.D., Gonzalez-Mondragon, E.G. and Vazquez-Tzompantzi, E. (2011) Chemical and pharmacological aspects of capsaicin. *Molecules* **16**(2), 1253–1270.

Riccioni, G., Bucciarelli, T., Mancini, B. *et al.* (2007) Plasma lycopene and antioxidant vitamins in asthma: the PLAVA study. *Journal of Asthma* **44**(6): 429–432.

Ried, K. and Fakler, P. (2011) Protective effect of lycopene on serum cholesterol and blood pressure: Meta-analyses of intervention trials. *Maturitas* **68**(4), 299–310.

Rios, J.L., Andujar, I., Escandell, J.M. *et al.* (2012) Cucurbitacins as inducers of cell death and a rich source of potential anticancer compounds. *Current Pharmaceutical Design* **18**(12), 1663–1676.

Saba, A.B., Oridupa, O.A., Oyeyemi, M.O. and Osanyigbe, O.D. (2009) Spermatozoa morphology and characteristics of male Wistar rats administered with ethanolic extract of *Lagenaria breviflora* Roberts. *African Journal of Biotechnology* **8**(7), 1170–1175.

Sabitha, V., Ramachandran, S., Naveen, K.R. and Panneerselvam, K. (2011) Antidiabetic and antihyperlipidemic potential of *Abelmoschus esculentus* (L.) Moench. in streptozotocin-induced diabetic rats. *Journal of Pharmacy and Bioallied Sciences* **3**(3), 397–402.

Saito, T., Matsukura, C., Sugiyama, M. *et al.* (2008) Screening for gamma-aminobutyric acid (GABA)-rich tomato varieties. *Journal of the Japanese Society for Horticultural Science* **77**(3), 242–250.

Sancho, R., Lucena, C., Macho, A. *et al.* (2002) Immunosuppressive activity of capsaicinoids: capsiate derived from sweet peppers inhibits NF-kappa B activation and is a potent antiinflammatory compound in vivole. *European Journal of Immunology* **32**(6), 1753–1763.

Sangh, P., Amit, K., Sharma, N. K. and Jha, K.K. (2012) *Luffa cylindrica*: an important medicinal plant. *Journal of Natural Product and Plant Resources* **2**(1), 127–134.

Sarkar, S. and Guha, D. (2008) Effect of ripe fruit pulp extract of *Cucurbita pepo* Linn. in aspirin induced gastric and duodenal ulcer in rats. *Indian Journal of Experimental Biology* **46**(9), 639–645.

Satyanarayana, M.N. (2006) Capsaicin and gastric ulcers. *Critical Reviews in Food Science and Nutrition* **46**(4), 275–328.

Schmidtlein, H. and Herrmann, K. (1975) On the phenolic acids of vegetables. II. Hydroxycinnamic acids and hydroxybenzoic acids of fruit and seed vegetables. *Zeitschrift fur Lebensmittel-Untersuchung und -Forschung* **159**(4), 213–218.

Seren, S., Lieberman, R., Bayraktar, U.D. *et al.* (2008a) Lycopene in cancer prevention and treatment. *American Journal of Therapeutics* **15**, 66–81.

Seren, S., Mutchnick, M., Hutchinson, D. *et al.* (2008b) Potential role of lycopene in the treatment of hepatitis C and prevention of hepatocellular carcinoma. *Nutrition and Cancer – an International Journal* **60**(6), 729–735.

Sharma, S.K., Puri, R., Jain, A. *et al.* (2012) Assessment of effects on health due to consumption of bitter bottle gourd (*Lagenaria siceraria*) juice. *The Indian Journal of Medical Research* **135**(1), 49–55.

Shimada M., Hasegawa T., Nishimura C. *et al.* (2009) Anti-hypertensive effect of gamma-aminobutyric acid (GABA)-rich chlorella on high-normal blood pressure and borderline hypertension in placebo-controlled double blind study. *Clinical and Experimental Hypertension* **31**, 342–354.

Silva, M.T.G., Simas, S.M., Batista, T.G.F.M. *et al.* (2005) Studies on antimicrobial activity, *in vitro*, of *Physalis angulata* L. (Solanaceae) fraction and physalin B bringing out the importance of assay determination. *Memorias do Instituto Oswaldo Cruz* **100**(7), 779–782.

Simonne, A.H., do Nascimento Nunes, C. and Brecht, J.K. (2011) Tomato and other Solananceous fruits. In: *Health-promoting Properties of Fruit and Vegetables* (ed. L.A. Terry), pp. 321–351. CAB International, Oxfordshire, UK.

Singh, S., Jarret, R., Russo, V. *et al.* (2009a) Determination of capsinoids by HPLC-DAD in *Capsicum* Species. *Journal of Agricultural and Food Chemistry* **57**, 3452–3457.

Singh, A.P., Luthria, D., Wilson, T. *et al.* (2009b) Polyphenols content and antioxidant capacity of eggplant pulp. *Food Chemistry* **114**(3), 955–961.

Singletary, K. (2011) Red pepper: overview of potential health benefits. *Nutrition Today* **46**(1), 33–47.

Slimestad, R. and Verheul, M. (2009) Review of flavonoids and other phenolics from fruits of different tomato (*Lycopersicon esculentum* Mill.) cultivars. *Journal of the Science of Food and Agriculture* **89**(8), 1255–1270.

Stevenson, D.E. and Hurst, R.D. (2007) Polyphenolic phytochemicals – just antioxidants or much more? *Cellular and Molecular Life Sciences* **64**, 2900–2916.

Story, E.N., Kopec, R.E., Schwartz, S.J. and Harris, G.K. (2010) An update on the health effects of tomato lycopene. *Annual Review of Food Science and Technology* **1**, 189–210.

Su, B.N., Gu, J.Q., Kang, Y.H. *et al.* (2004) Induction of the phase II enzyme, quinone reductase, by withanolides and norwithanolides from solanaceous species. *Mini-Reviews in Organic Chemistry* **1**(1), 115–123.

Sudheesh, S., Presannakumar, G., Vijayakumar, S. and Vijayalakshmi, N.R. (1997) Hypolipidemic effect of flavonoids from *Solanum melongena. Plant Foods for Human Nutrition* **51**(4), 321–330.

Sukranso, N. and Yeoman, M.M. (1993) Phenylpropanoid metabolism during growth and development of *Capsicum frutescens* fruits. *Phytochemistry* **32**(4), 839–844.

Szallasi, A. (2001) Vanilloid receptor ligands – Hopes and realities for the future. *Drugs and Aging* **18**(8), 561–573.

Szallasi, A. and Blumberg, P.M. (1999) Vanilloid (capsaicin) receptors and mechanisms. *Pharmacological Reviews* **51**(2), 159–211.

Szikszay, M., Obal, F., Jr. and Obal, F. (1982) Dose-response relationships in the thermoregulatory effects of capsaicin. *Naunyn-Schmiedeberg's Archives of Pharmacology* **320**(2), 97–100.

Tan, H.L., Thomas-Ahner, J.M., Grainger, E.M. *et al.* (2010) Tomato-based food products for prostate cancer prevention: what have we learned? *Cancer Metastasis Reviews* **29**, 553–568.

Tanaka, Y., Hosokawa, M., Otsu, K. *et al.* (2009) Assessment of capsiconinoid composition, nonpungent capsaicinoid analogues in *Capsicum* cultivars. *Journal of Agricultural and Food Chemistry* **57**, 5407–5412.

Taur, D.J. and Patil, R.Y. (2011) Mast cell stabilizing, antianaphylactic and antihistaminic activity of *Coccinia grandis* fruits in asthma. *Chinese Journal of Natural Medicines* **9**(5), 359–362.

Tiwari, A.K., Reddy, K.S., Radhakrishnan, J. *et al.* (2011) Influence of antioxidant rich fresh vegetable juices on starch induced postprandial hyperglycemia in rats. *Food and Function* **2**(9), 521–528.

Toor, R.K., Lister, C.E. and Savage, G.P. (2005) Antioxidant activities of New Zealand-grown tomatoes. *International Journal of Food Sciences and Nutrition* **56**(8), 597–605.

TPWC (Tomato Products Wellness Council) (2012) Nutrition Science Library [online]. Available: http://www.tomatowellness.com/research (last accessed 10 March 2013).

Trivedi, D., Mohanty, I., Srivastava, S. and Gupta, S.K. (2001a) Lycopene, a natural antioxidant protects against cataractous changes in lens epithelial cells. *Indian Journal of Pharmacology* **33**(3), 241–242.

Trivedi, D., Srivastava, S. and Gupta, S.K. (2001b) Prevention of cataract development by lycopene: an experimental study. *Indian Journal of Pharmacology* **33**(3), 241–242.

Tyagi, N., Sharma, G.N. and Hooda, V. (2012) Phytochemical and pharmacological profile of *Lagenaria siceraria*: an overview. *International Research Journal of Pharmacy* **3**(3) [online] Available: http://www.irjponline.com/admin/php/uploads/892_pdf.pdf (last accessed 10 March 2013).

Unlu, N.Z., Bohn, T., Francis, D. *et al.* (2007) Carotenoid absorption in humans consuming tomato sauces obtained from tangerine or high-beta-carotene varieties of tomatoes. *Journal of Agricultural and Food Chemistry* **55**(4), 1597–1603.

USDA (US Department of Agriculture) (2011) USDA National Nutrient Database for Standard Reference, Release 24. US Department of Agriculture, Agricultural Research Service, USDA Nutrient Data Laboratory (http://www.ars.usda.gov/nutrientdata; last accessed 10 March 2013).

van Breemen, R.B. and Pajkovic, N. (2008) Multitargeted therapy of cancer by lycopene. *Cancer Letters* **269**(2), 339–351.

Van Patten, C.L., de Boer, J.G. and Tomlinson Guns, E.S. (2008) Diet and dietary supplement intervention trials for the prevention of prostate cancer recurrence: a review of the randomized controlled trial evidence. *Journal of Urology* **180**(6), 2314–2321.

Velmurugan, V., Shiny, G. and Surekha, P.S. (2011) Phytochemical and biological screening of *Luffa cylindrica* Linn. Fruit. *International Journal of PharmTech Research* **3**(3), 1582–1585.

Wahlqvist, M.L. and Wattanapenpaiboon, N. (2001) Hot foods – unexpected help with energy balance? *Lancet* **358**(9279), 348–349.

Wahyuni, Y., Ballester, A.R., Sudarmonowati, E. *et al.* (2011) Metabolite biodiversity in pepper (*Capsicum*) fruits of thirty-two diverse accessions: Variation in health-related compounds and implications for breeding. *Phytochemistry* **72**(11-12), 1358–1370.

Watanabe, T., Ohnuki, K. and Kobata, K. (2011) Studies on the metabolism and toxicology of emerging capsinoids. *Expert Opinion on Drug Metabolism and Toxicology* **7**(5), 533–542.

Wertz, K. (2009) Lycopene effects contributing to prostate health. *Nutrition and Cancer-an International Journal* **61**(6), 775–783.

Wood, L.G., Garg, M.L., Powell, H. and Gibson, P.G. (2008) Lycopene-rich treatments modify noneosinophilic airway inflammation in asthma: proof of concept. *Free Radical Research* **42**(1), 94–102.

World Cancer Research Fund/American Institute for Cancer Research (2007) Food, Nutrition, Physical Activity, and the Prevention of Cancer: a Global Perspective. Washington, DC.

Yadav, M., Jain, S., Tomar, R. *et al.* (2010) Medicinal and biological potential of pumpkin: an updated review. *Nutrition Research Reviews* **23**(2), 184–190.

Yamasaki, M., Ebihara, S., Ebihara, T. *et al.* (2010) Effects of capsiate on the triggering of the swallowing reflex in elderly patients with aspiration pneumonia. *Geriatrics and Gerontology International* **10**(1), 107–109.

Yazawa, S., Suetome, N., Okamoto, K. and Namiki, T. (1989) Content of capsaicinoids and capsaicinoid-like substances in fruit of pepper (*Capsicum annuum*-L) hybrids made with CH-19 Sweet as a parent. *Journal of the Japanese Society for Horticultural Science* **58**(3), 601–607.

Yoshimura, M., Enomoto, T., Dake, Y. *et al.* (2007) An evaluation of the clinical efficacy of tomato extract for perennial allergic rhinitis. *Allergology International* **56**(3), 225–230.

Yoshimura, M., Toyoshi, T., Sano, A. *et al.* (2010) Antihypertensive effect of a gamma-aminobutyric acid rich tomato cultivar 'DG03-9' in spontaneously hypertensive rats. *Journal of Agricultural and Food Chemistry* **58**(1), 615–619.

Zaini, N.A.M., Anwar, F., Hamid, A.A. and Saari, N. (2011) Kundur *Benincasa hispida* (Thunb.) Cogn.: a potential source for valuable nutrients and functional foods. *Food Research International* **44**(7), 2368–2376.

FURTHER READING/RESOURCES

Bosland, P.W. and Votava, E.J. (eds) (2012) *Peppers: Vegetable and Spice Capsicums*, 2nd Edn. Crop Production Science in Horticulture #22, CABI Publishing, Wallingford, UK.

Heuvelink, E. (ed.) (2005) *Tomatoes*. Crop Production Science in Horticulture #12. CABI Publishing, Wallingford, UK.

Hodges D.M. and Lester G.E. (2011) Cucurbits [cucumber, melon, pumpkin and squash]. In: *Health-Promoting Properties of Fruit and Vegetables* (ed. L.A. Terry), pp. 118–134. CABI Publishing, Wallingford, UK.

Rao A.V. (ed.) (2006) *Tomatoes, Lycopene and Human Health. Preventing chronic diseases*. Caledonian Science Press Ltd.

Russo, V.M. (ed.) (2012) *Peppers: Botany, Production and Uses*. CABI Publishing, Wallingford, UK.

14 Potential Health Benefits of Avocados

Donald K. Ingram, Carrie M. Elks, Gary M. Davenport and George S. Roth

INTRODUCTION

The avocado fruit, *Persea americana* or *Persea gratissima*, also known as the alligator pear or butter fruit, has been cultivated for centuries, with its agricultural roots in Mexico; however, the fruit has become a major international food item only over the last few decades. Evidence of use of the avocado as a food item could date back to around 10 000 BC (Shepard and Bender, 2001). Spanish explorers reported Aztecs consuming the avocados, which they prized. English explorers were later introduced to avocados via Spanish contacts. In his 1519 book, Martín Fernández de Enciso provided the earliest known written description of the avocado in Europe. A more extensive description of avocado was written in 1526 by Gonzalo Fernandez de Oviedo, historian for the explorer Hernando Cortez, in his account, *Summario de la Natural Historia de las Indias*. In his 1696 index of Jamaican plants, Hans Sloane provided the first written record in English describing the avocado. In 1697 the word avocado first appeared in American print. Different accounts note the fruit was introduced to Indonesia in 1750, Brazil in 1809, and South Africa and Australia in the late 19th century. The horticulturist, Henry Perrine, was credited with planting the first avocado groves in the United States in Florida in 1833, with avocados reaching California as a crop around 1850. The US crop had little commercial value until the early 1900s, with cultivation confined to Florida, California and Hawaii. Little consumer interest existed until the 1950s when the avocado became a popular salad item, which increased consumption considerably. With greater consumer awareness and interest, worldwide avocado production has increased more than 300% since 1970 (FAOSTAT database produced by the Food and Agriculture Organization of the United Nations).

Regarding health benefits, the Aztecs considered the avocado to have properties as a sexual stimulant. This belief likely stemmed from the appearance of fruit on the tree as a male testicle. Indeed, the English name is derived from the Spanish word, *aguacate,* which is derived from the Aztec word, *ahuácatl,* which meant testicle tree. No modern research has supported the claim that avocado acts as a sexual stimulant. Indeed, research on the health benefits of the fruit has begun to emerge only recently.

Bioactives in Fruit: Health Benefits and Functional Foods, First Edition. Edited by Margot Skinner and Denise Hunter.

Table 14.1 Nutrition facts for avocado (as published by California Avocado Commission).

Serving Size 1/5 medium (30 g/1 oz)			
Servings per container 5			
Amount per serving			
Calories 50	Calories from fat 35		
			% Daily Value*
Total fat (g)	4.5		7%
Saturated fat (g)	0.5		
Trans fat (g)	0		
Polyunsaturated fat (g)	0.5		
Monounsaturated fat (g)	3		
Cholesterol (mg)	0		0%
Sodium (mg)	0		0%
Potassium (mg)	150		4%
Total carbohydrate (g)	3		1%
Dietary fibre (g)	2		8%
Sugars (g)	0		
Protein (g)	0		
Vitamin A 0%	• Vitamin C 4%		
Calcium 0%	• Iron 2%		
Vitamin E 4%	• Thiamin 2%		
Riboflavin 4%	• Niacin 4%		
Vitamin B6 4%	• Folate 6%		
Panotheric Acid 4%	• Phosphorus 2%		
Magnesium 2%	• Zinc 2%		
Copper 2%	• Manganese 2%		
***Percent Daily Values are based on a 2000 calorie diet. Your daily values may be higher or lower depending on your calorie needs:**			
	Calories	2000	2500
Total fat	Less than	65 g	80 g
Sat fat	Less than	20 g	25 g
Cholesterol	Less than	300 mg	300 mg
Sodium	Less than	2400 mg	2400 mg
Total carbohydrate		300 mg	375 g
Dietary fibre		25 g	30 g
Calories per gram: Fat 9 • Carbohydrate 4 • Protein 4			

With the well-established aversion to high fat foods during the 1970s and 1980s, avocados endured a poor reputation as a food due to their relatively high content of fat. Table 14.1 presents the nutrition label provided by the California Avocado Commission for their product. Given the serving size of one-fifth of a medium avocado, the whole fruit would contain about 250 kilocalories and 22.5 grams of fat; however, the types of fats should be noted, with monounsaturated fat accounting for two-thirds of total fat.

In this chapter the rather limited literature that has addressed the potential health benefits of avocados and avocado bioactives is reviewed. To this end, the focus is on studies in which the pulp or derivatives of the whole fruit have been used; however, there will be occasional references to other parts of the fruit, including the seed and the leaves. While the scientific literature yields some valuable insights into our topic, in general it must be concluded that, compared to research on many other fruits, studies supporting the health benefits of avocado have not been of the highest quality or quantity.

ANTIOXIDANT PROPERTIES

Regarding essential nutrients that act as exogenous antioxidants, avocados are relatively high in vitamins C (8.7 mg/100 g) and E (2.0 mg/100 g) compared to other fruits. Many fruits, particularly berries, have garnered scientific and consumer interest in their health benefits because of reports of their notable high levels of antioxidants related to polyphenolic content (Prior *et al.*, 2005). Technically, the avocado can be considered a large berry that contains a single seed.

While avocados possess a relatively high amount of polyphenols, their antioxidant rating is fairly low compared to other popular fruits, despite high levels of vitamins C and E (Wu *et al.*, 2006). For example, when considering antioxidant activity as measured by the widely applied oxygen radical absorbance capacity (ORAC) assay, cultivated blueberries and cranberries exhibit about five times the ORAC value of ripe avocados when compared on a wet weight basis (www.oracvalues.com/sort/food/, Wu *et al.*, 2006). Corral-Aguayo *et al.* (2008) used a variety of assays to assess antioxidant capacity, including ORAC, and found that avocados ranked well below guava, mango and strawberry when considering hydrophilic extracts; however, avocados ranked high when measuring lipophilic extracts (Wu *et al.*, 2006).

CARDIOVASCULAR AND METABOLIC DISEASE

Given the emphasis on reducing dietary fats to prevent cardiovascular disease (CVD), avocados were not a popular fruit among many Western consumers for a long time. Over the past couple of decades, however, a few preclinical and clinical studies have documented the potential benefits of avocados to mitigate some of the risk factors associated with cardiovascular and metabolic disease, such as diabetes.

Several constituents of the avocado fruit may function as bioactives and contribute to these health benefits. These include the relatively high levels of folate and vitamin B6 that have been associated with reduced CVD risk through their relationship to lower plasma homocysteine levels (McNulty *et al.*, 2008). On average, avocados contain about 0.3 mg B6 and 90 μg folate per 100 grams of fruit.

The consumption of 'healthy fats', such as monounsaturated fatty acids (MUFA) and polyunsaturated fatty acids (PUFA), has also been associated with lower CVD risk in epidemiological studies (Schwingshack *et al.*, 2011). Higher levels of MUFA are associated with less susceptibility of low density lipoproteins (LDL) to oxidation (Thijssen and Mensink, 2005). Specifically MUFA, mostly oleic acid, comprise 71% of the total fat content of ripe avocados while PUFA represent 13%, and saturated fatty acids (SFA) are about 16% of the fat. As it ripens, the fruit decreases in saturated fat and increases in MUFA and PUFA (Lu *et al.*, 2009).

Phytosterols are another major group of avocado bioactives that may have potential health benefits. Interestingly, phytosterols are similar to cholesterol in physical composition but contain an additional side chain. In general, avocados have among the highest concentrations of phytosterols of all popular fruits. Plant-derived sterols come in many forms, including campesterol and stigmasterol, but (β)-sitosterol is the most prevalent in avocados, with an estimated concentration of 0.76 mg/g in raw fruit (Duester, 2001). Compared to this concentration, campesterol represents the next highest sterol level at 0.05 mg/kg (Duester, 2001). In a wide range of studies, consumption of phytosterols has been associated with reduced risk of CVD (Gupta *et al.*, 2011; Rocha *et al.*, 2011). The most prominent effect has been reductions in circulating cholesterol levels, presumably acting through inhibition of fats in the intestine (Ostlund, 2004).

Preclinical studies

The health benefits of avocados and avocado-derived oil have been demonstrated in a number of animal models. Rats fed diets supplemented with avocado pulp exhibited increased plasma levels of high density lipoproteins (HDL) and lower triglycerides compared to rats fed an isocaloric control diet (Pérez-Méndez and García-Hernández, 2007). In an atherogenic rabbit model (high cholesterol diet), a diet enriched in avocado oil increased serum HDL compared to diets enriched in other oils (corn or coconut) while being comparable to an olive oil-containing diet. However, the diet containing avocado oil was associated with higher levels of serum cholesterol compared to the diets containing olive and corn oil but less than the diet with coconut oil (Kritchevsky *et al.*, 2003). These observations were repeated when rats were fed an atherogenic diet supplemented with defatted avocado pulp to enhance the dietary fibre concentration compared to a cellulose-based fibre diet. However, serum cholesterol did not differ between diets if no additional cholesterol was added to the diet (Naveh *et al.*, 2002). HDL was not measured, but liver fat levels were significantly lower in the rats fed the avocado diet. The authors noted that food intake was reduced in the rats fed the avocado diet, which may have contributed to the lower liver fat levels.

Other studies have investigated possible beneficial effects in animal models of metabolic disease, specifically diabetes. Using a streptozotocin (STZ) rat model of diabetes in which the pancreatic beta cells are destroyed, one study noted marked protection against the diabetic phenotype in rats given a daily oral dose of an avocado preparation derived from the whole fruit (Rao and Adinew, 2011). Specifically, the avocado-treated group exhibited significantly reduced levels of blood glucose, glycosylated haemoglobin, blood urea, and serum creatinine when compared to levels seen in the STZ hyperglycemic control rats. Another study reported similar protection against hyperglycaemia-related effects in an alloxan model of diabetes (similar to STZ models) in rats treated orally with a crude methanolic extract of avocado (Rao and Adinew, 2011).

Using a crude extract of avocado leaves, another group of investigators examined the effects of hypertension in a rat model (Ojewole *et al.*, 2007). Specifically, salt-sensitive Dahl rats exhibited reductions in arterial blood pressure following intravenous injections of the aqueous extract from the leaves.

Clinical studies

Compared to animal studies, relatively more clinical studies using avocado-enriched diets have been conducted to evaluate risk factors for CVD (mostly blood lipids) and diabetes. One well-controlled study, in which avocado pulp provided 75% of the fat content of a nutritious diet (30% total fat), confirmed the observations in animal studies of higher plasma HDL levels compared to a control diet with multiple fat sources (Carranza *et al.*, 1995). However, a follow-up study using vegetarian diets supplemented with avocado pulp provided to healthy volunteers reported reduced plasma HDL levels in an avocado-enriched diet (20% total fat with 75% of fat from avocados) with no difference in total cholesterol levels after four weeks on the diet (Carranza-Madrigal *et al.*, 1997). In contrast, triglyceride levels were significantly reduced in the subjects on the avocado-enriched diets. Similarly, in a study using a low-calorie diet (1600–2000 kcal) supplemented with avocado (200 mg/d) in healthy, but overweight or obese, volunteers, no differences relative to control diet were observed in plasma cholesterol, LDL, HDL, triglycerides, blood pressure or arterial compliance (Pieterse *et al.*, 2005). In another study using a more standard, but healthy, diet provided to both normal and hypercholesterolemic subjects, an avocado-enriched diet compared to an isocaloric control diet produced lower plasma levels of total cholesterol in both normal and hypercholesterolemic subjects after seven days and reduced LDL and triglycerides and increased HDL in the hypercholesterolemic subjects (López Ledesma, 1996). A well-controlled trial of normal females provided an avocado-enriched diet with 36% total fat showed reduced plasma cholesterol and LDL, without affecting HDL, compared to a low fat (21%) high complex carbohydrate diet (Colquhoun *et al.*, 1992). In a study of diabetics on a four-week, low-calorie diet enriched in both avocado and olive oils, reduced levels of total plasma cholesterol were noted in both diet groups without effects on LDL or HDL levels, while triglycerides were significantly lowered in the group that consumed the experimental diet enriched in avocado and olive oils (Lerman-Garber *et al.*, 1994).

In summary, while the results are mixed, a general conclusion can be drawn indicating that avocado-enriched diets can benefit risk factors of CVD measured as blood lipids. Based on this limited number of studies, however, it would appear that effects on lipids appear most reliably when the subjects are on diets with higher fat content or have other metabolic disorders. Most of the clinical studies cited above are dated and, given the indications of possible health benefits, it is interesting that more recent studies using avocado interventions for CVD and metabolic disorders have not emerged. The mechanisms of action for reducing blood lipids associated with avocado consumption also remain to be elucidated. Regarding the possibility that consumption of phytosterols may be involved in reducing cholesterol, a recent meta-analysis confirmed the generality of this effect; however, there was no overall evidence that avocado consumption reduced the incidence of CVD or related mortality (Genser *et al.*, 2012)

OSTEOARTHRITIS/BONE HEALTH

Regarding potential treatment of osteoarthritis, this subject features the highest number of papers related to the health benefits of avocados with the widest degree of well-controlled clinical trials to support the claims. Unfortunately, the reader is left uncertain regarding how specific the beneficial effects of avocado are because the great majority of studies have used a combination of avocado and soybean oils. Specifically, this combination has included what is known as ASU (avocado soy unsaponifiables). Saponification is the term used for the process of making soap from oil and lye. Unsaponifiables represent the small portions, less than 1%, of oil produced in this process that cannot be made into soap. ASU, as a brand product, represents a mixture of one part avocado oil unsaponifiables to two parts soybean oil unsaponifiables. It represents a product produced by the French company, Laboratoires Expanscience, where it is also known as A1S2 and Piasclédine 300®; the latter is the brand name of the prescription drug developed in the early 1990s for the treatment of arthritis (http://www.expanscience.com/en). ASU is sold over-the-counter in many other countries in similar compositions. Major components identified in ASU include β-sitosterol, campesterol and stimasterol, which are considered to provide therapeutic benefits via their anti-inflammatory and analgesic activities.

Since there are dozens of preclinical and clinical studies published applying this product or similar mixtures, it is not the intent to provide a comprehensive review of these papers in this section. Such reviews have already been published to which the reader is referred (Christensen *et al.*, 2008; Dinubile, 2010). Similarly, other studies and commentaries have been made in regard to the increased efficacy of ASU when mixed with other nutraceutically-based treatments for osteoarthritis, such as glucosamine/chondroitin sulfate (Henrotin, 2008) and epigallocatechin gallate, as found in green tea (Heinecke *et al.*, 2010); these studies are not covered in this section.

Preclinical studies

A large number of studies have demonstrated the efficacy of ASU in both *in vitro* and *in vivo* models (Dinubile, 2010). For *in vitro* studies the primary model has been bovine articular chondrocytes. Pretreatment of these cells with ASU following an inflammatory stimulus (lipopolysaccharide [LPS]) has been found to block the activation of several inducers and markers of inflammation, including cyclooxygenase-2 (COX-2), prostaglandin E2 (PGE2), inducible nitric oxide synthase (iNOS), IL-1α and tumour necrosis factor-α (TNF-α) (Au *et al.*, 2007; Dinubile, 2010). Similarly, using human chondrocytes stimulated with interleukin (IL)-1β to induce an inflammatory state, other investigators have also shown reductions in inflammatory markers in cultures treated with ASU, including IL6, IL8, and PGE2 (Henrotin *et al.*, 1998).

Other studies have used human chondrocytes to examine effects of ASU on metabolic and inflammatory events. When osteoblasts from sclerotic zones taken from human subjects with osteoarthritis were co-cultured with chondrocytes, one investigation reported significant inhibition in chondrocyte production of the matrix component, aggrecan, and decreased gene expression of collagen. When these cultures were pretreated with ASU, aggrecan production was not inhibited and gene expression indicated increased collagen production relative to control treated cultures (Henrotin *et al.*, 2006). In a similar study using human chondrocytes subjected to mechanical stress in addition to IL-1β stimulation, ASU treatment again showed protection against PGE2 release, as well as other markers of inflammation, including protection against degradation of IκB-α, which is an inhibitor of NFκB, a major promoter of inflammation (Gabay *et al.*, 2008).

Collectively, these *in vitro* studies demonstrate that ASU treatments can induce a wide range of therapeutic activities to counteract pathological processes involved in osteoarthritis, including collagen production and inflammation. Another study examined possible therapeutic effects of ASU on periodontal disease (Andriamanlijaona *et al.*, 2006). Specifically, this group noted reduced production of the transforming growth factor beta (TGF-β) in human periodontal ligament and human alveolar bone cells treated with IL-1β was rescued by treatment of the cultures with ASU. This finding was significant because TGF-β is involved in tissue regeneration and counteracts the effects of IL-1α.

Clinical studies

The preclinical findings regarding potential efficacy of ASU for treating osteoarthritis have been translated successfully into several clinical trials (Dinubile, 2010). In their meta-analysis of these studies, Christensen *et al.* (2008) considered only four to be well-controlled trials. Blotman *et al.* (1997) conducted a three-month trial treating osteoarthritis patients with either

300 mg/kg ASU per day or placebo. While pain scores developed to evaluate the condition did not differ significantly after treatment, subjective patient ratings were higher in the ASU group, and the need for treatment with non-steriodal anti-inflammatory drugs (NSAIDS) within this group was significantly reduced. As reported in the same paper, Blotman and colleagues also conducted a similar six-month trial recruiting patients with osteoarthritis in the knee or hip and found that measures of function and pain were significantly reduced in patients treated with ASU compared to placebo, in addition to reduced use of NSAIDs. Appelboom *et al.* (2001) conducted a three-month trial providing patients with either 300 or 600 mg ASU per day to confirm reduced use of NSAIDS and significant improvement in pain reports. Lequesne *et al.* (2002) reported on a two-year trial which provided patients with 300 mg ASU and focused on joint-space-narrowing as the primary measure, which was significantly improved; however, measures of pain were not.

In their meta-analysis of these studies, Christensen *et al.* (2008) concluded there was sufficient evidence to recommend ASU for treating osteoarthritis, particularly for those with disease in the knee compared to the hip. While side effects were notable, they appeared to be mild and did not affect a great portion of those treated successfully. In a more recent, well-controlled clinical trial, Pavelka *et al.* (2010) confirmed that daily treatment of patients with 300 mg ASU for two months was well tolerated and significantly reduced pain symptoms, increased mobility and decreased use of NSAIDS compared to baseline; however, this treatment did not differ significantly from treatment with chondroitin sulfate 400 mg provided three times daily. The authors concluded that patients would probably prefer ASU treatment because of the once daily versus thrice daily treatment regimen (Henrotin, 2008).

Studies evaluating the application of ASU for treating osteoarthritis in domestic animals have also been published. For example, Boileau *et al.* (2009) treated dogs with ASU (10 mg/kg per day) for eight weeks and observed reduced development of osteoarthritic cartilage and subchondral bone lesions in which the anterior cruciate ligament had been transected to model the disease. Kawcak *et al.* (2007) used a model of osteoarthritis in horses by surgically inducing osteochondral fragmentation. While treatment with ASU for 70 days did not appear to significantly affect pain or lameness, a significant reduction in severity of articular cartilage erosion and synovial haemorrhage as well as an increase in cartilage glycosaminoglycan synthesis were noted.

In summary, there is evidence of the efficacy of ASU for the treatment of osteoarthritis, but the specific role of the constituents of avocado cannot be determined from these studies. Thus, it remains uncertain what the impact of consumption of the whole fruit, although high in phytosterols, could have on this disease.

INFLAMMATION

Beyond the reports of the anti-inflammatory properties of ASU, there have been a few studies that have examined such properties in specific constituents derived from avocado. Some years ago, a Japanese group identified two novel compounds isolated from avocado, which they called persenone A and B (Kim *et al.*, 2000a). Using an *in vitro* cell system of mouse macrophages stimulated by lipopolysaccharide and interferon-γ, they demonstrated the potency of these compounds for inhibiting production of nitric oxide and superoxide radicals. However, they concluded that this activity was likely not due to their properties as radical scavengers, but rather to the ability to inhibit their production biochemically.

In follow-up studies, the group used cultures of mouse skin stimulated with a carcinogen to induce cancerous cells and found that treatment with persenone A inhibited hydrogen peroxide production as a marker of inflammation (Kim *et al.*, 2000b). Other analogues of persenone A were produced and confirmed to reduce nitric oxide and superoxide production in mouse leukocytes in which inflammation was induced (Kim *et al.*, 2000c). Given this progress, it is interesting that no recent reports using these compounds could be identified in the biomedical literature.

In a more recent application, Rosenblat *et al.* (2011) used an *in vitro* model of ultraviolet (UV) damage in skin cells to assess effects of several polyhydroxylated fatty alcohols (PFAs) derived from avocado seed and pulp. In human keratinocytes exposed to UV, treatment with the PFAs reduced production of IL-6 and PGE2 and also enhanced DNA repair. When human skin explants were exposed to UV, again treatment with the PFAs attenuated cellular damage. Thus, it appeared that constituents of the avocado could provide protection against UV damage, acting not only through anti-inflammatory properties but also through direct protection against damage and increased repair activities. These studies remain to be confirmed in *in vivo* animal studies or human clinical trials.

In summary, while anti-inflammatory properties of avocado have been proposed, research in this area, beyond that directed to evaluating ASU, has been sparse.

CANCER

Several studies have examined the potential anticancer effects of avocado and avocado constituents; however, nearly all of these have been *in vitro* studies. Screening a large number of edible Japanese plants in two cell lines, one study recorded significant activity of a chloroform extract from avocado pulp in inhibiting superoxide production from cells treated with a carcinogen (Kim *et al.*, 2002). Similarly, using an acetone extract from the pulp, other studies have reported reduced growth of precancerous cancer cells and shown apoptosis through arrested cell cycle (Lu *et al.*, 2005; Ding *et al.*, 2007). The former study focused on the high lutein content in the fruit cultivars but found that lutein, *per se*, was less effective in retarding cancer cell growth compared to the extract. Other studies have extended the observations to precancerous and cancerous oral cancer cell lines (Ding *et al.*, 2009; D'Ambrosio *et al.*, 2011) with some insight gained on possible molecular signalling pathways involved. Ironically, it was the ability of the chloroform extraction to increase reactive oxygen species, rather than decrease their production, that seemed to invoke the apoptotic process (Ding *et al.*, 2009). The follow-up study by the same group suggested that growth inhibitory effects of the extract involved the phosphorylation of important cancer signalling events, specifically, EGFR, c-RAFm and ERK1/2, and described two specific molecules derived from the extract that appeared to exert the antiproliferative effect (D'Ambrosio *et al.*, 2011). Interestingly, using a methanol extraction from avocado pulp, another investigation reported increased proliferation of lymphocytes as well as increased protection against cellular damage induced by cyclophosphamide treatment (Paul *et al.*, 2011). The significance of the lymphocyte proliferation would presumably increase immune response to tumours, while the latter indicated improved protection of normal cells against a chemotherapeutic agent.

Persin [(Z,Z)-1-(acetyloxy)-2-hydroxy-12,15-heneicosadien-4-1] is an oil-soluble compound derived from avocado that is higher in concentration in the plant leaves and that has been identified as a toxin (Oelrichs *et al.*, 1995). Some investigators have expanded on this discovery to examine its potential as an anticancer agent. Persin applied to human breast cancer cells can inhibit proliferation through suppression of the cell cycle (Butt *et al.*, 2006). The same group reported a possible synergy in suppression of growth of breast cancer cells when persin was combined with the cancer therapeutic drug, tamoxifen (Roberts *et al.*, 2007), and they have also identified several synthetic analogues of persin with pro-apoptotic actions in cancer cells (Brooke *et al.*, 2011).

In summary, while some promising results have emerged from *in vitro* studies indicating potential anticancer effects of avocado and avocado-derived compounds, these studies have been limited in number and do not appear to have expanded to any considerable degree. Clearly missing are animal studies to confirm efficacy *in vivo* followed by translation to clinical trials.

NEUROPROTECTION

A couple of studies have investigated the potential neuroprotective role of ASU in rat models of cerebral ischemia. Specifically, the ASU product was fed to rats for 10 days prior to inducing an ischemia reperfusion injury by manipulating occlusion of the cerebral arteries. Examining loss of neurons in the hippocampus, marked protection was noted in the group fed the ASU product compared to a control diet (Yaman *et al.*, 2007). Additionally, a marker of lipid peroxidation (MDA) and production of nitric oxide were also reduced in the hippocampi of the ASU group, while the antioxidant enzymes, superoxide dismutase (SOD) and catalase were increased. These results pertaining to neuronal damage and SOD levels were reproduced in a later study by the same group using similar methods but examining damage to the prefrontal cortex following ischemia reperfusion (Eser *et al.*, 2011). Additionally, in this study the rise in the inflammatory marker, TNFα, associated with neuronal damage, was also attenuated by ASU treatment.

Another study in mice has examined the anticonvulsive properties of a crude extract from avocado leaves (Ojewole and Amabeoku, 2006). Seizures were induced in mice by injecting either pentylenetetrazole, picrotoxin or bicuculline. Mice receiving injections of the extract showed significant reductions in seizure activity compared to placebo control treatment (distilled water) that were comparable to attenuation of seizures observed when two known anticonvulsants, phenobarbitone and diazepam, were provided to the animals before induction of seizures.

WOUND HEALING

One study examined the effects of avocado on wound healing in a rat model (Nayak *et al.*, 2008). Specifically, the time-course to healing of an excision wound was examined in rats receiving either oral or topical treatment of an avocado paste derived from the fruit pulp. Both treatment conditions showed faster wound healing (time to closure) compared to control treated animals.

AGING

D-Mannoheptulose (MH) is a rare seven-carbon sugar found in avocados that has garnered the interest of researchers investigating interventions that affect aging and age-related diseases

(Roth *et al.*, 2009; Ingram and Roth, 2011). Interest in MH by aging researchers has emerged from the relatively recent search for compounds that function as calorie restriction mimetics (CRM). CRM have been linked to the anti-aging and antidisease effects generally associated with low-caloric diets but without an actual reduction in calories consumed (Ingram *et al.*, 2006).

Avocados are relatively low in their content of conventional sugars (sucrose, glucose, and fructose) compared with other fruits, while MH and its reduced form, perseitol, contribute about 4 g per whole fruit (Shaw *et al.*, 1980). It has been well-established that the sugar content of avocados, including MH, decreases markedly as the fruit ripens (Liu *et al.*, 2002). Furthermore, it is known that MH is not confined solely to the avocado fruit, as the leaves, stems, trunks, buds and roots of the avocado tree also contain high levels of MH (Liu *et al.*, 2002).

Interest in CRM has been greatly heightened recently by studies touting the benefits of the phytochemical, resveratrol (Baur, 2010). Resveratrol is considered an activator of the SIRT1 gene that has been shown to be activated in animal and humans undergoing long-term calorie restriction. In contrast, MH has been proposed as a CRM because its functions as an inhibitor of glucose metabolism (Ingram and Roth, 2011). Specifically, MH inhibits the glycolytic enzyme, hexokinase, which is the first step in the cellular metabolism of glucose to adenosine triphosphate (ATP) and pyruvate. It is hypothesized that the inhibition of glucose metabolism creates a cellular response occurring during energy restriction. As a result, a number of genes are activated, including SIRT1, to increase the efficiency of cellular metabolism by up-regulating various cellular protective mechanisms (Ingram and Roth, 2011).

The biological activity of MH in humans was demonstrated in the 1970s as a potential treatment for hypoglycaemia. When doses of MH were delivered intravenously, a major increase in blood glucose was observed with a concomitant and marked decrease in blood insulin (Lev-Ran *et al.*, 1970). The mechanism for the enhanced blood glucose levels was considered to be MH-induced inhibition of the pancreatic beta cells to produce insulin. The high level of MH in avocado was known at that time, so additional experiments were conducted that involved the feeding of raw avocado to subjects (Viktora *et al.*, 1969). A similar endocrine response was observed that included a spike in blood glucose and a concomitant decrease in blood insulin. Results also revealed that MH could be detected in the urine following avocado consumption.

Subsequently the interest of investigators in developing an effective CRM turned to the potential benefit of avocados to reduce insulin levels, presumably through the actions of its constituent MH. As a CRM candidate, the MH-induced hyperglycaemia observed in the earlier studies was not a desirable outcome. Indeed, CR regimens in animals and humans are generally associated with lower plasma insulin levels and increased insulin sensitivity. In retrospect, it appears that the MH doses used in the early human studies may have been too high. When reduced doses of MH were delivered as a crude aqueous extract of avocado, insulin levels were reduced while insulin action improved without a concomitant rise in blood glucose (Roth *et al.*, 2009). Moreover, this effect was delivered without reducing food intake, which meets the standard definition of a CRM. Other studies conducted with dogs fed an aqueous extract of avocado fruit confirmed that MH is biologically available and biologically active based on increased circulating MH levels and reduced plasma insulin levels (Davenport *et al.*, 2010a; Davenport *et al.*, 2010b).

In theory, an effective CRM should attenuate the incidence and (or) delay the onset of a wide range of age- and metabolic-related diseases (Ingram and Roth, 2011). For example, several *in vitro* studies have used a variety of cancer cells to show that MH has the ability to retard cell proliferation (Board *et al.*, 1995; Rameriz *et al.*, 2001; Rasschaert *et al.*, 2001). Additionally, MH markedly retarded the growth of human tumour cells that were transplanted into mice (Board *et al.*, 1995). While cancer may be a strategic target for MH based on these studies, it should be noted that all of these studies used synthetic MH rather than naturally-occurring source of MH.

Several published reports from an Italian research group have described the use of an MH-containing extract for dermatological applications. The avocado-based product is called AV119 and is manufactured in France by Laboratoires Expanscience; the same company that produces ASU that was described previously. This group used cultured human keratinocytes as an *in vitro* model to demonstrate that common skin yeast was inhibited by AV119 (Donnarumma *et al.*, 2007). In a subsequent study, they demonstrated that treatment with AV119 increased the production of human beta-defensin 2 by keratinocytes. This compound exhibits antimicrobial activity when keratinocytes are attacked by bacteria, fungi and viruses (Paoletti *et al.*, 2010). AV119 has also been shown to attenuate the expression of pro-inflammatory genes (IL-1β, IL-6, IL-8, TNFα, ICAM-1) while up-regulating the expression of the protective heat-shock protein-70 when keratinocytes are stimulated with lipopolysaccharides (Donnarumma *et al.*, 2010). At this time, there are no *in vivo* studies involving AV119 in the literature.

In summary, avocados appear to be rather unique among fruits in their ability to produce high levels of the naturally-occurring seven-carbon sugar, MH. The fig is the only other fruit with detectable amounts of MH (Bevenue *et al.*, 1961). Emerging *in vivo* studies indicate that an aqueous extract of avocado fruit contains an appreciable level of MH that is bioavailable and bioactive in the dog model. Additional research will be needed to confirm the efficacy of this aqueous extract to address and alleviate specific metabolic and age-related health conditions.

TOXICITY

It is well accepted that some people are allergic to avocados. This allergy typically falls within a class of reactions known as latex-fruit syndrome (Blanco, 2003). A subset of individuals with an allergy to natural rubber latex may suffer from this broader syndrome. Avocados are not the only fruit implicated in this syndrome, as other fruits, like banana, pineapple, chestnut, kiwi, mango, strawberry and passion fruit, also contain a form of latex. Although a rare event, severe anaphylactic reactions have been reported following the consumption of avocados by this subset of people (Abrams *et al.*, 2011). Beyond the clear possibility of an allergic reaction, there appears to be no further evidence of toxicity of avocado to humans.

It should be noted that there currently persists a consensus that consumption of avocado-based components pose a toxic risk for some animals, such as birds, dogs, cats and horses (Craigmill *et al.*, 1989; McKenzie and Brown, 1991; Aregheore *et al.*, 1998; Hargis *et al.*, 1989). The toxic risk is primarily confined to the leaves and bark of the avocado tree and not the fruit itself, as no cases of persin toxicity have been reported due to the ingestion of avocado fruit (Burger *et al.*, 1994). Nevertheless, this belief was reinforced by a single report in 1994 regarding the deaths of two dogs that lived near an avocado grove in South Africa and regularly consumed avocado fruit and vegetative parts of the avocado trees (Buoro *et al.*, 1994). However, it has also been demonstrated that an aqueous extract of avocado fruit can be safely fed to adult dogs for six months with no adverse effects on their overall health and wellness (Davenport *et al.*, 2012).

In summary, there is literature indicating that the consumption of certain parts of the avocado plant may be toxic to certain animals. However, there are no reports of toxicity in humans with the exception of allergic reactions by persons suffering from the latex-fruit syndrome. Nonetheless, there would appear to be a need for further safety-based studies involving avocados and avocado-derived products with particular focus on select animal species.

CONCLUSIONS

Compared to the plethora of research demonstrating the health benefits of many fruits, similar research is far less developed for avocados, avocado-derived products or avocado-derived bioactive compounds. It is possible that one of the reasons for the limited research involving avocado may be its negative reputation as a high-fat food. However, research using *in vitro* and *in vivo* animal models has shown potential health benefits associated with avocado consumption that includes reductions in CVD risk factors, osteoporosis, cancer, and inflammation. Clinical trials have been limited primarily to studies of blood lipids and osteoporosis. While appearing highly promising, research studying the benefit on osteoporosis has been limited to a mixture of a product derived from avocados and soy products. As such, the sole contribution of avocado remains largely untested. The recent research that has focused on identifying calorie restriction mimetics is bringing new attention to this fruit (Ingram and Roth, 2011). Specifically, avocado-derived mannoheptulose continues to be investigated for its potential ability to address a wide range of health conditions. Given the potential for nutrition-based health benefits associated with avocado consumption, it is anticipated that research using avocados and avocado-derived products will greatly accelerate over the next decade.

REFERENCES

Abrams, E.M., Becker, A.B. and Gerstner, T.V. (2011) Anaphylaxis related to avocado ingestion: a case and review. *Allergy Asthma Clin Immunol* **7**, 12.

Andriamanalijaona, R., Benateau, H., Barre, P.E. *et al.* (2006) Effect of interleukin-1beta on transforming growth factor-beta and bone morphogenetic protein-2 expression in human periodontal ligament and alveolar bone cells in culture: modulation by avocado and soybean unsaponifiables. *J Periodontol* **77**, 1156–66.

Appelboom, T., Schuermans, J., Verbruggen, G. *et al.* (2001) Symptoms modifying effect of avocado/soybean unsaponifiables (ASU) in knee osteoarthritis. A double blind, prospective, placebo-controlled study. *Scand J Rheumatol* **30**, 242–7.

Aregheore, E.M. (1998) A review of implications of antiquality and toxic components in unconventional feedstuffs advocated for use in intensive animal production in Nigeria. *Vet Hum Toxicol* **40**, 35–9.

Au, R.Y., Al-Talib, T.K., Au, A.Y. *et al.* (2007) Avocado soybean unsaponifiables (ASU) suppress TNF-alpha, IL-1 beta, COX-2, iNOS gene expression and prostaglandin E-2 and nitric oxide production in articular chondrocytes and monocyte/macrophages. *Osteoarthr Cartil* **15**(11). 1249–55.

Baur, J.A. (2010) Resveratrol, sirtuins, and the promise of a DR mimetic. *Mech Ageing Dev* **131**, 261–9.

Bevenue, A., White, L.M., Secor, G.E. and Williams, K.T. (1961) The occurrence of two heptuloses in the fig plant. *J Assn Offic Agr Chemists* **44**, 265–266.

Blanco, C. (2003) Latex-fruit syndrome. *Curr Allergy Asthma Rep* **3**, 47–53.

Blotman, F., Maheu, E., Wulwik, A. *et al.* (1997) Efficacy and safety of avocado/soybean unsaponifiables in the treatment of symptomatic osteoarthritis of the knee and hip. A prospective, multicenter, three-month, randomized, double-blind, placebo-controlled trial. *Rev Rhum Engl Ed* **64**, 825–34.

Board, M., Colquhoun, A. and Newsholme, E.A. (1995) High Km glucose-phosphorylating (glucokinase) activities in a range of tumor cell lines and inhibition of rates of tumor growth by the specific enzyme inhibitor mannoheptulose. *Cancer Res* **55**(15), 3278–85.

Boileau, C., Martel-Pelletier, J., Caron, J. *et al.* (2009) Protective effects of total fraction of avocado/soybean unsaponifiables on the structural changes in experimental dog osteoarthritis: inhibition of nitric oxide synthase and matrix metalloproteinase-13. *Arthritis Res Ther* **11**, R41.

Brooke, D.G., Shelley, E.J., Roberts, C.G. *et al.* (2011) Synthesis and in vitro evaluation of analogues of avocado-produced toxin (+)-(R)-persin in human breast cancer cells. *Bioorg Med Chem* **19**, 7033–43.

Buoro, I.B., Nyamwange, S.B., Chai, D. and Munyua, S.M. (1994) Putative avocado toxicity in two dogs. *Onderstepoort J Vet Res* **61**, 107–9.

Burger, W.P., Naude, T.W., Van Rensburg, I.B. *et al.* (1994) Cardiomyopathy in ostriches (Struthio camelus) due to avocado (Persea americana var. guatemalensis) intoxication. *J S Afr Vet Assoc* **65**, 113–8.

Butt, A.J., Roberts, C.G., Seawright, A.A. *et al.* (2006) A novel plant toxin, persin, with *in vivo* activity in the mammary gland, induces Bim-dependent apoptosis in human breast cancer cells. *Mol Cancer Ther* **5**, 2300–9.

Carranza, J., Alvizouri, M., Alvarado, M.R. *et al.* (1995) Effects of avocado on the level of blood lipids in patients with phenotype II and IV dyslipidemias. *Arch Inst Cardiol Mex* **65**, 342–8.

Carranza-Madrigal, J., Herrera-Abarca, J.E., Alvizouri-Munoz, M. *et al.* (1997) Effects of a vegetarian diet vs. a vegetarian diet enriched with avocado in hypercholesterolemic patients. *Arch Med Res* **28**, 537–41.

Christensen, R., Bartels, E.M., Astrup, A. and Bliddal, H. (2008) Symptomatic efficacy of avocado-soybean unsaponifiables (ASU) in osteoarthritis (OA) patients: a meta-analysis of randomized controlled trials. *Osteoarthritis Cartilage* **16**, 399–408.

Colquhoun, D.M., Moores, D., Somerset, S.M. and Humphries, J.A. (1992) Comparison of the effects on lipoproteins and apolipoproteins of a diet high in monounsaturated fatty acids, enriched with avocado, and a high-carbohydrate diet. *Am J Clin Nutr* **56**, 671–7.

Corral-Aguayo, R.D., Yahia, E.M., Carrillo-Lopez, A. and Gonzalez-Aguilar, G. (2008) Correlation between some nutritional components and the total antioxidant capacity measured with six different assays in eight horticultural crops. *J Agric Food Chem* **56**, 10498–504.

Craigmill, A.L., Seawright, A.A., Mattila, T. and Frost, A.J. (1989) Pathological changes in the mammary gland and biochemical changes in milk of the goat following oral dosing with leaf of the avocado (Persea americana). *Aust Vet J* **66**, 206–11.

D'Ambrosio, S.M., Han, C., Pan, L. *et al.* (2011) Aliphatic acetogenin constituents of avocado fruits inhibit human oral cancer cell proliferation by targeting the EGFR/RAS/RAF/MEK/ERK1/2 pathway. *Biochem Biophys Res Commun* **409**, 465–9.

Davenport, G., Massimino, S., Hayek, M. *et al.* (2010a) Bioavailability of avocado-derived mannoheptulose in dogs. *FASEB J*, **24**, 725.3.

Davenport G, Massimino S, Hayek M *et al.* (2010b) Biological activity of avocado-derived mannoheptulose in dogs. *FASEB J*, **24**, 725.4.

Davenport, G., Shoveller, A. K., Hayek, M. *et al.* (2012) Tolerance and safety of an avocado-based ingredient for adult dogs. *FASEB J*, **26**, 825.1.

Ding, H., Chin, Y. W., Kinghorn, A. D. and D'Ambrosio, S. M. (2007) Chemopreventive characteristics of avocado fruit. *Semin Cancer Biol* **17**, 386–94.

Ding, H., Han, C., Guo, D. *et al.* (2009) Selective induction of apoptosis of human oral cancer cell lines by avocado extracts via a ROS-mediated mechanism. *Nutr Cancer* **61**, 348–56.

Dinubile, N.A. (2010) A potential role for avocado- and soybean-based nutritional supplements in the management of osteoarthritis: a review. *Phys Sportsmed* **38**, 71–81.

Donnarumma, G., Buommino, E., Baroni, A. *et al.* (2007) Effects of AV119, a natural sugar from avocado, on Malassezia furfur invasiveness and on the expression of HBD-2 and cytokines in human keratinocytes. *Exp Dermatol* **16**, 912–9.

Donnarumma, G., Paoletti, I., Buommino, E. *et al.* (2011) AV119, a natural sugar from avocado gratissima, modulates the LPS-induced proinflammatory response in human keratinocytes. *Inflammation* **34**, 568–75.

Duester, K.C. (2001) Avocado fruit is a rich source of beta-sitosterol. *J Am Diet Assoc* **101**(4), 404–5.

Eser, O., Songur, A., Yaman, M. *et al.* (2011) The protective effect of avocado soybean unsaponifilables on brain ischemia/reperfusion injury in rat prefrontal cortex. *Br J Neurosurg* **25**, 701–6.

Gabay, O., Gosset, M., Levy, A. *et al.* (2008) Stress-induced signaling pathways in hyalin chondrocytes: inhibition by Avocado-Soybean Unsaponifiables (ASU). *Osteoarthritis Cartilage* **16**, 373–84.

Genser, B., Silbernagel, G., De Backer, G. *et al.* (2012) Plant sterols and cardiovascular disease: a systematic review and meta-analysis. *Eur Heart J* **33**, 444–51.

Gupta, A.K., Savopoulos, C.G., Ahuja, J. and Hatzitolios, A.I. (2011) Role of phytosterols in lipid-lowering: current perspectives. *QJM* **104**, 301–8.

Hargis, A.M., Stauber, E., Casteel, S. and Eitner, D. (1989) Avocado (*Persea americana*) intoxication in caged birds. *J Am Vet Med Assoc* **194**, 64–6.

Heinecke, L.F., Grzanna, M.W., Au, A.Y. *et al.* (2010) Inhibition of cyclooxygenase-2 expression and prostaglandin E2 production in chondrocytes by avocado soybean unsaponifiables and epigallocatechin gallate. *Osteoarthritis Cartilage* **18**, 220–7.

Henrotin, Y. (2008) Avocado/soybean unsaponifiable (ASU) to treat osteoarthritis: a clarification. *Osteoarthritis Cartilage* **16**, 1118–9; author reply 1120.

Henrotin, Y.E., Labasse, A.H., Jaspar, J.M. *et al.* (1998) Effects of three avocado/soybean unsaponifiable mixtures on metalloproteinases, cytokines and prostaglandin E-2 production by human articular chondrocytes. *Clin Rheumatol* **17**(1), 31–9.

Henrotin, Y., Sanchez, C. and Balligand, M. (2005). Pharmaceutical and nutraceutical management of canine osteoarthritis: present and future perspectives. *Vet J* **170**, 113–23.

Henrotin, Y.E., Deberg, M.A., Crielaard, J.M. *et al.* (2006) Avocado/soybean unsaponifiables prevent the inhibitory effect of osteoarthritic subchondral osteoblasts on aggrecan and type II collagen synthesis by chondrocytes. *J Rheumatol* **33**, 1668–78.

Ingram, D.K. and Roth, G.S. (2011) Glycolytic inhibition as a strategy for developing calorie restriction mimetics. *Exp Geronto*. **46**, 148–54.

Ingram, D.K., Zhu, M., Mamczarz, J. *et al.* (2006) Calorie restriction mimetics: an emerging research field. *Aging Cell* **5**(2), 97–108.

Kawcak, C.E., Frisbie, D.D., McIlwraith, C.W. *et al.* (2007) Evaluation of avocado and soybean unsaponifiable extracts for treatment of horses with experimentally induced osteoarthritis. *Am J Vet Res* **68**, 598–604.

Kim, O.K., Murakami, A., Nakamura, Y. *et al.* (2000a) Novel nitric oxide and superoxide generation inhibitors, persenone A and B, from avocado fruit. *J Agric Food Chem* **48**(5), 1557–63.

Kim, O.K., Murakami, A., Takahashi, D. *et al.* (2000b) An avocado constituent, persenone A, suppresses expression of inducible forms of nitric oxide synthase and cyclooxygenase in macrophages, and hydrogen peroxide generation in mouse skin. *Biosci Biotechnol Biochem* **64**, 2504–7.

Kim, O.K., Murakami, A., Nakamura, Y. *et al.* (2000c) Inhibition by (+)-persenone [corrected] A-related compounds of nitric oxide and superoxide generation from inflammatory leukocytes. *Biosci Biotechnol Biochem* **64**, 2500–3.

Kim, H.W., Murakami, A., Nakamura, Y. and Ohigashi, H. (2002) Screening of edible Japanese plants for suppressive effects on phorbol ester-induced superoxide generation in differentiated HL-60 cells and AS52 cells. *Cancer Lett* **176**(1), 7–16.

Kritchevsky, D., Tepper, S.A., Wright, S. *et al.* (2003) Cholesterol vehicle in experimental atherosclerosis 24: avocado oil. *J Am Coll Nutr* **22**, 52–5.

Lequesne, M., Maheu, E., Cadet, C. and Dreiser, R.L. (2002) Structural effect of avocado/soybean unsaponifiables on joint space loss in osteoarthritis of the hip. *Arthritis Rheum* **47**(1), 50–8.

Lerman-Garber, I., Ichazo-Cerro, S., Zamora-Gonzalez, J. *et al.* (1994) Effect of a high-monounsaturated fat diet enriched with avocado in NIDDM patients. *Diabetes Care* **17**, 311–5.

Lev-Ran, A., Laor, J., Vins, M. and Simon, E. (1970) Effect of intravenous infusion of D-mannoheptulose on blood glucose and insulin levels in man. *J Endocrinol* **47**(1), 137–8.

Liu, X., Sievert, J., Arpaia, M.L. and Madore, M.A. (2002) Postulated Physiological Roles of the Seven-carbon Sugars, Mannoheptulose, and Perseitol in Avocado. *J Amer Soc Hort Sci* **127**, 108–14.

Lopez-Ledesma, R., Frati-Munari, A.C., Hernandez-Dominguez, B.C. *et al.* (1996) Monounsaturated fatty acid (avocado) rich diet for mild hypercholesterolemia. *Arch Med Res* **27**, 519–23.

Lu, Q.Y., Arteaga, J.R., Zhang, Q. *et al.* (2005) Inhibition of prostate cancer cell growth by an avocado extract: role of lipid-soluble bioactive substances. *J Nutr Biochem* **16**, 23–30.

Lu, Q.Y., Zhang, Y.J., Wang, Y. *et al.* (2009) California Hass Avocado: Profiling of carotenoids, tocopherol, fatty acid, and fat content during maturation and from different growing areas. *J Agric Food Chem* **57**(21), 10408–13.

McKenzie, R.A. and Brown, O.P. (1991) Avocado (*Persea americana*) poisoning of horses. *Aust Vet J* **68**, 77–8.
McNulty, H., Pentieva, K., Hoey, L. and Ward, M. (2008) Homocysteine, B-vitamins and CVD. *Proc Nutr Soc* **67**, 232–7.
Naveh, E., Werman, M.J., Sabo, E. and Neeman, I. (2002) Defatted avocado pulp reduces body weight and total hepatic fat but increases plasma cholesterol in male rats fed diets with cholesterol. *J Nutr* **132**, 2015–8.
Nayak, B.S., Raju, S.S. and Chalapathi Rao, A.V. (2008) Wound healing activity of *Persea americana* (avocado) fruit: a preclinical study on rats. *J Wound Care* **17**, 123–6.
Oelrichs, P.B., Ng, J.C., Seawright, A.A. *et al.* (1995) Isolation and identification of a compound from avocado (*Persea americana*) leaves which causes necrosis of the acinar epithelium of the lactating mammary gland and the myocardium. *Nat Toxins* **3**, 344–9.
Ojewole, J.A. and Amabeoku, G.J. (2006) Anticonvulsant effect of *Persea americana* Mill (Lauraceae) (Avocado) leaf aqueous extract in mice. *Phytother Res* **20**, 696–700.
Ojewole, J.A., Kamadyaapa, D.R., Gondwe, M.M. *et al.* (2007) Cardiovascular effects of *Persea americana* Mill (Lauraceae) (avocado) aqueous leaf extract in experimental animals. *Cardiovasc J Afr* **18**, 69–76.
Ostlund, R.E., Jr. (2004) Phytosterols and cholesterol metabolism. *Curr Opin Lipidol* **15**(1), 37.
Paul, R., Kulkarni, P. and Ganesh, N. (2011) Avocado fruit (*Persea americana* Mill) exhibits chemo-protective potentiality against cyclophosphamide induced genotoxicity in human lymphocyte culture. *J Exp Ther Oncol* **9**, 221–30.
Pavelka, K., Coste, P., Geher, P. and Krejci, G. (2010) Efficacy and safety of piascledine 300 versus chondroitin sulfate in a 6 months treatment plus 2 months observation in patients with osteoarthritis of the knee. *Clin Rheumatol* **29**, 659–70.
Perez Mendez, O. and Garcia Hernandez, L. (2007) High-density lipoproteins (HDL) size and composition are modified in the rat by a diet supplemented with "Hass" avocado (*Persea americana* Miller). *Arch Cardiol Mex* **77**, 17–24.
Pieterse, Z., Jerling, J. C., Oosthuizen, W. *et al.* (2005) Substitution of high monounsaturated fatty acid avocado for mixed dietary fats during an energy-restricted diet: effects on weight loss, serum lipids, fibrinogen, and vascular function. *Nutrition* **21**, 67–75.
Prior, R.L., Wu, X. and Schaich, K. (2005) Standardized methods for the determination of antioxidant capacity and phenolics in foods and dietary supplements. *J Agric Food Chem* **53**, 4290–302.
Ramirez, R., Rasschaert, J., Laghmich, A. *et al.* (2001) Uptake of D-mannoheptulose by normal and tumoral pancreatic islet cells. *Int J Mol Med* **7**, 631–8.
Rao, U.S. and Adinew, B. (2011) Remnant B-cell-stimulative and anti-oxidative effects of persea americana fruit extract studied in rats introduced into streptozotocin-induced hyperglycaemic state. *Afr J Tradit Complement Altern Med* **8**, 210–7.
Rasschaert, J., Kadiata, M.M. and Malaisse, W.J. (2001) Effects of D-mannoheptulose upon D-glucose metabolism in tumoral pancreatic islet cells. *Mol Cell Biochem* **226**, 77–81.
Roberts, C.G., Gurisik, E., Biden, T.J. *et al.* (2007) Synergistic cytotoxicity between tamoxifen and the plant toxin persin in human breast cancer cells is dependent on Bim expression and mediated by modulation of ceramide metabolism. *Mol Cancer Ther* **6**, 2777–85.
Rocha, M., Banuls, C., Bellod, L., Jover, A. *et al.* (2011) A review on the role of phytosterols: new insights into cardiovascular risk. *Curr Pharm Des* **17**, 4061–75.
Rosenblat, G., Meretski, S., Segal, J. *et al.* (2011) Polyhydroxylated fatty alcohols derived from avocado suppress inflammatory response and provide non-sunscreen protection against UV-induced damage in skin cells. *Arch Dermatol Res* **303**, 239–46.
Roth, G., Hayek, M., Massimino, S. *et al.* (2009) Mannoheptulose: glycolytic inhibitor and novel caloric restriction mimetic. *FASEB J*, **23**, 553.1.
Schwingshack, L., Strasser, B. and Hoffmann, G. (2011) Effects of monounsaturated fatty acids on cardiovascular risk factors: a systematic review and meta-analysis. *Ann Nutr Metab* **59**, 76–86.
Shaw, P.E., Wilson, C.W., 3rd and Knight, R.J., Jr. (1980) High-performance liquid chromatographic analysis of D-manno-heptulose, perseitol, glucose, and fructose in avocado cultivars. *J Agric Food Chem* **28**, 379–62.
Shepherd, J. and Bender, G. (2001) A History of the Avocado Industry in California. *California Avocado Society 2001 Yearbook*, **85**, 29–50.
Thijssen, M.A. and Mensink, R.P. (2005) Fatty acids and atherosclerotic risk. *Handb Exp Pharmacol*, 165–94.

Viktora, J.K., Johnson, B.F., Penhos, J.C. *et al.* (1969) Effect of ingested mannoheptulose in animals and mad. *Metabolism* **18**(2), 87–102.
Wu, X., Holden, J.M., Haytowitz, D.B. *et al.* (2006) Concentrations of anthocyanins in common foods in the United States and estimation of normal consumption. *J Agric Food Chem* **54**, 4069–75.
Yaman, M., Eser, O., Cosar, M. *et al.* (2007) Oral administration of avocado soybean unsaponifiables (ASU) reduces ischemic damage in the rat hippocampus. *Arch Med Res* **38**, 489–94.

15 Cardiovascular Benefits of Olive Oil: Beyond Effects of Fat Content

Elena M. Yubero-Serrano*, Antonio Camargo*, Jose Lopez-Miranda and Francisco Perez-Jimenez

INTRODUCTION

Olea europaea is one of the most important and widespread fruit trees in the Mediterranean countries and parts of Asia Minor (Loumou and Giourga, 2003). Its cultivation dates back to biblical, Greek and Roman times (Riley, 2002) where *Olea europea* products were used as a treatment for rheumatic and abdominal pain, paralysis and hypertension (Gilani *et al.*, 2005). Today, olives are one of the most extensively cultivated fruit crops in the world, and the production of virgin olive oil (VOO) from ripe olives still forms an essential part of farming practices in Mediterranean regions. A considerable part of olive production is aimed at the direct human consumption of different types of table olives, although most of the production is geared towards producing olive oil.

VOO is a real juice and it is the primary source of dietary fat in the countries where olives are grown (Visioli *et al.*, 2002; Wahrburg *et al.*, 2002), and constitutes part of what is known as the 'Mediterranean diet'. In addition to VOO, the Mediterranean diet is rich in fruit, fish, cereals, fibre and vegetables (6) and its consumption is associated with a decrease in the development of prevalent chronic diseases, particularly cardiovascular diseases (CVD) in Western countries and cancer (Martin-Moreno *et al.*, 1994; Keys *et al.*, 1991; Coni *et al.*, 2000). It is, therefore, important to establish which components of VOO contribute to these protective effects (Perez-Jimenez *et al.*, 2005). Historically, the healthy properties of VOO have been attributed to its high proportion of monounsaturated fatty acids (MUFAs), namely oleic acid, accounting for up to 80% of its total lipid composition (Tripoli *et al.*, 2005). This content in oleic acid has made it unique compared to most seed oils, which are composed mainly of polyunsaturated fatty acids (PUFA). MUFA have only one double bond, thus making it more resistant to oxidation and contributing to the antioxidant action, high stability and the long shelf life of olive oil compared to PUFA-enriched oils (Owen *et al.*, 2000a). Recent innovations in food production have led to other seed oils having levels of oleic acid similar to VOO, such as high-oleic sunflower and rapeseed (canola) oils. This means that these oils now share the biological effects linked to the fat content ((Perez-Jimenez *et al.*, 2007).

* These authors contributed equally to this work.

Bioactives in Fruit: Health Benefits and Functional Foods, First Edition. Edited by Margot Skinner and Denise Hunter.

Table 15.1 Principal components of virgin olive oil.

Triacylglycerols
Oleic acid (C18:1) (55–83% of olive oil)
Linoleic acid (C18:2) (3.5–21% of olive oil)
Palmitic acid (C16:0) (7.5–20% of olive oil)
Stearic acid (C18:0) (0.5–5% of olive oil)
Linolenic acid (C18:3) (specifically α-linolenic acid, 0–1.5% of olive oil)
Non-glyceric esters
Tocopherols
Sterols
Terpenic acids
Pigments
Chlorophyls
Carotenoids
Polyphenols
Hydroxytyrosol
Oleuropeine
Tyrosol
Caffeic acid
Ligstroside
Vanillic acid
Hydroxytyrosol esters

In addition to MUFA, VOO contains minor components, which make up 1–2% of the total content, each with their own biological properties. The minor components of VOO are classified into unsaponificable compounds (squalene, sitosterols, triterpenes, pigments, etc.), defined as the fraction extracted with solvents after the saponification of the oil (Lercker and Rodriguez-Estrada, 2000; Fito *et al.*, 2007), and soluble or hydrophilic compounds, which include phenolic compounds (Sanchez J, Harwood, 2002; Covas *et al.*, 2009) (Table 15.1). Such components also contribute to the unique flavour and taste of VOO, which distinguishes it from other oils derived from seeds. These seed oils need to be refined for human consumption, which means that they lose the vast majority of their original microcomponents, leaving them almost exclusively mere sources of fat. They are colourless, odour free and insipid, and for this reason of little gastronomic interest – the contrary of VOO, a real juice and one of the most highly appreciated edible oils both for its nutritional values and for its important gastronomic qualities.

The phenolic compounds of VOO have created much interest with regards to their health promoting properties. Different studies (in humans and in animals, *in vivo* and *in vitro*) have demonstrated that VOO phenolic compounds have positive, healthy effects in several physiological parameters, reducing pro-inflammatory, pro-oxidant and pro-thrombotic markers, and, thereby, reducing the risk of the development of chronic diseases (Carluccio *et al.*, 2003; Psaltopoulou *et al.*, 2004; Covas *et al.*, 2006a; Ruano *et al.*, 2007; Camargo *et al.*, 2010). The physical methods used to produce VOO preserve many of its antioxidant, anti-inflammatory and antimicrobial qualities.

As with other crops, the composition of the VOO and its components varies in relation to various factors. This review deals with the biological action and cardioprotective effects of VOO, with a special focus on phenolic compounds, which mainly account for their antiatherogenic and endothelium-protecting effects, and how these properties could be affected by the way the olives are ripened and processed and the VOO is stored.

FACTORS AFFECTING PHENOL COMPOUND CONTENT IN VOO

The composition and concentration of phenolic compounds in VOO can be modified according to several factors such as the cultivar, environmental growth conditions, maturation stage, processing techniques used, storage conditions of the olive oil and the cooking process employed (Rotondi *et al.*, 2004). A decrease in the concentration of phenolic compounds in VOO could reduce its positive effects on health. To help to preserve the quality and quantity of phenol compounds in VOO and maximize the health benefits associated with them, it is essential to identify the factors affecting the composition and concentration of these microcomponents (Figure 15.1).

Cultivar and geographic region play an important part in the composition and concentration of phenolic compounds (Gomez-Alonso *et al.*, 2002; Cerretani *et al.*, 2005; Vinha *et al.*, 2005). Several studies have shown differences in the phenolic content of olive fruit from different Italian (Esti *et al.*, 1998) and Spanish olive cultivars (Brenes *et al.*, 1999). Additionally, Gomez-Alonso and colleagues (Gomez-Alonso *et al.*, 2002) demonstrated that the concentration of various phenolic compounds changed significantly among different Spanish VOO varieties. Agronomic techniques and environmental factors, such as irrigation, have also been shown to affect the phenolic content of olive fruit. The application of different irrigation strategies to olive trees (Tovar *et al.*, 2001; Romero *et al.*, 2002) and the continuous exposure of olive fruit to low temperatures during the harvesting (Morello *et al.*, 2003) affect the phenolic profiles of the resulting VOO and, as a consequence, its oxidative stability and some related sensory attributes.

The degree of maturation of the olive fruit when harvested has also been shown to affect phenolic composition and concentration in VOO. Different studies carried out into how phenolic substances develop indicate that during olive ripening the concentration of phenols

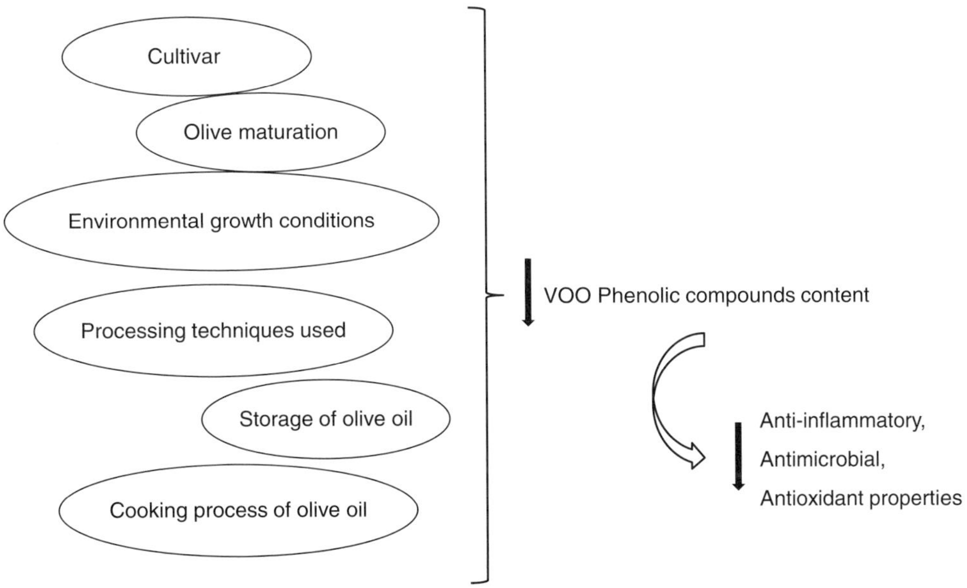

Fig. 15.1 Factors affecting the phenolic compounds content and health properties of virgin olive oil.

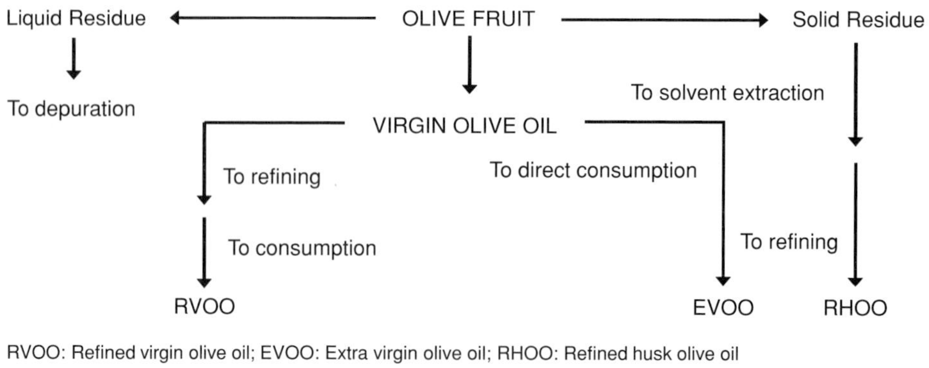

Fig. 15.2 Scheme for the production of various classes of olive oil.

progressively increases to a maximum level at the 'cherry' stage, and decreases as ripening progresses (Gimeno *et al.*, 2002; Rotondi *et al.*, 2004; Gomex-Rico *et al.*, 2006; Kalua *et al.*, 2006). Therefore, a suitable index of fruit ripening must be established specifically for each olive cultivar, because several chemical and enzymatic reactions may take place during ripening and processing which yield phenols of lower molecular weight (Amiot *et al.*, 1989). In general, the concentrations of hydroxytyrosol and tyrosol increase as the olive fruit ripens, in parallel with the breakdown of phenolic compounds of higher molecular weight (Rotondi *et al.*, 2004; Tripoli *et al.*, 2005).

To obtain VOO from olives, the fruits are crushed to create a *pomace*, which is homogenized before being pressed to produce oil. This mechanical and physical process involves collection, removal of leaves, washing, olive crushing, malaxition of the olive paste, centrifugation with or without added water, storage, filtration and bottling. The first oil extracted is the high quality virgin olive oil, which is produced using centrifugation and water only. Subsequent pressings and refining processes involving heat and chemical solvents produce lower-quality olive oils with the loss the most of the components (Figure 15.2) (Owen *et al.*, 2000b).

The presence of hydrophilic phenols in VOO depends primarily on different endogenous enzymes and composition of the olive fruits as well as extraction conditions. Several modifications, such as hydrolysis of glycerides by lipases, hydrolysis of glycosides and oligosaccharides by β-glucosidases, oxidation of phenolic compounds by phenoloxidases and polymerization of free phenols, can appear during these processes (Ryan and Robards, 1998).

The storage conditions of olive oil can affect the phenolic profile through oxidative stress and the consequent formation of oxidized phenols. Different studies have investigated the evolution of phenolic content during 21-month storage at room temperature and have found that the reduction of total phenolic compounds ranged from 43 to 73%, being higher in samples whose initial phenol content was greater (Gomez-Alonso *et al.*, 2002). Hydroxytyrosol increased linearly in most samples, whereas its complex forms decreased considerably (Baiano *et al.*, 2009). Brenes and others investigated the evolution of phenolic compounds in VOO of Arbequina, Hojiblanca and Picual varieties during storage after one year, in darkness at 30°C. The total concentration of the analysed phenolic compounds (hydroxytyrosol and tyrosol) fell slightly (<20% reduction), which had consequences for oil taste and antioxidant capacity (Brenes *et al.*, 2001). Changes occurring in the concentrations of α-tocopherol, total phenols and complex phenols during the storage of VOO under environmental conditions were also studied (Okogeri and Tasioula-Margari, 2002). Under diffused light, α-tocopherol decomposed by 79% in four months, whereas <45% of the phenols were lost during the same period. The levels of the above antioxidants were further related to peroxide formation. It has also been suggested that olive fruit storage at low temperatures may be beneficial, and could possibly increase oil yield and moderate the sensory quality of VOO (Kalua *et al.*, 2008).

The stability of VOO is primarily due to its fatty acid composition and the antioxidant activity of its polyphenols and tocopherols. Many olive oil phenolic compounds are subject to degradation on the application of heat during cooking, which leads to significant changes in the composition of the olive oil, affecting its biological properties (Andrikopoulos *et al.*, 2002; Brenes *et al.*, 2002; Gomez-Alonso *et al.*, 2003). For instance, the total phenolic concentration in a VOO decreased from 70–80% after the first frying operation to 20–30% after the eighth frying operation (Andrikopoulos *et al.*, 2002). Several studies have demonstrated that VOO composition is different depending on the method of heating. In the conventional procedure, a time-dependent effect is observed, causing a significant decrease in hydroxytyrosol and a rapid degradation of α-tocopherol and the glyceridic fraction of oils. In microwave heating, only minor losses in polyphenol content are observed, and oil degradation is lower than in thermoxidation assays (Brenes *et al.*, 2002). Similarly, the length of time the oil is heated in a microwave appears to have a dramatic effect on degradation of the oil (Caponio *et al.*, 2002). The heating methods assayed resulted in more severe polyphenol losses and oil degradation in Arbequina than in Picual oil, which could be related to the lower levels in PUFA in the Picual olive. These findings may be relevant to choice of cooking method and VOO cultivar in order to increase the intake of olive polyphenols (Brenes *et al.*, 2002; Perez-Herrera *et al.*, 2011).

THE HEALTH EFFECT OF VOO

Cardiovascular benefits of MUFA

In the past, the healthy properties of VOO were attributed to its high content of oleic acid (Perez-Jimenez *et al.*, 2007). It is now known that consumption of MUFA-rich diets promotes a healthy blood lipid profile, influencing the cholesterol metabolism, maintaining levels of HDL cholesterol and reducing those of LDL cholesterol, compared with consumption of saturated fatty acid (SFA)-rich diets. Moreover, MUFA-rich diets modulate favourably insulin sensitivity and glycemic control (Perez-Jimenez *et al.*, 2002; Thijssen and Mensink, 2005). There is also experimental and epidemiological evidence to suggest that a diet rich in MUFA could be potentially recommended for the prevention and control of hypertension (Salas *et al.*, 1999). In this way, Thomsen *et al.* observed that an olive oil rich diet (30% of total energy provided by MUFA) significantly reduced systolic and diastolic blood pressure (BP) compared with a PUFA-rich diet (27% of total energy intake) in normotensive type 2 diabetic subjects (Thomsen *et al.*, 1995). Moreover, the effect of two similar MUFA-rich diets (olive oil and high-oleic sunflower oil) in hypertensive female patients has been studied by Ruiz-Gutierrez *et al.* (1996), who observed that the diet rich in olive oil led to a significant reduction of blood pressure, which was also confirmed by Fito *et al.* (2005).

Additional studies have also supported the existence of the beneficial effects of MUFA on coagulation and fibrinolysis (Visioli *et al.*, 2005; Bogani *et al.*, 2007; Delgado-Lista *et al.*, 2011). Furthermore, consumption of a MUFA diet, compared with a SFA-rich diet, is also associated with maintenance of body weight and a decrease in risk of obesity, reducing central body fat adiposity (Bes-Rastrollo *et al.*, 2006, 2007) and leading to an anti-inflammatory obesity-linked gene expression profile in the adipose tissue of persons at risk of suffering metabolic syndrome (van Dijk *et al.*, 2009).

Besides having a high level of MUFA, olive oil, as a functional food, contains other minor components with biological properties, as shown (Covas *et al.*, 2006a, 2006b; Perez-Jimenez *et al.*, 2007 19, 59). The health benefits of this specific fraction of olive oil, which constitutes the real added value of this food, are discussed in more detail in the next sections.

Cardioprotective effects of olive oil phenolic compounds

Olive oil is the main source of fat in the Mediterranean diet (Med Diet) (Willett *et al.*, 1995). Although a few prospective cohort studies have shown that the adherence to a diet with a high MUFA/PUFA ratio, from an origin other than VOO, is associated with a significant improvement in health status, with a decreased rate of cardiovascular events and total mortality (Sofi *et al.*, 2008), the use of VOO has the additional advantage of the healthy effect of its minor components, in particular the phenolic compounds. In other words, a Med Diet with VOO as the main source of fat could bring advantages for health through the additional effects of phenolic compounds (De Lorgeril *et al.*, 1996; Willett, 2006). These microcomponents can be divided into three categories: simple phenolics, secoiridoids and lignans. Major phenolics include hydroxytyrosol, tyrosol, oleuropein (Perona *et al.*, 2006) and ligstroside (Owen *et al.*, 2000b). Tyrosol, hydroxytyrosol and their secoiridoid derivatives represent around 30%, and other conjugated forms, such as oleuropein, represent almost half the total phenolic content of VOO. Most of these microcomponents are lost during the oil refining process. Consequently, VOO is a real juice with a high content in these products (around 230 mg/ kg, common range 130–350 mg/ kg) (Owen *et al.*, 2000a). Phenolic compounds are partially responsible for the particular sensory properties of olive oil and for its high oxidative stability during storage, which makes it unique among other oils (Sanchez and Harwood, 2002; Rotondi *et al.*, 2004).

Atherosclerosis is an inflammatory disease and is the pathogenic substrate responsible for CVD. Atherosclerotic plaques are formed by the accumulation of lipid-rich macrophages, vascular smooth muscle cells, lipids and the extracellular matrix (Kong *et al.*, 2005); plaque rupture is the major mechanism for the development of acute coronary syndromes, strokes and sudden death (Lusis, 2000). This condition develops over years and VOO phenolic compounds have been shown to slow down several of the mechanisms underlying atheroma plaque formation, principally in its early stages.

A biologically plausible mechanism by which these components could exert antiatherogenic properties may be by increasing HDL cholesterol plasma concentrations. It has been shown that this lipoprotein particle slows down the process of atherogenesis at several key stages: removing cholesterol, inhibiting LDL oxidation and limiting the inflammatory component in atherosclerosis (Rader *et al.*, 2009). Thus, humans with low concentrations of HDL cholesterol are at greater risk of developing atherogenesis. Some information is available on the effect of the phenolic compounds of olive oil on plasma lipid levels, as has been demonstrated in two independent randomized, crossover, controlled human intervention studies, in healthy volunteers, with a significant increase in HDL cholesterol plasma concentrations after a three-week intake of phenolic compound-rich VOO (Covas *et al.*, 2006a; Marrugat *et al.*, 2004), and a decrease in the total cholesterol/HDL cholesterol ratio (Covas *et al.*, 2006a). This effect was observed after the consumption of 25 ml of VOO containing 366 mg of phenolic compounds/kg per day. Therefore, a relatively low dose of these microcomponents could have a major biological effect, that is by interacting with the regulatory mechanisms. However, there is still little experimental evidence and further studies are needed to establish a cause and effect relationship between olive oil phenolic compounds and the maintenance of normal blood HDL cholesterol concentrations.

High BP develops unnoticed over time and may compromise normal arterial and cardiac function. High BP is a CVD risk factor itself (Alexander, 1995), because it exerts an added strain on the artery walls, which over time damages the inner lining of the blood vessels, making them more vulnerable to narrowing and the build-up of plaque, and therefore more susceptible to the development of atherosclerosis (Onat *et al.*, 2011). The ability of olive oil phenolic compounds to modify arterial pressure is less well-known than other protective mechanisms. Animal studies suggest that olive oil phenolic compounds reduce BP. However, the limitation of these studies lies in the fact that either leaf extract was administered by oral administration (Khayyal *et al.*, 2002) or crude extract of the olive fruit was submitted intravenously (Gilani *et al.*, 2005), instead of testing a high consumption of phenolic compound-rich VOO. For this reason, the results on the effects of olive oil phenolic compounds, not administered in their natural matrix, are controversial, as previously has been shown (Acin *et al.*, 2006). Additionally, based on *in vitro* evidence, it has been proposed that the BP-lowering effect may be mediated by calcium channel blockade (Gilani *et al.*, 2005). In humans, the BP-lowering effect by olive oil phenolic compounds has been reproduced (Perrinjaquet-Moccetti *et al.*, 2008), again with a limitation of the consumption of olive leaf extract, instead of a high phenolic compound-rich VOO consumption and the use of a relatively high dose, ranging from 500 to 1000 mg of olive leaf extract per day.

Recent interest in VOO phenolic compounds has greatly increased because of their antioxidant and free radical scavenging abilities, which are both associated with potential benefits for human health. Olive oil contains a wide variety of antioxidant compounds that are not found in other oils, which make it a healthier alternative to the consumption of other oils. Several human studies have shown that plasma antioxidant activity increases after a short-term consumption of phenolic compounds-rich VOO (Visioli *et al.*, 2005; Bogani *et al.*, 2007). In fact, during the postprandial state, after the acute intake of a breakfast prepared with VOO (Visioli *et al.*, 2005; Bogani *et al.*, 2007), a decrease in the urinary excretion of isoprostanes has been shown (Visioli *et al.*, 2000), which is one of the most reliable ways of assessing oxidative stress and beneficially modulating the GSH turnover (Weinbrenner *et al.*, 2004; Covas *et al.*, 2006a). Therefore, VOO phenolic compounds could partly ameliorate the adverse effects of oxidative stress on endothelial dysfunction (Ruano *et al.*, 2005, 2007), which is one of the first steps in the development of arteriosclerosis (Davignon and Ganz, 2004).

Atherosclerosis is an inflammatory disease favoured by the infiltration of LDL particles in the vessel wall (Tedgui and Mallat, 2006). Once inside the tunica intima of the vessel, the lipoprotein is oxidized (oxLDL) by reactive oxygen species (ROS) and becomes more damaging to the arterial wall than native LDL cholesterol (Meisinger *et al.*, 2005). The uptake of oxLDL by macrophages through scavenger receptors leads to macrophage transformation into lipid-laden activated foam cells (Rose and Afanasyeva, 2003), which in turn triggers the immune response in the vessel responsible for atheroma plaque formation over time (Galkina and Ley, 2009). Otherwise, foam cells produce large quantities of ROS, thereby stimulating oxidation of LDL and further prolonging the vicious circle (Rose and Afanasyeva, 2003). Thus, oxLDL plasma level is considered a strong predictor of cardiovascular diseases (Witztum, 1994; Meisinger *et al.*, 2005).

The incorporation of phenolic compounds from VOO to the LDL particles, as a prerequisite to performing the protective effect of phenolic compounds on LDL oxidation, has been proposed as the mechanism by which VOO phenol compounds may protect LDL particles from peroxidation. Thus, the protective effect of olive oil phenolic compounds lies in their antioxidant properties, such as binding LDL (de la Torre-Carbot *et al.*, 2007) and protecting them against oxidation, as has been shown both with *in vitro* (Masella *et al.*, 2004; Franconi *et al.*, 2006) and *in vivo* studies (Marrugat *et al.*, 2004; Weinbrenner *et al.*, 2004; Covas *et al.*, 2006a; de la Torre-Carbot *et al.*, 2010). Randomized, crossover, controlled trials showed that phenolic compounds protect LDL against oxidation. This effect was observed by the consumption of 25 ml of olive oil per day with a phenolic compound content ranging from 150 to 685 mg/kg, with an intake of 4–15 mg of phenolic compounds per day during three weeks (Marrugat *et al.*, 2004; Weinbrenner *et al.*, 2004; Covas *et al.*, 2006a; de la Torre-Carbot *et al.*, 2010), which is a similar dose to a real-life daily intake of a commercial VOO. However, a shorter period of consumption (4 days) has also shown the protective effect of the consumption of phenol compound-rich VOO against LDL oxidation (Marrugat *et al.*, 2004; Weinbrenner *et al.*, 2004; Covas *et al.*, 2006a; de la Torre-Carbot *et al.*, 2010). Additionally, hydroxytyrosol monosulfate and homovanillic acid sulfate, two hydroxytyrosol metabolites (de la Torre-Carbot *et al.*, 2007; de la Torre-Carbot, 2008), seem to be responsible for this effect, as the concentration of these metabolites, but not of tyrosol sulfate, significantly increased in LDL particles after the ingestion of phenolic compounds-rich VOO (de la Torre-Carbot *et al.*, 2007). Thus, consequently, the consumption of phenolic compound-rich VOO may be used as a nutritional strategy against the development of atherosclerosis.

Endothelial dysfunction is one of the first stages in the development of arteriosclerosis (Davignon and Ganz, 2004), and it has been related to oxidative stress, which plays a critical role in the process of endothelial dysfunction (Heitzer *et al.*, 2001), as well as the thrombogenic state, caused by an imbalance between procoagulant and profibrinolytic activity (Nossel, 1981). Various constituents of the platelet and the coagulation system have been shown to be involved in the development of atherosclerosis (Borissoff *et al.*, 2011). The binding of platelets to each other and to circulating leukocytes produces coaggregates, which trigger leukocyte recruitment, and are critical for plaque progression (Rizvi *et al.*, 2003; Gleissner *et al.*, 2008; Lievens *et al.*, 2009). In addition, numerous coagulant proteins have been implicated in processes such as the disruption of the endothelial barrier, leukocyte recruitment and migration and proliferation of vascular smooth-muscle cells (Borensztajn *et al.*, 2008, 2009).

It has been shown that olive oil phenolic compounds can partly ameliorate the adverse effects of oxidative stress on endothelial dysfunction. In fact, the consumption of a breakfast prepared with a high-phenolic compound VOO improves ischemic reactive hyperemia during the postprandial state by the reduction in oxidative stress and the increase of nitric oxide metabolites (Ruano *et al.*, 2005, 2007). On the other hand, several studies have shown that the antithrombotic properties associated to VOO may be produced by olive oil phenol compounds (Perez-Jimenez *et al.*, 2007, Cicerale *et al.*, 2010; Granados-Principal *et al.*, 2010; Visioli and Bernardini, 2011). This effect has been shown by *in vitro* studies dealing with platelet aggregation, which was shown to be inhibited by several olive oil phenolic compounds (Petroni *et al.*, 1995; Dell'Agli *et al.*, 2008). Additionally, a randomized, sequential acute crossover study showed that the consumption of a VOO with a high content in phenols decreases the procoagulant factor plasminogen activator inhibitor-1 (PAI-1) and factor VII (FVII), both linked to the development of coronary heart disease (Ruano *et al.*, 2007).

Inflammation is a physiological response to tissue damage, and whether or not reduction of inflammatory response is beneficial to health depends on the context. Nevertheless, it has been acknowledged that high concentrations of serum inflammatory markers are associated with CVD (Packard and Libby, 2008), which is particularly important in several pathological conditions, such as obesity and MetS, which are characterized by a low-grade inflammation that favours the development of atherosclerosis (Subramanian and Chait, 2009). Both *in vitro* and *in vivo* studies have shown that some olive oil phenol compounds reduce the synthesis and release of lipid-derived inflammatory signals. *In vitro* studies showed that olive oil phenolic compounds decrease arachidonic acid synthesis and release (Moreno, 2003) as well as cyclooxygenases-1 (COX-1) and cyclooxygenase-2 (COX-2) activity inhibition (Beauchamp *et al.*, 2005). On the other hand, an *in vivo* study has shown a decrease in plasma thromboxane B_2 (TXB2) and leukotriene B_4 (LTB4) concentrations after the acute intake of a VOO rich in phenolic compounds (Bogani *et al.*, 2007).

Moreover, in terms of gene expression, two independent microarray studies in humans have shown that olive oil phenol compounds repress *in vivo* the gene expression of several pro-inflammatory genes (Khymenets *et al.*, 2009; Camargo *et al.*, 2010), many of which are involved in inflammatory processes mediated by the transcription factors NF-κB and AP-1, including cytokines such as *IL1B*, *IL6*, and *CXCL1*, or arachidonic acid pathways (*COX-2*) (Camargo *et al.*, 2010). Interestingly, these two studies showed that the anti-inflammatory properties arising from short-term consumption of a VOO rich in phenolic compounds (3 weeks) (Khymenets *et al.*, 2009; Camargo *et al.*, 2010), as well as the fast response of the genome to the acute intake of extra virgin olive oil rich in phenolic compounds (Khymenets *et al.*, 2009; Camargo *et al.*, 2010) demonstrated the pathways where phenolic compounds may directly interact.

CONCLUSION AND FUTURE PERSPECTIVES

In summary, some cardioprotective properties associated with the consumption of a Med Diet may be attributed, at least partially, to olive oil phenol compounds, the most important minor components of olive oil, and the main source of fat in the Med Diet. The mechanism of action described in this review, mainly through slowing down the earlier stages of the development of atherosclerosis, may explain the low rate of atherosclerosis and coronary heart disease in Mediterranean countries. Experimental studies have shown that olive oil phenol compounds cause a reduction in the cardiovascular risk factors HDL cholesterol and BP. Moreover, they also lead to an improvement in endothelial dysfunction, because of the proposed antioxidant, anti-inflammatory and antithrombotic properties associated with olive oil phenolic compounds, in addition to their antiatherogenic properties derived from the protection against LDL oxidation provided by olive oil phenolic compounds (Table 15.2).

The European Food Safety Authority (EFSA) has recently favourably assessed a cause and effect relationship between the consumption of phenolic compound-rich olive oil and protection of LDL particles from oxidative damage. However, the EFSA considers that the accumulated evidence for the other properties of olive oil phenolic compounds is still insufficient (EFSA, 2011). Although the current review provides evidence at the most basic level of healthy effects derived from phenolic compound-rich VOO consumption in humans, more studies are required to fully understand the effect of olive oil phenolic compounds

Table 15.2 Effects of olive oil phenol compounds on cardiovascular risk factors and other mechanisms related to atherogenesis.

Level of evidence	Kind of study	Type of effect
Demonstrated	Intervention studies in humans (Marrugat *et al.*, 2004; Weinbrenner *et al.*, 2004; Covas *et al.*, 2006a; de la Torre-Carbot*et al.*, 2010)	1. Increases resistance of LDL to oxidation
Probable	Intervention studies in humans (Visioli *et al.*, 2000, 2005; Weinbrenner *et al.*, 2004; Covas *et al.*, 2006a; Bogani *et al.*, 2007)	1. Antioxidant properties
	In vitro (Moreno 2003; Beauchamp *et al.*, 2005), intervention studies in humans (Bogani *et al.*, 2007; Khymenets *et al.*, 2009; Camargo *et al.*, 2010)	2. Anti-inflammatory properties
	Intervention studies in humans (Ruano *et al.*, 2005, 2007)	3. Improve endothelial dysfunction
Possible	Intervention studies in humans (Covas *et al.*, 2006a, Marrugat *et al.*, 2004)	1. Increases HDL cholesterol levels
	In animal models (Khayyal *et al.*, 2002; Gilani *et al.*, 2005), intervention study in humans (Perrinjaquet-Moccetti *et al.*, 2008)	2. Reduces arterial blood pressure
	In vitro (Petroni *et al.*, 1995; Dell'Agli *et al.*, 2008), intervention study in humans (Ruano *et al.*, 2007)	3. Reduces prothrombotic state

phenolic on human health. It would be particularly interesting to evaluate whether it is one or more particular phenolic compounds which cause these effects, or if they are the consequence of a synergic effect of the total phenolic fraction.

ACKNOWLEDGEMENTS

Source of support: This research was supported partly by research grants from the Ministerio de Ciencia e Innovación (AGL2006-01979, AGL2009-12270 to JL-M), ("CB06/03/0047-CIBER Fisiopatologia de la Obesidad y Nutrition" is an ISCIII grant awarded to FP-J), Consejería de Innovación, Ciencia y Empresa, Junta de Andalucía (P06-CTS-01425 and CVI-7450 to JL-M); Consejería de Salud, Junta de Andalucía (06/128, 07/43, PI0193/2009 to JL-M, 06/129 to FP-J).

REFERENCES

Acin S, Navarro MA, Arbones-Mainar JM *et al.* Hydroxytyrosol administration enhances atherosclerotic lesion development in apo E deficient mice. *J Biochem* 2006;**140**:383–91.

Alexander RW. Theodore Cooper Memorial Lecture. Hypertension and the pathogenesis of atherosclerosis. Oxidative stress and the mediation of arterial inflammatory response: a new perspective. *Hypertension* 1995;**25**:155–61.

Amiot M, Fleuriet A, Macheix J. Accumulation of oleuropein derivatives during maturation. *Phytochemistry* 1989;**28**:67–9.

Andrikopoulos NK, Dedoussis GV, Falirea A *et al.* Deterioration of natural antioxidant species of vegetable edible oils during the domestic deep-frying and pan-frying of potatoes. *Int J Food Sci Nutr* 2002;**53**:351–63.

Baiano A, Gambacorta G, Terracone C *et al.* Changes in phenolic content and antioxidant activity of Italian extra-virgin olive oils during storage. *J Food Sci* 2009;**74**:C177–83.

Beauchamp GK, Keast RS, Morel D *et al.* Phytochemistry: ibuprofen-like activity in extra-virgin olive oil. *Nature* 2005;**437**:45–6.

Bes-Rastrollo M, Sanchez-Villegas A, Gomez-Gracia E *et al.* Predictors of weight gain in a Mediterranean cohort: the Seguimiento Universidad de Navarra Study 1. *Am J Clin Nutr* 2006;**83**:362–70; quiz 394-5.

Bes-Rastrollo M, Sabate J, Gomez-Gracia E *et al.* Nut consumption and weight gain in a Mediterranean cohort: The SUN study. *Obesity (Silver Spring)* 2007;**15**:107–16.

Bogani P, Galli C, Villa M, Visioli F. Postprandial anti-inflammatory and antioxidant effects of extra virgin olive oil. *Atherosclerosis* 2007;**190**:181–6.

Borensztajn K, Peppelenbosch MP, Spek CA. Factor Xa: at the crossroads between coagulation and signaling in physiology and disease. *Trends Mol Med* 2008;**14**:429–40.

Borissoff JI, Spronk HM, Heeneman S, ten Cate H. Is thrombin a key player in the 'coagulation-atherogenesis' maze? *Cardiovasc Res* 2009;**82**:392–403.

Borissoff JI, Spronk HM, ten Cate H. The hemostatic system as a modulator of atherosclerosis. *N Engl J Med* 2011;**364**:1746–60.

Brenes M, Garcia A, Garcia P *et al.* Phenolic compounds in Spanish olive oils. *J Agric Food Chem* 1999;**47**:3535–40.

Brenes M, Garcia A, Garcia P, Garrido A. Acid hydrolysis of secoiridoid aglycons during storage of virgin olive oil. *J Agric Food Chem* 2001;**49**:5609–14.

Brenes M, Garcia A, Dobarganes MC *et al.* Influence of thermal treatments simulating cooking processes on the polyphenol content in virgin olive oil. *J Agric Food Chem* 2002;**50**:5962–7.

Camargo A, Ruano J, Fernandez JM *et al.* Gene expression changes in mononuclear cells in patients with metabolic syndrome after acute intake of phenol-rich virgin olive oil. *BMC Genomics* 2010;**11**: 253.

Caponio F, Pasqualone A, Gomes T. Effects of conventional and microwave heating on the degradation of olive oil. *Eur Food Res Technol* 2002;**215**:114–117.

Carluccio MA, Siculella L, Ancora MA *et al.* Olive oil and red wine antioxidant polyphenols inhibit endothelial activation: antiatherogenic properties of Mediterranean diet phytochemicals. *Arterioscler Thromb Vasc Biol* 2003;**23**:622–9.

Cerretani L, Bendini A, Rotondi A *et al.* Analytical comparison of monovarietal virgin olive oils obtained by both a continuous industrial plant and low-scale mill. *Eur J Lipid Sci Tech* 2005;**107**:93–100.

Cicerale S, Lucas L, Keast R. Biological activities of phenolic compounds present in virgin olive oil. *Int J Mol Sci* 2010;**11**:458–79.

Coni E, Di Benedetto R, Di Pasquale M *et al.* Protective effect of oleuropein, an olive oil biophenol, on low density lipoprotein oxidizability in rabbits. *Lipids* 2000;**35**:45–54.

Covas M, Nyyssonen K, Poulsen H *et al.* The effect of polyphenols in olive oil on heart disease risk factors: a randomized trial. *Annals Internal Medicine* 2006a;**145**:333–341.

Covas M, Ruiz-Gutierrez V, de la Torre R *et al.* Minor components of olive oil: evidence to date of health benefits in humans. *Nutr Rev* 2006b;**64**:20–30.

Covas MI, Konstantinidou V, Fito M. Olive oil and cardiovascular health. *J Cardiovasc Pharmacol* 2009;**54**:477–82.

Davignon J, Ganz P. Role of endothelial dysfunction in atherosclerosis. *Circulation* 2004;**109**:III27–32.

de la Torre R. Bioavailability of olive oil phenolic compounds in humans. *Inflammopharmacology* 2008;**16**:245–7.

de la Torre-Carbot K, Chavez-Servin JL, Jauregui O *et al.* Presence of virgin olive oil phenolic metabolites in human low density lipoprotein fraction: determination by high-performance liquid chromatography-electrospray ionization tandem mass spectrometry. *Anal Chim Acta* 2007;**583**:402–10.

de la Torre-Carbot K, Chavez-Servin JL, Jauregui O *et al.* Elevated circulating LDL phenol levels in men who consumed virgin rather than refined olive oil are associated with less oxidation of plasma LDL. *J Nutr* 2010;**140**:501–8.

De Lorgeril M, Salen P, Martin JL, *et al.* Effect of a mediterranean type of diet on the rate of cardiovascular complications in patients with coronary artery disease. Insights into the cardioprotective effect of certain nutriments. *J Am Coll Cardiol* 1996;**28**:1103–8.

Delgado-Lista J, Garcia-Rios A, Perez-Martinez P *et al.* Olive oil and haemostasis: platelet function, thrombogenesis and fibrinolysis. *Curr Pharm Des* 2011;**17**:778–85.

Dell'Agli M, Maschi O, Galli GV *et al.* Inhibition of platelet aggregation by olive oil phenols via cAMP-phosphodiesterase. *Br J Nutr* 2008;**99**:945–51.

EFSA Panel on Dietetic Products, Nutrition and Allergies (NDA) (2011) Scientific Opinion on the substantiation of health claims related to polyphenols in olive and protection of LDL particles from oxidative damage (ID 1333, 1638, 1639, 1696, 2865), maintenance of normal blood HDL cholesterol concentrations (ID 1639), maintenance of normal blood pressure (ID 3781), "anti-inflammatory properties" (ID 1882), "contributes to the upper respiratory tract health" (ID 3468), "can help to maintain a normal function of gastrointestinal tract" (3779), and "contributes to body defences against external agents" (ID 3467) pursuant to Article 13(1) of Regulation (EC) No 1924/2006. EFSA Journal 9(4):2033.(doi:10.2903/j.efsa.2011.2033).

Esti M, Cinquanta L, La Notte E. Phenolic compounds in different olive varieties. *J Agric Food Chem* 1998;**46**:32–35.

Fito M, Cladellas M, de la Torre R *et al.* Antioxidant effect of virgin olive oil in patients with stable coronary heart disease: a randomized, crossover, controlled, clinical trial. *Atherosclerosis* 2005;**181**:149–58.

Fito M, de la Torre R, Farre-Albaladejo M *et al.* Bioavailability and antioxidant effects of olive oil phenolic compounds in humans: a review. *Ann Ist Super Sanita* 2007;**43**:375–81.

Franconi F, Coinu R, Carta S *et al.* Antioxidant effect of two virgin olive oils depends on the concentration and composition of minor polar compounds. *J Agric Food Chem* 2006;**54**:3121–5.

Galkina E, Ley K. Immune and inflammatory mechanisms of atherosclerosis. *Ann Rev Immunol* 2009;**27**:165–97.

Gilani AH, Khan AU, Shah AJ *et al.* Blood pressure lowering effect of olive is mediated through calcium channel blockade. *Int J Food Sci Nutr* 2005;**56**:613–20.

Gimeno E, Castellote A, Lamuela-Raventos R *et al.* The effects of harvest and extraction methods on the antioxidant content (phenolics, alpha-tocopherol, and beta-carotene) in virgin olive oil. *Food Chem* 2002;**78**.

Gleissner CA, von Hundelshausen P, Ley K. Platelet chemokines in vascular disease. *Arterioscler Thromb Vasc Biol* 2008;**28**:1920–7.

Gomez-Alonso S, Salvador MD, Fregapane G. Phenolic compounds profile of cornicabra virgin olive oil. *J Agric Food Chem* 2002;**50**:6812–7.

Gomez-Alonso S, Fregapane G, Salvador MD, Gordon MH. Changes in phenolic composition and antioxidant activity of virgin olive oil during frying. *J Agric Food Chem* 2003;**51**:667–72.

Gomez-Rico A, Salvador MD, La Greca M, Fregapane G. Phenolic and volatile compounds of extra virgin olive oil (Olea europaea L. Cv. Cornicabra) with regard to fruit ripening and irrigation management. *J Agric Food Chem* 2006;**54**:7130–6.

Granados-Principal S, Quiles JL, Ramirez-Tortosa CL *et al.* Hydroxytyrosol: from laboratory investigations to future clinical trials. *Nutr Rev* 2010;**68**:191–206.

Heitzer T, Schlinzig T, Krohn K *et al.* Endothelial dysfunction, oxidative stress, and risk of cardiovascular events in patients with coronary artery disease. *Circulation* 2001;**104**:2673–8.

Kalua CM, Bedgood DR, Jr, Bishop AG, Prenzler PD. Changes in volatile and phenolic compounds with malaxation time and temperature during virgin olive oil production. *J Agric Food Chem* 2006;**54**:7641–51.

Kalua CM, Bedgood DR, Jr, Bishop AG, Prenzler PD. Changes in virgin olive oil quality during low-temperature fruit storage. *J Agric Food Chem* 2008;**56**:2415–22.

Khayyal MT, el-Ghazaly MA, Abdallah DM *et al.* Blood pressure lowering effect of an olive leaf extract (*Olea europaea*) in L-NAME induced hypertension in rats. *Arzneimittelforschung* 2002;**52**:797–802.

Keys A, Aravanis C, Van Buchem H *et al.* The diet and all-causes death rate in the Seven Countries Study. *Lancet* 1981;**2**:58–61.

Khymenets O, Fito M, Covas MI *et al.* Mononuclear cell transcriptome response after sustained virgin olive oil consumption in humans: an exploratory nutrigenomics study. *Omics* 2009;**13**:7–19.

Kong YZ, Yu X, Tang JJ *et al.* Macrophage migration inhibitory factor induces MMP-9 expression: implications for destabilization of human atherosclerotic plaques. *Atherosclerosis* 2005;**178**:207–15.

Lercker G, Rodriguez-Estrada MT. Chromatographic analysis of unsaponifiable compounds of olive oils and fat-containing foods. *J Chromatogr A* 2000;**881**:105–29.

Lievens D, Eijgelaar WJ, Biessen EA *et al.* The multi-functionality of CD40L and its receptor CD40 in atherosclerosis. *Thromb Haemost* 2009;**102**:206–14.

Loumou A, Giourga C. Olive groves: The life and identity of the Mediterranean. *Agric Human Values* 2003;**20**:87–95.

Lusis AJ. Atherosclerosis. *Nature* 2000;**407**:233–41.
Marrugat J, Covas MI, Fito M *et al.* Effects of differing phenolic content in dietary olive oils on lipids and LDL oxidation – a randomized controlled trial. *Eur J Nutr* 2004;**43**:140–7.
Martin-Moreno JM, Willett WC, Gorgojo L *et al.* Dietary fat, olive oil intake and breast cancer risk. *Int J Cancer* 1994;**58**:774–80.
Masella R, Vari R, D'Archivio M *et al.* Extra virgin olive oil biophenols inhibit cell-mediated oxidation of LDL by increasing the mRNA transcription of glutathione-related enzymes. *J Nutr* 2004;**134**:785–91.
Meisinger C, Baumert J, Khuseyinova N *et al.* Plasma oxidized low-density lipoprotein, a strong predictor for acute coronary heart disease events in apparently healthy, middle-aged men from the general population. *Circulation* 2005;**112**:651–7.
Morello J, Motilva M, Ramo T, Romero M. Effect of freeze injuries in olive fruit on virgin olive oil composition. *Food Chem Toxicol* 2003;**81**:547–53.
Moreno JJ. Effect of olive oil minor components on oxidative stress and arachidonic acid mobilization and metabolism by macrophages RAW 264.7. *Free Radic Biol Med* 2003;**35**:1073–81.
Nossel HL. Relative proteolysis of the fibrinogen B beta chain by thrombin and plasmin as a determinant of thrombosis. *Nature* 1981;**291**:165–7.
Okogeri O, Tasioula-Margari M. Changes occurring in phenolic compounds and alpha-tocopherol of virgin olive oil during storage. *J Agric Food Chem* 2002;**50**:1077–80.
Onat D, Brillon D, Colombo PC, Schmidt AM. Human vascular endothelial cells: a model system for studying vascular inflammation in diabetes and atherosclerosis. *Curr Diab Rep* 2011;**11**:193–202.
Owen RW, Mier W, Giacosa A *et al.* Phenolic compounds and squalene in olive oils: the concentration and antioxidant potential of total phenols, simple phenols, secoiridoids, lignansand squalene. *Food Chem Toxicol* 2000a;**38**:647–59.
Owen R, Giacosa A, Hull W, *et al.* Olive oil consumption and health: the possible role of antioxidants. *Lancet Oncol* 2000b;**1**:107–112.
Owen R, Haubner R, Wurtele G *et al.* Olives and olive oil in cancer prevention. *Eur J Cancer Prev* 2004;**13**:319–326.
Packard RR, Libby P. Inflammation in atherosclerosis: from vascular biology to biomarker discovery and risk prediction. *Clin Chem* 2008;**54**:24–38.
Perez-Herrera A, Delgado-Lista J, Torres-Sanchez LA *et al.* The postprandial inflammatory response after ingestion of heated oils in obese persons is reduced by the presence of phenol compounds. *Mol Nutr Food Res* 2011;**56**(3):510–514.
Perez-Jimenez F, Lopez-Miranda J, Mata P. Protective effect of dietary monounsaturated fat on arteriosclerosis: beyond cholesterol. *Atherosclerosis* 2002;**163**:385–98.
Perez-Jimenez F, Alvarez de Cienfuegos G, Badimon L *et al.* International conference on the healthy effect of virgin olive oil. *Eur J Clin Invest* 2005;**35**:421–4.
Perez-Jimenez F, Ruano J, Perez-Martinez P *et al.* The influence of olive oil on human health: not a question of fat alone. *Mol Nutr Food Res* 2007;**51**:1199–208.
Perona JS, Cabello-Moruno R, Ruiz-Gutierrez V. The role of virgin olive oil components in the modulation of endothelial function. *J Nutr Biochem* 2006;**17**:429–45.
Perrinjaquet-Moccetti T, Busjahn A, Schmidlin C *et al.* Food supplementation with an olive (*Olea europaea* L.) leaf extract reduces blood pressure in borderline hypertensive monozygotic twins. *Phytother Res* 2008;**22**:1239–42.
Petroni A, Blasevich M, Salami M *et al.* Inhibition of platelet aggregation and eicosanoid production by phenolic components of olive oil. *Thromb Res* 1995;**78**:151–60.
Psaltopoulou T, Naska A, Orfanos P *et al.* Olive oil, the Mediterranean diet, and arterial blood pressure: the Greek European Prospective Investigation into Cancer and Nutrition (EPIC) study. *Am J Clin Nutr* 2004;**80**:1012–8.
Rader DJ, Alexander ET, Weibel GL *et al.* The role of reverse cholesterol transport in animals and humans and relationship to atherosclerosis. *J Lipid Res* 2009;**50**(Suppl):S189–94.
Riley F. Olive oil production on Bronze Age Crete: nutritional properties, processing methods, and storage life of Minoan olive oil. *Oxford J Archaeol* 2002;**21**:63–75.
Rizvi Z, Rehman T, Malik S *et al.* An evaluation of topical and local anesthesia in phacoemulsification. *J Pak Med Assoc* 2003;**53**:167–70.
Romero MP, Tovar MJ, Girona J, Motilva MJ. Changes in the HPLC phenolic profile of virgin olive oil from young trees (Olea europaea L. Cv. Arbequina) grown under different deficit irrigation strategies. *J Agric Food Chem* 2002;**50**:5349–54.

Rose N, Afanasyeva M. Autoimmunity: busting the atherosclerotic plaque. *Nat Med* 2003;**9**:641–2.

Rotondi A, Bendini A, Cerretani L *et al.* Effect of olive ripening degree on the oxidative stability and organoleptic properties of cv. Nostrana di Brisighella extra virgin olive oil. *J Agric Food Chem* 2004;**52**:3649–54.

Ruano J, Lopez-Miranda J, Fuentes F *et al.* Phenolic content of virgin olive oil improves ischemic reactive hyperemia in hypercholesterolemic patients. *J Am Coll Cardiol* 2005;**46**:1864–8.

Ruano J, Lopez-Miranda J, de la Torre R *et al.* Intake of phenol-rich virgin olive oil improves the postprandial prothrombotic profile in hypercholesterolemic patients. *Am J Clin Nutr* 2007;**86**:341–6.

Ruiz-Gutierrez V, Muriana FJ, Guerrero A *et al.* Plasma lipids, erythrocyte membrane lipids and blood pressure of hypertensive women after ingestion of dietary oleic acid from two different sources. *J Hypertens* 1996;**14**:1483–90.

Ryan D, Robards, K. Phenolic compounds in olives. *Analyst* 1998;**123**:31R–44R.

Salas J, Lopez Miranda J, Jansen S *et al.* The diet rich in monounsaturated fat modifies in a beneficial way carbohydrate metabolism and arterial pressure. *Med Clin (Barc)* 1999;**113**:765–9.

Sanchez J, Harwood J. Biosynthesis of triacylglycerols and volatiles in olives. *Eur J Lipid Sci Technol* 2002;**104**:564–73.

Sofi F, Cesari F, Abbate R *et al.* Adherence to Mediterranean diet and health status: meta-analysis. *BMJ* 2008;**337**:a1344.

Subramanian S, Chait A. The effect of dietary cholesterol on macrophage accumulation in adipose tissue: implications for systemic inflammation and atherosclerosis. *Curr Opin Lipidol* 2009;**20**:39–44.

Tedgui A, Mallat Z. Cytokines in atherosclerosis: pathogenic and regulatory pathways. *Physiol Rev* 2006;**86**:515–81.

Thijssen MA, Mensink RP. Small differences in the effects of stearic acid, oleic acid, and linoleic acid on the serum lipoprotein profile of humans. *Am J Clin Nutr* 2005;**82**:510–6.

Thomsen C, Rasmussen OW, Hansen KW *et al.* Comparison of the effects on the diurnal blood pressure, glucose, and lipid levels of a diet rich in monounsaturated fatty acids with a diet rich in polyunsaturated fatty acids in type 2 diabetic subjects. *Diabet Med* 1995;**12**:600–6.

Tovar MJ, Motilva MJ, Romero MP. Changes in the phenolic composition of virgin olive oil from young trees (Olea europaea L. cv. Arbequina) grown under linear irrigation strategies. *J Agric Food Chem* 2001;**49**:5502–8.

Tripoli E, Giammanco M, Tabacchi G *et al.* The phenolic compounds of olive oil: structure, biological activity and beneficial effects on human health. *Nutr Res Rev* 2005;**18**:98–112.

van Dijk SJ FE, Bos MB, Hoelen DW *et al.* A saturated fatty acid–rich diet induces an obesity-linked proinflammatory gene expression profile in adipose tissue of subjects at risk of metabolic syndrome. *Am J Clin Nutr* 2009;**90**:1656–64.

Vinha A, Ferreres F, Silva B *et al.* Phenolic profiles of Portuguese olive fruits (*Olea europaea* L.): Influences of cultivar and geographical origin. *Food Chem* 2005;**89**:561–568.

Visioli F, Bernardini E. Extra virgin olive oil's polyphenols: biological activities. *Curr Pharm Des* 2011;**17**:786–804.

Visioli F, Caruso D, Galli C *et al.* Olive oils rich in natural catecholic phenols decrease isoprostane excretion in humans. *Biochem Biophys Res Commun* 2000;**278**:797–9.

Visioli F, Poli A, Gall C. Antioxidant and other biological activities of phenols from olives and olive oil. *Med Res Rev* 2002;**22**:65–75.

Visioli F, Caruso D, Grande S *et al.* Virgin Olive Oil Study (VOLOS): vasoprotective potential of extra virgin olive oil in mildly dyslipidemic patients. *Eur J Nutr* 2005;**44**:121–7.

Wahrburg U, Kratz M, Cullen P. Mediterranean diet, olive oil and health. *Eur J Lipid Sci Technol* 2002;**104**:698–705.

Weinbrenner T, Fito M, de la Torre R *et al.* Olive oils high in phenolic compounds modulate oxidative/antioxidative status in men. *J Nutr* 2004;**134**:2314–21.

Willett WC. The Mediterranean diet: science and practice. *Public Health Nutr* 2006;**9**:105–10.

Willett WC, Sacks F, Trichopoulou A, *et al.* Mediterranean diet pyramid: a cultural model for healthy eating. *Am J Clin Nutr* 1995;**61**:1402S–1406S.

Witztum JL. The oxidation hypothesis of atherosclerosis. *Lancet* 1994;**344**:793–5.

16 Cocoa, Blood Flow and the Brain

Crystal F. Haskell and Anthony W. Watson

INTRODUCTION

Research into the health benefits of cocoa was first stimulated by observations of Kuna Indians living off the coast of Panama. It was noted that incidence of hypertension was very low in this population and the common increase in blood pressure with age was not observed. However, examination of those Kuna who resided in Panama City revealed that this effect was lost upon migration, suggesting environmental, rather than genetic, factors are responsible for this effect (Hollenberg *et al.*, 1997). Diet was highlighted as a potential factor in these antihypertensive effects and the observation that island-dwelling Kuna consumed 10 times as much cocoa as their urban counterparts (McCullough *et al.*, 2006), as well as demonstrations of nitric oxide-dependent vasodilation following flavan-3-ol-rich cocoa consumption (Fisher *et al.*, 2003), led to the suggestion that cocoa may be responsible for these effects. In light of the importance of vascular function to cognition, and specifically cognitive decline (Brown and Thore, 2011; Gorelick *et al.*, 2011; Levine *et al.*, 2011), a number of recent studies have also explored the impact of cocoa on cognitive function. The purpose of this chapter is to explore the relationship between cocoa consumption, specifically flavan-3-ols, and brain activity as assessed by human intervention studies. Wherever possible, studies included will be conducted according to randomized, placebo-controlled, double-blind procedures. Evidence from animal studies and human cardiovascular studies will also be included, where appropriate, to provide potential mechanisms relevant to cognition. As the effects of caffeine and theobromine have also been the focus of research with regards the effects of cocoa on cognition, relevant studies of these compounds will also be included.

CULTIVATION AND EXTRACTION

Theobroma cacao, also known as the cocoa or cacao tree, is an evergreen tree, which although indigenous to tropical regions of the Americas has been cultivated in many other countries, including the Ivory Coast, Indonesia, Papua New Guinea and the Solomon Islands. There are three main types of cocoa; Criollo, Trinitario and Forastero, of which the latter comprises 95% of cocoa production. The fruit of the tree, termed the cocoa pod, is around 30 cm long.

Bioactives in Fruit: Health Benefits and Functional Foods, First Edition. Edited by Margot Skinner and Denise Hunter.

When ripe, this pod weighs around 500 g and contains approximately 20–60 seeds, known as cocoa beans, embedded in a white pulp. The cocoa bean contains around 40–60% fat and is abundant in phytochemicals. Early documentation of the use of cocoa depicts indigenous Mesoamericans using ground, roasted, cocoa beans mixed with chilli to make a cold, bitter drink. Presently, postharvest processing techniques are diverse and depend on the desired use of the cocoa, and ultimately the required taste. The main postharvest processes in the preparation of cocoa beans are drying (sun dried), roasting, fermentation and alkalisation (dutching). Once roasted or dried, cocoa beans are cracked out of their shells to create nibs. These nibs can be eaten alone or used in other foods; they are also ground and fat is extracted to create cocoa powder and cocoa butter.

ACTIVE COMPONENTS

Cocoa beans contain high levels of flavan-3-ols, a subclass of the larger group of plant compounds known as flavonoids, which are now widely accepted to be the main active component responsible for the health benefits observed following cocoa consumption. As with all polyphenols, there is large variability in content of flavan-3-ols as a consequence of cultivar, growing conditions, harvesting and processing; however, flavan-3-ols tend to make up at least 10% of cocoa powder (Hammerstone *et al.*, 2000). Of the flavan-3-ols, the monomers (+)–catechin and (–)–epicatechin are of particular interest. These monomers, along with their enantiomeric forms (–)–catechin and (+)–epicatechin, are often present in nature as subunits of the oligomers (procyanidins). Monomers have been shown to make up around 12% of the total procyanidin content, irrespective of whether measured in unsweetened or dark chocolate, whereas large polymers (> decamers) range from about 28% in dark chocolate to about 34% in unsweetened chocolate (Gu *et al.*, 2004).

Like raw cocoa beans, the flavan-3-ol content of cocoa-based products varies greatly with processing techniques. For example, milk chocolate contains about 41 mg of flavan-3-ols per 100 g whereas the flavan-3-ol content of dark chocolate can be up to 20 times higher than milk chocolate, ranging from about 93 mg to 793 mg/100 g of chocolate, dependent on the amount of cocoa solids (Langer *et al.*, 2011). Postharvest preparation techniques not only alter the quantity of phytochemicals in cocoa, but also their stereochemistry (Hurst *et al.*, 2011). For example, Kofink *et al.* (2007) quantified (–)–epicatechin, (+)–catechin and (–)–catechin (Figure 16.1) in roasted and unroasted cocoa beans, cocoa liquor, chocolate containing 85% cocoa solids and cocoa powder in both alkalised and non-alkalised form. As expected, (–)–epicatechin and (+)–catechin were present in all sources tested. However, only roasted cocoa beans and the alkalised cocoa powder additionally contained the atypical flavan-3-ol (–)–catechin. High temperatures during the cocoa bean roasting process and the alkalisation of the cocoa powder are cited as the main factors inducing the epimerisation reaction.

Although flavan-3-ols are widely accepted to be the main active component in regards to health benefits, cocoa has a diverse phytochemical profile. Methylxanthines such as theobromine and caffeine can be found in cocoa and, like flavan-3-ols, the quantities of these methylxanthines vary greatly between plant genotypes, with levels of theobromine ranging from 0.5 to 1.2% of dry cocoa weight and levels of caffeine ranging between 0.4 and 0.1% (Brunetto *et al.*, 2007). However, these values are considerably lower in chocolate with caffeine ranging from 0.03 and 0.04% and theobromine ranging from 0.2 to 0.4% (Brunetto *et al.*, 2007). Naturally occurring indole alkaloids Tetrahydro-β-carbolines are

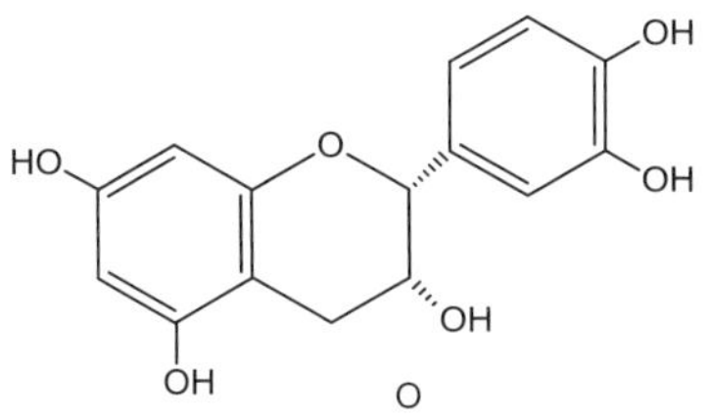

(−)-epicatechin

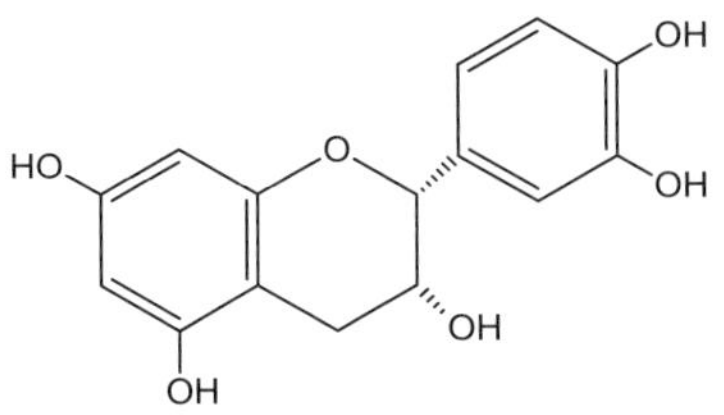

(+)-catechin

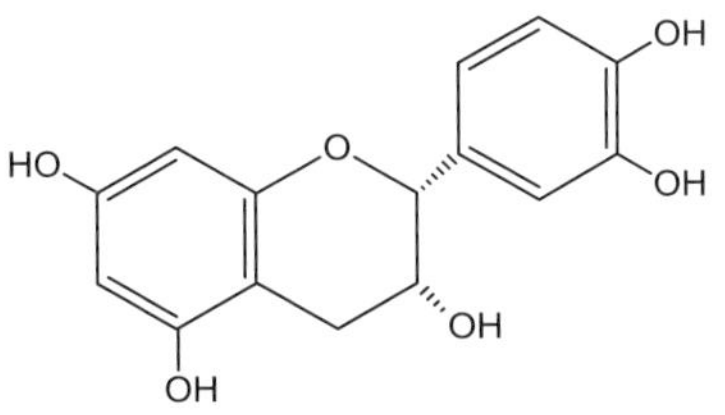

(−)-catechin

Fig. 16.1 Chemical structures of flavan-3-ol monomers most commonly found in cocoa.

also found in cocoa-based products along with their amino acid and amine precursors L-tryptophan, serotonin and tryptomine (Herraiz, 2000), tryptophan, phenylethylamine (Hurst and Toomey, 1981) and anandamide (di Tomaso *et al.*, 1996). Indole Tetrahydro-β-carbolines vary in quantity depending on the postharvest preparation of the food. Levels are low and range from 5.5 μg/g of dark chocolate, 3.3 μ/g of cocoa drinking powder and 2.3 μg/g of milk chocolate (Herraiz, 2000). Derivatives of the anthocyanin cyanidin and monomers and derivatives of the flavones quercetin, naringen, apigenin and luteolin have also been identified in cocoa bean, liquor and powder (Ortega *et al.*, 2008). However, the quantification of these phenolics has not been conducted in chocolate and it is possible they are hydrolysed by postharvest techniques. Cocoa beans also contain micronutrients such as phosphorus, potassium, magnesium, sodium and iron (Lockard and Burridge, 1965). These micronutrients withstand processing techniques and are found in processed cocoa powders; phosphorus 390 mg, potassium 2078 mg, magnesium 260 mg, sodium 1169 mg and iron 13 mg per 100 g of powder (CocoaVia^{TM}).

PHARMACOKINETICS AND ABSORPTION

The literature surrounding the pharmacokinetics of flavan-3-ols after consumption of cocoa based food is difficult to interpret due to varying methodologies and the inherent differences in phenolic content across different cocoa-based foods. This is exacerbated due to the lack of comparison between different cocoa products in the same controlled study, for example a cocoa drink vs. milk/dark chocolate. However, monomeric flavonoids can pass through mucosal cells in the small intestine and be absorbed without much hydrolysis or conjugation; therefore, epicatechin is found in plasma in its monomeric state. Polymeric procyanidins do not undergo depolymerization in the stomach (Rios *et al.*, 2002); they are, however, degraded in the microflora in the intestines and bowel. The produced metabolites can then be absorbed, or excreted in faeces. The monomeric flavan-3-ols and derivatives are modified on absorption during transfer across the small intestine and then by the liver, where they are mainly modified into glucuronide conjugates (Gonthier *et al.*, 2003). *In vitro* (Youdim *et al.*, 2003) and animal models (Manal *et al.*, 2002) have shown that flavan-3-ols can cross the blood–brain barrier and enter the brain after oral ingestion.

The amount of flavan-3-ols found in blood plasma after consumption of cocoa-based foods is observably affected by the postharvest preparation process and product matrix. With this in mind, the biological activity of these polyphenols, and therefore the health benefits, are largely dependent on their bioavailability (Ramiro-Puig *et al.*, 2009). Monomers and oligomers of (+)–catechin and (–)–epicatechin are found in plasma at the nanomolar or low micro molar levels after consumption of cocoa rich foods, with plasma concentration peaking between two and three hours post-supplementation (Serafini and Crozier, 2003). The absorption of (–)–epicatechin has been shown to be directly proportional to the concentration consumed in chocolate. For example, Richelle *et al.* (1999) supplemented healthy male volunteers with 40 or 80 g of dark chocolate along with bread and water with a one week washout period between each supplementation. Epicatechin intakes were 82 and 164 mg and plasma epicatechin concentrations were measured at baseline, 1, 2, 3, 4 and 8 hours post-supplementation via high performance liquid chromatography (HPLC). Results showed that epicatechin ^{T}max peaked at two hours with a maximum concentration of 111 ng/ml for the 40 g condition and 2.6 hours with a maximum concentration of 203 ng/ml for the 80 g condition, with maximal concentrations being proportional to the dose. In a study of the longer-term effects on plasma concentrations, Murphy *et al.* (2003) supplemented 32 healthy participants in parallel with either 234 mg of cocoa flavan-3-ols or a placebo in tablet form for 28 days. Both (+)–catechin and (–)–epicatechin were detectable in plasma at baseline but after 28 days supplementation plasma concentrations increased by 28% and 81% respectively with no significant post-supplementation changes in the control arm observed. To our knowledge, the atypical (–)–catechin in its monomeric or conjugate states has not been quantified in humans after oral consumption. It has, however, been shown to be present in plasma after oral consumption in rats, with (+)–catechin being more bioavailable than (–)–catechin (Donovan *et al.*, 2006).

Despite original findings by Serafini and Crozier (2003) outlining a distinct reduction in the absorption of cocoa polyphenols when consumed with milk, either as a drink or in milk chocolate, further studies have found that the food matrix, especially milk or milk proteins, does not affect the uptake of flavan-3-ols from cocoa-based products (Keogh *et al.*, 2007; Roura *et al.*, 2007; Mullen *et al.*, 2009). Although the effect of milk on the absorption of cocoa flavan-3-ols is somewhat unclear, Schramm *et al.* (2003) studied the effect of carbohydrates on the pharmacokinetic profile of cocoa flavan-3-ols. The authors supplemented six healthy volunteers with either cocoa alone or cocoa with sugar (8.75 kj/kg of body weight or 17.5 kj/kg body weight), providing all treatment groups with 1.53 mg of total flavan-3-ols per kilogram of body weight. The outcomes indicated a significant increase in area under the curve in the 8.75 kj and 17.5 kj groups when compared to cocoa alone and an increase in ^{C}max in the 17.5 kj group when compared to control. However, a decrease in the elimination half-life ($E_{\frac{1}{2}}$) of the cocoa flavan-3-ols in the 8.75 kj and 17.5 kj groups was also observed when compared to cocoa alone. This outlines a higher uptake of flavan-3-ols if consumed with sugar when compared to cocoa alone without any change in ^{T}max but with a shorter $E_{\frac{1}{2}}$. In a similar study Schramm *et al.* (2003) supplemented six healthy participants with 0.125 kg of cocoa per kilogram of body weight, either alone or with bread, steak or butter in a four-week crossover study design. Results showed that only the cocoa and bread treatment affected absorption of flavan-3-ols when compared to cocoa alone; as with the previous study, there was an increase in ^{C}max without an increase in ^{T}max when consumed with bread.

The methylxanthines caffeine and theobromine have been extensively investigated in terms of bioavailability and are both measurable in plasma in a dose-responsive fashion after oral ingestion (Eteng *et al.*, 1997). After oral administration of 72 mg of caffeine and 370 mg of theobromine as pure compound, the uptake of theobromine from the digestive tract is slow when compared to caffeine, with maximum concentrations of caffeine peaking after 0.5 hours at 1.93 μg/ml and maximum concentrations of theobromine peaking after about 3 hours at 6.72 μg/ml. When consumed in chocolate containing 72 mg of caffeine and 370 mg of theobromine, both methylxanthines are rapidly absorbed with maximum concentrations of caffeine peaking after 1.5 hours at 1.5 mg/ml. Interestingly, theobromine peaked after two hours at 8.5 mg/ml, one hour quicker than the compound ingested alone (Mumford *et al.*, 1996).

MECHANISM OF ACTION

In vitro

Flavan-3-ols and their metabolites have the potential to exert neuroprotective actions, including the capacity to protect neurons from damage induced neurotoxins, reduce neuroinflammation and promote cognitive function. Flavan-3-ols are known to possess a potent antioxidative capacity; however, it is unlikely that this antioxidant activity can fully explain their neuroprotective potential. *In vitro* studies have shown that the potential neuroprotective action of flavonoids is most likely via a cascade of cell signalling mechanisms, such as modulating mitogen-activated protein kinase (MAPK) activation (Ramiro-Puig *et al.*, 2009), inhibiting NF-kB, p53 and ERK activation (Huang *et al.*, 2005) and mediating central nervous system inflammation by suppressing microglia activation (Li *et al.*, 2004). (A comprehensive review is available elsewhere (Spencer, 2007).) Polymeric procyanidins extracted from cocoa have also been shown to cause endothelial relaxation in aortic rings of New Zealand rabbits. Furthermore, Ca21-dependent nitric oxide activity is also significantly increased in aortic endothelial cells when exposed to polymeric procyanidins without any effect from monomeric compounds (Karim *et al.*, 2000).

The main mechanism of action of theobromine is via antagonism of adenosine receptors, specifically A_1 and A_{2A} subreceptors (Fredholm *et al.*, 1999). As with caffeine, it has an equal affinity for the A_1 and A_2 receptors; however, theobromine has a much lower overall affinity when compared to caffeine (Fredholm and Lindstrom, 1999).

In vivo

If transferable *in vivo,* the reported *in vitro* mechanisms could promote synaptic elasticity and neuronal survival, and could therefore constitute a good source of exogenous neuroprotectants, particularly relevant to attenuating age-related neurocognitive decline. Indeed, *in vivo* animal models have revealed that fruit extracts high in flavonoids can interact with phosphor-c-Jun *N*-terminal kinase and the p53 signalling pathways, significantly reducing brain injury after focal cerebral ischaemic injury in male Sprague-Dawley rats after oral administration (Shin *et al.*, 2006). Epicatechin in isolation has been observed to aid the prevention of stroke damage through the Nrf2/HO1 pathway (Shah *et al.*, 2010). Flavonoid-rich extracts have also been shown to improve spatial learning and memory when fed orally to aged rats over a 10 week period. These increases in cognitive performance were directly correlated to flavonoid levels in the brains of the rats when compared to unfed controls (Andres-Lacueva *et al.*, 2005). Bisson *et al.* (2008) investigated the impact of 12 month's administration of a cocoa extract (Acticoa) containing 35 mg of polyphenols (89% procyanidins, 0.2% anthocyanins, 10% epicatechin 1% epicatechin gallate and 0.5% catechin) per 100 g of body weight versus pellet food only on cognition in 18 male Wistar-Unilever rats. Cognition was assessed by light extinction and water maze (adapted from Morris, 1984) prior to and at 0, 2, 6 and 10 months following intervention. The light extinction task involves the rat being placed in a highly illuminated cage equipped with an active and an inactive lever. By pressing the active lever, the rat switches off the light allowing access to a 30-second period of darkness, whereas the inactive lever has no effect on the environment. In the water maze task, rats were trained to find a hidden platform hidden 1 cm beneath the surface in a tank filled with water. After training, rats were placed into the tank and retested whilst time to find the platform was measured. Comparisons between groups across daily trials allowed an assessment of short-term memory, whereas comparisons over months represented long-term memory. Attenuation of cognitive decline in short- and long-term memory measures, as assessed by the water maze task, was observed after supplementation of the cocoa extract when compared to control. Rats supplemented with the cocoa extract also had more total lever pressing activity than control rats in the light extinction test along with significant discrimination between active and inactive levers at 17 and 21 months of age. However, Rozan *et al.* (2007) found no discrimination between active and inactive lever presses after 16 days' supplementation with 23 mg/kg of Acticoa when compared to control. Rats in the cocoa condition did, however, show significant reductions in oxidative stress when exposed to extreme heat stress when compared to control rats; this was also demonstrated after supplementation of 200 mg/kg/d of α-tocopherol (vitamin E). It is possible that the lack of effects on lever presses reflects the shorter time period of intervention used here but these findings, taken together with findings from Bisson *et al.* (2008) showing no effects until 17 months, may also suggest that memory benefits of cocoa-based products in animal models only occur with age.

HUMAN INTERVENTION TRIALS

Peripheral blood flow

In light of observations of Kuna Indians suggesting a relationship between cocoa consumption and blood pressure, a number of intervention studies have explored the effects of cocoa on blood flow-related parameters. Several of these studies have provided evidence to support a reduction in blood pressure in healthy participants (Fraga *et al.*, 2005; Grassi *et al.*, 2005a), as well as the overweight (Faridi *et al.*, 2008; Berry *et al.*, 2010) and those with evidence of glucose intolerance or cardiovascular risk factors (Taubert *et al.*, 2003; Grassi *et al.*, 2005b, 2008; Heiss *et al.*, 2010). However, others have failed to demonstrate these effects whether studying healthy participants or otherwise (Heiss *et al.*, 2003; Engler *et al.*, 2004; Balzer *et al.*, 2008; Davison *et al.*, 2008; Monagas *et al.*, 2009; Mellor *et al.*, 2010; Njike *et al.*, 2011). There are a number of methodological differences that may explain these inconsistencies in findings across studies but meta-analyses have supported a case for antihypertensive effects of cocoa (Taubert *et al.*, 2007; Desch *et al.*, 2010; Ried *et al.*, 2010). Specifically, Taubert *et al.* (2007) identified five studies that met their criteria for effects of

$\geq$7 days' cocoa intake, resulting in a combined total of 173 participants. The median duration of intervention was two weeks, with 64% male participants and 34% high–normal/hypertensive. Four of the five studies identified demonstrated a significant reduction in blood pressure, with the suggestion that the failure to find effects in the fifth study is due to below threshold levels of flavan-3-ols being administered. Pooled analyses support an antihypertensive effect, suggesting that younger participants with essential hypertension are more susceptible to these effects than younger normotensives or elderly hypertensives. Similar conclusions with regards the effects of cocoa on blood pressure were made in a more recent meta-analysis of 15 intervention trials examining effects of $\geq$14 days' intake (Ried *et al.*, 2010). Subgroup analyses demonstrated that significant blood pressure reductions following cocoa were only apparent in hypertensives/prehypertensives, whereas no significant effects were demonstrated in a normotensive subgroup. However, despite a range of flavan-3-ol doses from 30 to 1008 mg, no effects of dose were apparent, nor was age, duration or body mass index (BMI) shown to have a significant association.

Perhaps a more informative measure in the effects of cocoa on blood flow is Flow Mediated Dilation (FMD), as this indicates a specific effect of cocoa on endothelial function. A number of studies have demonstrated increased FMD across a similar range of populations as studied in relation to blood pressure (many of whom suffer endothelial dysfunction), whether studied chronically over periods of eight days to twelve weeks (Engler *et al.*, 2004; Grassi *et al.*, 2005b, 2008; Davison *et al.*, 2008; Heiss *et al.*, 2010), or acutely at two hours post-administration (Heiss *et al.*, 2003, 2005; Vlachopoulos *et al.*, 2005; Davison *et al.*, 2008; Faridi *et al.*, 2008; Berry *et al.*, 2010; Njike *et al.*, 2011). Studies have also demonstrated a significant acute increase superimposed upon a chronic increase, with effects being greatest when measured two hours post-ingestion at the final time point measured (whether this was following 8 or 30 days cocoa intake), suggesting a cumulative benefit of cocoa consumption that may extend beyond 30 days (Heiss *et al.*, 2007; Balzer *et al.*, 2008). Further support for a role of endothelial function in the relationship between cocoa and blood flow comes from studies showing an absence of modulation of endothelial-independent brachial artery diameter, as measured following sublingual administration of glycerol trinitrate (Heiss *et al.*, 2003, 2007; Balzer *et al.*, 2008; Grassi *et al.*, 2008). The acute effect of cocoa on FMD is correlated with increases in plasma nitric oxide species (Heiss *et al.*, 2003, 2005), with Heiss *et al.* (2005) demonstrating a reversal of the improvement to endothelial function when nitric oxide synthase was inhibited by intravenous infusion of L-NG-monomethyl-arginine nitrite (LNMMA). Support for the involvement of nitric oxide in the blood flow effects following longer term cocoa intake also comes from data showing increases in nitrite (an oxidative metabolite of nitric oxide) in parallel with FMD increases (Heiss *et al.*, 2007, 2010), with Heiss *et al.* (2010) also demonstrating an increase in circulating angiogenic cells involved in the maintenance and recovery of endothelial function, which have been shown to be linked to endothelial nitric oxide synthase (Ward *et al.*, 2011). Platelet function, which again relates to endothelial nitric oxide synthase, has also been observed to be reduced following cocoa consumption, both acutely (Hermann *et al.*, 2006; Flammer *et al.*, 2007) and chronically (Murphy *et al.*, 2003; Hamed *et al.*, 2008; Monagas *et al.*, 2009). Not all studies have demonstrated this effect (Farouque *et al.*, 2006), but it is possible that underlying group differences present at baseline may explain the lack of effects in this study. Other findings relevant to blood flow include reductions in insulin resistance (Grassi *et al.*, 2005a, 2008), cell adhesion molecule (Flammer *et al.*, 2007), and low density lipoprotein (LDL) cholesterol (Fraga *et al.*, 2005; Grassi *et al.*, 2005b; Balzer *et al.*, 2008). Recent meta-analyses of cocoa's impact on cardiovascular risk factors support robust improvements to FMD and insulin resistance, with lesser effects on blood pressure, fasting glucose, triglycerides and cholesterol (Shrime *et al.*, 2011; Hooper *et al.*, 2012). In terms of dose effects, analyses from Hooper *et al.* (2012) suggest reductions in blood pressure following ≥50 mg epicatechin, and benefits to fasting glucose and triglycerides following 50–100 mg. Similarly, Shrime *et al.* (2011) discovered optimal effects on FMD at about 500 mg total polyphenol dose, with decreasing effects at higher doses. There was also some indication of impact of duration with reductions in LDL and total cholesterol apparent in studies of shorter duration (3–4 weeks), whereas increases in high density lipoprotein (HDL) cholesterol are not observed unless measuring over a longer time-frame (Shrime *et al.*, 2011; Hooper *et al.*, 2012).

Although the studies above attribute blood flow-related improvements following cocoa to its flavan-3-ol content, theobromine has historically been used as a vasodilator (Kelly, 2005). van den Bogaard *et al.* (2010) also recently demonstrated, in a controlled intervention trial, that decreases to central and peripheral blood pressure were only apparent when comparing theobromine-enriched cocoa to placebo, and not when examining effects of cocoa with natural theobromine content. In support of this, when considering the role of theobromine in the antihypertensive effects of cocoa it is apparent that a number of studies that have employed a methylxanthine-matched placebo have failed to show these effects (Heiss *et al.*, 2003; Balzer *et al.*, 2008; Davison *et al.*, 2008), but this is not always the case (Berry *et al.*, 2008; Heiss *et al.*, 2010). However, when considering the same issue with regards FMD it is clear that the effects are consistent whether compared to a white chocolate control or a methylxanthine-matched control. Additionally, Schroeter *et al.* (2006) demonstrated significant increases in FMD following epicatechin in isolation, supporting a role for epicatechin in the endothelial effects of cocoa, but not eradicating the possibility of effects of theobromine on this measure; moreover, there may be a synergistic effect of cocoa components that requires further exploration.

Brain activity

Randomized controlled trials that have assessed the impact of cocoa or its major components on brain activity are summarized in Table 16.1.

Flavan-3-ols

In one of the first studies to extend the findings of improved peripheral vasodilation following cocoa consumption to cerebral blood flow (CBF), Sorond *et al.* (2008) assessed the impact, in elderly adults, of short-term flavan-3-ol-rich cocoa consumption on cerebral blood flow velocity and vasoreactivity as assessed by transcranial doppler (TCD). Beat-to-beat arterial pressure and cerebrovascular conductance were also assessed. Seven days' consumption of either flavan-3-ol-rich cocoa (900 mg) or flavan-3-ol-poor control (36 mg), matched for calories, macronutrients, micronutrients and methylxanthines, produced no significant effects on the outcomes measured. The lack of effects in this study may relate to effects of the known vasoconstrictor caffeine. Given that dietary control was simply limited to assessment taking place ≥2 hours post-meal, it is possible that dietary intake of caffeine prior to participation resulted in a reduction of CBF as previously demonstrated (e.g. Cameron *et al.*, 1990; Lunt *et al.*, 2004; Sigmon *et al.*, 2009; Kennedy and Haskell, 2011). Alternatively, caffeine withdrawal may have led to increases in CBF as previously shown (Field *et al.*, 2003; Addicott

Table 16.1 Randomized controlled trials assessing the effects of cocoa on brain function.

Reference	Sample	Design	Dose, duration	Placebo	Vehicle	Methods	Results
Mumford *et al*. (1994)	N = 7 (28–46 y, mean 45) 330 mg CA/d; 25.7 mg TB/d	Crossover	560 mg TB; 178 mg CA. 5 days	Lactose	Capsules	Subjective ratings (baseline; 0.25, 0.5, 0.75, 1, 1.5, 2, 4, 8 & 12 h PD)	No group effects of TB
Smit *et al*. (2004)	N = 20 (18–56 y; mean 32.6) 330 mg CA/day; 100 mg TB/day	Crossover	CP [47 mg CF* (23 mg monomers) + 250 mg TB + 19 mg CA]; 250 mg TB + 19 mg CA. Acute	Cellulose	Capsules	Mood/state VAS; SRT; RVIP; Tapping (baseline; 1, & 2 h PD)	SRT, energetic arousal & hedonic tone significantly improved by CP & TB + CA; TB + CA improved headache
Smit *et al*. (2004)	N = 22 (18–70 y; mean 35.4) 386 mg caffeine/day; 85 mg theobromine/day	Crossover	100 mg TB + 8 mg CA (low MX); 250 mg TB + 20 mg CA (high MX). Acute	Zero MX; 60ml water	'Chocolate' (icing sugar, milk powder, cocoa butter, emulsifier, food colouring, flavouring)	Mood/state VAS; SRT; RVIP; Tapping (baseline; 1, & 2 h PD & during 'extended intervention administration')	SRT significantly reduced by high MX; RVIP accuracy improved by high & low MX
Francis *et al*. (2006)	N = 4 (18–31 y)	Crossover	516 mg CF. Acute	39 mg CF, matched	Cocoa beverage	CBF measured with ASL MRI (baseline; 2, 4 & 6 h PD)	CF increased CBF (peak 2 h, return to baseline 6 h)
Francis *et al*. (2006)	N = 16 females (18–30 y) <120 mg caffeine/day	Crossover	172 mg CF. 5 days	13 mg CF, matched	Cocoa beverage	Switch task; BOLD; HR (1.5 h PD)	CF increased BOLD response. No effects on cognition
Crews *et al*. (2008)	N = 90 (≥60 y; mean 68.7) CI (≥24 MMSE)	Parallel groups	805 mg CF (51 mg monomers) + 481 mg TB + 36 mg CA. 6 weeks	41 mg CF + 1 mg TB + 2 mg CA	37 g chocolate + 237 ml cocoa beverage	BSRT; WMS-III Faces I & II; TMT; Stroop; WAIS-III DSST; total cholesterol (HDL, LDL, VLDL); triacylglycerol; C-reactive protein (baseline & 6 weeks). A-DACL General Activation; SBP; DBP; HR (baseline, & at 2 h PD at 3 & 6 weeks)	No positive effects. CF increased HR (3 & 6 weeks)
Sorond *et al*. (2008)	N = 21 (59–83 y; mean 72.2)	Parallel groups	900 mg CF + 673 TB + 37 mg CA. 1 week	36 mg CF, matched	Cocoa beverage	TCD of MCA to assess VR & MFV; CVR; ABP (baseline & PD)	No effects

(*continued*)

Table 16.1 *(Continued)*

Reference	Sample	Design	Dose, duration	Placebo	Vehicle	Methods	Results
Scholey *et al.* (2010)	N = 30 (18–35 years; mean 21.9)	Crossover	520 mg CF (129 mg monomers) + 429 mg TB + 44 mg CA; 994 mg CF (246 mg monomers) + 458 mg TB + 41 mg CA. Acute	46 mg CF, matched	Cocoa beverage	2 serial subtraction tasks (3 & 7 s); RVIP; VAS; STAI-Y1 (1.5 h PD)	CF increased correct serial 3 subtractions; 520 mg CF attenuated mental fatigue; 994 mg CF improved RVIP RT but increased serial 7 subtraction errors
Mitchell *et al.* (2011)	N = 24 females (18–70 years; mean 51.1) 420 mg caffeine/day	Crossover	700 mg TB; 120 mg CA; 700 mg TB + 120 mg CA. Acute	Cellulose	Capsules	Bond-Lader; DSST; ERTT; SBP; DBP; HR (baseline;1, 2 & 3 h PD). Implicit mood (1 h PD). Motivation & Workload Q (3 h PD)	CA increased alertness, contentment, interest in, & eagerness to perform tasks. TB reduced calmness & increased interest in tasks. ERRT data indicated negative mood was increased by TB but decreased by CA. CA increased SBP & DBP, & decreased HR. TB decreased SBP & increased HR
Field *et al.* (2011)	N = 30 (18–25 years)	Crossover	773 mg CF + 222 mg TB + 38 mg CA. Acute	White chocolate	35 g chocolate	CS; motion coherence threshold; motion integration time threshold; visual spatial WM; CRT (2 h PD)	CF improved CS, improved WM accuracy & speeded motion integration & CRT
Camfield *et al.* (2011)	N = 63 (40–65 years; mean 52.3)	Parallel groups	250 mg CF + 240 mg TB + 18 mg CA; 500 mg CF + 240 mg TB + 18 mg CA. 30 days	0 mg CF, matched	Cocoa beverage	SST-SSVEP; SWM (baseline & PD)	CF decreased SSVEP amplitude (250 mg) & increased latency (250 & 500 mg) in posterior parietal regions

ABP = Ambulatory Blood Pressure; A-DACL = Activation-Deactivation Adjective Check List; ASL = Arterial Spin Labelling; BOLD = Blood Oxygenation Level-Dependent response; BSRT = Buschke Selective Reminding Test; CA = caffeine; CBF = Cerebral Blood Flow; CF = Cocoa Flavanols; CI = Cognitively Intact; CP = cocoa powder; CS = Contrast Sensitivity; CVC = CerebroVascular Conductance; DBP = Diastolic Blood Pressure; ERTT = Emotive Reaction Time Task; g = gram; h = hours; HDL = High-Density Lipoprotein; HR = Heart Rate; LDL = Low-Density Lipoprotein; MCA = Middle cerebral Artery; MFV = Mean blood Flow Velocity; mg = milligram; ml = millilitre; MMSE = Mini Mental State Examination; MRI = Magnetic Resonance Imaging; MX = methylxanthine; N = Number of subjects; PD = post-dose; PM = Post-Meridian; Q = Questionnaire; RT = Reaction Time; RVIP = Rapid Visual Information Processing; SBP = Systolic Blood Pressure; SRT = Simple Reaction Time; SST = Steady State Topography; SSVEP = Steady State Visually Evoked Potential; STAI-Y1 = State Trait Anxiety Inventory state component; SWM = Spatial Working Memory; TB = theobromine; TCD = TransCranial Doppler; TMT = Trail Making Test; VAS = Visual Analogue Scales; VLDL = Very Low-Density Lipoprotein; VR = VasoReactivity; WAIS-III = Wechsler Adult Intelligence Scale-III; WM = Working Memory; WMS-III = Wechsler Memory Scale-III.

*Levels taken from Langer *et al.* (2011).

and Laurienti, 2009), making it difficult to demonstrate further superimposed increases. Another explanation is that effects of flavan-3-ols on CBF are only measurable acutely.

Acute effects of flavan-3-ols on CBF were explored by Francis *et al.* (2006) following either a flavan-3-ol-rich cocoa drink (516 mg) or matched control (39 mg) in healthy young adults. Significant increases in grey matter CBF were observed two hours post-consumption using arterial spin labelling, with a return to baseline by six hours post-consumption. These findings of a return to baseline levels at six hours support the suggestion that CBF increases would be unlikely to be observed in the Sorond *et al.* (2008) study, but do not rule out a cumulative effect with continued consumption, and the possibility that the intervention period of seven days was not long enough to demonstrate this in the aforementioned study. In a separate fMRI study Francis *et al.* (2006) examined effects of five days' consumption of flavan-3-ol-rich cocoa (172 mg) or a matched control (13 mg) in healthy young females. fMRI scanning took place during completion of a letter pair switching task at approximately 90 minutes post-consumption on the fifth day of consumption in each condition. BOLD signal intensity was shown to be significantly increased by flavan-3-ol rich cocoa as compared to the control. Given the previously demonstrated increase in grey matter CBF, it is not clear whether this increase in BOLD signal indicates an increase in neuronal activity or merely reflects modulation of CBF. However, as this increased activation was apparent in brain areas relevant to the task, it is perhaps surprising that no significant effects on cognitive performance were observed. One suggestion for the lack of behavioural effects is that the healthy young participants were performing close to ceiling and benefits were, therefore, unlikely to be observed. Another related explanation for this lack of effects is the use of neuropsychological tasks that may not be sensitive enough to detect small changes in cognition as a consequence of a nutritional intervention over the time-frame employed. This latter point is supported by findings from one of only two studies to date to detect an effect of cocoa flavan-3-ols on cognition. Scholey *et al.* (2010) demonstrated significant improvements to performance of an intense 60-minute Cognitive Demand Battery in healthy young adults at 90 minutes post-administration of 520 and 994 mg flavan-3-ols compared to matched control (46 mg). Of the two enriched drinks the 520 mg drink was more effective, leading to consistent improvements to serial threes subtraction performance and a robust attenuation of mental fatigue increases induced by performance of the demanding tasks.

In the only study to measure the cognitive effects of chronic cocoa supplementation in the absence of an acute load, Camfield *et al.* (2012) supplemented 40–65 year olds with cocoa at low ($\sim$0 mg), medium ($\sim$250 mg) and high ($\sim$500 mg) flavan-3-ol levels for 30 days. Steady State Visually Evoked Potentials (SSVEPs) were recorded using Steady State probe Topography (SST) (Silberstein *et al.*, 1990), whilst participants completed a spatial working memory task, at baseline and following the 30-day supplementation. SST is a form of electrophysiological brain imaging that measures steady state evoked potentials in response to task-irrelevant visual flickers, similar to Event Related Potentials (ERPs) but with greater temporal resolution. There were no effects on spatial memory. Surprisingly, the pattern of posterior-parietal SSVEP amplitude in the medium flavan-3-ol group was significantly lower than the pattern observed following the control. Latency was also decreased in the same region following both the medium and high flavan-3-ol interventions, indicative of increased neural processing speed. In light of the lack of effects on cognition and the reduction in latency, it is suggested by the authors that this decrease in amplitude may reflect an increase in neural efficiency, whereby participants are able to perform at the same level with reduced activation. This assertion is supported by previous data showing a positive association between increased posterior-parietal activation and task difficulty (Silberstein *et al.*, 2001).

Cocoa

Whilst the studies outlined above all employed a methylxanthine-matched placebo, other studies of cocoa have compared whole cocoa interventions to controls that are not matched for methylxanthine levels. The use of a non-matched placebo obviously presents a more ecologically valid study of the effects of cocoa and increases the likelihood of detecting effects that may otherwise be obscured by effects of caffeine/theobromine in the placebo condition, but has the disadvantage of not allowing any disentanglement of effects of potential active components. Crews *et al.* (2008) attempted to address the previously highlighted issue of ceiling effects in a parallel groups trial of 90 cognitively intact elderly participants. Participants received either 37 g dark chocolate plus 237 ml cocoa beverage (754 mg flavan-3-ols) or placebo products, matched only for sensory properties, to take daily over a six-week intervention period. Despite the older age of participants in this study and the long intervention period, effects on cognition were still not apparent. As previously, this lack of effects may relate to the use of insensitive neuropsychological tasks. Interestingly, this study also failed to find effects on cardiovascular and haematological measures that have previously been shown to be modulated by cocoa (Heiss *et al.*, 2003; Fraga *et al.*, 2005; Grassi *et al.*, 2005a, 2008); the only exception being an increase in heart rate, which has been shown following theobromine in isolation (Mitchell *et al.*, 2011). An alternative explanation for the lack of significant benefits may relate to the use of sugar to sweeten the placebo, rather than the sweetener employed to sweeten the cocoa intervention, resulting in threefold higher carbohydrate levels in placebo than active intervention. More recently, Field *et al.* (2011) demonstrated improvements to visual contrast sensitivity (as assessed by reading numbers that became progressively more similar in luminance to their background) and the time to detect motion direction, 90 minutes following acute consumption of dark chocolate (720 mg cocoa flavan-3-ols, 38 mg caffeine) as compared to white chocolate. This study also demonstrated improvements to a visual spatial memory task following dark chocolate. However, the use of a placebo that was not sensorially matched, thereby precluding blinding of participants, may have impacted upon data in this study, particularly given the use of a crossover design.

Theobromine

One potential component that may have impacted in the two studies outlined above is theobromine. Only four studies have been published to date that attempt to elucidate specific effects of theobromine on brain function; in the first of these, Mumford *et al.* (1994) failed to find group subjective effects of 560 mg of theobromine. However, four of the seven participants were able to discriminate this dose of theobromine from placebo. Smit *et al.* (2004) administered methylxanthine levels equivalent to that found in a 50 g bar of dark chocolate (250 mg theobromine, 19 mg caffeine) in two different encapsulated forms – cocoa powder, or theobromine + caffeine only. Both active treatments were observed to reduce simple reaction time and increase energetic arousal. Differential effects of cocoa powder and methylxanthines alone were observed on hedonic tone, with improvements following cocoa powder evident at two hours post-intervention compared to the peak observed at one hour following methylxanthines only. This would suggest a modulation of the effects of methylxanthines by other components within cocoa, at least in terms of the time-course of effects. In a second study, the effects of 60 g cocoa-free 'chocolate' with varying levels of methylxanthines (placebo = zero; low = 8 mg caffeine + 100 mg theobromine; high = 20 mg caffeine + 250 mg theobromine) were explored. Simple reaction time was again improved following the high methylxanthine intervention, whereas both the high and low methylxanthine interventions improved accuracy of Rapid Visual Information Processing. These data suggest psychoactive effects of methylxanthines at the doses tested but do not allow a disentanglement of the effects of theobromine from those of caffeine.

A study by Mitchell *et al.* (2011) addressed the issue of the differential contributions of caffeine and theobromine to behavioural effects observed following cocoa by examining the effects of encapsulated theobromine (700 mg), caffeine (240 mg) and their combination on mood, performance and autonomic measures. Caffeine increased alertness, contentment, interest in tasks and eagerness to do tasks. Theobromine decreased calmness and increased interest in tasks. These effects on mood were not reflected in performance measures and the combination of the two components appeared to eradicate any individual effects. As expected, caffeine increased blood pressure whilst decreasing heart rate; whereas theobromine decreased both systolic and diastolic blood pressure, and increased heart rate. Clearly, differential effects of these two methylxanthines exist with theobromine appearing to induce stimulation, whilst also improving blood flow. The lack of effects following the combination would seem to contradict Smit *et al.* (2004), but these differences are likely to reflect the higher doses employed as well as differences in the tasks used.

Caffeine

The doses of caffeine administered in studies of cocoa and brain activity (Table 16.1) range from 8 to 44 mg, which are lower than those doses typically employed in studies exploring the behavioural effects of caffeine. However, a limited number of studies has demonstrated effects of relevant doses; for instance, Lieberman *et al.* (1987) reported comparable effects on a visual reaction time task following 32, 128, and 256 mg of caffeine. All doses also improved speed of four-choice auditory reaction time, with comparable effects following 32 and 128 mg. Although it has been suggested that effects of caffeine merely reflect alleviation of withdrawal in habitual caffeine consumers (James and Rogers, 2005), these effects were independent of consumer status. Similarly, Smith *et al.* (1999) examined the effects of 40 mg of caffeine in a variety of beverages (tea, coffee, cola, diet cola, water, sparkling water) compared with decaffeinated equivalents. Caffeine improved attentional and mnemonic function and increased 'alertness' and 'anxiety' in low and high habitual caffeine consumers alike. The lowest reported psychoactive dose of caffeine in humans is 12.5 mg (Smit and Rogers, 2000), which, as measured by speeded simple reaction time, was equipotent to doses of 50 and 100 mg of caffeine in both high and low habitual caffeine consumers. However, improvements to vigilance working memory task following 12.5, 25, 50, and 100 mg of caffeine were only observed in high caffeine consumers. A number of studies have also demonstrated a reduction in cerebral blood flow following caffeine consumption using a variety of methods (Cameron *et al.*, 1990; Field *et al.*, 2003; Sigmon *et al.*, 2009) but the lowest dose explored (75 mg) is higher than that relevant to cocoa studies. This dose was shown to reduce cerebral blood flow as assessed by Near Infrared Spectroscopy, with significantly greater reductions in non-habitual consumers of caffeine (Kennedy and Haskell, 2011).

EPIDEMIOLOGICAL EVIDENCE

Support for intervention studies showing effects of cocoa relevant to blood flow and the brain comes from recent epidemiological research. In the first observational study of habitual cocoa intake and blood pressure or cardiovascular disease, Buijsse *et al.* (2006) demonstrated that intake of cocoa, as assessed by the cross-check dietary history method, was associated with blood pressure measured five years post-baseline. Specifically, in elderly Dutch men, the highest tertile of cocoa intake had lower systolic and diastolic blood pressure than the lowest tertile. Fifteen-year follow-up also revealed 45–50% lower risk of cardiovascular mortality, as well as all-cause mortality, in the highest tertile of cocoa intake; this was shown to be unrelated to blood pressure reductions, suggesting differing mechanisms for these effects. Buijsse *et al.* (2010) further explored the relationship of chocolate intake with cardiovascular risk and attempted to elucidate the impact of chocolate's blood pressure lowering effects on this relationship. In a prospective study of German adults, they again showed an inverse relationship of chocolate consumption with systolic and diastolic blood pressure; this was irrespective of whether milk or dark chocolate had been consumed. Chocolate consumption was also inversely related to cardiovascular disease, with 12% of this relationship explained by baseline blood pressure. These data are supported by a cross-sectional study of about 5000 participants showing an inverse relationship between chocolate consumption and coronary heart disease in men and women aged 25–93 years, with similar observations in those below and above 60 years (Djousse *et al.*, 2011). This relationship was robust in that it survived adjustment for age, sex, and traditional risk factors, as well as exclusion of those with diabetes or on weight loss diets. However, a similar inverse relationship between chocolate consumption and hypertension was lost when adjustments were made, suggesting that other factors are involved in mediating the relationship between coronary heart disease and chocolate. Cardiac mortality following acute myocardial infarction was also explored in a Swedish inception cohort study, demonstrating a strong inverse relationship with chocolate consumption (Janszky *et al.*, 2009). Similarly, Mostofsky *et al.* (2010) showed that moderate chocolate consumption was related to lowered rate of heart failure hospitalisation or death; a finding supported by observations by Lewis *et al.* (2010) in older Australian women. This latter study also showed that frequent chocolate consumers had a significantly lower prevalence of carotid atherosclerotic plaques.

Observational data from a study comparing death certificates of Kuna Indians living on mainland Panama with those still residing in their indigenous home, where cocoa is the main source of fluid, provide support for the findings of Lewis *et al.* (2010) in showing that as well as a lower incidence of death from ischaemic heart disease, death from stroke was also less frequent in islanders (Bayard *et al.*, 2007). Further evidence of the link between cocoa consumption and cerebrovascular disease is provided by a population-based study of 33 000 women aged 49–83 years old with a mean follow-up of 10.4 years, which demonstrated a significant inverse association between chocolate consumption and total stroke, cerebral infarction and haemorrhagic stroke, with a stronger association for haemorrhagic stroke than cerebral infarction (Larsson *et al.*, 2011). A recent systematic review of observational studies demonstrated that cocoa consumption was associated with a 29% reduction in risk of stroke, a relationship that survived adjustment for potential confounders (Buitrago-Lopez *et al.*, 2011). The role of flavonoid consumption in the development of dementia has also been explored in the Paquid study of 1367 French participants over the age of 65 years with a five-year follow-up (Commenges *et al.*, 2000). This study demonstrated a significant inverse relationship between flavonoid intake and incidence of dementia. This data set was further explored by Letenneur *et al.* (2007) who assessed mental state (Mini-Mental State Exam) and cognitive performance at baseline and at re-assessment four times over a 10-year period. A positive relationship between flavonoid intake and cognitive performance at baseline as well as an inverse relationship between intake and cognitive decline over the 10-year follow-up was demonstrated; however, chocolate was not explored as a specific source of flavonoids in either analysis. Nurk *et al.* (2009) did, however, study the specific relationship of chocolate with cognitive performance, as well as the association of performance with flavonoid-rich tea and wine in about 2000 elderly men and women living in Norway. In a multiple adjusted model, they found that consumers of chocolate (based on previous year's consumption) performed significantly better than non-consumers in five out of six cognitive tests. Increasing intake of chocolate was also associated with increased performance in five out of six tests with a plateau at 10 g per day, but this was reduced to association with only one task (semantic memory) in the fully adjusted model.

CONCLUSIONS

A wealth of intervention research exploring the impact of cocoa on blood flow parameters indicates strong support for recommending cocoa as a suitable dietary intervention to significantly alter cardiovascular risk factors such as hypertension and flow-mediated dilation. Whilst it has not been possible to include every study relevant to this area in the current chapter, the literature would seem to suggest a stronger relationship between cocoa consumption and FMD than blood pressure. Evidence from meta-analyses of blood pressure studies indicates that effects of cocoa (ranging from 30 to 1008 mg flavan-3-ols) are more likely to be observed in those with essential or pre-hypertension (Taubert *et al.*, 2007; Ried *et al.*, 2010). Although these meta-analyses were restricted to studies that explored effects of intervention periods of at least seven days, reductions in blood pressure have been shown acutely at two hours post-ingestion (e.g. Berry *et al.*, 2008; Faridi *et al.*, 2008). Effects of cocoa in increasing FMD have also been widely demonstrated whether measured following chronic cocoa intake, with intervention periods ranging from eight days to twelve weeks (Heiss *et al.*, 2007; Davison *et al.*, 2008), or following a single dose (Faridi *et al.*, 2008; Berry *et al.*, 2010). Peak acute effects have been demonstrated at two hours post-administration, whilst chronic effects in the absence of an acute dose have been observed as rapidly as three days following intervention initiation (Heiss *et al.*, 2007). However, when combined acute and chronic effects have been monitored, the greatest effects have been demonstrated when measured at two hours post-acute dose following repeated administration, indicating an additive acute effect despite an existing increase in dilation as a result of chronic effects (Heiss *et al.*, 2007; Balzer *et al.*, 2008).

Although studies of the impact of cocoa on blood flow parameters all attribute any effects to flavan-3-ols present, theobromine is also present in cocoa, which although often thought to be behaviourally inactive, has a long history of usage as a vasodilator. Studies have varied in whether they have employed a matched control or not, with very little reference to the importance of this. In those studies that have not employed a matched control (i.e. employing white chocolate as placebo), as well as the possible impact of the methylxanthines caffeine and, more importantly, theobromine, there is also the potential for confound from micronutrients such as potassium and magnesium, which have previously been shown to modulate blood pressure; although recent Cochrane reviews have indicated insufficient evidence in support of potassium (Dickinson *et al.*, 2006a), magnesium (Dickinson *et al.*, 2006b), or a combination of magnesium, potassium and calcium (Beyer *et al.*, 2006) in the treatment of hypertension. Evidence from a number of studies that have used a control matched for methylxanthine and micronutrient levels, which have demonstrated effects of cocoa on blood flow-related parameters, also illustrate the role of flavan-3-ols in these effects.

In terms of research investigating the impact of cocoa or its components on the brain, only 11 studies have been published to date and, with the exception of four studies that specifically explored the role of theobromine (Mumford *et al.*, 1994; Smit *et al.*, 2004; Mitchell *et al.*, 2011), these have all attributed any effects to flavan-3-ols. Despite the observation of behavioural effects of theobromine in three of these studies (Smit *et al.*, 2004) and the lack of any demonstration of greater effects of cocoa as compared to its methylxanthine components (Smit *et al.*, 2004; Mitchell *et al.*, 2011), this assertion seems warranted given that a number of studies have shown effects of enriched flavan-3-ol levels when compared to a control matched for such potential confounders as methylxanthines, micronutrients and energy content (Francis *et al.*, 2006; Scholey *et al.*, 2010; Camfield *et al.*, 2012). Despite studies demonstrating some evidence of modulation of either cerebral blood flow (Francis *et al.*, 2006) or brain activity (Francis *et al.*, 2006; Camfield *et al.*, 2012), only three studies have demonstrated improvements to cognition following cocoa consumption (Smit *et al.*, 2004; Scholey *et al.*, 2010; Field *et al.*, 2011). These studies were acute with an absorption period of 1.5 hours or, in the case of Smit *et al.* (2004), with assessments at one and two hours post-ingestion. In this latter study improvements to simple reaction time and energetic arousal were observed at both one and two hours post-dose but hedonic tone did not improve until two hours post-ingestion; this is in keeping with bioavailability data indicating a peak in epicatechin levels at two hours post-ingestion (Richelle *et al.*, 1999; Rein *et al.*, 2000) and evidence from FMD studies showing an increase in dilation at this time point (Berry *et al.*, 2008; Faridi *et al.*, 2008). Aside from improvements to hedonic tone, the time course and nature of the effects observed in Smit *et al.* (2004) would seem to indicate a methylxanthine effect, as is proposed by the authors. It is also likely that the 47 mg dose of flavan-3-ols administered in this study was below the threshold needed to elicit any effects over and above those produced by caffeine and theobromine. The improvements demonstrated by Field *et al.* (2011) are also broadly in keeping with effects observed following caffeine but it is not possible to pinpoint mechanism of action in this study.

It is important to note that all three studies that have demonstrated cognitive improvement have employed healthy, young adults and attempts to overcome potential ceiling effects in other studies by exploring effects in older adults have failed to show any improvements. In the only study to demonstrate improvements to cognition whilst controlling for methylxanthines and other potential actives in their control, Scholey *et al.* (2010) observed effects of cocoa flavan-3-ols using a mentally demanding paradigm intended to overcome any possible ceiling effects. Therefore, it is possible that improvements following flavan-3-ols can only be observed when the demands of the task or fatigue, as a result of prolonged processing, are sufficiently high. It is also possible that the observation of improvements in this study is specific to the processes involved in the task administered. The clearest and most robust effects were seen in a serial threes subtraction task following 520 mg flavan-3-ols. As the more difficult serial sevens subtraction task was unaffected by this dose, this would suggest that the improvement is not simply reliant upon increased demand. Both tasks have attentional components but the serial threes task has a larger psychomotor component, whereas the serial sevens task relies more heavily on working memory. Taken together with a robust attenuation of mental fatigue ratings associated with the paradigm, it is possible that the improvement to serial three subtractions following 520 mg flavan-3-ols simply reflects an increase in psychomotor speed as a consequence of an amelioration of fatigue. Further research is needed to allow a more comprehensive understanding of the conditions required to elicit enhancements to behaviour following cocoa consumption; however, it should be noted that a study examining the effects of the same interventions as employed by Scholey *et al.* (2010) across a range of domains with low levels of fatigue or demand, failed to elicit any significant effects in a healthy young sample (unpublished data).

As well as the importance of task selection in demonstrating behavioural improvements following cocoa, the dose employed is also likely to be an important factor. A wide range of cocoa flavan-3-ol doses have been employed in studies involving chronic administration (172–900 mg), whereas acute flavan-3-ol studies have used a smaller range (516–994 mg). Given that improvements have been demonstrated following 520–994 mg administered acutely, it seems unlikely that dose selection has contributed to null findings in these studies. However, it is possible that the dose employed (172 mg) in the Francis *et al.* (2006) study was simply too low to show effects. Referring to peripheral blood flow studies, a recent meta-analysis highlighted the optimum dose for FMD as about 500 mg total polyphenols with decreasing effects at higher doses (Shrime *et al.*, 2011), whereas Hooper *et al.* (2012) indicated that $\geq$50 mg epicatechin is necessary to induce antihypertensive effects. This latter reference to epicatechin levels is important in that it has been shown to be bioavailable (Rein *et al.*, 2000), is able to replicate the effects of cocoa on FMD (Schroeter *et al.*, 2006) and, of the behavioural studies that report epicatechin levels, only Scholey *et al.* (2010) exceed the 50 mg dose indicated as necessary to elicit antihypertensive effects, suggesting that this may be important in producing cognitive benefits. Further work is needed directed at assessing the cognitive effects of individual components of cocoa as well as assessing the contribution of each component to the synergistic effects of cocoa as a whole. It is also important to assess the contribution of flavan-3-ol metabolites to behavioural effects, as recent research indicates that the monomeric metabolite 3'-*O*-methyl-epicatechin-5-*O*-β-glucuronide promotes long-term potentiation and may improve synaptic plasticity in a mouse model of Alzheimer's disease (Wang *et al.*, 2012).

Only one study to date has assessed the impact of cocoa in the absence of an acute load. In their recent study Camfield *et al.* (2012) demonstrated that consumption of cocoa flavan-3-ols for 30 days may actually decrease brain activity without producing any decrement to performance, which they suggest may indicate an increase in neural efficiency. The impact of chronic consumption of cocoa clearly warrants further investigation both in the presence and absence of an acute load. Epidemiological studies provide some support for a beneficial health role of chocolate consumption; however, these studies tend to simply obtain information regarding chocolate consumption, irrespective of whether dark or milk, and since the majority of consumption of chocolate in the relevant countries would have been milk chocolate it is questionable how closely these data link to cocoa flavan-3-ols. Further work that differentiates dark chocolate from other chocolate consumption is needed in order to provide a fuller picture of the amount and frequency of cocoa consumption required to see benefits, particularly as the doses currently administered in intervention trials are far higher than would ordinarily be ingested as part of a healthy diet. Given the potential for this high energy food to promote weight gain it is important that any integration of chocolate or cocoa into a regular diet is coupled with advice to reduce caloric intake from other sources.

As with all dietary interventions, it may be more appropriate to integrate cocoa consumption into a healthy lifestyle as early as possible to avoid encountering unnecessary health problems as a consequence of ageing. However, evidence from cardiovascular research shows reversal of endothelial dysfunction following short-term cocoa intervention, suggesting the same reversal may be possible for cerebrovascular/cognitive impairment. Surprisingly, cocoa intervention trials specifically examining effects in those suffering, or at high risk for, stroke or dementia are lacking. Given the potential benefits of cocoa it is imperative that further research is carried out to assess the benefit that a simply dietary adjustment could have to those suffering vascular impairment.

REFERENCES

Addicott, M. A. and Laurienti, P. J. (2009) A comparison of the effects of caffeine following abstinence and normal caffeine use. *Psychopharmacology*, **207**, 423–431.

Andres-Lacueva, C., Shukitt-Hale, B., Galli, R. L. *et al.* (2005) Anthocyanins in aged blueberry-fed rats are found centrally and may enhance memory. *Nutritional Neuroscience,* **8**, 111–120.

Balzer, J., Rassaf, T., Heiss, C. *et al.* (2008) Sustained benefits in vascular function through flavanol-containing cocoa in medicated diabetic patients: a double-masked, randomized, controlled trial. *Circulation,* **118**, S1123–S1123.

Bayard, V., Chamorro, F., Motta, J. and Hollenberg, N. K. (2007) Does flavanol intake influence mortality from nitric oxide-dependent processes? Ischemic heart disease, stroke, diabetes mellitus, and cancer in Panama. *International Journal of Medical Sciences*, **4**, 53–8.

Berry, N., Davison, K., Buckley, J. D. *et al.* (2008) Flavanol-rich cocoa attenuates blood pressure responsiveness to exercise. *Journal of Hypertension,* **26**, S392–S393.

Berry, N. M., Davison, K., Coates, A. M. *et al.* (2010) Impact of cocoa flavanol consumption on blood pressure responsiveness to exercise. *British Journal of Nutrition,* **103**, 1480–1484.

Beyer, F. R., Dickinson, H. O., Nicolson, D. J. *et al.* (2006) Combined calcium, magnesium and potassium supplementation for the management of primary hypertension in adults. *Cochrane Database of Systematic Reviews* 3 (Art. No.: CD004805). doi: 10.1002/14651858.CD004805.pub2.

Bisson, J.-F., Nejdi, A., Rozan, P. *et al.* (2008) Effects of long-term administration of a cocoa polyphenolic extract (Acticoa powder) on cognitive performances in aged rats. *British Journal of Nutrition,* **100**, 94–101.

Brown, W. R. and Thore, C. R. (2011) Cerebral microvascular pathology in ageing and neurodegeneration. *Neuropathology and Applied Neurobiology*, **37**, 56–74.

Brunetto, M. D. R., Gutiérrez, L., Delgado, Y. *et al.* (2007) Determination of theobromine, theophylline and caffeine in cocoa samples by a high-performance liquid chromatographic method with on-line sample cleanup in a switching-column system. *Food Chemistry*, **100**, 459–467.

Buijsse, B., Feskens, E. J. M., Kok, F. J. and Kromhout, D. (2006) Cocoa intake, blood pressure, and cardiovascular mortality – The Zutphen Elderly Study. *Archives of Internal Medicine*, **166**, 411–417.

Buijsse, B., Weikert, C., Drogan, D. *et al.* (2010) Chocolate consumption in relation to blood pressure and risk of cardiovascular disease in German adults. *European Heart Journal,* **31**, 1616–1623.

Buitrago-Lopez, A., Sanderson, J., Johnson, L. *et al.* (2011) Chocolate consumption and cardiometabolic disorders: systematic review and meta-analysis. *British Medical Journal*, 343, Art. No.: d4488 (doi: 10.1136/bmj.d4488).

Cameron, O. G., Modell, J. G. and Hariharan, M. (1990) Caffeine and human cerebral blood-flow – a positron emission tomography study. *Life Sciences*, **47**, 1141–1146.

Camfield, D. A., Scholey, A., Pipingas, A. *et al.* (2012) Steady state visually evoked potential (SSVEP) topography changes associated with cocoa flavanol consumption. *Physiology and Behavior,* **105**, 948–957.

Commenges, D., Scotet, V., Renaud, S. *et al.* (2000) Intake of flavonoids and risk of dementia. *European Journal of Epidemiology,* **16**, 357–363.

Crews, W. D., Jr., Harrison, D. W. and Wright, J. W. (2008) A double-blind, placebo-controlled, randomized trial of the effects of dark chocolate and cocoa on variables associated with neuropsychological functioning and cardiovascular health: clinical findings from a sample of healthy, cognitively intact older adults. *American Journal of Clinical Nutrition*, **87**, 872–880.

Davison, K., Coates, A. M., Buckley, J. D. and Howe, P. R. C. (2008) Effect of cocoa flavanols and exercise on cardiometabolic risk factors in overweight and obese subjects. *International Journal of Obesity*, **32**, 1289–1296.

Desch, S., Kobler, D., Schmidt, J. *et al.* (2010) Low vs. higher-dose dark chocolate and blood pressure in cardiovascular high-risk patients. *American Journal of Hypertension*, **23**, 694–700.

di Tomaso, E., Beltramo, M. and Piomelli, D. (1996) Brain cannabinoids in chocolate. *Nature,* **382**, 677–678.

Dickinson, H. O., Nicolson, D. J., Campbell, F. *et al.* (2006a) Potassium supplementation for the management of primary hypertension in adults. *Cochrane Database of Systematic Reviews* 3 (Art. No.: CD004640). doi: 10.1002/14651858.CD004640.pub2.

Dickinson, H. O., Nicolson, D. J., Campbell, F. *et al.* (2006b) Magnesium supplementation for the management of essential hypertension in adults. *Cochrane Database of Systematic Reviews.*

Djousse, L., Hopkins, P. N., North, K. E. *et al.* (2011) Chocolate consumption is inversely associated with prevalent coronary heart disease: The National Heart, Lung, and Blood Institute Family Heart Study. *Clinical Nutrition,* **30**, 182–187.

Donovan, J. L., Crespy, V., Oliveira, M. *et al.* (2006) (+)–Catechin is more bioavailable than (–)–catechin: Relevance to the bioavailability of catechin from cocoa. *Free Radical Research*, **40**, 1029–1034.

Engler, M. B., Engler, M. M., Chen, C. Y. *et al.* (2004) Flavonoid-rich dark chocolate improves endothelial function and increases plasma epicatechin concentrations in healthy adults. *Journal of the American College of Nutrition,* **23**, 197–204.

Eteng, M., Eyong, E., Akpanyung, E. *et al.* (1997) Recent advances in caffeine and theobromine toxicities: a review. *Plant Foods for Human Nutrition (Formerly Qualitas Plantarum)*, **51**, 231–243.

Faridi, Z., Njike, V. Y., Dutta, S. *et al*L. (2008) Acute dark chocolate and cocoa ingestion and endothelial function: a randomized controlled crossover trial. *American Journal of Clinical Nutrition*, **88**, 58–63.

Farouque, H. M. O., Leung, M., Hope, S. A. *et al.* (2006) Acute and chronic effects of flavanol-rich cocoa on vascular function in subjects with coronary artery disease: a randomized double-blind placebo-controlled study. *Clinical Science,* **111**, 71–80.

Field, A. S., Laurienti, P. J., Yen, Y. F. *et al.* (2003) Dietary caffeine consumption and withdrawal: Confounding variables in quantitative cerebral perfusion studies? *Radiology,* **227**, 129–135.

Field, D. T., Williams, C. M. and Butler, L. T. (2011) Consumption of cocoa flavanols results in an acute improvement in visual and cognitive functions. *Physiology and Behavior*, **103**, 255–260.

Fisher, N. D. L., Hughes, M. and Hollenberg, N. K. (2003) Cocoa induces nitric-oxide-dependent vasodilation in healthy humans. *American Journal of Hypertension*, **16**, 72A.

Flammer, A. J., Hermann, F., Sudano, I. *et al.* (2007) Dark chocolate improves coronary vasomotion and reduces platelet reactivity. *Circulation,* **116**, 2376–2382.

Fraga, C. G., Actis-Goretta, L., Ottaviani, J. I. *et al.* (2005) Regular consumption of a flavanol-rich chocolate can improve oxidant stress in young soccer players. *Clinical and Developmental Immunology,* **12**, 11–17.

Francis, S. T., Head, K., Morris, P. G. and Macdonald, I. A. (2006) The effect of flavanol-rich cocoa on the fMRI response to a cognitive task in healthy young people. *Journal of Cardiovascular Pharmacology*, **47**, S215–S220.

Fredholm, B. B., Battig, K., Holmen, J. *et al.* (1999) Actions of caffeine in the brain with special reference to factors that contribute to its widespread use. *Pharmacological Reviews,* **51**, 83–133.

Fredholm, B. B. and Lindstrom, K. (1999) Autoradiographic comparison of the potency of several structurally unrelated adenosine receptor antagonists at adenosine A(1) and A(2A) receptors. *European Journal of Pharmacology*, **380**, 197–202.

Gonthier, M.-P., Donovan, J. L., Texier, O. *et al.* (2003) Metabolism of dietary procyanidins in rats. *Free Radical Biology and Medicine,* **35**, 837–844.

Gorelick, P. B., Scuteri, A., Black, S. E. *et al.* (2011) Vascular contributions to cognitive impairment and dementia a statement for healthcare professionals from the American Heart Association/American Stroke Association. *Stroke,* **42**, 2672–2713.

Grassi, D., Lippi, C., Necozione, S. *et al.* (2005a) Short-term administration of dark chocolate is followed by a significant increase in insulin sensitivity and a decrease in blood pressure in healthy persons. *American Journal of Clinical Nutrition,* **81**, 611–614.

Grassi, D., Necozione, S., Lippi, C. *et al.* (2005b) Cocoa reduces blood pressure and insulin resistance and improves endothelium-dependent vasodilation in hypertensives. *Hypertension*, **46**, 398–405.

Grassi, D., Desideri, G., Necozione, S. *et al.* (2008) Blood pressure is reduced and insulin sensitivity increased in glucose-intolerant, hypertensive subjects after 15 days of consuming high-polyphenol dark chocolate. *Journal of Nutrition,* **138**, 1671–1676.

Gu, L., Kelm, M. A., Hammerstone, J. F. *et al.* (2004) Concentrations of proanthocyanidins in common foods and estimations of normal consumption. *The Journal of Nutrition*, **134**, 613–617.
Hamed, M. S., Gambert, S., Bliden, K. P. *et al.* (2008) Dark chocolate effect on platelet activity, c-reactive protein and lipid profile: a pilot study. *Southern Medical Journal*, **101**, 1203–1208.
Hammerstone, J. F., Lazarus, S. A. and Schmitz, H. H. (2000) Procyanidin content and variation in some commonly consumed foods. *The Journal of Nutrition*, **130**, 2086S–2092S.
Heiss, C., Dejam, A., Kleinbongard, P. *et al.* (2003) Vascular effects of cocoa rich in flavan-3-ols. *JAMA: The Journal of the American Medical Association*, **290**, 1030–1031.
Heiss, C., Kleinbongard, P., Dejam, A. *et al.* (2005) Acute consumption of flavanol-rich cocoa and the reversal of endothelial dysfunction in smokers. *Journal of the American College of Cardiology*, **46**, 1276–1283.
Heiss, C., Finis, D., Kleinbongard, P. *et al.* (2007) Sustained increase in flow-mediated dilation after daily intake of high-flavanol cocoa drink over 1 week. *Journal of Cardiovascular Pharmacology,* **49**, 74–80.
Heiss, C., Jahn, S., Taylor, M. *et al.* (2010) Improvement of endothelial function with dietary flavanols is associated with mobilization of circulating angiogenic cells in patients with coronary artery disease. *Journal of the American College of Cardiology*, **56**, 218–224.
Hermann, F., Spieker, L. E., Ruschitzka, F. *et al.* (2006) Dark chocolate improves endothelial and platelet function. *Heart,* **92**, 119–120.
Herraiz, T. (2000) Tetrahydro-β-carbolines, potential neuroactive alkaloids, in chocolate and cocoa. *Journal of Agricultural and Food Chemistry*, **48**, 4900–4904.
Hollenberg, N. K., Martinez, G., McCullough, M. *et al.* (1997) Aging, acculturation, salt intake, and hypertension in the Kuna of Panama. *Hypertension*, **29**, 171–176.
Hooper, L., Kay, C., Abdelhamid, A. *et al.* (2012) Effects of chocolate, cocoa, and flavan-3-ols on cardiovascular health: a systematic review and meta-analysis of randomized trials. *American Journal of Clinical Nutrition,* **95**, 740–751.
Huang, Q., Wu, L.-J., Tashiro, S.-I. *et al.* (2005) (+)–Catechin, an ingredient of green tea, protects murine microglia from oxidative stress-induced DNA damage and cell cycle arrest. *Journal of Pharmacological Sciences*, **98**, 16–24.
Hurst, W. J. and Toomey, P. B. (1981) High-performance liquid chromatographic determination of four biogenic amines in chocolate. *Analyst*, **106**, 394–402.
Hurst, W. J., Krake, S., Bergmeier, S. *et al.* (2011) Impact of fermentation, drying, roasting and Dutch processing on flavan-3-ol stereochemistry in cacao beans and cocoa ingredients. *Chemistry Central Journal*, **5**, 53.
James, J. E. and Rogers, P. J. (2005) Effects of caffeine on performance and mood: withdrawal reversal is the most plausible explanation. *Psychopharmacology*, **182**, 1–8.
Janszky, I., Mukamal, K. J., Ljung, R. *et al.* (2009) Chocolate consumption and mortality following a first acute myocardial infarction: the Stockholm Heart Epidemiology Program. *Journal of Internal Medicine,* **266**, 248–257.
Karim, M., Mccormick, K. and Kappagoda, C. T. (2000) Effects of cocoa extracts on endothelium-dependent relaxation. *The Journal of Nutrition*, **130**, 2105S–2108S.
Kelly, C. J. (2005) Effects of theobromine should be considered in future studies. *American Journal of Clinical Nutrition*, **82**, 486–487.
Kennedy, D. O. and Haskell, C. F. (2011) Cerebral blood flow and behavioural effects of caffeine in habitual and non-habitual consumers of caffeine: A near infrared spectroscopy study. *Biological Psychology*, **86**, 298–306.
Keogh, J. B., Mcinerney, J. and Clifton, P. M. (2007) The effect of milk protein on the bioavailability of cocoa polyphenols. *Journal of Food Science*, **72**, S230–S233.
Kofink, M., Papagiannopoulos, M. and Galensa, R. (2007) (–)–Catechin in cocoa and chocolate: occurence and analysis of an atypical flavan-3-ol enantiomer. *Molecules*, **12**, 1274–1288.
Langer, S., Marshall, L. J., Day, A. J. and Morgan, M. R. A. (2011) Flavanols and methylxanthines in commercially available dark chocolate: a study of the correlation with nonfat cocoa solids. *Journal of Agricultural and Food Chemistry*, **59**, 8435–8441.
Larsson, S. C., Virtamo, J. and Wolk, A. (2011) Chocolate consumption and risk of stroke in women. *Journal of the American College of Cardiology*, **58**, 1828–1829.
Letenneur, L., Proust-Lima, C., Le Gouge, A. *et al.* (2007) Flavonoid intake and cognitive decline over a 10-year period. *American Journal of Epidemiology,* **165**, 1364–1371.
Levine, G. N., Bates, E. R., Blankenship, J. C. *et al.* (2011) 2011 ACCF/AHA/SCAI guideline for percutaneous coronary intervention: executive summary a report of the American College Of Cardiology

Foundation/American Heart Association task force on practice guidelines and the society for cardiovascular angiography and interventions. *Circulation,* **124**, 2574–2609.
Lewis, J. R., Prince, R. L., Zhu, K. *et al.* (2010) Habitual chocolate intake and vascular disease: a prospective study of clinical outcomes in older women. *Archives of Internal Medicine,* **170**, 1857–1858.
Li, R., Huang, Y.-G., Fang, D. and Le, W.-D. (2004) (−)-Epigallocatechin gallate inhibits lipopolysaccharide-induced microglial activation and protects against inflammation-mediated dopaminergic neuronal injury. *Journal of Neuroscience Research*, **78**, 723–731.
Lieberman, H. R., Wurtman, R. J., Emde, G. G. *et al.* (1987) The effects of low doses of caffeine on human performance and mood. *Psychopharmacology,* **92**, 308–312.
Lockard, R. G. and Burridge, J. C. (1965) The levels of macro- and micronutrients in the beans of cacao (theobroma cacao l.) in relation to shade, fertilizer, irrigation, and season. *Annals of Botany,* **29**, 377–382.
Lunt, M. J., Ragab, S., Birch, A. A. *et al.* (2004) Comparison of caffeine-induced changes in cerebral blood flow and middle cerebral artery blood velocity shows that caffeine reduces middle cerebral artery diameter. *Physiological Measurement*, **25**, 467–474.
Manal, A. E. M., Kuhnle, G., Rechner, A. R. *et al.* (2002) Uptake and metabolism of epicatechin and its access to the brain after oral ingestion. *Free Radical Biology and Medicine*, **33**, 1693–1702.
McCullough, M. L., Chevaux, K., Jackson, L. *et al.* (2006) Hypertension, the Kuna, and the epidemiology of flavanols. *Journal of Cardiovascular Pharmacology,* **47**, S103–S109.
Mellor, D. D., Sathyapalan, T., Kilpatrick, E. S. *et al.* (2010) High-cocoa polyphenol-rich chocolate improves HDL cholesterol in Type 2 diabetes patients. *Diabetic Medicine,* **27**, 1318–1321.
Mitchell, E. S., Slettenaar, M., Van Der Meer, N. *et al.* (2011) Differential contributions of theobromine and caffeine on mood, psychomotor performance and blood pressure. *Physiology and Behavior,* **104**, 816–822.
Monagas, M., Khan, N., Andres-Lacueva, C. *et al.* (2009) Effect of cocoa powder on the modulation of inflammatory biomarkers in patients at high risk of cardiovascular disease. *American Journal of Clinical Nutrition,* **90**, 1144–1150.
Morris, R. (1984) Developments of a water-maze procedure for studying spatial-learning in the rat. *Journal of Neuroscience Methods*, **11**, 47–60.
Mostofsky, E., Levitan, E. B., Wolk, A. and Mittleman, M. A. (2010) Chocolate intake and incidence of heart failure a population-based prospective study of middle-aged and elderly women. *Circulation-Heart Failure*, **3**, 612–616.
Mullen, W., Borges, G., Donovan, J. L. *et al.* (2009) Milk decreases urinary excretion but not plasma pharmacokinetics of cocoa flavan-3-ol metabolites in humans. *The American Journal of Clinical Nutrition,* **89**, 1784–1791.
Mumford, G., Evans, S., Kaminski, B. *et al.* (1994) Discriminative stimulus and subjective effects of theobromine and caffeine in humans. *Psychopharmacology,* **115**, 1–8.
Mumford, G. K., Benowitz, N. L., Evans, S. M. *et al.* (1996) Absorption rate of methylxanthines following capsules, cola and chocolate. *European Journal of Clinical Pharmacology,* **51**, 319–325.
Murphy, K. J., Chronopoulos, A. K., Singh, I. *et al.* (2003) Dietary flavanols and procyanidin oligomers from cocoa (Theobroma cacao) inhibit platelet function. *American Journal of Clinical Nutrition,* **77**, 1466–1473.
Njike, V. Y., Faridi, Z., Shuval, K. *et al.* (2011) Effects of sugar-sweetened and sugar-free cocoa on endothelial function in overweight adults. *International Journal of Cardiology,* **149**, 83–88.
Nurk, E., Refsum, H., Drevon, C. A. *et al.* (2009) Intake of flavonoid-rich wine, tea, and chocolate by elderly men and women is associated with better cognitive test performance. *Journal of Nutrition,* **139**, 120–127.
Ortega, N., Romero, M.-P., Macia, A. *et al.* (2008) Obtention and characterization of phenolic extracts from different cocoa sources. *Journal of Agricultural and Food Chemistry,* **56**, 9621–9627.
Ramiro-Puig, E., Casadesús, G., Lee, H.-G. *et al.* (2009) Neuroprotective effect of cocoa flavonids on *in vitro* oxidative stress. *European Journal of Nutrition,* **48**, 54–61.
Rein, D., Paglieroni, T. G., Wun, T. *et al.* (2000) Cocoa inhibits platelet activation and function. *The American Journal of Clinical Nutrition,* **72**, 30–35.
Richelle, M., Tavazzi, I., Enslen, M. and Offord, E. A. (1999) Plasma kinetics in man of epicatechin from black chocolate. *European Journal of Clinical Nutrition*, **53**, 22–26.
Ried, K., Sullivan, T., Fakler, P. *et al.* (2010) Does chocolate reduce blood pressure? A meta-analysis. *BMC Medicine,* **8**, 8–39.
Rios, L. Y., Bennett, R. N., Lazarus, S. A. *et al.* (2002) Cocoa procyanidins are stable during gastric transit in humans. *American Journal of Clinical Nutrition,* **76**, 1106–1110.

Roura, E., Andres-Lacueva, C., Estruch, R. *et al.* (2007) Milk does not affect the bioavailability of cocoa powder flavonoid in healthy human. *Annals of Nutrition and Metabolism*, **51**, 493–498.

Rozan, P., Hidalgo, S., Nejdi, A. *et al.* (2007) Preventive antioxidant effects of cocoa polyphenolic extract on free radical production and cognitive performances after heat exposure in Wistar rats. *Journal of Food Science,* **72**, S203–S206.

Scholey, A. B., French, S. J., Morris, P. J. *et al.* (2010) Consumption of cocoa flavanols results in acute improvements in mood and cognitive performance during sustained mental effort. *Journal of Psychopharmacology,* **24**, 1505–1514.

Schramm, D. D., Karim, M., Schrader, H. R. *et al.* (2003) Food effects on the absorption and pharmacokinetics of cocoa flavanols. *Life Sciences,* **73**, 857–869.

Schroeter, H., Heiss, C., Spencer, J. P. E. *et al.* (2006) Recommending flavanols and procyanidins for cardiovascular health: Current knowledge and future needs. *Molecular Aspects of Medicine,* **31**, 546–557.

Serafini, M. and Crozier, A. (2003) Nutrition: Milk and absorption of dietary flavanols. *Nature*, **426**, 788–788.

Shah, Z. A., Li, R. C., Ahmad, A. S. *et al.* (2010) The flavanol (–)–epicatechin prevents stroke damage through the Nrf2/HO1 pathway. *Journal of Cerebral Blood Flow and Metabolism: Official Journal of the International Society of Cerebral Blood Flow and Metabolism,* **30**, 1951–1961.

Shin, W.-H., Park, S.-J. and Kim, E.-J. (2006) Protective effect of anthocyanins in middle cerebral artery occlusion and reperfusion model of cerebral ischemia in rats. *Life Sciences*, **79**, 130–137.

Shrime, M. G., Bauer, S. R., Mcdonald, A. C. *et al.* (2011) Flavonoid-rich cocoa consumption affects multiple cardiovascular risk factors in a meta-analysis of short-term studies. *Journal of Nutrition,* **141**, 1982–1988.

Sigmon, S. C., Herning, R. I., Better, W. *et al.* (2009) Caffeine withdrawal, acute effects, tolerance, and absence of net beneficial effects of chronic administration: cerebral blood flow velocity, quantitative EEG, and subjective effects. *Psychopharmacology*, **204**, 573–585.

Silberstein, R. B., Schier, M. A., Pipingas, A. *et al.* (1990) Steady-state visually evoked potential topography associated with a visual vigilance task. *Brain Topography,* **3**, 337–347.

Silberstein, R. B., Nunez, P. L., Pipingas, A. *et al.* (2001) Steady state visually evoked potential (SSVEP) topography in a graded working memory task. *International Journal of Psychophysiology,* **42**, 219–232.

Smit, H. J. and Rogers, P. J. (2000) Effects of low doses of caffeine on cognitive performance, mood and thirst in low and higher caffeine consumers. *Psychopharmacology*, **152**, 167–173.

Smit, H. J., Gaffan, E. A. and Rogers, P. J. (2004) Methylxanthines are the psycho-pharmacologically active constituents of chocolate. *Psychopharmacology*, **176**, 412–419.

Smith, A., Sturgess, W. and Gallagher, J. (1999) Effects of a low dose of caffeine given in different drinks on mood and performance. *Human Psychopharmacology-Clinical and Experimental*, **14**, 473–482.

Sorond, F. A., Lipsitz, L. A., Hollenberg, N. K. and Fisher, N. D. L. (2008) Cerebral blood flow response to flavanol-rich cocoa in healthy elderly humans. *Neuropsychiatric Disease and Treatment*, **4**, 433–440.

Spencer, J. (2007) The interactions of flavonoids within neuronal signalling pathways. *Genes and Nutrition*, **2**, 257–273.

Taubert, D., Berkels, R., Roesen, R. and Klaus, W. (2003) Chocolate and blood pressure in elderly individuals with isolated systolic hypertension. *JAMA – Journal of the American Medical Association*, **290**, 1029–1030.

Taubert, D., Roesen, R., Lehmann, C. *et al.* (2007) Effects of low habitual cocoa intake on blood pressure and bioactive nitric oxide - A randomized controlled trial. *JAMA – Journal of the American Medical Association,* **298**, 49–60.

van den Bogaard, B., Draijer, R., Westerhof, B. E. *et al.* (2010) Effects on peripheral and central blood pressure of cocoa with natural or high-dose theobromine a randomized, double-blind crossover trial. *Hypertension*, **56**, 839–846.

Vlachopoulos, C., Aznaouridis, K., Alexopoulos, N. *et al.* (2005) Effect of dark chocolate on arterial function in healthy individuals. *American Journal of Hypertension,* **18**, 785–791.

Wang, J., Ferruzzi, M. G., Ho, L. *et al.* (2012) Brain-targeted proanthocyanidin metabolites for alzheimer's disease treatment. *Journal of Neuroscience,* **32**, 5144–5150.

Ward, M. R., Thompson, K. A., Isaac, K. *et al.* (2011) Nitric oxide synthase gene transfer restores activity of circulating angiogenic cells from patients with coronary artery disease. *Molecular Therapy,* **19**, 1323–1330.

Youdim, K. A., Dobbie, M. S., Kuhnle, G. *et al.* (2003) Interaction between flavonoids and the blood–brain barrier: in vitro studies. *Journal of Neurochemistry,* **85**, 180–192.

17 Breeding for Enhanced Bioactives in Berry Fruit

Alastair Currie, Jessica Scalzo and Bruno Mezzetti

INTRODUCTION

This chapter examines breeding for enhanced levels of bioactive compounds in berry fruit using the example of anthocyanin content in blackcurrant, blueberry and strawberry breeding programmes at The New Zealand Institute for Plant & Food Research Ltd (PFR) and Università Politecnica delle Marche, Ancona, Italy (UNIVPM).

Bioactive components have been defined as essential and non-essential compounds (e.g. vitamins or polyphenols) that occur in nature, are part of the food chain and can be shown to have an effect on human health (Biesalski *et al.*, 2009). Fruit and vegetable consumption has long been associated with healthy living and the World Health Organization and the Food and Agricultural Organization of the United Nations have recommended a daily intake of more than 400 g to promote good health (WHO, 2003). One of the major drivers in consumer spending is the pursuit of health and youth, so consumers are buying foods with added health- and youth-promoting functions, in particular foods that are naturally healthy with high levels of nutrients, additive free and high in bioactives (Sloan, 2005, 2011; Kearney, 2010; Kempster and Beverley, 2011).

One important group of bioactive compounds in berry fruit include anthocyanins, which are a group of over 500 compounds. Fruits and vegetables are a rich source of anthocyanins compared to other whole foods (Table 17.1). Anthocyanins provide berry fruit with their characteristic colours and their quantity and composition varies among fruit types (Table 17.2) (Beattie *et al.*, 2005; Wu *et al.*, 2006; Scalzo *et al.*, 2009a, 2009b).

Within the example crops at PFR and UNIVPM there is a considerable variation in anthocyanin content (Figure 17.1). A shift can be observed towards higher levels of anthocyanin content for selections from the blackcurrant and strawberry breeding programmes. The lack of progress in the blueberry breeding programme is discussed in the case studies in this chapter.

Anthocyanins have been shown to contribute to the antioxidant activity in the genus *Ribes, Vaccinium* L., *Fragaria* and *Rubus* (Wang *et al.*, 1997; Prior *et al.*, 1998; Kalt *et al.*, 1999b, 2001; Deighton *et al.*, 2000; Ehlenfeldt and Prior, 2001; Connor *et al.*, 2002c, 2005; McGhie *et al.*, 2002; Moyer *et al.*, 2002a, 2002b; Proteggente *et al.*, 2002; Wada and Ou, 2002; Scalzo *et al.*, 2005; Currie *et al.*, 2006). The value of anthocyanins has been reviewed (Kong *et al.*, 2003; Beattie *et al.*, 2005) and besides antioxidant capacity (Wang *et al.*, 1997;

Bioactives in Fruit: Health Benefits and Functional Foods, First Edition. Edited by Margot Skinner and Denise Hunter.

Table 17.1 Mean (range) of anthocyanin content for US whole foods (USDA, 2011).

Class	Mean (range) anthocyanin content (mg/100 g)	
Vegetables and vegetable products	25.7	(0.0–752.7)
Fruits and fruit juices	51.0	(0.0–558.2)
Beverages	90.3	(0.1–411.4)
Legumes and legume products	59.4	(0.0–262.5)
Cereal grains and pasta	25.9	—
Nut and seed products	4.0	(0.0–18.0)
Spices and herbs	0.7	—
Sweets	0.3	(0.2–0.3)

Matsumoto *et al.*, 2002; Stintzing *et al.*, 2002) anthocyanins are suggested to have a number of therapeutic properties, including:

- improved circulation (Matsumoto *et al.*, 2005; Shin *et al.*, 2006);
- benefits to vision (Matsumoto *et al.*, 2003, 2005; Nielsen *et al.*, 2003; Lee *et al.*, 2005);
- anti-inflammatory and antioxidative effects (Wang *et al.*, 1999; Hecht, 2004; Ghosh *et al.*, 2006; Kelley *et al.*, 2006; Lyall *et al.*, 2009);

Table 17.2 Mean anthocyanin content of a range of raw fruit (USDA, 2011).

Fruit type	Mean anthocyanin content (mg/100 g)
Illawara plum	558.2
Elderberries	485.3
Chokeberry	349.8
Bilberry	285.2
Blackcurrants	154.8
Blueberries	138
Blackberries	90.6
Cranberries	85.5
Grapes	56.6
Raspberries	40.6
Cherries	31.2
Strawberries	27.8
Pears	12.2
Gooseberries	9.5
Bananas	7.4
Plums	5.7
Peaches	1.3
Apples	1.1
Kiwi fruit	0.5
Figs	0.5
Avocados	0.3
Mangos	0.1
Pineapple	0.0
Watermelon	0.0
Melons	0.0
Oranges	0.0

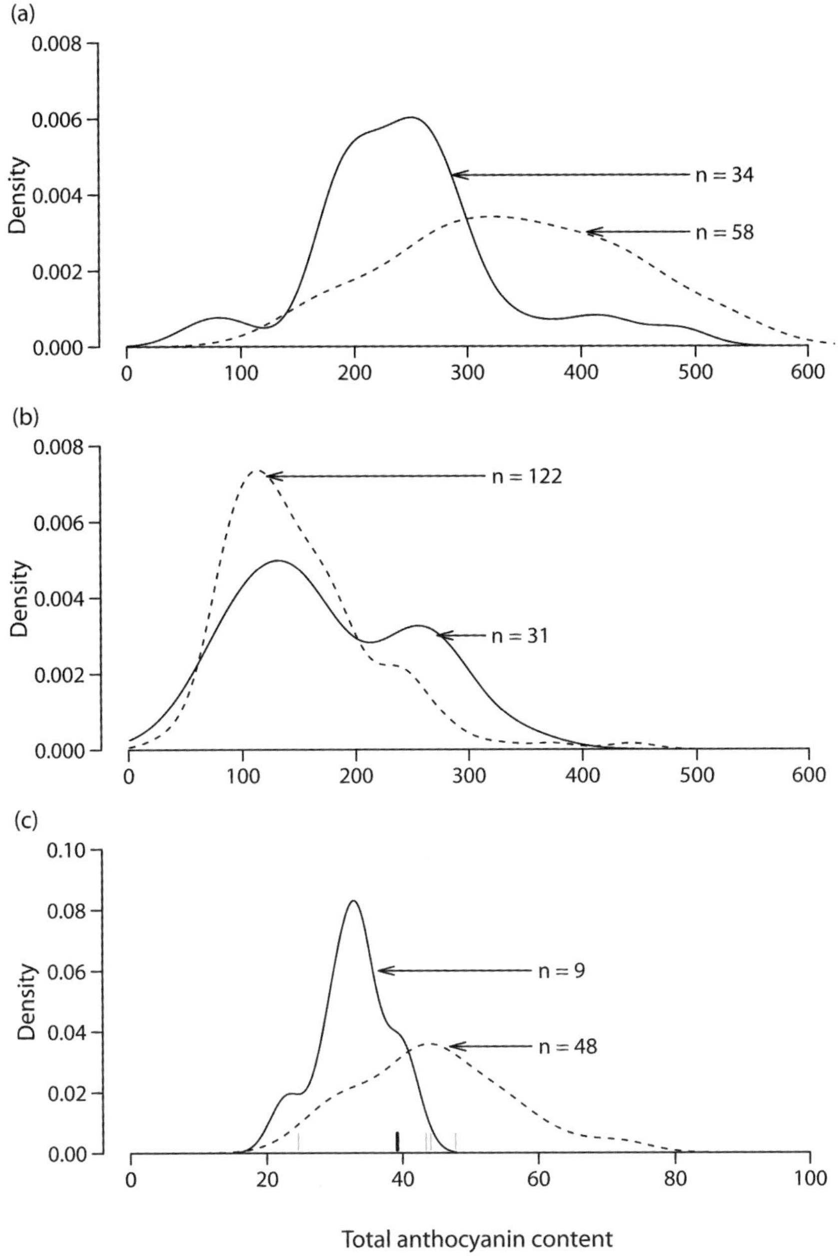

Fig. 17.1 Density plots for total anthocyanin content (mg/100 ml juice or 100 g fruit) for (a) blackcurrant, (b) blueberry and (c) strawberry. The solid line is for germplasm and the dashed line for selections. The number of genotypes in each category is indicated. Values for the four GMO strawberry selections are indicated by grey tick-marks above the horizontal axis, and that for 'Calypso' by a thicker black tick. Note the different scales for (c) (i.e. strawberry).

- preventing cardiovascular disease (Hertog *et al.*, 1993, 1997; Ness and Powles, 1997; Joshipura *et al.*, 2001);
- preventing stroke (Keli *et al.*, 1996);
- preventing some cancers (Steinmetz and Potter, 1996; Kumar *et al.*, 2008; Neyazaki *et al.*, 2008; Scalzo *et al.*, 2012);
- delaying dementia (Youdim and Joseph, 2001; Knekt *et al.*, 2002).

Fifteen different anthocyanins have been identified in blackcurrant (Moyer *et al.*, 2002a, 2002b; Slimestad and Solheim, 2002; Scalzo *et al.*, 2009a), of which the four main ones are the 3-glucosides and the 3-rutinosides of delphinidin and cyanidin.

The anthocyanins present in blueberry are galactosides, glucosides and arabinosides of the aglycones delphinidin, cyanidin, petunidin, peonidin and malvidin. Additionally, these glycosides may also be acetylated (Wu and Prior, 2005). According to Wu *et al.* (2006), blueberry is the richest source of malvidins and petunidins among the fruits and vegetables they analysed, while other authors (Kader *et al.*, 1996; Goiffon *et al.*, 1999) have reported that blueberries are rich in delphinidin-3-galactoside and petunidin-3-glucoside.

In strawberry the major anthocyanins are pelargonidin and cyanidin glucosides or acylated forms with a range of aliphatic acids (Goiffon *et al.*, 1999; Lopes-da-Silva *et al.*, 2002) and the presence of the main derivates seems to be constant in all varieties, but qualitative and quantitative variation have been observed among cultivars (Maatta-Riihinen *et al.*, 2004). Anthocyanin concentrations may strongly differ among the same variety, depending on the degree of ripeness, on climatic factors and on postharvest storage (Lopes-da-Silva *et al.*, 2007).

In addition to anthocyanins, berry fruit also contain a number of other bioactive compounds that are reported to be beneficial to health (Lyall *et al.*, 2009). One of the best known health effects of berry fruit is the protective effect of cranberries towards urinary tract infections (Howell, 2002), which is thought to be due to specific proanthocyanidins. Different *Vaccinium* spp. seem to be also rich in resveratrol (Rimando *et al.*, 2004) that has been shown to possess cancer chemopreventive activities (Jang *et al.*, 1997).

Berry fruit also contain ellagitannins (Amakura *et al.*, 2000; Maatta-Riihinen *et al.*, 2004; Koponen *et al.*, 2007) which according to Hakkinen *et al.* (1999) are responsible for more than 50% of total phenolics quantified for strawberries and raspberries.

Together with other berries, strawberry is also a relevant source of folate (Bailey and Gregory, 1999) and vitamin C, which is responsible for more than 20% of the total antioxidant activity of the fruit extracts (Carr and Frei, 1999).

BREEDING BERRY FRUIT FOR IMPROVED BIOACTIVES

Given the existence of a favourable market for foods with bioactive properties and a crop with ample bioactive compounds in the form of anthocyanins, it remains to put a case for breeding cultivars with more anthocyanins. Some people may argue that it is unnecessary to enhance levels of anthocyanins in fruits and that consumers only need to eat more berries to meet any dietary requirement for bioactive compounds. However, a berry bred with higher levels of health compounds may be preferentially purchased by consumers and so be a viable solution.

Breeding is a long-term process with repeated stages of evaluation, selection and vegetative propagation into increasingly rigorous selection trials.

In berry fruit breeding this process typically takes two to three stages over more than ten years. Success depends on a range of criteria, including knowledge of market trends, good access to a genetically diverse germplasm from which to select parents, knowledge of the properties of the crop and adequate resources to run the programme: land, laboratory and people with patience, perseverance and providence. Ron Beatson (personal communication) has compared progress towards a breeding objective similar to navigating an ocean super-tanker. Both require considerable effort to make progress and it is not easy if the path changes direction frequently. Breeders need to continue towards existing objectives and see if they can incorporate anthocyanins into the list of selection criteria. Knowledge of the genetic properties of anthocyanin expression and how it correlates with other selection criteria is an essential part of this process. Useful parameters include genetic variation, narrow sense heritability, breeding values and genetic correlations (trait–trait, trait–year and trait–environment correlations).

For there to be any response to breeding there has to be several components demonstrated by the breeders equation:

$$R = \frac{\sigma_A^2}{\sigma_P^2} S = h^2 S$$

Where response (R) is the product of S (selection differential), which is the average superiority of the parents from the base population, and the narrow sense heritability (h^2 referred to as heritability hereafter), which is the ratio of the additive genetic variation (σ_{A}^2) to the phenotypic variation (σ_{A}^2) (Falconer and Mackay, 1996; Lynch and Walsh, 1998).

Genetic variation is a key component. Without genetic variation in the genes controlling the trait there will be no opportunity to select genetically superior genotypes. The degree that the observed phenotype is controlled by the additive genetic effects is expressed by the *heritability*. Values of heritability range between zero and one; they can be changed by the source population and by manipulating the environment or the experimental design. For this reason heritability should be interpreted in conjunction with the experiment from which it was estimated. Despite this, estimates of heritability can be broadly comparable across a range of experiments with a crop.

Heritability greater than 0.5 indicates that the phenotype is a good predictor of the breeding value and that a good response to selection can be achieved by selecting individual phenotypes. Heritability below 0.3 means that the phenotype is a poor indication of the breeding value and information from relatives is needed to select good parents.

The *Breeding Value* (BV) of a plant is defined as twice the average deviation of its offspring from the population mean when mated randomly to an infinite population. The number is doubled as the progeny samples half of the genes from the parent of interest (Lynch and Walsh, 1998). Estimated breeding values (EBV) are used to rank and select superior parents for each selection trait, or they can be weighted and summed in a selection index for multitrait selection (Lynch and Walsh, 1998).

The response to multitrait selection depends on their *genetic correlation*. Selection for traits that have a positive genetic correlation will allow selection in one trait to also enable a good response to selection in the other traits and rapid progress is possible. If there is a strong negative correlation between traits, then response to selection will initially be slow. However, negative genetic correlations can change with multitrait selection, providing there is still genetic variation on the traits of interest (Falconer and Mackay, 1996; Kumar *et al.*, 2008).

If the correlation between the same genotypes replicated over different environments is poor, then the genotype by environment interaction (GxE) is said to be large and selection will have to be carried out in each different environment. A high correlation between genotypes across different environments (low GxE) means selection can be done at one site for a range of environments, so one breeding programme could deploy cultivars into several environments. A similar concept is repeated for the correlation between genotypes at different ages. A high correlation between ages allows a breeder to select at an early age.

Case studies for the three crops are provided here. The first two case studies look at using genetic parameters in the PFR blackcurrant and blueberry breeding programmes to predict the success of breeding new cultivars with higher levels of anthocyanins. The final case study looks at an application of genetic engineering in the strawberry breeding programme at UNIVPM. DNA markers are not available, so are not included in the discussion. However, a breeding programme that combines markers for anthocyanin content with quantitative genetics would increase the efficiency of the programme.

Case study: blackcurrant

Blackcurrant fruits (*Ribes nigrum* L.) are grown mainly for juice concentrate and are prized for their unique flavour and health properties. Four anthocyanin compounds are responsible for most of the colour in blackcurrants (Table 17.3).

Juice colour is a selection trait and is measured using juice buffered to pH 3 and recording absorbance at 515 nm (red) and 430 nm (brown) with a spectrophotometer. This measure is closely correlated to anthocyanin content. Anthocyanin content measured on PFR germplasm and selections ranged from 65 mg/100 ml juice to 484 mg/100 ml juice. Although anthocyanin content has only recently been routinely measured in the PFR breeding programme, selections are well represented among the genotypes with the top levels of anthocyanin due to indirect selection for anthocyanin content by juice colour (Figure 17.1 and Table 17.4).

Heritability

Currie *et al.* (2006) reported a heritability for total anthocyanin content of 0.64, which indicates that selecting parents on the basis of their observed anthocyanin content should be

Table 17.3 Mean (min–max) content of individual anthocyanins from over 92 blackcurrant genotypes in a New Zealand blackcurrant repository.

Individual anthocyanin	Content (mg/100 g)	(Range)
Delphinidin glucoside	45.7	(9.3–123.4)
Delphinidin rutinoside	130.0	(24.0–273.2)
Cyanidin glucoside	20.3	(4.7–47.7)
Cyanidin rutinoside	119.5	(21.1–261.5)
Minor anthocyanins	10.8	(1.2–27.7)
Total anthocyanin content	326.3	(64.9–618.4)

Table 17.4 Anthocyanin composition in blackcurrant juice of different genotypes collected from the PFR *Ribes* repository (sorted by anthocyanin content).

Cultivar	Total anthocyanin content (mg/100 ml)
Ben Lincoln*	483.9
Murchison*	468.5
Ben Rua*	457.5
Blackadder*	434.3
McWhite*	431.9
Ben Ard*	427.6
Ben Hope	355.6
Isobel*	328.8
Silvergeiters Zwarte	307.2
Cotswold Cross	285.4
Melina*	280.7
Millist	277.5
Black Down	273.5
Magnus	272.2
Baldwin	268.7
Tenah	266.6
EM1611/70	265.9
Kentish Hero	247.2
Laxtons Tinker	246.8
Linton	246.1
Levin Black	243.0
Constant	239.6
P7-7-11	230.0
EM1773/138	227.7
Goliath	222.9
Ben Lomond	206.9
EM1611/75	203.1
Tsema	199.6
EM1706/91	199.5
Hatton Black	196.7
Black Reward	186.1
EM1704/124	184.1
EM1704/125	181.4
Ben Mapua*	178.9
SAL SEL 6	178.5
Pilot Aleksandr Mamkin	96.4
Ben Nora	64.9

*Cultivars selected in New Zealand.

moderately successful. Experimental design can be used to increase the observed heritability, for example: replication, sample size and controlling sampling conditions.

Breeding values

We found a number of seedlings in each family had eBVs that exceeded their higher parental eBV for the total anthocyanin content (Figure 17.2) indicating that breeding for improved levels of anthocyanin could be successful.

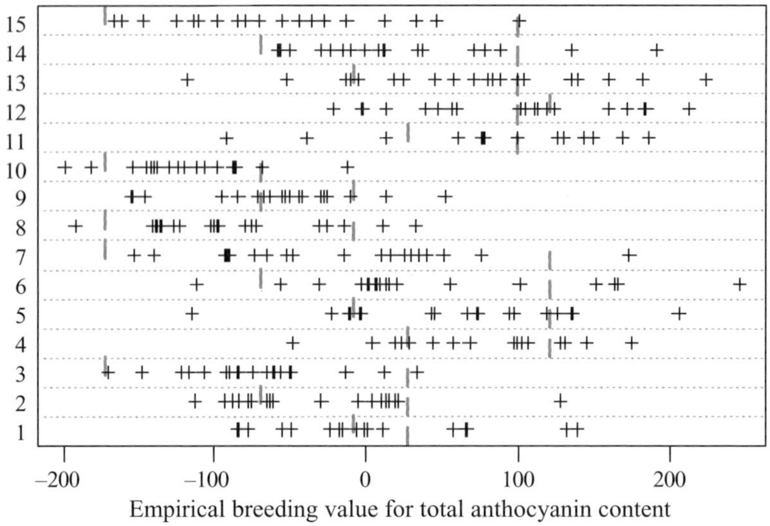

Fig. 17.2 Total anthocyanin content empirical breeding values from a 6 × 6 half-diallel study in blackcurrant. Crosses in each row represent the progeny from a family and the lines represent the male and female parents (male lines in upper portion, female lines in lower portion).

Genetic correlation

Anthocyanin content is reported to be highly correlated genetically (Currie *et al.*, 2006) and phenotypically (Moyer *et al.*, 2002a, 2002b; Currie *et al.*, 2006) to total phenolics and antioxidant activity, suggesting that either assay could be used as screening methodology.

We found no other genetic correlations reported between total anthocyanins and other traits. Phenotypic correlations are reported in Table 17.5.

Table 17.5 Phenotypic correlation between total anthocyanin content and fruit quality traits from 260 blackcurrant samples from the PFR blackcurrant repository.

Trait	Phenotypic correlation with total anthocyanin content
Colour absorbance 515 nm	0.93
Colour absorbance 420 nm	0.92
Flavonols	0.76
Brix	0.49
Vitamin C	0.30
Total sugars (mg/g*)	0.29
Total phenolics	0.18
Titratable acidity	−0.07
Juice content	−0.09
Total acids (mg/g)	−0.10
Firmness	−0.38
Berry weight	−0.42

*Estimated from a subset of 54 samples.

The negative correlation between anthocyanin content and fruit size is not unexpected, but Moyer and colleagues (2002b) reported a phenotypic correlation of 0.41 between total anthocyanin content and fruit size. We would expect a negative correlation because anthocyanins are produced in the skin of blackcurrants and not the flesh, so larger fruit could be expected to have lower levels of anthocyanin. Moyer *et al.* used a data set of 32 observations from cultivars that have been selected for both fruit size and colour and this may have influenced the relationship. Genetic correlations need to be estimated to confirm these correlations. If the genetic correlations have a similar magnitude as the phenotypic correlations the response to selection should be good for total anthocyanins and most fruit traits apart from fruit size and firmness.

Case study: blueberry

Blueberry fruits (*Vaccinium corymbosum*, *V. corymbosum* hybrids, *V. virgatum*, *V. angustifolium*) are true berries, generally blue–black or dark purple in colour, with a surface wax that provides the overall impression of a light blue 'bloom' on the fruit. Many anthocyanin compounds are responsible for the background colour in our blueberry collection (Table 17.6). In our study we could identify individual anthocyanins in two groups (acylated anthocyanin and minor anthocyanin). According to our study the most representative individual anthocyanin was Malvidin galactoside (Table 17.6); similar conclusions were found by Wu *et al.* (2006).

Blueberry fruit available from the market come from varieties that have been selected for specific agronomic and fruit traits (i.e. high crop, disease resistance, large fruit size and light blue fruit) and not necessarily for the concentration of healthy properties in the fruit. A survey of the variation of anthocyanin composition in blueberry fruit of different genotypes collected from the PFR *Vaccinium* repository was done to find sources of high levels of anthocyanins. We found that the lowest total anthocyanin content was found in *V. corymbosum* southern highbush type (Table 17.7), while the highest anthocyanin content was found in fruit of *V. virgatum* breeding lines. PFR selections were not always high in anthocyanin content (Table 17.7). The discussion on genetic correlations in the following section explains why this is the case.

Table 17.6 Mean content (range) of individual anthocyanins over 34 blueberry genotypes (Scalzo and Mezzetti, 2010).

Individual anthocyanin	Anthocyanin content (range) (mg/100 g)
Malvidin galactoside	31.0 (6.7–88.2)
Cyanidin galactoside + delphinidin arabinoside	22.9 (2.4–68.6)
Delphinidin galactoside	20.5 (8.0–67.4)
Malvidin arabinoside	18.3 (4.0–38.8)
Malvidin glucoside	14.1 (2.0–56.0)
Acylated Anthocyanin	13.1 (0.0–80.9)
Petunidin galactoside	12.4 (1.4–36.9)
Petunidin arabinoside	9.1 (2.1–30.4)
Minor nnthocyanins	7.3 (0.5–25.6)
Delphinidin glucoside	6.9 (0.1–36.4)
Petunidin glucoside	5.2 (0.1–23.2)
Cyanidin arabinoside	4.3 (0.1–27.3)
Cyanidin glucoside	2.8 (0.1–15.5)

Table 17.7 Anthocyanin composition in blueberry fruit of different genotypes collected from the PFR *Vaccinium* repository (grouped by species and sorted by anthocyanin content).

Cultivar	**Total anthocyanin content (mg/100 g)**
Northern highbush (*V. corymbosum*)	
Rubel	290.0
Darrow	286.8
Elliott	261.3
Northland	238.7
Duke	189.6
Jersey	184.5
Caroline Blue	153.6
Nui*	148.9
Hortblue Poppins*	138.9
Blue Moon*	138.9
Weymouth	120.2
Sunset Blue*	106.8
Brigitta	101.8
Nelson	99.9
Reka*	92.6
Elizabeth	89.8
Bluecrop	83.1
Puru*	45.9
Mean Northern Highbush	*154.0*
Rabbiteye (*V. virgatum*)	
Ono*	347.5
Centurion	275.5
Southland	266.2
Maru*	263.8
Brightblue	247.3
Dolce Blue*	238.2
Climax	234.1
Little Giant	230.7
Centra Blue*	212.0
Velluto Blue*	192.5
Ocean Blue*	189.0
Rahi*	168.5
Sky Blue*	167.2
Powderblue	165.3
Tifblue	145.3
Delite	111.9
Mean Rabbiteye	*215.9*
Southern highbush (*V. Corymbosum hybrids*)	
Island blue*	249.6
Marimba	148.6
Blue Bayou*	139.5
O'Neal	138.9
JU83	136.0
Misty	126.3
Ozarkblue	80.9
Mean Southern Highbush	*145.7*

*Cultivars selected in New Zealand.

Heritability

Connor *et al.* (2002b) reported a moderate heritability for total anthocyanin content of 0.56, which indicates that selecting parents on the basis of their observed anthocyanin content should be moderately successful. Experimental design to account for sources of variability in anthocyanin such as cultural conditions (Wang *et al.*, 2008), year (Connor *et al.*, 2002c), fruit maturity (Castrejon *et al.*, 2008) and storage conditions (Kalt *et al.*, 1999a; Connor *et al.*, 2002a) should increase the heritability.

Breeding values

We found a number of seedlings in each family that exceeded their higher parental eBV for the total anthocyanin content (Figure 17.3), showing that the progress with our breeding approach could be successful.

Genetic correlation

The genetic correlation between anthocyanin content and other fruit quality traits indicates how easy it will be to co-select for these traits. Genetic correlations were estimated by determining the correlations between breeding values for anthocyanin (Scalzo, personal communication) or total phenolics (Scalzo *et al.*, 2012) with fruit quality traits (Table 17.8). There is a strong and positive correlation reported between the anthocyanin content and total phenolics (Prior *et al.*, 1998; Ehlenfeldt and Prior, 2001; Connor *et al.*, 2002c; Moyer *et al.*, 2002b; Sellapan *et al.*, 2002) suggesting that either assay could be used as screening methodology.

In the PFR breeding programme the selection criteria has been for fruit weight, fruit firmness, fruit colour and pedicel scar. Fruit colour is genetically independent from anthocyanin

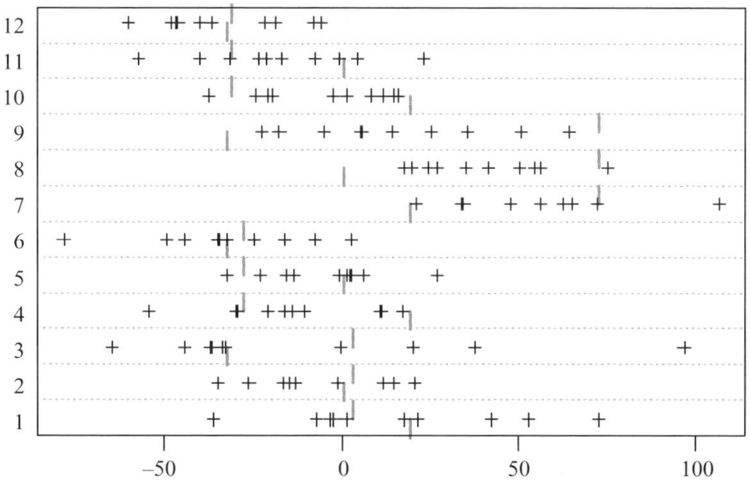

Fig. 17.3 Total anthocyanin content empirical breeding values from a 3 × 4 factorial study in blueberry. Crosses in each row represent the progeny from a family and the lines represent the male and female parents (male lines in upper portion, female lines in lower portion).

Table 17.8 Genetic correlations between anthocyanin content and total phenolics content with some blueberry fruit quality traits.

	Genetic correlation with:	
Trait	Anthocyanin content	Total phenolics content
Fruit colour	0.08	0.09*
Pedicel scar size	–0.16	–0.10*
Fruit firmness	–0.23	–0.33
Fruit weight	–0.48	–0.48*
Antioxidant activity	0.75	0.88
Total phenolics content	0.76	

*Scalzo *et al.* (2012).

content (Table 17.8). The other selection criteria are negatively correlated to anthocyanin content (Table 17.8), so we would expect to see lower anthocyanin content in selections from the PFR breeding programme, as is observed (Figure 17.1).

Response to selection would be initially slow for multitrait selection for anthocyanin content and fruit weight, fruit firmness, and pedicel scar size but with new germplasm and continued multitrait selection of parents, it is possible to change the correlations and improve response rate (Kumar *et al.*, 2008).

USE OF GENETIC ENGINEERING FOR IMPROVING BIOACTIVE COMPOUNDS

Genetic engineering is an alternative to hybridization as a means of genetic improvement and is favoured if the genes of interest are not available via hybridization or if the aim is to improve an existing cultivar without changing the other characteristics. In some countries (including New Zealand) transgenic crops are banned from the field, so transformation is limited to genetic experiments to prove concepts in a secure laboratory environment.

Transgenic approaches have been successfully used to increase the nutritional value of several worldwide important crops, for example rice (Paine *et al.*, 2005) and tomato (Davuluri *et al.*, 2005). Improving the fruit quality is a major breeding objective in strawberry breeding with the major objectives of improving flavour (sweetness, acidity and aroma), firmness and self-life. All these aspects are genetically controlled by the fruit ripening process (Manning, 1994). In climacteric fruits ripening is coordinated by the gaseous hormone ethylene, which is synthesized autocatalytically in the early stages of ripening. Non-climacteric fruits such as strawberry do not synthesize or respond to ethylene in this manner, yet undergo many of the same physiological and biochemical changes associated with the production of a ripe fruit. Wild strawberry (*Fragaria vesca*) is an attractive model system for studying ripening in non-climacteric fruit, because of its small diploid genome, its short reproductive cycle and its capacity for transformation. Strawberries may also serve as a model for other cultivated *Rosaceae,* such as fruit trees of which transgenic varieties have already been developed (e.g. plum and cherry trees) or are under development (e.g. apple or peach trees). Strawberries have the same basic flower structure and the same main pollen vectors (honey bees), but much shorter life cycles, and hence are much more easily amenable to experimentation.

A summary of the research in ripening and fruit quality:

- Eight ripening-induced cDNAs (Nam *et al.*, 1999) coding for acyl carrier protein, caffeoyl-CoA 3-*O*-methyltransferase, sesquiterpene cyclase, major latex protein, cystathionine gamma-synthase, dehydrin and an auxin-induced gene. Their putative functions are indicative of the wide range of processes up-regulated during fruit ripening.
- Wilkinson and colleagues (Wilkinson *et al.*, 1995) found five mRNAs with ripening-enhanced expression in strawberries using polymerase chain reaction (PCR) differential display. Three of the mRNAs appeared to be fruit-specific, with little or no expression detected in vegetative tissues. Sequence analysis of the cDNA clones revealed positive identities for three of the five mRNAs based on homology to known proteins. These results indicate that the differential display technique can be a useful tool to study fruit ripening and other developmental processes in plants at the RNA level.
- Civello and colleagues (Civello *et al.*, 1999) identified a cDNA sequence in strawberry that encoded a ripening-regulated expansin, FaExp2, part of the tissue softening process. A novel E-type endo-beta-1,4-glucanase with a putative cellulose-binding domain is highly expressed in ripening strawberry fruits of the octoploid cultivar Chandler (Trainotti *et al.*, 1999). Two full-length cDNA clones (faEG1 and faEG3, respectively) were isolated found to be down-regulated by treatment with the auxin analogue NAA.
- Antisense technology has been used to prevent strawberry fruit from softening by suppressing genes involved in fruit softening, without altering fruit quality (Woolley *et al.*, 2001; Palomer *et al.*, 2006; Sesmero *et al.*, 2007).
- Llop-Tous *et al.* (1999) isolated two cDNA clones (Cel1 and Cel2) from a cDNA library encoding divergent endo-beta-1,4-glucanases (EGases) (Trainotti *et al.*, 1999). Cel1 and Cel2 mRNA EGases play an important role in softening during ripening.
- Invertase (beta-fructofuranosidase) is responsible for catalysing the breakdown of sucrose to fructose and glucose during ripening. Invertase genes cloned from potato were integrated into two strawberry cultivars, Symphony and Senga Sengana, via *A. tumefaciens*-mediated transformation and are being assessed for fruit quality (Bachelier *et al.*, 1997).
- Park *et al.* (2006) generated transgenic plants with antisense cDNA of ADP-glucose pyrophosphorylase (AGPase) small subunit (FagpS), to evaluate the effects on carbohydrate contents during fruit development. Levels of *AGPase* mRNA were drastically reduced in the red stage of fruits and resulted in a 16–37% increment of total soluble sugar content and a 27–47% decrease of the starch content in mature fruit without significantly affecting other fruit characteristics such as colour, weight and hardness.
- DNA microarrays and appropriate statistical analyses were used to identify a novel strawberry alcohol acyltransferase (SAAT) gene that is part of the production of the fruity esters that are an important part of flavour development (Aharoni *et al.*, 2000).
- MADS-box genes encode transcription factors that are homeotic and control the spatial and temporal locations of specific organs. (Schwarz-Sommer *et al.*, 1992). AGAMOUS is a MADS-box gene from Aradopsis that is responsible for stamen and carpel identity and floral meristem determination (Mizukami and Ma, 1995). Homologs of this gene have been found in strawberry (Aharoni *et al.*, 2000).
- L-ascorbic acid in ripe strawberry fruit was increased in *A. thaliana* plants by over-expressing the strawberry gene *GaiUR,* encoding a D-galacturonic acid reductase (Agius *et al.*, 2003).

- Strawberries transformed with *DefH9-iaaM* and *rolC* genes to improve productivity (Mezzetti *et al.*, 2004a, 2004b). *DefH9-iaaM* increased auxin synthesis in the flowers and increased productivity but did not alter the antioxidant activity and the *rolC* gene increased the cytokinin metabolism producing plants with higher vigour, larger number of smaller fruit, higher sugar content and greater antioxidant activity of the fruit (Scalzo *et al.*, 2005).
- Recently the major enzymes and genes involved in flavonoid and proanthocyanidin biosynthesis during fruit development have been identified (Almeida *et al.*, 2007).
- The cloning and biochemical characterization of a glucosyltransferase involved in anthocyanin biosynthesis in strawberry fruit was reported by Griesser *et al.* (2008).
- *AmDFR* and *MiANS* transgenic lines of Calypso and Sveva and are being studied to elucidate the regulation of flavonoid pathway and improve the nutritional properties of strawberry (Montironi *et al.*, 2009).

Case study: Transgenic strawberry

Control and transgenic lines of 'Calypso' strawberry (*F. x ananassa*) were used to test the effect of transformation with the *rolC* genes in improving strawberry agronomic performance, fruit quality and the antioxidant attributes (Table 17.9). The *rolC* gene increased the cytokinin metabolism and the plant responded with higher vigour, larger number of smaller fruit, higher sugar content and greater antioxidant activity of the fruit (Scalzo *et al.*, 2005). Although the *rolC* gene has not been specifically targeted as a gene for improving anthocyanin content, it has improved anthocyanin levels and so shall be used as our example of genetic engineering to improve anthocyanins.

'Calypso' had 39.2 mg/100 g anthocyanins and the better transgenic line had 47.8 mg/100 g (Figure 17.1). The best transgenic line showed a considerable increase in anthocyanin content for 'Calypso'; however, there were parents in the repository with a 10-fold increase in anthocyanin content (485.5 mg/100 g FW) (Figure 17.1). With moderate heritability and positive genetic correlations a breeding programme could produce progeny with higher anthocyanin expression than the 'Calypso' transgenic lines.

There was some variation in the expression of anthocyanin content in the three transgenic lines suggesting that the insertion point may influence the phenotype. Given the relatively small effect the *rolC* gene has on anthocyanin production compared to the total range of anthocyanin content phenotypes in the germplasm, it is clear that there are other genes involved in the anthocyanin pathway, so the application of quantitative genetics to estimate genetic parameters like heritability, breeding value and genetic correlations would be useful in understanding how to incorporate the trait into a breeding programme.

Table 17.9 Antioxidant activity in transgenic *rolC* plants compared with 'Calypso' (Scalzo *et al.*, 2005).

Strawberry line	Total antioxidant capacity (μ moles Trolox Eq/g FW)	Total phenolics (mg GAE/g FW)
Calypso	13.26	2.26
rolC – line A	12.26	2.23
rolC – line B	14.33	2.41
rolC – line F	16.29	3.02

CONCLUSIONS

Breeding for bioactive compounds such as anthocyanin content is feasible and indications from genetic parameters and early results suggest good progress can be made in developing commercial cultivars with improved levels of anthocyanins by using conventional hybridization. Genetic transformation can provide gains in anthocyanin content but is limited to laboratory studies apart from in countries that allow genetic transformation in field crops.

REFERENCES

Agius, F., Gonzalez-Lamothe, R., Caballero, J.L. *et al.*, 2003. Engineering increased vitamin C levels in plants by overexpression of a D-galacturonic acid reductase. *Nat Biotechnol* **21**:177–181.

Aharoni, A., Keizer, L., Bouwmeester, H. *et al.*, 2000. Identification of the SAAT gene involved in strawberry flavor biogenesis by use of DNA microarrays. *Plant Cell* **12**:647–661.

Almeida, J., D'Amico, E., Preuss, A. *et al.*, 2007. Characterization of major enzymes and genes involved in flavonoid and proanthocyanidin biosynthesis during fruit development in strawberry (Fragaria x ananassa). *Arch Biochem Biophys* **465**:61–71.

Amakura, Y., Okada, M., Tsuji, S. and Tonogai, Y., 2000. High-performance liquid chromatographic determination with photodiode array detection of ellagic acid in fresh and processed fruits. *J Chromatogr A* **896**:87–93.

Bachelier, C., Graham, J., Machray, G. *et al.*, 1997. Integration of an invertase gene to control sucrose metabolism in strawberry cultivars. *Acta Hort* **439**:161–164.

Bailey, L. and Gregory, J., 1999. Folate metabolism and requirements. *J Nutr* **129**:779–782.

Beattie, J., Crozier, A. and Duthie, G.G., 2005. Potential health benefits of berries. *Current Nutrition & Food Science* **1**:71–86.

Biesalski, H.-K., Dragsted, L.O., Elmadfa, I. *et al.*, 2009. Bioactive compounds: Definition and assessment of activity. *Nutrition* **25**:1202–1205.

Carr, A. and Frei, B., 1999. Toward a new recommended dietary allowance for vitamin C based on antioxidant and health effects in humans. *Am J Clin Nutr* **69**:1086–1107.

Castrejon, A.D.R., Eichholz, I., Rohn, S. *et al.*, 2008. Phenolic profile and antioxidant activity of higbush blueberry (*Vaccinium corymbosum* L.) during fruit maturation and ripening. *Food Chem* **109**:564–572.

Civello, P., Powell, A., Sabehat, A. and Bennett, A., 1999. An expansin gene expressed in ripening strawberry fruit. *Plant Physiology* **121**:1273–1279.

Connor, A.M., Luby, J.J., Hancock, J.F. *et al.*, 2002a. Changes in fruit antioxidant activity among blueberry cultivars during cold-temperature storage. *J Agr Food Chem* **50**:893–898.

Connor, A.M., Luby, J.J. and Tong, C.B.S., 2002b. Variation and heritability estimates for antioxidant activity, total phenolic content, and anthocyanin content in blueberry progenies. *J Am Soc Hortic Sci* **127**:82–88.

Connor, A.M., Luby, J.J., Tong, C.B.S. *et al.*, 2002c. Genotypic and environmental variation in antioxidant activity, total phenolic content and anthocyanin content among blueberry cultivars. *J Am Soc Hortic Sci* **127**:89–97.

Connor, A.M., Finn, C.E., McGhie, T.K. and Alspach, P.A., 2005. Genetic and environmental variation in anthocyanins and their relationship to antioxidant activity in blackberry and hybridberry cultivars. *J Am Soc Hort Sci* **130**:680–687.

Currie, A., Langford, G., McGhie, T. *et al.* 2006. Inheritance of antioxidants in a New Zealand blackcurrant (*Ribes nigrum L.*) population. In: Mercer, C.F. (ed.), *Breeding for success: diversity in action*. Proceedings of the 13th Australasian Plant Breeding Conference, 18–21 April 2006, Christchurch, New Zealand, pp. 218-225.

Davuluri, G., van Tuinen, A., Fraser, P. *et al.*, 2005. Fruit-specific RNAi-mediated suppression of DET1 enhances carotenoid and flavonoid content in tomatoes. *Nat Biotechnol* **23**:890–895.

Deighton, N., Brennan, R.M., Finn, C.E. and Davies, H.V., 2000. Antioxidant properties of domesticated and wild *Rubus* species. *J Sci Food Agric* **80**:1307–1313.

Ehlenfeldt, M. and Prior, R., 2001. Oxygen radical absorbance capacity (ORAC) and phenolic and anthocyanin concentrations in fruit and leaf tissues of highbush blueberry. *J Agric Food Chem* **49**:2222–2227.

Falconer, D.S. and Mackay, T.F.C., 1996. *Introduction To Quantitative Genetics*. Longman Group, London.

Ghosh, D., McGhie, T.K., Zhang, J.L. *et al.*, 2006. Effects of anthocyanins and other phenolics of boysenberry and blackcurrant as inhibitors of oxidative stress and damage to cellular DNA in SH-SY5Y and HL-60 cells. *J Sci Food Agric* **86**:678.

Goiffon, J.-P., Mouly, P.P. and Gaydou, E.M., 1999. Anthocyanin pigment determination in red fruit juices, concentrated juices and syrups using liquid chromatography. *Anal Chim Acta* **382**:39–50.

Griesser, M., Hoffmann, T., Bellido, M. *et al.*, 2008. Redirection of flavonoid biosynthesis through the down-regulation of an anthocyanidin glucosyltransferase in ripening strawberry fruit. *Plant Physiology* **146**:1528–1539.

Hakkinen, S., Karenlampi, S., Heinonen, I. *et al.*, 1999. Content of the flavonols quercetin, myricetin, and kaempferol in 25 edible berries. *J Agric Food Chem* **47**:2274–2279.

Hecht, S., Kammhuber, K., Reiner, J. *et al.*, 2004. Biosynthetic experiments with tall plants under field conditions. O_2 incorporation into humulone from Humulus lupulus. *Elsevier/Science Direct* **65**:1057–1060.

Hertog, M., Feskens, E., Hollman, P. *et al.*, 1993. Dietary antioxidant flavanoids and risk of coronary heart-disease – The Zutphen Elderly Study. *Lancet* **342**:1007–1011.

Hertog, M., Feskens, E. and Kromhout, D., 1997. Antioxidant flavonols and coronary heart disease risk. *Lancet* **349**:699–699.

Howell, A.B., 2002. Cranberry proanthocyanidins and the maintenance of urinary tract health. *Crit Rev Food Sci Nutr* **42**:273–278.

Jang, M., Cai, E., Udeani, G. *et al.*, 1997. Cancer chemopreventive activity of resveratrol, a natural product derived from grapes. *Science* **275**:218–220.

Joshipura, K.J., Hu, F., Manson, J.E. *et al.*, 2001. Fruit and vegetable intake in relation to risk of coronary heart disease. *Annals of Internal Medicine* **134**:1–9.

Kader, F., Rovel, B., Girardin, M. and Metche, M., 1996. Fractionation and identification of the phenolic compounds of Highbush blueberries (*Vaccinium corymbosum*, L). *Food Chem* **55**:35–40.

Kalt, W., Forney, C.F., Martin, A. and Prior, R.L., 1999a. Antioxidant capacity, Vitamin C, phenolics, and anthocyanins after fresh storage of small fruits. *J Agr Food Chem* **47**:4638–4644.

Kalt, W., McDonald, J., Ricker, R. and Lu, X., 1999b. Anthocyanin content and profile within and among blueberry species. *Can J Plant Sci* **79**:617–623.

Kalt, W., Ryan, D., Duy, J. *et al.*, 2001. Interspecific variation in anthocyanins, phenolics, and antioxidant capacity among genotypes of highbush and lowbush blueberries (Vaccinium section cyanococcus spp.). *J Agric Food Chem* **49**:4761–4767.

Kearney, J., 2010. Food consumption trends and drivers. *Philosophical Transactions of the Royal Society B – Biological Sciences* **365**:2793–2807.

Keli, S., Hertog, M., Feskens, E. and Kromhout, D., 1996. Dietary flavonoids, antioxidant vitamins, and incidence of stroke – The Zutphen Study. *Arch Intern Med* **156**:637–642.

Kelley, D.S., Rasooly, R., Jacob, R.A. *et al.*, 2006. Consumption of Bing sweet cherries lowers circulating concentrations of inflammation markers in healthy men and women. *J Nutr* **136**:981–986.

Kempster, L. and Beverley, L., 2011. Leatherhead reveals 2012 food and drink trends. Leatherhead Food Research, Leatherhead, UK. (http://www.leatherheadfood.com/leatherhead-reveals-2012-food-and-drink-trends; last accessed 22 March 2013).

Knekt, P., Kumpulainen, J., Jarvinen, R. *et al.*, 2002. Flavonoid intake and risk of chronic diseases. *Am J Clin Nut* **76**:560–568.

Kong, J.M., Chia, L.S., Goh, N.K. *et al.*, 2003. Analysis and biological activities of anthocyanins. *Phytochemistry* **64**:923–933.

Koponen, J., Happonen, A., Mattila, P. and Torronen, A., 2007. Contents of anthocyanins and ellagitannins in selected foods consumed in Finland. *J Agric Food Chem* **55**:1612–1619.

Kumar, S., Burdon, R.D., Stovold, G.T. and Gea, L.D., 2008. Implications of selection history on genetic architecture of growth, form, and wood-quality traits in *Pinus radiata*. *Can J For Res* **38**:2372–2381.

Lee, J., Lee, H.K., Kim, C.Y. *et al.*, 2005. Purified high-dose anthocyanoside oligomer administration improves nocturnal vision and clinical symptoms in myopia subjects. *Br J Nutr* **93**:895–899.

Llop-Tous, I., Dominguez-Puigjaner, E., Palomer, X. and Vendrell, M., 1999. Characterization of two divergent endo-beta-1,4-glucanase cDNA clones highly expressed in the nonclimacteric strawberry fruit. *Plant Physiology* **119**:1415–1421.

Lopes-da-Silva, F., de Pascual-Teresa, S., Rivas-Gonzalo, J. and Santos-Buelga, C., 2002. Identification of anthocyanin pigments in strawberry (cv Camarosa) by LC using DAD and ESI-MS detection. *Eur Food Res Technol* **214**:248–253.

Lopes-da-Silva, F., Escribano-Bailón, M.T., Pérez Alonso, J.J. *et al.*, 2007. Anthocyanin pigments in strawberry. *LWT - Food Science and Technology* **40**:374–382.

Lyall, K., Hurst, S., Cooney, J. *et al.*, 2009. Short-term blackcurrant extract consumption modulates exercise-induced oxidative stress and lipopolysaccharide-stimulated inflammatory responses. *Am J Physiol Regul Integr Comp Physiol* **297**:R70–R81.

Lynch, M. and Walsh, B., 1998. *Genetics and Analysis of Quantitative Characters*. Sinauer Associates Inc., Sunderland.

Maatta-Riihinen, K.R., Kamal-Eldin, A. and Torronen, A.R., 2004. Identification and quantification of phenolic compounds in berries of Fragaria and Rubus species (family Rosaceae). *J Agric Food Chem* **52**:6178–6187.

Manning, K., 1994. Changes in gene-expression during strawberry fruit ripening and their regulation by Auxin. *Planta* **194**:62–68.

Matsumoto, H., Nakamura, Y., Hirayama, M. *et al.*, 2002. Antioxidant activity of black currant anthocyanin aglycons and their glycosides measured by chemiluminescence in a neutral pH region and in human plasma. *J Agric Food Chem* **50**:5034–5037.

Matsumoto, H., Nakamura, Y., Tachibanaki, S. *et al.*, 2003. Stimulatory effect of cyanidin 3-glycosides on the regeneration of rhodopsin. *J Agric Food Chem* **51**:3560–3563.

Matsumoto, H., Takenami, E., Iwasaki-Kurashige, K. *et al.*, 2005. Effects of blackcurrant anthocyanin intake on peripheral muscle circulation during typing work in humans. *Eur J Appl Physiol* **94**:36.

McGhie, T.K., Hall, H.K., Ainge, G.D. and Mowat, A.D., 2002. Breeding Rubus cultivars for high anthocyanin content and high antioxidant capacity. In: Brennan, R.M., Gordon, S.L., and Williamson, B. (eds), *Proceedings of the Eighth International Rubus and Ribes Symposium*, Vols 1 and 2. International Society Horticultural Science, Leuven, Belgium, pp. 495–500.

Mezzetti, B., Costantini, E., Chionchetti, F. *et al.*, 2004a. Genetic transformation in strawberry and raspberry for improving plant productivity and fruit quality. *Acta Hort* **649**:107–110.

Mezzetti, B., Landi, L., Pandolfini, T. and Spena, A., 2004b. The defH9-iaaM auxin-synthesizing gene increases plant fecundity and fruit production in strawberry and raspberry. *BMC Biotechnol* **4**:4.

Mizukami, Y. and Ma, H., 1995. Separation of Ag function in floral meristem determinacy from that in reproductive organ identity by expressing antisense Ag RNA. *Plant Mol Biol* **28**:767–784.

Montironi, E., Costantini, E., Mezzetti, B. *et al.*, 2009. Engineering strawberry anthocyanin levels by transformation with late flavonoid pathway genes. In: Lopez-Medina, J. (ed.), *Proceedings of the International Symposium on Cut Flowers in the Tropics*. Acta Horticulturae 482, International Society Horticultural Science, Leuven, Belgium, pp. 463–466.

Moyer, R., Hummer, K., Wrolstad, R.E. and Finn, C., 2002a. Antioxidant compounds in diverse Ribes and Rubus germplasm. In: Brennan, R.M., Gordon, S.L. and Williamson, B. (eds), *Proceedings of the Eighth International Rubus and Ribes Symposium*, Vols 1 and 2. International Society Horticultural Science, Leuven, Belgium, pp. 501–505.

Moyer, R.A., Hummer, K.E., Finn, C.E. *et al.*, 2002b. Anthocyanins, phenolics, and antioxidant capacity in diverse small fruits: Vaccinium, Rubus, and Ribes. *J Agric Food Chem* **50**:519–525.

Nam, Y., Tichit, L., Leperlier, M. *et al.*, 1999. Isolation and characterization of mRNAs differentially expressed during ripening of wild strawberry (*Fragaria vesca* L.) fruits. *Plant Mol Biol* **39**:629–636.

Ness, A. and Powles, J., 1997. Fruit and vegetables, and cardiovascular disease: A review. *Int J Epidemiol* **26**:1–13.

Neyazaki, K., Mizugami, T., Kim, J.G. *et al.*, 2008. Forcing culture of low-chilling Actinidia interspecific hybrid 'Kosui'. In: Zhang, D., Lee, J.M., and Tao, R. (eds), *Proceedings of the International Symposium on Asian Plants with Unique Horticultural Potential*. Acta Horticulturae 769, pp. International Society Horticultural Science, Leuven, Belgium, 271–275.

Nielsen, I.L.F., Haren, G.R., Magnussen, E.L. *et al.*, 2003. Quantification of anthocyanins in commercial black currant juices by simple high-performance liquid chromatography. Investigation of their pH stability and antioxidative potency. *J Agric Food Chem* **51**:5861–5866.

Paine, J., Shipton, C., Chaggar, S. *et al.*, 2005. Improving the nutritional value of Golden Rice through increased pro-vitamin A content. *Nat Biotechnol* **23**:482–487.

Palomer, X., Llop-Tous, I., Vendrell, M. *et al.*, 2006. Antisense down-regulation of strawberry endo-beta-(1,4)-glucanase genes does not prevent fruit softening during ripening. *Plant Science* **171**:640–646.

Park, J., Lee, Y., Chung, W. *et al.*, 2006. Modification of sugar composition in strawberry fruit by antisense suppression of an ADP-glucose pyrophosphorylase. *Mol Breed* **17**:269–279.

Prior, R., Cao, G., Martin, A. *et al.*, 1998. Antioxidant capacity as influenced by total phenolic and anthocyanin content, maturity, and variety of Vaccinium species. *J Agric Food Chem* **46**:2686–2693.

Proteggente, A., Pannala, A., Paganga, G. *et al.*, 2002. The antioxidant activity of regularly consumed fruit and vegetables reflects their phenolic and vitamin C composition. *Free Radical Research* **36**:217–233.

Rimando, A., Kalt, W., Magee, J. *et al.*, 2004. Resveratrol, pterostilbene, and piceatannol in Vaccinium berries. *J Agric Food Chem* **52**:4713–4719.

Scalzo, J. and Mezzetti, B., 2010. Biotechnology and breeding for enhancing the nutritional value of berry fruit. In: Bagchi, D., Lau, F.C. and Ghosh, D.K. (eds), *Biotechnology in Functional Foods and Nutraceuticals*, Taylor & Francis Group, pp. 61–79.

Scalzo, J., Battino, M., Costantini, E. and Mezzetti, B., 2005. Breeding and biotechnology for improving berry nutritional quality. *Biofactors* **23**:213.

Scalzo, J., Currie, A., Stephens, J. *et al.*, 2009a. The anthocyanin composition of different Vaccinium, Ribes and Rubus genotypes. *BioFactors* **34**:13–21.

Scalzo, J., Miller, S., Edwards, C. *et al.*, 2009b. Variation in phytochemical composition and fruit traits of blueberry cultivars and advanced breeding selections in New Zealand. *Acta Hort* **810**:823–830.

Scalzo, J., Sguigna, V., Mezzetti, B. *et al.*, 2012. Variation of fruit traits in Highbush blueberry seedlings from a factorial cross. *Acta Hort* **926**:79–83.

Schwarz-Sommer, Z., Saedler, H. and Sommer, H., 1992. Homeotic genes in the genetic control of flower morphogenesis in Antirrhinum majus. In: Russo, V.E.A., Brody, S., Cove, D. and Ottolenghi, S. (eds), *Development: The Molecular Genetic Approach*. Springer-Verlag, Berlin, Germany, pp. 242–256.

Sellapan, S., Akoh, C.C. and Krewer, G., 2002. Phenolic compounds and antioxidant capacity of Georgia-grown blueberries and blackberries. *J Agr Food Chem* **50**:2432–2438.

Sesmero, R., Quesada, M. and Mercado, J., 2007. Antisense inhibition of pectate lyase gene expression in strawberry fruit: Characteristics of fruits processed into jam. *J Food Eng* **79**:194–199.

Shin, W., Park, S. and Kim, E., 2006. Protective effect of anthocyanins in middle cerebral artery occlusion and reperfusion model of cerebral ischemia in rats. *Life Sci* **79**:130–137.

Slimestad, R. and Solheim, H., 2002. Anthocyanins from black currants (*Ribes nigrum* L.). *J Agric Food Chem* **50**:3228–3231.

Sloan, A., 2005. Top 10 global food trends. *Food Technology* **59**(4):20–22, 24, 26, 28, 30–32.

Sloane, A.E. 2011. Top 10 food trends show lasting influence of weak economy on consumers decisions. *Food Technology Magazine* **65**(4):24–26, 28, 30, 33–34, 36–38, 40–41.

Steinmetz, K.A. and Potter, J.D., 1996. Vegetables, fruit, and cancer prevention: a review. *J Am Diet Assoc* **96**:1027–1039.

Stintzing, F.C., Stintzing, A.S., Carle, R. *et al.*, 2002. Color and antioxidant properties of cyanidin-based anthocyanin pigments. *J Agric Food Chem* **50**:6172–6181.

Trainotti, L., Spolaore, S., Pavanello, A. *et al.*, 1999. A novel E-type endo-beta-1,4-glucanase with a putative cellulose-binding domain is highly expressed in ripening strawberry fruits. *Plant Mol Biol* **40**:323–332.

USDA, 2011. USDA Database for the Flavonoid Content of Selected Foods, Release 3. US Department of Agriculture, Agricultuarl research service (http://www.ars.usda.gov/Services/docs.htm?docid=6231; last accessed 22 March 2013).

Wada, L. and Ou, B., 2002. Antioxidant activity and phenolic content of oregon caneberries. *J Agric Food Chem* **50**:3495–3500.

Wang, H., Cao, G. and Prior, R., 1997. Oxygen radical absorbing capacity of anthocyanins. *J Agric Food Chem* **45**:304–309.

Wang, H., Nair, M., Strasburg, G. *et al.*, 1999. Antioxidant and antiinflammatory activities of anthocyanins and their aglycon, cyanidin, from tart cherries. *J Nat Prod* **62**:802.

Wang, S.Y., Chen, C.-T., Sciarappa, W. *et al.*, 2008. Fruit quality, antioxidant capacity, and flavonoid content of organically and conventionally grown blueberries. *J Agric Food Chem* **56**:5788–5794.

WHO, 2003. Population nutrient intake goals for preventing diet-related chronic diseases, Diet, nutrition and the prevention of chronic diseases. Report of the joint WHO/FAO expert consultation. World Health Organization, Geneva, Switzerland.

Wilkinson, J., Lanahan, M., Conner, T. and Klee, H., 1995. Identification of mRNAs with enhanced expression in ripening strawberry fruit using polymerase chain reaction differential display. *Plant Mol Biol* **27**:1097–1108.

Woolley, L., James, D. and Manning, K., 2001. Purification and properties of an endo-beta-1,4-glucanase from strawberry and down-regulation of the corresponding gene, cel1. *Planta* **214**:11–21.
Wu, X. and Prior, R., 2005. Systematic identification and characterization of anthocyanins by HPLC-ESI-MS/MS in common foods in the United States: Fruits and berries. *J Agric Food Chem* **53**:2589–2599.
Wu, X., Beecher, G., Holden, J. *et al.*, 2006. Concentrations of anthocyanins in common foods in the United States and estimation of normal consumption. *J Agric Food Chem* **54**:4069–4075.
Youdim, K.A. and Joseph, J.A., 2001. A possible emerging role of phytochemicals in improving age-related neurological dysfunctions: a multiplicity of effects. *Free Radical Biology and Medicine* **30**:583–594.

18 The Influence of Pre- and Postharvest Environmental Stress on Fruit Bioactives

David J. Burritt

INTRODUCTION

Plants are able to synthesize a wide range of organic molecules, including many that are unique to the kingdom Plantae, often generically referred to as phytochemicals. Many plant species are well known for the production of compounds with medicinal properties (Cravotto *et al.*, 2010) and in recent years there has been increasing interest in identifying phytochemicals that, when consumed in foods or when used as nutritional supplements, have health-promoting properties for humans (Cravotto *et al.*, 2010; Traka and Mithen, 2011). These health-promoting phytochemicals are often referred to as plant bioactives and there is an increasing body of evidence suggesting that bioactives could help to protect humans from a wide range of health problems (Traka and Mithen, 2011). Numerous studies have demonstrated the potential of plant bioactives as protective agents against chronic diseases such as cancer (Collet *et al.*, 2010; Lee *et al.*, 2011a; Weng and Yen, 2012), cardiovascular disease (Hansen *et al.*, 2005; Pase *et al.*, 2011; Vasanthi *et al.*, 2012) and diabetes (Borriello *et al.*, 2010; Lee *et al.*, 2011b). In addition to having potential protective roles some plant bioactives, due to their antioxidant (Korkina, 2007), anti-inflammatory capacity (Gonzalez-Gallego *et al.*, 2010; Korkina *et al.*, 2011), abilities to inhibit enzymes and cellular signalling cascades (Hou and Kumamoto, 2010; Kang *et al.*, 2011), or antiproliferative actions (Delmas *et al.*, 2011; Attoub *et al.*, 2011), also have the potential to function as therapeutic agents and could help to provide new and more effective treatments when used in combination with conventional therapies (Chalabi *et al.*, 2006).

Phytochemicals produced by plants can be classified as either primary or secondary metabolites. Primary metabolites are required for normal plant growth and development, and are mostly essential for survival. Secondary metabolites are generally not essential for survival, but have numerous important supplemental roles throughout a plants' life cycle and often enable plants to cope with periods of environmental stress. The production of secondary metabolites is generally at the expense of the production of primary metabolites, so can have a negative influence on plant growth and development, and crop yields (Capanoglu, 2010). Due to their sessile nature plants are continuously exposed to what can be a highly variable and stressful environment. As they grow and develop plants are exposed to a range of abiotic and biotic factors, including ultraviolet radiation (UVR), drought, fluctuations in temperatures, attack by insects and herbivores, and exposure to a variety of pathogens, all of which can

Bioactives in Fruit: Health Benefits and Functional Foods, First Edition. Edited by Margot Skinner and Denise Hunter.

impair cellular function and cause stress (Zhao *et al.*, 2007). To cope with these abiotic and biotic stressors plants have evolved complex defence mechanisms (Zhao *et al.*, 2005; Guo *et al.*, 2008), which involve major changes in a plants' metabolome (Debnath *et al.*, 2011; Nadella *et al.*, 2012). In particular, some abiotic and biotic stressors, and even some of the chemical and physical treatments used in crop production, can result in an enhancement of secondary metabolite production (Sudha and Ravishankar, 2002). As most bioactive phytochemicals are secondary metabolites, the levels of these molecules fluctuate in plant cells and tissues due to changes in environmental conditions; and while genetics determines the specific secondary metabolites that a plant species can produce, the environment is often the most important factor regulating tissue levels. This provides producers with the potential to manipulate environmental conditions, both pre- and postharvest, to maximize the yield of bioactives, using simple management practices or treatments, without having to resort to genetic manipulation. In this chapter an overview is given of the abiotic stressors that have, and potentially could be used as treatments, pre- and/or postharvest, to enhance the levels of bioactive compounds found in commonly cultivated fruit crops. Finally, the recently proposed concept of xenohormesis, a principle that attempts to explain how stressed plant produce could be of benefit to the animals that consume it, is briefly discussed with respect to crop production and human health.

THE MAIN GROUPS OF FRUIT BIOACTIVES SENSITIVE TO ENVIRONMENTAL STRESSORS AND THEIR ASSOCIATIONS WITH HUMAN HEALTH

Vitamins C and E

Ascorbic acid, commonly referred to as vitamin C, is an essential nutrient for humans. Three forms of ascorbic acid are found in cells: ascorbate and two oxidized forms, monodehydroascorbate and dehydroascorbate. Plants are a rich source of vitamin C, with high levels found in many fruits and vegetables. In plants, vitamin C functions as an electron donor and is the most abundant antioxidant found in plant cells. It is found in all cellular compartments and within the cell wall (Foyer and Noctor, 2012). The highest concentrations of ascorbate are found in chloroplasts, often in excess of 20 mM, where it plays an important role in the detoxification of hydrogen peroxide and act as a substrate of ascorbate peroxidase, a key enzyme of the ascorbate-glutathione cycle. Vitamin C levels in plant cells are controlled by the regulation of biosynthesis, catabolism, recycling and transport (Foyer and Noctor, 2012). Unlike plants, humans are unable to synthesize their own vitamin C, so must obtain it from the foods they consume. In humans vitamin C is required for the synthesis of collagen and functions as a cofactor for enzymes involved in many key cellular process, such as the synthesis of catecholamines, carnitine and cholesterol (Arrigoni and De Tullio, 2002). Vitamin C is also involved in controlling the expression of several redox-regulated genes, and acts as an important cellular antioxidant either as an electron donor or by directly scavenging reactive oxygen species (ROS) generated by normal cellular metabolism. As an antioxidant vitamin C plays an important role in protecting cells against the build-up of ROS that causes oxidative stress, which is thought to be a contributing factor leading to the development of neurodegenerative disorders such as Alzheimer's and Parkinson's diseases, and inflammatory responses, such as arteriosclerosis (Ames *et al.*, 1993). Studies have also concluded that increased vitamin C consumption, in combination with other therapies, could prevent chronic inflammatory and heart diseases, as well as some neurodegenerative disorders and cancers (Mandl *et al.*, 2009).

Vitamin E is a generic term that refers to the tocochromanols, a group of eight compounds that include four tocopherols and four tocotrienols. The tocopherols and tocotrienols are amphipatic molecules made up of lipid soluble, hydrophobic prenyl tails that often associate with membranes and polar chromanol head groups that are exposed to the cytosol. The difference between the tocopherols and tocotrienols is in the degree of saturation of their hydrophobic prenyl tails, while the α, β, γ and δ forms of the tocopherols and tocotrienols vary in both the position and number of methyl groups attached to the chromanol head, with the α form having three, the β and γ forms having two and the δ form having one (DellaPenna, 2005). α-Tocopherol is synthesized in plastid membranes and is stored in the thylakoid membranes and plastoglobuli of the chloroplast stroma (DellaPenna, 2005). In plants, the tocopherols function as lipid-soluble antioxidants and play an important role in limiting oxidative damage caused by environmental stressors, in particular by limiting lipid peroxidation. The tocopherols also have other important functions in plants, such as regulating carbon translocation from source tissues to sink tissues (Hofius *et al.*, 2004).

In humans, vitamin E is the major lipid-soluble chain-breaking antioxidant and is found in all cell membranes and plasma lipoproteins, with high concentrations found in red blood cells. As an antioxidant vitamin E helps to protect low density lipoproteins, nucleic acids, and polyunsaturated fatty acids from oxidative damage. It also plays important roles in the biosynthesis of haemoglobin, the modulation of immune responses and membrane stabilization (Brigelius-Flohe and Traber, 1999). Like vitamin C the antioxidant capacity of vitamin E has led to extensive research on its potential uses as a therapeutic agent. However, the results obtained to date have been mixed, with some studies suggested increasing the consumption of antioxidants like vitamin E could prevent disease while others have shown no preventative value (Firuzi *et al.*, 2011).

Plant phenolics

The term plant phenolic encompasses a diverse array of metabolites found throughout the plant kingdom that are characterized by having at least one aromatic ring (C6) with one or more hydroxyl groups. Plant phenolics include phenolic acids, flavonoids (flavonols, flavones and isoflavones, flavanols, proanthocyanidins), anthocyandins, tannins, stilbenes, lignans and other phenols (Tomas-Barberan and Espin, 2001). These phenolic compounds are produced from phenylalanine, via the shikimate, phenylpropanoid and flavonoid specific pathways. Plant phenolics are often found as esters or glycosides, conjugated with other molecules such as hydroxy-fatty acids, sterols or polymerized into larger molecules, such as condensed tannins and lignins. Plant phenolics have numerous functions in plants, ranging from providing structural support in thickened cell walls and antimicrobial activities, to acting as signal molecules (Bhattacharya *et al.*, 2010). There is also evidence that phenolics can act as antioxidants, neutralizing ROS and toxic free radicals. For example, hydroxycinnamic acid derivatives such as caffeic acid, chlorogenic acid, ferulic acid, *p*-coumaric acid and sinapic acid have strong antioxidant activities and can inhibit lipid peroxidation (Sroka and Cisowski, 2003; Cheng *et al.*, 2007). Flavonoids could also potentially act as antioxidants (Rice-Evans *et al.*, 1997); however, the importance of flavonoids as antioxidants in plants cells is still a matter of debate (Hernandez *et al.*, 2009).

Most plants synthesize phenolics constitutively, but their accumulation can be induced by abiotic and biotic stresses as detailed above. In humans, plant phenolics are also thought to function as antioxidants and it has been suggested that a diet high in plant phenolics can reduce the accumulation of free radicals and peroxides, which may cause oxidative damage to macromolecules such as DNA, lipids and proteins (Heim *et al.,* 2002). Oxidative damage has been linked to aging, atherosclerosis, cancer, inflammation and neurodegenerative disorders, and it has been suggested that the consumption of plant phenolics could help reduce susceptibility to these diseases/disorders (Burton-Freeman, 2010; Dai and Mumper, 2010; Kim, 2010; Tsao, 2010). However, the data regarding the absorption and bioavailability of dietary phenolics in humans are still controversial (de Oliveira and Bastos, 2011). In addition to acting as antioxidants, plant phenolics also have other potential health promoting functions. For example, quercetin, myricetin and fisetin inhibit the intestinal glucose transporter GLUT2, and thus can have antiglycaemic effects (Song *et al.,* 2002). Isoflavones such as daidzein and genistein are able to bind to mammalian oestrogen receptors and such anti-oestrogen actions may help to prevent breast cancer and osteoporosis in humans (Martinez, 2006).

Terpenes

The tetraterpene carotenoids are a group of lipophilic plant pigments derived from the isoprenoid pathway (Cazzonelli and Pogson, 2010). They are found in most of the above ground parts of plants. In chloroplasts they are essential photosynthetic pigments and are important for preventing photooxidative damage (Howitt and Pogson, 2006). In most plants the carotenoids are abundant in fruits and flowers, where they are responsible for the yellow, orange and red colours. Carotenoids are also important precursors for two important regulators of plant growth and development, the plant hormone abscisic acid (ABA) and the strigolactones (Cazzonelli and Pogson, 2010). Carotenoids are an important component of the human diet, with the most abundant consumed being β-carotene, lycopene, lutein, β-cryptoxanthin, zeaxanthin and astaxanthin (Riccioni, 2009). The carotenes, α-carotene, β-carotene, γ-carotene, and the xanthophyll β-cryptoxanthin are of particular importance as provitamin A. Vitamin A is required by the retina of the eye in the form of the light-absorbing diterpene retinol (Fernandez-Garcia *et al.,* 2012). In humans vitamin A is also important for hormone synthesis, immune response, and the regulation of cell growth and differentiation (Fernandez-Garcia *et al.,* 2012). Numerous studies have provided evidence that carotenoids, such as the carotenes and lycopene, are important dietary antioxidants with potential anticancer and antimutagenic properties (Beutner *et al.,* 2001; Riccioni, 2009).

The limonoids are a group of highly oxygenated modified triterpenes, derived from 4,4,8-trimethyl-17-furanylsteroid skeletons, with potential bioactive properties (Ejaz *et al.,* 2006). They are found in citrus fruits, mostly in the peel, and produce the bitter taste and zesty aroma (Ejaz *et al.,* 2006). Over 300 limonoids have been isolated, with large quantities found in citrus juices (350–400 μg/g in orange juice) as water-soluble limonoid glucosides, or in seeds as water-insoluble limonoidaglycones (Ejaz *et al.,* 2006). While the importance of limonoids for humans is unclear, initial studies have shown that they could be useful in preventing and/or treating cancer in humans (Ejaz *et al.,* 2006; Kim *et al.,* 2012).

Phytosterols

Phytosterols are sterols structurally similar to cholesterol, but with an extra methyl or ethyl group, and are found in the non-saponifiable fraction of plant oils (Jones and Abumweis,

2009). They play important roles in regulating the fluidity and permeability of plant membranes (Hartmann, 1998) and some phytosterols are precursors for the synthesis of the brassinosteroid group of plant hormones, which are important regulators of cell division, embryo development and growth rates in plants (Clouse and Sasse, 1998). Phytosterols are found in large amounts in vegetables, with the highest concentrations found in pea, cauliflower, broccoli and romaine lettuce, and at much lower levels in fruits, ranging from 1.6 to 32.6 mg/100 g, with the highest levels found in navel oranges, tangerines and mangos (Han *et al.,* 2009). In humans, because of their structural similarity to cholesterol, phytosterols can compete with cholesterol, limiting its absorption (Devaraj and Jialal, 2006). In addition, plant sterols may have anti-cancer, anti-heart disease, anti-inflammatory and antioxidant activities (Ostlund, 2002; Woyengo *et al.,* 2009; Genser *et al.,* 2012).

THE MAIN ENVIRONMENTAL STRESSORS THAT CAN INFLUENCE THE LEVELS OF BIOACTIVES IN FRUITS

Stress and secondary metabolite production

Elicitation has been widely used to induce *de novo* synthesis of secondary metabolites in plant cell cultures (Dicosmo and Misawa, 1985) and studies are now showing that manipulation of environmental conditions, either by changing crop management strategies or by direct exposure to stressors, can increase secondary metabolite production in whole plants. In addition, treatment with chemicals that mimic the hormonal signals associated with plant stress responses, or irradiation with UVR or ionizing radiation, can increase secondary metabolite production in whole plants (Karuppusamy, 2009).

Quantity versus quality

Climate change is become an increasing global problem. Studies have shown that abiotic stressors including drought, heat or cold, soil salinity, increased ozone levels and excess UVR are already causing crop losses globally, and such losses are predicted to increase in the future (Ortiz *et al.*, 2008; Wang and Frei, 2011). In many countries plant breeding programmes are focusing on the production of stress-resistant crops in order to mitigate potential yield losses caused by environmental stresses, and crop plants with reduced requirements for potentially limited resources such as water (Lobell *et al.,* 2008; Ortiz *et al.,* 2008; Wang and Frei, 2011).

While exposure to stress can reduce primary productivity, and hence yield, stress can also influence crop quality, a process about which much less is known. Exposure to abiotic stressors can cause numerous changes to a plants physiology, including changes in photosynthetic rates and the partitioning and translocation of assimilates (Morgan *et al.*, 2004), changes to water and nutrient uptake, usage and translocation (Katerji *et al.,* 2010; Sanchez-Rodriguez *et al.*, 2010; Rivelli *et al.*, 2002), significant biochemical changes, for example increased levels of antioxidants (Apel and Hirt, 2004), and altered gene expression and enzyme activity (Guo *et al.,* 2009; Ramakrishan & Ravishnar, 2011). The end result is major changes in a plants' metabolome (Debnath *et al.,* 2011; Nadella *et al.,* 2012).

For many staple crops, such as rice, wheat, maize (corn), millet and sorghum, and tubers, such as potatoes, cassava and taro, any reduction is yield is undesirable, but for fruit crops a reduction in yield may be acceptable if the quality is higher. Fruits are often consumed for their beneficial effects on health, mainly attributed to the bioactives they contain.

Pre-harvest, during harvest, and postharvest, fruits are often exposed to a range of stresses that can result in increased levels of secondary metabolites (Castellarin *et al.*, 2007; Capanoglu *et al.*, 2008; Capanoglu, 2010). For example, activation of various pathways leading to the accumulation of antioxidants and flavonoids has been associated with plant responses to drought, UVR, high-light, salinity, fluctuating or extreme temperatures, pathogen attack and wounding (Capanoglu, 2010). The manipulation of abiotic stress responses could provide growers with a simple and effective means to increase the bioactive contents, and hence potential marketability of many fruit crops. The main stressors that have potential to be used to increase the levels of fruit bioactives are briefly reviewed here.

Drought and salinity

Water deficit and salt stress are becoming increasing problems that threaten sustainable food production. Drought occurs when the available soil water is reduced to such low levels that plants loose more water than they can take up, hence cells become dehydrated, which results in osmotic stress and a reduction in both cytosolic and vacuolar volumes. Growth under saline conditions also leads to cellular dehydration and the cellular impacts of drought and exposure to moderate salinity are usually similar. Drought tolerance is seen in all plants, but the degree of tolerance varies with species. Tolerance involves numerous stress-induced changes in both primary and secondary metabolism (Krasensky and Jonak, 2012), which can influence the nutritional value of crop plants. For example, when grapevines (*Vitis vinifera* L.) are grown under drought conditions anthocyanin biosynthesis is strongly up-regulated in ripening fruit (Castellarin *et al.*, 2007), as is the total phenolic content of red peppers grown under moderately saline conditions (Navarro *et al.*, 2006).

Changes in temperature

Extreme or rapidly fluctuating temperatures are highly stressful for temperate plants. For example, many plant species adjust their metabolism during autumn and winter, increasing their cellular levels of a range of cryoprotective molecules in order to maximize cold tolerance (Janska *et al.*, 2010). Cold stress, caused by storage at low temperatures, can have positive effects on fruit bioactive levels, with increases in total phenolic levels in oranges and apples (Perez-Ilzarbe *et al.*, 1997; Rapisarda *et al.*, 2008), and anthocyanin levels in strawberries (Josuttis *et al.*, 2010), blueberries (Skrede *et al.*, 2000; Kalt *et al.*, 2001), grapes (Cantos *et al.*, 2000) and pomegranates (Holcroft *et al.*, 1998). However, extended storage at low temperatures often results in a decline in other quality attributes in many fruits (Tomas-Barberan and Espin, 2001). Thermal treatments (42–48°C for 3 hours) have also been used to prevent chilling injury in fruits such as tomato and grapefruit, and postharvest decay of strawberries, but such treatments generally results in reduce levels of bioactives (Tomas-Barberan and Espin, 2001).

Ultraviolet radiation

Ultraviolet radiation (UVR) is an important environmental factor that can stimulate the production of secondary metabolites and is often used in the food industry to reduce microbial growth. UVR between 280 and 320 nm (UV-B) plays an important role in regulating plant secondary metabolism, with relatively low doses triggering the accumulation of a wide range of secondary metabolites, including the flavonoids and other phenolic compounds, and the carotenoids. Phenylalanine ammonia lyase (PAL) and chalcone synthase (CHS) are key enzymes in the flavonoid biosynthetic pathway and the genes encoding these enzymes are well known to be UV-B inducible (Buer *et al.,* 2010).

Ionizing radiation

Postharvest irradiation of fruits and other foods is a technology commonly used to control spoilage and to eliminate microorganisms and insect pests. While pasteurization relies on heat, irradiation relies on the energy associated with ionizing radiation. Irradiation kills bacteria and fungi that could otherwise result in rot or spoilage and is usually supplied by gamma sources such as cobalt-60 or by electron beam generators (Arvanitoyannis *et al.,* 2009). Studies have shown that irradiation can induce oxidative stress in fruits and influence their nutritional composition (Girennavar *et al.,* 2008).

Electric currents

While few studies on the influence of exposure to electric fields on the bioactive contents of fruits have been conducted, exposure of garden cress sprouts and radish plants to different intensities of intermittent-direct-electric-current (IDC) have shown that IDC application can causes increases in the levels bioactives. For example, application of an IDC of 1400 mA to intact plants caused an increase in the total phenol content and other bioactive compounds in garden cress with no damage to the plants (Dannehl *et al.,* 2012). Similar results were found when radish plants were exposed to three different IDC treatments (200 mA, 600 mA and 1000 mA), with the total phenol content, anthocyanin levels and antioxidant capacity of harvested tubers increasing with increasing IDC (Dannehl *et al.,* 2009). Exposure of fruits to IDC is also possible, as studies have been conducted investigating the influence of continuous and IDC electric fields on postharvest respiration and climacteric patterns in apples in order to extend their shelf lives (Atungulu *et al.,* 2003, 2004). In addition, a high-voltage electrostatic field (HVEF) treatment postharvest has been shown to increase the levels of total phenols and antioxidants in tomato fruit (Zhao *et al.,* 2011).

Chemical priming

Numerous studies have demonstrated that plants can be treated with natural or synthetic elicitors, such as salicylic acid (SA), benzo (1,2,3) thiadiazole-7-carbothioicacid S-methyl ester (BTH), β-aminobutyric acid (BABA) or 2,6-dichloroisonicotinic acid (INA), to increase their tolerance to abiotic and biotic stresses (Capanoglu, 2010). This process is often referred to as priming and involves the chemically stimulated accumulation of secondary metabolites that help increase stress tolerance (Conrath, 2009; Capanoglu, 2010). While few studies have investigated chemical priming as a means to increase the levels of bioactives present in fruits, chemical priming is an attractive option, as unlike the abiotic stressors detailed above, spray application is simple and cost effective, with many of chemicals used for priming already used in the field (Ramakrishan and Ravishnar, 2011). Elicitor-treatment could be an effective strategy to enhance the production of plant bioactives, such as alkaloids, terpenoids, flavonoids and other phenolic compounds, both pre- and postharvest (Cao *et al.,* 2009; Conrath, 2009).

THE INFLUENCE OF ENVIRONMENTAL STRESSORS AND OTHER TREATMENTS ON THE LEVELS OF BIOACTIVES IN SOME COMMERCIALLY IMPORTANT FRUITS

Blueberries and raspberries

In recent years the health-promoting properties of small berry fruits has received increased attention due to their potential to promote human health and help to prevent some cancers, heart disease and neurodegenerative disorders (Seeram, 2010). In addition to containing high levels of vitamins, minerals and dietary fibre, small berry fruits often contain high levels of a diverse array of bioactives (Seeram, 2010). Many small berry fruits have high concentrations of anthocyanins, up to four times more than non-berry fruits, 10 times more than vegetables and 40 times more than cereals (Halvorsen *et al.*, 2002). Both pre- and postharvest treatments and cultural practices can be used to increase the levels of bioactives found in both blueberries and raspberries.

Blueberries (*Vaccinium sp.*) are rich in antioxidant compounds (Wu *et al.*, 2004), containing high levels of phenolics, in particular anthocyanins (Kalt *et al.*, 2001). It has been suggested that the consumption of wild blueberries (*Vaccinium angustifolium* Ait.) reduces the risk of cancer (Smith *et al.*, 2000) and stroke (Sweeney *et al.*, 2002), and help maintain cardiovascular health (Norton *et al.*, 2005). A major aim in blueberry breeding programmes is to enhance tolerance to winter freezing and spring frosts. To this end, studies have been conducted using cDNA microarrays to monitor changes in gene expression during cold acclimation of blueberry plants under field and cold room conditions (Dhanaraj *et al.*, 2007; Naik *et al.*, 2007). These studies have showed that genes associated with stress tolerance are rapidly up-regulated to very high levels in stressed plants and demonstrate the potential for short duration cold treatments, and/or endogenous growth regulator treatments that up-regulated genes associated with cold tolerance, to increase bioactive levels in fruits. In a study of the *V. angustifolium* Percival and MacKenzie (2007) used plant growth regulators to increase the polyphenolic contents and antioxidant properties of fruit. Four plant growth regulators (methyl jasmonate, ABA, riboflavin and Retain®) were applied pre-harvest to maturing berries and the berries were harvested three weeks after treatment. The application of methyl jasmonate resulted in a small increase in the levels of bioactives, but at the expense of harvestable yields. The authors suggested while some plant growth regulator treatments may increase bioactive levels, the value of this must be assessed against possible yield reductions.

Postharvest treatment of blueberries with UV-C (2.15, 4.30, or 6.45 kJ/m^2) has been shown to be a more effective way to increase bioactive levels in fruit, with increases in resveratrol, myricetin-3-arabinoside, quercetin-3-galactoside, quercetin-3-glucoside, kaempferol-3-glucuronide, delphinidin-3-galactoside, cyanidin-3-galactoside, delphinidin-3-arabinoside, petunidin-3-galactoside, petunidin-3-glucoside, petunidin-3-arabinoside, malvidin-3-galactoside, malvidin-3-arabinoside and chlorogenic acid, and a significantly higher antioxidant capacity found in UV-C treated fruit compared to control fruit (Wang *et al.*, 2009).

Raspberries (plant species of genus *Rubus*, mostly of subgenus *Idaeobatus*) are highly valued by European and North American consumers. Despite their high commercial value raspberries have relatively low levels of bioactive phenolics compared to other berry fruits and breeders and growers are working to add further value to this fruit using both breeding and modified management practices. Kassim *et al.* (2009) investigated the influence of seasonal climatic variables and growth conditions on the levels of eight anthocyanins, cyanidin-3-sophoroside (C3S), cyanidin-3-glucoside (C3G), cyanidin-3-glucosylrutinoside (C3GR), cyanidin-3-rutinoside (C3R), pelargonidin-3-sophoroside (P3S), pelargonidin-3-glucoside (P3G), pelargonidin-3-glucosylrutinoside (P3GR) and pelargonidin-3-rutinoside (P3R), in progeny of a cross between two Rubus subspecies, *Rubus idaeus* (cv. Glen Moy) and *Rubus strigosus* (cv. Latham). While differences in anthocyanin levels between fruit from plants grown in the field versus polytunnel houses were low, significant seasonal variations in anthocyanin levels were observed. Fruit harvested from the 2006 growing season, which was hotter, drier and had more hours of sunshine hours than the 2007 season, contained much higher levels of anthocyanins. These results suggest that a more stressful pre-harvest environment could result in increased levels of bioactive anthocyanins. Further evidence of the importance of pre-harvest stress on the levels of bioactives in raspberries comes from a study on the influence of salinity on bioactive levels. Neocleous and Vasilakakis (2008) showed that a mild salinity treatment, increased the antioxidant capacity, vitamin C content and total phenolic content of fruit.

Citrus fruits

Citrus fruits contain a wide range of bioactive phytochemicals, including high concentrations of vitamin C, the phenolic acids caffeic acid, q-coumaric acid, ferulic acid, and sinapic acid, the flavonoids narirutin, naringin, eriocitrin, neoeriocitrin, rutin, diosmin, poncirin and nobiletin, and the flavanone hesperidin (Jourdan *et al.*, 1985; Kawaii *et al.*, 1999). In addition, numerous carotenoids have been identified in citrus fruits, including lutein, zeaxanthin and β-cryptoxanthin, with the orange colour of the peel and flesh largely due to carotenoids (Mangels *et al.*, 1993) and the red coloration of Red Navel and Valencia oranges mainly due to the presence of lycopene and cryptoxanthin, respectively (Lee, 2001). Many citrus fruits also contain triterpene limonoids (Ejaz *et al.*, 2006). Attempts to increase the levels of bioactives found in commercially important citrus crops have largely involved postharvest treatments, as prolonged postharvest storage of citrus fruits is often required when fruits are shipped long distances, and fruit are often chemically treated or irradiated to control insect pests or prevent fungal rots. The effects of postharvest treatment and storage on the nutritional value of citrus fruits have been widely studied and there is some potential for the careful use of such treatments to increase the bioactive contents of certain fruits.

Grapefruit (*Citrus* × *paradisi*) are rich in vitamin C, flavonoids, carotenoids, furocoumarins and limonoids (Girennavar *et al.*, 2008). Girennavar *et al.* (2008) investigated the influence of electron-beam (E-beam) irradiation (0, 1.0, 2.5, 5.0 and 10.0 kGy) on the bioactive compounds found in 'Rio Red' and 'Marsh White' grapefruits. They found the vitamin C content was unchanged at 1 kGy, but decreased with higher radiation doses. Lycopene levels decreased slightly, while β-carotene levels increased slightly, with increasing dose. In contrast, the flavoniod naringin increased significantly, compared to controls, at 10 kGy in both cultivars. These results demonstrate that low dose irradiation can induce changes in the levels of grapefruit bioactives, but the overall health benefit to the consumer may be small.

Lemons (*Citrus* × *limon*) are grown worldwide with the fruit's juice, pulp and peel both consumed fresh and used in the manufacture of foods. Interdonato *et al.* (2011) studied the influence of postharvest exposure of lemon fruits (*Citrus limon*, cv. Limoneira 8A) to UV-B on soluble carbohydrate metabolism and secondary metabolite accumulation. Fruits were harvested in winter, when natural UV-B levels are low, and treated with 0.43 W/m^2 (22 $kJ/m^2/d$ UV-BBE) of UV-B radiation for various times up to five minutes and then stored at

25°C for 24 hours. Peel and flesh samples were analysed separately and the data obtained showed that two or three minutes of UV-B exposure increased the levels of UV-B absorbing compounds, mostly flavonoids, and total phenolics in peels without causing any detrimental changes in peel colour compared with non-irradiated lemons. Exposure for more than three minutes did not result in a further increase in the levels of UV-B absorbing or phenolic compounds.

The sweet orange (*Citrus sinensis* L) is the most commonly grown tree fruit globally. Oranges are eaten whole, processed for juice and also used in the production of processed foods. Dubery (1992) induced oxidative stress in citrus peel by exposure to a gamma radiation dose of 2 kGy. While this treatment did induce the formation of small necrotic lesions, a transient increase in PAL activity was observed, which was accompanied by the accumulation of phenols, including coumarins and flavanols. Due to a growing demand to reduce the postharvest use of pesticides and concerns associated with gamma irradiated foods, oranges are increasingly being heat treated, 48 hour at 37°C and 90% relative humidity, followed by 24 hours at 20°C, before storage at 5°C (Perotti *et al.,* 2011). In a study of the proteome and the biochemistry of *Citrus sinensis* cv Valencia late, Perotti *et al.* (2011) found that heat treatment increased the levels of antioxidant enzymes, with a large increase in superoxide dismutase (SOD) observed in both the flavedo and juice sacs. While not classified as a traditional bioactive, recent research suggests that fruit juices and/or extracts rich in SOD have potential health benefits, slowing the development of atherosclerosis and liver steatosis (Decorde *et al.,* 2010). The results of these studies clearly show the potential for postharvest treatments as a means to manipulate the levels of bioactives in citrus fruits.

Grapes

Grapes (*Vitis vinifera* L.) are one of the most important fruit crops globally, being eaten raw or used to make jams, juice and wine. Berry quality is highly dependent upon the environmental conditions in which the plants are grown and the berries mature, with high light, temperature, UVR and drought all potentially influencing secondary metabolite composition and quantity (Downey *et al.,* 2006). Manipulation of certain environmental factors pre-harvest could, therefore, be used to increase the levels of health-promoting bioactives. For example, Deluc *et al.* (2009) conducted a study on the effects of long-term water deficit on the berries of Cabernet Sauvignon, a red wine grape, and Chardonnay, a white wine grape. They found that water deficit affected the phenylpropanoid, ABA, isoprenoid, carotenoid, amino acid and fatty acid metabolic pathways. In Cabernet Sauvignon increased anthocyanin levels were observed, while in Chardonnay increased levels of antheraxanthin, flavonols and aroma volatiles were detected. In addition, Deluc *et al.* (2011) found that water deficit increased stilbene metabolism in Cabernet Sauvignon berries. These changes in metabolism not only influence berry flavour and wine quality, but could also contribute to increased human-health benefits.

Both visible light and UVR exposure, both of which can easily be manipulated in the field by the selective removal of leaves from around bunches of berries, have also been shown to influence the levels of grape bioactives. Price *et al.* (1995) compared the anthocyanin and flavonol contents of Pinot Noir berry skins from sun exposed and shaded berries, from the same clusters. While the anthocyanin content was not influenced by sun exposure, levels of quercetin glycoside were ten times higher in sun-exposed skin compared to shaded skin. Wines made from Pinot Noir berries either shaded, moderately sun exposed or highly sun exposed, contained 4.5, 14.8, and 33.7 mg/l of quercetin glycosides respectively. The levels of quercetin aglycone were also found to increase with sun exposure and the authors concluded that the primary factor determining quercetin levels in grapes and wine was sun exposure, a finding now supported by several other studies (Cortell and Kennedy, 2006; Downey *et al.*, 2006, 2008).

Grapes are also often subjected to postharvest treatments. For example, postharvest withering or wilting of grape berries is used to increase the concentration of simple sugars in berries used for the production of dessert and fortified wines (Zamboni *et al.*, 2008). This treatment induces a dehydration stress response and increases the expression of genes involved in drought stress protection, including genes associated the phenolic and terpene compound pathways (Zamboni *et al.*, 2008). Other studies have also shown that postharvest dehydration increases the synthesis of selected phenolic compounds including the trihydroxystilbene resveratrol (Versari *et al.*, 2001). In addition, postharvest exposure of grapes berries to UVR has been shown to have similar effects on bioactive contents as dehydration (Cantos *et al.*, 2000; Versari *et al.*, 2001; Guerrero *et al.*, 2010), as have short anoxic treatments (Jimenez *et al.*, 2007). Both pre- and postharvest abiotic stress treatments have great potential for manipulating the bioactive contents of grapes.

Strawberries

Strawberries (*Fragaria ananassa* Duch) are a commercially important fruit, widely consumed fresh or processed to produce jams, juices and jellies (Giampieri *et al.*, 2012). Strawberries have relatively high levels of bioactive compounds, including vitamin C and folate, and contain more than 40 phenolic compounds, including anthocyanins, proanthocyanidins, ellagitannins, phenolic acids and flavonols (Proteggente *et al.*, 2002), giving fresh strawberries and extracts relatively high antioxidant capacities (Wang and Lin, 2000; Scalzo *et al.*, 2005). Because of their economic importance the nutritional value of strawberries has been studied extensively and while cultivars differ greatly in their phenolic composition (Buendia *et al.*, 2010), environmental factors such as drought, salinity, temperature and UVR can all greatly influence the levels of strawberry bioactives (Wang and Zen 2001; Terry *et al.*, 2007; Keutgen and Pawelzik, 2007, 2008a, 2008b, 2008c; Crespo *et al.*, 2010; Josuttis *et al.*, 2010, 2011). Therefore, the potential exists for bioactive levels to be maximized using simple management strategies.

For example, in a study of the influence of pre-harvest cultivation temperature on the levels of bioactive compounds in strawberry fruit (cv. Elsanta), Josuttis *et al.* (2011) grew plants under three temperature treatments: cool (day/night: 16/12°C), warm (day/night: 22/16°C) and warm with soil heating to 26°C during the day. While increasing the air or the soil temperature decreased the average fruit weight, fruit quality attributes (dry matter and soluble solids contents, and titratable acidity) were not significantly influenced. With respect to bioactivity, strawberries from the cool environment had a lower antioxidant capacity than strawberries produced in the warmer environment, due to lower concentrations of total monomeric anthocyanins, total pro-anthocyanidins, total ellagitannins and kaempferol 3-glucoside-malonate. However, the ascorbic acid levels in fruits grown in the cooler environment were higher than those from the warmer environment. Increased soil temperature had little additional influence on the accumulation of fruit phenolics, except for *p*-coumaroyl glucose, and the concentrations of pro-anthocyanidins, which declined. These results show the importance of growth temperatures on the phenolic content of strawberry fruit.

Josuttis *et al.* (2010) grew *F. ananassa* cvs. Everest and Elsanta plants in tunnel houses covered with films that blocked UV-B or transmitted 70% of UV-B, as well as under open-field conditions and evaluated the nutritional and quality parameters of the harvested fruits. While pre-harvest exposure to UV-B did not influence the total anthocyanin or ascorbate contents, or the total antioxidant capacity, an HPLC analysis of extracted phenolics showed that individual phenolics were affected by the presence or absence of UV-B. The content of the anthocyanin cyanidin 3-glucoside and the flavonols quercetin 3-glucuronide and kaempferol 3-glucoside decreased in the fruits grown under UV blocking films, compared to those grown in the open-field. The authors suggested that in order for tunnel house grown strawberries to retain the bioactive properties of field grown strawberries tunnel houses should be glad with UV-B transparent plastics.

In addition to temperature and UV-B, salinity and drought have also been shown to influence the bioactive levels in strawberry fruit. Keutgen and Pawelzik (2007, 2008a, 2008b, 2008c) exposed two strawberry cultivars to pre-harvest salt stress (0, 40 or 80 mmol NaCl/l in the nutrient solution) for four months. While there was a reduction in mean fruit weight, dry matter and in the total soluble carbohydrates, salt stress significantly increased the ascorbic acid, anthocyanin and SOD levels in both cultivars. In addition, the contents of free and essential amino acids and minerals also increased. An increase in reduced glutathione and a better fruit taste was also observed for one cultivar, but not the other. Drought has also been shown to have a positive influence on the content of bioactives in strawberry fruits. Fruit from strawberry plants grown under mild water deficit generally had increased monosaccharide levels, sugar/acid ratios and higher levels of total phenolics, and hence a greater antioxidant capacity than watered controls plants (Terry *et al.*, 2007).

Tomatoes

Tomatoes (*Solanum lycopersicum*) are one of the most widely cultivated fruits, either grown in fields or in greenhouses in colder parts of the world. Most fruits are relatively high in vitamin C, lycopene, carotene and other antioxidants, but varieties have been breed with higher than normal amounts of lycopene (Lenucci *et al.*, 2012), and purple anthocyanin producing cultivars have also been developed (Li *et al.*, 2011).

Both pre- and postharvest stressors have been shown to increase the levels of bioactives found in tomatoes and could readily be used to produce fruit with greater health promoting properties. Both moderate salinity and dehydration stress have been shown to significantly increase the levels of bioactives in tomatoes. Sgherri *et al.* (2007) found that the fruits of standard sized tomatoes harvested from plants treated with diluted seawater had increased levels of ascorbic acid, lipoic acid, and α-tocopherol, but that the effect was genotype specific, with some cultivars showing reduced levels. The leaves of cherry tomato plants treated with dilute seawater also have been shown to contained higher levels of bioactives, with increased levels of ascorbic acid, lipoic acid, tocopherols and phenolic acids when compared to control plants (Sgherri *et al.*, 2008), although in this study fruit levels were not determined. In a similar study, Borghesi *et al.* (2011) investigated the influence of salinity on carotenoid and anthocyanin levels in one non-anthocyanin accumulating and three anthocyanin accumulating tomato genotypes. They found that salinity had mixed effects, increasing carotenoid levels in some cultivars compared to controls and not in others. Anthocyanins levels in the anthocyanin accumulating genotypes also showed variable responses to salinity, with a twofold increase in total anthocyanins in fruits of the cultivar Sun Black and a tenfold reduction in fruits of another cultivar. The authors suggested specific cultivars of tomatoes could be grown on saline soils to produce fruit with higher levels of bioactives.

In addition to pre-harvest stress, exposure to stress postharvest has also been shown to increase the levels of bioactives in the fruits of tomato plants. Jagadeesh *et al.* (2011)

exposed mature green tomato fruit to a dose of UV-C (3.7 kJ/m^2) and compared the levels of bioactive in UV-C treated fruits with untreated fruits. Fruit was stored at 13°C and 95% relative humidity for 10, 20, and 30 days and transferred to room temperature (23°C) to allow ripening. Ascorbic acid and total the phenolic content was higher in the UV-C treated fruit, but the lycopene content was less. Postharvest thermal processing has also been shown to increase the bioaccessible lycopene content and total antioxidant activity of tomato fruit (Dewanto *et al.*, 2002). For tomatoes exposure to pre- and postharvest abiotic stressors has the potential to increase the bioactive contents of fruit, but cultivar selection is critical.

Other fruits

Numerous others studies have also shown the potential of pre- or postharvest abiotic stress to increase the levels of bioactives in fruit. For example, Schmitz-Eiberger and Blanke (2012) grew five early maturing cherry (*Prunus avium L.*) varieties under transparent plastic covers that reduced incident radiation and increased daytime temperatures without additional heating. The aim of the study was to produce fruit that would be ripe before field-grown German fruit became available. However, the authors found that not only could fruit be produced earlier but that the fruit produced were larger, and the fruit of some cultivars had significantly higher anthocyanin, ascorbic acid and total phenolic contents. The authors attributed the increase in bioactives to heat stress.

Mangos (*Mangifera indica* L) are tropical fruits, high it vitamin C, carotenoids, quercetin, kaempferol, gallic acid, caffeic acid, catechins, tannins and the xanthonoid, mangiferin (Singh *et al.*, 2004), but fruit are very susceptible to microbial action during storage, so postharvest UV-C irradiation is commonly used to increase the shelf life. Gonzalez-Aguilar *et al.* (2007) found that UV-C treatment not only reduced fruit losses due to microbial action but that treated fruit had higher total phenol and flavonoid levels than untreated fruit. Other treatments used to control potential microbial action in mangos are salicylic acid (SA) and ionizing radiation. While SA has been shown to increase the activity of PAL in mango fruits (Zeng *et al.*, 2006), no measurements of flavonoid levels in SA treated fruits were taken; however, as PAL activity increased there is the potential for SA priming to increase flavonoid levels. Reyes and Cisneros-Zevallos (2007) treated harvested mango fruit with electron-beam ionizing radiation and found that this treatment caused an increase in flavonol levels during storage.

In addition to the above, postharvest heat treatments, visible light and UV-R have been shown to be effective methods for increasing the bioactive contents of a wide range of other fruits including apples and pears (Hagen *et al.*, 2007; Zhang *et al.*, 2012). From the studies detailed above it appears that pre- and/or postharvest stress treatments have the potential to be simple and effective ways to increase the levels of bioactives found in fruit. However, no single stressor appears to work for all fruit crops and significant differences in responses between cultivars are highly likely.

XENOHORMESIS: AN ANCIENT RESPONSE WITH MODERN APPLICATIONS

The xenohormesis hypothesis as proposed by Howitz and Sinclair (2008) attempts to explain why many phytochemicals influence mammalian physiology in ways that promote health and well-being. Howitz and Sinclair (2008) proposed that animals (and fungi) have co-evolved with plants in such a way that they are able to sense the bioactive compounds produced by environmentally stressed plants and that these molecules act as warning signals of potentially detrimental environmental conditions, thus allowing animals to prepare before the environment becomes too unfavourable. Hooper *et al.* (2010) extended the original hypothesis and suggested that animals have evolved to 'piggyback off the products of plants' sophisticated stress responses' and that bioactive compounds, when consumed by animals, confer stress resistance and survival benefits. If the concept of xenohormesis is true, more research on the phytochemicals contained within the fruits of plants adapting to stressful environments should lead to the discovery of more health-promoting plant bioactives. In addition, in a world undergoing climate change a better understanding of the effects of abiotic stressors on the nutritional value and bioactive contents of plant produce could lead to the adaption of crop management strategies, such as planting of specific crop plants in saline soils, that could maximize bioactive concentrations and shift the focus of crop production, at least in part, to quality rather that quantity. In addition, the development of smart postharvest technologies that activate and use plant stress responses to increase the levels of bioactives should also enable the production of plant-based products with even greater heath promoting benefits.

REFERENCES

Ames, B.N., Shigenaga, M.K. and Hagen, T.M. (1993) Oxidants, antioxidants, and the degenerative diseases of aging. *Proceedings of the National Academy of Sciences of the United States of America*, **90**, 7915–7922.

Apel, K. and Hirt, H. (2004) Reactive oxygen species: Metabolism, oxidative stress, and signal transduction. *Annual Review of Plant Biology*, **55**, 373–399.

Arrigoni, O. and De Tullio, M.C. (2002) Ascorbic acid: Much more than just an antioxidant. *Biochimica Et Biophysica Acta – General Subjects*, **1569**, 1–9.

Arvanitoyannis, I.S., Stratakos, A.C. and Tsarouhas, P. (2009) Irradiation applications in vegetables and fruits: A review. *Critical Reviews in Food Science and Nutrition*, **49**, 427–462.

Attoub, S., Hassan, A.H., Vanhoecke, B. *et al.* (2011) Inhibition of cell survival, invasion, tumor growth and histone deacetylase activity by the dietary flavonoid luteolin in human epithelioid cancer cells. *European Journal of Pharmacology*, **651**, 18–25.

Atungulu, G., Nishiyama, Y. and Koide, S. (2003) Use of an electric field to extend the shelf life of apples. *Biosystems Engineering*, **85**, 41–49.

Atungulu, G., Nishiyama, Y. and Koide, S. (2004) Respiration and climacteric patterns of apples treated with continuous and intermittent direct current electric field. *Journal of Food Engineering*, **63**, 1–8.

Beutner, S., Bloedorn, B., Frixel, S. *et al.* (2001) Quantitative assessment of antioxidant properties of natural colorants and phytochemicals: Carotenoids, flavonoids, phenols and indigoids. The role of carotene in antioxidant functions. *Journal of the Science of Food and Agriculture*, **81**, 559–568.

Bhattacharya, A., Sood, P. and Citovsky, V. (2010) The roles of plant phenolics in defence and communication during agrobacterium and rhizobium infection. *Molecular Plant Pathology*, **11**, 705–719.

Borghesi, E., Gonzalez-Miret, M.L., Escudero-Gilete, M.L. *et al.* (2012) Effects of salinity stress on carotenoids, anthocyanins, and color of diverse tomato genotypes. *Journal of Agricultural and Food Chemistry*, **59**, 11676–11682.

Borriello, A., Cucciolla, V., Della Ragione, F. and Galletti, P. (2010) Dietary polyphenols: Focus on resveratrol, a promising agent in the prevention of cardiovascular diseases and control of glucose homeostasis. *Nutrition Metabolism and Cardiovascular Diseases*, **20**, 618–625.

Brigelius-Flohe, R. and Traber, M.G. (1999) Vitamin e: Function and metabolism. *FASEB Journal*, **13**, 1145–1155.

Buendia, B., Gil, M.I., Tudela, J.A. *et al.* (2010) HPLC-MS analysis of proanthocyanidin oligomers and other phenolics in 15 strawberry cultivars. *Journal of Agricultural and Food Chemistry*, **58**, 3916–3926.

Buer, C.S., Imin, N. and Djordjevic, M.A. (2010) Flavonoids: New roles for old molecules. *Journal of Integrative Plant Biology*, **52**, 98–111.

Burton-Freeman, B. (2010) Postprandial metabolic events and fruit-derived phenolics: A review of the science. *British Journal of Nutrition*, **104**, S1–S14.

Cantos, E., Garcia-Viguera, C., De Pascual-Teresa, S. and Tomas-Berberan, F.A. (2000) Effect of postharvest ultraviolet irradiation on resveratrol and other phenolics of cv. Napoleon table grapes. *Journal of Agricultural and Food Chemistry*, **48**, 4606–4612.

Cao, S.F., Hu, Z.C. and Wang, H.O. (2009) Effect of salicylic acid on the activities of antioxidant enzymes and phenylalanine ammonia-lyase in cucumber fruit in relation to chilling injury. *Journal of Horticultural Science and Biotechnology*, **84**, 125–130.

Capanoglu, E. (2010) The potential of priming in food production. *Trends in Food Science and Technology*, **21**, 399–407.

Capanoglu, E., Beekwilder, J., Boyacioglu, D. *et al.* (2008) Changes in antioxidant and metabolite profiles during production of tomato paste. *Journal of Agricultural and Food Chemistry*, **56**, 964–973.

Castellarin, S.D., Pfeiffer, A., Sivilotti, P. *et al.* (2007) Transcriptional regulation of anthocyanin biosynthesis in ripening fruits of grapevine under seasonal water deficit. *Plant Cell and Environment*, **30**, 1381–1399.

Cazzonelli, C.I. and Pogson, B.J. (2010) Source to sink: Regulation of carotenoid biosynthesis in plants. *Trends in Plant Science*, **15**, 266–274.

Chalabi, N., Delort, L., Le Corre, L. *et al.* (2006) Gene signature of breast cancer cell lines treated with lycopene. *Pharmacogenomics*, **7**, 663–672.

Cheng, J.-C., Dai, F., Zhou, B. *et al.* (2007) Antioxidant activity of hydroxycinnamic acid derivatives in human low density lipoprotein: Mechanism and structure-activity relationship. *Food Chemistry*, **104**, 132–139.

Clouse, S.D. and Sasse, J.M. (1998) Brassinosteroids: Essential regulators of plant growth and development. *Annual Review of Plant Physiology and Plant Molecular Biology*, **49**, 427–451.

Collett, N.P., Amin, A.R.M.R., Bayraktar, S. *et al.* (2010) Cancer prevention with natural compounds. *Seminars in Oncology*, **37**, 258–281.

Conrath, U. (2009) Priming of induced plant defense responses. Plant innate immunity. *Advances in Botanical Research*, **51**, 361–395.

Cortell, J.M. and Kennedy, J.A. (2006) Effect of shading on accumulation of flavonoid compounds in (*Vitis vinifera* L.) pinot noir fruit and extraction in a model system. *Journal of Agricultural and Food Chemistry*, **54**, 8510–8520.

Cravotto, G., Boffa, L., Genzini, L. and Garella, D. (2010) Phytotherapeutics: An evaluation of the potential of 1000 plants. *Journal of Clinical Pharmacy and Therapeutics*, **35**, 11–48.

Crespo, P., Bordonaba, J.G., Terry, L.A. and Carlen, C. (2010) Characterisation of major taste and health-related compounds of four strawberry genotypes grown at different Swiss production sites. *Food Chemistry*, **122**, 16–24.

Dai, J. and Mumper, R.J. (2010) Plant phenolics: Extraction, analysis and their antioxidant and anticancer properties. *Molecules*, **15**, 7313–7352.

Dannehl, D., Huyskens-Keil, S., Eichholz, I. *et al.* (2009) Effects of intermittent-direct-electric-current (idc) on polyphenols and antioxidant activity in radish (*Raphanus sativus* L.) during growth. *Journal of Applied Botany and Food Quality – Angewandte Botanik*, **83**, 54–59.

Dannehl, D., Huyskens-Keil, S., Wendorf, D. *et al.* (2012) Influence of intermittent-direct-electric-current (IDC) on phytochemical compounds in garden cress during growth. *Food Chemistry*, **131**, 239–246.

De Oliveira, D.M. and Bastos, D.H.M. (2011) Phenolic acids bioavailability. *Quimica Nova*, **34**, 1051–1056.

Debnath, M., Pandey, M. and Bisen, P.S. (2011) An omics approach to understand the plant abiotic stress. *Omics – a Journal of Integrative Biology*, **15**, 739–762.

Decorde, K., Ventura, E., Lacan, D. *et al.* (2010) An SOD rich melon extract extramel (r) prevents aortic lipids and liver steatosis in diet-induced model of atherosclerosis. *Nutrition Metabolism and Cardiovascular Diseases*, **20**, 301–307.

DellaPenna, D. (2005) Progress in the dissection and manipulation of vitamin e synthesis. *Trends in Plant Science*, **10**, 574–579.

Delmas, D., Solary, E. and Latruffe, N. (2011) Resveratrol, a phytochemical inducer of multiple cell death pathways: Apoptosis, autophagy and mitotic catastrophe. *Current Medicinal Chemistry*, **18**, 1100–1121.

Deluc, L.G., Quilici, D.R., Decendit, A. *et al.* (2009) Water deficit alters differentially metabolic pathways affecting important flavor and quality traits in grape berries of cabernet sauvignon and chardonnay. *BMC Genomics*, **10**, 212 (doi: 10.1186/1471-2164-10-212).

Deluc, L.G., Decendit, A., Papastamoulis, Y. *et al.* (2011) Water deficit increases stilbene metabolism in Cabernet Sauvignon berries. *Journal of Agricultural and Food Chemistry*, **59**, 289–297.
Devaraj, S. and Jialal, I. (2006) The role of dietary supplementation with plant sterols and stanols in the prevention of cardiovascular disease. *Nutrition Reviews*, **64**, 348–354.
Dewanto, V., Wu, X.Z., Adom, K.K. and Liu, R.H. (2002) Thermal processing enhances the nutritional value of tomatoes by increasing total antioxidant activity. *Journal of Agricultural and Food Chemistry*, **50**, 3010–3014.
Dhanaraj, A.L., Alkharouf, N.W., Beard, H.S. *et al.* (2007) Major differences observed in transcript profiles of blueberry during cold acclimation under field and cold room conditions. *Planta*, **225**, 735–751.
Dicosmo, F. and Misawa, M. (1985) Eliciting secondary metabolism in plant-cell cultures. *Trends in Biotechnology*, **3**, 318–322.
Downey, M.O., Dokoozlian, N.K. and Krstic, M.P. (2006) Cultural practice and environmental impacts on the flavonoid composition of grapes and wine: A review of recent research. *American Journal of Enology and Viticulture*, **57**, 257–268.
Downey, M.O., Harvey, J.S. and Robinson, S.P. (2008) The effect of bunch shading on berry development and flavonoid accumulation in Shiraz grapes. *Australian Journal of Grape and Wine Research*, **10**, 53–65.
Dubery, I.A. (1992) Elicitation of enhanced phenylpropanoid metabolism in citrus flavedo by gamma-radiation. *Phytochemistry*, **31**, 2659–2662.
Ejaz, S., Ejaz, A., Matsuda, K. and Lim, C.W. (2006) Limonoids as cancer chemopreventive agents. *Journal of the Science of Food and Agriculture*, **86**, 339–345.
Fernandez-Garcia, E., Carvajal-Lerida, I., Jaren-Galan, M. *et al.* (2012) Carotenoids bioavailability from foods: From plant pigments to efficient biological activities. *Food Research International*, **46**, 438–450.
Firuzi, O., Miri, R., Tavakkoli, M. and Saso, L. (2011) Antioxidant therapy: Current status and future prospects. *Current Medicinal Chemistry*, **18**, 3871–3888.
Foyer, C.H. and Noctor, G. (2012) Managing the cellular redox hub in photosynthetic organisms. *Plant Cell and Environment*, **35**, 199–201.
Genser, B., Silbernagel, G., De Backer, G. *et al.* (2012) Plant sterols and cardiovascular disease: A systematic review and meta-analysis(dagger). *European Heart Journal*, **33**, 444–U28.
Giampieri, F., Tulipani, S., Alvarez-Suarez, J.M. *et al.* (2012) The strawberry: Composition, nutritional quality, and impact on human health. *Nutrition*, **28**, 9–19.
Girennavar, B., Jayaprakasha, G.K., Mclin, S.E. *et al.* (2008) Influence of electron-beam irradiation on bioactive compounds in grapefruits (*Citrus paradisi* macf.). *Journal of Agricultural and Food Chemistry*, **56**, 10941–10946.
Gonzalez-Aguilar, G.A., Zavaleta-Gatica, R. and Tiznado-Hernandez, M.E. (2007) Improving postharvest quality of mango 'Haden' by UV-C treatment. *Postharvest Biology and Technology*, **45**, 108–116.
Gonzalez-Gallego, J., Garcia-Mediavilla, M.V., Sanchez-Campos, S. and Tunon, M.J. (2010) Fruit polyphenols, immunity and inflammation. *British Journal of Nutrition*, **104**, S15–S27.
Guerrero, R.F., Puertas, B., Fernandez, M.I. *et al.* (2010) Induction of stilbenes in grapes by UV-C: Comparison of different subspecies of vitis. *Innovative Food Science and Emerging Technologies*, **11**, 231–238.
Guo, J., Han, W. and Wang, M.H. (2008) Ultraviolet and environmental stresses involved in the induction and regulation of anthocyanin biosynthesis: A review. *African Journal of Biotechnology*, **7**, 4966–4972.
Guo, W.M., Kong, E. and Meydani, M. (2009) Dietary polyphenols, inflammation, and cancer. *Nutrition and Cancer – an International Journal*, **61**, 807–810.
Hagen, S.F., Borge, G.I.A., Bengtsson, G.B. *et al.* (2007) Phenolic contents and other health and sensory related properties of apple fruit (*Malus domestica* borkh., cv. Aroma): Effect of postharvest UV-B irradiation. *Postharvest Biology and Technology*, **45**, 1–10.
Halvorsen, B.L., Holte, K., Myhrstad, M.C.W. *et al.* (2002) A systematic screening of total antioxidants in dietary plants. *Journal of Nutrition*, **132**, 461–471.
Han, J.H., Yang, Y.X., Feng, M.Y. and Wang, G.D. (2009) The analysis of phytosterols contents in plant food and the estimation of intake in chinese people. *Annals of Nutrition and Metabolism*, **55**, 595–595.
Hansen, A.S., Marckmann, P., Dragsted, L.O. *et al.* 2005. Effect of red wine and red grape extract on blood lipids, haemostatic factors, and other risk factors for cardiovascular disease. *European Journal of Clinical Nutrition*, **59**, (3) 449–455.
Hartmann, M.A. (1998) Plant sterols and the membrane environment. *Trends in Plant Science*, **3**, 170–175.
Heim, K.E., Tagliaferro, A.R. and Bobilya, D.J. (2002) Flavonoid antioxidants: Chemistry, metabolism and structure-activity relationships. *Journal of Nutritional Biochemistry*, **13**, 572–584.

Hernandez, I., Alegre, L., Van Breusegem, F. and Munne-Bosch, S. (2009) How relevant are flavonoids as antioxidants in plants? *Trends in Plant Science*, **14**, 125–132.

Hodges, D.M. and Tolvonen, P.M.A. (2008) Quality of fresh-cut fruits and vegetables as affected by exposure to abiotic stress. *Postharvest Biology and Technology*, **48**, 155–162.

Hofius, D., Hajirezaei, M.R., Geiger, M. *et al.* (2004) RNAi-mediated tocopherol deficiency impairs photoassimilate export in transgenic potato plants. *Plant Physiology*, **135**, 1256–1268.

Holcroft, D.M., Gil, M.I. and Kader, A.A. (1998) Effect of carbon dioxide on anthocyanins, phenylalanine ammonia lyase and glucosyltransferase in the arils of stored pomegranates. *Journal of the American Society for Horticultural Science*, **123**, 136–140.

Hooper, P.L., Hooper, P.L., Tytell, M. and Vigh, L. (2010) Xenohormesis: Health benefits from an eon of plant stress response evolution. *Cell Stress and Chaperones*, **15**, 761–770.

Hou, D.X. and Kumamoto, T. (2010) Flavonoids as protein kinase inhibitors for cancer chemoprevention: Direct binding and molecular modeling. *Antioxidants and Redox Signaling*, **13**, 691–719.

Howitt, C.A. and Pogson, B.J. (2006) Carotenoid accumulation and function in seeds and non-green tissues. *Plant Cell and Environment*, **29**, 435–445.

Howitz, K.T. and Sinclair, D.A. (2008) Xenohormesis: Sensing the chemical cues of other species. *Cell*, **133**, 387–391.

Interdonato, R., Rosa, M., Nieva, C.B. *et al.* (2011) Effects of low UV-B doses on the accumulation of UV-B absorbing compounds and total phenolics and carbohydrate metabolism in the peel of harvested lemons. *Environmental and Experimental Botany*, **70**, 204–211.

Jagadeesh, S.L., Charles, M.T., Gariepy, Y. *et al.* (2011) Influence of postharvest UV-C hormesis on the bioactive components of tomato during post-treatment handling. *Food and Bioprocess Technology*, **4**, 1463–1472.

Janska, A., Marsik, P., Zelenkova, S. and Ovesna, J. (2010) Cold stress and acclimation – what is important for metabolic adjustment? *Plant Biology*, **12**, 395–405.

Jimenez, J.B., Orea, J.M., Urena, A.G. *et al.* (2007) Short anoxic treatments to enhance trans-resveratrol content in grapes and wine. *European Food Research and Technology*, **224**, 373–378.

Jones, P.J.H. and Abumweis, S.S. (2009) Phytosterols as functional food ingredients: Linkages to cardiovascular disease and cancer. *Current Opinion in Clinical Nutrition and Metabolic Care*, **12**, 147–151.

Josuttis, M., Dietrich, H., Treutter, D. *et al.* (2010) Solar UVB response of bioactives in strawberry (*Fragaria x Ananassa* duch. L.): A comparison of protected and open-field cultivation. *Journal of Agricultural and Food Chemistry*, **58**, 12692–12702.

Josuttis, M., Dietrich, H., Patz, C.D. and Kruger, E. (2011) Effects of air and soil temperatures on the chemical composition of fruit and agronomic performance in strawberry (*Fragaria x Ananassa* duch.). *Journal of Horticultural Science and Biotechnology*, **86**, 415–421.

Jourdan, P.S., Mcintosh, C.A. and Mansell, R.L. (1985) Naringin levels in citrus tissues. 2. Quantitative distribution of naringin in *Citrus paradisi* MacFad. *Plant Physiology*, **77**, 903–908.

Kalt, W., Ryan, D.A.J., Duy, J.C. *et al.* (2001) Interspecific variation in anthocyanins, phenolics, and antioxidant capacity among genotypes of highbush and lowbush blueberries (*Vaccinium section* cyanococcus spp.). *Journal of Agricultural and Food Chemistry*, **49**, 4761–4767.

Kang, N.J., Shin, S.H., Lee, H.J. and Lee, K.W. (2011) Polyphenols as small molecular inhibitors of signaling cascades in carcinogenesis. *Pharmacology and Therapeutics*, **130**, 310–324.

Karuppusamy, S. (2009) A review on trends in production of secondary metabolites from higher plants by *in vitro* tissue, organ and cell cultures. *Journal of Medicinal Plants Research*, **3**, 1222–1239.

Kassim, A., Poette, J., Paterson, A. *et al.* (2009) Environmental and seasonal influences on red raspberry anthocyanin antioxidant contents and identification of quantitative traits loci (QTL). *Molecular Nutrition and Food Research*, **53**, 625–634.

Katerji, N., Rana, G. and Mastrorilli, M. (2010) Modelling of actual evapotranspiration in open top chambers (OTC) at daily and seasonal scale: Multi-annual validation on soybean in contrasted conditions of water stress and air ozone concentration. *European Journal of Agronomy*, **33**, 218–230.

Kawaii, S., Tomono, Y., Katase, E. *et al.* (1999) Quantitation of flavonoid constituents in citrus fruits. *Journal of Agricultural and Food Chemistry*, **47**, 3565–3571.

Keutgen, A.J. and Pawelzik, E. (2007) Modifications of strawberry fruit antioxidant pools and fruit quality under nacl stress. *Journal of Agricultural and Food Chemistry*, **55**, 4066–4072.

Keutgen, A.J. and Pawelzik, E. (2008a) Contribution of amino acids to strawberry fruit quality and their relevance as stress indicators under NaCl salinity. *Food Chemistry*, **111**(3), 642–647.

Keutgen, A.J. and Pawelzik, E. (2008b) Quality and nutritional value of strawberry fruit under long term salt stress. *Food Chemistry*, **107**, 1413–1420.

Keutgen, A.J. and Pawelzik, E. (2008c) Influence of pre-harvest ozone exposure on quality of strawberry fruit under simulated retail conditions. *Postharvest Biology and Technology*, **49**, 10–18.

Kim, Y.C. (2010) Neuroprotective phenolics in medicinal plants. *Archives of Pharmacal Research*, **33**, 1611–1632.

Kim, J., Jayaprakasha, G.K., Uckoo, R.M. and Patil, B.S. (2012) Evaluation of chemopreventive and cytotoxic effect of lemon seed extracts on human breast cancer (mcf-7) cells. *Food and Chemical Toxicology*, **50**, 423–430.

Korkina, L.G. (2007) Phenylpropanoids as naturally occurring antioxidants: From plant defense to human health. *Cellular and Molecular Biology*, **53**, 15–25.

Korkina, L.G., Kostyuk, V., De Luca, C. and Pastore, S. (2011) Plant phenylpropanoids as emerging anti-inflammatory agents. *Mini-Reviews in Medicinal Chemistry*, **11**, 823–835.

Krasensky, J. and Jonak, C. (2012) Drought, salt, and temperature stress-induced metabolic rearrangements and regulatory networks. *Journal of Experimental Botany*, **63**, 1593–1608.

Lee, H.S. (2001) Characterization of carotenoids in juice of red navel orange (*Cara Cara*). *Journal of Agricultural and Food Chemistry*, **49**, 2563–2568.

Lee, K.W., Bode, A.M. and Dong, Z.G. (2011) Molecular targets of phytochemicals for cancer prevention. *Nature Reviews Cancer*, **11**, 211–218.

Lee, S.M., Yang, H., Tartar, D.M. *et al.* (2011) Prevention and treatment of diabetes with resveratrol in a non-obese mouse model of type 1 diabetes. *Diabetologia*, **54**, 1136–1146.

Lenucci, M.S., Serrone, L., De Caroli, M. *et al.* (2012) Isoprenoid, lipid, and protein contents in intact plastids isolated from mesocarp cells of traditional and high-pigment tomato cultivars at different ripening stages. *Journal of Agricultural and Food Chemistry*, **60**, 1764–1775.

Li, H.Y., Deng, Z.Y., Liu, R.H. *et al.* (2011) Characterization of phytochemicals and antioxidant activities of a purple tomato (*Solanum lycopersicum* l.). *Journal of Agricultural and Food Chemistry*, **59**, 11803–11811.

Lobell, D.B., Burke, M.B., Tebaldi, C. *et al.* (2008) Prioritizing climate change adaptation needs for food security in 2030. *Science*, **319**, 607–610.

Mandl, J., Szarka, A. and Banhegyi, G. (2009) Vitamin C: Update on physiology and pharmacology. *British Journal of Pharmacology*, **157**, 1097–1110.

Mangels, A.R., Holden, J.M., Beecher, G.R. *et al.* (1993) Carotenoid content of fruits and vegetables – an evaluation of analytic data. *Journal of the American Dietetic Association*, **93**, 284–296.

Martinez, J.M.G. (2006) Isoflavones in menopause women. *Medicina Clinica*, **127**, 352–356.

Merzlyak, M.N. and Chivkunova, O.B. (2000) Light-stress-induced pigment changes and evidence for anthocyanin photoprotection in apples. *Journal of Photochemistry and Photobiology B – Biology*, **55**, 155–163.

Morgan, P.B., Bernacchi, C.J., Ort, D.R. and Long, S.P. (2004) An in vivo analysis of the effect of season-long open-air elevation of ozone to anticipated 2050 levels on photosynthesis in soybean. *Plant Physiology*, **135**, 2348–2357.

Nadella, K.D., Marla, S.S. and Kumar, P.A. (2012) Metabolomics in agriculture. *Omics – a Journal of Integrative Biology*, **16**, 149–159.

Naik, D., Dhanaraj, A.L., Arora, R. and Rowland, L.J. (2007) Identification of genes associated with cold acclimation in blueberry (*Vaccinium corymbosum* l.) using a subtractive hybridization approach. *Plant Science*, **173**, 213–222.

Navarro, J.M., Flores, P., Garrido, C. and Martinez, V. (2006) Changes in the contents of antioxidant compounds in pepper fruits at different ripening stages, as affected by salinity. *Food Chemistry*, **96**, 66–73.

Neocleous, D. and Vasilakakis, M. (2008) Fruit quality of red raspberry as affected by salinity. *European Journal of Horticultural Science*, **73**, 131–137.

Norton, C., Kalea, A.Z., Harris, P.D. and Klimis-Zacas, D.J. (2005) Wild blueberry-rich diets affect the contractile machinery of the vascular smooth muscle in the Sprague-Dawley rat. *Journal of Medicinal Food*, **8**, 8–13.

Ortiz, R., Sayre, K.D., Govaerts, B. *et al.* (2008) Climate change: Can wheat beat the heat? *Agriculture Ecosystems and Environment*, **126**, 46–58.

Ostlund, R.E. (2002) Cholesterol absorption. *Current Opinion in Gastroenterology*, **18**, 254–258.

Pase, M.P., Grima, N.A. and Sarris, J. (2011) The effects of dietary and nutrient interventions on arterial stiffness: A systematic review. *American Journal of Clinical Nutrition*, **93**, 446–454.

Percival, D. and Mackenzie, J.L. (2007) Use of plant growth regulators to increase polyphenolic compounds in the wild blueberry. *Canadian Journal of Plant Science*, **87**, 333–336.

Perez-Ilzarbe, J., Hernandez, T., Estrella, I. and Vendrell, M. (1997) Cold storage of apples (cv Granny Smith) and changes in phenolic compounds. *Zeitschrift Fur Lebensmittel-Untersuchung Und-Forschung a – Food Research and Technology*, **204**, 52–55.

Perotti, V.E., Del Vecchio, H.A., Sansevich, A. *et al.* (2011) Proteomic, metabalomic, and biochemical analysis of heat treated Valencia oranges during storage. *Postharvest Biology and Technology*, **62**, 97–114.

Price, S.F., Breen, P.J., Valladao, M. and Watson, B.T. (1995) Cluster sun exposure and quercetin in pinot-noir grapes and wine. *American Journal of Enology and Viticulture*, **46**, 187–194.

Proteggente, A.R., Pannala, A.S., Paganga, G. *et al.* (2002) The antioxidant activity of regularly consumed fruit and vegetables reflects their phenolic and vitamin c composition. *Free Radical Research*, **36**, 217–233.

Ramakrishan, A. and Ravishnar, G.A. (2011) Influence of abiotic stress signals on secondary metabolites in plants. *Plant Signaling & Behavior*, **6**, 1720–1731.

Rapisarda, P., Lo Bianco, M., Pannuzzo, P. and Timpanaro, N. (2008) Effect of cold storage on vitamin c, phenolics and antioxidant activity of five orange genotypes [*Citrus sinensis* (L.) osbeck]. *Postharvest Biology and Technology*, **49**, 348–354.

Reyes, L.F. and Cisneros-Zevallos, L. (2007) Electron-beam ionizing radiation stress effects on mango fruit (*Mangifera indica* L.) antioxidant constituents before and during postharvest storage. *Journal of Agricultural and Food Chemistry*, **55**, 6132–6139.

Riccioni, G. (2009) Carotenoids and cardiovascular disease. *Current Atherosclerosis Reports*, **11**, 434–439.

Rice-Evans, C.A., Miller, J. and Paganga, G. (1997) Antioxidant properties of phenolic compounds. *Trends in Plant Science*, **2**, 152–159.

Rivelli, A.R., James, R.A., Munns, R. and Condon, A.G. (2002) Effect of salinity on water relations and growth of wheat genotypes with contrasting sodium uptake. *Functional Plant Biology*, **29**, 1065–1074.

Sanchez-Rodriguez, E., Del Mar Rubio-Wilhelmi, M., Miguel Cervilla, L. *et al.* (2010) Study of the ionome and uptake fluxes in cherry tomato plants under moderate water stress conditions. *Plant and Soil*, **335**, 339–347.

Scalzo, J., Mezzetti, B. and Battino, M. (2005) Total antioxidant capacity evaluation: Critical steps for assaying berry antioxidant features. *Biofactors*, **23**, 221–227.

Schmitz-Eiberger, M.A. and Blanke, M.M. (2012) Bioactive components in forced sweet cherry fruit (*Prunus avium* L.), antioxidative capacity and allergenic potential as dependent on cultivation under cover. *LWT – Food Science and Technology*, **46**(2), 388–392.

Seeram, N.P. (2010) Recent trends and advances in berry health benefits research. *Journal of Agricultural and Food Chemistry*, **58**, 3869–3870.

Sgherri, C., Kadlecova, Z., Pardossi, A. *et al.* (2008) Irrigation with diluted seawater improves the nutritional value of cherry tomatoes. *Journal of Agricultural and Food Chemistry*, **56**, 3391–3397.

Sgherri, C., Navari-Izzo, F., Pardossi, A. *et al.* (2007) The influence of diluted seawater and ripening stage on the content of antioxidants in fruits of different tomato genotypes. *Journal of Agricultural and Food Chemistry*, **55**, 2452–2458.

Singh, U.P., Singh, D.P., Singh, M. *et al.* (2004) Characterization of phenolic compounds in some Indian mango cultivars. *International Journal of Food Sciences and Nutrition*, **55**, 163–169.

Skrede, G., Wrolstad, R.E. and Durst, R.W. (2000) Changes in anthocyanins and polyphenolics during juice processing of highbush blueberries (*Vaccinium Corymbosum* L.). *Journal of Food Science*, **65**, 357–364.

Smith, M.A.L., Marley, K.A., Seigler, D. *et al.* (2000) Bioactive properties of wild blueberry fruits. *Journal of Food Science*, **65**, 352–356.

Song, J., Kwon, O., Chen, S.L. *et al.* (2002) Flavonoid inhibition of sodium-dependent vitamin C transporter 1 (svct1) and glucose transporter isoform 2 (glut2), intestinal transporters for vitamin c and glucose. *Journal of Biological Chemistry*, **277**, 15252–15260.

Sroka, Z. and Cisowski, W. (2003) Hydrogen peroxide scavenging, antioxidant and anti-radical activity of some phenolic acids. *Food and Chemical Toxicology*, **41**, 753–758.

Sudha, G. and Ravishankar, G.A. (2002) Involvement and interaction of various signaling compounds on the plant metabolic events during defense response, resistance to stress factors, formation of secondary metabolites and their molecular aspects. *Plant Cell Tissue and Organ Culture*, **71**, 181–212.

Sweeney, M.I., Kalt, W., Mackinnon, S.L. *et al.* (2002) Feeding rats diets enriched in lowbush blueberries for six weeks decreases ischemia-induced brain damage. *Nutritional Neuroscience*, **5**, 427–431.

Terry, L.A., Chope, G.A. and Bordonaba, J.G. (2007) Effect of water deficit irrigation and inoculation with botrytis cinerea on strawberry (*Fragaria x Ananassa*) fruit quality. *Journal of Agricultural and Food Chemistry*, **55**, 10812–10819.
Tomas-Barberan, F. and Espin, J.C. (2001) Phenolic compounds and related enzymes as determinants of quality in fruits and vegetables. *Journal of the Science of Food and Agriculture*, **81**, 853–876.
Traka, M.H. and Mithen, R.F. (2011) Plant science and human nutrition: Challenges in assessing health-promoting properties of phytochemicals. *Plant Cell*, **23**, 2483–2497.
Tsao, R. (2010) Chemistry and biochemistry of dietary polyphenols. *Nutrients*, **2**, 1231–1246.
Vasanthi, H.R., Shrishrimal, N. and Das, D.K. (2012) Phytochemicals from plants to combat cardiovascular disease. *Current Medicinal Chemistry*, **19**, 2242–2251.
Versari, A., Parpinello, G.P., Tornielli, G.B. *et al.* (2001) Stilbene compounds and stilbene synthase expression during ripening, wilting, and UV treatment in grape cv. Corvina. *Journal of Agricultural and Food Chemistry*, **49**, 5531–5536.
Wang, Y.X. and Frei, M. (2011) Stressed food – the impact of abiotic environmental stresses on crop quality. *Agriculture Ecosystems and Environment*, **141**, 271–286.
Wang, S.Y. and Lin, H.S. (2000) Antioxidant activity in fruits and leaves of blackberry, raspberry, and strawberry varies with cultivar and developmental stage. *Journal of Agricultural and Food Chemistry*, **48**, 140–146.
Wang, S.Y. and Zheng, W. (2001) Effect of plant growth temperature on antioxidant capacity in strawberry. *Journal of Agricultural and Food Chemistry*, **49**, 4977–4982.
Wang, C.Y., Chen, C.T. and Wang, S.Y. (2009) Changes of flavonoid content and antioxidant capacity in blueberries after illumination with UV-C. *Food Chemistry*, **117**, 426–431.
Weng, C.J. and Yen, G.C. (2012) Chemopreventive effects of dietary phytochemicals against cancer invasion and metastasis: Phenolic acids, monophenol, polyphenol, and their derivatives. *Cancer Treatment Reviews*, **38**, 76–87.
Woyengo, T.A., Ramprasath, V.R. and Jones, P.J.H. (2009) Anticancer effects of phytosterols. *European Journal of Clinical Nutrition*, **63**, 813–820.
Wu, X.L., Beecher, G.R., Holden, J.M. *et al.* (2004) Lipophilic and hydrophilic antioxidant capacities of common foods in the United States. *Journal of Agricultural and Food Chemistry*, **52**, 4026–4037.
Zamboni, A., Minoia, L., Ferrarini, A. *et al.* (2008) Molecular analysis of post-harvest withering in grape by AFLP transcriptional profiling. *Journal of Experimental Botany*, **59**, 4145–4159.
Zeng, K.F., Cao, J.K. and Jiang, W.B. (2006) Enhancing disease resistance in harvested mango (*Mangifera indica* L. Cv. 'Matisu') fruit by salicylic acid. *Journal of the Science of Food and Agriculture*, **86**, 694–698.
Zhang, D., Yu, B., Bai, J.H. *et al.* (2012) Effects of high temperatures on uv-b/visible irradiation induced postharvest anthocyanin accumulation in 'Yunhongli no. 1' (*Pyrus pyrifolia* nakai) pears. *Scientia Horticulturae*, **134**, 53–59.
Zhao, J., Davis, L.C. and Verpoorte, R. (2005) Elicitor signal transduction leading to production of plant secondary metabolites. *Biotechnology Advances*, **23**, 283–333.
Zhao, T.J., Liu, Y., Yan, Y.B. *et al.* (2007) Identification of the amino acids crucial for the activities of drought responsive element binding factors (Drebs) of *Brassica napus*. *FEBS Letters*, **581**, 3044–3050.
Zhao, R.P., Hao, J.X., Xue, J. *et al.* (2011) Effect of high-voltage electrostatic field pretreatment on the antioxidant system in stored green mature tomatoes. *Journal of the Science of Food and Agriculture*, **91**, 1680–1686.

19 Recovery of Valuable Bioactives from Residues Arising from Fruit Processing

Zaid Saleh, Reinhold Carle and Dietmar Rolf Kammerer

INTRODUCTION

Recently, phenolics have been extensively studied because of their antioxidant activity and associated health benefits, including reducing the risk of various cancers (Tsao *et al.*, 2005). Additionally, they help with the retention of high density lipoproteins and removal of low density lipoproteins within the body (Nawaz *et al.*, 2006), as well as reducing the risk of cardiovascular diseases (Soler *et al.*, 2009). Consequently, apple phenolics as well as polyphenols from grape and sunflower were focused on in a series of studies aiming at their selective recovery both by membrane separation and adsorption/ion exchange technology.

Current efficient phenolic extraction methods using adsorption technology involve organic solvents including ethanol, methanol and hexane, which are less desirable because they create potential health and safety challenges for production and impart harmful impurities into the extract (Nawaz *et al.*, 2006; Saleh *et al.*, 2006). Also these methods unfortunately co-extract sugar with the phenolics. From a health perspective, sugar is an undesired admixture to the ingredients of foods enriched with phenolic phytochemicals, and hence an economical separation technology is required (Nawaz *et al.*, 2006; Saleh *et al.*, 2006). Membrane filtration is a technology with several advantages compared to other filtration technologies through, eliminating the use of additives and organic solvents since being a physical process operating under mild conditions (Nawaz *et al.*, 2006), lowering the labour costs and energy usage, reducing amounts of waste to be disposed of (De Bruijn *et al.*, 2003), and easy to model and scale up (Nawaz *et al.*, 2006). Furthermore, membrane technology is cost effective and already commercially available in the food and dairy industries for the clarification and concentration of fruit and vegetables juices, and for the recovery of proteins from dairy processing side streams. In nanofiltration, the hydrophobic and lipophilic compounds are mainly retained and the hydrophilic compounds, such as the sugars, pass through the membrane (Mantarri *et al.*, 2006).

Studies have shown that membrane technology is effective in the separation of phenolics and sugars from clear apple juice and possibly other clear juices (Saleh *et al.*, 2006). This study noted that a possible secondary membrane, caused by fouling, might be formed during the filtration when the feed concentration and the transmembrane pressure were increased.

Bioactives in Fruit: Health Benefits and Functional Foods, First Edition. Edited by Margot Skinner and Denise Hunter.

Fouling, which leads to a decline in the permeate flux over time, decreases membrane yield and performance. During membrane filtration of apple juice, fouling is mainly caused by the retention and interactions of carbohydrates (pectin, starch, hemicellulose and cellulose), phenolics (tannins and other) and/or proteins (De Bruijn *et al.*, 2002; De Bruijn and Borquez, 2006; Saleh *et al.*, 2006). The main fouling mechanisms are concentration polarization, formation of a secondary or cake layer, and internal pore blocking/constriction (De Bruijn *et al.*, 2002; De Bruijn and Borquez, 2006). Concentration polarization is an accumulation of particles near the surface of the membrane and is, therefore, easily overcome by water flushing. Internal pore blocking and cake layer formation are often the result of stronger adsorption and deposition of foulants, and can only be partially removed by water flushing. Complete removal requires intensive chemical cleaning (De Bruijn *et al.*, 2002; Saleh *et al.*, 2006).

Several studies have analysed the mechanisms of fouling that occur during membrane filtration, using different models. These include the resistances-in-series model, which looks at the different fouling resistances (Jiraratananon and Chanachai, 1996; De Bruijn *et al.*, 2002; Cassano *et al.*, 2007), a model that considers variations in permeate flux and filtration time (Wiesner and Aptel, 1996; Cassano *et al.*, 2008), and a model that predicts the fouling mechanism over time using the initial and critical/steady state permeate fluxes (Field *et al.*, 1995; Rai *et al.*, 2006).

The effects of different operating parameters, including temperature, initial juice concentration, transmembrane pressure (TMP) and pH, on the permeate flux have been investigated to determine the optimal operating conditions to mitigate fouling. The concentrations of total phenolics were determined, and a phenolic profile was created using high performance liquid chromatography (HPLC) (Stevenson *et al.*, 2006). A series of investigations have been performed aiming to characterize and compare the type of fouling occurring during the nanofiltration of apple juice concentrate using existing models, and to recommend a fouling mitigation strategy.

Even though resin-based technologies are characterized by some drawbacks, such as the use of organic solvents, applied among others for the recovery and purification of plant secondary metabolites, adsorption and ion exchange processes are valuable tools for the enrichment and fractionation of certain target compounds and, thus, were thoroughly investigated in a series of studies reported here. Adsorption and ion exchange processes have been exploited for millennia, for example by the Egyptians and Sumerians, who used charcoal around 4000 years BC for bronze manufacturing, and by the Phoenicians, who applied sorption processes for the preparation of potable water. Such applications were rather empirical, and more systematic and quantitative experiments were not performed before the eighteenth century. Similarly, ion exchange processes have also been applied for a long time, and first reports on ion exchange phenomena may probably be found in the Old Testament. Around a thousand years later, Aristotle also dealt with ion exchange processes. However, it is assumed that the principles behind these phenomena were still unknown at that time. Subsequently, this separation technique has progressed very rapidly, which is due to more detailed investigations and the development of novel synthetic ion exchange and adsorber resins with well-defined physicochemical properties. Consequently, the application areas of sorption and ion exchange phenomena are highly diverse nowadays and form part of a wide range of industrial production processes, which are based on activated carbon, natural and synthetic zeolites as well as synthetic resins based on polystyrene, polyacrylic esters or phenols (Dorfner, 1970; Helfferich, 1995; Inglezakis and Poulopoulos, 2006; LeVav and Carta, 2007; Zagorodni, 2007).

Water purification is by far the most important application of adsorption technology, which aims at the preparation of highly purified water, not only for food production but also for further industrial branches, such as the pharmaceutical and cosmetics industry, for paper and fibre production as well as for the production of electronic components. In addition, sorptive wastewater treatment is gaining increasing importance (Inglezakis and Poulopoulos, 2006; Zagorodni, 2007; Shannon *et al.*, 2008).

In the food industry, adsorber resins are commonly applied for the standardization and stabilization of fruit juices by removing undesired browning compounds or by reducing the contents of components which contribute to the formation of hazes and sediments upon storage. Thus, adsorption technology may be applied as an alternative to conventional fining procedures using, for example, bentonite or gelatin (Lyndon, 1996; Weinand, 1996; Weiss, 2001). Moreover, debittering of citrus juices is frequently achieved by adsorption resins. This process is based on selective sorptive removal of naringin and limonin from grapefruit juices (Johnson and Chandler, 1982) and of hesperidin and limonin from navel orange juices (Chandler *et al.*, 1968).

Ion exchangers are also commonly applied in industrial practice, mainly for water softening and purification, as described for adsorber resins (Melear and Zoltek, 1982; Inglezakis and Poulopoulos, 2006; Zagorodni, 2007). In addition, ionic interactions with synthetic resins are exploited for the purification and enrichment of desired compounds, such as of lactic acid from fermentation broths (Tong *et al.*, 2004; Ataei and Vasheghani-Farahani, 2008). Moreover, ion exchange is successfully applicable for the removal of unwanted components, such as the Maillard reaction or hexose degradation products formed upon alkaline treatment in the course of sucrose production (Serpen *et al.*, 2007; Coca *et al.*, 2008). Further applications comprise the decaffeination of coffee, tea and cocoa products (Dawson-Ekeland and Stringfield, 1991) as well as the stabilization of beverages such as beer and wine by removing proteins, which may cause the formation of hazes and sediments (Bellamy and Pease, 2009).

More recently, adsorber and ion exchange resins have attracted increased interest, since they may serve as an ideal tool for the recovery of valuable compounds from the byproducts of food processing (Kammerer *et al.*, 2011a). Among others, resin-based processes for such purposes have been described for the purification of proteins from waste streams (Wasswa *et al.*, 2007; Weisz *et al.*, 2010) or for the recovery of phenolic compounds from press cakes, such as hesperidin from orange peels (Di Mauro *et al.*, 1999), anthocyanins and colourless phenolics from grape pomace (Kammerer *et al.*, 2004; Maier *et al.*, 2008, 2009) and the simultaneous recovery of pectins and polyphenols from apple pomace (Carle *et al.*, 2001; Schieber *et al.*, 2003).

Despite the fact that adsorption and ion exchange have already been implemented in a wide range of industrial processes, and even though there are a number of models and mathematical approaches for theoretically describing sorption and ion exchange phenomena, and thus predicting optimal sorption conditions, the underlying principles are mostly still largely unknown. Consequently, such processes have been developed empirically and have not been systematically optimized (Bretag *et al.*, 2009a). Furthermore, there are only a few systematic studies evaluating sorption and ion exchange processes for the enrichment of plant secondary metabolites (Kammerer *et al.*, 2013). Among these, only a few compounds have been studied in detail, such as the sorption behaviour of cyanidin 3-*O*-glucoside upon contact with resins of differing physicochemical characteristics (Scordino *et al.*, 2004). Similarly, the sorption of hesperidin has been thoroughly investigated, varying pH and temperature conditions (Scordino *et al.*, 2003). Further attention has been paid to the sorptive recovery of anthocyanins from the byproducts of citrus and grape processing (Di Mauro *et al.*, 2002;

Scordino *et al.*, 2005). However, further more detailed studies of sorption and ion exchange phenomena allowing the selective and cost efficient recovery of plant secondary metabolites are still widely lacking. For this reason, a series of investigations have been performed aiming at the evaluation of crucial process parameters and their effects on polyphenol binding applying adsorber and ion exchange resins to systematically enrich polyphenols and fractionate complex phenolic extracts. Such investigations contribute to the optimization of the selectivity and cost-efficiency of industrial sorption processes, which will play an even more important role in the future with regard to sustainable agricultural and food production.

MEMBRANE SEPARATION AND FILTRATION

Membrane technology

Membrane is used in a pressure-driven separation process, which selectively passes components through membrane while rejecting the other components. The passing stream is called permeate and the rejecting stream is called retentate. The maximum size of the component that can permeate through the membrane is called the cut-off value. Membrane technology is classified into four types: Microfiltration (MF), Ultrafiltration (UF), Nanofiltration (NF) and Reverse Osmosis (RO).

Membrane module design and configurations

Membrane is only one component of the whole filtration system, which also includes pumps, feed vessels, tubing, associated connections, necessary monitors and controllers for pressure, temperature, flow rate and the membrane module. Among these, the membrane module is of vital importance. It is a physical unit which houses membrane in an appropriate configuration and satisfies mechanical, hydrodynamic and economic requirements (Zeman and Zydney, 1996). In any module design, one of the critical aspects is the mass transfer characteristic.

Only when the rate of solute mass transfer in the feed is sufficiently fast to avoid concentration polarization, can the intrinsic selectivity and membrane flux characteristics be obtained. Concentration polarization refers to accumulation of a highly concentrated boundary layer of retained solutes at the membrane surface. It is often the dominant factor that controls the membrane performance (Zeman and Zydney, 1996).

To obtain a high rate of mass transfer, high tangential velocity and shear forces are required. This can be achieved by tangential flow, which means flow parallel to the surface of the filtration membrane and perpendicular to the pressure.

Conventional filtration processes operate in normal flow, also known as dead-end flow, in which fluid flow is normal to the membrane surface and the entire fluid volume passes through the membrane. Particulates that are larger than the membrane pores are retained while smaller molecules pass through to the downstream. In tangential flow filtration, also known as cross-flow filtration, fluid is pumped tangentially to the membrane surface, and only a portion of the fluid is forced through the membrane. The larger particulates, which cannot pass through the membrane, are swept along upstream by the tangential force, providing a self-cleaning effect. Dead-end flow is employed in sterile filtration of clean streams, clarifying pre-filtration, MF sterilization cartridges in the pharmaceutical industry. Ultrafiltration is conventionally performed in cross-flow mode, which is advantageous when the retained material is likely to plug the membrane (Noble and Stern, 1995).

Commercial membrane modules can be classified into four groups: hollow fibre, tubular, plate-and-frame and spiral wound. Hollow fibre and spiral wound membranes have the highest packing density; however, the trade-off is more susceptible to particulate plugging and difficult cleaning (Noble and Stern, 1995).

Ultrafiltration of polyphenols

Research on the ultrafiltration of polyphenolics in the fruit industry is either to extract polyphenolics from fruit for health-related function or remove them from juice to prevent browning.

The recovery of phenolic compounds from apple juice has been studied on a laboratory pilot-scale using a commercially viable filtration process (Saleh *et al.*, 2010). The work was carried out to understand how the processing parameters during nanofiltration processing can be manipulated to optimize the production of apple phenolics as functional food ingredients. The fouling mechanism was characterized and modelled using a polysulfone SelRO® spiral-wound membrane with different molecular weight cut-offs (MWCO). The retention of various compounds, including total and individual phenolics, proteins and certain types of sugars, were determined analytically. The effects of temperature, feed concentration, transmembrane pressure and pH on permeate flux, fouling and membrane performance were determined. Despite the increase in polyphenolics in the retentate there was little difference in the phenolic composition between retentate and permeate solutions. This indicated that the rejection was related to the formation of a secondary membrane because of fouling. A mass balance of polyphenolics in the final retentate and permeate compared with the initial feed solution indicated that a lot of polyphenolics were bound to the membrane. Mitigation of fouling could be achieved by controlling the operating conditions and pre-filtering larger molecular weight polyphenols. It was concluded that membrane separation represents a potentially efficient and cost-effective technology to concentrate the phenolic fraction of fruit juice for use as a functional ingredient.

Different types of fouling mechanisms, including reversible and irreversible concentration polarization, cake formation, pore blocking, or a combination of the aforementioned, were investigated. Modelling of the permeate flux using available predictive fouling models in literature gave comparable results. Analyses of the resistances found that reversible cake formation was greater at higher feed concentration and lower at higher pH values. Furthermore, performing diafiltration increased the permeate flux significantly by reducing the concentration polarization effect. The filtration process was optimized and a fouling mitigation strategy was recommended.

In another study (Saleh *et al.*, 2006) ultrafiltration and nanofiltration were used to separate polyphenolics and sugar from apple juice, by using 1 and 0.25 kDa molecular weight cut-off, spiral wound membranes. For 1 kDa membrane, the maximum concentration factors of polyphenolics and sugar in retentate achieved were up to 4 and 1.5, respectively, under the optimum process conditions, which were lower temperature (30°C), acidic pH (2), lower transmembrane pressure (5 bar) and higher initial sugar concentration (20°Brix). The permeate solution was then filtered using a 0.25 kDa membrane, and the polyphenolics concentration in retentate increased by a factor of up to two.

The removal of polyphenol oxidase (PPO) from banana extracts was also investigated (Tanada-Palmu *et al.*, 1999). PPO was extracted by ultrafiltration using polysulfone membranes with 20 kDa MWCO. Two studies were carried out at different operating pressures,

600 and 800 kPa (UF 600 and UF 800). The operating temperature was maintained between 25 and 30°C. It was found that the permeate flux decreasing rate was about 4.7 $l/h/m^2$ for UF 600 and 3.5 $l/h/m^2$ for UF 800. Therefore, a transmembrane pressure of 600 kPa was favoured over 800 kPa due to more stable permeate fluxes.

In another study (Borneman *et al.*, 2001), colour and polyphenol components were selectively removed by custom membrane (polyethersulfone (PES)/polyvinylpyrrolidone (PVP)). The membranes were prepared with PES (MW 4300) and polyvinylpyrrolidone PVP (MW 360 000). The pore size of custom membrane varied according to different polymer compositions. The commercial regenerated cellulose membranes used were from Millipore with 10 and 100 kDa MWCO. Operating pressure was set at one bar. Temperature was 50°C. The concentration of polyphenols increased as a function of permeates volume, but permeate flux decreased with time. It has been found that PES/PVP membranes reduced more polyphenols and yellowish-brown colour pigments in apple juice than commercial regenerated cellulose membranes. The average removal of polyphenols by PES/PVP membrane was 30%.

The removal of polyphenols from pomegranate juice was investigated by ultrafiltration and UF-laccase combinations (Alper and Acar, 2004). Ultrafiltration was performed with a laboratory device which was cross-flow, stirred cell, Amicon 8200. The membrane was made of cellulose acetate with 10 kDa MWCO. The operating pressure was 2 bar, and the flux rate was 12.7 $l/h/m^2$. By using UF-laccase combinations, 9.05% of phenolic compounds and 97.58% of cloudy particles were removed. Only 8.82% of phenolic compounds could be removed by the UF method. Although laccase can improve the removal of phenolic compounds from fruit juice, it can cause discoloration of the fruit juice, resulting in poor acceptance of the final product.

The polyphenols from grape seed have also been shown to be concentrated by ultrafiltration (Nawaz *et al.*, 2006). Ultrafiltration was performed by a Millipore stirred ultrafiltration cell with a volume of 50 ml. Two membranes were selected, Millipore type GS 0.22 μm and Millipore type HA 0.45 μm. The operating pressure was five bar and temperature was 23 ± 1°C. Sample was prepared by extraction with ethanol solvent. Double extraction referred to sample extraction by ethanol solvents twice prior to ultrafiltration. It has been found that double extraction with the smaller membrane pore size (0.22 μm) could achieve phenol extraction of 11.4% of the total seeds weight from the grape seed. It was concluded from the aforementioned studies that optimum operating parameters need to be determined specifically for the feed material.

Membrane fouling

Membrane characteristics

Membranes are normally characterized by flux, pure water flux and flux decline (Abbas, 2004).

Flux

Flux is directly related to the driven-force. It is measured in unit volume of permeates per unit time per unit membrane area. The forces are mostly generated by the difference in pressure, concentration or electrical potential. Flux can be the results of combinations of the forces, as in the case of osmosis process in which volume flow is obtained by applied pressure but also induced by a difference in concentration (Abbas, 2004).

Pure water flux

Pure water flux (PWF) is used to characterize membrane by the permeate volume of deionized water per unit area of membrane per unit time ($m^3/m^2/s$). It is the maximum permeation of a new membrane and normally used to assess fouling conditions after each run. Furthermore, PWF is often taken as the base line to evaluate the starting condition of the new membrane. It is of note that used membrane can never return to the original PWF value, even after repeated cleaning. The reason is that membrane was configured in a delicate form that exhibits irreversible deformation when subjected to the whole process, including filtration, cleaning and installation (Abbas, 2004).

Flux decline

Flux decline caused by fouling is the major problem in membrane technology. During filtration processes, molecules or impurities precipitate in the feed side on the membrane surface or inside membrane pores, reducing the effective membrane area of filtration. Fouling can also occur when the concentration of solute is increased near the membrane surface and forms a thick layer, thus adding extra resistance for filtration. This phenomenon is called concentration polarization or gel formation. Fouling can be reduced by increasing temperature and flow turbulence within an acceptable range (Abbas, 2004).

Fouling characterization

Determination of fouling resistances using the resistance-in-series model

The resistance-in-series model (Jiraratananon and Chanachai, 1996; De Bruijn *et al.*, 2002; Cassano *et al.*, 2007) determines the individual resistances that lead to a decline in permeate flux. The permeate flux is calculated in terms of transmembrane pressure, viscosity and total resistance.

The cake layer resistance, R_c, is dominated by the concentration polarization effect plus a thin film deposit of larger, more water-soluble macromolecules (Jiraratananon and Chanachai, 1996). This resistance is removed by diafiltration of the membrane with water. The fouling resistance, R_f, is a strongly adsorbed fouling layer that can only be removed by chemical cleaning.

Modelling of permeate flux variations over time

The decline of the permeate flux was evaluated by three models reported in the literature that considered a relationship between the variation of the permeate flux and filtration time (Wiesner and Aptel, 1996; Cassano *et al.*, 2008).

Model I assumes the dominant type of fouling is the formation of a cake layer due to the characteristics of the membrane. In Model II, it is also assumed that the dominant type of fouling is the formation of a cake layer with a thin layer covering the whole membrane surface. In Model III, the dominant type of fouling is assumed to be internal pore blocking/constriction.

Modelling the fouling mechanism using modified cross-flow filtration equations

Using the constant pressure dead-end filtration equations modified for cross-flow filtration by Field and colleagues (Field *et al.*, 1995; Rai *et al.*, 2006), the flux decline can be modelled.

When particles are similar in size to the membrane MWCO, they may become absorbed and completely block the pores. Partial pore blocking occurs when the particles are smaller than the pores, adsorbing on the inside of the pores and partially blocking them.

Particles that are larger than the membrane pores are deposited on the membrane surface forming a cake or accumulate close to the surface leading to concentration polarization. It was assumed that the decline in flux occurs due to one of the three fouling mechanisms.

Effect of operating conditions on fouling

Effect of temperature

Temperature is an important variable in membrane fouling. Most fouling involving organic systems has shown a strong dependence on temperature (Watkinson and Wilson, 1997) because of changes to the fluid characteristics which produce deposits. In cases where fouling is due to chemical reaction or solubility effects in the bulk solution, bulk temperatures can play a major role in the formation of fouling precursors (Eaton and Lux, 1984). The bulk temperature and viscosity of fruit juice are inversely related. Therefore, the effect of temperature on viscosity, and hence flow regime through the membrane, is very important.

The recovery of phenolic compounds from apple juice was studied on a laboratory pilot-scale using a commercially viable filtration process. The work was carried out to understand how the processing parameters during batch nanofiltration processing can be manipulated to optimize the production of apple phenolics as functional food ingredients. Temperature had an effect on both the permeate flux and the recovery of phenolic compounds. The concentration of phenolic compounds in the retentate decreased by 26% with increasing temperature from 20 to 40°C (Saleh *et al.*, 2010, 2011). The higher temperatures resulted in an increased solubility of the phenolic compounds and could also have increased the membrane pore size, as polysulfone is sensitive to temperature variations (Goosen *et al.*, 2002; Saleh *et al.*, 2006). Possible denaturation and aggregation of proteins, and interactions between the hydrophilic oligo- and polysaccharides, may have formed a gel layer that acts as a secondary membrane that resulted in reduced pore size and, thus, restriction in the size of particles that could pass through the membrane. The hydrophilic compounds, such as the sugars, are expected to pass through the membrane. Previously reported research on the filtration of passion fruit juice demonstrated that fouling increased at 50°C because of the formation of a gel layer (Jiraratananon and Chanachai, 1996).

The effect of different feed solution temperatures on the permeate flux was notable. The permeate flux dropped sharply at the beginning of the experiments until it reached steady state. The initial decline in permeate flux was most severe at lower temperatures, with the initial value being higher with increasing temperatures. This was expected given that higher temperatures decrease the viscosity and increase the Reynold's number, therefore effectively increasing the cross-flow flow rates and turbulence. Higher temperatures may also change the flow regime by increasing the turbulence and the cross-flow velocity removing fouling material with higher shear forces (Tanada-Palmu *et al.*, 1999). This is consistent with the findings of Goosen *et al.* (2002), that the porosity of polysulfone membranes can be very sensitive to changes in the feed temperature. The lower temperatures could also increase the formation of a secondary membrane that restricted the flow and reduced the pore size of the membrane. The flux declined immediately after the start of each run, and the decrease was faster at lower temperatures. The improvement in flux with increasing temperature was primarily due to viscosity effects on the water (Goosen *et al.*, 2002).

An HPLC analysis was carried out on the feed, retentate and permeates. The phenolic profile shows quite clearly the noticeable increase in the retentate concentration from the feed concentration. The final permeate shows that most of the phenolics were retained. However, the overall composition of the phenolic profile did not change. The HPLC results demonstrated a clear difference in the concentrations of phenolic compounds.

Effect of feed concentration

The effect of concentration is very important in membrane fouling. Membrane processes are very sensitive to a critical concentration of moderately high molecular weight molecules (Karode *et al.*, 2000). The phenolic concentration from apple juice on the retentate side of the membrane increased significantly with increasing feed concentration (Saleh *et al.*, 2010, 2011).

The large increase in fouling materials indicates that concentration polarization was more dominant at higher concentrations due to larger molecules forming a denser layer next to the membrane. The higher feed concentrations had higher viscosity and lower Reynold's numbers, which resulted in lower cross-flow flow rates and less movement at the surface and, therefore, a higher fouling rate (Saleh *et al.*, 2010, 2011).

The permeate flux declined more rapidly at the beginning of the trial with higher feed concentrations. At higher concentrations, the steady state flux was very low, which resulted in increasing the length of the experiment considerably (Saleh *et al.*, 2010, 2011).

Effect of transmembrane pressure

It was found that fouling increased considerably at higher transmembrane pressures using apple juice as model solution (Saleh *et al.*, 2010, 2011), which is in agreement with the literature (Tanada-Palmu *et al.*, 1999). This can be explained by higher pressure compressing the rejected solutes into a thicker and denser fouling layer and lower cross-flow velocity removing retained materials because of lower shear forces.

Reproducibility was addressed by running the experiment with the optimum operating conditions again. Figures 19.1 and 19.2 show the permeate flux declines with time. The decline for both membranes were very similar, with the 1 kDa membrane having an R^2 value of 0.99 compared to the 0.97 R^2 value of the 0.25 kDa membrane. The concentration of phenolics across the membrane was also measured and found to be very similar for both membranes.

Effect of pH

The pH of the apple juice solution was adjusted with 10% phosphoric acid or 5% sodium hydroxide. Decreasing the pH from five to two increased the concentration of phenolics on the retentate side by a factor of 1.2 (Saleh *et al.*, 2010, 2011). This may have been due to the tightening of the polysulfone membrane pores with decreasing pH, in effect making the membrane more efficient. This finding is in agreement with our previous research on concentrating phenolics (Saleh *et al.*, 2006), but in contrast to other research on fruit juice, where the most effective filtration was found to occur at a pH of 7.5 (Balakrishnan *et al.*, 2000). Another effect of decreasing the pH is an increase in the protonation of phenolic compounds, rendering them more hydrophobic (Kammerer *et al.*, 2007). This will make these compounds less soluble and more easily retained in the retentate.

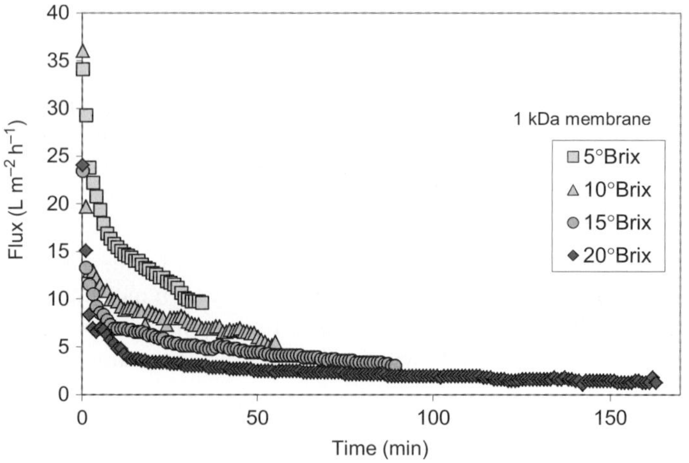

Fig. 19.1 Decline of the permeate flux with time for the duplicate runs using the 1 kDa membrane. (Operating conditions were: 30°C, 15 bar, pH 4.)

The permeate flux decline was more severe at lower pH values. The tightening of the pores and increased hydrophobicity of the phenolic compounds provide conditions that favour the formation of a secondary membrane, thus decreasing the flux. Because of the lower flux, the runs at lower pH values were longer (Saleh *et al.*, 2010, 2011).

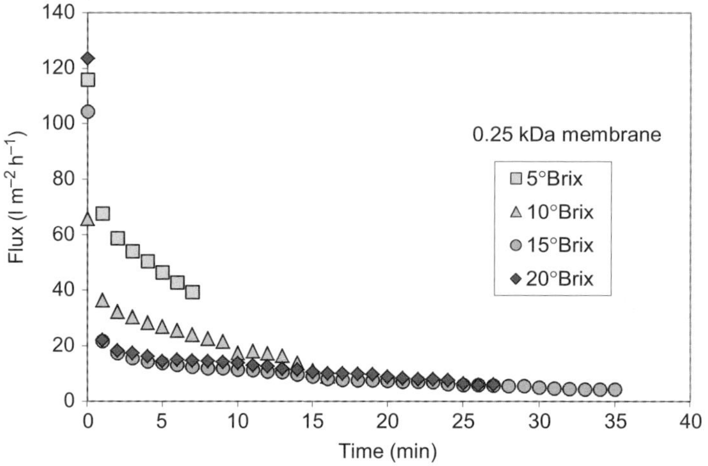

Fig. 19.2 Decline of the permeate flux with time for the duplicate runs using the 0.25 kDa membrane. (Operating conditions were: 30°C, 15 bar, pH 4.)

Protein and sugar content

The amounts of protein present in the feed, retentate and permeate from apple juice were analyzed for the experiments with the best retention of phenolics, to determine whether proteins were also contributing to fouling. The operating conditions had a very noticeable effect on the efficiency of the membrane. At a temperature of 20°C, some retention of

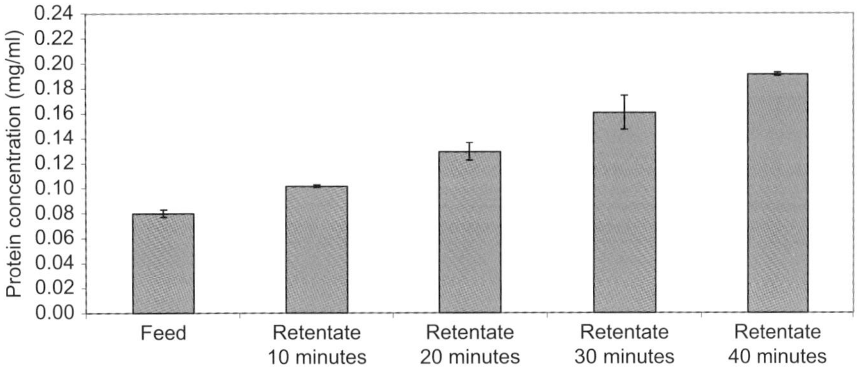

Fig. 19.3 Protein concentration (mg/ml) in the feed and retention over the duration of the experiment. (Operating conditions: 50°C, 10°Brix, 10 bar, pH 4.)

proteins occurred; however, a remarkable amount still passed through the membrane. It is possible that at 20°C the proteins are more tightly folded, so able to pass through the membrane. No proteins passed through the membrane at 50°C, indicating that they were aggregated on the surface of the membrane, and that a gel layer of polysaccharides restricted the movement of protein, as well as other constituents including polyphenols, through the membrane (Bradford, 1976).

Unexpectedly, no protein passed through when the feed concentration was increased to 20°Brix. This indicates that the high concentration created a high concentration polarization effect and reversible or irreversible fouling, which effectively created a separate secondary membrane, restricting protein movement through the membrane. At pH 2, no proteins passed through the membrane, further indicating constriction of the pores due to the acidic conditions and protonation of the phenolic–protein complexes. The retention of proteins may have also been influenced by their isoelectric points at the various feed pHs. Overall, most of the proteins were retained because of the larger molecular weight of proteins compared with those of phenolics (Saleh *et al.*, 2010, 2011). The protein concentration on the retentate side of the membrane during the experimental run at 50°C is shown in Figure 19.3. The steady increase in protein concentration indicates that the protein is being concentrated rather than fouled on the membrane surface.

The amounts and types of sugars present in the feed, retentate and permeate were analyzed for the optimum experiment and are shown in Table 19.1. Of the smaller molecular weight

Table 19.1 The concentration of various sugars in the feed, retentate and permeate and the ratio between the sugar and phenolic concentration. (Operating conditions: 30°C, 10°Brix, 15 bar, pH 4.)

Sample	Sorbitol (mg/ml)	Inositol (mg/ml)	Glucose (mg/ml)	Fructose (mg/ml)	Sucrose (mg/ml)
Feed 1 kDa	2.23	0.21	12.57	51.03	11.49
Retentate 1 kDa	2.11	0.20	13.83	58.73	19.60
Permeate 1 kDa and Feed 0.25 kDa	1.28	0.14	7.05	29.46	4.15
Retentate 0.25 kDa	2.02	0.29	13.20	56.08	13.33
Permeate 0.25 kDa	1.32	0.10	6.43	25.87	0.60

sugars (sorbitol, inositol, glucose and fructose), about 50% of the sugars passed through the membranes. The molecular weight of sucrose is higher and, therefore, more retention and fouling occur, with the most of the sucrose being retained by the membrane. The ratios between the phenolic and sugar concentration shows clearly that sugar concentration increased and phenolic concentration decreased in the permeate, indicating separation of sugars from phenolics (Saleh *et al.*, 2010, 2011).

Diafiltration

The effect of diafiltration on the permeate flux was evaluated under various operating conditions using apple juice as a model solution (Saleh *et al.*, 2010, 2011). It was observed that the permeate flux increased noticeably with each diafiltration. The first diafiltration was slightly more effective than the second diafiltration. In this particular case, diafiltration restored the permeate flux to close to the flux at the beginning of the experiment. This suggests that concentration polarization was the dominant fouling mechanism, and that most of the fouling was reversible. The remaining permeate flux was recovered by the chemical cleaning procedure. Samples of the retentate and permeate were collected at the end of each diafiltration. Diafiltration was effective in diluting and removing the concentration polarization effect and also in diluting and removing the secondary membrane. The diafiltration diluted the concentration of compounds on the membrane surface and remarkably increased the permeate flux.

Determination of resistances using the resistance-in-series model

The fouling resistances, including the total resistance, membrane resistance, cake layer resistance and reversible fouling resistance, were calculated for different feed concentrations and pH values. The percentage values of the resistances for different feed concentrations and pH values were calculated (Saleh *et al.*, 2010, 2011). The membrane resistance, R_m, decreased with increasing feed concentration. The cake layer resistance, R_c, increased noticeably with increasing feed concentration. This indicates a thicker and denser cake layer and a more evident concentration polarization effect with increasing feed concentration.

The cake layer resistance decreased with increasing pH, indicating less concentration polarization, possibly due to larger pore openings. Interestingly, the reversible fouling remained reasonably constant (Table 19.2). In contrast, the contribution of the membrane resistance to the total resistance increased proportionally with increasing pH, because of increasing pore size thus reducing the conditions that favour the formation of a cake layer (Saleh *et al.*, 2010, 2011).

Analyses of membrane fouling using the permeate flux variation over time model

Models I, II and III reported in the literature (Wiesner and Aptel, 1996; Cassano *et al.*, 2008) were used to analyze the dominant types of membrane fouling (Saleh *et al.*, 2010, 2011) for both 0.25 and 1 kDa MWCO membranes at the same operating conditions. Using the 1 kDa membrane, the applicability of model II was evident, with an R^2 value of 0.97. This indicates

Table 19.2 The various resistance values at different feed concentrations and different pH values.

Parameters	5°Brix	10°Brix	15°Brix	20°Brix	pH 2	pH 3	pH 4	pH 5
R_t ($\times 10^{12}$) (m^{-1})	2.83	3.45	4.37	6.88	4.19	3.57	3.45	2.46
R_m ($\times 10^{12}$) (m^{-1})	0.99	0.99	0.99	0.99	0.99	0.99	0.99	0.99
R_c ($\times 10^{12}$) (m^{-1})	1.06	1.43	2.52	4.77	2.18	1.68	1.43	0.809
R_{frev} ($\times 10^{12}$) (m^{-1})	0.36	0.57	0.36	1.68	0.617	0.759	0.57	0.35
R_m/R_t (%)	35.01	28.71	22.66	14.39	23.63	27.74	28.71	40.21
$(R_c + R_{frev})/R_t$ (%)	64.99	71.29	77.34	85.61	76.37	72.26	71.29	59.79
R_c/R_t (%)	37.40	41.32	57.61	69.27	51.92	47.11	41.32	32.87
R_{frev}/R_t (%)	27.59	29.97	19.73	16.34	24.45	25.15	29.97	26.92

Total resistance R_t, membrane resistance, R_m, cake layer resistance R_c, reversible fouling resistance, R_{frev}.

that the dominant fouling was the formation of a cake layer covering the entire surface of the membrane, which confirms our earlier results. This is in agreement with a previous study on the ultrafiltration of kiwifruit juice, which found that the most dominant fouling mechanism was the formation of a cake layer (Cassano *et al.*, 2008).

The three models were applied again for a typical run using the 0.25 kDa membrane. With an R^2 value of 0.99, Model III was the best fit. This indicates that the dominant fouling mechanism for the 0.25 kDa membrane was internal pore plugging, rather than the formation of a cake layer as observed with the 1 kDa membrane. This may be because the 1 kDa membrane filtered out most of the macromolecules, such as proteins and tannins, which might have caused the formation of a cake layer and resulted in reduced pore size, leaving only the smaller molecules which could then plug the 0.25 kDa membrane pores. In most cases, diafiltration recovered the permeate flux markedly, which indicates that the particles are weakly bound to the membrane. Similar trends and fouling mechanisms were obtained when applying Models I, II and III on both membrane cut-offs using all operating conditions.

The permeate flux decline was also modelled and the results show a less clear fouling mechanism trend. For both membrane cut-offs (1 and 0.25 kDa), Model I is the better fit with an R^2 value of 0.93 and 0.97, respectively. Model I indicates the fouling mechanism is cake formation due to the characteristics of the membrane. As previously mentioned, at 50°C a possible secondary layer of oligo-polysaccharides and aggregated proteins has been formed (Saleh *et al.*, 2010, 2011). The results obtained from this model suggest that using polysulfone membranes may not be particularly suited to filtering apple juice at high operating temperatures.

Analyses of fouling mechanism and prediction of flux decline

As the complete pore blocking, partial pore blocking and cake formation models reported in literature by Field *et al.* (1995) and Jiraratanon and Chanachai (1996) are quite similar, a coefficient of determination (R^2) was calculated for each model and applied on both membrane cut-offs (0.25 kDa and 1 kDa). Table 19.3 shows the optimized values of the filtration constants (k) and the corresponding R^2 value for each model and both membrane cut-offs (Saleh *et al.*, 2010, 2011).

Table 19.3 The membrane molecular weight cut-off (MWCO), k value and corresponding R^2 value. (Operating conditions 30°C, 10°Brix, 15 bar, pH 4.)

Membrane MWCO (kDa)	Complete blocking		Partial pore blocking		Cake formation	
	k (s^{-1})	R^2	k (s^{-1})	R^2	k (s/m^2)	R^2
1	0.189	0.87	0.00865	0.93	0.000487	0.97
0.25	0.196	0.97	0.00662	0.94	0.000243	0.88

For the 1 kDa membrane, the R^2 values range from 0.87 to 0.97, with the cake formation model having an R^2 closest to 1. This is in agreement with the previous fouling model (permeate flux variations over time) (Wiesner and Aptel, 1996), which indicated the dominant type of fouling was the formation of a cake layer, without considering partial pore blocking. The partial pore blocking and complete pore blocking curves at 0.93 and 0.87 are not such a good fit against the experimental data and, therefore, are not applicable.

In the case of the 0.25 kDa membrane, the R^2 values range from 0.88 to 0.97 in reverse order to the 1 kDa membrane. The R^2 value of 0.97 for complete pore blocking is the best fit, indicating the more prevailing fouling mechanism was complete pore blocking. This is again in agreement with the findings of the previous models (permeate flux variation over time). This is because small molecular weight compounds are being deposited onto the membrane surface or into membrane pores and are blocking them (Jiraratananon and Chanachai, 1996). The mineral analyses of the feed showed that calcium, iron, magnesium, phosphorus and potassium were present, which may have contributed to the fouling. The free minerals would pass through the 1 kDa membrane unless they were bound to proteins, and thus may have been responsible for some pore blocking in the 0.25 kDa membrane (Saleh *et al.*, 2010, 2011).

Table 19.4 shows the values of k and R^2 for the run at 50°C with the model results for the 1 and 0.25 kDa membrane plotted in Figures 19.4 and 19.5, respectively. For the 1 kDa membrane, the most dominant fouling mechanism was cake formation with an R^2 value of 0.96. For the 0.25 kDa membrane the best R^2 value (0.98) was for the complete pore blocking and the partial pore blocking model. In contrast, the findings of the permeate flux variation over time model indicated the dominant fouling mechanism was cake formation due to the characteristics of the membrane for both the 1 and 0.25 kDa membranes (Saleh *et al.*, 2010, 2011).

Table 19.4 The membrane molecular weight cut-off (MWCO), k value and corresponding R^2 value. (Operating conditions 50°C, 10°Brix. 10 bar, pH 4.)

Membrane MWCO (kDa)	Complete blocking		Partial pore blocking		Cake formation	
	k (s^{-1})	R^2	k (s^{-1})	R^2	k (s/m^2)	R^2
1	0.182	0.81	0.00674	0.89	0.000312	0.96
0.25	0.166	0.98	0.00966	0.98	0.000692	0.88

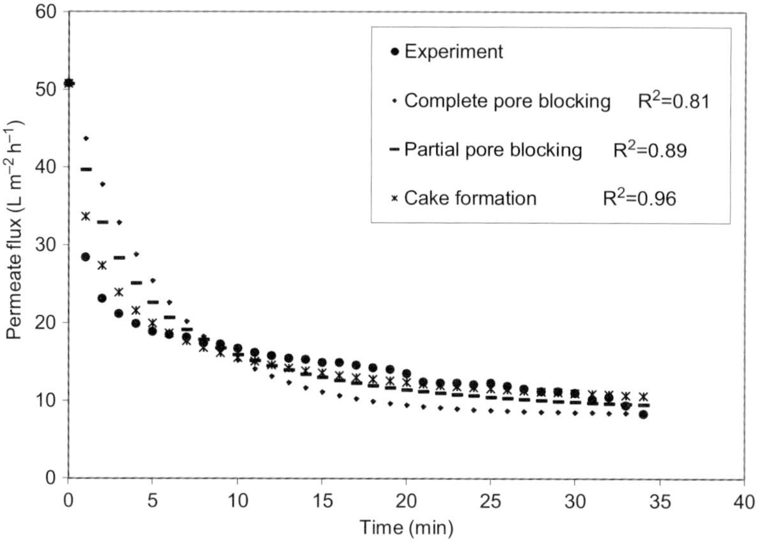

Fig. 19.4 The experimental data of the 1 kDa membrane plotted against the predicted flux declines. (Operating conditions were: 50°C, 10°Brix, 10 bar, pH 4.)

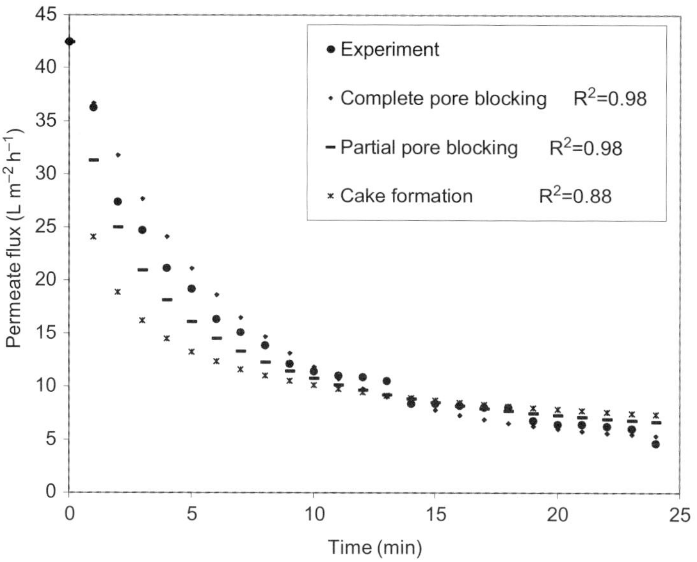

Fig. 19.5 The experimental data of the 0.25 kDa membrane plotted against the predicted flux decline. (Operating conditions were: 50°C, 10°Brix, 10 bar, pH 4.)

Conclusions

The aforementioned studies of the separation, concentration and fractionation of polyphenolics from different juices and the fouling mechanism during membrane filtration have led to a number of conclusions.

Results from the aforementioned studies will help ingredient manufacturers to adapt the membrane filtration technology for the separation of phenolic phytochemicals as functional ingredients for foods for the well-being market. Membrane filtration can potentially be used to recover and enrich polyphenols from apple juice and, by extension, would work for other clear juices.

Protein retention was strongly affected by different operating conditions. Increasing temperature, pressure and feed concentration, and decreasing the pH all increased the retention of proteins. The analyses of sugars showed that a significant amount of the simple sugars passed through both membranes and that the disaccharide sucrose was found to be largely retained. The membrane filtration process was shown to be significantly affected by temperature with the phenolics concentration on the retentate side decreasing significantly with increasing temperature. The concentration of polyphenolics in the retentate side increased with lowering feed pH and increasing initial sugar concentration. Increasing the transmembrane pressure increased fouling and indicated an attachment controlled mechanism due to the formation of a secondary membrane. The increase in fouling paralleled decreases in the amount of polyphenols found in the permeate solution. The importance of particulate fouling was proven by a significant decrease in flux and increase of fouling material over each run. Therefore, higher molecular weight polyphenolics in the feed are expected to have a significant role in membrane fouling. Increasing the feed concentration severely reduced the permeate flux and caused a noticeable increase in the phenolic fouling material accumulated on the membrane. The dominant type of fouling resulting from this is the polarization concentration phenomenon and the formation of a weak cake layer.

The effect of diafiltration was remarkable, with the permeate flux having been shown to recover significantly the original water flux; the phenolic contents on the retentate side after diafiltration was reduced remarkably. The recommended operating conditions for similar feed streams have been identified and will help ingredient manufacturers to adapt the membrane filtration technique reliably at a much lower cost. The 'resistance-in-series' model found that cake layer resistance increased with increasing feed concentration and decreasing pH values.

ADSORPTION AND ION EXCHANGE PROCESSES

In addition to the aforementioned process strategies, resin adsorption and ion exchange were systematically evaluated and optimized in further trials as a means to enrich, purify and fractionate plant secondary metabolites, such as polyphenols, which may be exploited both for their techno- and biofunctional features. For this purpose, isolated phenolic compounds were brought into contact by stirring phenolic model solutions with defined amounts of adsorber and ion exchange resins. The application of D-optimal experimental designs evaluating the effects of pH value, temperature, solute concentration and resin amount on the sorption and ion exchange behaviour of individual compounds, as well as mathematical modelling, allowed prediction of the recovery rates of phenolic compounds under any given experimental condition. Furthermore, the experimental data were applied to common adsorption isotherms, for example the Langmuir and Freundlich isotherms, demonstrating that these theories are not generally applicable to all sorption systems, even in relatively simple experimental settings, presumably due to deviations from the simplifying assumptions of these isotherms. The sorption and ion exchange systems were further characterized in terms of their binding kinetics and thermodynamic features, allowing the establishment of affinity orders of individual compounds upon contact with the resins, which is of particular importance for the successful fractionation of complex crude plant extracts in order to obtain preparations with well-defined functional properties. Sorption and ion exchange phenomena were further assessed in the presence of non-phenolic constituents, such as saccharides and amino acids, which were also shown to have an impact on phenolic recovery rates and which, thus, may not be neglected upon systematic optimization of sorption and ion exchange processes. Finally, this technology was demonstrated to be successfully applicable for the recovery of phenolic compounds from grape and apple pomace as well as sunflower expeller, thus valorizing these underused byproducts of plant food processing and contributing to sustainable agricultural production.

Characterization of sorption processes by evaluating sorption kinetics

To more systematically characterize sorption processes, the principles behind them are commonly described by assessing sorption kinetics and sorption isotherms. Sorption kinetics describes the time-dependent course of sorption processes until equilibrium conditions are reached. The mass transport going along with this process is usually subdivided into four consecutive phases: firstly, the sorptive is transported from the fluid phase into a layer which is formed on the sorbent surface and the concentration of which differs from that of the surrounding fluid phase. The second kinetic step consists of the transport of the target molecules through this liquid layer, which is also referred to as film diffusion. The third step is characterized by the transport of the components into the pores of the sorbent material, a process called pore diffusion, which also covers the diffusion of the components along the inner surface of the sorbent. Finally, the fourth step is dominated by the interaction of the target molecule with the active sites of the sorbent surface (Kümmel and Worch, 1990; Bathen and Breitbach, 2001).

The Lagergren equation (Equation 19.1) is a pseudo-first-order model commonly applied to describe the kinetics of sorption processes:

$$\frac{dN(t)}{dt} = k_1 \left(N^{(e)} - N(t)\right) \tag{19.1}$$

One major drawback of this model is the fact that it is only applicable to experimental data that are not too far from equilibrium conditions. In this case, $N^{(e)}$ represents the amount of the target compound bound on the resin surface under equilibrium conditions, whereas $N(t)$ represents the corresponding amount bound after time t. The linearized form of the Lagergren regression (Equation 19.2) yields a linear graph when plotting $\ln(N^{(e)} - N(t))$ values against time, t:

$$\ln \left(N^{(e)} - N(t)\right) = \beta - k_1 t \tag{19.2}$$

Consequently, this linearized form only yields linear graphs if experimental data of systems close to equilibrium conditions are evaluated. k_1 is a constant which, together with ß, can be deduced from the slope and intercept of the linear graph of Equation 19.2 (Kammerer *et al.*, 2011a).

Based on the four-step kinetic model described, the initial phase of the sorption kinetics requires particular consideration. Consequently, the Lagergren equation is not appropriate to describe the entire sorption process. Therefore, a modified Lagergren equation has been proposed recently, which is also referred to as the modified-pseudo-first-order model (MPFO, Equation 19.3) (Rudzinski and Plazinski, 2008a).

$$\ln \left(N^{(e)} - N(t)\right) + \frac{N(t)}{N^{(e)}} = \ln N^{(e)} - K_M t \tag{19.3}$$

This model ($N^{(e)}$ and $N(t)$ as described above, K_M : constant) may be described as a hybrid of the square root dependence, which is particularly appropriate to describe the initial phase of the sorption kinetics, and the Lagergren dependence, which is applicable to mathematically describe the sorption phase close to equilibrium. This equation further allows differentiation between the different phases of the adsorption process (Rudzinski and Plazinski, 2007; Rudzinski and Plazinski, 2008a, 2008b).

In most cases, adsorption and ion exchange phenomena are not differentiated in practical applications. Most theories and models have been developed based on sorption processes. The results obtained with these models have been transferred to ion exchange phenomena, even though in the latter case processes are much more complicated compared to sorption processes due to the involvement of ionic forces (Inglezakis and Poulopoulos, 2006).

Characterization of sorption processes at equilibrium: sorption isotherms

A large number of isotherms have been described in the literature for characterizing sorption and ion exchange systems at equilibrium (Kammerer *et al.*, 2011a). Among these models, the Langmuir and Freundlich isotherms are the most well-known and most frequently applied models. Originally, both isotherms had been developed to describe the sorption of gases onto solid surfaces. However, these models are also frequently applied to evaluate solid–liquid systems.

Langmuir isotherm

The Langmuir model is applied to characterize sorption systems with a monomolecular coverage of the sorbent surface. Furthermore, this model assumes energetically homogeneous sorption sites; mutual interactions of the molecules bound by the sorbent are excluded. The mathematical form of the Langmuir isotherm is represented by Equation 19.4:

$$q_s = \frac{Q_m a_L C_S}{1 + a_L C_S} = \frac{K_L C_S}{1 + a_L C_S} \tag{19.4}$$

with Q_m being the sorption capacity at complete monolayer coverage of the sorbent surface under equilibrium conditions, which means the maximum amount of the target compound that may be bound, given the aforementioned premises are fulfilled, and q_s and C_s, being the amount bound per gram of sorbent and the concentration of the target compound in the liquid phase under equilibrium conditions, respectively. Upon application of the linearized form of the Langmuir isotherm (Equation 19.5) to experimental data, K_L and Q_m can be deduced from the slope and intercept of the linear graphs that are obtained when plotting $1/q_s$ against $1/C_s$ (Kümmel and Worch, 1990; Allen *et al.*, 2003; Liu *et al.*, 2007).

$$\frac{1}{q_S} = \frac{1}{K_L C_S} + \frac{1}{Q_m} \tag{19.5}$$

Freundlich isotherm

In contrast to the Langmuir isotherm, the Freundlich model is applicable to describe monolayer sorption systems with sorption sites of different energy levels. This model is characterized by C_s and q_s, which are defined as described for the Langmuir isotherm, and by the adsorption capacity, K_F, of the sorbent and b_F as a measure of the adsorption intensity (Equation 19.6). The latter two parameters are empirical values and depend on the experimental conditions.

$$q_S = K_F C_S^{b_F} \tag{19.6}$$

Using the linearized form of the Freundlich isotherm, K_F and b_F can be deduced from the slope and intercept of the linear graph which is obtained when plotting log q_s against log C_s (Equation 19.7) (Kümmel and Worch, 1990; Liu *et al.*, 2007).

$$\log q_S = \log K_F + b_F \log C_S \tag{19.7}$$

However, it must be kept in mind that both the Langmuir and Freundlich isotherms are not generally applicable, since in practice conditions often significantly deviate from the premises described above. Thus, the applicability of these models to experimental data strongly depends on the target compounds as well as further experimental conditions. If the linearized form of both models (Equations 19.5 and 19.7) does not yield linear graphs upon evaluation of the data, the isotherms are not applicable to describe the corresponding sorption system (Bretag *et al.*, 2009a, 2009b).

Characterization of polyphenol adsorption in single compound model systems

For a systematic evaluation of the sorption behaviour of individual phenolic compounds, batch adsorption experiments were performed first using isolated polyphenols in single compound systems. A first study was performed with rutin (quercetin 3-*O*-rutinoside) using Amberlite® XAD 16 HP, a food-grade styrene-divinylbenzene copolymeric resin (Bretag *et al.*, 2009a). For this purpose, the phenolic compound was brought into contact with the synthetic adsorber resin by stirring rutin solutions in batch experiments with pre-defined resin amounts. A D-optimal experimental design was applied to study the effects of pH value, temperature, polyphenol concentration and resin amounts on the binding behaviour. Polyphenol concentrations in the solutions contacted with the resin were determined by HPLC-DAD (high performance liquid chromatography with diode array detection), whereas the amounts bound onto the adsorber surface were calculated by difference, considering the polyphenol amounts in solution prior to and after the adsorption experiment.

This statistical design proved to be highly appropriate for evaluating rutin adsorption under any experimental condition with only a limited number of experimental settings. Regression analyses allowed deduction of a mathematical model for predicting adsorption rates depending on the experimental conditions if rutin is present in isolated form. This further allowed calculating optimal sorption conditions for rutin. The data revealed a temperature-dependence for rutin sorption, with this effect also being dependent on the sorptive concentration and resin amounts. Furthermore, acidity of the solutions contacted with the adsorber resin played a dominant role, since phenolic hydroxyl functions are less dissociated under more acidic conditions rendering the compounds more lipophilic (Soto *et al.*, 2011). This favoured the interaction between the target compound and the apolar resin. Maximum sorption of 94.3% of initial rutin amounts was achieved at pH 1 and a temperature of 313 K, when 0.15 g resin/100 ml were stirred in a rutin solution of 50 mg/l. Figure 19.6 illustrates rutin adsorption rates

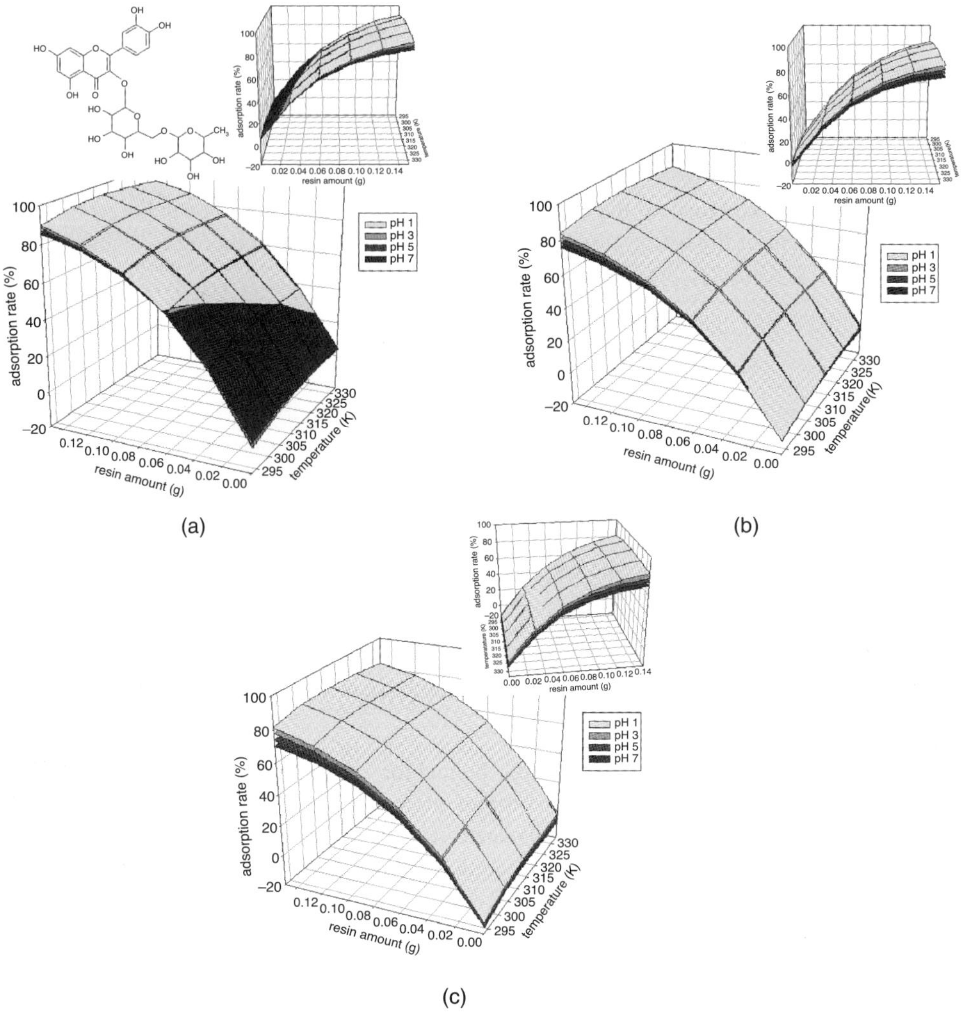

Fig. 19.6 Relative amounts of rutin adsorbed onto Amberlite® XAD 16HP depending on pH value, resin amount and temperature at a solute concentration of (a) 50 mg/l, (b) 100 mg/l and (c) 150 mg/l. (Reprinted from *Food Chemistry* (2009) **114**, J. Bretag, D.R. Kammerer, U. Jensen, R. Carle, Adsorption of rutin onto a food-grade styrene-divinylbenzene copolymer in a model system, 151–160 (Figure 1), Copyright 2008, with permission from Elsevier.)

depending on the solute concentration, the pH value, temperature and resin amounts calculated from the mathematical model obtained through regression analysis of the experimental data.

The Langmuir and Freundlich isotherms were successfully applied to the experimental data, which allowed thorough characterization of the rutin sorption system. As an example, the application of these models yielded maximum sorption capacity Q_m values when the resin surface is covered by a monomolecular layer of rutin. This parameter revealed a strong dependence on the initial sorptive concentration. Further, the sorption system at equilibrium was evaluated thermodynamically, thus calculating values of the changes of free energy

(Δ G), of enthalpy (Δ H) and of entropy (Δ S) caused by the adsorption process. These parameters were calculated both based on equilibrium constants, which were deduced from equilibrium rutin concentrations in the liquid phase and on the sorbent surface, and based on the constants derived from the Langmuir and Freundlich isotherms. Positive values of Δ S were interpreted as an increase of disorder (Liu *et al.*, 2007; Mittal *et al.*, 2007). However, the data calculated according to these two options differed significantly, which is due to the fact that major parameters deduced from the Langmuir and Freundlich isotherms were dependent on the resin amounts. Consequently, the thermodynamic data generated from the equilibrium constants appeared more reliable, which was also corroborated by the fact that these were widely in accordance with the experimental data. Nevertheless, Δ G and Δ H values reflected the temperature dependence of the sorption process, which can also be seen from Figure 19.6 (Bretag *et al.*, 2009a).

To further assess the impact of differing compound structures, which go along with differences in their physicochemical characteristics, on the sorption behaviour of individual phenolics upon their contact with the aforementioned styrene-divinylbenzene copolymeric resin, a number of further components representing different subclasses of the structurally highly diverse polyphenols were covered in a subsequent study. These included caffeic acid (hydroxycinnamic acid) and chlorogenic acid (5-*O*-caffeoylquinic acid; esterified hydroxycinnamic acid), catechin (flavanol) and phloridzin (dihydrochalcone). This study also aimed at evaluating the sorption behaviour of individual compounds when separately contacted with the sorbent by applying a D-optimal statistical design and mathematical modelling (Bretag *et al.*, 2009b). Since the studies were performed covering a broad pH range (pH 1.0–7.0), the stability of individual compounds under the experimental conditions had to be verified. Only chlorogenic acid revealed pH-dependent isomerization reactions. Experiments performed at pH 7 and some of the trials at pH 5 showed the formation of further components merely by stirring isolated chlorogenic acid in the presence of the adsorber resin. By mass spectrometric analyses these compounds were identified as neochlorogenic acid (3-*O*-caffeoylquinic acid) and cryptochlorogenic acid (4-*O*-caffeoylquinic acid), the formation of which due to isomerization in aqueous solution and under mild acidic and neutral pH conditions has been reported in the literature (Clifford, 2003). Similar effects were observed in experiments performed with catechin, where the formation of epicatechin was found. These isomers formed during the sorption experiments were quantitated and considered when calculating sorption rates.

As for rutin adsorption, regression analyses enabled mathematical modelling of phenolic compound sorption depending on temperature, pH value, resin amount and polyphenol concentration. Despite significant differences in the compound structures, all four compounds were most effectively bound under almost the same sorption conditions. Temperature was found to be one of the most decisive process parameters affecting sorption rates. Furthermore, pH conditions also markedly affected the binding behaviour, especially of caffeic and chlorogenic acids. The latter two compounds were particularly susceptible to pH variations due to their carboxylic functions exhibiting much higher acidity than phenolic hydroxyl functions and which are, thus, more easily deprotonated upon pH increase, rendering the compounds more hydrophilic, which weakens the interaction with the non-functionalized adsorber resin. This can also be clearly seen from the graphs in Figure 19.7, which illustrate the adsorption rates of caffeic acid, catechin, chlorogenic acid and phloridzin. Thus, optimal sorption of all phenolics was observed at pH 1 and a temperature of 293 and 303 K, respectively. Nevertheless, the phenolics differed in their affinity towards the resin. Under their respective optimal sorption conditions, the phenolics revealed the following order of

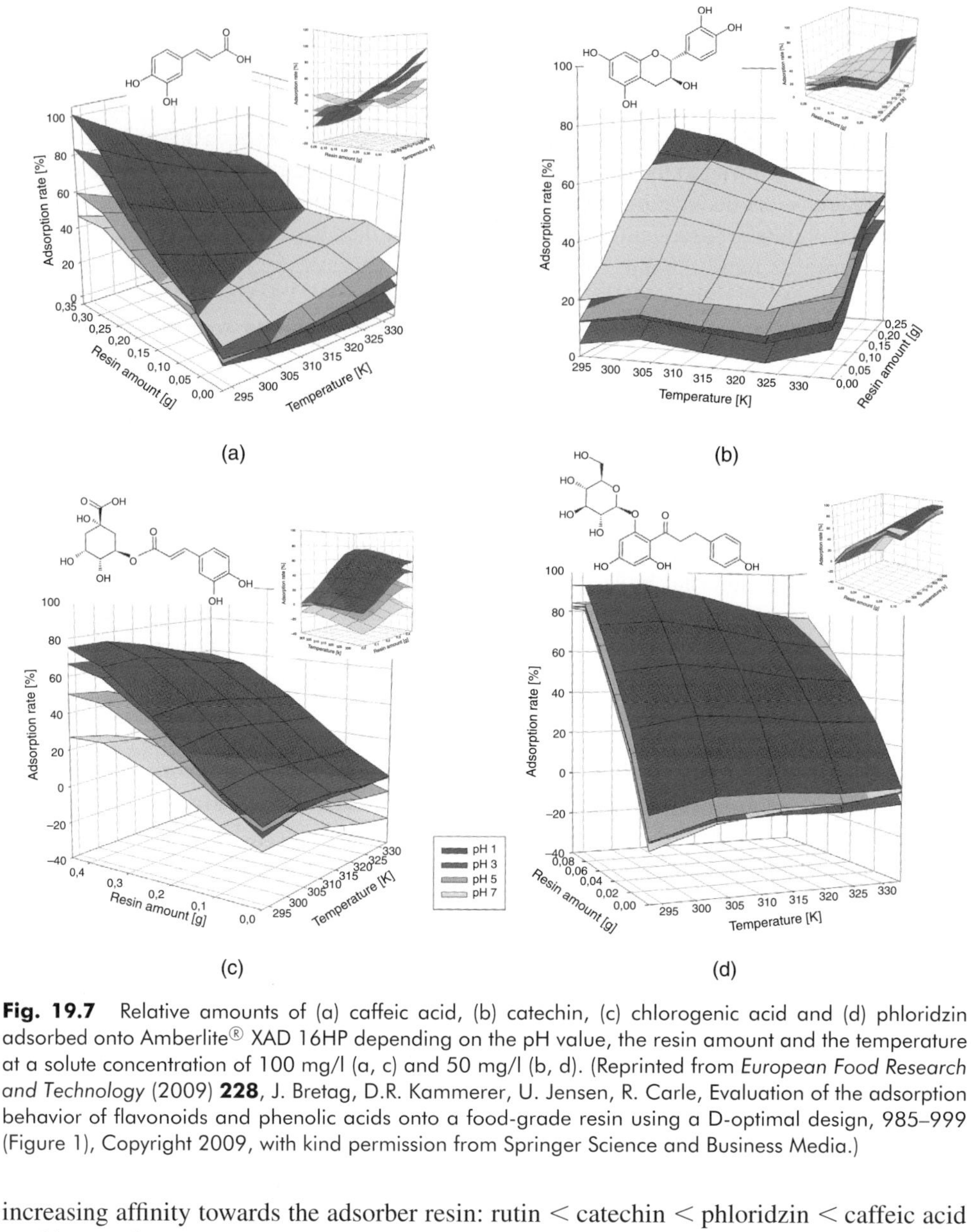

Fig. 19.7 Relative amounts of (a) caffeic acid, (b) catechin, (c) chlorogenic acid and (d) phloridzin adsorbed onto Amberlite® XAD 16HP depending on the pH value, the resin amount and the temperature at a solute concentration of 100 mg/l (a, c) and 50 mg/l (b, d). (Reprinted from *European Food Research and Technology* (2009) **228**, J. Bretag, D.R. Kammerer, U. Jensen, R. Carle, Evaluation of the adsorption behavior of flavonoids and phenolic acids onto a food-grade resin using a D-optimal design, 985–999 (Figure 1), Copyright 2009, with kind permission from Springer Science and Business Media.)

increasing affinity towards the adsorber resin: rutin < catechin < phloridzin < caffeic acid < chlorogenic acid.

Interestingly, data evaluation applying the Langmuir and Freundlich models showed that these two isotherms are only applicable for describing rutin and catechin adsorption. Both models exclude mutual interactions of the compounds bound onto the resin surface (Scordino *et al.*, 2004; Kammerer *et al.*, 2011a), which obviously is not generally fulfilled in practice. Instead, upon sorption of phenolic compounds, interactions via hydrogen bridges of the hydroxyl and carboxylic functions and, accordingly, the formation of multilayer coverages of the sorbent surface are even likely. Thus, it is not astonishing that the aforementioned isotherms are not generally applicable to describe all sorption systems. Rutin and catechin are presumably an exception, which is possibly due to the size of their molecular structure which aggravates the accumulation of higher amounts of phenolics on the sorbent surface and, thus, the formation of multilayer coverages.

The changes of free energy ΔG of the adsorption process revealed minima under conditions which were well in agreement with the optimal sorption conditions determined via the model equations. In contrast, ΔH and ΔS values could not be deduced from the experimental data of the four phenolics, which is probably also due to deviation from the ideal conditions as required by the Langmuir and Freundlich isotherms (Bretag *et al.*, 2009b).

Interaction of phenolic compounds upon adsorption from multicomponent model solutions

Single compound experiments allow optimal sorption conditions to be deduced exactly; however, they do not allow the behaviour upon adsorption from multicomponent solutions to be predicted. Consequently, such systems were assessed in further experiments. For this purpose, the five aforementioned phenolic compounds were used, which were applied as mixtures of two, three, four and five compounds, respectively, and contacted with the aforementioned food-grade adsorber resin in batch systems (Kammerer *et al.*, 2010a).

If, for example, conditions optimal for catechin adsorption were applied, this compound could still not be selectively enriched by resin adsorption, since all other phenolics were bound in comparable amounts. By direct comparison it could be demonstrated that hydrophilicity of the molecules is obviously not exclusively decisive with regard to phenolic affinity towards the resin. This was exemplified for the conditions optimal for caffeic acid adsorption. Under these conditions, more hydrophilic chlorogenic acid was bound in even higher amounts than caffeic acid.

The multicompound systems revealed complex mutual interactions of the phenolic compounds; these were observed both as synergistic and antagonistic effects. It was found that none of the experimental conditions was appropriate for isolating individual compounds from complex mixtures. However, certain compounds or compound classes may be enriched, either in the fluid phase or on the sorbent surface, by careful adjustment of the process conditions. It was further observed that these interactions may modulate the affinities of individual compounds. Thus, in contrast to the affinity order mentioned above, which was deduced from single compound systems, the following order of increasing affinity to the styrene-divinylbenzene copolymeric resin was deduced from multicompound systems: caffeic acid $<$ catechin $<$ chlorogenic acid $<$ phloridzin $<$ rutin.

For a better explanation of these observations, differences in the sorption kinetics of individual compounds were evaluated as well. Due to some major drawbacks of the Lagergren equation, which has been commonly applied to describe sorption kinetics, the so-called modified-pseudo-first-order (MPFO) model was applied in the present study. This model describes sorption as a two-step kinetic process, with the first phase being dominated by processes occurring at the sorbent surface, while intraparticle diffusion dominates the second phase (Rudzinski and Plazinski, 2007, 2008a, 2008b, 2008c). The successful application of this model confirmed that, in most cases, intraparticle diffusion presents the rate-determining step of adsorption in the phase close to equilibrium conditions. On the contrary, the kinetics in the initial phase of the sorption process confirmed the aforementioned affinity order, with rutin and phloridzin being characterized by faster binding in this important phase of adsorption.

In summary, the optimal conditions of adsorption as deduced from single compound sorption systems are not necessarily transferable to more complex multicompound systems. Thus, the impact of further molecules competing with the target compounds for sorption sites must be carefully considered when systematically optimizing selectivity and cost efficiency of secondary plant metabolite recovery and fractionation from crude plant extracts (Kammerer *et al.*, 2010a).

Impact of saccharides and amino acids on the interaction of phenolic compounds with adsorber and ion exchange resins

To approach the more complex systems of crude plant extracts, the effects of non-phenolic constituents on the sorption behaviour of polyphenols was assessed in batch systems. For this purpose, the impact of glucose, fructose and sucrose, which are usually present in much higher amount in plant matrices, on polyphenol sorption was studied. In addition, the L-amino acids alanine, glutamic acid, phenylalanine, lysine and histidin were supplemented to study their effects on polyphenol binding, especially since for the aromatic acids a significant interaction with phenolic compounds and apolar adsorber resins was expected. To further improve the selectivity of polyphenol recovery processes and provide opportunities to reproducibly fractionate complex plant extracts, several adsorber resins as well as cation and anion exchange resins were compared (Kammerer *et al.*, 2010b).

Synthetic ion exchangers were considered in this study to evaluate ionic interactions of the phenolics with the resins. Furthermore, it should be studied to what extent interactions with the apolar backbone of such functionalized resins decrease the selectivity procured by the functional groups. As an example, the binding rates of catechin, caffeic acid and chlorogenic acid were significantly lower when polyacrylamide-based anion exchangers were applied compared to equivalent resins based on polystyrene. This indicates the polymeric matrix of ion exchangers to also contribute to polyphenol binding via hydrophobic interactions, in addition to ionic forces imparted by the functional groups of the resins. Expectedly, an increase of binding rates using anion exchangers was observed at elevated pH values due to deprotonation of phenolic compounds, thus enabling ionic interactions with the functional groups of the resins. This effect was most pronounced for phenolic acids due to higher acidity of their carboxylic functions. In summary, a systematic comparison of ion exchangers and adsorber resins under varied experimental conditions revealed the potential to fractionate complex phenolic mixtures based on the physicochemical properties of individual compounds.

Sugar supplementation generally caused lower polyphenol binding rates when using anion exchange resins, which is probably due to saccharide interactions with the functional groups of the exchangers, thus interfering the interaction of phenolic compounds with the resins (Ghim and Chang, 1982). In a few cases, even unexpected effects were observed, for example an enhanced sorption of caffeic acid by a styrene-divinylbenzene copolymer in the presence of saccharides. This may be explained by the formation of multilayer coverages of the sorbent surface, which may be enhanced in the presence of saccharides through hydrogen bridges. As a consequence, common sorption isotherms, such as the Langmuir and Freundlich isotherms, are no longer applicable, since these models exclude such mutual interactions (Kammerer *et al.*, 2010a). In general, the sorption and ion exchange behaviour of phenolic compounds was markedly affected in numerous cases upon sugar supplementation, with ion exchange processes being more affected than adsorber processes. Furthermore, differences in the effects of different sugars were observed, which is presumably attributed to differing steric orientations of the hydroxyl functions (Vente *et al.*, 2005).

The supplementation with amino acids did not cause any effects that may be generalized. In some model systems polyphenol binding was not affected at all, while in others markedly reduced polyphenol recovery rates, due to competitive interaction of the amino acids with the resins, were observed. The degree of interaction strongly depended on the type of resin material. Interestingly, in a few cases even an amino acid induced increase of polyphenol binding rates was found.

These findings demonstrate that both saccharides and amino acids may not be neglected when polyphenol purification and fractionation must be systematically optimized. However, careful selection of process parameters and resin types provides the opportunity to selectively purify and fractionate phenolic compounds from extracts characterized by complex profiles of phenolic and non-phenolic constituents (Kammerer *et al.*, 2010b).

Concentration and fractionation of phenolic compounds using anion exchange resins

To demonstrate the potential to fractionate complex phenolic extracts based on the physico-chemical features of individual compounds and produce phenolic preparations with well-defined functional properties, in a further study phenolic compounds were enriched using anion exchangers in a column system. Ion exchange processes via the ammonium functions of the resin were distinguished from sorption processes via the apolar resin backbone (Inglezakis and Poulopoulos, 2006; Rudzinski and Plazinski, 2006, 2008b) by performing experiments at pH 1 and pH 7. In addition, the effects of different resin backbones on polyphenol binding under these conditions were evaluated in column experiments. For this purpose, one anion exchanger with a styrene-divinylbenzene structure and one based on polyacrylamide were used (Kammerer *et al.*, 2011b).

Under acidic conditions phenolic compounds mainly exist in non-deprotonated form. This only allows π–π interactions with the aromatic backbone of the polystyrene resin and van der Waals forces between the phenolics and the apolar resin matrix, respectively. In contrast, at pH 7 ionic interactions of the deprotonated phenolics with the quaternary ammonium functions of the resins play a dominant role. Interactions between the target compounds in solution and of those bound on the resin surface are mediated via π–π and van der Waals forces as well as dipole dipole forces and hydrogen bridges and generally increase binding rates (Cheng *et al.*, 2006).

At pH 1 progressive resin loading went along with increasing leakage rates in the column effluents. Upon column loading with phenolic solutions at pH 7, leakage rates were generally lower due to the additional ionic interactions described above. Especially, caffeic acid and chlorogenic acid revealed higher affinity towards the resin compared to the other phenolics, which is due to the higher acidity of their carboxylic functions (Figure 19.8). Furthermore, the binding behaviour of phloridzin and rutin could also be markedly modulated by pH variation, since around 50% of the molecules of both compounds are present as anions at pH 7. In contrast, catechin binding was hardly affected by a change of solvent acidity due to its comparatively high pk_a value. Further, the simultaneous application of several polyphenols to the ion exchange resins revealed significant interactions of the compounds, which must be considered when complex phenolic extracts are applied to sorbent and ion exchange resins.

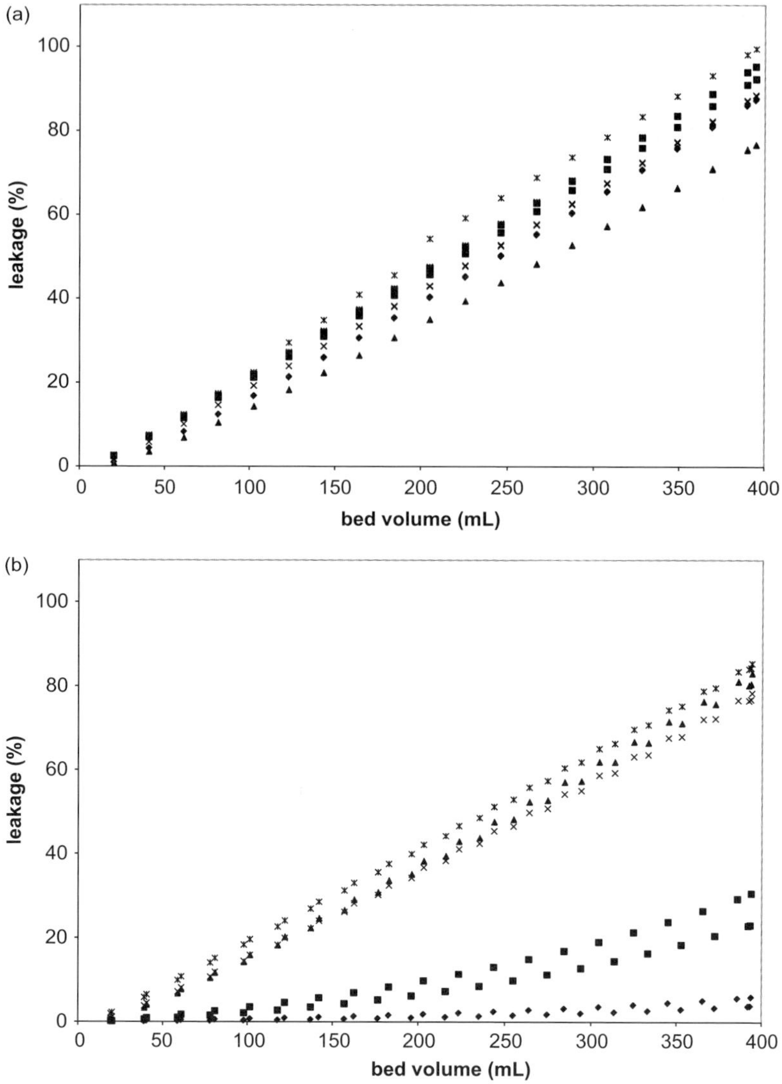

Fig. 19.8 Relative leakage of individual phenolic compounds upon application of a solution containing catechin (▲), caffeic acid (♦), chlorogenic acid (■), phloridzin (x) and rutin (⋇) onto Lewatit® S 6328 A at (a) pH 1 and (b) pH 7 (experiments performed in duplicate). (Reprinted from *LWT – Food Science and Technology* (2011) **44**, J. Kammerer, J. Boschet, D.R. Kammerer, R. Carle, Enrichment and fractionation of major apple flavonoids, phenolic acids and dihydrochalcones using anion exchange resins, 1079–1087 (Figures 1 and 2), Copyright 2010, with permission from Elsevier.)

Both ion exchangers with comparable exchanger capacities were directly compared to elucidate the impact of their matrix structures on binding rates and selectivity of the recovery process. Both resins were suitable for selectively enriching phenolic acids under neutral conditions. Under acidic conditions, significantly higher binding rates were observed with the polystyrene-based resin, which is ascribed to interactions of phenolic compounds with the resin backbone structure. Consequently, resin selection must also be carefully considered for selectively enriching and fractionating phenolic compounds (Kammerer *et al.*, 2011b).

Transfer of results from model experiments to the recovery of phenolic compounds from apple and grape pomace

To assess the transferability of results obtained from experiments with model solutions, further studies were performed directly comparing the sorption and ion exchange behaviour of phenolic compounds from crude extracts of apple pomace and red grape peels with those from model solutions, the compositions of which were adapted to these plant extracts with regard to their phenolic profile and amino acid composition (Kammerer *et al.*, 2011c).

The solvent type used for the preparation of the plant extracts proved to be decisive with regard to their composition and the proportion of phenolic compounds and amino acids. Thus, extract preparation must be carefully optimized prior to the application of ion exchange and adsorber resins, since this markedly contributes to the success of resin-based purification and fractionation processes. As expected from their structures, amino acids showed significant interactions both with cation and anion exchangers, which were characterized by marked pH optima. Moreover, the structure of individual amino acids, their initial concentrations and resin structures were important parameters determining binding rates. However, in mixtures composed of amino acids and polyphenols, the comparatively low recovery rates of amino acids were even reduced as a result of higher affinities of the phenolics towards the resins.

In the acidic pH range polyphenol binding from model solutions was hardly affected by the presence of amino acids. In contrast, under neutral conditions even enhanced interactions of the phenolics with the resins were observed in the presence of low amino acid amounts, which can be explained by synergistic effects as a result of co-adsorption. However, a specific prediction of possible interactions of both compound classes upon their contact with adsorber and ion exchange resins proved to be highly complex. This is further aggravated by additional compounds present in plant matrices, such as organic acids, minerals, pectins and other high molecular components, which are also expected to competitively interact with the sorbents.

Due to the presence of further compounds not considered in the model experiments, such as polymeric polyphenols, resin capacity was exhausted very quickly in experiments performed with pomace extracts. Thus, greater column dimensions and lower extract concentrations had to be used. Nevertheless, binding rates that were comparable to those of the model studies were determined in these studies. Expectedly, a connection of several adsorber and ion exchange columns in series contributed to markedly reduced leakage rates and, thus, may assist in significantly increasing the efficiency of such processes and in fractionating crude phenolic extracts (Kammerer *et al.*, 2011c).

Recovery of preparations rich in phenolic compounds from apple extracts

A re-diluted apple juice concentrate was used as a model substrate in further investigations of the systematic optimization of polyphenol recovery from a plant extract using a polymethyl-methacrylate resin. For this purpose, sorption and desorption rates of individual phenolic compounds were monitored varying the phenolic concentrations of the model solutions as well as temperature, pH value and resin amounts in batch experiments (Kammerer *et al.*, 2007).

As demonstrated in previous single compound experiments, this study proved the proportion of resin amounts and initial polyphenol concentration to have a marked impact on sorption kinetics and on phenolic amounts bound per mass unit of resin. With constant resin amounts, its loading increased almost linearly with the phenolic solute concentration. However, phenolic losses, that is the proportion of phenolics which remained unbound in solution

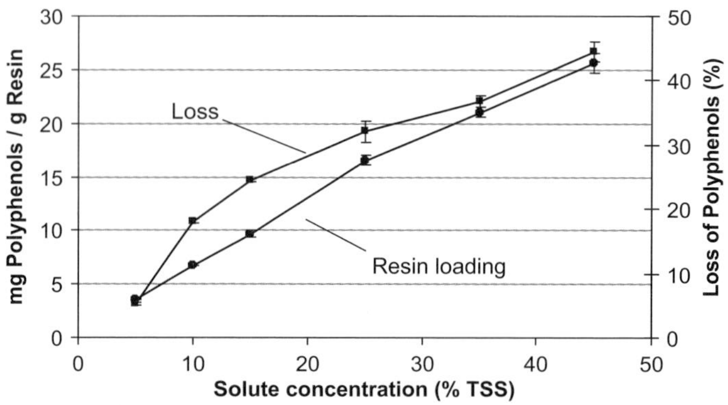

Fig. 19.9 Amounts of polyphenols per unit of weight of resin and losses of phenolic compounds depending on solute concentrations for experiments performed at 40°C. (Reprinted from *European Food Research and Technology* (2007) **224**, D.R. Kammerer, Z.S. Saleh, R. Carle, R.A. Stanley, Adsorptive recovery of phenolic compounds from apple juice, 605–613 (Figure 2), Copyright 2006, with kind permission from Springer Science and Business Media.)

at the end of the sorption experiment, showed similar trends, as illustrated in Figure 19.9. Both effects need to be carefully considered when optimizing the cost efficiency of polyphenol recovery processes. Interestingly, the contents of polymeric phenolic compounds, which are responsible for the brown colour of processed apple products, were more efficiently reduced than observed for total phenolic contents. This is presumably due to a higher hydrophobicity of polymeric compounds, and, for this reason, adsorber and ion exchange processes are successfully applied in industrial processing for the removal of browning compounds (Zaganiaris, 2011).

HPLC analyses of low molecular weight phenolic compounds revealed their structures and hydrophilicity to be a key parameter determining sorption rates. Thus, compounds of different polyphenol subclasses could be well differentiated based on their differing sorption rates. Further, even the glycosylation pattern of quercetin glycosides was reflected in differences of the affinities towards the resin, and steric effects were also noticeable, as could be deduced from the sorption behaviour of various caffeoylquinic acid isomers. Even though in some of the single compound batch experiments the Langmuir and Freundlich isotherms were found to be not applicable (Bretag *et al.*, 2009b), both models could be successfully applied in this study to describe sorption under equilibrium conditions. Mutual interactions of the phenolics in solution as well as the interaction with non-phenolic compounds may have caused that the preconditions for the applicability of these isotherms were fulfilled.

Desorption studies demonstrated an increase in temperature to yield higher polyphenol desorption rates. The application of hydroalcoholic solvents was the most efficient way for polyphenol elution. Maximum yields were obtained at an ethanol concentration of 60 vol-% and a methanol content of 70–80 vol-%, respectively. Under these conditions, around 70% of total phenolics bound onto the resin surface were eluted, whereas HPLC analyses revealed almost quantitative desorption of low molecular weight phenolic compounds. Accordingly, oligomeric and polymeric components are less efficiently recovered from adsorber resins, thus contributing to the aforementioned losses and incomplete compound elution. Depending on the solvent type, the desorption of individual phenolics differed, which again correlated with their lipophilicity. Consequently, a fractionation of polyphenols bound by adsorber resins is possible through consecutively applying different solvent systems (Kammerer *et al.*, 2007).

Recovery and fractionation of apple polyphenols by adsorption technology on pilot-plant scale

To test the transferability of these results obtained on laboratory scale to production on a larger scale, the same resin type was used in a subsequent study on pilot-plant scale. For this purpose, column sorption and desorption experiments were performed evaluating the effects of temperature, pH value, flow rate, polyphenol concentration of the feed solution and solvent type on the binding and elution behaviour of total and individual phenolic compounds (Kammerer *et al.*, 2010c).

Naturally, the conditions under which phenolic losses may jeopardise the cost effectiveness of the recovery process were significantly determined by the concentration of the solutions applied to the adsorber column. Thus, continuous control of the service life of adsorber columns is a prerequisite to optimize the cost efficiency of the entire process. The column experiments corroborated the findings of the laboratory trials, since more hydrophobic components exhibited higher affinity towards the resin material. Thus, highly hydrophilic compounds such as hydroxycinnamic acids showed higher leakage rates upon sample application to the column, which can be clearly seen in Figure 19.10. In contrast, dihydrochalcones and most quercetin glycosides were almost quantitatively bound onto the resin surface. At higher initial polyphenol concentrations, losses of hydroxycinnamates increased markedly, and, interestingly, such effects were also observed for dihydrochalcones and flavonols, which were bound almost quantitatively from more diluted solutions. Obviously, competitive effects, presumably caused by oligomeric and polymeric phenolics as well as non-phenolic compounds, become more important at higher concentrations, as could be seen from a marked reduction of binding rates.

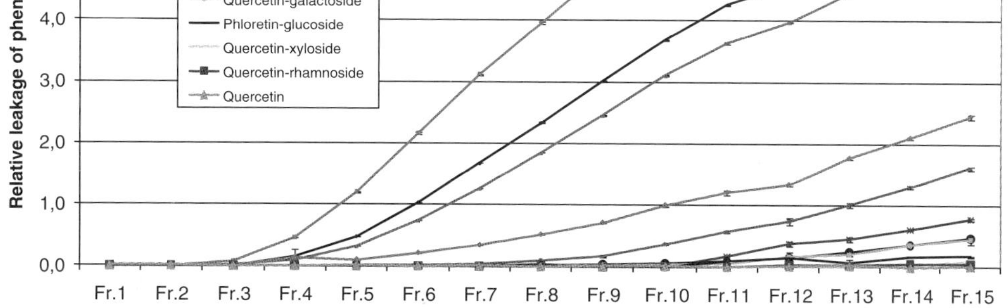

Fig. 19.10 Relative leakage of individual phenolic compounds quantified in individual fractions (V = 2 l each) collected during adsorption; mean ± standard deviation of the mean. (Reprinted with permission from *Journal of Agricultural and Food Chemistry* (2010) **58**, D.R. Kammerer, R. Carle, R.A. Stanley, Z.S. Saleh, Pilot-scale resin adsorption as a means to recover and fractionate apple polyphenols, 6787–6796 (Figure 2), Copyright 2010, American Chemical Society.)

Whereas flow rate during the column loading step only moderately affected polyphenol binding, a temperature increase from 20 to 60 and 80°C, respectively, caused a significant reduction of binding rates, which was in accordance with the results of the corresponding laboratory trials (Kammerer *et al.*, 2007). Hydroxycinnamic acids, which are obviously only bound to the resins via weak forces, were particularly susceptible to a temperature increase, as could be deduced from markedly reduced binding even at slightly increased temperatures. At the same time, this polyphenol subclass was most affected by acidification of the sample solutions, since protonation of their carboxylic function yielding neutral molecules went along with an enhanced interaction with the adsorber resin.

Desorption trials were performed with ethanol and methanol as well as their mixtures with water. Recovery rates and elution profiles were strongly dependent on the type of alcohol and on its concentration applied for desorption of the phenolics. In accordance with the laboratory experiments, total phenolic determinations using the Folin Ciocalteu assay revealed only partial recovery, not exceeding 80% of bound phenolics, which was due to strong binding of oligomeric and polymeric compounds, since HPLC analyses showed quantitative elution of low molecular weight phenolics with the pure alcohols. At lower alcohol concentrations, differences in the hydrophilicity of individual compounds could be deduced from their elution behaviour. Thus, even on a pilot-plant scale quercetin derivatives differing in their glycosidic substitution showed different desorption rates. Accordingly, these findings demonstrate the opportunity to fractionate complex phenolic crude extracts even on an industrial scale by applying adsorber resins and exploiting the different physicochemical characteristics of various polyphenol subclasses that go along with differences in their sorption and desorption behaviour (Kammerer *et al.*, 2010c).

Combined recovery of phenolic compounds and pectin from apple pomace

In a further approach, apple pomace arising in large amounts from apple juice production should be valorized by exploiting both phenolics and pectins. So far, apple pomace is commonly used for industrial production of pectin, which is used in foods as gelling agent and stabilizer. Since apple pomace is a particularly rich source of pectin, amounting to around 10–15% on a dry weight basis (Endreß, 2000), pectin recovery can be performed cost efficiently. However, due to co-extraction of oxidized phenolic compounds apple pectins exhibit brown hues, which may lead to limitations regarding their use in light-coloured products, even though gelling properties are mostly superior to citrus pectins. Therefore, this study aimed at the establishment of a process for the combined recovery of pectin and phenolic compounds based on polyphenol adsorption onto a food-grade adsorber resin (Carle *et al.*, 2001; Schieber *et al.*, 2003).

For this purpose, an acidic pomace extract was prepared, which is commonly used for direct precipitation of apple pectins, and applied to a pilot-scale column filled with Amberlite XAD 16 HP at a flow rate of 10 bed volumes/h. After sample application, the resin was washed with water to remove residual pectins, whereas the phenolics bound onto the resin were recovered by elution with two bed volumes of methanol.

This process allows simultaneous recovery of both compound classes, since pectins passed the resin without significant interaction with the styrenic polymer (Figure 19.11). The acidic conditions of the pectin extraction process further enhanced polyphenol binding by the adsorber resin, since the phenolic profile of apples and apple pomace is characterized by high

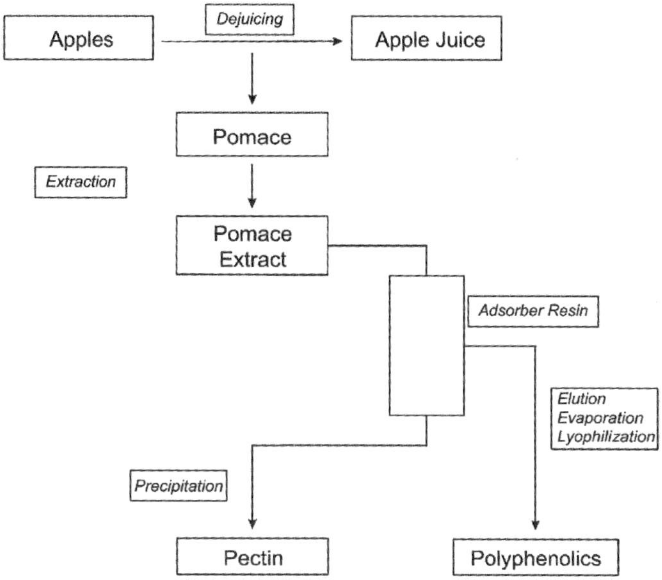

Fig. 19.11 Process for the combined recovery of pectin and phenolic compounds from apple pomace. (Reprinted from *Innovative Food Science and Emerging Technologies* (2003) **4**, A. Schieber, P. Hilt, P. Streker, H.-U. Endreß, C. Rentschler, R. Carle, A new process for the combined recovery of pectin and phenolic compounds from apple pomace, 99–107 (Figure 1), Copyright 2002, with permission from Elsevier.)

proportions of phenolic acids, which are rendered more hydrophobic under low pH conditions due to protonation of their carboxylate functions. Thus, the resin proved to efficiently retain phenolic compounds, and only very low levels of chlorogenic acid were co-eluted with the pectin-containing column effluent. Furthermore, the adsorber treatment caused a significant increase of lightness of the extract, and the colour turned to more yellow hues. By methanolic elution from the column, a phenolic preparation rich in phloridzin, chlorogenic acid and a number of quercetin glycosides was obtained. Since this process can be easily integrated in commercial pectin production, the industrial scale-up is a straightforward process (Schieber *et al.*, 2003).

Application of adsorber and ion exchange resins for the valorization of plant food processing byproducts: recovery of proteins and phenolic compounds from sunflower expeller

The potential of adsorber and ion exchange resins for the recovery of valuable plant constituents and the valorization of so far underused byproducts of plant food processing was demonstrated by developing a process for the combined recovery of proteins and polyphenols from sunflower expeller. Sunflower seeds are a particularly rich source of proteins with promising nutritional and technofunctional properties. Nevertheless, these proteins have not been exploited for human consumption so far. This is mainly due to the high phenolic contents of the press residues originating from sunflower oil recovery, amounting to 1–4% (Pickardt *et al.*, 2009, 2011). These compounds mainly belong to hydroxycinnamic acids (Weisz *et al.*, 2009) and are known to be prone to spontaneous oxidation under the alkaline conditions which are commonly applied for protein extraction. The quinones resulting from this oxidation are highly reactive, thus forming covalent bonds to reactive side chains of the proteins, which affects both their nutritional value and functional properties. Furthermore, protein isolates recovered under conditions conventionally applied for protein recovery exhibit greyish-brown shades, which is due to the presence of co-extracted polymerized phenolic compounds (Pickardt *et al.*, 2009, 2011; Weisz *et al.*, 2010).

In an innovative approach a process for protein extraction under mild acidic conditions (pH 6.0) was developed, adding sodium chloride (1.3 mol/L) to partly compensate for decreased protein extractability, when compared to alkaline extraction, by shifting the isoelectric point of the proteins. At the same time, these conditions efficiently prevent polyphenol oxidation and the unwanted effects associated therewith, such as covalent binding to the proteins (Pickardt *et al.*, 2009). In a subsequent step, co-extracted polyphenols were recovered by adsorption and ion exchange, thus providing the opportunity to valorize sunflower expeller by recovering both proteins for human nutrition and phenolic compounds as plant secondary metabolites with putative health-beneficial properties.

For this purpose, several food-grade adsorber and anion exchange resins were compared with regard to their potential to decrease phenolic contents in the protein extracts. The ion exchangers were more efficient in binding sunflower phenolics, which is due to the predominance of phenolic acids in sunflowers and their deprotonation at the extraction pH value (6.0). Vice versa, adsorber resins were much better suited for the sorption of polymeric polyphenols, which was deduced from a strong decoloration of the brown protein crude extracts (Weisz *et al.*, 2010).

Optimal sorption conditions in column experiments using an apolar adsorber resin were deduced applying a D-optimal experimental design. According to these results, temperature during adsorption should be kept as low as possible. Furthermore, a low flow rate of three bed volumes/h and a low extract concentration were found favourable for maximum polyphenol binding. However, the mathematical model established for predicting sorption rates under any experimental condition revealed a single-step sorption process to be insufficient for complete polyphenol removal from the extracts. Thus, in a scale-up experiment extracting 100 kg of sunflower expeller with 1500 l of a 1.3 M NaCl solution, both an ion exchange (90 L) and an adsorber column (90 L) were combined in series to reduce the contents of low molecular weight phenolic acids and of polymeric browning compounds. Subsequent isoelectric precipitation of the proteins from the extracts purified through ion exchange and adsorption as well as washing of the precipitated proteins allowed the removal of 99.4% of the low molecular weight phenolic compounds (Weisz *et al.*, 2010). This process yielded almost colourless protein isolates with low residual phenolic contents, which are, thus, applicable for human nutrition.

To also valorize the phenolic compounds removed from the protein crude extracts as natural antioxidants and for the production of functional and enriched foods, their recovery by desorption from the resins was optimized in a further study. For this purpose, an adsorber column with bound sunflower phenolics was treated with methanol, ethanol and 2-propanol as well as with mixtures of these alcohols with water. For modelling the desorption step, a D-optimal design was applied (Weisz *et al.*, 2013). Initial screening experiments comparing these three alcohols demonstrated 2-propanol to be a suitable alternative to toxic methanol and ethanol, which is not allowed for food production in a number of countries. Almost quantitative desorption of bound phenolics was achieved with 70% 2-propanol. In contrast, the flow rate of the solvent as well as the temperature during desorption did not significantly affect recovery rates. Only at low flow rates did an increase in temperature appear beneficial.

Even with only one bed volume of 70 and 50% 2-propanol, around 99 and 86%, respectively, of mono- and dimeric polyphenols bound onto the resin surface were eluted, while with the second bed volume only further 0.2 and 0.5% were recovered. Thus, the bound phenolics can be eluted with very low solvent volumes and be strongly concentrated (Weisz *et al.*, 2013).

SUMMARY

The present chapter has demonstrated membrane separation and filtration as well as adsorption and ion exchange to be valuable tools for the enrichment, purification and fractionation of phenolic compounds. The studies discussed in this context have shown that these methodologies may be precisely described applying mathematical modelling, allowing prediction of recovery rates under any experimental condition. However, it has also been found that the complex composition of plant extracts comprising a wide range of different compound classes aggravates systematic prediction of recovery rates. Nevertheless, the methods presented are an indispensable tool for optimizing polyphenol recovery, since most processes realized in industrial practice are based on empirical investigations and have not been systematically optimized so far. Thus, such methods also contribute to sustainable agricultural and food production, since they provide the opportunity to valorize byproducts of food processing which mostly have been underused so far. Novel developments have revealed potential combinations of membrane and ion exchange processes for the fractionation of crude phenolic extracts by using, for example, sulfonated membrane materials, thus allowing separations both based on molecular size and ionic charge (Juadjur and Winterhalter, 2012). This allows production of phenolic preparations with well-defined functional properties, even though a wide range of biofunctional characteristics still needs to be proven in future in human trials.

NOMENCLATURE

K_F	Adsorption capacity
$N^{(e)}$	Amount of target compound bound on the resin surface under equilibrium conditions
$N(t)$	Amount of target compound bound on the resin surface after time t
t	Filtration / adsorption time (min)
μ_w	Viscosity of pure water (Pa.s)
P	Transmembrane pressure (kPa)
Q_m	Sorption capacity at complete monolayer coverage of the sorbent surface under equilibrium conditions
q_s	Amount of target compound bound per gram of sorbent under equilibrium conditions
R_r	Membrane filter resistance (m^{-1})
R_t	Total resistance (m^{-1})
R_c	Cake layer resistance (m^{-1})
R_{frev}	Reversible fouling resistance (m^{-1})
K and k	Filtration constants (m^2/l) or (s^{-1}) or (m^{-1}) or (s/m^2)
n	Filtration constant

REFERENCES

Abbas, H. (2004) Anhydrous milkfat fractionation using membrane. Thesis, Chemical and Materials Engineering, The University of Auckland, New Zealand.

Allen, S.J., Gan, Q., Matthews, R. and Johnson, P.A. (2003) Comparison of optimized isotherm models for basic dye adsorption by kudzu. *Bioresource Technology* **88**, 143–152.

Alper, N. and Acar, J. (2004) Removal of phenolic compounds in pomegranate juice using ultrafiltration and laccase-ultrafiltration combinations. *Nahrung – Food* **48**(3), 184–187.

Ataei, S.A. and Vasheghani-Farahani, E. (2008) In situ separation of lactic acid from fermentation broth using ion exchange resins. *Journal of Industrial Microbiology and Biotechnology* **35**, 1229–1233.

Balakrishnan, M., Dua, M. and Bhagat, J.J. (2000) Effect of operating parameters on sugarcane juice ultrafiltration: Results of a field experience. *Journal of Separation and Purification Technology* **19**(3), 209–220.

Bathen, D. and Breitbach, M. (eds) (2001) *Adsorptionstechnik*. Springer, Berlin, Heidelberg.

Bellamy, S.A. and Pease, S. (2009) Stabilization of liquid food and beverages. *US Patent 2009291171*.

Borneman, Z., Gökmen, V. and Nijhuis, H.H. (2001) Selective removal of polyphenols and brown colour in apple juices using PES/PVP membranes in a single ultrafiltration process. *Separation and Purification Technology* **22–23**, 53–61.

Bradford, M.M. (1976) A rapid and sensitive method for the quantitation of microgram quantities of protein utilizing the principle of protein-dye binding. *Analytical Biochemistry* **72**(1–2), 248–254.

Bretag, J., Kammerer, D.R., Jensen, U. and Carle, R. (2009a) Adsorption of rutin onto a food-grade styrene-divinylbenzene copolymer in a model system. *Food Chemistry* **114**, 151–150.

Bretag, J., Kammerer, D.R., Jensen, U. and Carle, R. (2009b) Evaluation of the adsorption behavior of flavonoids and phenolic acids onto a food-grade resin using a D-optimal design. *European Food Research and Technology* **228**, 985–999.

Carle, R., Keller, P., Schieber, A. *et al.* (2001) Method for obtaining useful materials from the by-products of fruit and vegetable processing. *Patent WO 01/78859*.

Cassano, A., Donato, L. and Drioli, E. (2007) Ultrafiltration of kiwifruit juice: Operating parameters, juice quality and membrane fouling. *Journal of Food Engineering* **79**(2), 613–621.

Cassano, A., Donato, L., Conidi, C. and Drioli, E. (2008) Recovery of bioactive compounds in kiwifruit juice by ultrafiltration. *Innovative Food Science and Emerging Technologies* **9**(4), 556–562.

Chandler, B.V., Kefford, J.K. and Ziemelis, G. (1968) Removal of limonin from bitter orange juice. *Journal of the Science of Food and Agriculture* **19**, 83–86.

Cheng, S., Tang, H. and Yan, H. (2006) Effects of multiple weak interactions on the binding of phenolic compounds by polymeric adsorbents. *Journal of Applied Polymer Science* **102**, 4652–4658.

Clifford, M.N. (2003) The analysis and characterization of chlorogenic acids and other cinnamates. In: *Methods in Polyphenol Analysis* (eds C. Santos-Buelga, G. Williamson), pp. 314–447. The Royal Society of Chemistry, Cambridge, UK.

Coca, M., García, M.T., Mato, S. *et al.* (2008) Evolution of colorants in sugarbeet juices during decolorization using styrenic resins. *Journal of Food Engineering* **89**, 429–434.

Dawson-Ekeland, K.R. and Stringfield, R.T. (1991) Decaffeination of liquid medium. *US Patent 5,021,253*.

De Bruijn, J. and Borquez, R. (2006) Analysis of the fouling mechanisms during cross-flow ultrafiltration of apple juice. *Food Science and Technology* **39**(6), 861–871.

De Bruijn, J., Venegas, A. and Borquez, R. (2002) Influence of crossflow ultrafiltration on membrane fouling and apple juice quality. *Desalination* **148**(1–3), 131–136.

De Bruijn, J., Venegas, A., Salazar, F.N. and Borquez, R. (2003) Ultrafiltration performance of Carbosep membrane for the clarification of apple juice. *Food Science and Technology* **36**(4), 397–406.

Di Mauro, A., Fallico, B., Passerini, A. *et al.* (1999) Recovery of hesperidin from orange peel by concentration of extracts on styrene-divinylbenzene resin. *Journal of Agricultural and Food Chemistry* **47**, 4391–4397.

Di Mauro, A., Arena, E., Fallico, B. *et al.* (2002) Recovery of anthocyanins from pulp wash of pigmented oranges by concentration on resins. *Journal of Agricultural and Food Chemistry* **50**, 5968–5974.

Dorfner, K. (ed.) (1970) *Ionenaustauscher*. Walter de Gruyter and Co., Berlin.

Eaton, P. and Lux, R. (1984) Laboratory fouling test apparatus for hydrocarbon feedstocks: Fouling in heat exchange equipment. *ASME HTD*, **35**, 33–42.

Endreß, H.-U. (2000) High quality resulting from products integrated environment protection - PIUS. *Fruit Processing* **10**, 273–276.

Field, R.W., Wu, D., Howell, J.A. and Gubta, B.B. (1995) Critical flux concept for microfiltration. *Journal of Membrane Science* **100**(3), 259–272.

Ghim, Y.S. and Chang, H.N. (1982) Adsorption characteristics of glucose and fructose in ion-exchange resin columns. *Industrial and Engineering Chemistry Fundamentals* **21**, 369–374.

Goosen, M.F.A., Sablani, S.S., Al-Maskari, S.S. *et al.* (2002) Effect of feed temperature on permeate flux and mass transfer coefficient in spiral-wound reverse osmosis systems. *Desalination* **144**(1–3), 367–372.

Helfferich, F. (ed.) (1995) *Ion exchange*. Dover Publications, New York.

Inglezakis, V. and Poulopoulos, S. (eds) (2006) *Adsorption, ion exchange and catalysis: design of operations and environmental applications*. Elsevier, Amsterdam, The Netherlands.

Jiraratananon, R. and Chanachai, A. (1996) A study of fouling in the ultrafiltration of passion fruit juice. *Journal of Membrane Science* **111**(1), 39–48.

Johnson, R.L. and Chandler, B.C. (1982) Reduction of bitterness and acidity in grapefruit juices by adsorption process. *Journal of the Science of Food and Agriculture* **33**, 287–293.

Juadjur, A. and Winterhalter, P. (2012) Development of a novel adsorptive membrane chromatographic method for the fractionation of polyphenols from bilberry. *Journal of Agricultural and Food Chemistry* **60**, 2427–2433.

Kammerer, D., Gajdos Kljusuric, J., Carle, R. and Schieber, A. (2004) Recovery of anthocyanins from grape pomace extracts (*Vitis vinifera* L. cv. Cabernet Mitos) using a polymeric adsorber resin. *European Food Research and Technology* **220**, 431–437.

Kammerer, D.R., Saleh, Z.S., Carle, R. and Stanley, R.A. (2007) Adsorptive recovery of phenolic compounds from apple juice. *European Food Research and Technology* **224**, 605–613.

Kammerer, J., Kammerer, D.R., Jensen, U. and Carle, R. (2010a) Interactions of apple polyphenols in a multi-compound system upon adsorption onto a food-grade resin. *Journal of Food Engineering* **96**, 544–554.

Kammerer, J., Kammerer, D.R. and Carle, R. (2010b) Impact of saccharides and amino acids on the interaction of apple polyphenols with ion exchange and adsorbent resins. *Journal of Food Engineering* **98**, 230–239.

Kammerer, D.R., Carle, R.,Stanley, R.A. and Saleh, Z.S. (2010c) Pilot-scale resin adsorption as a means to recover and fractionate apple polyphenols. *Journal of Agricultural and Food Chemistry* **58**, 6787–6796.

Kammerer, J., Carle, R. and Kammerer D.R. (2011a) Adsorption and ion exchange: Basic principles and their application in food processing. *Journal of Agricultural and Food Chemistry* **59**, 22–42.

Kammerer, J., Boschet, J., Kammerer, D.R. and Carle, R. (2011b) Enrichment and fractionation of major apple flavonoids, phenolic acids and dihydrochalcones using anion exchange resins. *LWT – Food Science and Technology* **44**, 1079–1087.

Kammerer, J., Schweizer, C., Carle, R. and Kammerer, D.R. (2011c) Recovery and fractionation of major apple and grape polyphenols from model solutions and crude plant extracts using ion exchange and adsorbent resins. *International Journal of Food Science and Technology* **46**, 1755–1767.

Kammerer, D.R., Kammerer, J. and Carle, R. (2013) Anreicherung und Fraktionierung phenolischer Verbindungen mittels Adsorber- und Ionenaustauscherharzen. *Deutsche Lebensmittel-Rundschau*, **109**, 25–32.

Karode, S.K., Courtois, T., and Gupta, B.B., (2000) Ultrafiltration of raw Indian sugar solution using polymeric and mineral membranes. *Separation Science and Technology* **35**(15), 2473–2483.

Kümmel, R. and Worch, E. (eds) (1990) *Adsorption aus wässrigen Lösungen*. VBE Deutscher Verlag für Grundstoffindustrie, Leipzig.

LeVav, M.D. and Carta, G. (eds) (2007) *Perry's Chemical Engineers' Handbook: Adsorption and Ion Exchange*. McGraw-Hill Professional, New York.

Liu, X., Xu, Z., Gao, Y. *et al.* (2007) Adsorption characteristics of anthocyanins from purple-fleshed potato (*Solanum tuberosum* Jasim) extract on macroporous resins. *International Journal of Food Engineering* **3**, 1–16.

Lyndon, R. (1996) Kommerzialisierung der Adsorbertechnologie in der Fruchtsaftindustrie. *Flüssiges Obst* **63**, 499–503.

Maier, T., Göppert, A., Kammerer, D.R. *et al.* (2008) Optimization of a process for enzyme-assisted pigment extraction from grape (*Vitis vinifera* L.) pomace. *European Food Research and Technology* **227**, 267–275.

Maier, T., Schieber, A., Kammerer, D.R. and Carle, R. (2009) Residues of grape (*Vitis vinifera* L.) seed oil production as a valuable source of phenolic antioxidants. *Food Chemistry* **112**, 551–559.

Mantarri, M., Pihlajamaki, A. and Nystrom, M. (2006) Effect of pH on hydrophilicity and charge and their effect on the filtration efficiency of NF membranes at different pH. *Journal of Membrane Science* **280**(1–2), 311–320.

Melear, E. and Zoltek Jr, J. (1982) Physical and chemical methods. *Journal of the Water Pollution Control Federation* **54**, 584–592.

Mittal, A., Kurup, L. and Mittal, J. (2007) Freundlich and Langmuir adsorption isotherms and kinetics for the removal of tartrazine from aqueous solutions using hen feathers. *Journal of Hazardous Materials* **146**, 243–248.

Nawaz, H., Shi, J., Mittal, G.S. and Kakuda, Y. (2006) Extraction of polyphenols from grape seeds and concentration by ultrafiltration. *Separation and Purification Technology* **48**(2), 176–181.

Noble, R.D. and Stern S.A. (1995). *Membrane Separations Technology Principles and Applications*. Elsevier, New York.

Pickardt, C., Neidhart, S., Griesbach, C. *et al.* (2009) Optimization of mild-acidic protein extraction from defatted sunflower (*Helianthus annuus* L.) meal. *Food Hydrocolloids* **23**, 1966–1973.

Pickardt, C., Hager, T., Eisner, P. *et al.* (2011) Isoelectric protein precipitation from mild-acidic extracts of de-oiled sunflower (*Helianthus annuus* L.) press cake. *European Food Research and Technology* **233**, 31–44.

Rai, P., Majumdar, G.C., Sharma, G. *et al.* 2006. Effect of various cutoff membranes on permeate flux and quality during filtration of Mosambi (*Citrus sinensis* (L.) Osbeck) juice. *Food and Bioproducts Processing* **84**(3), 213–219.

Rudzinski, W. and Plazinski, W. (2006) Kinetics of solute adsorption at solid/solute interfaces: A theoretical development of the empirical pseudo-first and pseudo-second order kinetic rate equations, based on applying the statistical rate theory of interfacial transport. *Journal of Physical Chemistry B* **110**, 16514–16525.

Rudzinski, W. and Plazinski, W. (2007) Studies of the kinetics of solute adsorption at solid/solution interfaces: On the possibility of distinguishing between the diffusional and the surface reaction kinetic models by studying the pseudo-first order kinetics special features of the initial adsorption kinetics. *The Journal of Physical Chemistry C* **111**, 15100–15110.

Rudzinski, W. and Plazinski, W. (2008a) Kinetics of solute adsorption at solid/aqueous interfaces: Searching for the theoretical background of the modified-pseudo-first order kinetic equation. *Langmuir*, **24**, 5393–5399.

Rudzinski, W. and Plazinski, W. (2008b) Kinetics of solute adsorption at solid/solution interfaces: On the special features of the initial adsorption kinetics. *Langmuir* **24**, 6738–6744.

Rudzinski, W. and Plazinski, W. (2008c) Kinetics of dyes adsorption at the solid-solution interfaces: A theoretical description based on the two-step kinetic model. *Environmental Science and Technology* **42**, 2470–2475.

Saleh, Z.S., Stanley, R.A. and Wibisono, R. (2006) Separation and concentration of health compounds from fruit juice using membrane technology. *International Journal of Food Engineering* **2**(3), 1556–3758.

Saleh, Z.S., Farr, J.M. and Friedrich, S.A. (2010) Mitigating and modeling membrane fouling by bioactive mixtures. Proceedings of The American Separation and Filtration Society Conference, San Antonio, Texas, USA, 21–26 March 2010.

Saleh, Z.S., Farr, J.M. and Friedrich, S.A. (2011) Mitigating and modelling membrane fouling by bioactive mixtures. *Journal of Filtration*, **11**(1), 34–45.

Schieber, A., Hilt, P., Streker, P. *et al.* (2003) A new process for the combined recovery of pectin and phenolic compounds from apple pomace. *Innovative Food Science and Emerging Technologies* **4**, 99–107.

Scordino, M., Di Mauro, A., Passerini, A. and Maccarone, E. (2003) Adsorption of flavonoids on resins: Hesperidin. *Journal of Agricultural and Food Chemistry* **51**, 6998–7004.

Scordino, M., Di Mauro, A., Passerini, A. and Maccarone, E. (2004) Adsorption of flavonoids on resins: Cyanidin 3-glucoside. *Journal of Agricultural and Food Chemistry* **52**, 1965–1972.

Scordino, M., Di Mauro, A., Passerini, A. and Maccarone, E. (2005) Selective recovery of anthocyanins and hydroxycinnamates from a byproduct of citrus processing. *Journal of Agricultural and Food Chemistry* **53**, 651–658.

Serpen, A., Ataç, B. and Gökmen, V. (2007) Adsorption of Maillard reaction products from aqueous solutions and sugar syrups using adsorbent resin. *Journal of Food Engineering* **82**, 342–350.

Shannon, M., Bohn, P.W., Elimelech, M. *et al.* (2008) Science and technology for water purification in the coming decades. *Nature* **452**, 301–310.

Soler, C., Soriano, J.M. and Manes, J. (2009) Apple-products phytochemicals and processing: a review. *Natural Product Communications* **4**(5), 659–670.

Soto, M.L., Moure, A., Domínguez, H. and Parajó, J.C. (2011) Recovery, concentration and purification of phenolic compounds by adsorption: A review. *Journal of Food Engineering* **105**, 1–27.

Stevenson, D.E., Wibisono, R., Jensen, D.J. *et al.* (2006) Direct acylation of flavonoid glycosides with phenolic acids catalysed by Candida antarctica lipase B (Novozym 435.5®). *Enzyme and Microbial Technology* **36**(6), 1236–1241.

Tanada-Palmu, P., Jardine, J., Matta, V. (1999) Production of a banana (*Musa cavendishii*) extract containing no polyphenol oxidase by ultrafiltration. *Journal of the Science of Food and Agriculture* **79**(5), 643–647.

Tong, W.-Y., Fu, X.-Y., Lee, S.-M. *et al.* (2004) Purification of L(+)-lactic acid from fermentation broth with paper sludge as a cellulosic feedstock using weak anion exchanger Amberlite IRA-92. *Biochemical Engineering Journal* **18**, 89–96.

Tsao, R., Yang, R., Xie, S. *et al.* 2005. Which polyphenolic compounds contribute to the total antioxidant activities of apple? *Journal of Agricultural and Food Chemistry* **53**(12), 4989–4995.

Vente, J.A., Bosch, H., de Haan, A.B. and Bussmann, P.J.T. (2005) Comparison of sorption isotherms of mono- and disaccharides relevant to oligosaccharide separations for Na, K, and Ca loaded cation exchange resins. *Chemical Engineering Communications* **192**, 23–33.

Wasswa, J., Tang, J. and Gu, X.-H. (2007) Desalting fish skin protein hydrolysates using macroporous adsorption resin. *American Journal of Food Technology* **2**, 406–413.

Watkinson, A.P. and Wilson, D.I., (1997) Chemical reaction fouling: a review. *Experimental Thermal and Fluid Science*, **14**, 361–374.

Weinand, R. (1996) Apple juice stabilization and decoloration by modern adsorption techniques. *Flüssiges Obst* **96**, 495–496, 498.

Weiss, J. (2001) Fruchtsaftstabilisierung. In: *Frucht- und Gemüse-Säfte: Technologie, Chemie, Mikrobiologie, Analytik, Bedeutung, Recht* (ed. U. Schobinger), pp. 163–178. Ulmer, Stuttgart.

Weisz, G.M., Kammerer, D.R. and Carle, R. (2009) Identification and quantification of phenolic compounds from sunflower (*Helianthus annuus* L.) kernels and shells by HPLC-DAD/ESI-MS^n. *Food Chemistry* **115**, 758–765.

Weisz, G.M., Schneider, L., Schweiggert, U. *et al.* (2010) Sustainable sunflower processing – I. Development of a process for the adsorptive decolorization of sunflower [*Helianthus annuus* L.] protein extracts. *Innovative Food Science and Emerging Technologies* **11**, 733–741.

Weisz, G.M., Carle, R. and Kammerer, D.R. (2013) Sustainable sunflower processing – II. Recovery of phenolic compounds as a by-product of sunflower protein extraction. *Innovative Food Science and Emerging Technologies*, **17**, 169–179.

Wiesner, M.R. and Aptel, P. (1996) Mass transport and permeate flux and fouling in pressure-driven processes. In: *Membrane Processes in Water Treatment* (eds J. Mallevialle, P.E. Odendaal, M.R. Wiesner). McGraw-Hill, New York, pp. 4.1–4.30.

Zaganiaris, E.J. (ed) (2011) *Ion Exchange Resins and Synthetic Adsorbents in Food Processing*. Books on Demand, Paris.

Zagorodni, A.A. (ed.) (2007) *Ion exchange materials – properties and applications.* Elsevier, Amsterdam, The Netherlands.

Zeman, L.J. and Zydney, A.L. (1996) *Microfiltration and Ultrafiltration Principles and Applications*. Marcel Dekker, Inc., New York.

20 Stability and Bioaccessibility of Fruit Bioactives in Foods: Food Component Interactions and Matrix Effect

Dongxiao Sun-Waterhouse

INTRODUCTION

Consumers increasingly demand natural healthy foods containing nutrients and bioactive components with demonstrated health benefits. This has created a huge opportunity in developing novel foods enhanced with various plant-derived nutrients and bioactives. Fruits contain a wide range of health-promoting components including vitamins and antioxidants. However, people today do not consume sufficient fruits; for example, on average, an American consumer eats only about three servings of fruits and vegetables a day (US National Cancer Institute, http://progressreport.cancer.gov/). Therefore, developing different formats of fruit-derived foods can help to increase the total consumption of fruits.

Whilst the potential health-promoting properties of individual bioactive components, including polyphenols, vitamins and carotenoids, have been progressively proven, the ultimate effects of these bioactives must be delivered in the form of edible products, and preferably foods instead of supplements. The delivery efficiency, bioavailability and ultimate bioactivity of bioactive compounds in the diet/foods are determined by the interactions between bioactives and other co-existing food matrix substances during food processing, storage, consumption and bioabsorption, where exposure to various chemical and physical detriments occurs (Chaudhari, 2009; Sun-Waterhouse *et al.*, 2009a, 2010a, 2010b, 2010c; Sun-Waterhouse, 2011). Such interactions can be exclusive, additive or synergistic. The benefits of a finished food come from both the individual bioactive components and the food matrix (Gidley, 2004). Stability of bioactives is influenced by food format, formulation, processing and storage conditions (Duhard *et al.*, 1997; Dietrich *et al.*, 2003; Renard, 2005; Wegrzyn *et al.*, 2008; Sun-Waterhouse *et al.*, 2009b, 2010a). The stability, extractability and bioaccessibility of bioactives within the food matrix is an important prerequisite for overcoming their loss during food processing and handling, and ensuring their bioavailability and bioabsorption through the digestion/absorption and blood–brain barriers (Vattem *et al.*, 2005; Sun-Waterhouse *et al.*, 2009a; Wibisono *et al.*, 2009; Scheepens *et al.*, 2010).

Consumers demand proven health benefits, without compromising flavour, taste and colour (Bhaskaran and Hardley, 2002; Tuorila, 2007). The incorporation of 'functional' plant-based ingredients, either individually or in combinations, into a food may increase a product's health benefit but also introduces undesirable sensory attributes. Bitterness and astringency

Bioactives in Fruit: Health Benefits and Functional Foods, First Edition. Edited by Margot Skinner and Denise Hunter.

have been identified as the key sensory issues for the products containing a high polyphenol content, because of the interaction between polyphenols and the glycoproteins in saliva (Kallithraka *et al.*, 1998; Peleg *et al.*, 1999; Shi *et al.*, 2005; Jaeger *et al.*, 2009; McRae and Kennedy, 2011). Flavour formulation and development is one of the key requirements in processing a successful functional food (Milo Ohr, 1999; Sun-Waterhouse, 2011). The addition of food additives such as sugars or salts can significantly influence the stability and extractability of polyphenols such as anthocyanins (de Rosso and Mercadante, 2007; Sivam *et al.*, 2012). The stability of bioactives such as anthocyanins also directly affects the colour and flavour of foods. The use of fruit extracts can modify the flavour and mouthfeel of foods, for example gluten-free bread enhanced with green kiwifruit aqueous extract had a strong fruity flavour and softer and smoother crumb texture, compared with traditional gluten-free bread (Sun-Waterhouse *et al.*, 2009b).

This chapter provides insight into the bioaccessibility of bioactive compounds from fruits in a range of model food systems. The selected model foods showcase the recent innovations in the food development area. Some key technical aspects, for example sensory issues resulting from incorporating bioactives in functional or health-promoting foods, will be explored and addressed. This chapter also presents methodologies to examine and quantify the effects of interactions among food components and matrices on bioactive stability and bioaccessibility.

KEY COMPONENTS PRESENT IN FRUIT-BASED FUNCTIONAL FOODS

Food legislation usually requires food products to display a list of ingredients and specification of food additives. In a finished fruit-based functional food, three main types of ingredients are included: functional ingredients containing target bioactives, protective ingredients that can improve the stability and bioavailability of target bioactives, and other food ingredients or additives that are essential for a specific food product (Sun-Waterhouse *et al.*, 2009a, 2009b, 2010b, 2011a, 2011b, 2011c). The food ingredients and additives used for manufacturing fruit-based functional foods are mainly composed of carbohydrates, proteins, lipids, phytochemicals and water. Their chemical properties and processing functionalities are summarized in the following sections.

Phytochemicals

Fruits, including cranberry, blueberry, pomegranate, apple, blackcurrant, acai, acerola, guarana, mango, bilberries, grapes, cherries, kiwifruits, strawberries, feijoa, peach, plums cranberry, blueberry and pomegranate, contain a wide range of health-promoting nutrients and bioactive components, such as polyphenols, carotenoids, vitamins and minerals (Kriventsov and Karakhanova, 1972; Vattem *et al.*, 2005; Cevallos-Casals *et al.*, 2006; Starling, 2007; Masibo and Qian, 2009; Adyanthaya *et al.*, 2010; Hoffelner, 2010; Weston, 2010). Not surprisingly, these native bioactives in fruits are valuable to the food industries because they can be used as functional ingredients to exert beneficial biological effects on human health (such as phytochemical antioxidants) and/or as food additives to facilitate food manufacturing (such as pigments, flavours, preservatives, stabilizers, emulsifiers and thickeners).

Due to the high risk of consumption of synthetic food additives (such as artificial colorants and flavouring agents) in the long term (Nair *et al.*, 2006), there exists a remarkably increased interest in natural food additives from fruits and vegetables, for example novel apple and kiwifruit cultivars with various coloured flesh and skin (Sun-Waterhouse *et al.*, 2011a, 2011b). Huge innovation opportunities exist in using these fruit components alone or in combinations to create various new food processing or biological functionality for consumer foods. Multiple benefits can be achieved through combining health-beneficial phytochemicals and other fruit components that possess a specific food additive function into one single fruit-based ingredient for subsequent food manufacturing. The use of these fruit materials or ingredients in novel combinations of different intrinsic components during food processing can deliver beneficial synergies and also reduce the need to use non-natural food additives. For example, aqueous kiwifruit extracts can extend food shelf life, impart anti-inflammatory activities and improve food texture, because the extracts intrinsically contain antioxidative, antimicrobial, colouring, flavouring and texturing properties (Philpott *et al.*, 2007; Sun-Waterhouse *et al.*, 2009a).

A nutritive food would contain a good mix of desirable phytochemicals. Phytochemicals are chemicals of plant origin including polyphenols, carotenoids, organosulfur compounds, phytosterols, vitamins and minerals. Certain phytochemicals are found in all plants and derived foods, whereas others only occur in some particular foods, for example phloridzin in apples. Polyphenols are the most numerous and widely distributed group of bioactive molecules with different polarity, including simple molecules such as phenolic acids, biphenyls and flavonoids, and polyphenols (Rice-Evans *et al.*, 1996; Haslam, 1998). Chemically, these compounds possess a number of active functional groups for reactions, such as phenol structural unit (e.g. ortho-diphenol/catechol aromatic rings), hydroxyl, carbonyl and carboxylic acid groups. These functional groups are usually unstable when exposed to heat, oxygen and light. The number of functional groups and characteristics of chemical structures underlie the unique physical, chemical and biological properties of the phytochemicals ('structure–activity relationship'). Polyphenols are sensitive to non-enzymatic redox reactions and enzymatic reactions, and can undergo oxidation either directly or mediated by oxidative enzymes (e.g. polyphenol oxidase, PPO). Most polyphenols are weak organic acids with pKa of 8–12 (Antolovich *et al.*, 2000; Tura and Robards, 2002; Luthria, 2006). Polyphenols can also undergo a series of condensation reactions to produce high molecular weight complex molecules. Another important group of phytochemicals is carotenoids, which are naturally occurring pigments in fruits and vegetables (Bartley and Scolnik, 1995). They are oxygenated or non-oxygenated hydrocarbons containing at least forty carbons and an extensive conjugated double bond system. Non-polar carotenoids include beta-carotene, alpha-carotene and lycopene, and polar carotenoids are dominated by lutein. They are found in fruits like mango, apricot, peach, orange, papaya, persimmon, guava, pineapple, plum, kiwifruit, raspberry, blueberry, strawberry, apple and watermelon. Anthocyanins, tannins and carotenoids significantly contribute to the visual appeal of whole fruits (e.g. tamarillo), exposed flesh (e.g. cherry) or extracted juice (e.g. guava), as well as to flavour characteristics (mild and pleasing or bitter and astringent). Anthocyanins impart different colours at different pH and can be stabilized through complexing with other flavonoids (co-pigmentation) (Brouillard, 1988; Dangles and Elhajji, 1994; Boulton, 2001). Condensed tannins can complex with salivary proteins, causing the astringent sensation of fruit products and finished foods from grapes, peaches, apples, pears and berries (Baxter *et al.*, 1997; Dinnella *et al.*, 2009).

Carbohydrates

Carbohydrates are one of the major food components and used in the form of starch, sugars, dietary fibre, natural sweeteners or gummy polysaccharide additives for various food products. In addition to satisfying energy, nutrition and metabolic requirements, carbohydrates also impart structural, rheological and sensory properties to foods. Carbohydrates can be divided into simple and complex carbohydrates. Simple carbohydrates are single or two sugar molecules that have joined together (termed mono- and disaccharides, respectively). Complex carbohydrates are made up of many interconnected sugar molecules, including starch and other polysaccharides. Fruits are rich in 'simple' carbohydrates, which possess little nutritional value, except for calories. Melons, peaches, pineapples, mangos and tangerines contain 6–12% sucrose, and dates contain up to 58% sucrose. Some simple sugars may impart desired food processing properties, for example sucrose aids the structure of ice cream (Schmidt, 2004). Most fruits contain soluble reducing sugars, mainly in the form of fructose and glucose mixtures. 'Maillard' browning reactions induced by reducing sugars (such as inulin), contribute to the antioxidant activity and colour of foods (Damodaran, 1996).

Dietary fibre is listed as a subcomponent under total carbohydrate in the Nutrition Information Panel (NIP) of food. Dietary fibres are an important group of bioactive substances, providing structure to the plant cell walls and consist of many undefined polymers that are held together by bonds that cannot be broken by human digestive enzymes. Fruit fibres have balanced soluble and insoluble dietary fibre content, therefore may be suitable for dietary fibre enrichment (López *et al.*, 1996; Grigelmo-Miguel *et al.*, 1999; McKee and Latner, 2000). Fruit cell wall polysaccharide networks contain components such as cellulose, hemicellulose and noncellulose polysaccharides, glycoproteins, water and other smaller molecules like phenolic acids (Fry, 1988; Sun-Waterhouse *et al.*, 2008a, 2008b, 2011a). Apples, cranberries, mango, oranges, persimmon, feijoa and figs are rich sources of water-soluble dietary fibres. Pectic polysaccharides (also called pectins) are the predominant soluble fibres in fruits, occurring in the middle lamella and primary cell walls of different plant tissues (Fry, 1988; De Man, 1990). Water-insoluble dietary fibres such as hemicelluloses, lignin and pentosans occur in a significant amount in apples, pear, persimmon, feijoa, figs, pineapples, bananas and berries. The composition, structure and funcitonality of dietary fibre preparations depend largely on their origin and also the extraction method used (Bacic *et al.*, 1988). There exists a wide range of laboratory techniques (Fry, 1988; Percy *et al.*, 1997; Sun-Waterhouse *et al.*, 2008a, 2008b) and processing technologies (Fleury and Lahaye, 1993; Nawirska and Kwaniewska, 2004; Choudhari and Ananthanarayan, 2007) for the production of fibre ingredients, including various milling processes, simple extraction and drying procedures, and enzyme and additive treatments. Each method has limitations.

The diversity of dietary fibre, ranging from plant-derived oligo- and polysaccharides through to microbial exudates and to amylase-resistant starch, and their consequent different physicochemical properties, has led to new dietary fibre-enriched products being marketed. Dietary fibre enters the food market mostly as a natural food component, an added ingredient or a polysaccharide food additive (also referred to as 'a hydrocolloid' or 'a gum'). Food legislation enables dietary fibres to be valuable ingredients in functional foods. The dietary fibre's functionality affects both the organoleptic and nutritional aspects of a finished food. The functionality of plant cell walls depends on the polysaccharide composition, as well as the location and orientation of polysaccharides in the cell-wall networks. The physical properties of 3D plant cell-wall polymer networks are of direct relevance to their behaviour as dietary fibre. The porosity and water-holding capacity of primary plant cell walls will vary as the pH and ionic strength vary during gut transit. The polysaccharides from plant cell walls, such as pectic polysaccharides, xyloglucans, glucomannans and galactoglucomannans, possess ester, ether and cyclic acetal functional groups for various chemical reactions, and a number of free carboxyl groups (e.g. uronic acid) (Bacic *et al.*, 1988; Kaplan *et al.*, 1988; David, 2001).

Pectic polysaccharides are versatile and health beneficial biopolymers, with backbones of galacturonic acid connected by α-1,4-linkages. These polysaccharides contain distinct monosaccharide composition, degree of methyl esterification and acetylation, araban, galactan and amide contents. Pectin substances occur mainly in four forms: (i) all polygalacturonic acid-containing materials; (ii) water-insoluble protopectins that can yield pectins upon hydrolysis; (iii) partly esterified polygalacturonic acids with different degrees of methyl esterification and acetylation, such as low methoxy and high methoxy pectins; (iv) pectinic acids in which all carboxyl groups are in the free form and water insoluble. Pectins exhibit various physicochemical properties (viscosity, gelation, water-holding capacity and ion exchange) depending on their chemical structure during food formulation and processing, and exert positive nutritional and physiological properties to the human body, for example their specific effects in the gastrointestinal tract in relation to lipid, glucose, mineral and vitamin metabolism (Fernandez, 2001). Commercial pectins are generally obtained through processing the parenchymatic tissues of plant materials. Sun-Waterhouse *et al.* (2008a) reported that apple cell wall preparations with high uronic acid content offered remarkable protection against ascorbic acid degradation. Commercial pectins also significantly stabilized the antioxidant activity of ascorbic acid.

Starch is the other polysaccharide that is widely used as a food ingredient or additive. Starch is also used as a thickening, stabilizer or processing aid for protecting colour, flavour, odour and texturizing component for applications in beverages, powdered mixes and emulsified foods (Whistler and Daniel, 1984; Ellis *et al.*, 1998; Sandhu and Singh, 2007; Singh and Nath, 2009; Chung *et al.*, 2011). Starch is a heterogeneous polymer of α-D-glucopyranosyl units. There are two basic forms of the polymer in starch, one with α-1→4 links, called amylose, and one with α-1→4 links and α-1→6 links, called amylopectin (Whistler and Daniel, 1984, 1985; Myers *et al.*, 2000). Starch granules from different plants (and animals) differ in size, appearance, proportions of amylopectin and amylose, chain lengths and branching patterns (Hoseney, 1994; Dziedzic and Kearsley, 1995). Amylose mostly consists of about 17–30% of total starch (e.g. in corn, rice and potato starches), but can be up to 75% of total starch (e.g. pea and corn starches). High amylose starches (with amylose content >30%) have very distinct properties; for example, barely gelatinized above 100°C, form films and fibres, retard granule swelling and have more helical structure (which can entrap molecules like fatty acids and phenols). To improve food processing properties, sensory attributes and health benefits, modifications of the hydroxyl groups available on the starch polymer are made singly or in combination via chemical or enzyme-catalysed reactions ('modified starch'). Resistant starch is defined from a nutritional perspective, as starch and starch degradation products that escape digestion in the small intestine of healthy individuals (Bird *et al.*, 2000; Alsaffar, 2011). It can be native or modified and soluble or insoluble starch.

Various physical types and chemical modifications of starch are marketed for industrial use. For example, the introduction of carboxyl and carbonyl groups to oxidized starches, through treatments with alkaline hypochlorite, leads to minimal retrogradation in starch, low viscosity and increased protective effects on colloid action. The introduction and increase of ionic phosphate groups in the cross-bonded waxy starch helps to tackle water separation issues and improve freeze–thaw stability. Starch exhibits different rheological and physicochemical properties depending upon process conditions (Sandhu and Singh, 2007; Lu *et al.*, 2008; Oh *et al.*, 2008; Singh and Nath, 2009). Starch is not a highly efficient emulsifying agent. The use of starch in emulsifying fats and oils depends merely on the mechanical separation of oil droplets by swollen granules. Syneresis is commonly found in chilled and frozen products, for example sauces and dips, which are formulated with unmodified amylose-containing starch. Retrogradation is especially evident when amylose-containing starches are cooled. When food processing requires heating at high temperature, such as during sterilization of canned goods by autoclaving or retorting, the stability of starch thickeners decreases remarkably. During processing, the starch thickener must not thin excessively under conditions of agitation and shear. Starch paste pumped through the jet of a spray dryer or through the orifice of a homogenizer may undergo rather extensive hydrolytic cleavage of molecules due to the enormous shear action.

Proteins

Proteins are a key food component. The functionality of proteins is governed by their primary, secondary, tertiary and quaternary structures. Primary structure is the sequential order of the 20 different amino acid residues connected via covalent peptide bonds. Secondary structure is the spatial configuration of the polypeptide backbone stabilized largely by hydrogen bonds. Tertiary structure is the three-dimensional organization of secondary structures, via disulfide, hydrophobic and hydrogen bonds, and electrostatic, dipole–dipole and charge–dipole interactions. Quaternary structure is the aggregation of protein 'subunits'. Despite the complexity of protein structure, most of the protein properties, including cross-linking, are associated with the following functional groups: primary amines ($-NH_2$), carboxyls (–COOH), sulfhydryls (–SH) and carbonyls (–CHO). Sulfhydryl groups mainly exist in the side chain of cysteines in the form of disulfide bonds (–S–S–). The primary amine group exists at the N-terminus of each polypeptide chain and in the side chain of lysine residues. These positively charged groups are usually on the outer surface of proteins, thereby accessible for conjugation without denaturing protein structure, and in the presence of reducing sugars could be involved in Maillard reactions. Carboxyl groups exist at the C-terminus of each polypeptide chain and in the side chains of aspartic acid and glutamic acid. The nature and magnitude of covalent and non-covalent bonds determine the size, shape, surface charge and properties of proteins. For example, charged amino acids enhance electrostatic interactions, stabilizing globular proteins and in enhancing water-binding and, thereby, affect food properties such as hydration, solubility, gelation and surface activity. Changes in protein structure that occur in acidic media are irreversible. The isoionic pH region of 4.0–6.0 normally facilitates greater amounts of non-ionized amino and carboxyl groups. Food formulation and processing can be tailored to use or avoid this isoionic pH region, for example for ice cream making that involves a large amount of freshly prepared fruit juices (Sun-Waterhouse *et al.*, 2011b). Conformational changes can be reversible or irreversible: those generated at alkaline pH values can be reversed by increasing ionic strength, whilst conformational changes are irreversible when increased levorotation, dissociation or aggregation occurs.

Change in protein structure is normally associated with changes in physiochemical properties, shelf life and nutritional value. The correlation between protein hydration and viscosity of a food system is affected by protein concentration and environmental factors such as pH, ionic strength and temperature. Proteins with high solubility have low gelation properties and low viscosities. Proteins may modify food flavour directly due to their composition, or indirectly because of their involvement in flavour development after heating and their interactions with other food components like flavours during food processing and storage. The off flavours of many food proteins are often associated with lipids, amino acids (for Strecker degradation), bitter peptides or nucleotides present in the products. During food processing, heat treatments such as boiling, baking and roasting can cause amino acid side chains to degrade and/or interact with other food components (i.e. lysine and reducing sugars). Excessive heating could cause protein decomposition and a loss of protein nutritional value.

Protein denaturation is a modification of the secondary, tertiary or quaternary structure of the protein molecule that does not involve breaking covalent bonds. Denaturation of gluten proteins occur during pasta cooking. Cooking pasta for as short as 10 minutes can cause a decrease in solubility of protein such as gliadins (Autran and Berrier, 1984; Sun-Waterhouse and Jin, 2011). Upon heating, pasta protein subunits exhibit different abilities to aggregate and form a firm and viscoelastic network, involving hydrophobic interactions, disulfide bond formation, disulfide interchange reactions and non-polar group interactions. These changes would be influenced by reducing agents like polyphenols, indicated by the decreased firmness of the polyphenol-enhanced pasta dough and cooked pasta (Sun-Waterhouse and Jin, 2011). The ability to aggregate decreases in the order of high molecular weight glutenins > low molecular weight glutenins > α-β-γ-gliadins > ω-gliadins (Autran and Berrier, 1984). Heat denaturation of protein is not necessarily undesirable. For example, heat denaturation of fermented wheat dough is important for baking processes. Heating milk would enable a major shift of soluble to colloidal phosphate, and complexation of denatured β-lactoglobulin with the κ-casein component of casein micelles providing adequate heat stability during sterilization (Singh *et al.*, 1989). Heating milk also causes interaction of proteins and lactose, mainly through the Maillard browning reaction, resulting in reduced availability of lysine (Morr, 1983, 1984). Addition of polyphenols to milk modifies these reaction processes and then the physicochemical properties of dairy products, due to the interactions between polyphenols and proteins (Wegrzyn *et al.*, 2008; Sun-Waterhouse *et al.*, 2010b, 2010c, 2011c).

Enzymes

Enzymes are proteins and catalyse chemical reactions in living cells. They are generally composed of about 200–1000 amino acid residues covalently linked in a sequence and adopt a 3D conformation connected through non-bonded covalent interactions such as hydrogen bonds. The uniqueness of the conformation enables an enzyme's catalytic activity and specificity. The enzyme's active site is directly responsible for catalytic action through binding substrates. Altering the environmental pH or temperature can disrupt an enzyme's conformation, causing loss or reduction of enzyme activity or stability. The presence of inhibitors may result in decreasing enzyme activity or retard enzyme-catalysed reactions. There is a wide variety of enzymes that function inside living organisms and offer potential for the improvement of foods. Food enzymes are currently used under different legislation systems as food additives [Regulation (EC) No 1333/2008] or processing aids [Regulation (EC) No 1332/2008]. Enzymes are used in food processing to facilitate bioconversions, desired viscosity and flavour, and compound synthesis, extraction and separation. Industrially significant enzymes for fruit processing and derived food development include pectinolytic enzymes, proteases, polyphenoloxidase (PPO), lipoxygenase (LOX) and peroxidase (POD).

Pectinolytic enzymes are a heterogeneous group of related enzymes that are widely distributed in higher plants and microorganisms and hydrolyse the pectic substances (Whitaker, 1990). They play important roles in plant cell wall extension, and plant maturation and storage (Ward and Moo-Young, 1989; Aguilar and Huirton, 1990; Sakai, 1992). Pectinolytic enzymes include protopectinases that degrade the insoluble protopectin and increase highly polymerized soluble pectin, esterases that catalyse the de-esterification of pectin through removing methoxy esters, and depolymerases that catalyse the hydrolysis of the pectic substances through breaking the α-1$\rightarrow$4 glycosidic bonds of D-galacturonic acid units (Sakai, 1992; Palomäki and Saarilahti, 1997). Clarification or stabilization steps are often required to tackle the discoloration and haze formation caused by certain polyphenols during fruit juice processing. The pectinolytic enzymes are used to hydrolyse the esters of hydroxycinnamic acid for this purpose (Macheix and Fleuriet, 1998). A protease (also known as peptidase, proteinase or proteolytic enzymes) refers to a group of enzymes, such as fungal protease, pepsin, trypsin, chymotrypsin, papain, bromelain and actinidin, which hydrolyse the peptide bonds between amino acids within the polypeptide chain of protein. The ability of proteases for proteolysis differs, and each type of the proteases only breaks down a specific kind of peptide bonds. Proteases such as actinidin in kiwifruits break down proteinaceous substances in the food matrix, which may impart beneficial or detrimental effects on food attributes, for example improved texture of green kiwifruit extract-enhanced gluten-free bread (Sun-Waterhouse *et al.*, 2009b) or severe curdling issues during pasteurization or ageing of ice cream mix containing fresh kiwifruit juices (Sun-Waterhouse *et al.*, 2011b). PPO oxidizes phenolic substrates in the presence of oxygen to produce *o*-quinones that can be further polymerized into melanin (Lamikanra, 2002). PPO-catalysed oxidative browning occurs commonly when certain fruits, such as apples and green kiwifruits, are sliced (Yemenivioğlu *et al.*, 1997; Sun-Waterhouse *et al.*, 2009b; Falguera *et al.*, 2011). LOX catalyses the oxidation of polyunsaturated fatty acids like oleic (18:1), linoleic (18:2) and linolenic (18:3) acids in the presence of oxygen (Williams *et al.*, 1986; Boyes *et al.*, 1992), and would contribute to the sourness and grassy flavour of kiwifruit puree (Chen *et al.*, 1999). POD contributes to enzymatic browning due to its affinity to accept hydrogen donors such as polyphenols (Prestamo and Manzano, 1993; Richard-Forget and Gauillard, 1997).

Lipids

Lipids are one of the major macronutrients required by the human body, offering many essential functions, such as energy supply and transport of nutrients and metabolites. However, overconsumption of lipids results in various health issues including obesity, heart disease and diabetes. Good lipids enhance good health whilst bad lipids contribute to chronic illness.

Food lipids are principally triacylglycerols, phospholipids and sterols, occurring in two forms at room temperature: liquid oil or solid fat form. In nature, triacylglycerols contain a glycerol moiety with each hydroxyl group esterified to a fatty acid. There are four types of fats or fatty acids that are found in foods and often claimed on the nutrition labels of processed foods: Monounsaturated fats (mostly found in olive oil, canola oil, avocado oil), polyunsaturated fats (including essential fats like omega-3 fatty acids in the flaxseed, walnuts and salmon, and omega-6 fatty acids in corn, sunflower and soy), saturated fats (found at high concentrations in coconuts, whole milk, cream cheese), and trans fats (the worst lipids, found in hydrogenated fat products, fried foods and shortening). Lipids can also be grouped into two classes: Simple lipids (which produce at most two types of primary product per mole upon hydrolysis) and complex lipids (which yield three or more primary hydrolysis products per mole). Complex lipids include glycerophospholipids (also termed as phospholipids, containing a polar phosphorus moiety and a glycerol backbone) and glycolipids (which contain a polar carbohydrate moiety such as glycoglycerolipids and glycosphingolipids).

The role of lipids in a food product varies considerably. Oils and fats have a melting range rather than a discrete melting point which enables distinct physicochemical properties of foods. Lipids give calories, carry fat-soluble vitamins and nutrients, and impart flavour, colour, taste, texture and volume. These roles partly explain the strong preference and market advantage of fat-rich foods. However, lipids are chemically unstable and readily undergo free-radical chain reactions, including undesirable chemical oxidative or enzymatic (i.e. lipoxygenase) modifications, during raw material collection, processing, storage, handling and distribution (Kolakowska, 2003). The instability of lipids would not only cause deterioration of the lipid itself, but also lead to generation of oxidation products including off-flavour compounds, degradation of other food components such as proteins, vitamins and pigments, and interactions between lipids and other macromolecules via cross-links into non-nutritive polymers. Owing to the sensory and structural importance of lipids in foods, approaches for preserving lipids using efficient and nutritionally desired antioxidants (e.g. plant-derived polyphenol antioxidants), and replacing fat with carbohydrate-based, protein-based and oil-based substitutes, are highly sought. In the functional food and pharmaceutical sectors, lipids are likely used to formulate various food emulsion systems to deliver desired bioactives such as polyphenols with improved stability and bioavailability (Sørensen *et al.*, 2008). It is worth noting the distinct roles of lipids as liquid medium or means of heat and mass transfer during food processing and digestion. At physiologic pH, the hydroxyl groups of most polyphenols may interact with the polar head groups of phospholipids at the membrane surface via hydrogen bonds (Verstraeten *et al.*, 2003). The differences in solubility and polarity of bioactives including polyphenols in various lipids would lead to different stability, extractability and bioaccessibility of target bioactives, and ultimately their different bioavailability. Furthermore, interactions between polar lipids and polyphenols may result in active compounds with greater stability and bioaccessibility due to the changes in amphiphilic properties. Glucosylceramide may interact with polyphenols between the hydroxy functions of the polyphenols and the polar head of the sphingomyelin, although the fatty acid moieties may not be involved in the complexation between glucosylceramide and polyphenols (Bombardelli *et al.*, 1989).

Water

Water is the essential component of all life, accounting for 50–90% of the weight of living things including fruits. Some fresh fruits contain water greater than 90% in their edible portion, for example, berries contain 80–90% water and citrus fruits have 85–90% water. Water transports, combines and aids the metabolic breakdown of the substances in protoplasm such as ions, amino acids, monosaccharides, nucleic acids, proteins, lipids and polysaccharides (Guyton and Hall, 2005). Water contains no calories. Pure water is an odourless, tasteless and clear liquid.

Water, together with other fluids, is responsible for food component mobility, changes of state, mass transfer and flow behaviour of food products. Water is the solvent of hydrophilic constituents, affecting food structure, appearance, taste and stability. Water determines the pattern or structure of food macromolecules such as proteins and carbohydrates. Part of the water is absorbed on the surface of macromolecular colloids (starches, pectins, cellulose, and proteins). In addition to serving as a solvent and reactant, water can participate directly in hydrolytic cleavage (i.e. in the formation of invert sugar from sucrose under acidic conditions), and act as a plasticizer of amorphous materials. Water influences the properties of macronutrients, such as the sorption, viscosity, gelation, solubility, emulsification, and rheological, textural and sensory properties of polysaccharides and proteins. Imbibition of water in proteins and polysaccharides is important in many foods, for example wheat flour dough and starch dispersions, affecting characteristics such as texture/mouth feel, thickening and viscosity. Water governs food processing, product properties, quality, and stability; and the significance of non-equilibrium glassy and rubbery states to the time-dependent textural, rheological and thermomechanical properties. Absorption of water and swelling change the hydrodynamic properties of a food system, which are reflected in thickening, increased viscosity and flow characteristics.

Water content and water activity (a_w) are important parameters for food evaluation. The stability of foods is greatly governed by the interactions between water and other food components and the ability of water as a medium to facilitate the interactions among the food components (Labuza, 1971). Water activity indicates the availability of water from a medium for chemical reactions and is affected by temperature (Troller and Christian, 1978; Sperber, 1983). A change of 10°C can cause a change in a_w from 0.03 to 0.2, depending on the type of product. When $a_w < 0.3$, water is virtually unavailable as a solvent or reagent, as it is strongly linked to the food components via covalent bonds and cannot be separated by even severe techniques. When $0.3 < a_w < 0.7$, water is available and moderately reactive, as it is weakly bound in the form of a multilayer partially covering the surface of the substrate. When $a_w > 0.7$, water is 'free' and available as a reactive medium, as it is on the surface of the substrate via hydrogen bonds. Optimum stability is generally obtained when a_w is between 0.2 and 0.3, a water activity of 0.7 is considered a lower limit that prevents microbial growth.

Given the characteristics of the above food components, their usefulness in functional food applications depends on their native origin (e.g. type of fruit, fruit cultivars or fruit tissue), the approaches for isolating and extracting these components from source materials, and the subsequent food processing methods. Both the preparation of food ingredients or additives and the production of consumer processed foods are complex processes, involving various chemical and physical reactions induced by treatments using acid, alkali, enzymes, heat, pressure and so on (Carle *et al.*, 2001; Choudhari and Ananthanarayan, 2007; O'Connor *et al.*, 2007; Tsuzuki *et al.*, 2008; de Carvalho and Maubois, 2009; Sun-Waterhouse *et al.*, 2008b, 2008c, 2009b, 2011c, 2012a; Lokuruka, 2011). These processes can modify the nature of the native components through mechanisms such as polymerization or depolymerization, reduction or oxidation, degradation, binding, complexation, hydrolysis, dissolution, precipitation, enzyme activation or deactivation, thus resulting in distinct functionality.

The preparation of food ingredients should adopt a finished food-oriented approach. This process can make a significant contribution to finished food processing. The bioactive components, along with matrix substances that are desirable as nutrients of final foods or as 'protective agents' to facilitate bioactive stability and beneficial processing synergies, should be maximized during ingredient preparation. The conditions of ingredient preparation determine the composition and processing functionality of obtained ingredients; for example, choosing an extraction medium with appropriate polarity, viscosity and surface tension can optimize the transfer of electrons/protons and the extractability of phytochemicals (Duttaroy and Jørgensen, 2004; Perez-Jimenez and Saura-Calixto, 2006). Some treatments may impart multiple effects on the ingredient preparation, for example an alkali pretreatment of fruit materials could increase the yield of extracted compounds through loosening fruit tissues, and alter the polyphenol profile of the final ingredient (Sun-Waterhouse *et al.*, 2009b, 2012a). The existing form of bioactive compounds (e.g. the aglycone or glycoside form) could be altered after a specific extraction process. Some nutrients that intrinsically occur in fruits, such as proteins, could be altered to undesired impurities by the extraction medium. Unwanted substances may also be brought in the final ingredient extracts, causing negative effects on the subsequent finished food processing.

Processing is expected to affect the food composition and nutritional value. The bioactives, nutrients, processing aids and undesired impurities present in the ingredients, all interact with one another during processing operations, which determines the ultimate properties of food products. In support of this fact, examples of some innovative concept foods that are enhanced with phytochemcials are provided in the next section. Possible changes in the stability and extractability of polyphenols and some key nutrients, as well as the chemical and physical properties of the finished foods, during and after processing are demonstrated.

INTERACTIONS BETWEEN FRUIT BIOACTIVES AND OTHER FOOD COMPONENTS IN VARIOUS FOOD SYSTEMS

There is a growing trend toward the incorporation of plant-based ingredients into popular consumer processed foods, because of the positive consumer perception on the 'naturalness' attributes and diverse nutrient profiles of fruits and vegetables (Mares-Perlman *et al.*, 2002; Starling, 2005, 2007; Lauren *et al.*, 2009; Rahavi and Kapsak, 2010; Weston, 2010). Bioactive ingredients such as fruit polyphenol extracts, fruit juice, fruit dietary fibres, either individually or in combination with other bioactive ingredients, can conveniently be incorporated into processed foods to deliver the positive health functionality of these active ingredients.

Challenges occur during the translation of consumer interests into real products. Both the biological activities and sensory attributes of a functional food result from individual compounds as well as chemical and physical interactions within the food matrix during processing, storage, ingestion and digestion. The effects of the food matrix on the bioaccessibility of polyphenols, depend on the nature of the direct interactions between polyphenols and food components such as proteins and polysaccharides, or indirect effects such as food processing, handling and digestion parameters (Hollman et al., 1997, 2001; van het Hof et al., 1998; Manach *et al.*, 1998; Goldberg *et al.*, 2003; Butts *et al.*, 2010; Parkar *et al.*, 2010). Attention should be paid to the important characteristics of bioactives and food matrix components amenable to changes during raw material/ingredient processing and food processing, such as changes in molecular structure, bonding and transformation, as well as particle and bulk characteristics and microstructure.

Intrinsic functional properties represent the composite properties of the components of a food system as related to the amino acid composition, primary sequence, conformation, molecular shape and size, charge distribution and intramolecular and intermolecular bonding. Different types of bonds that govern structural forces in foods include:

- Covalent bonds (via which the atoms bound by a common electron pair, such as C–C, C–N, C=O, C–H, C–N–C and S–S) for bridging and ordering function within a food matrix. Reducing agents, such as sulfite and mercaptoethanol, can disrupt these bonds.
- Ionic bonds (attraction between opposite charges, such as $-NH_3^+$ and $-COO^-$) for solvent interactions and salt effects within a food matrix.

- Hydrogen bonds (hydrogen shared between two electronegative atoms, such as N–H···O=C or hydroxyl, amide and phenol groups) for bridging and stabilizing. Urea solutions, guanidine hydrochloride and dimethylformamide can disrupt these bonds.
- Hydrophobic bonds (via apolarity, involving apolar groups such as long aliphatic chains and aromatic groups) for strand thickening, strengthening and stabilizing. Detergents and organic solvents can disrupt these bonds.
- Electrostatic repulsion (via coulombic repulsion between particles with same charges, involving polar groups of side chains like carboxyl, amino, imidazole and guanido groups). Salt solutions and high or low pH can disrupt these bonds.
- van der Waals repulsion (repulsion of apolar groups that are close, like steric hindrance between side chain groups).

The co-existence of added fruit bioactives and other food components such as protein, carbohydrates and lipids suggests that fruit bioactives will interact extensively with other food components. For example, proteins occur in foods quite often in combination with carbohydrates or lipids such as glycoproteins, glycolipids and lipoproteins. The stability of protein structure in food systems containing lipids could be largely affected by protein–lipid interactions via hydrophobic interactions. Protein hydration is influenced by the presence of hydrophilic polysaccharides, lipids and salts, as well as the pH of food and food processing and storage conditions. The addition of polyphenols would further influence protein macromolecular structure and increase the entropy of a food system, through interfering with intra/intermolecular interactions mediated by non-covalent forces such as hydrogen bonds, van der Waals forces and hydrophobic effects. Binding between polyphenols and other food components does not necessarily lead to the loss of antioxidant activity and, in some cases, such binding (e.g. albumin-bound quercetin) can still exert antioxidant activity due to the accessibility of antioxidant functional groups (e.g. the catechol moiety), and may enable additional desired biological properties (Dangles *et al.*, 1999). Protein and starch polysaccharides would interact and form complexes via attraction of their opposite charges during gelatinization. Polyphenols with varied polarity and molecular weight/size may influence the conformation of proteins and starch polymers, subsequently their hydrophobic domains/pockets that exert inclusion mechanisms and, finally, the stability and extractability of polyphenols and proteins after food processing, for example after bread making (Table 20.1, Figure 20.1) (Sun-Waterhouse *et al.*, 2011c).

There are quite a few factors influencing bioactive functionality and food product properties, such as ingredient composition, matrix characteristics, functional groups of target

Table 20.1 Recovery of added polyphenols for control and fortified breads.

Type of added polyphenol ingredient	Recovery of added polyphenols (%)
Control	N/A
Kiwifruit polyphenol extract	28–35^{ab}
Blackcurrant polyphenol extract	30–39^{a}
Apple polyphenol extract	9–12^{b}
Epicatechin	7–8^{c}

Different lowercase superscript letters $^{a-c}$(within the same entire column) indicate statistically significant differences at $P < 0.05$.

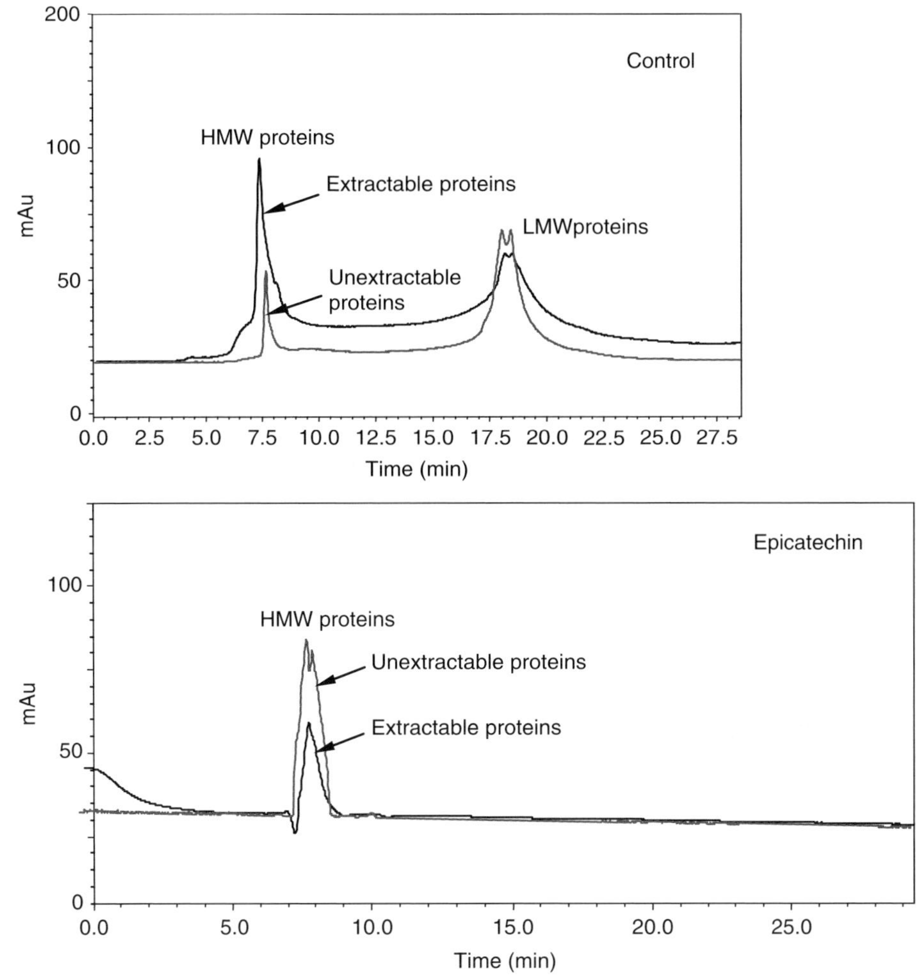

Fig. 20.1 SE-HPLC profiles ($\lambda = 214$ nm) of extractable and unextractable fractions from the control and epicatechin-spiked breads.

bioactives, crystallinity, surface characteristics, molecular chemical structure, intermolecular bonding and polymer conformation of food components. Exposure to light has a major effect on the content of most flavonoids (Macheix et al., 1990). Such changes in colour and organoleptic characteristics may be beneficial (as for black tea) or harmful (browning of fruit) to consumer acceptability. Storage under different conditions would affect the content of polyphenols that are easily oxidized, generating degradation products or polymerized substances, which lead to changes in food quality (Spanos et al., 1990; Miller et al., 1995).

This section presents examples to demonstrate the effects of incorporated fruit bioactives on the organoleptic, microstructural, physical and chemical properties of several food systems, and the food matrix effects on the stability, extractability, bioaccessibility and antioxidant activity of added fruit bioactives. The outcomes and associated technical challenges that result from the interactions between added fruit bioactives and other food components are discussed.

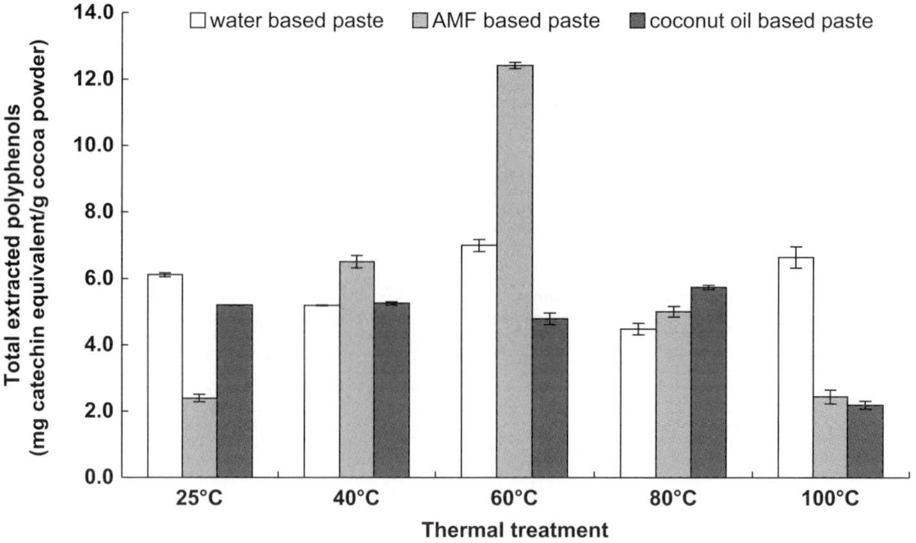

Fig. 20.2 Total extracted polyphenol contents of the unheated and heated cocoa pastes prepared in water or lipids.

Interactions between polyphenols and lipids

The polarity of polyphenols enables their functions at water–lipid interfaces. Modification of the structure of polyphenols such as glucuronidation, sulfation and esterification would render polyphenols more hydrophilic or hydrophobic, influencing their intrinsic reductive capacity (O'Leary *et al.*, 2001; Spencer *et al.*, 2001; Cren-Olive *et al.*, 2003). Wibisono *et al.* (2009) investigated the stability and extractability of poplyphenols in water- and lipid-based cocoa pastes after subjected to heating at different temperatures (Figure 20.2). The total extracted polyphenol contents of cocoa pastes formulated using water, anhydrous milk fat (AMF) or coconut oil were determined by the Folin-Ciocalteu assay, after heating the cocoa pastes at different temperatures and, subsequently, extracting polyphenols from the cocoa pastes. At 25°C, the unheated cocoa paste prepared in water had the highest amount of retained polyphenols, followed by the pastes prepared in coconut oil and AMF. However, heat treatment changed the trend. The influence of heat treatment on the formulated cocoa pastes varied with different media. There are intramolecular or intermolecular interactions among these bioactive compounds and medium components (e.g. water and triglycerides) (Perez-Jimenez and Saura-Calixto, 2006). Such interactions may be influenced by factors like the partition coefficient of the functional compounds, heat treatment temperature and viscosity of the cocoa paste matrix. Catechins are very hydrophilic and polar polyphenols. Caffeic acid is a more lipophilic phenolic acid. The structural feature responsible for the antioxidative and free radical scavenging activity of caffeic acid is the ortho-dihydroxyl functionality in the catechol ring. The presence of the electron-donating hydroxyl group at the ortho-position would lead to intermediate oxidation and increase the rate of hydrogen atom transfer to peroxyl radicals (Lucarini *et al.*, 2002), and generate a stable phenoxy radical (Chen and Ho, 1997; Wright *et al.*, 2001).

Increasing fat content might decrease the partition coefficient of the most lipophilic compound, whereas addition of fat would increase the partition coefficient of hydrophobic or amphiphilic phenolics (Relkin *et al.*, 2004). AMF and coconut oil are in the form of di- and triglycerides, which can bind to lipophilic and partly lipophilic compounds via weak van der Waals bonds and hydrophobic interactions (Plug and Harring, 1993). The binding may be affected by the chain length and degree of unsaturation of the triglycerides (Hyvonen *et al.*, 2003). The difference in the viscosity of the water and lipid matrices may affect the rate of heat and electron/hydrogen atom transfer among compound molecules (Bakker and Mela, 1996). The partitioning of polyphenols between aqueous and lipid phases is largely dependent on their hydrophilicity and hydrophobicity. A higher resistance to mass transfer in fat and oil than in water is possible (DeRoos, 2006). For a highly hydrophobic compound, an increased extractability from a lipid-based matrix is possible, compared with a water-based matrix, because of the greater solubility of this hydrophobic compound in the hydrophobic lipid matrix (Druaux and Voilley, 1997). Hydrophilic compounds exhibit solubility in water and are partitioned in the water phase (Druaux and Voilley 1997). Theoretically, a more hydrophobic compound may be more extractable from less saturated and polar triacylglycerols (e.g. AMF), and a hydrophilic compound may behave in the opposite manner (being more extractable from more saturated triacylglycerols, for example coconut oil) (Harrison *et al.*, 1997).

Preventing or minimizing lipid oxidation of edible oils is a major focus of research. Fortification of antioxidant(s) into edible oils has long been used as an approach for improving oil stability. Polyphenol antioxidants are the subject of increasing attention because of their health-promoting properties, including the ability to intercept free radicals (Kaul and Khanduja, 1998; Kampa *et al.*, 2004). Polyphenols vary in the carbon skeleton and hydroxylation of the phenolic rings (Boudet, 2007), and may suppress lipid oxidation via donating hydrogen atoms to lipid peroxyl radicals to interfere with the initiation or propagation of primary oxidation (Cheung *et al.*, 2007; Tsuzuki *et al.*, 2008). The differences in the protective effect of polyphenols on edible oils also demonstrate the interactions between polyphenols and lipids. Tsuzuki *et al.*, Sun-Waterhouse *et al.* (2011d) reported that the addition of caffeic acid or *p*-coumaric acid to avocado or coconut oil had influenced differently oil oxidation during the primary oxidation stage (Figure 20.3).

Unsaturated fatty acids are more easily oxidized, because of their lower activation energy to form fatty acid radicals, than saturated fatty acids (Przybylski *et al.*, 1993). Oxidation of fatty acids takes place primarily on methylene groups activated by adjacent double bonds,

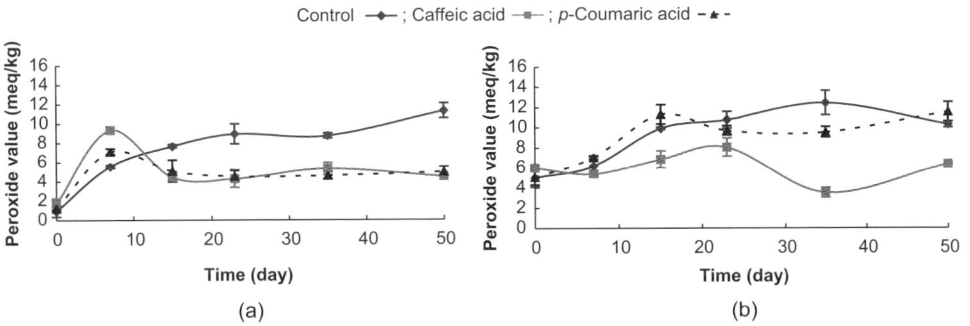

Fig. 20.3 Peroxide values of control and polyphenol-fortified (a) avocado and (b) coconut oils stored at 60°C.

and the oxidation rate increases with the number of double bonds and the number of adjacent double bonds. The affinity between fatty acid moieties to triacylglycerol molecules could be different in lipids containing different concentrations of unsaturated fatty acids (Zhang *et al.*, 2007). The bond dissociation enthalpies of the O–H bond and the capacity to donate electrons of polyphenols are affected by the matrix medium (Pino *et al.*, 2006; Pazos *et al.*, 2007). Avocado oil contains a greater amount of unsaturated fatty acids than coconut oil (De Leon and Delores, 2005). *p*-Coumaric acid and caffeic acid are a monohydroxy and dihydroxy derivative of hydroxycinamic acids, respectively, possessing different ability to scavenge lipid peroxidation-derived radicals (peroxyl, alkoxyl and alkyl radicals) (Murkovic, 2003). They possess different polarity and solubility in the oils containing varied ratios of saturated fatty acids/unsaturated fatty acids, for example avocado and coconut oils, which leads to different protection of polyphenols against oil oxidation. Furthermore, the phenoxy radicals derived from caffeic acid or *p*-coumaric acid had different stability, which would have caused different rates of propagation and subsequent oil oxidation reactions (Gordon, 1990). This result suggests the feasibility of using polyphenols from natural sources like fruits and vegetables to replace synthetic antioxidants such as butylated hydroxyanisole (BHA), butylated hydroxytoluene (BHT) and *tert*-butylhydroquinone (TBHQ) for improving oil stability, which may offer dual benefits through increasing the nutritional value of the final oil product.

Interactions among polyphenols, proteins and starch polysaccharides

Proteins are one of the most reactive food components and can react with polyphenols, fibre polypsaccharides, sugars, oil, water and other food components such as salt in the same food system (Arai and Watanabe, 1989; Sivam *et al.*, 2011a, 2011b; Sun-Waterhouse *et al.*, 2011c). These reactions may modify nutritive value and sensory attributes of foods as well as the extractability and bioaccessibility of polyphenols. Sivam *et al.* (2012) explored the interactions between blackcurrant polyphenols, pectin, wheat flour, sugar (sucrose), canola oil and salt in model bread systems. Increased nutritional value was found in the bread fortified with 3% wet weight dough of pectin fibre and 3% wet weight dough of blackcurrant polyphenol extract (containing 27% polyphenols): the fortified model bread contained 26 times higher total extractable polyphenols than control model bread. The polyphenols retained in the fortified bread were mainly anthocyanin antioxidants, delphinidin-3-*O*-glucoside, delphindin-3-*O*-rutinoside, cyanidin-3-*O*-glucoside and cyanidin-3-*O*-rutinoside. Adding blackurrant polyphenol extract and pectin also influenced the extractability of gluten proteins, for example it decreased the proportion of unextractable high molecular weight proteins (Figure 20.4) (Sivam *et al.*, 2012). Ingredients used for bread making, including blackcurrant polyphenols, pectins, oil, salt, yeast and sugar, all interact with one another during bread development and baking. Changes were detected in the gluten protein secondary structure, due to the removal of the breadmaking ingredients oil and salt, evidenced from the associated alterations in gluten protein's amide I and II bands (Figure 20.5). The presence of oil or salt affects the surface area and properties of proteins, which are possibly associated with the adsorption of lipid by proteins, primarily through physical entrapment and changes of ionic strength induced by salt. The removal of oil facilitates interactions among hydrophilic molecules, for example water molecules move more freely and can more easily approach gluten, causing increased ß-sheets which in turn stabilizes ß-turns. The removal of salt reduced the concentration of ions in the bread system, which influenced the reactivity of polyphenols with

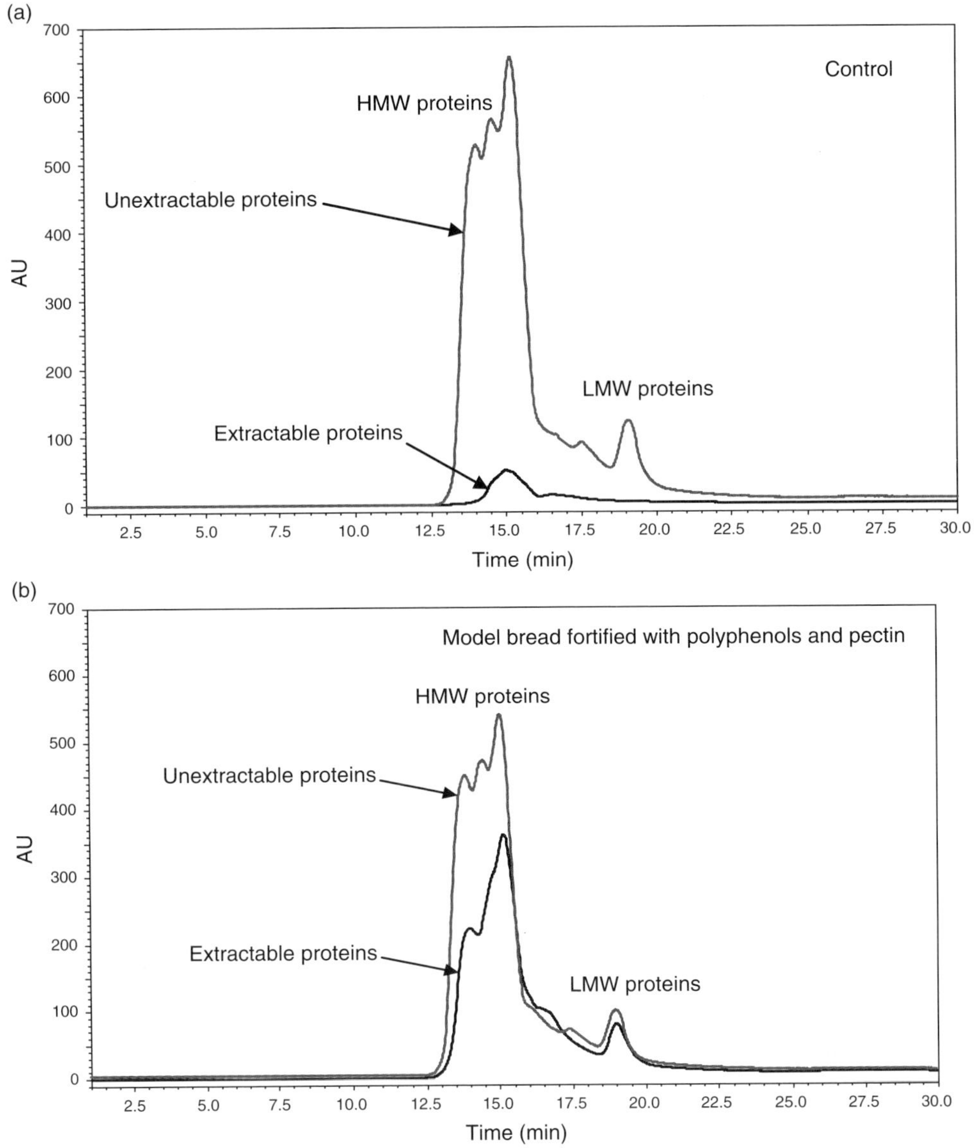

Fig. 20.4 Extractable and unextractable protein fractions for (a) control and (b) fortified model breads.

different polarity, gelation of pectin, solubility and conformation of gluten proteins (Sivam *et al.*, 2012).

Polar groups in proteins contribute to the hydrogen bonding, which affects protein's secondary structure, for example wheat flour protein α-helix and β-sheet conformations, and subsequently food product attributes. Hydrophobic interactions play important roles in the tertiary folding of proteins, contributing to emulsification and flavour-binding functions (which influences the solubility and extractability of polyphenols as well as sensory attributes of final product). The thiol group normally has a pKa of 8.3, although this value may vary due

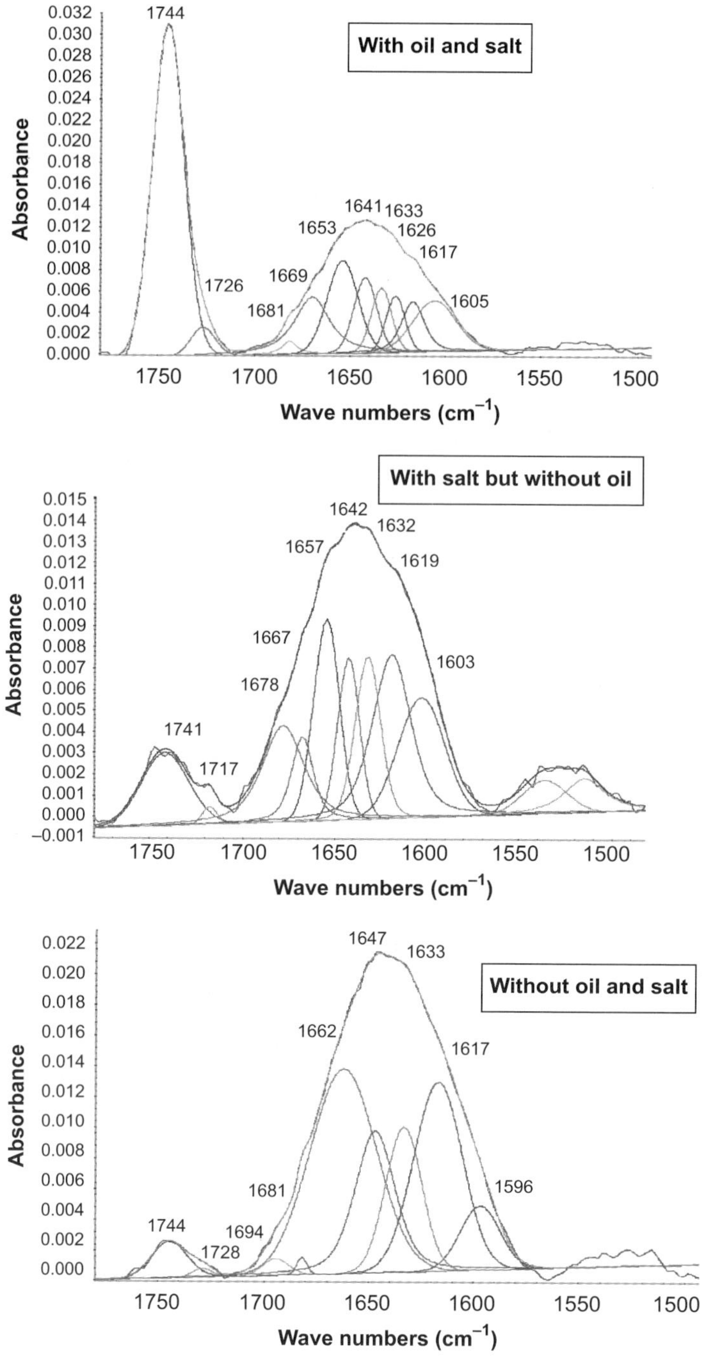

Fig. 20.5 FTIR spectra with curve fitting results for model bread systems with/without canola oil and salt. Curve fitting applied to bands at 1745 cm^{-1} (CO stretch), 1700–1595 cm^{-1} (Amide I), 1580–1500 cm^{-1} (Amide II) (reduced $\psi^2 = 2.77–9.99$; $r^2 = 0.004–0.007$; standard error = 0.003–0.008).

to the surrounding food matrix. The interchange reactions between thiol (–SH) groups and disulfide bonds (–S–S–) greatly influence the physical properties of proteins (Sun-Waterhouse *et al.*, 2011c). Proteins are surface active, which is generally affected by hydrogen ions or hydroxyl ions. Carboxyl groups also occur on the surface of protein structure. Carbonyl group in ketone or aldehyde structure can actively interact with phytochemicals and other co-existing substances in the food matrix.

Heating of protein solutions, such as baking the flour and water mixture during bread making, increases protein surface hydrophobicity (because of the unfolding effects) and changes the solubility of heated protein (e.g. increase of protein molecular size through hydrophobic interaction and disulfide formation upon heating). The gluten proteins are largely responsible for the viscoelastic properties of wheat dough (Tatham and Shewry, 1985). Large protein molecules are more heat sensitive than smaller ones. The gliadins in wheat gluten proteins are soluble in aqueous alcohols and are monomers associated by hydrogen bonding and hydrophobic interactions (Tatham and Shewry, 1985). The glutenins in wheat gluten proteins can aggregate into high molecular weight complexes which may not be soluble in aqueous alcohols, but the individual subunits, after reduction, are soluble at low pH. After heatingfor example to about 90°C, the extractability of gluten proteins decreases substantially (Sun-Waterhouse *et al.*, 2011c). The protein species unaffected by the heat are ω-gliadins, α-, β- and γ-gliadins and low molecular weight alcohol-soluble glutenins. The latter were less extractable, particularly at the higher temperatures (Tatham and Shewry, 1985; Sun-Waterhouse *et al.*, 2011c). Although other types of bonding, such as hydrogen bonds and hydrophobic bonds, contribute to the rheological and functional properties of gluten, increased covalent disulfide bonding could be the most significant factor that is responsible for the irreversible effects of heat on gluten functionality (Xu *et al.*, 1999; Sun-Waterhouse *et al.*, 2011c). The addition of a reducing agent such as polyphenols to the bread system was found to introduce changes in the disulfide–sulfydryl interchange, including the oxidation of the S–H groups in bread, especially the water-soluble S-H groups, which leads to their removal from the bread system (Cauvain *et al.*, 1992; Sun-Waterhouse *et al.*, 2011c). Introducing an antioxidant like polyphenols into the bread system affects the redox status of the S–H groups (Cauvain *et al.*, 1992; Collins, 1994; Frei, 1994; Xu *et al.*, 1999; Sun-Waterhouse *et al.*, 2011c), causing direct oxidation of an S–H group in a glutenin molecule to a stable form that will not be involved in further disulfide–sulfydryl interchange reactions, or formation of an –S-S– bond between a water-soluble protein –S–H group and a glutenin –S–H group and/or between two glutenin –S–H groups during bread dough development. Moreover, polyphenols may act as reducing agents and use some available oxygen, a process that competes the oxygen uptake by the yeast. The molecular size of polyphenols may also play a role in the aforementioned interactions with the thiol groups. Smaller polyphenol molecules may approach the –S–H groups easier, due to less steric hindrance from macromolecules like gluten and/or some protection by these macromolecules via an entrapment mechanism (Schofield and Chen, 1995).

CP/MAS ^{13}C NMR analyses (Sivam *et al.*, 2011b) showed that the relative crystallinity of the control bread (24.6%) was greater than the bread fortified with 3% blackcurrant polyphenol extract and 3% high methoxyl pectin (10.6%), suggesting a loss of crytallinity upon the fortification of polyphenols and pectin. Moreover, the fortified bread had greater amorphous material (evidenced by the C4 resonance of the α-1→4 glucans in starch around 80–84 ppm, 14.1%), than the control bread (the C4 resonance 9.42%). Starch is a key component of wheat flour (63–72% of wheat flour) and shows strong structure–functionality relationships (Atwell, 2001). Starch granules consist of well-organized crystallites embedded in less-organized amorphous regions (Liu *et al.*, 2006; Tahir *et al.*, 2011). The crystal structure forms include Type A, Type B, Type C (a combination of types A and B) and Type V (Tang and Hills, 2003; Bertoft, 2004). The starch conformation in the fortified and control breads differed: in addition to their common V-type structure or amorphous starch structure, the control bread could also contain A-type starch polymorphic structures. These detected changes in wheat flour starch induced by added polyphenols and pectin were associated with the changes in wheat flour proteins; the fortified breads contained greater amounts of unextractable proteins than extractable proteins, but the control bread had an opposite trend (Sun-Waterhouse *et al.*, 2011c). The wheat flour contains 66% w/w starch (amylopectin:amylose = 3:1) and 4.8% w/w non-starch polysaccharides. Thus, it was possible for the flour polysaccharides to be involved in the detected changes of bread upon the addition of polyphenols and pectin. Considering the high concentrations of starch in the flour, it is feasible to examine the changes in starch conformation, which are correlated with properties of chromatographic mobilities, solubility, crystallization, nuclear magnetic resonance and so on. Heating of aqueous dispersions of starch leads to gelatinization, which is a disruption of the molecular order within granules. A practical consequence of granule swelling is paste viscosity (Atwell *et al.*, 1988). The behaviour of starch during pasting, and especially during cooling, is determined to a large extent by the affinity of hydroxyl groups. The linear amylose chains in particular are hydrogen bonded. Linear amylose molecules have a greater tendency to re-associate and form hydrogen bonds than the large and branched amylopectin molecules. Conformation of α-1,4-linkages in amylose and amylopectin causes these polymers to assume tightly coiled helical structures which possess a hydrophobic inner surface to include a variety of substances, for example lipids and phenols, whilst conformation of β-1,4-linked homopolymers leads to flat, ribbon-like structures stabilized by intra- and interchain hydrogen bonds (Rees, 1977). The complexation of polyphenol by bread starch that results in starch adopting different conformation (Solms *et al.*, 1973) could also contribute to the detected difference in polyphenol extractability in this study. The interactions among bread components led to an increase in bread firmness from 0.22 kgf to 0.34 kgf was detected after the fortification of blackcurrant polyphenols and pectin (Sun-Waterhouse *et al.*, 2011c).

Interactions between polysaccharide polymers in the absence and presence of fruit polyphenol extracts or fruit juices

Food gels are important food vehicles that offer a structural matrix for holding water and other food ingredients. There is a general interest in the use of natural and plant-based polysaccharide additives, such as polysaccharide gums, to provide a means to increase total dietary fibre in the diet or popular consumer foods and modify the rheological properties of digesta for specific gut health benefits. Each polysaccharide gum has its limitations in food applications, for example carboxymethylcellulose (CMC) loses viscosity and may precipitate at low pHs. The combined use of gums, for example mixed gels of pectin, locust bean gum xanthan gums or CMC, can avoid such drawbacks and exhibit unique properties or cost-process advantages, for example increased fibre polysaccharide content without making a beverage too viscous, or adding a small amount of a gum additive to modify viscosity for a desired mouth feel.

Synergistic effects on the rheological behaviour and cloud point of the combined hydrocolloid dispersions are possible and could be affected by temperatures (Sun-Waterhouse and

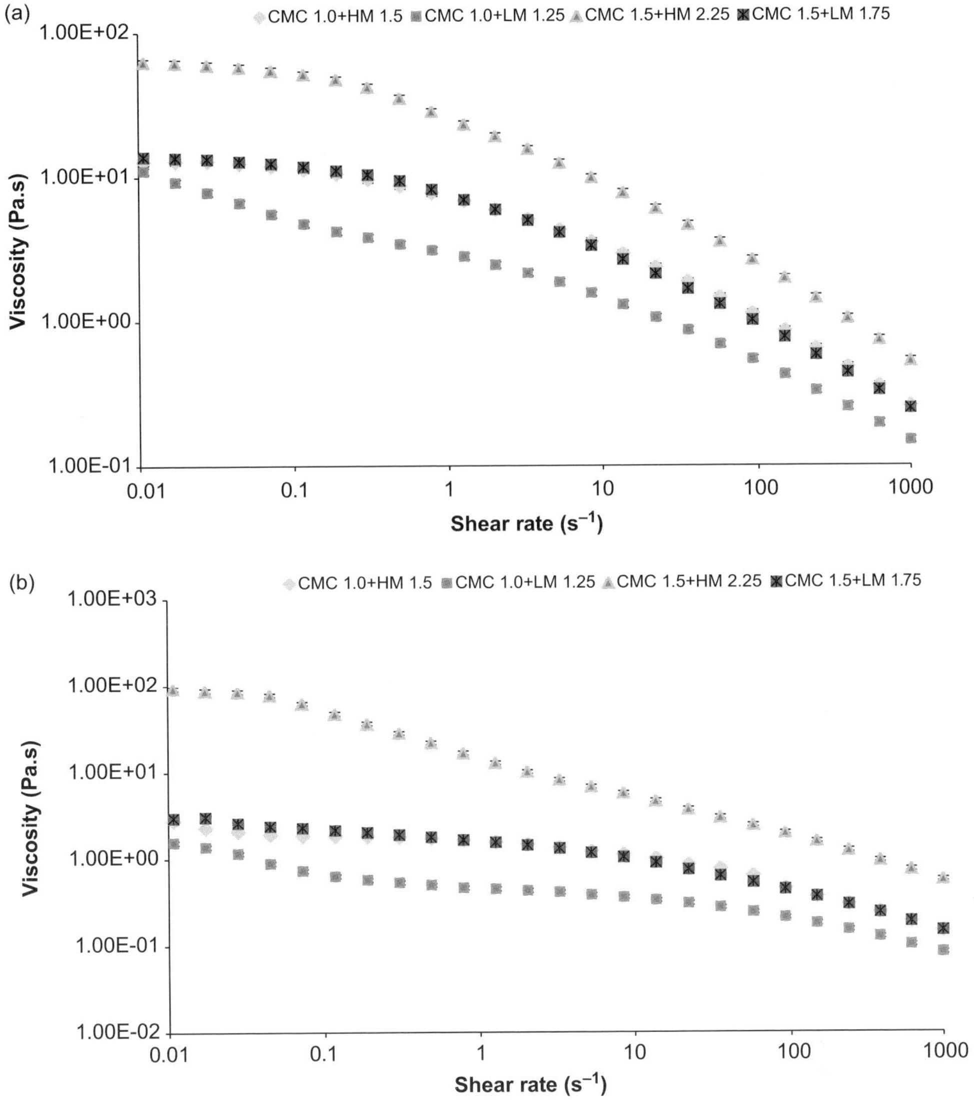

Fig. 20.6 Viscosity of carboxymethylcellulose (CMC)–HM pectin and CMC–LM pectin dispersions at (a) 20°C and (b) 60°C.

Wadhwa, 2012). It is possible to achieve the same viscosity profile through different combinations of hydrocolloid additives, such as 1% CMC–1.5% HM pectin and 1.5% CMC–1.75% LM pectin at 20 or 60°C (Figure 20.6). The combination of 1% CMC–1.25% LM pectin generally had lower viscosity at 60°C than at 20°C, but the trend reversed for 1.5% CMC–2.25% HM pectin. However, their elasticity could be quite different. Figure 20.7 shows that the combinations of 1% CMC–1.25% LM pectin and 1% CMC–1.5% HM pectin had similar G′ (storage modulus which quantifies the amount of elastic energy 'stored' in the system) or G″ (loss modulus which quantifies the mechanical energy lost in the form of heat due to the

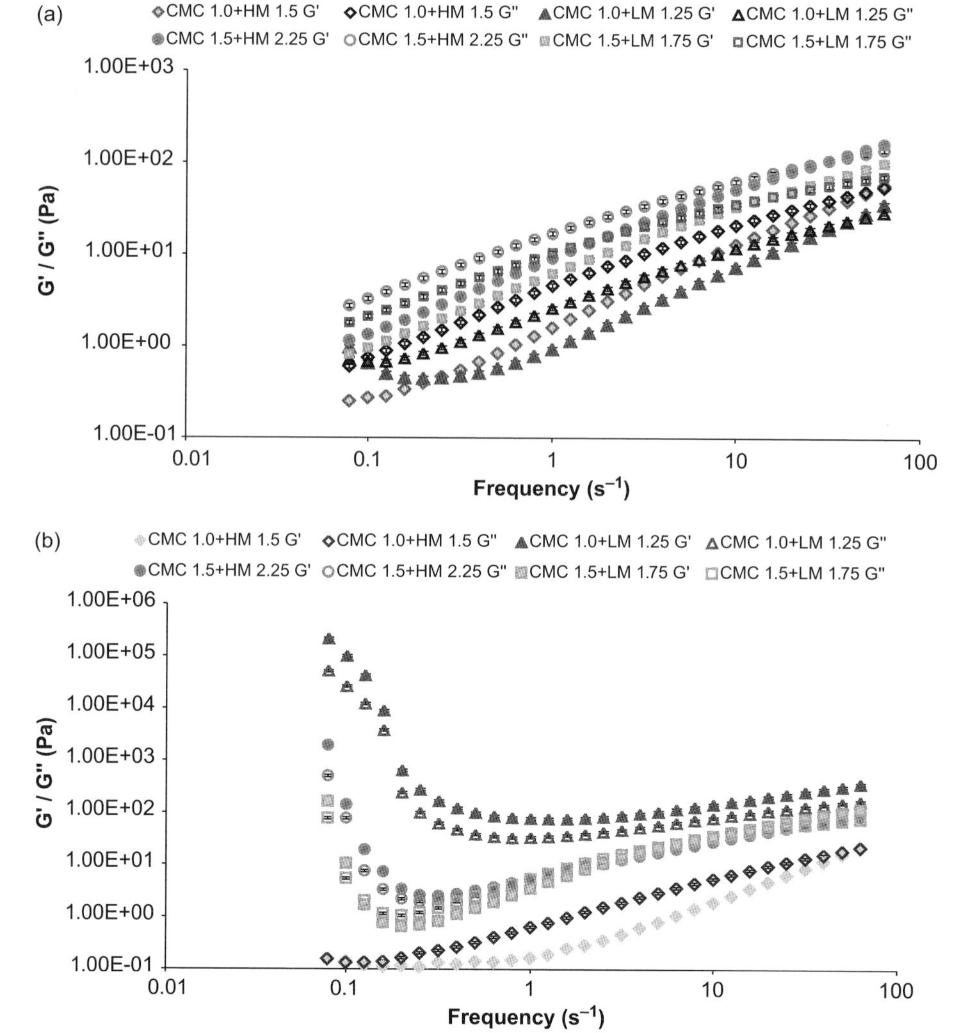

Fig. 20.7 Storage and loss moduli of CMC–HM and CMC–LM dispersions analysed at (a) 20°C and (b) 60°C.

viscous forces acting on the gel) at 20°C, but at 60°C had dramatically different G′ or G″. At both testing temperatures, G′ > G″ for the 1% CMC–1.25% LM pectin combination, but the trend reversed for the 1% CMC–1.5% HM pectin combination. Pectins are quite often bound to calcium and exhibit peculiar hydration, ion exchange and rheological properties which are sensitive to pH, temperature, enzymes and other matrix factors. The solubility of pectins increases with an increase in the degree of esterification and with a decrease in molecular weight. The less soluble a pectin is in water, the easier it can be precipitated by adding an electrolyte. HM pectins require high concentrations of soluble solids (> 55% w/w) and a low pH (<3.5) for gelation. LM pectins can gel via the same mechanism as that of HM pectins over a wide range of soluble solid contents and pH, but require sufficient Ca^{2+} for gelation (Imeson, 1997; Phillips and Williams, 2000).

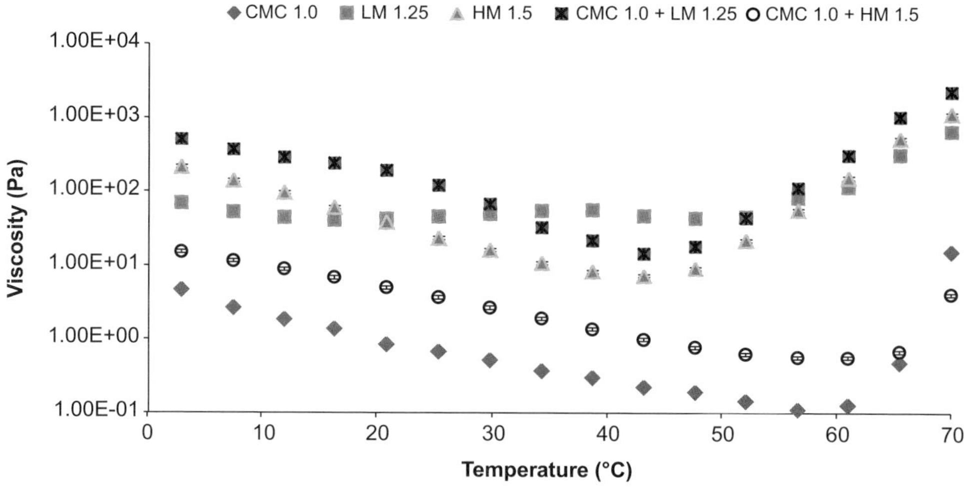

Fig. 20.8 Dynamic temperature ramp plots of CMC alone (1.00%), LM pectin alone (1.25%), HM pectin alone (1.50%), and CMC–HM and CMC–LM dispersions.

For the dispersions that contain combinations of hydrocolloid additives, the synergistic effects can also be reflected by the cloud points of the resultant dispersions. Figure 20.8 shows that the cloud points of 1% CMC, 1.25% LM pectin, 1.5% HM pectin, 1% CMC–1.25% LM pectin and 1% CMC–1.5% HM pectin dispersions occurred at 58, 52, 43, 43 and 61°C, respectively. The cloud point of 1% CMC–1.25% LM pectin occurred at a much lower temperature than individuals (1% CMC alone or 1.25% LM pectin alone). CMC appears to play a more dominant role than HM pectin for the cloud point of 1% CMC–1.5% HM pectin dispersions. The degree of rheological synergy varied depending on the type and concentration of hydrocolloids. The selection of hydrocolloid additives for combined use in food applications needs to consider all the possible changes in synergism of viscosity, elasticity and cloud point during food processing, storage and consumption (especially the temperature variable). Any changes in the polymer network, for example polymer chain length and entanglement complexity, may require different energies for separating or aggregating hydrocolloid chains (Larson, 1999).

Adding fruit polyphenol extracts to various hydrocolloid polymer networks would lead to different amounts of extractable polyphenols retained after processing. Figure 20.9 shows that for the sample fruit polyphenol ingredient, that is kiwifruit or blackcurrant extract, using different hydrocolloid polymers as the gelling ingredient results in different amounts of extracted polyphenols in the final gel products. Different methoxyl content of pectin also caused a difference (up to 3.5-fold) in the total extracted polyphenol content, with lower methoxylation facilitating a greater amount of extracted polyphenols.

Furthermore, adding fruit juices to hydrocolloid gel systems would cause changes in physicochemical properties after processing of fruit juice products due to enzymatic and non-enzymatic reactions (Richard-Forget and Gauillard, 1997). The nature of fruit juices and final processed foods greatly affect the polyphenol stability and extractability as well as food attributes, thus should be taken into account during food formulation design and food processing method selection (e.g. pasteurization methods). For example, freshly pressed apple and pineapple juice contain significantly different levels of protease activity. When

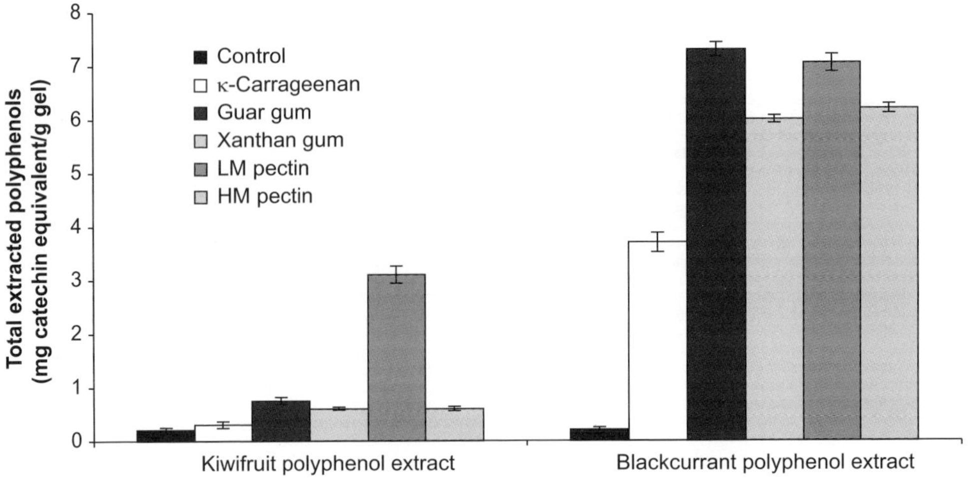

Fig. 20.9 Total extracted polyphenol contents of food gels enhanced with kiwifruit or blackcurrant polyphenol extract.

these juices were used to prepare pouch- or sealed-container-type of fruit juice jellies, which involves pasteurization such as retorting, Ultra High Temperature treatment and High Pressure Processing, the effects of the nature of fruit juices, such as acidity and intrinsic enzymes, on the nutritional values and sensory attributes of the final products should be considered. Modification of proteins via enzymes such as papain (Arai and Watanabe, 1989), or chemical reactions such as mild acid hydrolysis (Ma and Khanzada, 1987), can lead to significant changes in functionality of the proteins. Hydrophobic groups could be introduced into hydrophilic proteins using papain, which would alter the surface activities and emulsifying properties of α_{s1}-casein, and emulsion stability (Arai and Watanabe, 1989). Deamidation of proteins could be a possible mechanism under mildly acidic conditions, which cause greater solubility and emulsifying properties, and water- and lipid-binding capacities of proteins (Ma and Khanzada, 1987).

Sun-Waterhouse *et al.* (2011e) reported the differences in the rheological properties, total polyphenol content and protease activities of the fruit juice gels prepared using apple or pineapple juices and two gelling ingredient combinations, citrus fibre (AQ plus) and Locus bean gum (LBG) or citrus fibre (AQ plus) and gelatine. Both the type of fruit juices and gelling ingredients were responsible for the detected differences in the viscosity (Figure 20.10) and the amount of extracted polyphenols (Figure 20.11) of the fruit juice gels after food processing. For the same gelling agents AQ plus fibre and gelatine, adding apple juice resulted in much higher viscosity than did the pineapple juice of the same concentration. The extracted polyphenols retained after processing was the greatest for the apple juice-gel formulated with AQ plus fibre and LBG or gelatine, but such an enhancement in the stability and extractability of polyphenol was less profound in the case of pineapple juice. Thus, the combination of AQ Plus fibre and LBG could be used for apple or pineapple juice-gel applications, and apple juice-gel could be a good food format to deliver apple polyphenols with increased stability and extractability.

Polysaccharides are both nutrients and dietary fibres. A lot of functional properties of polysaccharides, such as structural function and water-binding properties, come from

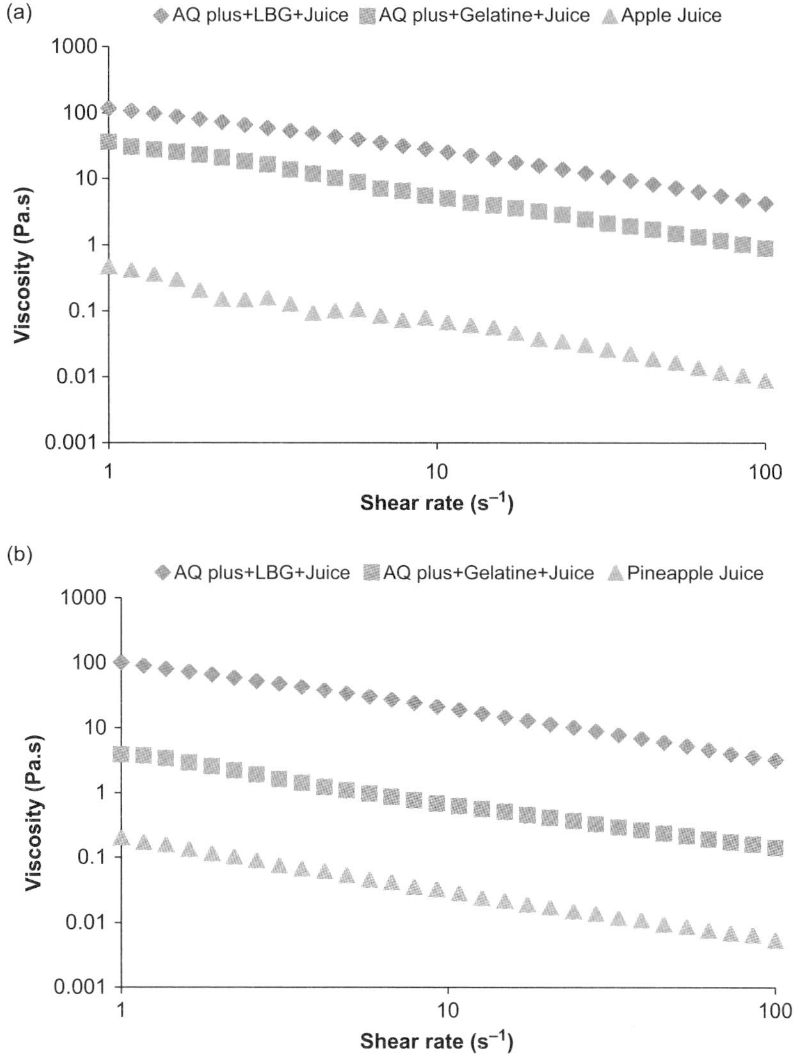

Fig. 20.10 Viscosity of (a) apple or (b) pineapple juice and their juice gels as a function of shear rate.

non-covalent bonds, especially hydrogen bonds, van der Waals forces, electrostatic forces, steric repulsive forces and hydrophobic interactions (e.g. complexation with lipid and polyphenol molecules). Although individual non-covalent bonds are weak, their cumulative effect in a large polysaccharide molecule can be very strong. Stable conformations require large numbers of non-covalent bonds (hydrogen, dipolar, ionic). The food environment surrounding each polysaccharide polymer also imparts an influence on proton exchanges and the polymer structure of polysaccharides via chemical reactions and physical effects. Solvent effects are one kind of the phenomena observed; for example, polysaccharides can interact with up to 100 times their own weight of water. Gummy polysaccharides, such as pectin, locust bean gum or guar gum, are used as additives in foods to add textural and other sensory and processing attributes. Rheological properties of these polysaccharides are affected by

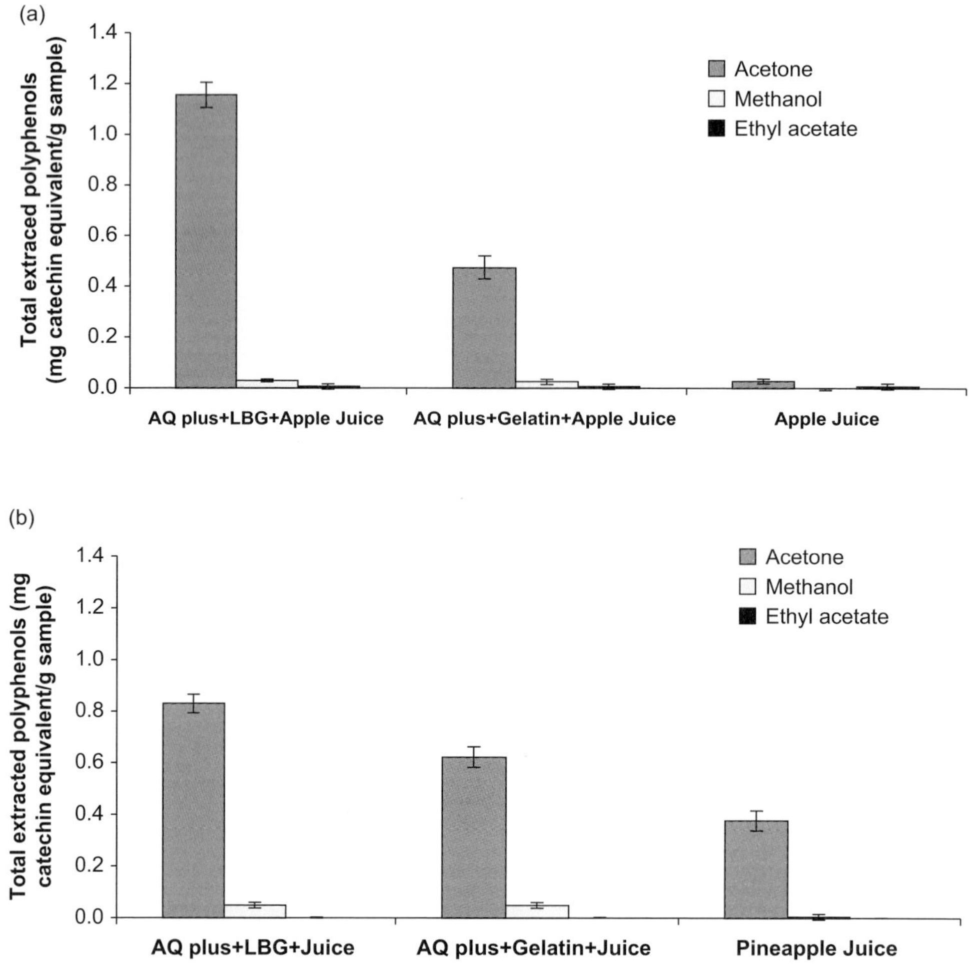

Fig. 20.11 Total extracted polyphenol contents of the (a) apple or (b) pineapple juice and their processed fruit gels.

molecular size and conformation, orientation of molecules, association of molecules, water binding and swelling, concentration, particle size, particle shape and particle specific surface, degree of dispersion and viscosity of the dispersing medium (Nottelmann and Kulicke, 1991; Grigorescu and Kulicke, 2000; Storz *et al.*, 2010). For the viscosity of hydrocolloidal polysaccharides, the concentration effect is not linear. Low concentrations facilitate unfolding and high mobility due to excess water. High concentrations cause increases in viscosity due to interactions with water and between polysaccharide particles. A comparable concentration and relative molecular weight, linear polymers normally exhibit higher viscosity than branched polymers. Other co-existing substances can directly affect rheological behaviours, for example salt inhibits hydration and reduces viscosity; acids and bases have minimal effect on viscosity of non-ionic colloids at pH 3–10.

Intrinsic enzymes in the fruit juices including PPO, protease, or pectinase play a role in the polyphenol extractability and physical properties. For example, higher quantity and/or

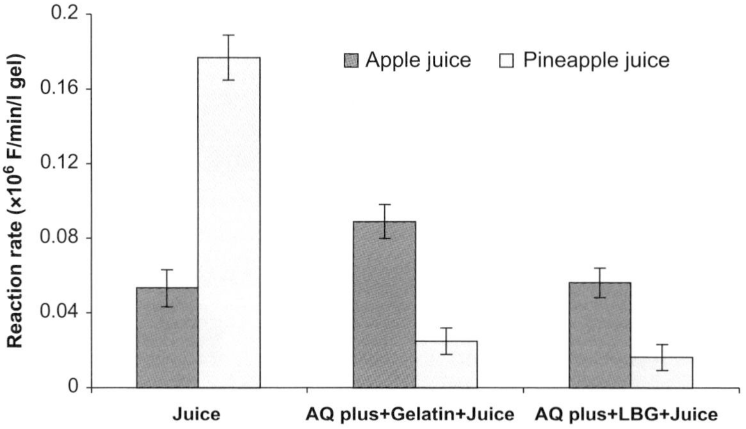

Fig. 20.12 Protease activities in the apple or pineapple juice and derived gels.

greater enzymatic activity of PPO would have led to greater loss of polyphenols. The activity of protease and pectinase would influence the cross-linking and microstructure of the polymer network associated with proteins from gelatin and pectic polysaccharide components from AQ Plus ingredient. Figure 20.12 shows that, for the same quantity, pineapple juice had approximately 3.5 times greater protease activity than apple juice, which accounts for the rate and extent of enzymatic reactions in the fruit juice gels, for example protease-catalysed hydrolysis of gelatine. However, the presence and type of gelling ingredients also determines the protease activity and reaction rate. For either juice, a greater reaction rate of protease was observed in the gel containing gelatine (a proteinaceous substance) than in the non-protein gel system. Several modes of action might contribute to the above phenomena. The presence of gelling agents would have generated physical barriers and steric hindrance via their gel network, slowing down the reaction rate of enzyme from fruit juice. The two combinations of gelling agents led to different viscosity (i.e. AQ plus + LBG > AQ plus + gelatin in both juices), with the viscosity of AQ plus + gelatin + apple juice being much greater than that of AQ plus + gelatin + pineapple juice. Thus, the degree of the reduced reaction rate caused by the gel network depended on the juice type and the gelling agent combinations. The presence of gelatine would have provided additional substrate for protease, causing an increased rate of enzyme activity. Greater disruption of the gel network was visually observed in the formulation with gelatin than that with LBG, which was possibly due to the presence of protease (Tochi *et al.*, 2008; Bhattacharya and Bhattacharyya, 2009) and other enzymes such as pectinase (polygalacturonase and pectin esterase) (Ben-Arie *et al.*, 1979; Whitaker, 1990). It was observed that the consistency of apple or pineapple juice gel made with AQ plus and LBG was similar, but varied remarkably between the apple and pineapple juice gels prepared using AQ plus and gelatine (Sun-Waterhouse *et al.*, 2011e). Greater disruption on the gel network was visually observed in the formulation with gelatin than that with LBG, which was possibly due to the presence of protease (Tochi *et al.*, 2008; Bhattacharya and Bhattacharyya, 2009) and other enzymes such as pectinase (polygalacturonase and pectin esterase)(Whitaker, 1990; Ben-Arie *et al.*, 1979). The gel matrix played an important role in these detected total extractable polyphenols. Polysaccharides such as pectins (the key component of AQ plus) and galactomannans (the key component of LBG), as well as proteins/proteinaceous substances (the key component in gelatine), were found to form complexes with phenolics including

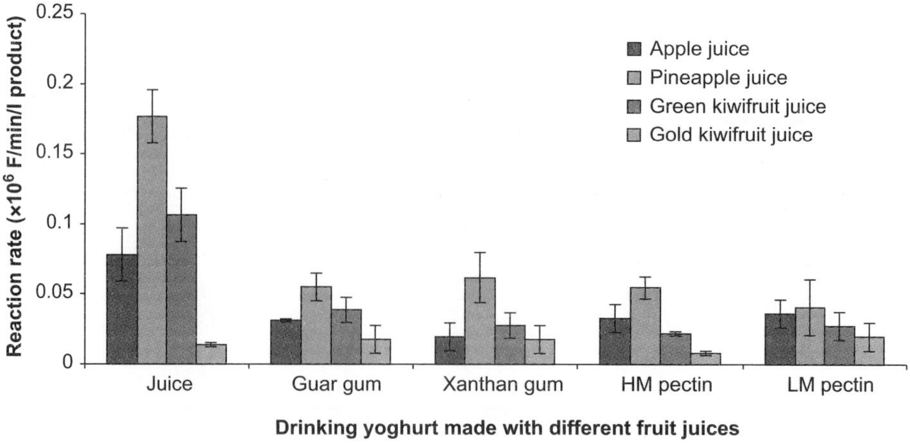

Fig. 20.13 Protease activities in the apple, pineapple, green and gold kiwifruit juice and derived drinking yoghurts.

ferulic acid, coumaric acid and quercetin (Haslam, 1998; Rawel *et al.*, 2001; Rohn *et al.*, 2004; Perez-Jimenez and Saura-Calixto, 2006). Pectins, galactomannans and proteins may or may not possess the same affinity to various polyphenols that have different chemical structure (Rice-Evans *et al.*, 1996).

For the case of fruit juice-enhanced yoghurts, difference in enzyme reaction rate was detected even in the absence of proteinaceous gelling ingredients or stabilizer additives (Figure 20.13). Extractability of fruit juice polyphenols from the final food gel products varied with gelling agent/stabilizer, nature of fruit juice (e.g. intrinsic enzymes, acidity, polyphenol type) and nature of food matrix (e.g. protein-based versus non-protein foods). For example, green kiwifruit is a different species to gold kiwifruit, although both kiwifruit cultivars contain similar types of cell wall polysaccharides and a high proportion of cellulose. Gold kiwifruit has a higher proportion of hemicellulosic polysaccharides and lower proportion of pectic polysaccharides (Sauvageau *et al.*, 2010). The variations in the composition, acidity and intrinsic enzyme profiles of the two types of kiwifruit juices (Dawes and Keene, 1999; Nieuwenhuizen *et al.*, 2007; Nishiyama, 2007; Sauvageau *et al.*, 2010), led to differences in milk coagulation kinetics (i.e. the gel time and coagulum firming rates) (Fagan *et al.*, 2006), the complexation between kiwifruit polyphenols and milk proteins and polysaccharides during yoghurt processing (Rawel *et al.*, 2001; Renard *et al.*, 2001; Rohn *et al.*, 2004) and the physical form of proteins (Goff *et al.*, 1995; Marshall and Arbuckle, 1996; Adapa *et al.*, 2000), which ultimately cause a change in yoghurt microstructure and the stability and extractability of polyphenols. Formation of denatured protein gels requires an initial balance between attractive and repulsive forces, allowing native protein to turn into unfolded polypeptides that may associate gradually to form the gel matrix. Greater disruption on the gel network was visually observed in the formulation with gelatin than that with LBG, which was possibly due to the presence of protease (Tochi *et al.*, 2008; Bhattacharya and Bhattacharyya, 2009) and other enzymes such as pectinase (polygalacturonase and pectin esterase)(Ben-Arie *et al.*, 1979; Whitaker, 1990). The protease activities would break down the protein network, increase the incompatibility between the two gel components and could lead to different degrees of phase separation, which can be detected as the difference in product consistency and mouth feel. Thus, it is essential to monitor gel network to maximize the delivery efficiency and bioaccessibility of targeted fruit bioactives such as polyphenols. Pretreatments to suppress the enzyme activities in fruit juice which are detrimental to gelation may be required, depending on the selection of gelling ingredients. The use of self-assembled food ingredients and complementary antioxidants can protect bioactives during processing and storage, and control the release of bioactives at the desired site of action in the body, without compromising the sensory properties of the food, by reducing contact with the environment and other food components in the product (Ruben and Larsson, 1985; Robins and Wilson, 1994; Kilcast and Clegg, 2002). Food microstructure and macrostructure can be modified to alter volatility and retention (controlled release) of aroma compounds, or slow down or reduce the contact between bitter compounds and taste receptors (Lafarge *et al.*, 2008).

Effects of beverage matrix containing dietary fibres on fruit bioactives

The positive roles of dietary fibres and polyphenols in health enhancement and disease prevention encourage attempts to add these two types of fruit bioacitve ingredients (including insoluble fibres) at higher concentrations into convenient consumer foods and beverages. Smoothies are a delivery format with high consumer popularity. In addition to product sensory issues, the stability and bioaccessibility of incorporated fruit bioactives are other challenges associated with beverage formulation, processing and storage (Sun-Waterhouse *et al.*, 2010a, 2012b). Monitoring the interactions among polyphenols, vitamin C and fibre polysaccharides during formulation, processing and storage is the key to a successful beverage to deliver relatively high concentrations of fruit bioactives.

The fibre concentration in a smoothie beverage affects the extracted polyphenols retained after beverage processing over time (Sun-Waterhouse *et al.*, 2010a). Smoothies were produced under the same processing conditions, using apple fibre (containing total fibre content 67.4%, of which 16% is soluble) at concentrations of 2.5, 4.5 and 7.5 g per 300 ml per serve, in the presence of a 250 mg per 300 ml serve of polyphenols (blackcurrant and apple polyphenol extracts at a fixed ratio of 20:80). It was found that at the same added polyphenol concentration, the reduction in total extracted polyphenol content after 14 days was the greatest for the smoothie containing 2.5 g fibre, followed by that with 4.5 g fibre and then that with 7.5 g fibre (Figure 20.14). Moreover, the rates of losing vitamin C (derived from ingredients like apple puree) appeared to be different during a 14-day storage for the smoothies that contained the same amount of fibre (7.5 g per 300 ml serve) but different polyphenol concentrations (250, 500 and 750 mg per 300 ml serve) (Figure 20.15). Rapid degradation of vitamin C was observed in smoothies during a 14-day storage period at 4°C. After eight days, vitamin C was not detectable in the smoothie containing 750 mg polyphenols, while about 33% of the original vitamin C level was retained in smoothies containing either 250 or 500 mg polyphenols. The degradation of vitamin C was associated with the concentration of polyphenols. Vitamin C is susceptible to degradation even during refrigerated storage and the degradation could be affected by the matrix environment, for example the soluble solids content (Robertson and Samaniego-Esguerra, 1990; Torregrosa *et al.*, 2006). Vitamin C initially present in the smoothie would have been used up more quickly when the smoothie contained greater amounts of polyphenols, as it would be used to slow the degradation of these polyphenols and regenerate polyphenols during storage (Liao and Seib, 1988; Evans and Shahidi, 1997).

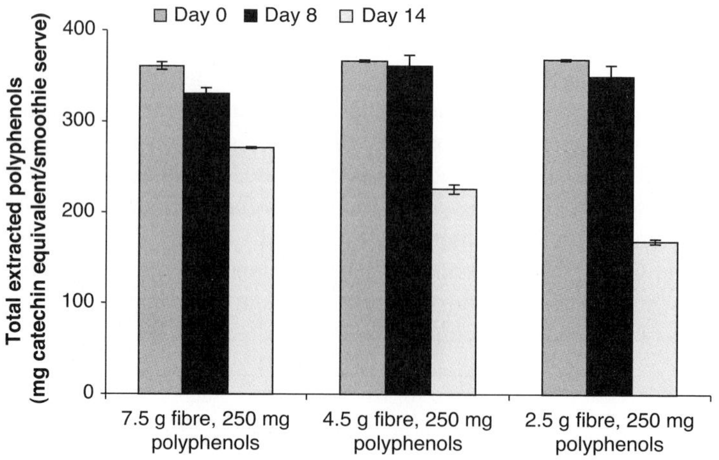

Fig. 20.14 Total extracted polyphenol content of smoothies (250 mg polyphenols, and 2.5, 4.5 or 7.5 g fibre) during storage at 4°C for 14 days.

The differences in the fruit fibre used (such as the particle size) and the stabilizer (such as the type or concentration) also influence the extracted polyphenol content and viscoelastic properties of this type of smoothie after food processing (Sun-Waterhouse *et al.*, 2012b). Smoothies were produced under the same processing conditions using the same concentration of apple polyphenols (500 mg catechin equivalent/300 ml serve), two types but the same concentration (12.5 g/300 ml serve) of apple fibre ingredients (both contain soluble and insoluble fibres but have different particle size: AF01 250 μm, AF09 500–700 μm) and a stabilizer carboxymethylcellulose (CMC) at different concentrations (0.05–2.00%). The total extracted polyphenol content (Figure 20.16) and viscosity (Figure 20.17) of these smoothies after food processing varied. For the smoothie formulated with AF09 (coarser fibre), the highest extracted polyphenol content (351 mg catechin equivalent/300 ml beverage) occurred at 1% CMC (Figure 20.16). The viscosity of the AF09 and AF01 smoothies changed, different

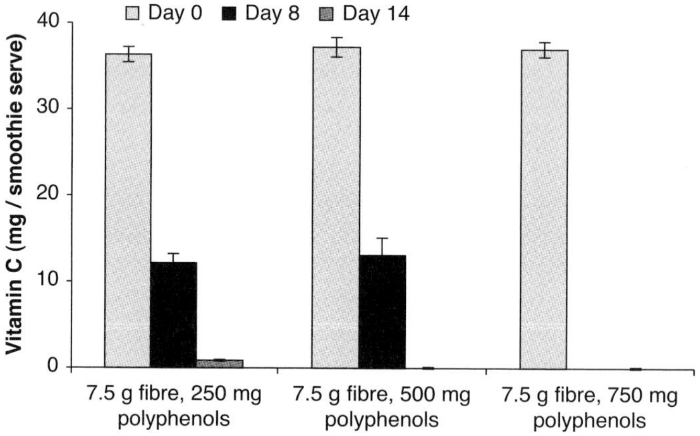

Fig. 20.15 Vitamin C content of smoothies (250, 500 or 750 mg polyphenols, and 7.5 g fibre) during storage at 4°C for 14 days.

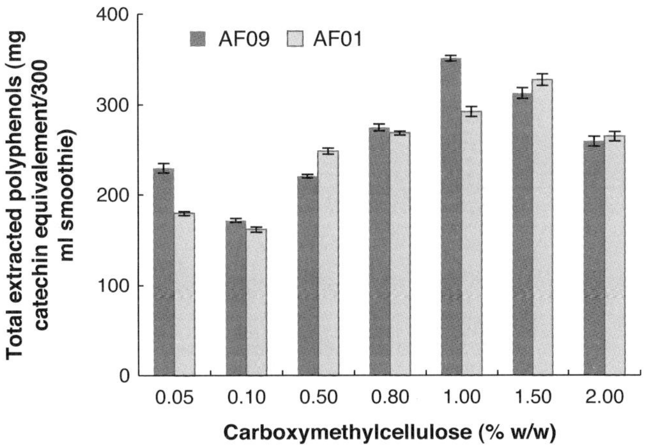

Fig. 20.16 Total extracted polyphenol contents of AF09 and AF01 smoothies with different concentrations of carboxymethylcellulose.

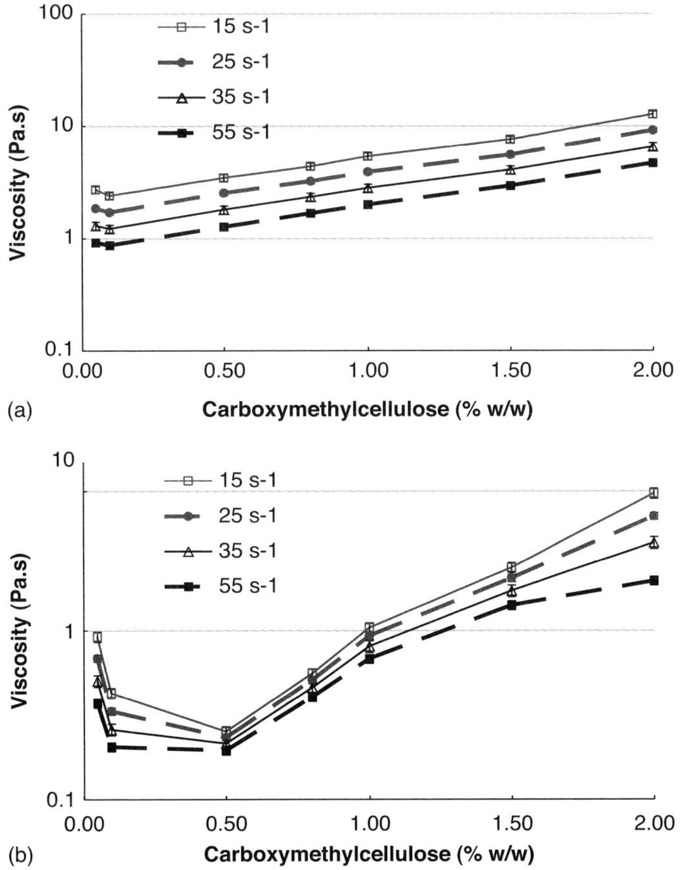

Fig. 20.17 Viscosity of the (a) AF09 and (b) AF01 smoothies at 37°C and different shear rates (15, 25, 35 and 55 s^{-1}).

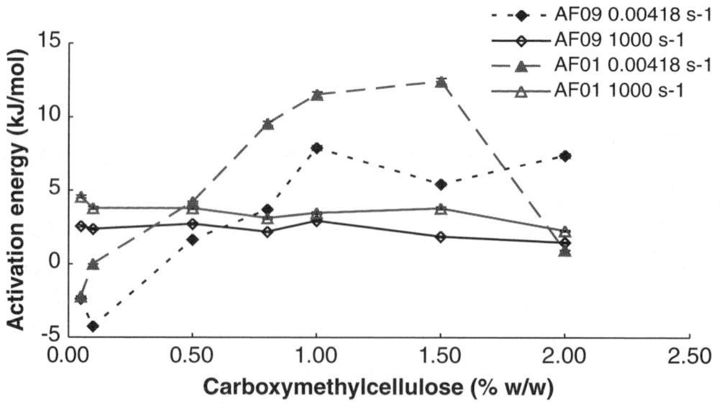

Fig. 20.18 Variation of the calculated activation energies of the AF09 and AF01 beverage systems as a function of CMC concentrations at low (0.00418 s^{-1}) and high (1000 s^{-1}) shear rates.

patterns with the increasing CMC concentration (Figure 20.17). Except for the 0.05% CMC, the viscosity of the AF09 smoothie steadily increased with an elevated CMC concentration, whilst a valley value of viscosity was detected for the AF01 beverage at the 0.5% CMC. The viscosity difference of the AF01 beverage at different shear rate became larger at 2% CMC, suggesting that it is easier to detect the viscosity difference when the shear rate of the in-mouth processing is altered from 15 to 55 s^{-1} (Wood and Goff, 1973). In this shear rate range, the viscosity of both the AF09 and AF01 smoothies containing CMC $\geq$0.10% decreased when shear rate was increased.

There were interactions between stabilizer molecules, between stabilizer molecules and the fibre network, between stabilizer molecules and polyphenols, between water and soluble or insoluble polymers, and between the fibre network and polyphenols. The amount of stabilizer polymer contributes to the phase stability of smoothie products, especially at low shear rates. The smoothies prepared using fibres of different particle size (AF01 or AF09) had different activation energy profiles (Figure 20.18), especially at low shear rates. The broader microstructure of the AF09 fibre material (compared with much finer AF01 fibre) would have caused a stiffer structure with larger-sized aggregates and stronger intermolecular interactions, with more energy being required for the breakdown of the aggregates. The alignment, orientation and breakdown of aggregates and disentanglement of stabilizer polymer coils (induced by the shear) all contribute to the shear-thinning behaviours of a smoothie. A lower calculated activation energy at a stabilizer concentration and shear rate indicates less energy required to generate the viscous flow.

In this type of fruit bioactive enhanced beverage system, interactions occur not only between an individual fruit bioactive and other food matrix components, but also among all the added fruit bioactives, such as polyphenols, fibres and vitamin C. Fibres in the water-soluble and water-insoluble forms and with different particle sizes could significantly affect the physical stability and rheological properties of beverages. Consumer acceptability decreased with elevated fibre and polyphenol contents, with fibre concentration playing a more dominant role (Sun-Waterhouse *et al.*, 2010a).

CONCLUSION AND FUTURE OUTLOOK

Novel functional foods containing natural plant-based bioactives are achieving increased market share in the food sector. Such foods provide a vehicle for delivering the health benefits of fruits and vegetables to consumers. Developing functional food products is a consumer-oriented process. While selecting appropriate processing methods to produce raw materials or ingredients containing target bioactives is important, it does not promise that sufficient amounts of bioactives in the required forms will be 'preserved' or incorporated in the final finished foods and enable a labelling claim with regard to a desired biological function to be made. The stability, bioaccessibility and bioavailability of bioactives in a suitable food matrix need to be addressed during food or beverage formulation and processing.

The incorporation of active or natural ingredients into foods presents technical challenges related to food sensory issues, bioactive stability and delivery efficiency. Desirable biological function, product type (food format), processing feasibility, technological constraints and consumer acceptability all relate to the interactions among the components in a functional food throughout processing, storage, handling and digestion. From the product development perspective, there exist challenges due to inherent contradictions that can exist between a desired biological function end-point and a commercially acceptable product, which requires reconciliation of the physicochemical properties to the desired biological functionality and the sensory attributes of finished products. Foods including functional foods must be enjoyable to consume. Furthermore, 'dose/concentration' influences the food format selection and food formulation design. Bitterness and astringency associated with foods with high polyphenol content, and grittiness associated with drinks containing high insoluble fibre content, are obstacles that need to be addressed during product development.

This chapter highlights how the interactions among components within the matrices of consumer foods and beverages can be positive or negative to food properties. Adding fruit bioactive singly or in combinations to a food system would cause the structural changes in key food components such as protein and carbohydrate macromolecules. Such changes would have impacts on the stability and extractability of added fruit bioactives such as polyphenols and vitamin C. Soluble fibre, including various gum polysaccharides, and insoluble fibre with different particle sizes, greatly affect the physical stability and textural characteristics of a beverage, as well as the polyphenols retained after processing. Food additives such as sugar, salt and stabilizer also influence the changes of bioactive properties and food attributes. Careful consideration of the synergies between a bioactive ingredient and other food or beverage matrix components is key to successful product development. To maximize the beneficial synergies derived from food component interactions, future work should include investigation into the food matrix effect on the bioavailability and biological activity of each type of fruit bioactive throughout the whole metabolic process in the human body; for example, improved penetration of fruit polyphenols through the gut wall, by monitoring their polarity with a desired hydrophilic–hydrophobic balance during food formulation and processing. Such knowledge will guide the industrial manufacturing and domestic cooking of foods containing desirable fruit bioacitves, as well as the optimization of diets for various consumer populations.

REFERENCES

Adapa, S., Dingeldein, H., Schmidt, K. A. and Herald, T. J. (2000) Rheological properties of ice cream mixes and frozen ice creams containing fat and fat replacers. *J Dairy Sci* **83**, 2224–2229.

Adyanthaya, I., Kwon, Y.I., Apostolidis, E. and Shetty, K. (2010) Health benefits of apple phenolics from postharvest stages for potential type 2 diabetes management using in vitro models. *J Food Biochem* **34**, 31–49.

Aguilar, G. and Huirton, C. (1990) Constitutive exo-pectinase produced by Aspergillus sp. CH-Y-1043 on different carbohydrate source. *Biotechnol Lett* **12**, 655–660.

Alsaffar, A.A. (2011) Effect of food processing on the resistant starch content of cereals and cereal products-a review. *Int J Food Sci Technol* **46**, 455–462.

Antolovich, M., Prenzler, P., Robards, K. and Ryan, D. (2000) Sample preparation in the determination of phenolic compounds in fruits. *Analyst* **125**, 989–1009.

Arai, S. and Watanabe, M. (1989) Modification to change functional properties of proteins. In: *Protein Quality and the Effects of Processing* (eds R.D. Phillips and J.W. Finley), pp. 315–329. Marcel Dekker, New York.

Atwell, W.A. (2001) *Composition of commercial flour in wheat flour*, pp. 27-45. American Association of Cereal Chemists, St Paul, MN.

Atwell, W.A., Hood, L.F., Lineback, D.R. *et al.* (1988) The terminology and methodology associated with basic starch phenomena. *Cereal Food World* **33**(3), 306–311.

Autran, J.-C. and Berrier, R. (1984) Durum wheat functional protein subunits revealed through heat treatments. Biochemical and genetic implications. In: *Proceedings of the 2nd International Workshop on Gluten Proteins* (eds A. Graveland and J.H.E. Moonen), pp. 175–183. Institute for Cereals, Flour and Bread, TNO, Wageningen, The Netherlands.

Bacic, A., Harris, P.J. and Stone, B.A. (1988) Structure and function of plant cell walls. In: *The Biochemistry of Plants* (ed. J. Preiss), Vol. **14**, pp. 297–371. Academic Press, San Diego, CA.

Bakker, J. and Mela, D.J. (1996) Effect of emulsion structure on flavor release and taste perception. In: *Flavor-Food Interactions.* (eds R.J. McGorrin and J.V. Leland), pp. 36–47. American Chemical Society, Washington, DC.

Bartley, G.E. and Scolnik, P.A. (1995). Plant carotenoids: Pigments for photoprotection, visual attraction, and human health. *Plant Cell* **7**, 1027–1038.

Baxter, N.J., Lilley, T.H., Haslam, E. and Williamson, M.P. (1997) Multiple interactions between polyphenols and a salivary proline-rich protein repeat result in complexation and precipitation. *Biochemistry* **36**, 5566–5577.

Ben-Arie, R., Kislev N. and Frenkel, C. (1979) Ultrastructural changes in the cell walls of ripening apple and pear fruit. *Plant Physiol* **64**(2), 197–202.

Bertoft, E. (2004) Analysis of starch structure. In: *Starch in Food, Structure, Function and Applications* (ed. A. Eliasson), 1st edn, pp. 57–81. CRC Press, Boca Raton, FL.

Bhaskaran, S. and Hardley, F. (2002). Buyer beliefs, attitudes and behaviour: foods with therapeutic claims. *J Consum Market* **19**, 591–606.

Bhattacharya, R. and Bhattacharyya, D. (2009) Preservation of natural stability of fruit. "Bromelain" from *Ananas Comosus* (pineapple). *J Food Biochem* **33**, 1–19.

Bird, A.R., Brown, I.L. and Topping, D.L. (2000) Starches, resistant starches, the gut microflora and human health. *Current Issues Intest Microbiol* **1**(1), 25–37.

Bombardelli, R., Curri, S.B., Della Loggia, R. *et al.* (1989) Complexes between phospholipids and vegetal derivatives of biological interest. *Fitoterapia* **60**(Suppl N.1), 1–9.

Boudet, A.M. (2007). Evolution and current status of research in phenolic compounds. *Phytochemistry* **68**, 2722–2735.

Boulton, R. (2001) The copigmentation of anthocyanins and its role in the color of red wine: A critical review. *Am J Enol Viticul* **52**(2), 67–87.

Boyes, S., Perera, C. and Young, H. (1992) Kiwi fruit lipoxygenase: Preparation and characteristics. *J Food Sci* **57**(6), 1390–1394.

Brouillard, R. (1988) Flavonoids and flower colour. In: *The Flavonoids, Advances in Research Since 1980*, (ed. J.B. Harborne), pp. 525–538. Chapman and Hall, London.

Butts, C., Paturi, G., de Guzman, C.E. *et al.* (2010) The effect of dietary vegetable and fruit fibres on gut health in healthy rats. *Proceedings Nutr Soc NZ* **34**, 24–29.

Carle, R., Keller, P., Schieber, A. *et al.* (2001) Method for obtaining useful materials from the by-products of fruit and vegetable processing. Patent application, WO 01/78859 A1.

Cauvain, S.P., Collins, T.H. and Pateras, I. (1992) Effects of ascorbic acid during processing. *Chorleywood Digest* **121**, 111–114.

Cevallos-Casals, B.A., Byrne, D., Okie, W.R. and Cisneros-Zevallos, L. (2006) Selecting new peach and plum genotypes rich in phenolic compounds and enhanced functional properties. *Food Chem* **96**, 273–280.

Chaudhari, R. (2009) Satiety ingredients: losing fats not nutrients. *Asia Pac Food Ind* **42**, 42–44.

Chen, K.S., Xu, C.J., Lou, J. *et al.* (1999) Lipoxygenase in relation to the ripening and softening of Actinidia fruit. *Acta Phytophysiol Sin* **25**, 138–144.

Chen, J.H. and Ho, C.-T. (1997) Antioxidant activities of caffeic acid and its related hydroxycinnamic acid compounds. *J Agric Food Chem* **45**, 2374–2378.

Cheung, S., Szeto, Y. and Benzie, I. (2007). Antioxidant protection of edible oils. *Plant Food Human Nutr (Formerly Qualitas Plantarum)* **62**, 39–42.

Choudhari, S.M. and Ananthanarayan, L. (2007) Enzyme aided extraction of lycopene from tomato tissues. *Food Chem* **102**(1), 77–81.

Chung, H., Liu, Q., Lee, L. and Wei, D. (2011) Relationship between the structure, physicochemical propertis and *in vitro* digestibility of rice starches with different amylose contents. *Food Hydrocoll* **25**, 968–975.

Collins T. H. (1994) Mixing, moulding and processing of bread doughs in the UK. In: *Breeding to Baking*, Proceedings of an International Conference at FMBRA (15–16 June). Chorleywood, Chipping Campden, UK, pp.77–83.

Cren-Olive, C.C., Teissier, E., Duriez, P. and Rolando, C. (2003) Effect of catechin O-methylated metabolites and analogues on human LDL oxidation. *Free Radic Biol Med* **34**, 850–855.

Damodaran, S. (1996) Amino acids, peptides and proteins. In: *Food chemistry* (ed. O.R. Fennema), pp. 322–429. Marcel Dekker Inc., New York.

Dangles, O. and Elhajji, H. (1994) Synthesis of 3-methoxy- and 3-β-Dglucopyranosy1oxy)flavylium Ions. Influence of the flavylium substitution pattern on the reactivity of anthocyanins in aqueous solution. *Helvetica Chimica Acta* 77, 1595–1610.

Dangles, O., Dufour, C. and Bret, S. (1999) Flavonol–serum albumin complexation. Two-electron oxidation of flavonols and their complexes with serum albumin. *J Chem Soc (Perkin 1)* **2**, 737–744.

David, O. (2001) Physicochemical properties of dietary fiber: Overview. In: *Handbook of Dietary Fiber* (eds. S.S. Cho and M.L. Dreher), p. 199. Marcel Dekker, Inc., New York.

Dawes, H. M. and Keene, J. B. (1999) Phenolic composition of kiwi fruit juice. *J Agric Food Chem* **47**, 2398–2403.

de Carvalho, A.F. and Maubois J.L. (2009). Applications of membrane technologies in the Dairy Industry. In: *Engineering Aspects of Milk and Dairy Products*, (eds J.S.R. Coimbra and J.A. Teixeira), pp. 33–56, CRC Press, Boca Raton, FL.

De Leon, S.Y. and Delores, M.I. (2005) Coconut. In: *Processing Fruits: Science and Technology* (eds D.M. Barrett, L. Somogyi and H. Ramaswamy). CRC Press Ltd, New York.

de Rosso, V.V. and Mercadante, A.Z. (2007). HPLC-PDA-MS/MS of anthocyanins and carotenoids from dovyalis and tamarillo fruits. *J Agric Food Chem* **55**, 9135–9141.

DeRoos, K.B. (2006) How lipids influence flavor perception. In: *Food Lipids Chemistry, Flavor, and Texture* (eds F. Shahidi and H. Weenan), pp. 145–158. American Chemical Society, Washington DC.

De Man, J.M. (1990) Carbohydrate. In: *Principles of Food Chemistry*, 2nd edn, pp. 173–175. AVI Publishers: Van Nostrand Reinhold, New York.

Dietrich, H., Rechner, A., Patz, C.-D. *et al.* (2003). Influence of processing on the phenolic antioxidants of apple juices. *Deutsche Lebensmittel-Rundschau*, **99**, 1–11.

Dinnella, C., Recchia, A., Fia, G. *et al.* (2009) Saliva characteristics and individual sensitivity to phenolic astringent stimuli. *Chem Senses* **34**, 295–304.

Druaux, C. and Voilley, A. (1997) Effect of food composition and microstructure on volatile flavour release. *Trends Food Sci Tech* **8**, 364–368.

Duhard, V., Garnier, J.C. and Megard, D. (1997). Comparison of the stability of selected anthocyanin colorants in drink model systems. *Agro Food Industry Hi-Tech*, **8**, 28–34.

Duttaroy, A.K. and Jørgensen, A. (2004) Effects of kiwi fruit consumption on platelet aggregation and plasma lipids in healthy human volunteers. *Platelets* **15**, 287–292.

Dziedzic, S.Z. and Kearsley, M.W. (1995) The technology of starch production. In: *Handbook of Starch Hydrolysis Products and their Derivatives* (eds M.W. Kearsley and S.Z. Dziedzic), 1st edn, pp. 1–25. Chapman and Hall, Cambridge.

Ellis, R.P., Cochrane, M.P., Dale, M.F.B. *et al.* (1998) Starch production and industrial use. *J Sci Food Agric* **77**, 289–311.

Evans, R.J. and Shahidi, F. (1997) Optimizing lipid stability with natural inhibitors. In: *Natural Antioxidants: Chemistry, Health Effects and Applications*, pp. 225–244. AOCS Press, Champaign, IL.

Fagan, C.C., O'Donnell, C.P., Cullen, P.J. and Brennan C.S. (2006). The effect of dietary fibre inclusion on milk coagulation kinetics. *J Food Eng* **77**, 261–268.

Falguera, V., Sánchez-Riaño, A.M., Quintero-Cerón, J.P. *et al.* (2011) Characterization of polyphenol oxidase activity in juices from 12 underutilized tropical fruits with high agroindustrial potential. *Food Bioprocess Tech,* doi: 10.1007/s11947-011-0521-y.

Fernandez, M.L. (2001) Pectin: composition, chemistry, physicochemical properties, food applications, and physiological effects. In: *Handbook of Dietary Fibre* (eds S.S. Cho and M.L. Dreher), pp. 583–602. Marcel Dekker, Inc., New York.

Fleury, N. and Lahaye, M. (1993) Studies on by-products from the industrial extraction of alginate. *J Appl Phycol* **5**(1), 605–611.

Frei, B. (1994) Reactive oxygen species and antioxidant vitamins: Mechanisms of action. *Am J Med* **97**(3, Suppl 1), S5–S13.

Fry, S.C. (1988) Chemical and metabolic analysis. In: *The Growing Plant Cell Wall*, pp. 1–333. Longman and Scientific Technical, New York.

Gidley, M.J. (2004). Naturally functional foods – challenges and opportunities. *Asia Pac J Clin Nutr* **13**, S31.

Goff, H.D., Freslon, B., Sahagian, M.E. *et al.* (1995). Structural development in ice cream B dynamic rheological measurements. *J Text Stud* **26**, 517–536.

Goldberg, D.M., Yan, J. and Soleas, G.J. (2003) Absorption of three wine-related polyphenols in three different matrices by healthy subjects. *Clin Biochem* **36**, 79–87.

Gordon, M.H. (1990) The mechanisms of antioxidant action in vitro. In: *Food antioxidants* (ed B.J.F. Hudson), pp.1–18. Elsevier Applied Science, London.

Grigelmo-Miguel, N., Gorinstein, S. and Martín-Belloso, O. (1999) Characterisation of peach dietary fibre concentrate as a food ingredient. *Food Chem* **65**, 175–181.

Grigorescu G. and Kulicke W.-M. (2000) Prediction of viscoelastic properties and shear stability of polymers in solution. *Adv Polym Sci* **152**, 1–45.

Guyton A.C. and Hall J.E. (2005) Protoplasm is composed mainly of five basic substances: water, electrolytes, proteins, lipids and carbohydrates. In: *Textbook of Medical Physiology*, 11th edn (eds A.C. Guyton and J.E. Hall) (ebook). Saunders.

Haslam, E. (1998). *Practical Polyphenolics: From Structure to Molecular Recognition and Physiological Action*. Cambridge University Press, Cambridge.

Harrison, M., Hills, B.P., Bakker, J. and Clothier, T. (1997) Mathematical models of flavor release from liquid emulsions. *J Food Sci* **62**, 653–664.

Hoffelner, S. (2010). Functional constituents – what is allowed under the Health Claim Regulations? *Fluessiges Obst* **77**(2), 70–72.

Hollman, P.C.H., van Trijp, J.M.P., Buysman, M.N.C.P. *et al.* (1997) Relative bioavailability of the antioxidant flavonoid quercetin from various foods in man. *FEBS Lett* **418**, 152–156.

Hollman, P.C., van het Hof, K.H., Tijburg, L.B. and Katan, M.B. (2001). Addition of milk does not affect the absorption of flavonols from tea in man. *Free Radic Res* **34**, 297–300.

Hoseney, R.C. (ed.). (1994) *Principles of Cereal Science and Technology*, 2nd edn. American Association of Cereal Chemists, St Paul, MN.

Hyvonen, L., Linna, M., Tuorila, H. and Dijksterhuis, G. (2003) Perception of melting and flavor release of ice cream containing different types and contents of fat. *J Dairy Sci* **86**, 1130–1138.

Imeson, A. (1997) *Thickening and Gelling Agents for Food*, 2nd edn. Blackie Academic and Professional, London, UK.

Jaeger, S.R., Axten, L.G., Wohlers, M.W. and Sun-Waterhouse, D. (2009) Polyphenol-rich beverages: insights from sensory and consumer science. *J Sci Food Agric* **89**(14), 2356–2363.

Kallithraka, S., Bakker, J. and Clifford, M.N. (1998). Evidence that salivary proteins are involved in astringency. *J Sens Stud* **13**, 29–43.

Kampa, M., Alexaki, V. I., Notas, G. *et al.* (2004) Antiproliferative and apoptotic effects of selective phenolic acids on T47D human breast cancer cells: potential mechanisms of action. *Breast Cancer Res* **6**, 63–74.

Kaul, A. and Khanduja, K.L. (1998) Polyphenols inhibit promotional phase of tumorigenesis: relevance of superoxide of superoxide radicals. *Nutr Cancer* **32**, 81–85.

Kaplan, D., Christiaen, D. and Arad, S. (1988) Binding of heavy metals by algal polysaccharides. In: *Algal Biotechnology*, pp. 179–187. Elsevier, London.

Kilcast, D. and Clegg, S. (2002). Sensory perception of creaminess and its relationship with food structure. Food *Qual Pref* **13**, 609–623.

Kolakowska, A. (2003) Lipid oxidation in food systems. In: *Chemical and Functional Properties of Food Lipids* (eds Z.E. Sikorski and A. Kolakowska), pp. 133–166. CRC Press Ltd, Boca Raton, FL.

Kriventsov, V.I. and Karakhanova, S.V. (1972) Biologically active substances of several subtropical fruit crops of the Crimea. In: *Trudy Vsesoyuznogo Seminara po Biologicheski Aktivnym (Lechebnym) Veshchestvam Plodov i Yagod* (16–20 March 1970, Michurinsk, USSR) (ed. E.P. Franchuk), 4th edn, pp. 82–86. Vsesoyuznyi Nauchno-Issledovatel'skii Institut Sado-vodstva, Michurinsk, USSR.

Labuza, T.P. (1971) Properties of water as related to the keeping quality of foods. In: *Proceedings of the International Congress of Food Science and Technology SOS/70*, pp. 618–635. Institute of Food Technologists, Chicago, IL.

Lafarge, C., Bard, M.H., Breuvart, A. *et al.* (2008) Influence of the structure of cornstarch dispersions on kinetics of aroma release. *J Food Sci* **73**, S104–S109.

Lamikanra, O. (2002) Enzymatic effects on flavor and texture of fresh-cut fruits and vegetables. In: *Fresh cut fruits and vegetables: Science,technology, and market* (ed. O. Lamikanra), pp. 127–147. CRC Press, Boca Raton, FL.

Larson, R.G. (1999) *The Structure and Rheology of Complex Fluids*. Oxford University Press, New York.

Lauren, D.R., Smith, W.A., Adaim, A. *et al.* (2009) Chemical composition and in vitro anti-inflammatory activity of apple phenolic extracts and of their sub-fractions. *Int J Food Sci Nutr* **60**, 188–205.

Liao, M.L. and Seib P.A. (1988) Chemistry of L-ascorbic acid related to foods. *Food Chem* **30**, 293–312.

Liu, H., Yu, L., Xie, F. and Chen, L. (2006) Gelatinization of cornstarch with different amylose/amylopectin content. *Carbohydr Polym* **65**, 357–363.

Lokuruka, M.N.I. (2011) Effects of processing on soybean nutrients and potential impact on consumer health: An Overview. *African J Food Agric Nutr Devel* **11**(4) (http://www.bioline.org.br/request?nd11045; last accessed 19 March 2013).

López, G., Ros, G., Rincón, F. *et al.* (1996) Relationship between physical and hydration properties of soluble and insoluble fibre of artichoke. *J Agric Food Chem* **44**, 2773–2778.

Lu, T., Duh, C., Lin, J. and Chang, Y. (2008) Effect of granular characteristics on the viscoelastic properties of composites of amylose and waxy starches. *Food Hydrocoll* **22**, 164–173.

Lucarini, M., Mugnaini, V. and Pedulli, G.F. (2002) Bond dissociation enthalpies of polyphenols: The importance of cooperative effects. *J Org Chem* **67**, 928–931.

Luthria, D.L. (2006) Significance of sample preparation in developing analytical methodologies for accurate estimation of bioactive compounds in functional foods. *J Sci Food Agric* **86**, 2266–2272.

Ma, C.-Y. and Khanzada, G. (1987) Functional properties of deamidated oat protein isolates. *J Food Sci* **52**, 1583–1587.

Macheix, J.-J., Fleuriet, A. and Billot, J. (1990) *Fruit phenolics*. CRC Press, Boca Raton, FL.

Macheix, J.J. and Fleuriet, A. (1998) Phenolic acids in fruits. In: *Flavonoids in health and disease*. (eds C. Rice-Evans and L. Packer), pp. 35–59. Marcel Dekker, Inc, New York.

Manach, C., Morand, C., Crespy, V. *et al.* (1998) *Quercetin is recovered in human plasma as conjugated derivatives which retain antioxidant properties*. *FEBS Lett* **426**, 331–336.

Mares-Perlman, J.A., Millen, A.E., Ficek, T.L. and Hankinson, S.E. (2002) The body of evidence to support a protective role for lutein and zeaxanthin in delaying chronic disease. *Overview. J Nutr* **132**, 518S–524S.

Marshall, R.T. and Arbuckle, W.S. (1996) *Ice Cream*, 5th edn. Chapman and Hall, New York.

Masibo, M. and Qian, H. (2009) Mango bioactive compounds and related nutraceutical properties – a review. *Food Rev Int* **25**, 346–370.

McKee, L.H. and Latner, T.A. (2000) Underutilized sources of dietary fibre: A review. *Plant Food Human Nutr* **55**, 285–304.

McRae J.M. and Kennedy J.A. (2011) Wine and grape tannin interactions with salivary proteins and their impact on astringency: A review of current research. *Molecules* **16**, 2348–2364.

Miller, N.J., Diplock, A.T. and Rice-Evans, C.A. (1995) Evaluation of the total antioxidant activity as a marker of the deterioration of apple juice on storage. *J Agric Food Chem* **43**, 1794–801.

Milo Ohr, L. (1999). Functional foods put flavor up front. *Prep Foods*, **168**, 59–60.

Morr, C.V. (1983) Physico-chemical basis for functionality of milk proteins. *Kiel Milchwirtsch Forschungsber* **35**, 333–344.

Morr, C.V. (1984) Production and use of milk proteins in food. *Food Tech* **38**(7), 39–42, 44, 46–48.

Myers, A.M., Morell, M.K., James, M.G. and Ball, S.G. (2000) Recent progress toward understanding biosynthesis of the amylopectin crystal. *Plant Physiol* **122**, 989–997.

Murkovic, M. (2003) Phenolic compounds. In: *Encyclopaedia of Food Sciences and Nutrition*. (eds B. Caballero, L.C. Trugo and P.M. Finglas), 2nd edn, pp. 4507–4514. Elsevier Science Ltd, Oxford.

Nair, J., Ehimare, U., Beitman, B.D. *et al.* (2006) Clinical review: Evidence-based diagnosis and treatment of ADHD in children. *Mol Med* **103**, 617–621.

Nawirska, A. and Kwaniewska, M. (2004) Dietary fibre fractions from fruit and vegetable processing waste. *Food Chem* **91**, 221–225.

Nieuwenhuizen, N.J., Beuning, L.L., Sutherland, P.W. *et al.* (2007). Identification and characterisation of acidic and novel basic forms of actinidin, the highly abundant cysteine protease from kiwifruit. *Funct Plant Biol* **34**, 946–961.

Nishiyama, I. (2007) Fruits of the actinidia genus. *Adv Food Nutr Res* **52**, 293–324.

Nottelmann H. and Kulicke W.-M. (1991) Preparation, characterisation and rheological behaviour of water-swellable polymer networks. In: *Polymers as Rheology Modifiers* (eds D.N. Schulz and J.E. Glass), ACS Book Series vol 462, pp. 62–87. American Chemical Society, Washington, DC.

O'Connor, C.J., Lal, S.N.D. and Eyres, L. (eds) (2007) *Handbook of Australasian Edible Oils*. Oils and Fats Specialist Group of NZIC, Auckland, New Zealand.

Oh, H.E., Hemar, Y., Anema, S.G. *et al.* (2008) Effect of high-pressure treatment on normal rice and waxy rice starch-in-water dispersions. *Carbohydr Polym* **73**, 332–343.

O'Leary, K.A., Day, A.J., Needs, P.W. *et al.* (2001) Flavonoid glucuronides are substrates for human liver beta-glucuronidase. *FEBS Lett* **503**, 103–106.

Palomäki, T. and Saarilahti, H.T. (1997) Isolation and characterization of new C-terminal substitution mutation affecting secretion of polygalacturonases in Erwinia carotovora ssp. carotovora. *FEBS Lett* **400**, 122–126.

Parkar, S., Trower, T., Stevenson, D. *et al.* (2010) Food, fibre and satiety – how does fibre make you feel full. In: *Proceedings of the 2nd TNO Beneficial Microbes Conference* (15-17 March 2010). TNO, Noordwijkerhout, The Netherlands.

Pazos, M., Andersen, M.L., Medina, I. and Skibsted, L.H. (2007) Efficiency of natural phenolic compounds regenerating R-tocopherol from R-tocopheroxyl radical. *J Agric Food Chem* **55**, 3661–3666.

Peleg, H., Gacon, K., Schlich, P. and Noble, A.C. (1999). Bitterness and astringency of flavan-3-ol monomers, dimers and trimers. *J Sci Food Agric* **79**, 1123–1128.

Percy, A.E., Melton, L.D. and Jameson, P.E. (1997) Xyloglucan and hemicelluloses in the cell wall during apple fruit development and ripening. *Plant Sci* **125**, 31–39.

Perez-Jimenez, J. and Saura-Calixto, F. (2006) Effect of solvent and certain food constituents on different antioxidant capacity assays. *Food Res Int* **39**, 791–800.

Phillips, G.O. and Williams, P.A. (2000) *Handbook of Hydrocolloids*. CRC Press, Boca Raton, FL.

Philpott, M., Mackay, L., Ferguson, L.R. *et al.* (2007) Cell culture models in developing nutrigenomics foods for inflammatory bowel disease. *Mutat Res* **622**, 94–102.

Pino, E., Campos, A.M., Lopez-Alarcon, C. *et al.* (2006) Free radical scavenging capacity of hydroxycinnamic acids and related compounds. *J Phys Org Chem* **19**, 759–764.

Plug, H. and Harring, P. (1993) The role of ingredient-flavor interactions in the development of fat-free foods. *Trends Food Sci Tech* **4**, 150–152.

Prestamo, G. and Manzano, P. (1993) Peroxidases of selected fruits and vegetables and the possible use of ascorbic acid as an antioxidant. *Hort Sci* **28**, 48–50.

Przybylski, R., Malcolmson, L.J., Eskin, N.A.M. *et al.* (1993) Stability of low linolenic acid canola oil to accelerated storage at 60°C. *Food Sci Tech* (Lebens Wiss U Technol), **26**, 205–209.

Rahavi, E.B. and Kapsak, W.R. (2010) Health and wellness product development. Prepared Foods Network: http://www.preparedfoods.com/articles/print/article-health-and-wellness-product-development-february-2010 (last accessed 22 March 2013).

Rawel, H., Kroll, J. and Hohl, U. (2001) Model studies on reactions of plant phenols with whey proteins. *Nahrung* **45**, 72–78.

Rees, D.A. (1977) *Polysaccharide Shapes*. Chapman and Hall, London.

Relkin, P., Fabre, M. and Guichard, E. (2004) Effect of fat nature and aroma compound hydrophobicity on flavor release from complex food emulsions. *J Agric Food Chem* **52**, 6257–6263.

Renard, C.M.G.C. (2005). Effects of conventional boiling on the polyphenols and cell walls of pears. *J Sci Food Agric* **85**, 310–318.

Renard, C.M.G.C., Baron, A., Guyot, S. and Drilleau, J.-F. (2001) Interactions between apple cell walls and native apple polyphenols: Quantification and some consequences. *Int J Biol Macromol* **29**(2), 115–125.

Rice-Evans, C.A., Miller, N.J. and Paganga, G. (1996) Structureantioxidant activity relationships of flavonoids and phenolic acids. *Free Radic Biol Med* **20**, 933–956.

Richard-Forget, F.C. and Gauillard, F.A. (1997) Oxidation of chlorogenic acid, catechins, and 4-methylcatechol in model solutions by combinations of pear (*Pyrus communis* cv. Williams) polyphenols oxidase and peroxidase: A possible involvement of peroxidase in enzymatic browning. *J Agric Food Chem* **45**, 2472–2476.

Robertson, G.L. and Samaniego-Esguerra, C.M. (1990) Effect of soluble solids and temperature on ascorbic acid degradation in lemon juice stored in glass bottles. *J Food Qual* **13**(5), 361–374.
Robins, M.M. and Wilson, P.D.G. (1994). Food structure and microbial growth. *Trends Food Sci Tech* **5**, 289–293.
Rohn, S., Rawel, H. and Kroll, J. (2004) Antioxidant activity of proteinbound quercetin. *J Agric Food Chem* **52**, 4725–4729.
Ruben, C. and Larsson, K. (1985) Relations between antioxidant effect of alpha-tocopherol and emulsion structure. *J Disp Sci Tech* **6**, 213–221.
Sakai, T. (1992) Degradation of pectins. In: *Microbial degradation of natural products* (ed. G. Winkelmann), pp. 57–81. VCH, Weinheim.
Sandhu, K.S. and Singh, N. (2007) Some properties of corn starches II: physicochemical, gelatinization, retrogradation, pasting and gel textural properties. *Food Chem* **101**, 1499–1507.
Sauvageau, J., Hinkley, S.F., Carnachan, S.M. and Sims, I.M. (2010) Characterisation of polysaccharides from gold kiwi fruit (Actinidia chinensis Planch. 'Hort16A'). *Carbohydr Polym* **82**, 1110–1115.
Scheepens, A., Kee, T. and Paxton, J.W. (2010). Improving the oral bioavailability of beneficial polyphenols through designed synergies. *Genes Nutr* **5**, 75–87.
Schmidt, K. A. (2004). Dairy: Ice cream. In: *Food Processing – Principles and Applications*. pp. 287–296. Blackwell Publishing, Ames, IA.
Schofield, J. and Chen, X. (1995) Analysis of free reduced and free oxidised glutathinoe in wheat flour. *J Cereal Sci* **21**, 127–136.
Shi, J., Nawaz, H., Pohorly, J. *et al.* (2005). Extraction of polyphenolics from plant material for functional foods – engineering and technology. *Food Rev Int*, **21**, 139–166.
Singh, A.V. and Nath, L.K. (2009) Evaluation of physicochemical character and pasting behavior of *Phaseolus Acontifolius Jacq*. Starch. *Electr J Environm Agric Food Chem* **8**(10), 984–990.
Singh, H., Creamer, L.K. and Newstead, D.F. (1989) Effects of heat on the proteins of concentrated milk systems. *Bull Intern Dairy Fed* **238** (Chap. 12), 94–104.
Sivam, A.S., Sun-Waterhouse, D., Perera, C.O. and Waterhouse, G.I.N. (2011a) Application of FT-IR and Raman spectroscopy for the study of biopolymers in breads fortified with fibre and polyphenols. *Food Res Int* doi: 10.1016/j.foodres.2011.03.039.
Sivam, A.S., Waterhouse, G.I.N., Zujovic, Z.D. *et al.* (2011b) Structure and dynamics of wheat starch in breads fortified with polyphenols and pectin: An ESEM and solid-state CP/MAS ^{13}C NMR spectroscopic study. *Food Bioprocess Tech.* doi: 10.1007/s11947-011-0699-z.
Sivam, A.S., Sun-Waterhouse, D., Perera, C.O. and Waterhouse, G.I.N. (2012) Exploring the interactions between blackcurrant polyphenols, pectin and wheat biopolymers in model breads; a FT-IR and HPLC investigation. *Food Chem* **131**(3), 802–810.
Solms, J., Osman-Ismail, F. and Beyler, M. (1973) The interaction of volatiles with food components. *Can Inst Food Sci Tech J* **6**, A10-A16.
Sørensen, A.-D.M., Haahr, A.-M., Becker, E.M. *et al.* (2008) Interactions between iron, phenolic compounds, emulsifiers, and pH in omega-3-enriched oil-in-water emulsions. *J Agric Food Chem* **56**, 1740–1750.
Spanos, G.A., Wrolstad, R.E. and Heatherbell, D.A. (1990) Influence of processing and storage on the phenolic composition of apple juice. *J Agric Food Chem* **38**, 1572–1579.
Spencer, J.P., Schroeter, H., Crossthwaithe, A.J. *et al.* (2001) Contrasting influences of glucuronidation and O-methylation of epicatechin on hydrogen peroxide–induced cell death in neurons and fibroblasts. *Free Radic Biol Med* **31**, 1139–1146.
Sperber, W.H. (1983) Influence of water activity on foodborne bacteria – A review. *J Protection* **46**, 142–145.
Starling, S. (2005) Fruit dons 'healthy' halo. *Func Food Nutrac* **6**, 6.
Starling, S. (2007) Superfruits – superheroes of functionality. *Func Food Nutrac* **64**, 22–26.
Storz, H., Zimmermann, U., Zimmermann H. and Kulicke W.-M. (2010) Viscoelastic properties of ultra-high viscosity alginates. *Rheol Acta* **49**, S155–167.
Sun-Waterhouse, D. (2011). The development of fruit-based functional foods targeting the health and wellness market: A review. *Int J Food Sci Tech* **46**(5), 899–920
Sun-Waterhouse, D. and Jin, D. (2011) Pasta enriched with fruit phenolics and pectin fibre – a convenient food in a healthy diet? In: *Proceedings of the 2011 NZIFST Conference 'Science to Reality: New Zealand and Beyond'* (29 June-1 July 2011, Rotorua, New Zealand). The New Zealand Institute of Food Science and Technology (NZIFST), Palmerston, New Zealand.
Sun-Waterhouse, D. and Wadhwa, S.S. (2012) Synergistic effects on the rheological behaviour and cloudy-point of aqueous hydrocolloid dispersions. In: *Proceedings of the 2012 NZIFST Conference* (26-28 June

2012, Hamilton, New Zealand). The New Zealand Institute of Food Science and Technology (NZIFST), Palmerston, New Zealand.
Sun-Waterhouse, D., Melton, L.D., O'Connor, C.J. *et al.* (2008a) Effect of apple cell walls and their extracts on the activity of dietary antioxidants. *J Agric Food Chem* **56**, 289–295.
Sun-Waterhouse, D., Farr, J., Wibisono, R. and Saleh, Z. (2008b) Fruit-based functional foods I: production of novel food grade apple fibre ingredients. *Int J Food Sci Tech* **43**, 2113–2122.
Sun-Waterhouse, D., Smith, B.G., O'Connor, C.J. and Melton, L.D. (2008c) Effect of raw and cooked onion dietary fibre on the antioxidant activity of ascorbic acid and quercetin. *Food Chem*, **111**, 580–585.
Sun-Waterhouse, D., Chen, J., Chuah, C. *et al.* (2009a) Kiwi fruit -based polyphenols and related antioxidants for functional foods: Kiwi fruit extract-enhanced gluten-free bread. *Int J Food Sci Nutr* **60**(S7), 251–264.
Sun-Waterhouse, D., Wen, I., Wibisono, R. *et al.* (2009b) Evaluation of the extraction efficiency for polyphenol extracts from by-products of green kiwi fruit juicing. *Int J Food Sci Tech* **44**, 2644–2652.
Sun-Waterhouse, D., Nair, S., Wibisono, R. *et al.* (2010a) Insights into smoothies with high levels of fibre and polyphenols: Factors influencing chemical, rheological and sensory properties. *World Acad Sci Eng Tech* **65**, 276–285.
Sun-Waterhouse, D., Zhou, J. and Wadhwa, S.S. (2010b) Effects of adding apple polyphenols before and after fermentation on the properties of drinking yoghurt. *Food Bioprocess Tech.* doi: 10.1007/s11947-011-0563-1.
Sun-Waterhouse, D., Teoh, A., Massarotto, C. *et al.* (2010c). Comparative analysis of fruit-based functional snack bars. *Food Chem* **119**, 1369–1379.
Sun-Waterhouse, D., Luberriaga, C., Jin, D. *et al.* (2011a) Juices, fibres and skin waste extracts from white, pink or red fleshed apple genotypes as potential food ingredients: A comparative study. *Food Bioprocess Tech.* doi: 10.1007/s11947-011-0692-6.
Sun-Waterhouse, D., Edmonds, L., Wadhwa, S.S. and Wibisono, R. (2011b) Producing ice cream using a substantial amount of juice from kiwi fruits with green, gold or red flesh. *Food Res Int.* doi: 10.1016/j.foodres.2011.05.030.
Sun-Waterhouse, D., Sivam, A.S., Cooney, J. *et al.* (2011c). Effects of added fruit polyphenols and pectin on the properties of finished breads revealed by HPLC/LC-MS and Size-Exclusion HPLC. *Food Res Int* **44**(9), 3047–3056.
Sun-Waterhouse, D., Thakorlal, J. and Zhou, J. (2011d) Effects of added phenolics on the storage stability of avocado and coconut oils. *Int J Food Sci Tech* **46**(8), 1575–1585.
Sun-Waterhouse, D., Wang, S.-Y., Wadhwa, S. and Zhou, J. (2011e) Comparative studies on the rheological and chemical properties of food gels incorporating freshly pressed fruit juice. *J Adv Food Hosp Tour* **1**(4), May (on-line).
Sun-Waterhouse, D., Wang, W. and Wadhwa, S.S. (2012a) Utilisation potential of feijoa fruit waste as ingredients for functional foods. In: *Proceedings of the 2012 NZIFST Conference* (26-28 June 2012, Hamilton, New Zealand). The New Zealand Institute of Food Science and Technology (NZIFST), Palmerston, New Zealand.
Sun-Waterhouse, D., Wadhwa, S.S. and Bekkour, K. (2012b) Complex rheological behaviours for predicting the stability of beverages containing high concentrations of polyphenols and soluble and insoluble fibres. In: *Proceedings of the 2012 NZIFST Conference* (26-28 June 2012, Hamilton, New Zealand). The New Zealand Institute of Food Science and Technology (NZIFST), Palmerston, New Zealand.
Tahir, R., Ellis, P.R., Bogracheva, T.Y., Meares-Taylor, C. and Butterworth, P.J. (2011) Study of the structure and properties of native and hydrothermally processed wild-type, *lam* and *r* variant pea starches that affect amylolysis of these starches. *Biomacromolecules* **12**, 123–133.
Tang, H. and Hills, B.P. (2003) Use of 13C MAS NMR to study domain structure and dynamics of polysaccharides in the native starch granules. *Biomacromolecules* **4**, 1269–1276.
Tatham, A.L. and Shewry, P.R. (1985) The conformation of wheat gluten proteins. The secondary structures and thermal stabilities of α-, β-, γ-, and ω-gliadins. *J Cereal Sci* **3**, 103–113.
Tochi, B.N., Wang, Z., Xu, S.–Y. and Zhang, W. (2008) Therapeutic application of pineapple protease (bromelain): A Review. *Pak J Nutr* **7**(4), 513–520.
Torregrosa, F., Esteve, M.J., Frígola, A. and Cortés, C. (2006) Ascorbic acid stability during refrigerated storage of orange-carrot juice treated by high pulsed electric field and comparison with pasteurized juice. *J Food Eng* **73**(4), 339–345.
Troller, J.A. and Christian, J.H.B. (1978) *Water Activity and Food*, pp. 13–47. Academic Press, New York.
Tsuzuki, W., Nagata, R., Yunoki, R. *et al.* (2008) Cis/trans-isomerisation of triolein, trilinolein and trilinolenin induced by heat treatment. *Food Chem* **108**, 75–80.

Tuorila, H. (2007). Sensory perception as a basis for food acceptance and consumption. In: *Consumer-led Food Product Development* (ed. H. MacFie), pp. 34–80. Woodhead Publishing, Cambridge.
Tura, D. and Robards K. (2002) Sample handling strategies for the determination of biophenols in food and plants. *J Chromatogr A* **975**, 71–93.
van het Hof, K.H., Kivits, G.A., Weststrate, J.A. and Tijburg, L.B. (1998) Bioavailability of catechins from tea: the effect of milk. *Eur J Clin Nutr* **52**, 356–359.
Vattem, D.A., Ghaedian, R. and Shetty, K. (2005). Enhancing health benefits of berries through phenolic antioxidant enrichment: focus on cranberry. *Asia Pac J Clin Nutr* **14**, 120–130.
Verstraeten, S.V., Keen, C.L., Schmitz, H.H. *et al.* (2003) Flavan-3-ols and procyanidins protect liposomes against lipid oxidation and disruption of the bilayer structure. *Free Radic Biol Med* **34**, 84–92.
Ward, O.P. and Moo-Young, M. (1989) Enzymatic degradation of cell wall and related plant polysaccharides. *CRC Crit Rev Biotechnol* **8**, 237–274.
Wegrzyn, T.F., Farr, J.M., Hunter, D.C. *et al.* (2008) Stability of antioxidants in an apple polyphenol-milk model food system. *Food Chem* **109**, 310–318.
Weston, R.J. (2010) Bioactive products from fruit of the feijoa (Feijoa sellowiana, Myrtaceae): A review. *Food Chem* **121**, 923–926.
Whistler, R.L. and Daniel, J.R. (1984) Molecular structure of starch. In: *Starch: Chemistry and Technology* (eds R.L. Whistler, J.N. BeMiller and E.F. Paschall), 2nd edn, pp. 153–178. Academic Press, FL.
Whistler, R.L. and Daniel, J.R. (1985) Carbohydrates. In: *Food Chemistry* (ed O.R. Fennema), 2nd edn, pp. 69–137. Marcel Dekker Inc., New York.
Whitaker, J.R. (1990) Microbial pectinolytic enzymes. In: *Microbial enzymes and biotechnology*, (eds W.M. Fogarty and C.T. Kelly), 2nd edn, pp. 133–176. Elsevier Science Ltd, London.
Wibisono, R., Zhou, J. and Sun-Waterhouse, D. (2009) Properties of water- and lipid-based cocoa pastes before and after heat treatment. *J Adv Food Hosp Tour* **1**(1), December (on-line).
Williams, D.C., Lim, M.H., Chen, A.O. *et al.* (1986) Blanching of vegetables for freezing – which indicator enzyme to use. *Food Tech* **40**(6), 130–140.
Wood, F.W. and Goff, T.C. (1973). The determination of the effective shear rate in the Brabender Viscograph and in other systems of complex geometry. *Starch* **25**(3), 89–91.
Wright, J.S., Johnson, E.R. and DiLabio, G.A. (2001) Predicting the activity of phenolic antioxidants: Theoretical method, analysis of substituent effects, and application to major families of antioxidants. *J Am Chem Soc* **123**, 1173–1183.
Xu, F., Brown, K.M., Dybdal, L. *et al.* (1999) Controlled stepwise reduction of disulfide bond and heat induced modification of wheat dough protein. *Cereal Chem* **76**(6), 931–937.
Yemenivioğlu, A., Özkan, M. and Cemeroğlu, B. (1997) Heat inactivation kinetics of apple polyphenoloxidase and activation of its latent form. *J Food Sci* **62**(3), 508–510.
Zhang, W.H., Shi, B. and Jeff, S. (2007) A theoretical study on autoxidation of unsaturated fatty acids and antioxidant activity of phenolic compounds. *J Am Leather Chem Assoc* **102**(3), 99–105.

Index

Bioactives in Fruit: Health Benefits and Functional Foods, First Edition. Edited by Margot Skinner and Denise Hunter.